Chemistry of Plant Natural Products

Sunil Kumar Talapatra • Bani Talapatra

Chemistry of Plant Natural Products

Stereochemistry, Conformation, Synthesis, Biology, and Medicine

Volume 1

With a Foreword by Professor K.C. Nicolaou

 Springer

Sunil Kumar Talapatra
Bani Talapatra
formerly Professors
Dept. Chemistry
University of Calcutta
Kolkata
India

ISBN 978-3-642-45409-7   ISBN 978-3-642-45410-3 (eBook)
DOI 10.1007/978-3-642-45410-3
Springer Heidelberg New York Dordrecht London

Library of Congress Control Number: 2014959101

Printed on acid-free paper

Springer is part of Springer Science+Business Media (www.springer.com)

Dedicated
To the loving memories of our mothers
Chapala-Basanti Talapatra and Sushama Rani Chaudhuri

And

A gift with love
To our students
Who inspired us through years
And
To our children
Anupam and Sharmishtha
Who ungrudgingly cooperated during our early professional years and
Helped in all possible ways during the preparation of this manuscript

Foreword

Striving to understand nature provided the impetus for much of our philosophical ideas, scientific discoveries, and technological advances whose beneficial results we came to enjoy as a species throughout history. Understanding the molecules of life—nucleic acids, proteins, and secondary metabolites, commonly referred to by organic chemists as natural products—is arguably one of the most important human endeavors, one in which the benefits reaped by science and society are both immense and undisputed. Indeed, the emergence of the structure of the molecule and the art of its synthesis in the nineteenth century set the stage for the development of all of organic chemistry and biochemistry and much of modern biology and medicine. Natural products played a protagonist role in these developments. Their isolation, structural elucidation, and synthesis challenged and stimulated analytical and purification techniques and instrumentation, method development, and synthetic strategy design. Today, because of the enormous advances made over the last two centuries in the chemistry, biology, and medicine of natural products, people around the world enjoy untold benefits with regard to healthcare, nutrition, cosmetics, and fashion, just to name a few areas.

I was honored by the invitation of Professors Sunil and Bani Talapatra to provide a Foreword to their book and enjoyed reading various versions of their prepublication manuscripts. Much to my delight, I found them to be stunningly illuminating in terms of breadth and depth of coverage of essentially the entire field of chemistry, biology, and medicine of plant-derived natural products. Their accomplishment in putting together this tome is beyond the normal boundaries of most books written on the subject, and they deserve our admiration and respect for bringing together the various aspects of this important field.

In the chapters that follow, the authors focus on the molecules of life, particularly those natural products derived from plants, dwelling on such wide ranging aspects as enzymatic transformations, biosynthetic pathways, isolation, and structural elucidation (including techniques and instrumentation), conformation, total synthesis, biological activities and functions, symbiotic relationships, medical applications, biochemistry, and stereochemistry. The latter topic is discussed in considerable detail and its impact on the chemical and biological properties of the molecule and its asymmetric synthesis are articulated and explained. The special emphasis on conformation is particularly informative and educational and provides useful insights and understanding of the nature of the molecule to students of organic chemistry. The various aspects of the science of natural products are elaborated upon by using selected natural product classes and individual compounds, such as terpenoids and alkaloids, to underscore the influential impact of these endeavors in advancing the discipline of organic chemistry, including theory, experimental methods and techniques, chemical synthesis, biology, medicine, and nutrition.

Three appendices at the end of the book provide interesting and useful biographical information on selected pioneers, major discoveries and inventions, and extra educational material for learning. Indeed, this opus is intended as both a reference book and a teaching text where one can find important facts, inspiration, and pedagogy. The structure of the book allows for easy selection of separate topics for reading and teaching. Congratulations and many thanks to Professors Sunil and Bani Talapatra for an outstanding treatise which will remain a classic in the field and hopefully serve to maintain its momentum and proliferation with new admirers, recruits, and supporters.

Houston, TX   K.C. Nicolaou

Preface

Linus Pauling [NL 1954 (Chemistry), 1962 (Peace)] once said *"A good book or a speech should be like a Lady's dress. It should be long enough to cover the essentials and short enough to reveal the vital statistics."*

With this caveat in mind, it becomes a challenging task to author an adequately balanced textbook which shall capture the essentials and address the leading frontiers, while staying focused within the broader scope of the subject area. During our long postgraduate-level teaching career spanning more than three decades, we felt the unmet need of a textbook on Plant Natural Products, the Secondary Metabolites, that covered the fundamental aspects of the relevant chemistry, stereochemistry, biosynthesis, and bioactivity in sufficient depth, elaborated using well-known representative members.

Here we have made an attempt to write a book intended to serve as a textbook for advanced undergraduate and graduate/postgraduate students as well as a reference guide for researchers and practicing organic and pharmaceutical chemists who may wish to gain a better understanding of various aspects of the chemistry of natural products. It is almost an impossible task to write a comprehensive textbook dealing with all facets of natural products chemistry of even some of the selected classes of molecules due to the rapidly expanding literature in this area. Hence, we have deliberately not roamed too far to make this a comprehensive textbook, but rather tried to create one that will provide sufficient background in the general area of natural products chemistry and serve as an adequate launching pad for an in-depth and specialized investigation.

This book grew primarily out of our 68 years of combined postgraduate teaching experience in natural products chemistry and stereochemistry at the Calcutta University and at some other universities. We have limited our discussions to some well-studied natural products of plant origin, which have contributed significantly to the advancement of organic chemistry. The choice from the endless variations of natural products is entirely personal, driven by our perception of model natural products for our target readers. Our coverage of the material is illustrative of the vastness, diversification, and continued growth of this particular area of chemistry.

The distinguishing feature of this book is the discussion on the stereochemical aspects, the hallmark of chiral natural products, and the inclusion of a chapter exclusively dedicated to stereochemistry (Chap. 2). This chapter will be helpful in understanding various updated nomenclatures, stereostructures, asymmetric synthesis, biosynthesis, and bioactivities (chiral recognition in vivo) of natural products that appear in subsequent chapters. Furthermore, wherever possible, we have given plausible mechanistic rationalizations of product formations which are less obvious, during the portrayal of reactions and synthesis of natural products. Original literature has been referenced, with titles of the articles (with the exception of titles of some very old references, which could not be procured), to give interested readers an idea of the subject matter contained in such articles.

Chapter 1 offers a preliminary introduction to enzymes, coenzymes, and primary and secondary metabolites (natural products), illustrates various biological functions of the metabolites, and summarizes the identified metabolic pathways leading to the different types of natural products.

Chapter 3 deals with few important biochemical events related to the discussions in the subsequent chapters on the biogenesis and biosynthesis of major skeletal patterns formed sometime during three billion years of evolution.

Various synthetic methodologies, separation techniques, and instrumental analysis have been invented and developed in connection with the natural products chemistry research. In Chap. 4, we have discussed some widely used isolation procedures and separation techniques (Sect. 4.1). The chapter also touches upon traditional and state-of-the-art methods for structural elucidation of naturally occurring molecules (Sect. 4.2).

During the discussions on the "Biosynthesis of Terpenoids: the oldest Natural Products" (Chap. 5) or on the biosynthesis of other natural products (other Chapters), no extensive description of the enzymes and their processes has been provided. The biosynthetic conversions are explained in terms of the currently accepted mechanisms of organic reactions, with special attention to the stereochemical features of the processes.

Mother Nature uses organic molecules as the building blocks for the natural product framework. Barton (NL 1969) said, "*if we assume that enzymatically-induced reactions follow the same mechanistic principles as ordinary organic reactions we can at least make an approach to the subject*" [1]. Likewise, R. Kluger wrote in an article [2] "*It is useful if the reactions can be systematically divided into mechanistic types (such as Ingold formulation [Ingold, 1953]. The most common examples of this type of classifications are the two general nucleophilic substitution mechanisms, S_N1 and S_N2.*" The synthetic methodologies used by Nature are amazingly simple, as exemplified by aldol, acyloin, and Claisen condensations, olefin-cation addition/cyclization, Markonikov additions, simple S_N1, S_N2, S_N2', E1, and E2 reactions, and 1,2-*trans* migrations for creating carbon–carbon bonds—the life string of organic molecules. Nature's substrates for the biosynthesis of complex natural products are also simple (e.g., small carbonyl compounds, L-amino acids, olefins, etc.).

Chapters 6–29, including a number of sections (sub-chapters) of Chaps. 6–8, 10, 13, and 14, deal with the chemistry of individual compounds classified according to the major biogenetic pathways. Many basic concepts and ideas have been discussed. Though the tone of discussions on natural products has changed during the last few decades, we have given adequate attention to the work of past great chemists in dealing with the individual compounds, because we believe that their work will never go out of date; rather it will exist to impart the spirit of enquiry and forms the foundation to our current understanding. In this context we quote from the abstract of Barton's talk entitled "Oxygen and I" in the 10th Johnson Symposium 1995 held at Stanford University: "*In Chemical Sciences, the distant past, the near past and the present join together in continuous harmony..... The lecture will illustrate how past chemistry should not be forgotten.*"

Chapter 15 is a general introduction to alkaloids. Chapters 16–29 include detailed discussions on some well-known alkaloids derived from various biogenetic precursors. Chapter 30 contains structures with relevant references of a number of alkaloids of diverse skeletal patterns (not discussed due to space constraint).

Important concepts such as conformational analysis, stereochemistry, biomimetic synthesis, retrosynthetic approach, pericyclic reactions, Fischer and Newman projection formulas, and many organic reactions owe their genesis and/or refinement to the natural products chemistry. With his understanding of squalene-2,3S-oxide (a natural product), the phenomenon of catalysis, and his passion for the transition metals of the Periodic Table, "the most elegant organizational chart ever devised" [3], Sharpless (NL 2001) invented Sharpless epoxidation—a near chemical substitute for an enzymatic reaction. The trisubstituted olefinic alcohol, geraniol, a natural product with an attractive smell that was a great favorite of Sharpless [4], was used by him to successfully implement the most challenging asymmetric epoxidation and dihydroxylation. Chapter 31 attempts to bring together such important outcomes of natural products chemistry research.

The contribution of natural products in asymmetric synthesis is immense. Natural chiral auxiliaries take the place of enzymes in some chemical reactions, and the chemical literature is flooded with reports on such chiral auxiliaries. Chapter 32 has been dedicated to chiral recognition in biological systems and the use of natural products and their derivatives as chiral auxiliaries. This chapter illustrates the differences in biological properties exhibited by each component in a pair of stereochemically nonequivalent enantiomeric twins, i.e., enantiomeric stereoselectivity.

The medicinal values of many natural products are now evaluated scientifically to give them the drug status. Further, many of them served as the lead molecules for the synthesis of cost-effective therapeutic molecules with promising medicinal values. This is one of the important value-added dimensions of natural products chemistry. Chapter 33 deals with *The Natural Products in the Parlor of the Pharmaceuticals*. Chapter 34, entitled *Organic Phytonutrients, Vitamins and Antioxidants*, presents the multitude of useful biological functions of nutraceuticals, as well as natural products that are consumed daily through diet in the form of spices, vegetables, fruits, and drinks. This chapter may also stimulate interest of the uninitiated readers.

Carbohydrates form a very important class of bioorganic molecules whose chemical and biochemical properties revolve around their stereochemical and conformational features. The chemistry of carbohydrates has grown enormously during the last few decades and involves their use as templates in the synthesis of complex chiral natural products. However, since their biosynthesis and biodegradation are routed through primary metabolic/catabolic pathways, they are referred to as primary metabolites according to conservative definition and do not fall within the scope of secondary metabolites and hence have not been included.

Brief biographical sketches of 31 pioneering chemists, who we believe immensely enriched the realm of natural products chemistry, appear in Appendix A. We hope that these great personalities will inspire the young readers and motivate them in their professional pursuits. In the absence of space constraint, several more such life sketches could have been included, since concepts and ideas often develop through the collaborative and collective intellectual engagement and may not be attributed to a single discoverer. We sincerely regret our inability to be all-inclusive in this regard.

Appendix B chronologically lists some landmark inventions/discoveries in the field of natural products which may be interesting and inspiring to readers.

Appendix C contains miscellaneous information geared towards students and some often-overlooked conventions. For convenience, a list of Abbreviations has been provided before the Table of Contents.

Both in the Preface and in the body of the text, we took the liberty of quoting from books, articles, lectures, and letters (original references provided wherever appropriate), in order to impart deeper insights and broader perspectives related to the topics discussed.

We hope that the present version of the text will be well adapted to the teaching of natural products chemistry at the senior undergraduate and graduate/postgraduate levels. Students should get a flavor of the subject and be able to comfortably move forward with more advanced topics both within the domain of natural products chemistry and beyond. The teachers with limited teaching time and resources should be able to pick and choose topics of interest to them, aligned with the level of the course they intend to offer.

Acknowledgments: Acknowledgments and thanks are not casual words—they have their origin in our hearts. The job of writing this book could not have been accomplished without the help of our supporters and well-wishers from various walks of life.

We express our deepest gratitude and heartfelt thanks to Professor K. C. Nicolaou first for kindly going through the Sect. 8.4 on **Taxol**® and then the whole of the first draft of the Chapters and Appendices, at our requests from time to time, for enriching them with helpful suggestions and valuable comments on the overall content, and finally for kindly writing the Foreword with his candid opinion, in spite of his heavy academic and administrative commitments.

Informational support received from Professors A. R. Battersby, C. Djerassi, K. Nakanishi, K. C. Nicolaou, and R. Noyori during drafting of their life sketches is acknowledged with deep gratitude. We are grateful to Professor Alex Nickon for his encouraging comments on Sect. 7.5 (on caryophyllene) and to Professor Sukh Dev for his encouragement in our endeavor. We are sincerely thankful to Dr. B. N. Roy (Lupin Limited, Pune) for several academic correspondences and constructive suggestions. Our affectionate thanks go to Dr. Sanchita Goswami [Calcutta University (CU)] for continual supply of PDF files of research papers needed as references for the last 3 years and to Mr. Debajyoti Saha [Indian Association for the Cultivation of Science (IACS), Kolkata] for the past 1 year.

The inputs received from our professional colleagues, friends, and former students during the preparation of the manuscripts have been extremely valuable. Special mention should be made of Dr. Bhupesh C. Das (CNRS, Gif-sur-Yvette, France), Professor Manoj K. Pal (SINP, Kolkata), Professor Amitabha Sarkar (IACS), Professors Anil K. Singh and S. Kotha (IIT Bombay), Professor Kankan Bhattacharyya (IACS), Professor Swadesh R. Roychaudhuri (Jadavpur University, Kolkata), Dr. Arun K. Shaw (CDRI, Lucknow), Dr. Debashis Samanta (CLRI, Chennai), Professor Asutosh Ghosh, and Dr. Dilip K. Maiti (CU), Dr. Rajyasri Ghosh (Scottish Church College, Kolkata), Mr. Nandadulal Biswas (Library, CU), Mr. Asoke Mukherjee (CU, College Street Campus), Mr. Amar Chand De (Sarat Book House, Kolkata), Mr. Harry Ridge (Sheffield), and Dr. Bimal R. Bhattacharyya (Market Harborough, England). The library facilities of the Calcutta University, IIT Bombay, IACS (Kolkata), and UC Berkeley are acknowledged with gratitude.

We are solely responsible for errors, inconsistencies, and infelicities that may persist in the text and Figures, despite our most sincere efforts to eliminate them. Our endeavors will be meaningful if the book benefits the target readers. We welcome any corrective suggestions and recommendations for the future edition of this book.

Kolkata, India   Sunil Kumar Talapatra <talapatrask@gmail.com>
July, 2013   Bani Talapatra <banitalapatra@yahoo.co.in>

References

1. D. H. R. Barton, A Region of Biosynthesis, *Chem. Brit*, **1967**,August Issue, 330-336 (Pedlar Lecture, Chemical Society, London, **1967**).
2. Ronald Kluger, The Mechanistic Basis of Enzyme Catalysis in *Enzyme Chemistry. Impact and Applications*, Ed., Colin J. Suckling, Chapman and Hall, London, **1984**, p. 9.
3. Robert E. Krebs in 'The History and Use of Our Earth's Chemical Elements', *vide* Bill Bryson, *A Short History of Nearly Everything*, Black Swan, **2004**, p. 143.
4. K. B. Sharpless, Searching for New Reactivity (Nobel Lecture), *Angew. Chem., Int. Ed.* **2002**, *41*, 2024-2032 (pertinent pages. 2025-2026).

About the Images

(−)-Quinine

$3(R), 4(S), 8(S), 9(R)$

Cinchona calisaya

Cinchona officinalis

Quinine has been one of the most celebrated natural products since its first isolation in 1820 from the *Cinchona* bark by J. Pelletier and J. Caventou. It remained as the only effective drug against malaria for more than a century. The pictures of *C. calisaya* and *C. officinalis* are shown. Malaria is thus one of the first few diseases to be treated by a pure organic compound. The French Government issued a postal stamp (shown) in 1970 in honor of the discoverers of quinine to mark the 150th year of its discovery and also to introduce this molecule of immense health importance to the general public.

Quinine also served as the lead compound for a number of effective synthetic antimalarial drugs. Its chiral and basic character allowed its use for the first chiral resolution of a racemate acid, and a new horizon called stereochemistry appeared in the sky of chemical sciences. *Cinchona* alkaloids with some chemical manipulations have been extensively used as chiral auxiliaries and chiral catalysts in asymmetric synthesis and also as phase-transfer catalysts. For the multifarious importance of quinine along with its inspiring structure with four chiral centers, it has served as a target molecule for its first elegant synthesis in 1944. Its stereoselective synthesis appeared thrice even in the early twenty-first century. In fact, quinine serves as an interesting illustrative example of the (*R*,*S*)-designation of its four chiral centers. Thus, quinine deserves the front matter recognition of a book on the chemistry of bioactive plant natural products.

List of Abbreviations

α	Observed optical rotation in degrees
$[\alpha]$	Specific rotation [expressed without units; the actual units, deg mL/g dm) are understood].
Ac	Acetyl
ACP	Acyl carrier protein
acac	Acetylacetonate, acetylacetonyl
AD	Asymmetric dihydroxylation
ADP	Adenosine 5′-diphosphate
AIBN	2,2′-Azobisisobutyronitrile
aka	Also known as
AMP	Adenosine-5′-monophosphate
AO	Atomic orbital
anhyd	Anhydrous
Ar	Aryl
atm	Atmosphere(s)
ATP	Adenosine 5′-triphosphate
ATPase	Adenosine triphosphatase
aq.	Aqueous
–B⟩	9-Borabicyclo[3.3.1]nonyl
BBEDA	N,N'-bis(benzylidene)ethylenediamine
9-BBN	9-Borabicyclo[3.3.1]nonane
BINAL-H	2,2′-Dihydroxy-1,1′-binaphthylaluminum hydride
BINAP	(2R,3S),2,2′-bis(diphenylphos- phino)1,1′-binaphthyl
Bn	Benzyl
BOC, t-BOC, Boc	tert-Butoxycarbonyl
BOM	Benzyloxymethyl
bp	Boiling point
bpy, bipy	2,2′-Bipyridyl

BTMSA	Bis(trimethylsilyl)acetylene
Bu	*n*-Butyl
s-Bu	*sec*-Butyl
t-Bu	*tert*-Butyl
Bz	Benzoyl
18-C-6	18-Crown-6
c	Cyclo-
°C	Degrees Celsius
calcd	Calculated
CAM	Carboxamidomethyl
CAN	Ceric ammonium nitrate
cat	Catalyst
Cbz, CBZ	Benzyloxycarbonyl
CD	Circular dichroism
CHD	Coronary heart disease
CI	Chemical ionization (in MS)
CIP	Cahn-Ingold-Prelog
cm	Centimeter (s)
CoA	Coenzyme A
COD/cod	1,5-Cyclooctadiene (legand)
concd	Concentrated
COSY	Correlation spectroscopy (NMR)
cot, (COT)	Cyclooctatetraene (ligand)
Cp	Cyclopentadienyl
CSA	10-Camphorsulfonic acid
CTP	Cytidine triphosphate
Cy Cy-Hex	Cyclohexyl
δ	Chemical shift in parts per million downfield from TMS (NMR)
d	Day(s); doublet (spectral)
DABCO	1,4-Diazabicyclo[2.2.2]octane
DAIB	3-*exo*-(dimethylamino)isoborneol
DAST	Diethylaminosulfur trifluoride
dba	*trans,trans*-Dibenzylideneacetone
DBE	1,2-Dibromoethane
DBN	1,5-Diazabicyclo[4.3.0]non-5-ene
DBS	5-Dibenzosuberyl
DBU	1,8-Diazabicyclo[5.4.0]undec-7-ene
DCBI	*N,N'*-dicyclohexyl-*O*-benzylisourea
DCC	*N,N*-dicyclohexylcarbodiimide
DCE	1,2-Dichloroethane
DDQ	2,3-Dichloro-5,6-dicyano-1,4-benzoquinone
de	Diastereomer(ic) excess
DEA	Diethylamine
DEAD	Diethyl azodicarboxylate

DEIPS	Diethylisopropylsilyl
DEPT	Distortionless enhancement by polarization transfer (NMR)
DET	Diethyl tartrate
DHP	3,4-Dihydro-2H-pyran
DHQ	Dihydroquinine
DHQD	Dihydroquinidine
DIAD	Diisopropyl azodicarboxylate
Dibal-H, DIBALH	Diisobutylaluminum hydride
DIOP	2,3-O-isopropylidene-2,3-dihydroxy-1,4-bis (diphenylphosphino)butane
DiPAMP	1,2-Bis(o-anisylphenylphosphino)ethane
DIPT	Diisopropyl tartrate
DMA	N,N-dimethylacetamide
4-DMAP, DMAP	4-(dimethylamino)pyridine
DMB	3,4-Dimethoxybenzyl
Dmca	(S) and (R)-6,7-dimethoxy-4-coumaryl) alanines
DMDO	Dimethyldioxane
dppe	Bis(diphenylphosphino)ethane $Ph_2PCH_2CH_2PPh_2$
dppf	Bis(diphenylphosphino)ferrocene
dppm	Bis(diphenylphosphino)methane $(Ph_2P)_2CH_2$
dppp	1,3-Bis(diphenylphosphino)propane
dr	Diastereomer ratio
dvd	Divenylbenzene
DME	1,2-dimethoxyethane
DMF	N,N-dimethylformamide
DMPU	1,3-Dimethyl-3,4,5,6-tetrahydro-2(1H)-pyrimidinone
DMS	Dimethyl sulfide
DMSO	Dimethyl sulfoxide
DNA	Deoxyribonucleic acid
L-DOPA	3-(3,4-dihydroxyphenyl)-L-alanine
DPC	Dipyridine chromium(VI) oxide
DTBMS	Di($tert$-butyl)methylsilyl
EDC (EDCI)	1-(3-dimethylaminopropyl)-3-ethylcarbodiimide hydrochloride
E_1	Unimolecular elimination
E_2	Bimolecular elimination
ED_{50}	Dose that is effective in 50 % of test subjects
EDTA	Ethylenediaminetetraacetic acid
e	Electron
ee	Enantiomeric excess
EE	1-Ethoxyethyl
EI	Electron impact (in MS)
Et-DuPHOS	1,2-Bis($2',5'$-diethylphospholano)ethane

ESR	Electron spin resonance
Et	Ethyl
equiv	Equivalent(s)
FAB	Fast atom bombardment (in MS)
FAD	Flavin adenine dinucleotide (oxidized form)
$FADH_2$	Flavin adenine dinucleotide (reduced form)
FD	Field desorption (in MS)
FID	Flame ionization detection
Fmoc	9-Fluorenylmethoxycarbonyl
FPP	Fernesyl pyrophosphate
FT	Fourier transform
g	Gram (s)
GC	Gas chromatography
GFPP	Geranylfarnesyl pyrophosphate
GPP	Geranyl pyrophosphate
GGPP	Geranylgeranyl pyrophosphate
GSSG	Oxidized glutathione
h	Hour (s)
HETE	Hydroxyeicosatetraenoic
hfc	3-(heptafluoropropylhydroxymethylene)-D-camphorato
HDL	High-density lipoprotein
His	Histidine
HMPA	Hexamethylphosphoramide
HMG	Hydroxymethylglutaryl
HPLC	High-pressure liquid chromatography
HWE	Horner-Wadsworth-Emmons
Im (imid.)	Imidazole
IND	Indoline
Ipc	Isopinocampheyl
KHMDS	Potassium hexamethyldisilazide, potassium bis(trimethylsilyl) amide
L	Liter(s)
LASER	Light Amplification by Stimulated Emission of Radiation
LDA	Lithium diisopropylamide
LDL	Low-density lipoprotein
LHMDS	Lithium hexamethyldisilazide, lithium bis(trimethylsilyl)amide
LTMP	Lithium 2,2,6,6-tetramethylpiperidide
*m*CPBA	3-Chloroperoxybenzoic acid
Me	Methyl
MEM	(2-Methoxyethoxy)methyl
Mes	Mesityl,2,4,6-trimethylphenyl (not methanesulfonyl)
MHz	Megahertz (NMR)
min	Minute(s)
mM	Millimoles per liter

MO	Molecular orbital
MOM	Methoxymethyl
MS	Mass spectrometry
Ms	Methanesulfonyl(mesyl)
mol	Mole(s)
(+)-MTPA	(R)-(+)-α−Methoxy-α−trifluoromethyl- phenyl acetate / acetyl chloride
m/z	Mass to charge ratio (in MS)
NAD	Nicotinamide adenine dinucleotide
NADH	Reduced NAD
NaHMDS	Sodium bis(trimethylsilyl)amide
NB	2-Nitrobenzyl
nbd	Norbornadiene
*p*NB	*p*-Nitrobenzyl
NBS	*N*-bromosuccinimide
NCS	*N*-chlorosuccinimide
NIS	*N*-iodosuccinimide
NMM	4-Methylmorpholine
NMO	4-Methylmorpholine *N*-oxide
NMP	1-Methyl-2-pyrrolidinone
NMR	Nuclear magnetic resonance
NOE	Nuclear Overhauser effect
NOESY	NOE spectroscopy
Nu	Nucleophile
OD	Optical density
ORD	Optical rotatory dispersion
op	Optical purity (discouraged, see *ee*)
ORTEP	Oak Ridge thermal ellipsoid plot
PCC	Pyridinium chlorochromate
PDC	Pyridinium dichromate
PDF	Portable document transfer format
PG	Prostaglandin
Ph	Phenyl
PHAL	Phthalazine
Phth	Phthalimido (phthalate)
Piv	Pivaloyl
PMB	4-Methoxybenzyl
PNB	4-Nitrobenzyl
PMP	4-Methoxyphenyl
PPA	Poly(phosphoric acid)
ppm	Parts per million (in NMR)
PNNP	*N*,*N'*-bis(1-phenylethyl)-*N*,*N'*-bis-(diphe-nylphosphino) ethylenediamine
PPTS	Pyridinium 4-toluenesulfonate

Pr	Propyl
i-Pr	Isopropyl
psi	Pounds per square inch
py (pyr,Py)	Pyridine
PYR	Diphenylpyrimidine
q	Quartet (spectral)
Ra-Ni	Raney nickel
Red-Al	Sodium bis(2-methoxyethoxy)aluminum hydride
Rf	Retention factor (in chromatography)
rt	Room temperature
rf	Radio frequency
s	Second(s)
SEM	$2'$-(trimethylsilyl)ethoxymethyl
SET	Single electron transfer
S_N^1	Unimolecular nucleophilic substitution
S_N^2	Bimolecular nucleophilic substitution
$S_N^{2'}$	Bimolecular nucleophilic substitution with rearrangement
Sia	Siamyl
TADDOL	$\alpha,\alpha,\alpha',\alpha'$-Tetraaryl-4,5-dimethoxy-1,3-dioxolane
TASF	*Tris*-(diethylamino)sulfonium difluorotrimethyl silicate
TBAF	Tetra-*n*-butylammonium fluoride
TBAI	Tetra-*n*-butylammonium iodide
TBDMS	*tert*-Butyldimethylsilyl
TBDPS	*tert*-Butyldiphenylsilyl
TBHP	*tert*-Butylhydroperoxide, (*t*-BuOOH), l
TBS	*tert*-Butyldimethylsilyl
TCNE	Tetracyanoethylene
TEBA	Triethylbenzylammonium $Bu(Et)_3N^+$
TEMPO	Tetramethylpieridinyloxy free radical
TEOC	2-(trimethylsilyl)ethoxycarbonyl
TES	Triethylsilyl
Tf	Trifluoromethanesulfonyl (triflyl)
Tf(OTf)	Triflate $-SO_2CF_3(-OSO_2CF_3)$
TFA	Trifluoroacetic acid
TFAA	Trifluoroacetic anhydride
thexyl	1,1,2-Trimethylpropyl
THF	Tetrahydrofuran
THP	Tetrahydropyran-2-yl
TIPDS	1,1,3,3-Tetraisopropyldisiloxane-1,3-diyl
TIPS	Triisopropylsilyl
TMEDA	N,N,N',N'-tetramethyl-1,2-ethylenediamine
TMG	1,1,3,3-Tetramethylguanidine
TMS	Trimethylsilyl (tetramethylsilane)
TMS N_3	Trimethylsilylazide

Tol	4-Methylphenyl
Torr	1 mg Hg, 1/760 atm
TPAP	Tetra-*n*-propylammonium perruthenate
TPS	*tert*-Butyldiphenylsilyl
Tr	Triphenylmethyl (trityl)
Triton B	Benzyltrimethylammonium hydroxide
Ts (Tos)	Tosyl, 4-toluenesulfonyl
TS	Transition state
T_R	Retention time in chromatography
UV	Ultraviolet
X_c	Chiral auxiliary

Common Latin Abbreviations

De novo (L)	A new, a fresh, again from the beginning
et al. (L)	And others
Ibid (Ibidim L)	In the same book, Chapter, pages etc.
Idem (L)	The same as previously given
infra	Below
in situ	Existing place or position
in vitro (L)	Controlled experimental environment rather than within the living system
in vivo (L)	Made to occur within a living system
modus operandi (L)	Mode of operation of working
supra	Above
tincture	Medicinal solution of drug in alcohol
	For more terms see Appendix C, item (**9**)

LIST 2

Compound Names

At the beginning of sentences or in headings, the **first** letter of the chemical name after the prefix is capitalized.

Plant Names in Italics

See Sect. 4.1.2, p. 245 for details

Element Symbol Locants in Italics

N,*N*′-dimethylurea
N-ethylaniline
O,*O*,*S*-triethyl-3*H*-fluorene

Positional and Structural Prefixes in Italics

o, m, p, n, 1°, 2°, 3°, sec, t, tert, peri

Configurational Prefixes in Italics

(R), (S), (Z), (E), cis, trans, cisoid, transoid, rel, **d, l,** *meso, sn, endo, exo, sym, syn, anti, amphi, erythro, threo, altro, ribo, xylo, vic, gem, Pref, Parf*

Contents

Contents for Volume 2

About the Authors

Sunil Kumar Talapatra's scholastic association with his mentor, a renowned natural products chemist Professor (Mrs.) Asima Chatterjee at the Calcutta University (CU) for over 6 years, and with an outstanding organic chemist Professor Michael P. Cava at the Ohio State University (OSU), USA, for over 4 years and his close association with Professor M. S. Newman at OSU served as inspiration for him to pursue the career of teaching and fundamental research. During his active career as Lecturer (1965–1971), Reader (1971–1975), and UGC Professor of Chemistry (1975–1998), CU, he taught stereo-

chemistry, conformational analysis, reaction mechanism, and alkaloids in the postgraduate classes of CU and Presidency College (now University), Kolkata, for more than three decades and has been teaching stereochemistry for the last three decades in some other universities of W. Bengal and Tripura as a Guest teacher. He has been a pioneer in teaching reaction mechanism in CU and stereochemistry and conformational analysis in the country. His research contributions include the discovery and structure elucidation of more than a hundred new bioactive secondary metabolites, mostly alkaloids, terpenoids, polyphenolics, etc., of different classes and cardiac glycosides from medicinal plants and one new C_{26} tetracyclic diamine from a marine sponge (with Professor P. Crews at UCSC). In 1968 he proclaimed for the first time from chemical and model studies a double skew-boat conformation of rings D/E of multiflorenol and friedelin (confirmation appeared in the literature much later in early 1980s by X-ray studies) and a skew-boat conformation of ring E in bauerenol. His more than two dozens of PhD students and several postdoctoral fellows have served in universities, IITs, research institutes, and colleges and many of them have carried out high standard research works. He published more than 175 papers, is a recipient of Premchand Roychand Scholarship

(CU) and Fulbright Travel Fellowship, and was selected for an 1851 Exhibition Scholarship of UK (but did not avail it since he was already offered a postdoctoral fellowship of OSU by Professor Cava). He extensively visited many major laboratories, universities, and institutions of USA, UK, Europe, Japan, Singapore, Bangkok, Hong Kong, Seoul, Kuala Lumpur, Bandung, Hanoi, Karachi, Kathmandu, Colombo, etc., on Exchange Programs, as a British Council Guest Visitor, as the FACS General Assembly Member (1981–1987), and as Executive Committee Member (1981–1983 and 1987–1988) from India, for International Conferences, participated in more than 30 International Conferences, and lectured on his research findings. As a Visiting Scientist he was associated with Professors N. R. Farnsworth and G. A. Cordell (Illinois U, Chicago) (19740, P. Crews (UCSC) (1995) and T. C. McMorris (UCSD) (2000) for 3 months at each place and lectured. As a UNESCO Senior Visiting Scientist in 1982, he was associated with Professors Koji Nakanishi (Columbia U), J. Meinwald (Cornell U), E. J. Corey (Harvard U), H Rapoport and C. H. Heathcock (UC Berkeley), N. Takahashi and K. Mori (Tokyo Agri. U), and Y. Kanaoka (Hokkaido U) for 2 weeks at each place. He has been elected C.Chem., FRIC (London) in 1974, President of the Indian Chemical Society (2002 and 2003), Vice President, Indian Science News Association (2008-), and an Ex officio Vice President, Indian Chemical Society since 2004, and was a recipient of Mouat Gold Medal of the CU. He is interested in photography, visiting museums and art galleries, watching cricket, tennis, and football, and reading books, and he loves traveling and making friends.

Bani Talapatra Professor of Chemistry (1983–2002), Calcutta University (CU), taught natural products chemistry covering their all aspects in the postgraduate classes for more than three decades and in some other universities of W. Bengal and Tripura as a Guest teacher for more than a decade. She also taught general chemistry in a local undergraduate college for several years during her early professional career. She obtained the D.Sc. degree of the CU working in the laboratory of Professor (Mrs) Asima Chatterjee. She was offered the Smith-Mund Scholarship by USEFI and also a Postdoctoral Fellowship at the Vanderbilt University (VU) by a renowned microbiologist, Professor V. Najjar, and she accepted the latter offer. At VU very fortunately she came in personal contact in 1964 with Sir Christopher K. Ingold, who was completing there the second edition of his famous book "Structure and Mechanism in Organic Chemistry." She learnt from Professor Ingold much about the fundamental concepts on organic reaction mechanism. As a UNESCO Senior

Visiting Scientist (in 1976 and 1983) she extensively traveled and visited many major universities and research institutes of USA, Europe, Japan, and South East Asia and was briefly associated with Professors W. Parker (Stirling U), W. Kraus. (Hohenheim U, Stuttgart), F.Bohlmann (Tech. Univ. Berlin), H. Wagner (Munich U), S. Masamune (MIT), W. Herz (Florida State U), H. Rapoport (UC Berkeley), and N. Takahashi (Agri. U, Tokyo). She participated in many international conferences and presented her work. Her research interest includes the various aspects of natural products chemistry with special reference to terpenoids, alkaloids, polyphenolics, and reaction mechanism. She published more than 130 research papers; 15 students obtained PhD degree of CU under her guidance. She received Premchand Roychand Scholarship of CU. She is fond of books, friends, music, paintings, museums, art gallery visits, and traveling. She authored two books, one of poems in her mother tongue Bengali (1998) and the other one mostly a collection of articles and poems (2013).

Chapter 1
Introduction: Enzymes. Cofactors/ Coenzymes. Primary and Secondary Metabolites. Natural Products and their Functions. Plant Chemical Ecology. Biosynthesis. Metabolic Pathways

The evolution of natural products (secondary metabolites) in plants, their functions, environmental interactions with living organisms, and some participating components involved in these dynamic biological processes are briefly presented in this chapter.

1.1 Nature. Life. Cells. Molecules. Self-Replication

Nature is the mysteriously integrated network of incredibly simple to immensely complex phenomena that are vastly unknown and very little known and will remain forever in the thoughts and quests of scientists and philosophers as the childhood rhyme, "How I wonder what you are!" [1].

In absence of any definitive evidence, it is believed that ours is the only planet that is blessed with the supreme gift of Mother *Nature*, the life often referred to as carbon–liquid water life [2] enjoying the environmental conditions of this planet. Anything living should grow, respond to external stimuli, be capable of self-replication, and eventually die; and for that, all living organisms need to do an enormous variety of actions. These biological actions include reactions within organisms by way of assimilation, synthesis, transformations, interconversions, etc. involving the participation of simple to complex organic molecules and thus within the cell "... these diverse materials can take part in the amazing dance that we call life" [3]. In biochemical as well as energy sense, a continuous process of bond breaking and bond formation is going on during the biological processes, as long as life is there. And as a consequence, an enormously large number of simultaneous, parallel and recurring chemical events are going on too. Thus, the *fundamental biological processes of life involve* (1) the *biosynthesis* of small molecules and their further conversion to complex molecules (*anabolism*), (2) *self-replication* by which life is sustained through generations, and (3) *catabolism*, i.e., the breaking down of molecules into smaller ones [4]. These reactions proceed with profound stereospecificity and at enormously rapid rates in the cells.

S.K. Talapatra and B. Talapatra, *Chemistry of Plant Natural Products*,
DOI 10.1007/978-3-642-45410-3_1, © Springer-Verlag Berlin Heidelberg 2015

"The cell has been compared to many things, from 'a complex chemical refinery' (by the physicist James Trefil) to 'a vast teeming metropolis' (the biochemist Guy Brown). A cell is both of those things and neither. It is like a refinery in that it is devoted to chemical activity on a grand scale, and like a metropolis in that it is crowded and busy and filled with interactions that seemed confused and random but clearly have some system to them" [5].

1.2   Enzymes [6–16]

1.2.1   Nature of Enzymes

Enzymes are highly evolved biological catalysts that virtually catalyze almost all metabolic reactions with high specificities at ambient temperature at very rapid rates (with few exceptions, e.g., *rubisco,* see Chap. 3). Jack bean urease was the first enzyme to be recognized as a protein, which was crystallized in 1926 by James B. Sumner (NL 1946). It was shown to catalyze the hydrolysis of urea to carbon dioxide and ammonia [9]. Sometimes, they also serve as targets for useful therapeutics (Chap. 5, footnote 1). Enzymes are natural proteins, made up of 20 amino acids {of L-configuration at the α-carbon atom (save achiral glycine) carrying both NH_2 and COOH groups, see Appendix C item (7)}, and *obviously, exist in only one enantiomeric form*; thus, their optical antipodes do not occur in nature. They elaborate their primary structures by way of sequencing the amino acids. Ribonuclease A (having 124 amino acids, molecular mass 13,680 Da) is the first enzyme to be sequenced by Christian B. Anfinsen, Stanford Moore, and William. H. Stein in 1960. They shared the Nobel Prize in 1972 for the methodology they developed for this purpose and for their contribution to the chemistry of enzymes. The other aspects of enzyme structures [6, 7, 16] include the tetrahedral configuration of sp^3 carbons, planar amide (–CO–NH–) bonds, appropriate stereochemically comfortable hydrogen bondings, hydrophobic character, and various elements of protein folding, playing chemical origami. The ability of polypeptide chains to fold into a great variety of topologies, combined with the enormous variety of sequences that can be derived from 19 different L-amino acids and achiral glycine (Appendix C), confers on proteins their prowess of recognition and selectivity of catalysis, and of specificity as structure forming elements. With all the armory currently available in experimental and theoretical methods, we still cannot predict reliably or even comprehend rationally how the sequence of amino acids in a given polypeptide chain would determine and control the ultimate spatial arrangement of the protein.

1.2.2 Functions of Enzymes

Catalysis and specificity are the central tenets of enzymatic reactions, i.e., "stereo-specificity is inherent in the catalytic actions of enzymes, and that enzymes would not be equally efficient as catalysts of nonstereospecific reactions" [7]. Enzymes can have molecular masses of several thousands to several million daltons; still they can catalyze conversions on molecules as small as carbon dioxide and nitrogen. At least 21 different hypotheses for the mechanisms of enzyme-catalyzed reactions have been proposed [9, 10]. It is an accepted proposition that an enzyme recognizes the substrate and catalyzes the reaction by formation of an *enzyme–substrate* (or *E–S*) *complex*. In general, an enzyme accelerates the reaction rate by lowering the transition state energy and thus decreasing the free energy of activation. They provide the reaction surface and suitable environment for bringing the reactants together in a favorable configuration to attain the transition state (TS) and weaken the bonds of the reactants. The loss of entropy on binding the enzyme is compensated by the favorable enthalpic interactions; the substrates come in the productive alignment from the randomly moving individuals; thus, though a considerable part of the entropy of reaction is removed, a favorable situation of intramolecular reaction results [10]. The enzymes may also raise the ground state energy (*cf.* steric acceleration). The structure of the TS of an enzyme catalyzed reaction cannot be detected by any spectroscopic method. The reasons for the specificities of enzymes as well as their profound catalytic power are not fully understood. Enzymes, being composed of only 20 amino acids, pose a limited repository of functional groups. Thus, enzymes by themselves, with their limited built-in resources, are quite often incapable of performing complex biochemical reactions entrusted upon them by the need of the complexity of living organisms.

1.2.3 Enzyme Classification and Nomenclature [12–14]

The name of an enzyme reveals the type of reaction it catalyzes. The names of enzymes usually carry the suffix "ase," while proteolytic enzymes, e.g., pepsin, chymotrypsin, trypsin, etc. are exceptions. Thus, an oxidase enzyme catalyzes an oxidation reaction. It should be noted that enzymes can catalyze the forward and backward reactions of an equilibrium reaction. An oxidase enzyme can catalyze oxidations as well as reductions, depending on the nature of the substrate, i.e., whether it is in the reduced or oxidized state. So, this type of enzymes may be classified as oxidoreductases (Class 1, stated below).

Enzymes are classified according to the general class of the reaction they catalyze and are coded with an EC number [12]. EC stands for Enzyme Commission, a body set up by the International Union of Biochemistry. In the system of nomenclature recommended and accepted, each enzyme is designated by a *four digit number separated by dots*. The *first digit* indicates the main class to which the enzyme belongs, and, *spells out the function the enzyme performs, viz.*

1. *Oxidoreductase* (oxidase/oxygenase/dehydrogenase/reductase) (causing oxidation or reduction)
2. *Transferase* (causing transfer of functional groups like acyl, glycosyl, etc. or one-carbon compound)
3. *Hydrolase* (causing hydrolysis)
4. *Lyase*
5. *Isomerase* (causing isomerization and intramolecular group transfer)
6. *Ligases* (*Synthetases*)

Thus, the names of the classes excepting *lyase* and *ligase* indicate the functions they perform. A *lyase* effects removal of a group (not by hydrolysis) from the substrate leaving a double bond. Conversely, an enzyme which adds a group to a double bond is also a lyase. As stated above, enzymatic reactions are reversed depending upon the substrate, although the same name is given to the enzyme, which sometimes is confusing. A *ligase* catalyzes the joining of two substrates, where energy is provided by the hydrolysis of ATP, etc. (see Chap. 3). *Ligase* is also classified as *synthetase* or *synthase*, causing synthesis.

The second digit constitutes a subclass, since it elaborates the class by specifying the substrate or reaction. The third digit or subsubclass gives further details of the reaction. The fourth digit is a specific number allocated to only one enzyme. Thus, there may be several enzymes with the same first three figures but a different fourth figure. However, these codes for enzymes will be understood only from a Table of the codes of classes, subclasses, and subsubclasses, and hence the codes are not handy. The interested reader may consult reference [13] for the complete rules of enzymatic nomenclature. Generally, the simplified way to classify an enzyme is to *spell out the function indicated by the first figure.* Some other specific functions of enzymes, which may be included in one of the six main classes (first digit) mentioned, are usually expressed by the specific names, e.g., a *cyclase* effecting cyclization, a *kinase* effecting phosphorylation, a *carboxylase* affecting carboxylation, a *transaminase* effecting transfer of amino group, etc. Further, these functional words, which fall in class 1, are tagged with the substrate/product to confer specificity. Some examples of nomenclature of enzymes in vogue are *phenylalanine transaminase, tyrosine aminotransferase, acetylcoenzyme synthetase, hydroxyphenyl-pyruvate reductase, squalene:hopene cyclase, taxadiene synthase, strictosidine synthase, geraniol 10-hydroxylase, secologanin synthase, abietadiene synthase, chorismate lyase, 4-amino-4-deoxychorismate lyase, anthranilate-CoA ligase, etc.* Sometimes, a single enzyme can perform more than one function, e.g., *rubisco* the most abundant enzyme on this planet, functions both as a *carboxylase* and as an *oxygenase* (see Chap. 3).

1.3 Cofactors/Coenzymes

Enzymes are Nature's catalysts. The catalytic activity of many enzymes depends on the presence of small organic molecules, or metals or, metal ions. They are termed **cofactors** (Chap. 3). They are Nature's version of common organic reagents. Their precise role varies with the enzyme. In general, these cofactors are able to perform complex biochemical reactions by activating the enzymes. An enzyme without its cofactor is called *apoenzyme*. The apoenzyme along with the cofactor becomes catalytically active and is called *holoenzyme*. Some coenzymes possess innate catalytic activity.

Cofactors can be divided into two groups: (1) elemental cofactors, e.g., polar metal ions like K^+, Mg^{++}, Ca^{++}, etc., transitional metal ions like Fe^{++}/Fe^{+++}, Ni^{++}, Cu^+/Cu^{++}, Zn^{++}, and metals like Mn, Co, Se, Mo, etc. and (2) small organic molecules called **coenzymes** (mostly derived from vitamins), e.g., *coenzyme A* (CoA), *nicotinamide adenine dinucleotide* (NAD^+) and *its hydride* (NADH), *nicotinamide adenine dinucleotide phosphate* ($NADP^+$) and *its hydride* (NADPH), *flavin adenine dinucleotide* (FAD+), *thiamine pyrophosphate(diphosphate)* (TPP/TDP), *pyridoxal phosphate* (PLP), *adenosine triphosphate* (ATP), *biotin, tetrahydrofolate*, etc. A tightly bound (covalently bonded) coenzyme is called a *prosthetic* group, while a loosely associated (noncovalently bonded) coenzyme behaves like a *cosubstrate*; like substrates and products, they bind to the enzyme to come to a compatible conformation and are finally released from it. Enzymes that use the same coenzyme usually execute catalysis by similar mechanism.

Arthur Harden (1865–1940, an English biochemist, shared 1929 Nobel Prize with Hans August Simon von Euler-Chelpin of Germany) first recognized the existence of coenzymes and their assistance to coenzyme-dependent enzymes [16, 17]. Statistics suggests that the coenzyme-dependent enzymes are more in number than the enzymes capable of doing work by themselves. The involvement of some coenzymes in various enzymatic reactions has been discussed in Chap. 3 and other chapters in appropriate cases.

In summary, we quote Birch [15], "Enzymes selectively assemble molecules (often chemically activated by combination with coenzymes), arrange them structurally, and promote, sterically and chirally, bond-forming processes within an asymmetric molecularly engineered cavity." In essence, enzymes catalyze reactions that cells need to live and replicate. *In the Nature's laboratory and in a chemical laboratory, the same chemistry exists. The difference is that Nature is superb and unparallel at chemistry (e.g., biosynthesis), and chemists are only learning little bit and trying to emulate to some extent even at this 21st century. The name enzyme was given by Kuhne in 1878/1881 to some active components present in yeast which caused fermentation. The two words **in yeast** are expressed in **Greek as en(in) and zyme (yeast)** and hence the name.*

1.4   Metabolism. The Vital Biological Processes. Primary Metabolites

Each organism is carrying out within itself an integrated network of enzyme-catalyzed cellular reactions, sum total of which is regarded as *metabolism*. It includes anabolism, reactions concerned with an increase in molecular complexity, and catabolism, a reverse process, i.e., reactions concerned with molecular degradation. However, the extent of such network varies from one organism to another because of the differences in the diversifications and individual characteristics. The paths of these reactions constitute *metabolic pathways.* But, whatever may be the diversification—the mode of formation of the most important and basic chemical ingredients of life, *viz.*, proteins, fats, nucleic acids, carbohydrates, etc. follows more or less the paths where a fundamental commonality is observed. They are formed from a small number of fundamental units (*building blocks*), which, in turn, are formed by *photosynthesis*, directly or indirectly.

Of all living organisms, plants are most efficient in elaborating organic compounds with superb molecular diversity, initiated by the pivotal natural synthetic tool *photosynthesis* (Sect. 3.1), by which light energy of the sun is utilized by the green plants to produce organic compounds from carbon dioxide in presence of water. The initial products of photosynthesis are carbohydrates which upon subsequent metabolic alterations give rise to a series of simple and universally distributed organic compounds of low molecular weights, like common sugars and sugar derivatives, the 20 or so amino acids that make the majority of the proteins, and the low molecular weight aliphatic carboxylic acids which form common fats and lipids. These small molecules are needed for sustaining life and are present in almost all living systems, and hence they are known as *primary metabolites*. In plants they all have the same metabolic functions. The pathways involved in their formations and transformations during anabolism and catabolism are described as *primary metabolic pathways* and constitute the *primary metabolism* of plants. Unlike the plants, the animals and the microorganisms derive these raw materials from their food and diet, and hence animals develop mechanisms for locomotion and perception of external objects.

1.5   Metabolism. Secondary Metabolites (Natural Products)

There is also another area of metabolism that falls outside the domain of *primary metabolism* where simple *primary metabolites* enter into further reactions to form more complex molecules. *These reactions are genetically controlled and enzymatically catalyzed.* The resulting products are of *limited or restricted distribution, i.e., unlike primary metabolites* they are found to occur in certain groups of organisms or plant genera and species only. These compounds are classified as *secondary metabolites* and the pathways involved in their formation fall in the domain of *secondary metabolism.*

> "The term 'secondary metabolites' was coined by the biochemist Albrecht Kossel in 1891 to characterize cell components that contrast with 'primary metabolites', and are not found in any developing cell. The former term was adopted by Friedrich Czapek in his book *Biochemistry of Plants* and has been used ever since." [18]

Incidentally, it may be mentioned that Christophersen [19] commented, "An exact definition of primary and secondary metabolites is not very useful since it is deemed to be artificial. ...From a purely formal point of view this distinction is meaningless and best abandoned."

In view of the above comment, the present authors opine that such definition as well as classification helps in harmonizing the chemical chaos of Nature. It serves as the useful working proposition for understanding the purpose of primary and secondary metabolisms and evokes serious interest to unveil the cryptic functions of secondary metabolites, which are distinctly different from those of primary metabolites.

1.5.1   Natural Products

The secondary metabolites are termed by the chemists as *"natural products"* to distinguish them from those compounds which traditionally have been the province of biochemists. They are unique in the sense of being of restricted occurrence and are more characteristic of specific botanical sources (i.e., a family or a genus and sometimes a single species of plants). Most of the natural products, usually of relatively complex structures, possess biological activities. Many of them find their way to the parlor of pharmaceuticals (Chap. 33) and also are used as insecticides, pesticides, plant growth hormones, etc. Their interesting structural variations, stereochemical configurations, conformations, molecular shapes, etc. claim the academic attention of chemists all over the globe.

1.5.2   Biomacromolecules

It should be pointed out that *plants elaborate the vital life-essential biopolymers like proteins, nucleic acids, polymeric carbohydrates, lipids,* etc. Though *like primary metabolites* they occur in all living organisms, but unlike the same they possess complex structures—a characteristic of many *secondary metabolites.* But *unlike secondary metabolites* they are not of restricted occurrence, and the purpose of their formation by the organisms is well understood. Hence, they fall outside the conservative definitions of both primary and secondary metabolism. They have their own domain—a defused overlapping area of metabolism that cannot be associated with either primary or secondary metabolites, as long as the conservative definition of the latter is in vogue.

However, all metabolic products may be looked upon as biological entities whose origin, function, transformation, and fate constitute a vast unlimited area of research toward our understanding of their purpose in the living systems.

1.6 Functions of the Natural Products: Chemical Ecology—Plant Chemical Ecology

The functions of the secondary metabolites in the organisms producing them are yet to be deciphered in most cases. However, in some cases, the functions have been revealed through observations and experiments.

1.6.1 Chemical Ecology

Chemical ecology means chemically mediated interactions between organisms and their living (biotic) and nonliving (abiotic) environments [18]. It includes a broad range of *chemical interactions/communications and signaling processes as follows:*

(a) Chemical defenses of organisms, plant defenses against herbivores
(b) Chemical communication with insects and plant–insect interactions (*phero-mones, ecdysones, and pollination*)
(c) Mutualistic interactions of plants and fungi (*endophytes*)
(d) Plant–plant interactions (*allelopathy*)
(e) Plant–microorganism interactions (*phytoalexins*)
(f) Natural products and human being

Ecological factors or geographical parameters (e.g., altitude, season, climate, etc.) influence the nature, yields, and the chemical ecology of the secondary metabolites (Sect. 4.1.2). The above interactions involving essential ecological role of natural products may be regarded as *chemical ecology of natural products or plant chemical ecology.* Such interactions will be briefly discussed with a few illustrations. *However, this classification is not mutually exclusive; there is some overlapping.* The pioneering work of A. Kerner (1831–1898), L. Errera (1858–1906), and E. Stahl (1848–1919) on the ecological role of plant secondary metabolism in interactions between plants and a mostly hostile environment finally seems to be broadly accepted [18]. These interactions reveal enormously amazing facts. The chemical ecological research is now regarded as an important area of natural products research, since it is possible at present to isolate and do the structures of the responsible secondary metabolites even in microgram scales using the available sophisticated instruments. The intensive works of Schildknecht, Meinwald, Rembold, Towers, and others have significantly enriched the chemical ecology area.

1.6.2   Chemical Defenses of Organisms. Plant Defenses Against Herbivores

Plants elaborate natural products for their own benefit. Kerner emphasized the importance of chemical defenses for plants' survival [18].

1.6.2.1   Antifeedants. Repelling Insects

The bitter principles produced in some Meliaceae plants (e.g., *Melia azadirachta, neem*/*nim*) might repel the insects and protect the plants. Some toxic materials produced in certain organisms prevent the attack on them by *predators*. Many secondary metabolites act as *antifeedants to insects*, i.e., they inhibit feeding but do not kill the insects directly and the insects perhaps die of starvation [20]. Azadirachtin, a complex terpenoid constituent of *Azadirachta indica* (syn. *Melia azadirachta*) (neem/*nim* tree) (Fam. Meliaceae), and polygodial (*Polygonum hydropiper*) (Fam. Polygonaceae) (Fig. 1.1) are perhaps *two most promising antifeedants of* plant origin, known till date [20]. In case of polygodial, the enedial functionality probably blocks the insect chemoreceptor by reacting with an amino group to form a pyrrole derivative. The homodrimanes (**I**) and (**II**), synthesized from the diterpene (−)-sclareol, showed significantly more antifeedant activity, perhaps since they may interact with an amino group of the insect chemoreceptor more strongly yielding pyridine derivatives [21]. Interestingly, it has been mentioned that if the predator eats the opium plant/exudate, it will soon feel drowsy and will not be able to do any further harm to the plant.

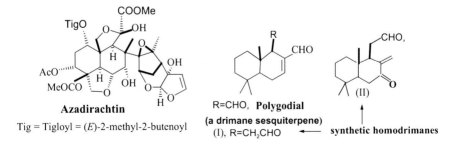

Fig. 1.1 Antifeedants azadirachtin and polygodial, and two synthetic homodrimanes

1.6.2.2 Wounded Plants Emitting Prussic Acid

To protect themselves from herbivores many plants, when wounded by animals emit defense chemicals. Almost 10 % of all plants use *prussic acid* (HCN), a very potent poison, as defense against being eaten by animals. It is an inhibitor of cytochrome *c* oxidase affecting the final step in the respiratory electron transport chain and is present widely in plants in the bound form as *cyanogenic glycosides* [22, 23]. *When wounded, the compartmentalization of the glycoside and the enzyme glycosidase gets disturbed and their contact results in the hydrolysis of the glycoside liberating prussic acid* [22, 23] (Fig. 1.2). Many cyanogenic glycosides have been isolated from several flora, most of them are O–β-D-glucosides. The old name of hydrocyanic acid is prussic acid since it was prepared by Scheele in 1782 by heating Prussian blue(ferric ferrocyanide) with sulfuric acid.

1.6.2.3 Wounded Plants Emitting Volatile Mustard Oils

Likewise, an isothiocyanate, R–N=C=S, poisonous at high concentration, is generated when radish, cabbage, and mustard plants, containing a thioglycoside, *glycosinolate,* also named *mustard oil glycoside,* are wounded; then the thioglycoside comes in contact with the enzyme *thioglycosidase,* located in separate compartments of the plant tissues and gets hydrolyzed and rearranged to form the isothiocyanate [22, 23] (Fig. 1.3). When cells of these plants get damaged, the pungent smell of mustard oil is detected indicating the formation of the isothiocyanate. Several dozens of glycosinolates are known to occur in Capparaceae and Cruciferae families. Most of them are S–β-D-glucosides.

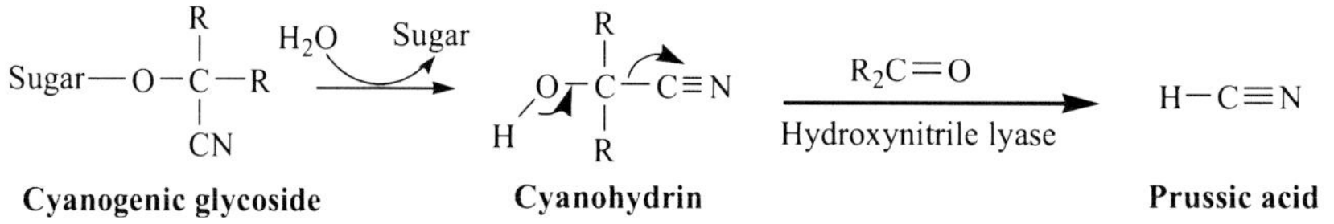

Fig. 1.2 Enzymatic hydrolysis of cyanogenic glycoside to prussic acid (R=Me); two substituents may be different, e.g., Me, Et; Ph, H (both enantiomers); *p*-hydroxyphenyl, H (both enantiomers). Most of them are O–β-D-glucosides

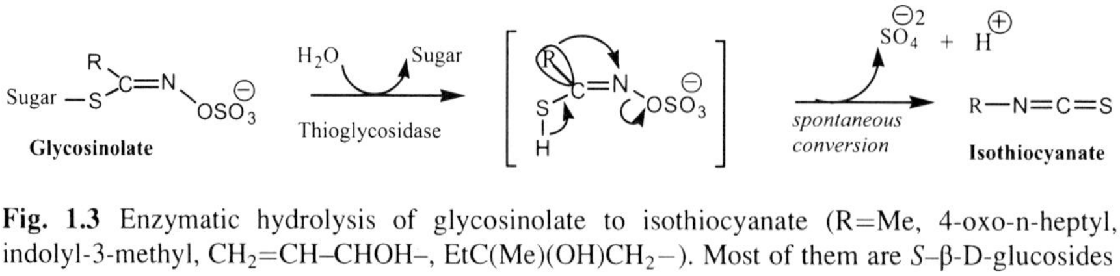

Fig. 1.3 Enzymatic hydrolysis of glycosinolate to isothiocyanate (R=Me, 4-oxo-n-heptyl, indolyl-3-methyl, CH_2=CH–CHOH–, EtC(Me)(OH)CH_2–). Most of them are S–β-D-glucosides

1.6.2.4   Plants Toxic to Animals

A number of secondary metabolites also provide plants protective means from herbivores. Ruminants (e.g., cows, buffaloes, and other animals living on plants) generally do not browse plants containing alkaloids. For example, pyrrolizidine alkaloids present in some Leguminosae (*Crotalaria* species) and Boraginaceae (*Heliotropium* species) plants cause severe liver damage. They are hepatotoxic. They have attracted a lot of attention because of the heavy loss of livestock in many countries. Even human food (cereal grains) contaminated with the seeds of these plants, when consumed, may cause food poisoning and liver damage.

Molded (fermented) sweet clover (*Melilotus officinalis*, Leguminosae) contains the toxic principle dicoumarol (**1**) (a bis-4-hydroxycoumarin derivative) (Fig. 1.4) with pronounced blood anticoagulant properties which can cause the death of livestock (e.g., cows) by severe internal bleeding (hemorrhage). Dicoumarol interferes with the effects of vitamin K in blood coagulation. This observation led to the synthesis of some medicinal anticoagulants, having 3-substituted-4-hydroxycoumarin structures, e.g., warfarin (Sect. 13.2, Chap. 33) used in thrombosis. Kerner realized that plants toxic for some animals may not be toxic for others, e.g., *Belladonna* berries are toxic to ruminant animals but harmless to many birds [18].

1.6.2.5   Plants Deceiving Herbivores with False Amino Acids

Many plants elaborate unusual amino acids, e.g., (*S*)-L-canavanine (**2**) (Fig. 1.4) (occurring in *Canavalia ensiformis, jack bean*) which is a structural analogue of (*S*)-L-arginine (**3**), a protein building amino acid. Herbivores consume canavanine from their food. This false amino acid cannot be distinguished by the arginine-transfer RNAs of those herbivores during protein biosynthesis. The wrong incorporation changes the desired three-dimensional structure of proteins

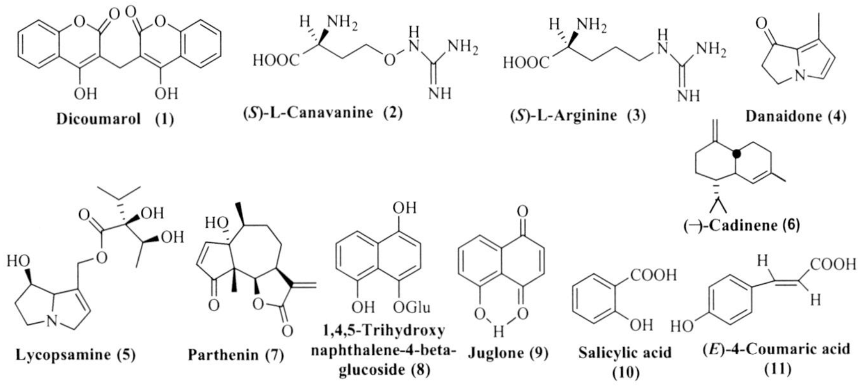

Fig. 1.4   Some natural products involved in plant chemical ecology

which lose their biological function partially or completely, and hence become toxic to those herbivores. The arginine-transfer RNAs do not react with canavanine in plants which synthesize it, so it is nontoxic for such plants. Plants contain many false amino acids, which are toxic for herbivores in an analogous manner.

1.6.3   Chemical Communication with Insects and Plant–Insect interactions

1.6.3.1   Formation of Courtship Pheromones

Chemical interactions between plants and insects constitute a type of signal exchange. Animals or insects of the same species quite often release some second-ary metabolites or chemicals, broadly known as *pheromones*, which serve to influence the physiology and behavior of other members of the same species so as to attract them and effect courtship. While feeding on plants some insects receive the precursor chemicals of their courtship pheromones. For an example, pyrrolizidine alkaloids occurring in the *Crotalaria* (Leguminosae) and *Heliotropium* (Boraginaceae) plants, though poisonous to livestock as stated above (see Sect. 1.6.1.4), are useful as precursor pheromones to some insects such as danaid butterflies and arctiid moths. The dual role of these plant alkaloids as defensive chemicals and as the precursors of male courtship pheromones in danaid butterflies and arctiid moths have been thoroughly studied by Meinwald [24–27] and coworkers during several decades. They isolated a heterocyclic ketone danaidone (**4**) (Fig. 1.3) from the hairpencils' secretion of Trinidad butterfly, *Lycorea ceres*. The ketone is the courtship pheromone in the Florida queen butterfly (*Danaus berenice*). It was found to be *biosynthesized* in African monarch, *Danaus chrysippus* from the pyrrolizidine alkaloid lycopsamine (**5**), delivered to their system from the plant food *Heliotropium steudneri*. This observation establishes a deep relationship between arctiid moth and its *Crotalaria* plant food elaborating the pyrrolizidine alkaloid that the moth thus utilizes on becoming an adult butterfly in making the courtship pheromone.

1.6.3.2   Formation of Ecdysones (Molting Hormones)

Insects require **ecdysones** *having molting hormone properties for their larval development and metamorphosis.* Biosynthetically, ecdysones are derived from steroids (Sects. 11.2 and 11.3). However, insects are unable to synthesize steroids. They use the plant food as the source of ecdysones. Since ecdysones are not widely elaborated by plants, they use more abundantly occurring plant steroids (e.g., sitosterol, stigmasterol, etc.) as well as less abundant cholesterol from plant food as the precursors of ecdysones biosynthesis. The structural features 14a-OH, *a*,β-unsaturated ketone system and *cis*-A/B rings are responsible for their bioactivity. The involvement of secondary metabolites in plant–insect interactions, their utilization as signaling agents, the biosynthetic conversion of a plant alkaloid to an insect sex pheromone, and of plant steroids to molting hormones and various other interesting findings are significant extensions of natural products chemistry research.

1.6.3.3 Pollination

The beautiful colors of the flowers, caused by the presence of flavonoids, anthocyanins, etc. (Chap. 14) and other pigments *and their smells* (due to monoterpenes like geraniol, linalool, β-ionone, etc.; aromatic compounds like vanillin, eugenol, etc.; aliphatic compounds like isooctanol; various amines; etc.) and nutritious exudates (e.g., honey) might attract the insects like bees, butterflies, flies, etc. to help in the *pollination*—the first step in the fertilization in flowering plants. This step is performed unknowingly by all these visitors (including birds and bats) collecting male gametes from the flowers while in search for food in nectar and pollens. Different pollinators are attracted to different colors, e.g., butterflies like bright colors, while flies prefer green and brown colors. These colors are due to various plant pigments belonging to the classes of secondary metabolites, which include quinones, chalcones, flavonoids, anthocyanins, carotenoids, etc., to be discussed in appropriate chapters. Pollinators visit the same plant species as they are capable of distinguishing the species.

Interestingly, it may be mentioned that the fragrance of *Ophrys* flowers (orchids) resembles the odor coming from the odor glands of *Andrena* female bees. Further, one of the components of the flower fragrance, identified as (−)-cadinene (**11**) (Fig. 1.4), has been the sex pheromone of the female bees. The deceived *Andrena* male bees are attracted, and this attraction results in the pollination.

1.6.3.4 Plant–Insect Interactions: Another Example

Insects are sometimes attracted by the nutritious exudates of some plants. Many ants use the nutritious exudates of some *Acacia* species as their food and in turn protect the host plants from herbivores by forming ant colonies on them. If the ant colonies are removed from the plants, the herbivores damage the plants. This proves the mutualistic beneficial relationship between insects and plants. Some *Acacia* species, where ants do not form colonies to protect the plants from herbivores, produce cyanogenic glycosides, toxic to herbivores (see Sect. 1.6.1.2).

Both pollination (Sect. 1.6.2.3) and plant–insect interaction (Sect. 1.6.2.4) stated above may be grouped under mutualistic interactions of plants and insects, since they both are benefitted. In case of the former, propagation of plants is caused while the insects, etc., are getting their food.

1.6.4   Mutualistic Interactions of Plants and Fungi: Endophytes [28]

1.6.4.1   Endophyte Fungus and Host Plant

Endophyte refers to a situation when one organism lives inside another organism. Endophyte fungus is a beneficial fungus and holds a symbiotic relation with the host plant in the cells of which the former grows. When it feeds on the host plant, it produces toxic chemicals (e.g., alkaloids) that protect the host plants from environmental influences (e.g., pests, diseases, etc.). Thus, in this phenomenon, the endophyte fungus can grow and at the same time the host plant gets a natural protection. Hence, a mutualistic beneficial relationship exists.

An example: the turf grass has a built-in protection provided by some specific endophytes, present in high percentage in the seeds. It protects the turf grass from the pests eating the surface of the leaves, and sometimes enhances drought tolerance; chemical pesticides are not needed to be added. Endophytes are transferred from plant to plant through seeds. Many perennial rye grass and tall fescue varieties (*Festuca* genus) contain endophytes. There are specific endophytes that can be used for forage grasses, which produce alkaloids toxic to pests and not to livestock.

An important as well as interesting study revealed that the endophytic novel fungus *Taxomyces andreanae* isolated from the inner bark of a yew tree (*Taxus brevifolia*) growing in northwestern Montana elaborates, though in small amount, the same compound taxol as its host plant [29]. Taxol is an extremely important drug for cancer. Hopefully, microbial synthesis of taxol [30] will help in procuring the drug (Sect. 8.4). Likewise, many endophytic fungi have been shown to produce a number of interesting biologically active natural products like their host plants. Endophytic fungus can thus potentially serve as an alternative source for the bioactive natural product/s.

1.6.5   Plant–Plant Interactions: Allelopathy [31]

The chemical communicative phenomenon between some plants where their chemical constituent/s (natural product/s) disallow or inhibit the growth of other plants in their vicinities is termed as allelopathy. The phenomenon takes place through the soil. *Parthenium hysteroporus* (Asteraceae), a troublesome weed of India and notorious for its virulent growth, contains a sesquiterpene lactone, parthenin (**7**) (Fig. 1.4), identified as a growth inhibitor for other plants. The leaves of the walnut tree, *Juglans nigra* (Juglandaceae), elaborate juglone (**9**) (=5-hydroxynaphthaquinone) and 1,4,5-trihydroxynaphthalene-4-glucoside (**8**). The latter being a glucoside is washed down by rainwater and dew. On coming in contact with the microorganisms in the wet soil, it gets hydrolyzed to the aglycone which is readily oxidized to juglone (**9**). Juglone inhibits the growth of other plants under this walnut tree. However, some species like Kentucky blue grass, *Poa pratensis* can grow under a walnut tree, showing that such plants somehow neutralize the adverse effect of juglone and can compete for soil nutrients. Some allelopathic agents (allomones) are produced by some plants, e.g., salicylic acid (**10**) by the oak tree, *Quercus falcate*; *E*-4-coumaric acid (**11**) (*E*-4-hydroxycinnamic acid) and some quinones by some plants including the shrubs *Adenostoma fasciculatum and Arctostaphylos glandulosa,* native to southwest California where they inhibit the growth of other shrubs in the neighborhood.

Incidentally, it may be mentioned that parthenins/parthenolides possess fairly good insecticidal properties. They affect the nervous system of the insect. Flowers containing parthenins as the chemical constituents [*Chrysanthemum cinerariaefolium* (Asteraceae)] are processed as insecticidal spray for domestic and agricultural purposes. Parthenins are biodegradable and nontoxic to mammals but toxic to fish and amphibians. Various synthetic parthenins/parthenolides with improved insecticidal properties and decreased toxicity are in use.

The undergrowth in groves of different trees generating allomones varies considerably. The bare belt area around such plants thus differs. *Salvia leucophylla* (mint) and *Artemisia californica* thickets do not allow other herbs to grow in their vicinity. The allomones from these trees have been identified as volatile monoterpenes like camphor and cineol. When these thickets are removed by way of burning, grass and annual herbs start growing on the bare patches. Allelopathic effect is caused by E-cinnamic acid or its derivatives. Allelopathic effect has been studied in vitro with some compounds, which inhibit the growth of some seeds.

1.6.6   Plant–Microorganism Interactions: Phytoalexins [32, 33]

The loss of crops by microbial attack is a global problem and thus a matter of great concern. Much needed intense research in this area revealed that the microorganisms—the pathogens—are host specific; they multiply in the plant tissues and produce toxic metabolism by generating compounds termed as *Elicitors. The latter* activate the genes in the plant so as to produce chemicals which are capable of detoxifying the microbial-generated toxins by way of oxidation, glycosylation, or counteractingly by the host–plant-generated defense chemicals. The defense chemicals becomes of maximum concentration immediately after the microbial attack on the host plants. In course of studies on plants exposed to microorganisms and various infected tubers, hundreds of defense chemicals of diverse skeletal patterns have been discovered. These defense chemicals are termed as phytoalexins, e.g., Orchidaceae plants elaborate phenanthrene derivatives, while Compositae plants elaborate acetylenes as phytoalexins.

Plants and Fungi/Bacteria/Pathogens Secondary metabolites often protect plants from pathogenic microorganisms. The latter are provided by fungi and or bacteria while they are using the plant sources for their nutritional purposes. Varied classes of secondary metabolites comprising *isoprenoids* (Chaps. 6–12), *phenylpropanoids* (Chap. 13), *polyketides* (Chap. 14), and *alkaloids* (Chaps. 15–30) include natural pesticides that protect plants against herbivores and pathogenic microorganisms.

For more information on different aspects of chemical ecology, as discussed in Sect. 1.6, as well as sequestration of natural products by insects, etc. two excellent reviews by J. J. Harborne [32, 33] may be consulted.

1.6.7 Natural Products and Human Being

Many natural products, being remarkably bioactive, serve as the cradle for pharmaceuticals. A number of molecules not only saved many lives in the past but also continue to do so even after the advent of modern day pharmaceutical sciences consisting of various scientific disciplines. Their activities have largely been studied in relation to human benefit or compatibility resulting into the discovery of many important drugs, namely morphine, quinine, reserpine, vincristine, vinblastine, taxol, artemisinin, dicoumarol, camptothecin, etc. (Chap. 33)—to name only a few well-known ones, and various crude drugs as well. However, the bioactivity of most natural products still remains unknown. The occurrence of many natural products in fruits, vegetables, and spices are responsible for our liking the odors and tastes and finally for our health benefit (Chap. 34).

Some poisonous natural products like curare alkaloids (quaternary bisbenzylisoquinoline alkaloids) of calabash-curare [22, 23] are strong dart poisons, causing the voluntary muscle relaxation (paralysis) to a great extent. It was used in hunting in South America, particularly in Amazon and Orinoco basins. Taking advantage of the muscle relaxation property of the curare alkaloids, by the judicial use of curare preparations along with light anesthesia, some surgical operations could be performed without recourse to deep anesthesia.

Some plants (leaves), exudates, fruits, etc. elaborate natural products that cause addiction when taken and immensely affect the human behavior. Cannabinoids (*charas*), morphine, acetyl morphine, cocaine, and nicotine (tobacco leaves) are a few prominent examples of such types of natural products.

The beautiful *colors* of the flowers, due to the presence of flavonoids, anthocyanins, etc. (Chap. 14) and other pigments are visual delight to human being from time immemorial. The *fragrance* of flowers and natural perfumes (prepared from plant sources) satisfy our olfactory nerves to a great extent. However, pollen volatiles of sweet-scented flowers often act as vital aeroallergens (generally allergens are mostly proteins) and cause seasonal fever to susceptible people, and bring to them a lot of physical miseries.

These are few of the many relations of natural products with human being, and the relationship or interactions are directly understood.

However, in defining the role of natural products in the internal economy of the organisms producing them, William et al. [34] suggested that biosynthesis of natural products is programmed by many kilobytes of DNA, and the energy expenditure involved is allowed, since they serve to improve the fitness of the natural product forming organisms by acting at specific receptors in the competing organisms. This is an important aspect of natural product formation under the pressure of the survival; otherwise, Darwin's natural selection would have precluded this energy expenditure, if necessary.

In this context, we may quote [35] Prelog (NL 1975), "Every natural product, which has appeared sometime during the three billion years of evolution and survived, carries in its structure a message and many of them have a yet unrevealed function. To decipher the message and to find out their function will remain for a long time one of the most challenging tasks of chemistry."

1.7 Biosynthesis: Studies with Isotopically Labeled Precursors[1] [36–38]

With the advent of isotope chemistry and advanced technology, it has been possible to synthesize molecules by putting isotopic (mostly ^{14}C, ^{13}C, ^{15}N, ^{2}H, or ^{3}H) labels into the assumed precursors. In the feeding experiments, the labeled precursors are administered into the plants. After a suitable period of growth, the expected incorporation of the precursors with radiolabels could be detected by trapping the intermediate/s or by degradation of the product/s, which appear with isotopic signature. In cases of ^{2}H-labeled precursors mass shifts in the products are observed in mass spectroscopy.

In the last few decades of the twentieth century, ^{13}C NMR spectroscopy has become an important tool for biosynthetic studies of natural products, as evident from the accumulation of a large body of data in the literature. ^{13}C Rich precursor is given in an adequate culture media during a biosynthetic study. The isolation of the specifically ^{13}C-labeled product/s showing larger intensity for some carbons in its ^{13}C NMR spectra will allow the identification of the site of incorporation, and hence help in elucidating the biosynthetic pathway of the compound under study. Generally, enrichment of 0.5 % ^{13}C above the natural abundance (1.1 %) is sufficient to locate the labeled site/s [39]. The biosynthesis of camptothecin [40] demonstrated

---

[1] Since the discovery of carbon14 by Martin Kamen and his fellow chemist Sam Ruben in 1940 while working at E.O. Lawrence's famous radiation laboratory at the UC, Berkeley, this radioactive isotope has been used as a very important and useful tracer in the study of various biological pathways including photosynthesis. It has profoundly influenced the studies of biochemistry and biology. Using $^{14}CO_2$ Melvin Calvin deciphered the photosynthetic pathway and received the Nobel Prize in 1961. William Libby also used ^{14}C decay in radioactive dating technique and received the Nobel Prize in 1960.

for the first time the suitability of ^{13}C NMR spectroscopy for biosynthetic studies in higher plants.

Thus, labeled compounds used as probable precursors help to map the various biological events with stereochemical implications involved in the biosynthesis of natural products. This provides an insight regarding the synthetic strategies of *Nature*. However, one has to be very careful about the biosynthetic experiments because sometimes the labeled precursor gets degraded prior to its incorporation, and also measurements of the various parameters are to be done with much repetitive care. The results are to be carefully interpreted to avoid inadvertent misleading conclusions. The negative results should also be rationalized with caution, keeping in mind that the experiment might not have been done at the opportune time for the natural biosynthesis of the concerned compound.

Despite the unbelievable diversity occurring in natural products, the plants utilize only a few amazingly simple molecules [e.g., acetate, C_5-units (isoprenoids), and amino acids (phenylalanine, tyrosine, ornithine, etc.)] as the building blocks. More amazingly, plants use very simple reactions at ambient temperature revolving around carbonyl chemistry, cation-olefin addition/cyclization, 1,2-alkyl or H migration, Michael addition, Diels–Alder reaction, Mannich condensation and a few other reactions to create carbon–carbon bonds during the construction of organic molecules of all conceivable structural patterns.

1.8   Metabolic Pathways: Mevalonic Acid Pathway, 1-Deoxy-D-xylulose Phosphate Pathway, Shikimic Acid Pathway, and Polyketide Pathway

Metabolic pathways provide definite sequence of biological reactions and elaborate discrete and controlled steps. Some of the identified well-known metabolic pathways are mevalonic acid (MVA), 1-deoxy-D-xylulose phosphate (DXP), shikimic acid, and polyketide pathways. It is helpful to classify the compounds based on their precursors from which the natural products are formed, as well as based on the metabolic pathways. Identification of the metabolic pathways brings a harmony in the chaotic chemical richness of *Nature*.

Mevalonic acid pathway and deoxy-D-xylulose phosphate pathway are responsible for the biosynthesis of terpenoids, steroids, and related natural molecules in the plant cells, while shikimic acid pathway forks out at the point of formation of chorismic acid, and the latter then participates in different pathways leading to the formations of natural products with structural carbon contents C_6–C_3, C_6–C_3–C_6, C_6–C_3–C_3–C_6, etc., and various aromatic amino acids and related products. The polyketide pathway allows the formation of simple to complex phenolic compounds (Table 1.1). There are many compounds which are of mixed biogenetic origin, i.e., certain fragment of the molecule originates following a route, while the

Table 1.1   Metabolic pathways and natural products[a]

Metabolic pathways	Class of compounds
Mevalonic acid (MVA)/deoxy-D-xylulose P (DXP)	Terpenoids, steroids, carotenoids, etc. and also the prenyl (C_5, C_{10}, C_{15}, C_{20}, etc.) units as such or in their modified forms as part/s of other compounds of different biogenetic origin, e.g., monoterpene indole alkaloids and other strictosidine-derived alkaloids, etc.
Shikimic acid (via chorismic acid)	Coumarins, lignans, etc., aromatic amino acids, aromatic phenolic acids, etc. ↓ Different classes of alkaloids
Polyketide	Aromatic phenolic compounds, polyphenolic compounds, flourenones, phenanthraquinones, antibiotics (mostly in microorganisms), etc.
Shikimic acid + polyketide	C_6–C_3–C_6 Compounds which include flavonoids, isoflavonoids, anthocyanins, kava pyrones, etc.
Shikimic acid + MVA/DXP	Coumarins with prenyl (C_5, C_{10}, C_{15}, C_{20}, etc.) units or, their modified forms
Shikimic acid + polyketide + MVA/DXP	Flavonoids, isoflavonoids, and anthocyanins carrying prenyl (C_5, C_{10}, C_{15}, C_{20}, etc.) units or, their modified forms

[a]Many other natural products fall outside these classifications and will not be discussed

other part is formed from another route, and then they get united to form the natural products, e.g., flavonoids having carbon content C_6–C_3–C_6 are formed via polyketide and shikimic acid pathways: C_6 (from polyketide) and C_3–C_6 (cinnamic acid equivalent) (from shikimic acid) (Table 1.1). *The above pathways will be dealt in fairly detail with the stereochemical implications while discussing the chemistry of the compounds derived from these pathways.*

The L-amino acids serve as the source of the nitrogen atom in the molecules of a major class of natural products called *alkaloids*. In fact, in most cases the amino acids as such get incorporated in the alkaloidal framework (decarboxylation takes place at some stage of biosynthesis). Rarely nitrogen is delivered to the alkaloidal skeleton by way of direct amination as in case of coniine (Chap. 17). The associated pathways and the derived natural products are summarized in Table 1.1 in a very general fashion.

Biosynthesis of the simple molecules like 3-phosphoglyceraldehyde, 3-phosphoglycerate, erythrose phosphate, ribose 1,5-diphosphate, fructose 1,6-diphosphate, pyruvic acid, phosphoenol pyruvate, acetone dicarboxylic acid, dihydroxyacetone diphosphate, etc. (Calvin cycle) (fixation products of C_3-plants), and of the C_4-acids like aspartate and malate (the primary fixation products in cases of C_4-plants) will be discussed briefly in Chap. 3. These are the *key primary metabolites,* which enter into the formation of natural products.

References

1. Jane Taylor (1783-1827), "The Star" (a poem).
2. Barry W. Jones, *Life in the Solar System and Beyond,* Springer-Praxis, **2000**, p. 29.
3. Bill Bryson, *A Short History of Nearly Everything,* Black Swan, **2004**, p. 353.
4. Barry W. Jones, *Life in the Solar System and Beyond,* Springer-Praxis, **2000**, p.34.
5. Bill Bryson, *A Short History of Nearly Everything,* Black Swan, **2004**, p. 456.
6. Linus Pauling, *The Nature of the Chemical Bond,* Cornell University Press, Ithaca, New York, **1939**, 3rd ed. **1960**, Oxford and IBH Publishing Co., Calcutta, Indian ed. **1967**.
7. Alan Fersht, *Enzyme Structure and Mechanisms,* 2nd ed., W. H. Freeman and Co., New York, **1985**.
8. J. W. Cornforth, Enzymes and Stereochemistry, *Tetrahedron,* **1974**, *30*, 1515-1524.
9. Richard B. Silverman, *The Organic Chemistry of Drug Design and Drug Action*, 2nd Edn., Elsevier, Amsterdam, **2004**, pp. 173-225.
10. M I. Page, in *Enzyme Mechanisms*, (Eds.: M. I. Page and A. Williams), Royal Society of Chemistry, London, **1987**, p.1.
11. Ronald Kluger, The Mechanistic Basis of Enzyme Chemistry in *Enzyme Chemistry: Impact and Applications,* Ed. Colin J . Suckling Chapman and Hall, London, New York, **1984**, pp. 8-31.
12. G. Brown, *An Introduction to Biochemistry,* The Royal Institute of Chemistry, **1971**; pertinent pages 107-109, Appendix 1.Enzyme Classification and Nomenclature.
13. *Enzyme Nomenclature, Recommendation 1964 of the International Union of Biochemistry*, Amsterdam: Elsevier, **1965**.
14. Jeremy M. Berg, John L. Tymoczko, and Lubert Stryer, *Biochemistry*, W. H. Freeman & Co., New York, Sixth Ed. **2007**, pp 236-237.
15. Arthur J. Birch, *To See the Obvious* (Profiles, Pathways, and Dreams, Autobiographies of Eminent Chemists), (Series Ed.: Jeffrey I. Seeman), Am. Chem. Soc, Washington DC, **1995**, pp. 177-178 .
16. Trimothy D. H. Bugg, The Development of Mechanistic Enzymology in the 20th Century, *Nat. Prod. Rep.*, **2001**, *18*, 465-493.
17. William D. McElroy, *Cellular Physiology and Biochemistry,* Prentice Hall of India, New Delhi, **1963**, p. 36.
18. Thomas Hartmann, The Lost Origin of Chemical Ecology in the Late 19th Century, *Proc. Natl. Acad. Sci.,* USA, **2008**, *105*, 4541-4546.
19. Carsten Christophersen, Theory of The Origin, Function and Evolution of Secondary Metabolites in *Studies of Natural Products Chemistry*, Elsevier, Amsterdam, (Ed.: Atta-ur-Rahaman), Vol. 18, **1996**, pp. 677-737.
20. T. A. van Beek and Ae. de Groot, Terpenoid Antifeedants, Part I, An Overview of Terpenoid Antifeedants of Natural Origin, *Recl. Trav. Chim. Pays-Bas,* **1986**, *105*, 513-527 and references cited.
21. Alejandro F. Barrero, Enrique A. Manzaneda, Joaquin Altarejos, Sofia Salido, Jose M. Ramos, M. S. J. Simmonds, and W.M. Blaney, Synthesis of Biologically Active Drimanes and Homodrimanes from (−)-Sclareol, *Tetrahedron,* **1995**, *51*, 7435-7450 and references cited.
22. Hans-Walter Heldt, *Plant Biochemistry and Molecular Biology*, Oxford University Press, Oxford, New York, **1997**, Chapter 16, pp. 352-359; *Idem,* Plant Biochemistry, Academic Press (An Imprint of Elsevier), San Diego, 3rd ed., **2005**, Chapter 16, pp. 403-412.
23. A. M. Rizk, The Phytochemistry of the Flora of Qatar, Kingprint of Richmond, UK (on behalf of the University of Qatar), **1986**, pp 447-450.
24. Jerrold Meinwald, Personal Reflections on Receiving the Roger Adams Award in Organic Chemistry, *J. Org. Chem.,* **2005**, *70*, 4903-4909.
25. J. Meinwald, Y. C. Meinwald, J. W. Wheeler, T. Eisner and L. P. Brower, Major Components in the Exocrine Secretion of a Male Butterfly (Lycorea), *Science,* **1966**, *151*, 583-585.

26. J. Meinwald, Y C. Meinwald and P. H. Mazzocchi, Sex Pheromone of the Queen Butterfly: Chemistry, *Science,* **1969**, *164*, 1174-1175.

27. S. Schultz, W. Francke, M. Boppre', T. Eisner and J. Meinwald, Insect Pheromone Biosynthesis: Stereochemical Pathway of Hydroxydanaidal Production from Alkaloidal Precursors in *Creatonotos transiens* (Lepidoptera, Arctiidae) *Proc. Natl. Acad.* Sci. *USA,* **1993**, *90*, 6834-6838.

28. John R. Porter, Plant Fungal Endophytes: Interactions, Metabolites and Biosyntheses in *Selected Topics in The Chemistry of Natural Products* (Ed. Raphael Ikan), World Scientific, **2007**, pp. 503-580.

29. Andrea Stierle, Gary Strobel, Donald Stierle, Paul Grothaus, and Gary Bignani, The Search for a Taxol-producing Microorganism Among the Endophytic Fungi of the Pacific Yew, *Taxus Brevifolia, J. Nat. Prod.,* **1995**, *58*, 1315-1324.

30. Andrea Stierle, Gary Strobel, and Donald Stierle, Taxol and Taxane Production by *Taxomyces andreanae,,* an Endophytic Fungus of Pacific Yew, *Science,* **1993**, *260*, 214-217.

31. E. L. Rice, *Allelopathy, Academic Press, New York, 2nd ed.,* **1984.**

32. Jeffrey B. Harborne, Twenty-five Years of Chemical Ecology, *Nat. Prod. Rep.* (Millennium Review), **2001**, *18*, 361-379.

33. Jeffrey B. Harborne, Plant Chemical Ecology in *Comprehensive Natural Products Chemistry,* Vol. 8, Vol. Editor Kenji Mori, Elsevier, Pergamon Press, **1999**, 137-196.

34. Dudley H. Williams, Martin J. Stone, Peter R. Hauck and Shirley K. Rahman, Why are Secondary Metabolites (Natural Products) Biosynthesized? *J. Nat. Prod.,* **1989**, *52*, 1189-1208.

35. V. Prelog, Why Natural Products? *Croatica Chemica Acta,* **1985**, *58*, 349-351.

36. A. R. Battersby, Alkaloid Biosynthesis, *Quart. Rev.,* **1961**, *15*, 259-286.

37. D. H. R. Barton, A Region of Biosynthesis, *Chem. Brit,* **1967**, 330-337.

38. J. W. Cornforth, Exploration of Enzyme Mechanisms by Asymmetric Labelling,, *Quart. Rev.,* **1969**, 125-140.

39. Eberhard Breitmaiser and Wolfgang Voelter, Carbon-13 NMR Spectroscopy, VCH Third Revised Edition, **1987**, p. 451; see also Appendix **B, 1966.**

40. C. R. Hutchinson, A. H. Heckendorf, P. E. Daddona, E. Hagaman, and E. Wenkert, Biosynthesis of Camptothecin. I. Definition of the Overall Pathway Assisted by Carbon-13 Nuclear Magnetic Resonance, *J. Am. Chem. Soc.,* **1974**, *96*, 5609-5611, and reference 16 cited therein.

Further Reading

T.A. Geissman and D.H.G. Crout, Organic Chemistry of Secondary Plant Metabolism, Freeman, Cooper, San Francisco, 1969.

James B. Hendrickson, The Molecules of Nature, W. A. Benzamin, Inc. New York, 1965.

Paul M. Dewick, Medicinal Natural Products, John Wiley & Sons, 3rd edition, 2009.

E.A. Bell and B.V. Charlwood (Eds.), Secondary Plant Products, Encyclopedia of Plant Physiology, Vol. 8, Springer Verlag, Berlin, 1980.

Kurt B. G. Torssell, Natural Product Chemistry: A mechanistic, biosynthetic and ecological approach, 2nd ed., Taylor and Francis, Chapter 2, 1997.

J. Mann, R. S. Davidson, J. B. Hobbs, D. V. Benthrop and J. B. Harborne, Natural Products, Their Chemistry & Biological Significance, Longman, Science & Technology, 1994.

Hans-Walter Heldt, Plant Biochemistry and Molecular Biology, Oxford University Press, Oxford, New York, 1997.

A. M.Rizk, The Phytochemistry of the Flora of Qatar, Kingprint of Richmond, UK (on behalf of the University of Qatar), 1986.

T. Eisner and J. Meinwald, in Insect Pheromone. Biochemistry and Molecular Biology, (Eds.: G. J. Blomquist and R. C. Vogt), Elsevier Academic Press, 2003.

Enzyme Chemistry: Impact and Applications, Editors Colin J. Suckling, Colin L. Gibson, and Andrew R. Pill, 2nd Edition, Blackie Academic & Professional, 1998.

Reference 9 and other relevant Chapters of the book.

Chapter 2
Fundamental Stereochemical Concepts and Nomenclatures

2.1 Introduction

In this chapter some essential stereochemical concepts associated with organic molecules (natural or synthetic), as reflected in course of their many reactions, their asymmetric synthesis, biosynthesis, and biological activities, have been discussed. This treatment is expected to be quite handy, advantageous, and helpful to the readers to understand the chiral/achiral designations (nomenclatures), the stereochemical features, and related properties of the natural products dealt in the chapters that follow. Without having adequate stereochemical concepts, it may not be possible to understand and appreciate properly the stereochemistry of the natural or synthetic products. Thus, some essential static and dynamic aspects of stereo-chemistry will be dealt with sufficient illustrative examples in figures along with discussions in the text to reasonable extents. Further, this treatment will give an idea of some basic stereochemical concepts as applied to organic molecules in general.

The chirality of the natural products imparts specific remarkable medicinal properties (in vivo interactions) (Chap. 32) and opened up in vitro two new chemically important avenues:

1. Synthesis of chiral molecules using chiral templates
2. Asymmetric synthesis of chiral molecules, based on stereoselectivity, in profitable enantiomeric or diastereomeric excess (involving development of chirality from a prochiral system).

Optically active natural products are synthesized in the plant cells in genetically controlled and enzymatically catalyzed processes with well-defined stereochemistry and with definite orientations (Z/E) of double bond/s, if present.

S.K. Talapatra and B. Talapatra, *Chemistry of Plant Natural Products*,
DOI 10.1007/978-3-642-45410-3_2, © Springer-Verlag Berlin Heidelberg 2015

2.2   Chirality. Symmetry Elements. Optical Rotation

In a lecture in 1893 Lord Kelvin introduced the term *chirality*. The word *chiral* is from the Greek name *cheir* (for hand or pertaining to hand). Chirality means handedness (topological). It is a purely geometrical property. An object, for that matter, a molecule is chiral if it is not superposable on its mirror image. Organic molecules may be conformationally mobile. *A chiral molecule must be nonsuperposable on its mirror image even after operation of rotation* around single bond/s or translation, *.i.e.,* it must be chiral in all conformations. Thus, in order to ascertain whether a flexible molecule is chiral or achiral (not chiral), all possible conformers must be analyzed.

Chirality of a particular molecular conformation of known stereostructure may be decided by

(i)   Intuition (not useful for beginners),
(ii) Constructing the molecular models of the stereostructure of the compound and its mirror image and examining their probable superposability (not always possible)
(iii) Looking for *symmetry elements* of that structure by *symmetry operation(s).*

Molecules (a particular conformation in case of a flexible molecule) which have a plane of symmetry (σ or S_1), a center of symmetry (i or S_2), or other alternating axis of symmetry (S_n, n even) are said to have reflection symmetry. Such molecules are superposable on their mirror images and are termed *achiral*. Molecules lacking any reflection symmetry are nonsuperposable on their mirror images and are termed *chiral. Thus, nonsuperposability on the mirror image of a molecule is the necessary and sufficient condition for chirality and for displaying optical activity.* The principal symmetry elements and symmetry operations are summarized in Table 2.1 for their convenient application and for subsequent finding out their point groups, which will be illustrated in the sequel.

A preliminary introduction to the different symmetry elements and their operations follow.

Table 2.1 Symmetry elements (symbols expressed in italics) and symmetry operations

Symbol	Symmetry elements	Symmetry operations
(i) C_n	Simple or proper axis of symmetry	Rotation about an axis through $360°/n$
(ii) σ (S_1)	Plane of symmetry	Reflection in a plane
(iii) i (S_2)	Center of symmetry	Inversion through a center
(iv) S_n	Alternating or improper, or rotation-reflection or reflection-rotation axis of symmetry	*Rotation-Reflection*: Rotation about an axis by $360°/n$, followed by reflection in a plane orthogonal ($\perp$r) to the axis. *Reflection-Rotation*: The order of the above two operations may be reversed to give the same result.

2.2.1   Simple or Proper Axis of Symmetry

A simple or proper axis of symmetry of order (multiplicity) n is such that rotation of the model or structure of the molecule around an axis by $360°/n$ leads to a structure indistinguishable from the original. Such an axis is denoted by Cn, expressed in *italics*. Some examples are shown in Fig. 2.1.

Principal Axis If a molecule possesses several simple axes of symmetry, the axis with maximum multiplicity (order) is designated as the principal axis of symmetry. If the simple axes of symmetry are of the same order, the simple axis passing through the atoms is regarded as the principal or main axis. For an example see allene (Fig. 2.3).

   C_1 *Axes. Any object* or *molecule* contains an infinite number of C_1 axes since its rotation by $360°$ around any axis passing through the object or molecule results in the original three-dimensional orientation.

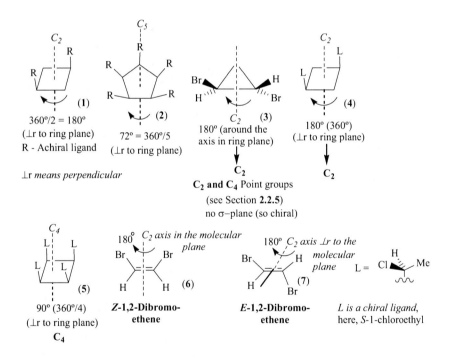

Fig. 2.1 Examples of molecules with C_2, C_4, and point groups $\mathbf{C_2}$ and $\mathbf{C_4}$

2.2.2   Plane of Symmetry

A plane of symmetry or a sigma plane (**σ plane**) is defined as a *mirror plane* which bisects a rigid object or a molecule so that one-half of it coincides with the reflection of the other half in the mirror, i.e., one-half of it reflects its enantiomeric or identical half.

Examples of some common objects with only one plane of symmetry are chairs, cups, file cabinets, spoons, and tooth brushes. Some examples of molecules with one σ-plane are shown in Fig. 2.2.

σ_v and σ_h C_n axes and σ planes often occur together in molecules. The principal proper axis of symmetry is conventionally taken as vertical. The symbol σ_v is used to designate a vertical plane of symmetry containing the principal axis; σ_h is used to designate a horizontal plane of symmetry perpendicular to the principal axis of symmetry; and σ_d is used to represent a diagonal plane bisecting the angle between two C_2 axes. The three types of planes are illustrated with examples in Fig. 2.3.

A plane of symmetry is equivalent to onefold alternate axis of symmetry S_1 (*vide infra*).

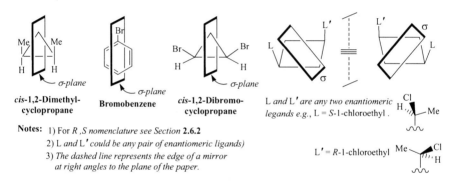

cis-1,2-Dimethyl-cyclopropane   **Bromobenzene**   *cis*-1,2-Dibromo-cyclopropane   L *and* L' *are any two enantiomeric legands e.g.,* L = S-1-chloroethyl .

L' = R-1-chloroethyl

Notes: 1) For *R ,S nomenclature see Section* **2.6.2**
2) L *and* L' *could be any pair of enantiomeric ligands)*
3) *The dashed line represents the edge of a mirror at right angles to the plane of the paper.*

Fig. 2.2 Achiral molecules with only one symmetry plane

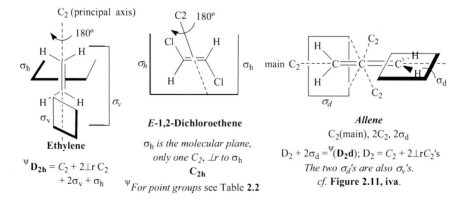

Ethylene   **E-1,2-Dichloroethene**   *Allene*

$^\Psi \mathbf{D_{2h}} = C_2 + 2\perp r\, C_2$
$+ 2\sigma_v + \sigma_h$

σ_h *is the molecular plane, only one* C_2, *⊥r to* σ_h
C_{2h}
$^\Psi$*For point groups see Table* **2.2**

C_2(main), $2C_2$, $2\sigma_d$
$D_2 + 2\sigma_d = ^\Psi(\mathbf{D_2d})$; $D_2 = C_2 + 2\perp rC_2$'s
The two σ_d's are also σ_v's.
cf. **Figure 2.11, iva.**

Fig. 2.3 Examples of σ_v, σ_h, and σ_d planes

2.2.3 Center of Symmetry or Inversion Center

The center of symmetry designated i is a point such that all imaginary straight lines that can be drawn through it meet identical atoms at equal distance from the point. The objects including molecules with a center of symmetry (point symmetry) are termed *centroasymmetric*.

A center of symmetry is equivalent to twofold alternating axis of symmetry S_2, as illustrated in Fig. 2.4 (see Sect. 2.2.4).

2.2.4 Alternating or Improper or Rotation-Reflection Axis (S_n)

An alternating or an improper or a rotation-reflection axis of order n (S_n) present in a molecule is such that it can be rotated about the axis by an angle of $360°/n$ and then reflected across a plane perpendicular to the axis to provide a structure indistinguishable from the original. The order of the two operations may be reversed (reflection-rotation) to give the same result. It has been exemplified for *meso*-2,3-butanediol and *meso*-tartaric acid, each having an S_2 axis (Fig. 2.4), and also for substituted cyclobutane with two mirror image (enantiomeric) ligands at 1,3 positions having an S_4 axes (Fig. 2.5). Another example with an S_4 axis is illustrated in Fig. 2.5.

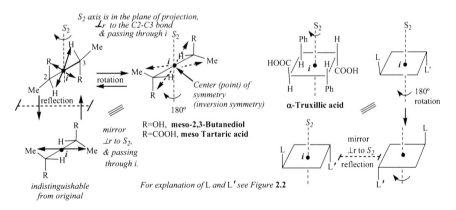

Fig. 2.4 Achiral molecules with center of symmetry (S_2) (inversion symmetry, i)

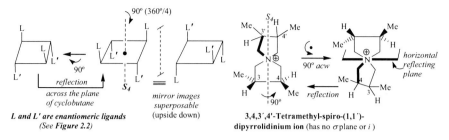

Fig. 2.5 Achiral molecules with S_4 axis explained

2.2.5 Dissymetric and Asymmetric Molecules. Chiral and Achiral Point Groups. Central Chirality

All chiral molecules are *dissymmetric* since they lack the symmetry elements S_n, i, and σ. A dissymmetric molecule may have one axis or more axes of symmetry (C_n). On the other hand, the chiral or dissymmetric molecules which lack even C_n axis (save C_1 axis) is termed *asymmetric*. Thus, a molecule can be classified as shown in the following Fig. 2.6.

A chiral or dissymmetric compound having at least one or more (a) chiral centers (central chirality), (b) chiral axis (axial chirality), (c) chiral plane (planar chirality), or (d) helicity (helical chirality) is nonsuperposable on its minor image (enantiomer) and hence is optically active. Axial chirality, planar chirality, and helicity—lacking any chiral center—will be discussed in Sects. 2.16, 2.17, and 2.18, respectively.

The *point group* defines the symmetry class to which a molecule belongs. The symmetry operations (or symmetry elements) are combined to form a point group (notation by Schönfliess), since each operation *leaves a point*, the center of gravity of the molecule, *unchanged*. The C_n and S_n operations organize a point group which is by convention expressed in **bold letters**. The chiral point groups (Fig. 2.6) are illustrated by examples in Figs. 2.7, 2.8, 2.9, and 2.10).

Some Examples of Chiral Point Groups

(i) **Point group C_1**

All organic molecules with only one stereogenic center or axis or plane belong to C_1 point group.

C_n axis and C_n operation

The compounds (**3**) and (**4**) of Fig. 2.1 are chiral and belong to the point groups C_2 and C_4. Compounds (**1**), (**2**), (**6**), and (**7**) possess σ-planes and hence achiral (Fig. 2.1). A few more compounds belonging to C_2 point group are illustrated in Figs. 2.8 and 2.9.

(ii) **Point group C_2** (*possesses the only symmetry element, C_2*)

(iii) **Point group C_3**

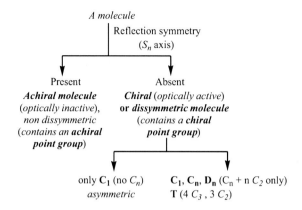

Fig. 2.6   Chiral and achiral molecules based on reflection symmetry. Chiral point groups

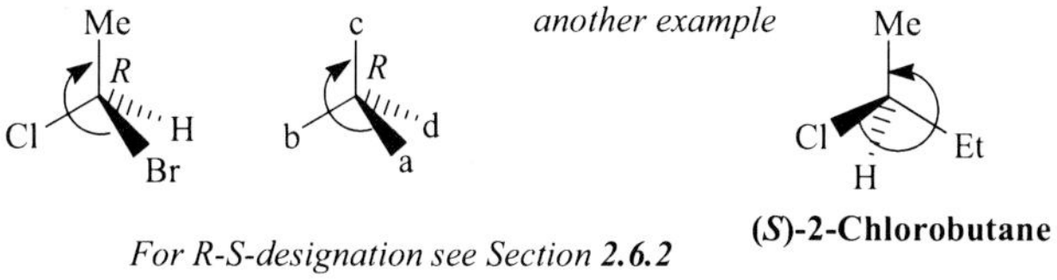

For R-S-designation see Section 2.6.2

(*S*)-2-Chlorobutane

Fig. 2.7   Asymmetric compounds with **C₁** point group

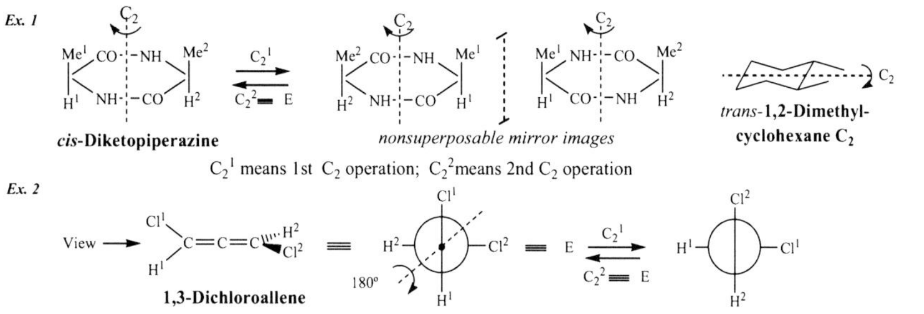

$C_2{}^1$ means 1st C_2 operation; $C_2{}^2$ means 2nd C_2 operation

cis-Diketopiperazine   *nonsuperposable mirror images*   *trans*-**1,2-Dimethyl-cyclohexane C₂**

Ex. 2

View →   **1,3-Dichloroallene**

- The C_2 axis passes through the central carbon and bisecting the dihedral angle between the two Cl's and two H's (taking the three allene carbons as a C-C bond in the Newman projection formulas.
- Any operation like $C_2{}^2$ in the above examples, producing a shape identical with the original one is called identity or 'E' or 'I' operation.
- The C_2 symmetry of 1,3-dichloroallene is best seen in its Newman projection, as shown.

Fig. 2.8   Asymmetric compounds belonging to **C₂** point group

(iv) **Point groups Dₙ and T.** The chiral point group **Dₙ** consists of a principal C_n axis of maximum multiplicity (or passing through some atoms of the molecule) and n C_2 axes orthogonal to the principal axis (Fig. 2.10).

  T point group. Tetrahedral (**T**) point group is also chiral, possessing four C_3 and three C_2 axes, but no σ-plane. The four chiral ligands are necessarily having same structure and absolute configuration (Fig. 2.10).

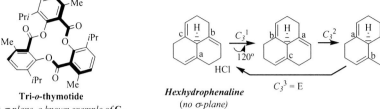

Tri-*o*-thymotide
(*no σ-plane, a known example of* **C₃**
point group having the shape of a
3-bladed propellar)

Hexhydrophenaline
(*no σ-plane*)

Imaginary hydrocarbon (propellar type) (C₃ axis ⊥r to the paper
plane, passing through central H-C bond)

Fig. 2.9   Asymmetric compounds belonging to **C₃** point group

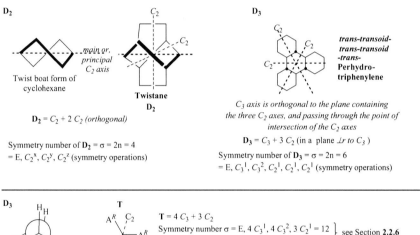

D₂

Twist boat form of
cyclohexane

main or,
principal
C_2 axis

Twistane
D₂

$D_2 = C_2 + 2 C_2$ *(orthogonal)*

Symmetry number of $D_2 = \sigma = 2n = 4$
$= E, C_2^x, C_2^y, C_2^z$ (symmetry operations)

D₃

trans-transoid-
trans-transoid
-trans-
Perhydro-
triphenylene

C₃ axis is orthogonal to the plane containing
the three C₂ axes, and passing through the point of
intersection of the C₂ axes

$D_3 = C_3 + 3 C_2$ (in a plane ⊥r to C_3)
Symmetry number of $D_3 = \sigma = 2n = 6$
$= E, C_3^1, C_3^2, C_2^1, C_2^1, C_2^1$ (symmetry operations)

D₃

Ethane
(*neither staggered*
nor eclipsed)

T

$T = 4 C_3 + 3 C_2$
Symmetry number $\sigma = E, 4 C_3^1, 4 C_3^2, 3 C_2^1 = 12$ } see Section **2.2.6**
Symmetry order also 12

For distribution of all simple axes of symmetry *vide infra* (point group T_d)

Fig. 2.10   Few molecules belonging to **D₂**, **D₃**, and **T** chiral point groups

2.2.6   Symmetry Number, Order of Point Groups, Achiral Point Groups

Point group notations of *achiral* molecules are enumerated in Table 2.2.

Symmetry number σ of a molecule is the number of indistinguishable but nonidentical positions into which the molecule can be turned by rigid rotation (around all simple axes of symmetry), taking the parent position as 1.

Order of the point group is defined as the total number of symmetry operations that can be performed in that point group. Table 2.3 lists the order and symmetry number (σ) of various chiral and achiral point groups.

Achiral Point Groups: Several examples of achiral molecules belonging to different point groups (Table 2.2) are displayed in Fig. 2.11. The point groups could be allocated to each molecule by finding out the C_n, S_1 (σ-plane), and S_2 (center of symmetry) possessed by the molecule.

Table 2.2 Point group notations of achiral molecules

C_s	S_n	C_{nv}	C_{nh}	D_{nd}		D_{nh}	T_d
One σ only	(no σ) n even	$Cn + n\,\sigma_v$ only	$C_n + \sigma_h$ no σ_v	$C_n + n\,C_2$ ($\underline{1}$r) + $n\,\sigma_v$ but no σ_h		$C_n + n\,C_2$ ($\underline{1}$r) + n $\sigma_v + \sigma_h$	$4\,C_3 + 3$ C_2 $+\,6\,\sigma$

Table 2.3 Symmetry number and order of point groups

Pt. group	C_1	C_n	D_n	C_s	$S_n^{\,\psi}$	C_{nv}; C_{nh}	$C_{\infty v}$	D_{nd}; D_{nh}	T_d	O_h
Order	1	n	$2n$	2	n	$2n$	∞	$4n$	24	48
Symmetry number, σ	1	n	$2n$	1	$n/2$	n	1	$2n$	12	24

$S_n^{\,\psi}$ includes $S_2 = C_i$, σ = 1, order 2

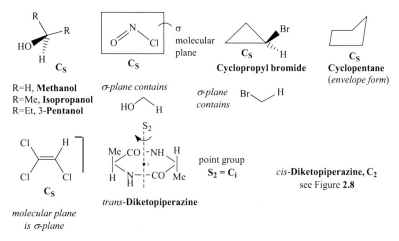

R=H, **Methanol**
R=Me, **Isopropanol**
R=Et, **3-Pentanol**

C_s

σ-plane contains

C_s

σ-plane contains

C_s Cyclopropyl bromide

C_s Cyclopentane (*envelope form*)

molecular plane is σ-plane C_s

*trans-*Diketopiperazine point group $S_2 = C_i$

*cis-*Diketopiperazine, C_2 see Figure **2.8**

S_4 see Figure **2.5**
One more example :

Here S_4 includes S_2 (C_i). For S_4 order is 4
Operators: E, $S_4^{\,1}$, $S_4^{\,2}$ (C_2), $S_4^{\,3}$

Symmetry number σ = 2

Fig. 2.11 (continued)

(ii) $C_{nv} = C_n + n\ \sigma_v$; no σ_h   *Each example contains a C_2 (shown) and two σ_v planes, as shown in the 1st example.*

   (a) $C_{2v} = C_2 + 2\ \sigma_v$ *Other examples:*

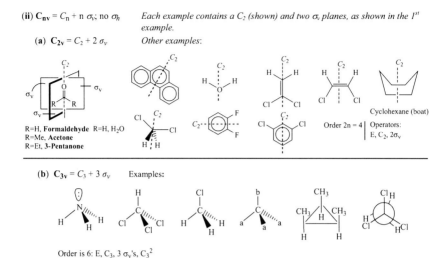

R=H, **Formaldehyde** R=H, H₂O
R=Me, **Acetone**
R=Et, **3-Pentanone**

     Order 2n = 4 | Operators:
                E, C₂, 2σᵥ

Cyclohexane (boat)

   (b) $C_{3v} = C_3 + 3\ \sigma_v$ Examples:

Order is 6: E, C₃, 3 σᵥ's, C₃²

(c) $C_{4v} = C_4 + 4\ \sigma_v$ all *cis*   **(d) $C_{5v} = C_5 + 5\ \sigma_v$**   the ring is taken as planar

     (e) $C_{\infty v} - C_\infty + \infty \sigma_v$
         Conical symmetry.   H-Cl; C=O; H-C≡N; H-C≡C-Cl

(iii) $C_{nh} = C_n + \sigma_h$; no σ_v.

   (a) $C_{2h} \rightarrow$ order is 4: E, C₂¹ σₕ, i
       σ = 2

               Examples of point group **$C_{2h} = C_2 + \sigma_h$**

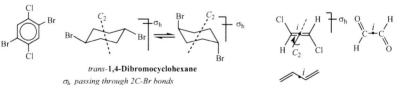

trans-**1,4-Dibromocyclohexane**
σₕ *passing through 2C-Br bonds*
C₂ *passing through the mid points of opposite bonds shown*

s-trans-**1,3-Butadiene**

(b) $C_{3h} = C_3 + \sigma_h$

order 6 :
E, C₃¹, C₃², σ, S₃¹, S₃²
σ = 3
This particular conformation
has point group **C_{3h}**

Cl—Pt⟨Cl / Cl, F, F

PtCl₃F₂

Fig. 2.11 (continued)

(iv) $\mathbf{D_{nd}} = C_n + n\,C_2\,(\perp r) + n\,\sigma_v\,(\sigma_d)$; no σ_h

(a) $\mathbf{D_{2d}} = C_2 + 2\perp r\,C_2 + 2\,\sigma_d$
also S_4 (prin. C_2)

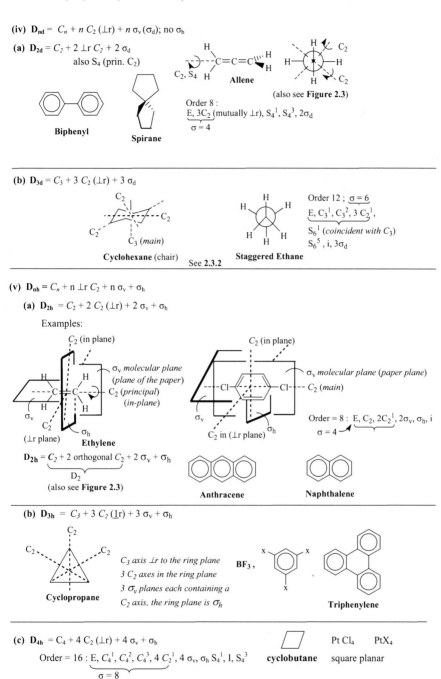

Biphenyl

Spirane

Allene

(also see **Figure 2.3**)

Order 8 :
E, $3C_2$ (mutually $\perp r$), $S_4{}^1$, $S_4{}^3$, $2\sigma_d$

$\sigma = 4$

(b) $\mathbf{D_{3d}} = C_3 + 3\,C_2\,(\perp r) + 3\,\sigma_d$

C_3 (main)

Cyclohexane (chair)   See **2.3.2**

Staggered Ethane

Order 12 ; $\sigma = 6$
E, $C_3{}^1$, $C_3{}^2$, $3\,C_2{}^1$,
$S_6{}^1$ (coincident with C_3)
$S_6{}^5$, i, $3\sigma_d$

(v) $\mathbf{D_{nh}} = C_n + n\perp r\,C_2 + n\,\sigma_v + \sigma_h$

(a) $\mathbf{D_{2h}} = C_2 + 2\,C_2\,(\perp r) + 2\,\sigma_v + \sigma_h$

Examples:

C_2 (in plane)

σ_v molecular plane
(plane of the paper)
C_2 (principal)
(in-plane)

($\perp r$ plane)   **Ethylene**

$\mathbf{D_{2h}} = \underbrace{C_2 + 2\ \text{orthogonal } C_2}_{D_2} + 2\,\sigma_v + \sigma_h$

(also see **Figure 2.3**)

C_2 (in plane)

σ_v molecular plane (paper plane)
C_2 (main)

C_2 in ($\perp r$ plane)

Order = 8 : E, C_2, $2C_2{}^1$, $2\sigma_v$, σ_h, i
$\sigma = 4$

Anthracene   **Naphthalene**

(b) $\mathbf{D_{3h}} = C_3 + 3\,C_2\,(\underline{1}r) + 3\,\sigma_v + \sigma_h$

Cyclopropane

C_3 axis $\perp r$ to the ring plane
3 C_2 axes in the ring plane
3 σ_v planes each containing a
C_2 axis, the ring plane is σ_h

BF_3 ,

Triphenylene

(c) $\mathbf{D_{4h}} = C_4 + 4\,C_2\,(\perp r) + 4\,\sigma_v + \sigma_h$
Order = 16 : E, $C_4{}^1$, $C_4{}^2$, $C_4{}^3$, $4\,C_2{}^1$, $4\,\sigma_v$, $\sigma_h\,S_4{}^1$, I, $S_4{}^3$

$\sigma = 8$

cyclobutane   **square planar**

$Pt\,Cl_4$ PtX_4

Fig. 2.11 (continued)

(d) D_{6h}

Benzene

order = 24, σ = 12

sym. no.

$$D_{6h} = C_6 + 6\,C_2 + 6\,\sigma_v + \sigma_h$$

$$\sigma = E,\ C_6{}^1,\ C_6{}^2,\ C_6{}^3,\ C_6{}^4,\ C_6{}^5,\ 6\,C_2{}^1 = 12$$

order $= \sigma + 6\,\sigma_v,\ \sigma_h,\ S_6{}^5,\ S_6{}^4,\ S_6{}^3,\ S_6{}^2,\ S_6{}^1 = 24$

sym. no. $(i = S_6{}^3)$

(vi) $D_{\infty h}$ E, C_∞, $\infty\,C_2$ ($\perp$ r C_∞ axis), $\infty\,\sigma_v$, σ_h cylindrical   H–H, O=C=O, HC≡CH

(vii) T_d

Pt gr. of regular
of tetrahedron
A = achiral group like
H, Me, Cl etc.

E, 3 C_2 (mutually $\perp$r) passing through pairs of opposite edges
4 $C_3 \rightarrow$ passing though each apex and the center
of opp. face (containing A-C bond)

(i)

6 $\sigma_d \rightarrow$ Each σ_d contains each A–C–A plane;
$\therefore$ 4 C_2 = 6 combinations
$3S_4 \rightarrow$ coincident with C_2 axes

$\}$ ----- C_2 or S_4

$\sigma = 12;\ E,\ 4\,C_3{}^1,\ 4\,C_3{}^2,\ 3\,C_2{}^1 = 12.$ order $= 12 + \underline{6\,\sigma_d + 3\,S_4{}^1 + 3\,S_4{}^3} = 24$

Underlined ones are additional operations

(ii)

Adamantane
Chiral center in the
center, not at any carbon

(iii)

R = *t*-Bu, synthesised by Maier in 1978
R = H, could not be synthesised as yet

(viii) O_h octahedral symmetry

cubic pt. gr.

Sym. operations : E, 3 C_4 (mutually perp.) + 4 C_3, 3 S_4, 3 C_2
(coincident with 3 C_4 axes) 6 C_2, 9 σ, 4 S_6 (coincident with $4C_3$ axes)

Other examples:

$\sigma = 1 + 9 + 8 + 6 = 24$
order $= 24 \times 2 = 48$
(double by the existence of symmetry planes

$[PtCl_6]^-$; $[Co(NH_3)_6]^{3+}$

first synthesised
by Eaton in 1964

Cubane

Fig. 2.11 Examples of some achiral compounds belonging to different achiral point groups

2.2.7   Local Symmetry (or Site Symmetry). Desymmetrization

Local symmetry shows the symmetry properties of atoms or groups within a molecule (Fig. 2.12).

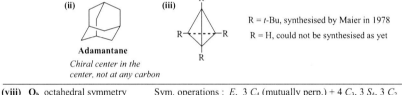

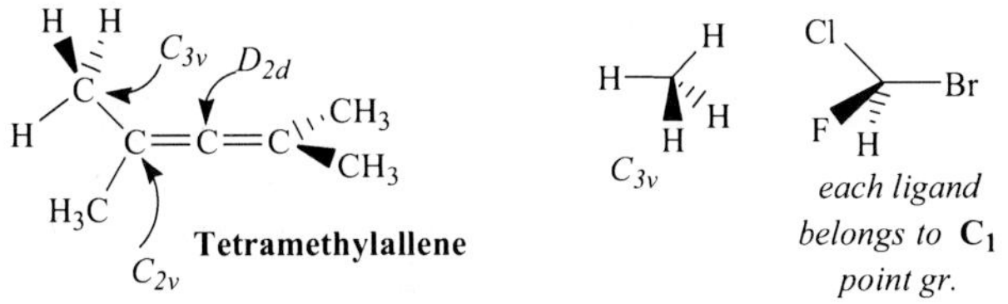

Tetramethylallene

each ligand
belongs to C_1
point gr.

Desymmetrization results from successive substitution on a symmetrical lattice framework

Ex.1 $\boxed{\begin{array}{c} CH_4 \\ T_d \end{array}}$ → $\begin{array}{c} CH_3Cl \\ C_{3v} \end{array}$ → $\begin{array}{c} CH_2Cl_2 \\ C_{2v} \end{array}$ → $\begin{array}{c} CH_2ClBr \\ C_S \end{array}$ → $\begin{array}{c} CHFClBr \\ C_1 \end{array}$

Desymmetrization results in distortion of the framework, small but real, as demanded by the symmetry of the system.

Ex.2 $\begin{array}{c} Cl_2C=C=CH_2 \\ C_{2v} \end{array}$ ← $\begin{array}{c} H_2C=C=CH_2 \\ D_{2d} \\ \textit{(Elongated tetrahedron)} \end{array}$ → $\begin{array}{c} ClCH=C=CHBr \\ C_1 \end{array}$ ← $\begin{array}{c} ClCH=C=CH_2 \\ C_S \end{array}$ → $\begin{array}{c} ClCH=C=CHCl \\ C_2 \end{array}$

Ex.3

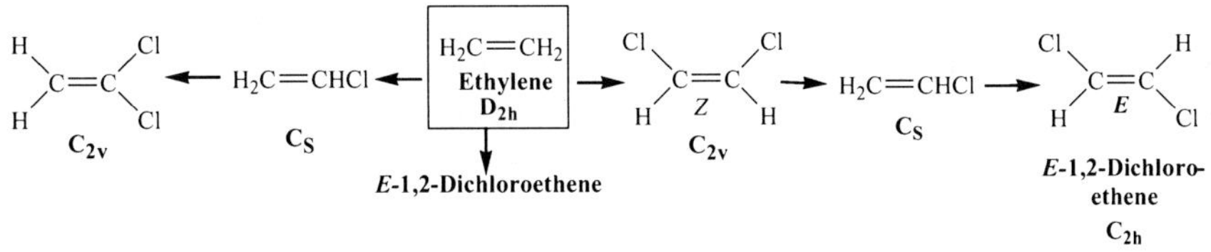

Fig. 2.12   Local symmetry and desymmetrization illustrated

2.2.8   *Optical Isomerism. Optical Rotation*

2.2.8.1   Optical Activity Due to Chiral Molecular Structure

A chiral compound in the solid, fused, gaseous (rarely) liquid or dissolved state, *e.g.*, lactic acid, tartaric acid, glucose etc., exhibits optical activity, *i.e.*, rotates the plane of polarization of an incident beam of polarized monochromatic light. During the early nineteenth century the French physicist Jean Baptiste Biot (1774–1862) discovered that many natural organic compounds exhibit optical activity. Since this optical rotation occurs in the liquid, dissolved or gaseous phase, it must be a molecular phenomenon. Here, the optical activity is entirely due to the *dissymmetry of the molecular structure*. The original molecule and its nonsuperposable mirror images are known as *enantiomers, enantiomorphs* (this name is taken from crystallography), or optical antipodes, one of which is dextrorotatory (rotates the plane of polarization in the clockwise direction by a positive (+)-angle) and the other levorotary (rotates in the anti-clockwise direction by a negative (−)-angle). For the measurement of optical rotation, earlier polarimeters have now been replaced by digital polarimeters.

2.2.8.2   Optical Activity Due to Crystalline Structure

Some crystals, *e.g.*, *quartz, sodium chloride, benzil*, etc. are optically active in the crystalline state only, *i.e.*, rotate the polarized light. Quartz is the first substance shown by Biot in 1812 to be optically active. It exists in two hemihedral nonsuperposable mirror image forms, one of which is *dextrorotatory* and the other *levorotary*. Here, the optical activity is a property of the chiral crystal and not of the molecule, as the rotation disappears when the crystal is melted or dissolved to give the achiral molecules. Such pairs of nonsuperposable crystals are said to be *enantiomorphous*.

   (See also Appendix B, 1821/1822, 1948, 1950)

2.2.8.3   Dependence of Rotation (α) on Concentration and Cell Length. Value of α

The observed angle of rotation of the plane of polarization is denoted by α, which can be recorded in a range of $-90°$ and $+90°$. For example, distinction can't be made between α and $\alpha \pm n$ $180°$; the plane when rotated by $\pm180°$ or its integer n, the new plane will coincide with the old one. Thus theoretically, no difference appears between $+40°$, $+220°$ ($40° + 180°$), $+400°$ ($40° + 2 \times 180°$), or $-140°$ ($40° - 180°$). Hence, rotation must be measured at least at two different concentrations to get α unequivocally. Since α is proportional to concentration, if the original solution is diluted to $1/10$, α would be $+4°$, $+22°$, $+40°$, $-14°$, respectively. Again α being proportional to length (discovered by Biot) can be unambiguously determined by using smaller cells, *e.g.*, of 0.25 dm length for the original concentrations, when observed α will be $+10°$, $+55°$, $+100°$ (equivalent to $-80°$), or $-35°$, respectively, again all clearly distinguishable.

2.2.8.4   Dependence of Sign of [α] of Polar Compounds on Solvent, Concentration, and pH

The rotation of compounds, especially the polar ones, is affected by the solvent because of its participation in solvation and association phenomena. It has been reported by Winther [1, 2] in 1907 that the specific rotation [α]$_D$ (Sect. 2.2.9) of (+)-nicotine (Chap. 18) is positive in polar solvents like formamide (maximum positive specific rotation) and water, [α]$_D$ decreasing with increase of concentration. It remains almost same in MeOH and EtOH and increases in benzene, *o-*, *m-*, and *p*-xylenes, and mesitylene with increase of concentration. Whereas, the rotation of natural nicotine in ethylene bromide and chloroform at lower concentrations (>0.6 g/g) was observed to be negative. Specific rotation of nicotine at infinite dilution of the solvent (meaning no solvent) obtained by extrapolation of the specific rotations in solvents of different polarities has been found to be ~8 units

(Sect. 2.2.9); this is known as *intrinsic rotation* $\{\alpha\}$. As the concentration of the solute increases, solute–solute interactions are maximized, which can differ greatly from one solvent to another. In polar solvents solute–solute association effect may be suppressed by competition with concentration-independent solvent association, which appears to increase the rotation.

Reversal of the sign of the rotation of (+)-nicotine takes place in less polar solvents like $CHCl_3$ and ethylene bromide at much less concentrations. Other examples: 2-methyl-2-ethylsuccinic acid displays positive rotation in $CHCl_3$ containing 0.7 % EtOH at >6.3 %, no rotation (null) at ~6.3 %, and negative rotation at <6.3 % concentration; but no reversal is observed in alcohol solvents, pyridine, diglyme, and acetonitrile.

Rotation of acids and bases is dependent upon the pH, *e.g.*, (*S*)-(+)-lactic acid is dextrorotatory in water, but its sodium salt is levorotatory. Another example: L-Leucine is levorotatory in water but dextrorotatory in aqueous hydrochloric acid.

2.2.9 Specific Rotation. Molecular Rotation. Units (Fig. 2.13)

The rotation per unit length in dm and unit concentration in g/ml is called *specific rotation* expressed as follows:

Specific rotation

$$[\alpha]_D^t = \frac{\alpha}{l.c}$$

where α is the observed rotation, l = length of the cell in dm,
c = conc. in g/ml, t = temperature, c' = conc. in g/100 ml,
d = density of the neat liquid

(Rotation is measured at the wavelength of Na-D line (589 nm))

$$= \frac{\alpha.\,100}{l.\,c'}$$

For liquids $[\alpha]_D^t = \frac{\alpha}{l.d}$

A new term "molecular rotation" [M] or [ϕ] also called molar rotation is defined to compensate for the effect of differing mol. wts. as follows :

$$[M] = [\varphi] = \frac{[\alpha]_D^t.M}{100^a} = \frac{\alpha.M}{l.c.100^a} = \frac{\alpha}{l.\frac{c}{M}(100\ ml)^{-1}} = \frac{\alpha}{l.c''}$$

where M = mol. wt., c'' = mole per 100 ml

[a]The division by 100 is arbitrary in order to keep its numerical value on the same approximate scale as that of specific rotationl.

Note: *Thus, optical rotation is proportional to the number of molecules encountered. Hence, for a compound with mol. wt. 100, [ϕ] = [α]*

Units or dimensions of [α] and [φ]

$$[\alpha] = \frac{\alpha}{l.\,c}$$ Thus *units of [α]* are $$\frac{degree}{dm.g\,cm^{-3}} = \frac{degree}{100\,cm.g.cm^{-3}} = \frac{degree}{100\,cm^{-2}\,g} = 10^{-1}\ \text{deg. cm}^2\,\text{g}^{-1}$$

So, while α is given in degrees, [α] should not be expressed in degree alone and should always be given without the units (understood to be 10^{-1} deg $cm^2\ g^{-1}$), as has been done in this book [*cf.* references in further reading (iii) and (v)].

$$Again, [\varphi] = \frac{[\alpha]M}{100} = \frac{\alpha}{l.mole\ per\ 100\ ml}$$

∴ Units of molar rotation are $$\frac{degree}{10\,cm.mole.\,10^{-2}cm^{-3}} = \frac{degree}{10^{-1}cm^{-2}\,mole} = 10\ deg\ cm^2\ mole^{-1}$$

[ϕ] is also expressed **without any unit**, understood to be 10 deg $cm^2\ mole^{-1}$

Fig. 2.13 Specific and molecular rotations and their units

2.2.10 Fischer Projection. Flying Wedge Formulas. Tetrahedral Representations of Cabcd

A common representation of a tetracoordinated compound with one chiral center was first proposed by Emil Fischer [3] in 1891, which is much easier than tetrahedral representation. While writing a Fischer projection formula of an organic compound having one or more chiral center/s, each sp^3 carbon atom in the chain is written as a cross +. Thus, the *agreed modes of projection* of the three-dimensional molecule on two-dimensional plane (projection plane, PP) to give the Fisher projection (FP) formula are as follows:

(1) The asymmetric (or dissymmetric) carbon should be in the projection plane.

The main chain must be in the vertical plane and pointing away from the observer, so that the vertical or up and down bonds are below the projection plane extending toward rear. The other two sideways substituents (ligands) of each carbon must be in the horizontal plane (above the PP) and pointing forward toward the observer.

Corollaries

1. One pair of ligands when exchanged gives the enantiomer.[1]
2. Two exchanges give another FP formula of the same enantiomer.
3. Ligands may be rotated in groups of three.

Limitations

1. FP formula being two-dimensional cannot be lifted out of the P.P. and turned over.
2. FP formula may be rotated in the projection plane by even multiple of 90° (*i.e.*, 180°), but not by odd multiple of 90° (*i.e.*, 90° or 270°).

Three-dimensional (flying wedge) and tetrahedral representations of enantiomers with one chiral center, C_{abcd}, and their two-dimensional representations as Fischer projection formulae are exemplified in Fig. 2.14.

There are 12 ways of writing the Fischer projection formula of an enantiomer of a compound with one chiral center, as depicted in Fig. 2.16. For detailed discussion on (D,L) and (R,S) nomenclatures, see Sect. 2.6.1 and 2.6.2.

The best way to confirm whether the two-dimensional (Fischer projection) or three-dimensional (flying wedge or tetrahedral) representations are identical or mirror images is to specify their (R,S) nomenclature (Fig. 2.6.2). In Fig. 2.14 all

---

[1] It may be noted that a tetrahedron is the only skeleton in which every transposition of ligands is equivalent to reversal of the ligated assembly, in which the tetrahedral atom is at the center of the tetrahedron and the four bonds of that atom are directed toward the four vertices of the tetrahedron carrying the four ligand atoms.

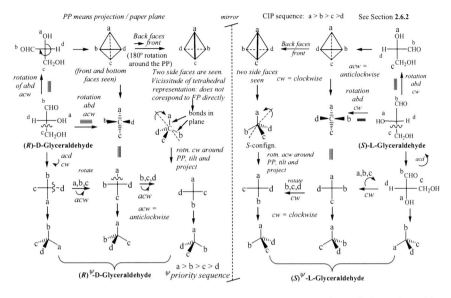

Fig. 2.14 Fisher projection, flying wedge, and tetrahedral representations of (*R*)-D-glyceralde-
hyde and (*S*)-L-glyceraldehyde

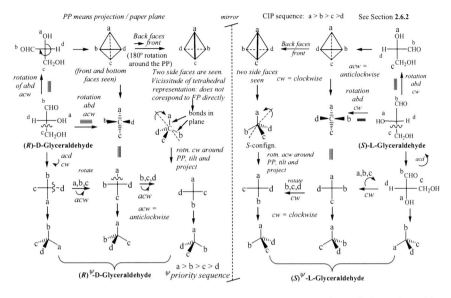

Fig. 2.15 A different tetrahedral representation of a few compounds, Cabcd (one enantiomer)

representations on the left side or right side of the mirror are of (*R*)-glyceraldehyde
or (*S*)-glyceraldehyde. The names are given under the Fischer projection formulas
specifying the (D,L) nomenclature. Each of the 12F.P. formulae of (*R*)-C$_{abcd}$
(Fig. 2.16) can give many more flying wedge formulas by tilting and rotating to
different extents and can be confirmed by specifying the absolute configuration
according to CIP rules (Sect. 6.2.1).

The tetrahedral representation of (*R*)-C$_{abcd}$ can also be written as shown in
Fig. 2.15, tilting the bottom side of the F.P. formula downwards so that the bottom
ligand goes to the rear side of the triangle (front face) made by the other three
ligands; the other three faces of the tetrahedron goes to the rear side. A few other
common compounds with one chiral center are also shown in this figure.

For a chiral compound with one chiral center there are factorial four (4!) = 4
× 3 × 2 = 24 ways of writing the Fischer projection formula, this being the
permutations of four ligands among four sites. Of these, 12 FP formulas will
correspond to the *R*-enantiomer and the remaining 12 to the (*S*)-enantiomer.

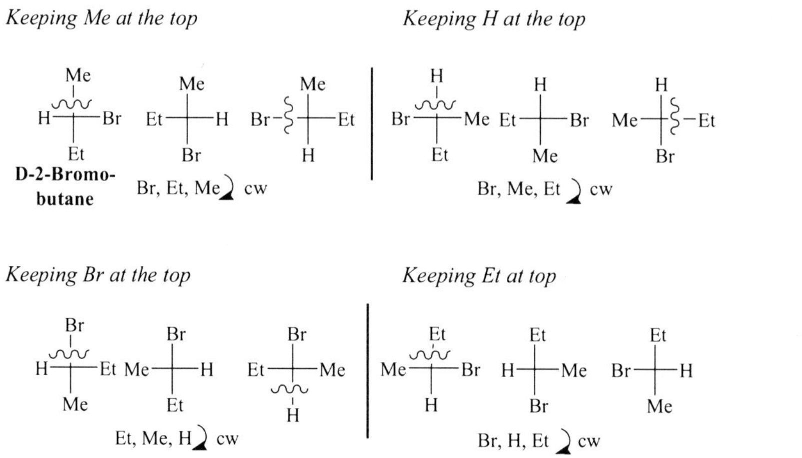

Keeping Me at the top   *Keeping H at the top*

D-2-Bromo-butane   Br, Et, Me ⟩ cw   Br, Me, Et ⟩ cw

Keeping Br at the top   *Keeping Et at top*

Et, Me, H ⟩ cw   Br, H, Et ⟩ cw

Note: Each F.P. is having (S-) and D-absolute configuration. The D-configuration is revealed only from the first F.P. formula.

Fig. 2.16 12 Fischer projection formulae of (S)-D-2-bromobutane

The 12 Fischer projection formulas for (S)-D-2-bromobutane are depicted in Fig. 2.16.

2.3 Conformation of Simple Acyclic Molecules

The term *conformation* of a molecule (excepting the diatomic ones) signifies any one of the infinite number of momentary arrangements of its atoms in space that result from rotation around single bonds and also from twisting around bonds. The conformations of a molecule that correspond to the minimum energy in its potential energy diagram are known as conformers or *conformational isomers*. Thus, any point on the curves (Figs. 2.20, 2.21, 2.22) corresponds to some conformation of the concerned molecule. Conformations are not superposable upon each other.

Conformational analysis involves the interpretation or prediction of the physical (including spectral) properties, thermodynamic stabilities, and reactivities of substances in terms of the conformation or conformations of their molecules.

2.3.1 Dihedral Angle. Torsion Angle. Torsional Strength

The *dihedral angle* θ (*theta*) denotes the angle between two planes containing A–B–C and B–C–D, respectively, in a nonlinear molecule, A–B–C–D, as shown in Fig. 2.17. It is best seen in a Newman projection formula (Fig. 2.17) in which the

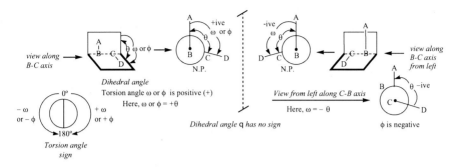

Fig. 2.17 Torsion angles and dihedral angles for two enantiomers

molecule is viewed from left along B–C axis; the dot in the front indicates the front atom B, and the circle indicates the back atom C. Here if B and C are tetracoordinated atoms, the remaining two bonds of each one are not shown. Thus, dihedral angle is a three-dimensional parameter of a molecule involving four atoms. The dihedral angle if given *a directional sense* becomes torsion angle denoted by Greek letter ω (omega) or ϕ (phi). If looking along the B–C axis *in either direction* the turn from A to D or D to A is clockwise, ω is positive; if the turn is anticlockwise, ω is negative. The torsion angle ω is best represented by Newman projection (N.P.) formula for each enantiomer (Fig. 2.17).

Torsional strain or Pitzer strain (V_ϕ) or potential energy (E) is caused by the rotational motion around the bond axis. Torsional strain is represented by the following equation:

$$V_\phi = \tfrac{1}{2} V_0 (1 + \cos n\Delta\phi) \; \text{kcal/mol}$$

where $\Delta\phi$ is the displacement of the dihedral angle (torsional displacement), V_0 is the torsional energy barrier or Pitzer strain, and n is the periodicity, *i.e.*, the number of times that a given conformation recurs during a complete revolution ($\Delta\phi = 360°$).

2.3.2 Klyne–Prelog Nomenclature for Torsion Angles. Conformational Chirality

A general method of nomenclature has been worked out by Klyne and Prelog [4] in 1960 to describe the steric relationship across a single bond in a molecule or a part of a molecule. The following rules are followed.

1. It has been mentioned earlier that in contrast to dihedral angle, torsion angle has a *directional property*, being mentioned as (+) when measured in a clockwise direction, and as (−) when measured in an anticlockwise direction. In a molecule

of the type A−B−C−D, the measurement is to be started from the *front substituent* A at 0° ending at the *rear substituent D* at 180° after rotation of the back substituent *D* around *B−C* bond.

2. *Conformation Selection Rule* (CSR): The two fiducial (reference) groups A and D are specified according to the *Conformation Selection Rule* by Cahn, Ingold, and Prelog [5] in 1966 as follows:

(i) If all the atoms or groups (ligands) of the set on both carbons (front and back) are different, the ligand most preferred by the standard *sequence sub-rules* (Sect. 2.6.2) is fiducial.

(ii) If two ligands of a set on one or both carbon/s are identical, the nonidentical (unique) group is fiducial, irrespective of the sequence rule.

(iii) If all or two ligands of the set on carbons are identical, that which provides the smallest torsion angle is fiducial.

3. The fiducial group A at the front atom is preferably (but not necessarily) placed at the top of the Newman projection formula, and the torsion angle is named in terms of the three designations **a, b,** and **c** (Fig. 2.18). The circle is divided into six segments by combination of **a, b,** and **c,** as shown in **d** in the figure. The conformation of any molecule bears the designation of the torsion angle. *Synperiplanar* and *antiperiplanar* conformations are expressed with (±) sign because of probable variation or libration of ω around 0° and 180°, respectively, by upto +30° or −30°

The sign of a torsion angle in any conformation remains unchanged whether the molecule is viewed from the front or from the rear. Since the exact values of torsion angles are often not known, the (+) and (−) signs immediately show the direction of a torsion angle, and the symbols *sp, sc, ac, ap* in the Klyne–Prelog system show the range of a torsion angle. The system is applicable to any molecule A−B−C−D, whether B and C are tetrahedral or trigonal atom; this system may also be used to describe partial conformation of ring compounds and polymer chains. Figure 2.19 depicts a few examples of different types. **Example 1** illustrates the torsion angle nomenclatures of the conformers and also the higher dipole moment of active stilbene dichloride than that of its meso isomer based on the analysis of the three conformers in each case.

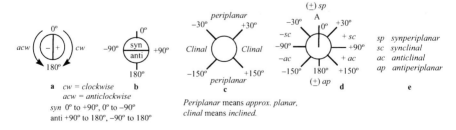

Fig. 2.18 Designation of conformation based on torsion angle ω or ϕ

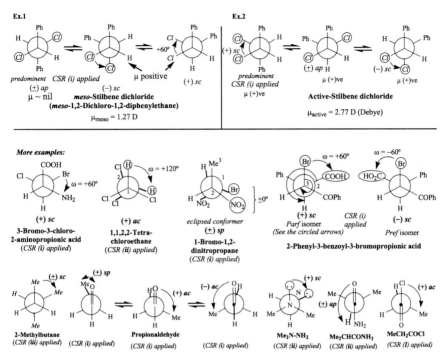

Fig. 2.19 A few examples of Klyne–Prelog nomenclature of torsion angles (*conformational chirality*)

2.3.3   Torsional Strain Curve (Potential Energy Diagram) of Ethane

The change of V_ϕ with ϕ or better with $\Delta\phi$ in case of ethane with the conformations having maximum and minimum energies is represented graphically in Fig. 2.20. The energy difference between the eclipsed conformer of maximum torsional strain or potential energy and the staggered conformer of minimum energy, the so-called *torsional or rotational energy barrier*, is only about 3 kcal/mol (12.5 kJ/mol). Hence, rotation around C–C single bond, though not free, is quite facile. Here, the greater the periodicity, the smaller the torsional barrier. For example, in case of nitromethane ($MeNO_2$) periodicity $n = 6$; the torsional energy barrier is only 0.006 kcal/mol.

Recent studies have shown that the high torsional energy of the eclipsed form of ethane and hence *the torsional energy barrier* is not due to van der Waals repulsive steric interaction between the two eclipsed H atoms (the distance between them is 2.3 Å = 0.23 nm, whereas the van der Waals radii of two H atoms is 1.83 Å), or not due to electrostatic interaction between weakly polarized C–H bonds, but it *is due to torsional strain caused by unfavorable overlap interaction between the C–H bond orbitals*. In case of the staggered form (the distance between two nearest H

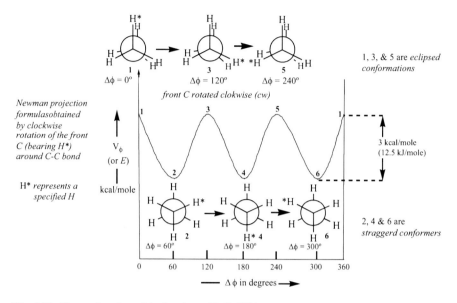

Fig. 2.20 V_ϕ as a function of Δ_ϕ in ethane (H_3C-CH_3)

atoms on C1 and C2 is 2.5 Å) favorable interaction between the bonding–antibonding orbitals makes its torsional strain 3 kcal/mol less than the eclipsed conformation and makes it more stable. This energy difference leads at 25 °C to the fact that for each 160 staggered ethane molecules, there is only one molecule of eclipsed ethane, *i.e.*, in negligible proportion.

2.3.4 Torsional Strain Curve of Propane

The torsional strain curve of propane is similar to that of ethane with a slightly higher torsional strain barrier (3.36 kcal/mol = 14 kJ/mol). The small difference between the energy barriers of ethane (Fig. 2.20) and propane (Fig. 2.21) indicates clearly that the torsional energy does not originate from steric effects, as already mentioned. The eclipsing of H with CH_3 in propane is hardly more unfavorable than the eclipsing of H and H in ethane, despite the bulkiness of the methyl group. In case of propane also the periodicity n is 3.

2.3.5 Torsional Strain Curve of Molecules ACX_2CX_2B, n-Butane

Torsional strain curve (V_ϕ as a function of $\Delta\phi$) of a molecule $AC(XY)C(XY)B$ becomes more complex when a given conformation does not recur within a complete torsional displacement ($\Delta\phi = 360°$), *i.e..*, n (periodicity) = 1. The

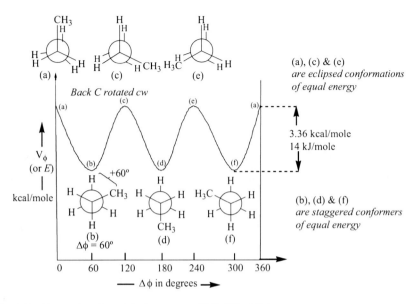

Fig. 2.21 V_ϕ as a function of Δ_ϕ in propane ($H_3C–CH_2–CH_3$)

torsional strain curve of the molecules of butane, $CH_3CH_2CH_2CH_3$, (where A = B = CH_3 and X = Y = H) giving the Newman projection formula of the maximum and minimum energy conformations, is displayed schematically in Fig. 2.20. The diagram shows two types of conformations having maxima: (**a**) and (**c**) = (**e**). The conformation (**d**), called *anti*, represents the energy minimum corresponding to the lowest valley (minimum); the energies of the other conformers (**b**) and (**f**), called *gauche*, corresponding to the other valleys (minima) are measured relative to that of this conformer (**d**), as shown in Fig. 2.22. The conformers, (**c**) and (**e**), representing the steric interaction energy of two CH_3 groups at $\Delta\phi = 60°$ or $300°$ are destabilized by 0.8-0.9 kcal/mol (~ 3.3 kJ/mol) at room temperature relative to the *anti* conformer (**a**).

This energy difference corresponds to about 1 mol of butane as the *gauche* conformer for every 2 moles of butane as the *anti* conformer. This may be calculated for the equilibrium [gauche] ⇆ [anti], as follows:

$\Delta G° = \Delta H° - T\Delta S° = -0.8$ kcal/mol$-(-TR\ln2) = (0.8-0.41 = 0.39$ kcal/mol at 27 °C, T = 300 °K, and

$R\ln2 = 1.38$ cal/deg/mol.

Putting the value of $\Delta G°$ in the equation $\Delta G = -RT\ln K$

K = [anti]/[gauche] ≈ 2

Thus, it is possible to predict the most stable conformer of a long chain ($-CH_2)_n$, which possesses repeated butane chain, as a *zigzag* planar arrangement of the chain.

The torsional strain curve of other molecules having the same molecular formula ACX_2CX_2B will be of same pattern as that of butane. Only the energy barriers will be different. In *n*-propyl chloride, $CH_3CH_2CH_2Cl$ (A = CH_3, B = Cl, and X = H),

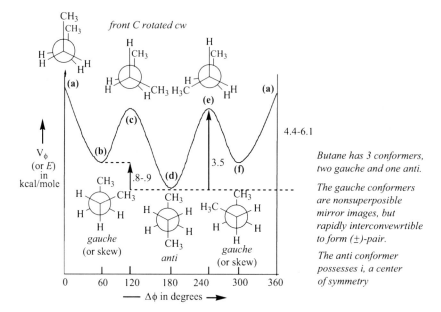

Fig. 2.22 V_ϕ or E of butane as function of Δ_ϕ (displacement of torsion angle)

and 1,2-dichloroethane ($A = B = Cl$, $X = H$) the *anti* and gauche conformers have approximately the same energy, whereas in 1,1,2,2-tetrachloroethane and 1,1,2,2-tetrabromoethane ($A = B = H$, $X = Cl$ or Br), the *gauche* conformer is known to be more stable than the *anti* conformer by ~1.0 kcal/mol, perhaps due to the stabilizing *gauche effect of the two halogen atoms*.

2.4 Configuration. Relative Configuration. Absolute Configuration

Molecules with the same molecular formula and same constitution or bonding connectivities may still be different. The arrangement of the ligands (atoms or groups) in space around the dissymmetric part of a molecule—in the simplest case around an asymmetric center or around the rigid part of a molecule like a double bond—is termed as *configuration*. If there is any dissymmetry in the molecule (lacking a reflection symmetry)—possessing a chiral point group (Sect. 2.2.5)—the molecule will be nonsuperposable on its mirror image and the two mirror image forms (enantiomers) will exhibit opposite optical rotation. When molecules differ only in relative orientation of ligands in space, stereoisomerism results. Stereoisomers having same bond connectivity but different absolute configuration are often called configurational isomers. Most of the natural products and biologically active molecules occur in specific stereoisomeric forms; their chemical and biological behaviors are controlled by their absolute configuration and also conformation.

In 1780 Scheele isolated $(-)$-lactic acid [= $(-)$-α-hydroxypropionic acid] from sour milk. Berzelius isolated $(+)$-lactic acid from muscle tissues in 1807 (Fig. 2.23). Actually, in 1848 Berzelius established that although they have same structure and same properties, they were not same, but enantiomers (nonsuperposable mirror images) exhibit opposite optical rotation. *Configuration is a theoretical concept* related to the molecular architecture expressed by a three-dimensional or projection formula, *while rotation is an experimental property*, measured in a polarimeter. Configuration is more fundamental than rotation and remain unchanged as long as any bond to the chiral center is not broken; but specific rotation changes with concentration and with solvent; even the sign of rotation may sometimes change with some solvent at different concentrations (*vide* Sect. 2.2.8). Again $(+)$-lactic acid when dissolved in aqueous sodium hydroxide, the resulting solution is levorotatory and would have to be called a solution of $(-)$-sodium lactate, although the configuration remains unchanged.

The simplest way to explain relative and absolute configuration is to state that two enantiomers (*e.g.*, of lactic acid) have got the same relative configuration, but have different absolute configurations.

Absolute configuration is specified by (D,L) notation (Sect. 2.6.1) in some cases and by (*R,S*) notation universally (Sect. 2.6.2), to be discussed in the respective sections. In a compound with several asymmetric centers, configuration at each center is to be specified in order to specify its configuration completely.

The arrangement of groups (or ligands) about the rigid double bond or ring (*i.e.* the configuration) is expressed as *cis* or *trans* (*Z* or *E*) for double bond compounds (Sect. 2.6.2.6) and *cis* or *trans* for cyclic compounds (Sect. 2.12.4).

The Fischer projection and tetrahedral representations of two common chiral molecules are shown in Fig. 2.23).

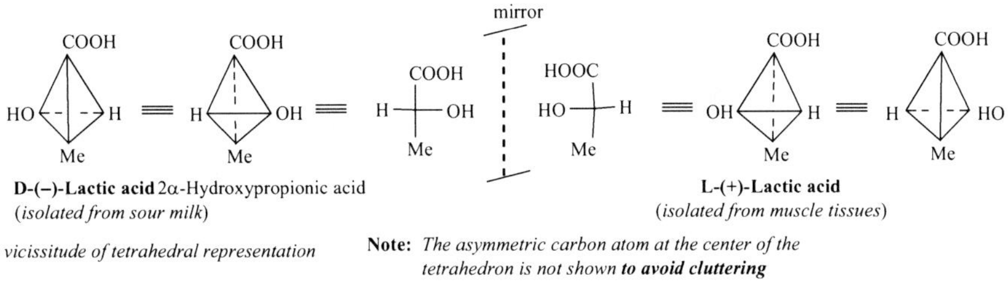

D-(−)-Lactic acid 2α-Hydroxypropionic acid
(*isolated from sour milk*)

L-(+)-Lactic acid
(*isolated from muscle tissues*)

vicissitude of tetrahedral representation

Note: *The asymmetric carbon atom at the center of the tetrahedron is not shown to avoid cluttering*

Another example: mandelic acid (α-hydroxy-α-phenylacetic acid).

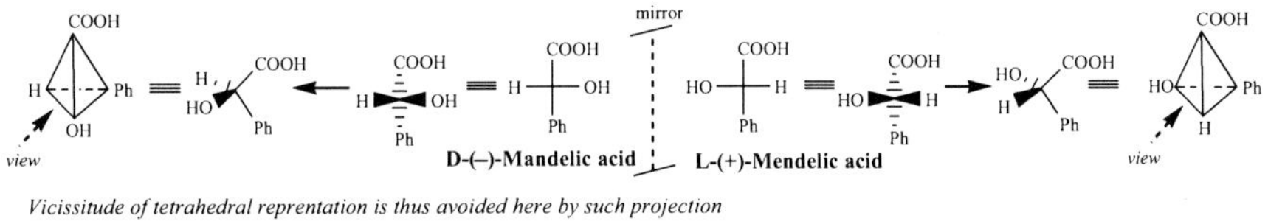

D-(−)-Mandelic acid L-(+)-Mandelic acid

view

Vicissitude of tetrahedral reprentation is thus avoided here by such projection

Fig. 2.23 The enantiomers of lactic acid and mandelic acid. Each pair has same relative configuration but different absolute configuration

2.5   Relationship Between Two Molecules of Same Molecular Formula. Homomers, Constitutional Isomers, Stereoisomers, Enantiomers, Diastereomers, Configurational/Conformational Enantiomers/ Diastereomers

Relationship between two molecules of same molecular formula giving rise to different stereochemical or structural terms is delineated schematically with examples in Fig. 2.24. The definition of each term, expressed in bold face, is understood from this figure.

2.6   Configurational Nomenclature

Two different methods are in use for completely specifying the absolute configuration of a chiral center in a molecule.

2.6.1   Fischer's D,L Nomenclature

This oldest system of nomenclature of chiral compounds was introduced by Emil Fischer [3] in 1891, while working with carbohydrates. Rosanoff [6] modified the system in 1906 and suggested the *following conventions* for a *projection* nomenclature of D,L system:

(i) As in Fischer's system, the molecule is written with the longest carbon chain placed vertically.

(ii) The C1 carbon or the most highly oxidized end of the chain is placed at the top, following Fischer's convention, *e.g.*, COOH > CHO > CH_2OH (according to IUPAC also).

Some examples of the use of D,L nomenclature and its extension in some cases are delineated in Fig. 2.25).

(a) The system works well for compounds $RCHXR^1$, when X is a hetero atom (Fig. 2.25). The molecule is D if X is on the right, and L if X is on the left. While writing the Fischer projection phenyl or aryl group should be at the bottom.

(b) In compounds of the type RR^1CXR^2, this system is applicable; D means that X is on the right, and the small alkyl group R^1 is to the left. The enantiomer of D is L.

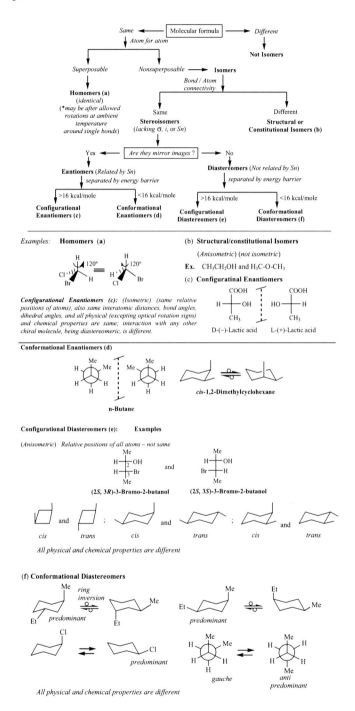

Fig. 2.24 Relationship between two molecules of same molecular formula: different terms with examples

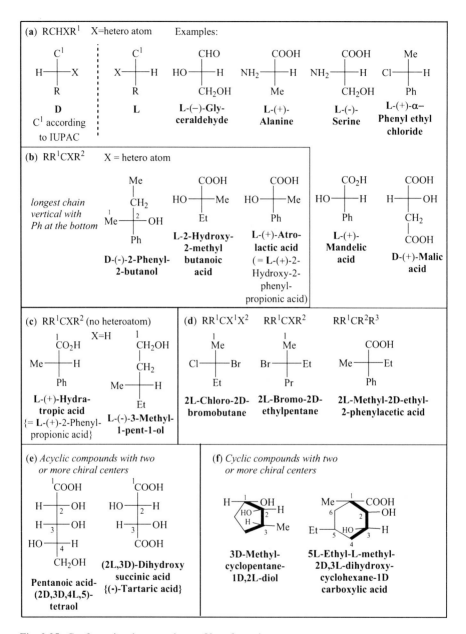

Fig. 2.25 Configurational nomenclature. Use of D and L

(c) If there is no hetero atom, then the smallest alkyl group (usually Me) becomes the fiducial group; if it is on the right it is D, if on the left it is L.

(d) When the chiral center carries two different substituents of comparable electronegativity, D- and L- may be used separately for each substituent to specify the configuration. Thus, the system may be applicable to compounds of the types $RR^1CR^2R^3$, having asymmetric quaternary carbon, *e.g.*, 2L-methyl-2D-ethyl-2-phenylacetic acid.

(e) For compounds containing more than one asymmetric carbon and (f) for compounds possessing asymmetric carbons in rings the *D,L* nomenclature has been extended by Klyne [7], *e.g.*, 3D-methylcyclopentane-1D,2L-diol. Rings are oriented with the edge of the lowest numbers toward the viewer, the main chain being numbered from the top bearing the more oxidized carbon [6] or C1 (Fig. 2.25).

(g) In carbohydrate chemistry, D,L nomenclature is assigned on the basis of configuration of the last chiral center of the chain written vertically counting from the top in the Fischer projection, more oxidized carbon being at the top (Fig. 2.26) [6].

It is to be noted that D,L *(like R,S) nomenclature has no genetic relationship.* Thus, the COOH group of D-hydratropic acid when converted to NH_2 gives L-1-phenylethylamine, although no bond to the chiral center is broken (Fig. 2.27).

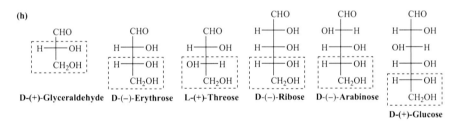

(h)

D-(+)-Glyceraldehyde   D-(-)-Erythrose   L-(+)-Threose   D-(-)-Ribose   D-(-)-Arabinose   D-(+)-Glucose

Fig. 2.26 Configurational (D,L) nomenclature, Glyceraldehyde, tetroses, pentoses, and Glucose

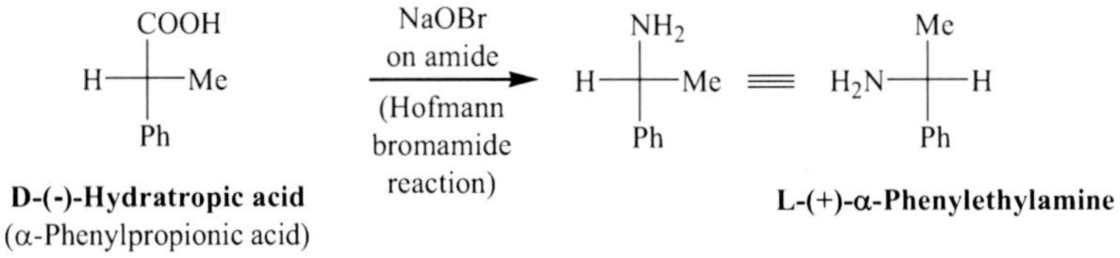

D-(-)-Hydratropic acid
(α-Phenylpropionic acid)

NaOBr on amide
(Hofmann bromamide reaction)

L-(+)-α-Phenylethylamine

Fig. 2.27 D,L nomenclature has no genetic relationship. Illustration

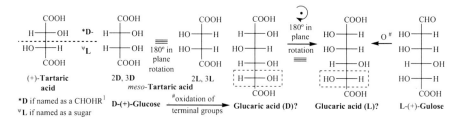

Fig. 2.28 A few ambiguities of D,L nomenclature

Ambiguities/Shortcomings of D,L Nomenclature (Fig. 2.28)

(i) As already mentioned, too many arbitrary conventions are to be remembered for D,L nomenclature of different types of compounds.

(ii) This system cannot be conveniently used for polycyclic compounds.

(iii) Ambiguity arises for D and L designation of tartaric acid diastereomers (Fig. 2.28).

(iv) The oxidation product of D-(+)-glucose may be termed both D-glucaric acid and L-glucaric acid which is also obtained by oxidation of L-(+)-gulose.

2.6.2   R,S Nomenclature for Absolute Configuration

The D,L nomenclature of Fischer (Sect. 2.6.1) applies to the Fischer projection. In spite of the shortcomings/ambiguities already discussed, the Fischer nomenclature has proved particularly useful for sugars or amino acids. However, one should avoid confusion between d, l (dextrorotatory or levorotatory) and D,L (absolute configuration). A d compound may have the absolute configuration either D or L, e.g., (d)-L-alanine and (d)-D- glyceraldehyde, which are now expressed as L-(+)-alanine and D-(+)-glyceraldehyde.

As D,L nomenclature is not generally useful to some types of compounds with chiral centers mentioned earlier. Cahn, Ingold, and Prelog (CIP) developed R-S nomenclature, [5, 8, 9] generally applicable to chiral compounds of all three different types having (i) *a center of chirality*, (ii) *an axis of chirality*, and (iii) *a plane of chirality*. Each of these chirality elements requires a particular method of nomenclature. We will now discuss the CIP (Cahn–Ingold–Prelog) rules for compounds containing one or more chiral centers. Compounds of the types (ii) and (iii) will be discussed in Sects. 2.16 and 2.17, respectively.

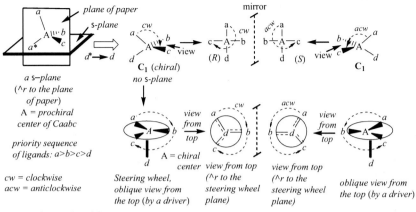

Fig. 2.29   Prochiral center. *R,S*-Designation of a chiral center (*cf.* steering wheel)

2.6.2.1   *R,S* Nomenclature. Center of Chirality

A center of chirality is usually associated with a nonplanar tetracoordinated atom "A" bonded (or ligated) to four different ligands (atoms or groups) or an asymmetric carbon atom, A*abcd*. The latter could be derived from an *achiral* or a *prochiral* precursor A*aabc* by changing a ligand *a* to a new ligand *d* (different from the existing ligands). This precursor possesses a plane of symmetry containing A–b and A–c bonds (Fig. 2.29); the chirality of A*abcd* (point group C_1) is the consequence of the destruction of the plane of symmetry of A*aabc*.

2.6.2.2   Specification of Center/s of Chirality

Two CIP rules are involved: (i) CIP *sequence rule* and (ii) CIP *chirality rule.*

(i)   **CIP Sequence rule:** Ligands are sequenced by comparing them at each step in bond-to-bond exploration in branched ligands *along the branch path of highest precedence*. This rule will be illustrated by examples.

   The following "Standard Sub-rules" are used for finding out the priority sequence of all possible types of compounds—*each to exhaustion in turn* (Fig. 2.30).

Sub-rule 0:	Nearer end of axis or side of plane precedes the further end (applicable to chiral axis or chiral plane).
Sub-rule 1:	Higher atomic number precedes lower.
Sub-rule 2:	Higher atomic mass number precedes lower, *e.g.*, T > D > H $^{14}C > ^{13}C$
	Two ligands varying only by isotopes produce marginal chirality and hence small optical rotation
Sub-rule 3:	*Seq cis* (*Z*) precedes *seq trans* (*E*)
Sub-rule 4:	*Like pair precedes unlike pair,* > stands for 'precedes' or 'has priority over'; M = minus helicity
	RR or *SS* > *SR* or *RS*; P = plus helicity
	MM or *PP* > *MP* or *PM*; *RM* or *SP* > *RP* or *SM*
	MR or *PS* > *MS* or *PR*; *r* > *s* (in case of achirotopic but stereogenic center)
	(Like and unlike pairs are gleaned from the sub-rule 5)
Sub-rule 5:	*R* > *S*; *M* > *P*

The priority sequence in the decreasing order, a > b > c > d is assigned to the four ligands of the chiral center.

Fig. 2.30 CIP standard sub-rules for ascertaining the priority sequence of ligands

(ii) **CIP Chirality rule** (Fig. 2.29): The path of the sequence of precedence/priority $a \rightarrow b \rightarrow c$ is followed from the preferred side of the model (containing the three preferred ligands a, b, and c), *i.e.*, remote from the ligand of the lowest precedence *d*. If the path turns *right* (traces clockwise, which is the entire sense of direction of abc) the element is assigned chiral label "*R*" (*rectus,* Latin for right). If the said path turns *left* (traces anticlockwise, which is the entire sense of direction of *abc*) then the stereogenic center is assigned the chiral label "*S*" (*sinister,* Latin for left).

One can imagine that *a, b*, and *c* are placed on a steering wheel of a car, and *d* is placed on the shaft (the axis of the wheel is A-d); if from a position above the wheel one rotates the wheel on the right side (in the clockwise sense $a \rightarrow b \rightarrow c$), the car moves to the *right*, and the absolute configuration is called *R*, whereas in the opposite case, the car moves to the *left* and the absolute configuration is called *S*.

It is evident from Fig. 2.29 that in the Fischer projection formula, if *d* (the least priority group) is at the *bottom* or *top*: $a \rightarrow b \rightarrow c$ makes a clockwise movement in the *R*-configuration; $a \rightarrow b \rightarrow c$ makes an anticlockwise movement in the *S*-configuration.

R,S-configurational specification of some common compounds is shown in Fig. 2.31.

Corollaries of the Chirality Rule

1. If *d* is on the *right* or *left* side of the Fischer projection formula, or *above* the plane containing the two in-plane bonds in *the flying wedge* formula, the

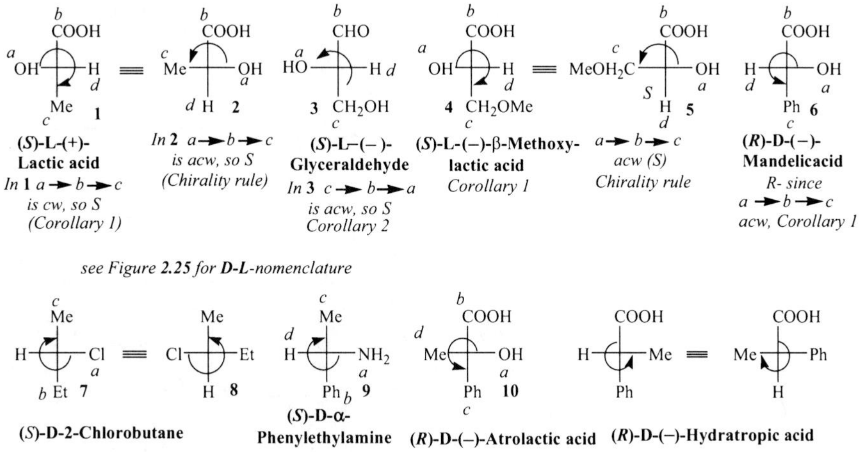

Fig. 2.31 R-S-configurational designation of some common chiral compounds. Use of corollaries (**1**) and (**2**) also

anticlockwise movement from $a \to b \to c$ (or $1 \to 2 \to 3$) will define R-configuration, and *clockwise* movement will define S-configuration.

2. In the above case, clockwise movement from $c \to b \to a$ (or $3 \to 2 \to 1$) will define R-configuration, and anticlockwise movement from $c \to b \to a$ (or $3 \to 2 \to 1$) will define S-configuration.

The sequence sub-rules are applied considering different aspects, as needed:

1. Sub-rule 0 gets the topmost priority but is applicable to axially dissymmetric compounds to be discussed later (Sect. 2.16).
2. If the atoms of two ligands attached to the asymmetric tetracovalent atom (like C, N, P, Si, etc.) are the same, their respective states of substitution are considered. If the second atom gives no choice, third, etc., is to be considered, following *along the branches with closer atoms of higher atomic number*.
3. If the two atoms directly linked to the asymmetric atom are different, the atoms of higher atomic number gets precedence, *e.g.*, $I > Br > Cl > S > P > Si > F > O > N > C$
4. If the atoms of two ligands attached to the asymmetric atom are same, that with more substituents of higher atomic number gets precedence

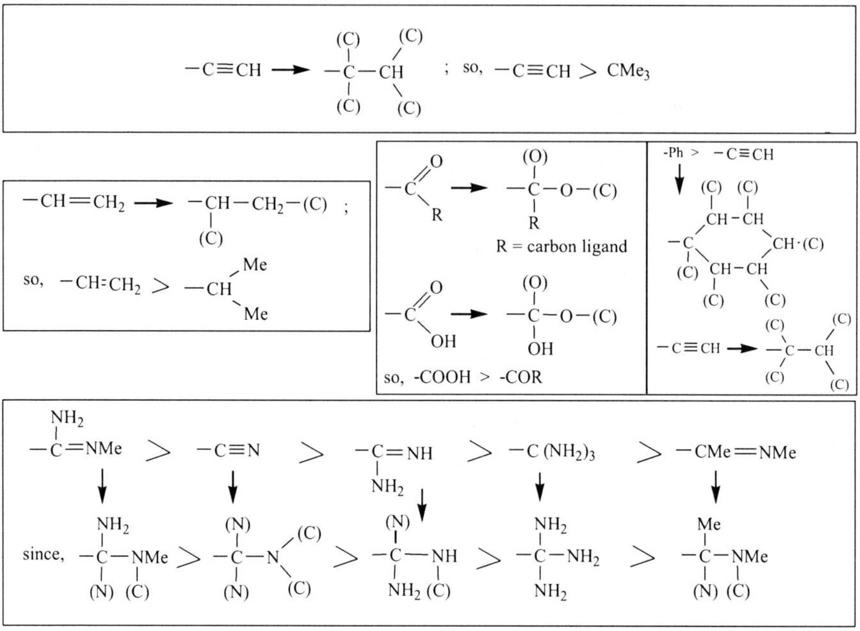

Fig. 2.32 Multiple bonded atoms treated as 4-coordinated ones for determining priority sequence

5. If the ligands attached to the asymmetric center are composed of only C and H atoms, then *the ligand with less number of H atom gets priority.*

For example: $-CR^{1R2R3} > -CHR^{1R2} > -CH_{2R}^{1} > CH_3$

(here R^1, R^2, R^3 may be same or different alkyl or aryl substituent)

6. *Multiple bonded atoms* are treated as four-coordinated ones by adding replica atoms of the same type (duplications or triplications) which are bracketed to signify that they are surrounded by phantom atoms of atomic number zero. This is illustrated by some examples given in Fig. 2.32.

Table 2.4 contains some of the most common groups in the decreasing order of precedence obtained by applications (2–6) of the sequence sub-rules 1 and 2.

7. A few more examples of *R-S* designation of compounds with several asymmetric atoms (written as Fischer projection or flying wedge formulas) have been included in Fig. 2.33.

Table 2.4 Descending order of priority of some common ligands according to the sequence rules

1. $-I$	25. $-NH_2$	49. $-Ph$
2. $-Br$	26. $-CX_3$	50. $-C\equiv CH$
3. $-Cl$	27. COX	51. $-CMe_3$
4. $-PR_2$	28. $-CO_2CMe_3$	52. $-CH=CH-Me$
5. $-SO_3H$	29. $-CO_2Ph$	53. $-C_6H_{11}$
6. $-SO_2R$	30. $-CO_2CH_2Me$	54. $-CH(Me)Et$
7. $-S(O)R$	31. $-CO_2Me$	55. $-CH=CH_2$
8. $-SR$	32. $-CO_2H$	56. $-CHMe_2$
9. $-SH$	33. $CONH_2$	57. $-CH_2Ph$
10. $-F$	34. $-COPh$	58. $-CH_2C\equiv CH$
11. $-OSO_2R$	35. $-COMe$	59. $-CH_2-CMe_3$
12. $-OCOR$	36. $-CHO$	60. $-CH_2CH=CH_2$
13. $-OR$	37. $-CR_2OH$	61. $-CH_2CHMe_2$
14. $-OH$	38. $-CH(OH)R$	62. $-CH_2CH_2Et$
15. $-NO_2$	39. $-CH_2OR$	63. $-CH_2CH_2Me$
16. $-NO$	40. $-CH_2OH$	64. $-CD_2Me$
17. $-N^+R_3$	41. $-CN$	65. $-CHDMe$
18. $-NR_2$	42. $-CH_2NH_2$	66. $-CH_2Me$
19. $-NHCOPh$	43. $-C_6H_4Me(o)$	67. $-CD_3$
20. $-NHCOR$	44. $-C\equiv CMe$	68. $-Me$
21. $-NHCH_2Me$	45. $-C_6H_4NO_2(m)$	69. $-T$
22. $-NMe_2$	46. $-C_6H_4Me(m)$	70. $-D$
23. $-NHR$	47. $-C_6H_4NO_2(p)$	71. $-H$
24. $-N^+H_3$	48. $-C_6H_4Me(p)$	72. Electron pair

2.6.2.3 Priority Sequence of the Application of the CIP Sub-rules

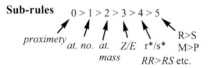

Sub-rule O (proximity rule) enjoys the topmost priority in cases of axial chirality.

Few *examples of the priorities of the sub-rules are provided in Fig.* 2.34.

2.6.2.4 Modification of Sub-rule 3

The sub-rule 3 has been modified by Prelog and Helmchen [10] in 1982 by an alternative proposal. The modified sub-rule and some examples illustrating it are given in Fig. 2.35.

2(S), 3(R), 4(R)-3-Amino, 3-bromo-2,4-dihydroxypentanoicacid
Zigzag formula
D-(+)-Glucose (open chain form)
2 (R) 3 (S) 4 (R) 5 (R)
Zigzag formula
b-D-glucose ψ
1(R), 2(R), 3(S), 4(S), 5(R)
ψ Although the absolute configuration of C4 remains unchaged after ring formation, its specification has changed due to the reversal of the CIP p. s. of its ligands, viz. C5>C3.
Notes: At any t-carbon, presence of a dot or an a-substituent indicates b-H, and whose absence indicates a -H.
*Arrows start from the priority a bond and end at the priority c bond of a chiral center.
3(S), 5(S), 8(R), 9(S), 10(S), 13(R), 14(S), 17(R), 20(R)
Cholestan-3b-ol

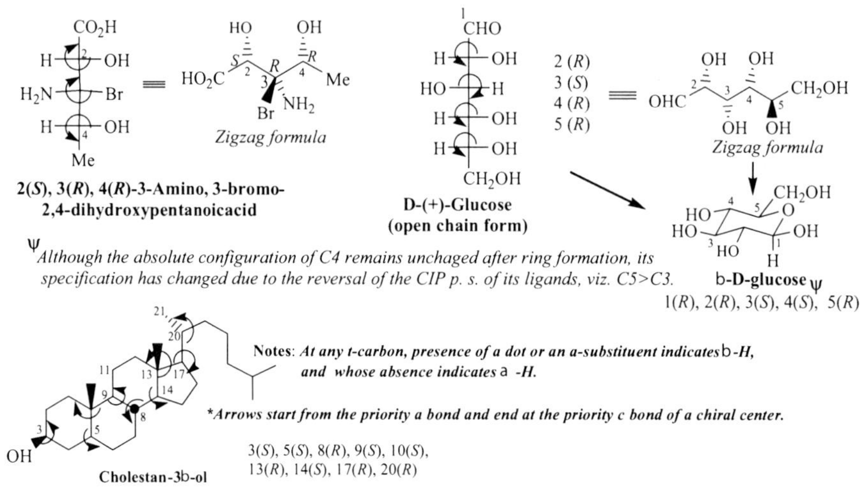

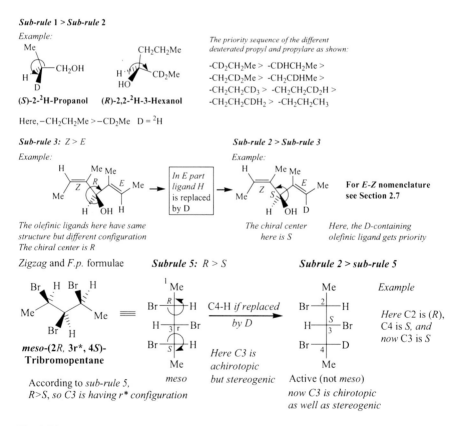

Fig. 2.33 *R-S* Designation of compounds with multi-chiral centers and monocyclic and polycyclic compounds

Sub-rule 1 > Sub-rule 2
Example:
(S)-2-²H-Propanol
(R)-2,2-²H-3-Hexanol
The priority sequence of the different deuterated propyl and propylare as shown:
-CD2CH2Me > -CDHCH2Me > -CH2CD2Me > -CH2CDHMe > -CH2CH2CD3 > -CH2CH2CD2H > -CH2CH2CDH2 > -CH2CH2CH3
Here, –CH2CH2Me > –CD2Me D = ²H
Sub-rule 3: Z > E
Example:
In E part ligand H is replaced by D
Sub-rule 2 > Sub-rule 3
Example:
For E-Z nomenclature see Section 2.7
The olefinic ligands here have same structure but different configuration
The chiral center is R
The chiral center here is S
Here, the D-containing olefinic ligand gets priority
Zigzag and F.p. formulae
Subrule 5: R > S
Subrule 2 > sub-rule 5
meso-(2R, 3r*, 4S)-Tribromopentane
C4-H if replaced by D
Here C3 is achirotopic but stereogenic
meso
Example
Here C2 is (R), C4 is S, and now C3 is S
Active (not meso)
now C3 is chirotopic as well as stereogenic
According to sub-rule 5, R>S, so C3 is having r* configuration

Fig. 2.34 Priority of some sub-rules over others. *R*-*S*-nomenclature of compounds with multi-chiral (here 3) centers from zigzag formula

Modified sub-rule 3 (alternative proposal) by *Prelog & Helmchen* [8]

The olefinic ligand with the higher priority substituent on the terminal olefinic carbon on the same side as the chiral centre will get priority over the other olefinic carbon and is designated R_n. The other olefinic carbon is designated S_n

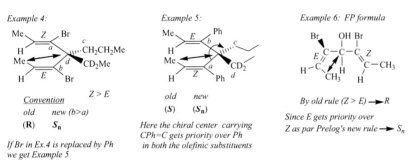

Example 1:

Here, by old and new conventions the configuration is **R** and **R_n** respectively

Examples 2:

Convention
old   new (c>b)
(S)   (**R_n**)

Examples 3:

Here, by old convention (b>c) the chiral center is S but by new convention it is **R_n** since c>b

Notes: Old convention: Z>E
New convention: In either example 2 or example 3, the lower part of the olefinic substituent carrying Cl in example 2 and Me in example 3 on the same side of the double bond as the chiral center, gets priority over the upper olefinic substituent, which has Z-configuration.

Example 4:

Convention   Z > E
old   new (b>a)
(R) S_n

If Br in Ex.4 is replaced by Ph we get Example 5

Example 5:

old   new
(S)   (S_n)

Here the chiral center carrying CPh=C gets priority over Ph in both the olefinic substituents

Example 6: FP formula

By old rule (Z > E) ⟶ R

Since E gets priority over Z as par Prelog's new rule ⟶ S_n

Note: *Intervening CH_2 or CH_2's will not change the chirality specification*

Fig. 2.35 Modification of sub-rule 3 by Prelog and Helmchen. Several examples to show chirality specifications

2.6.2.5   R* and S* Nomenclature

When *R-S* nomenclature is applied, in a molecule with multiple chiral centers both the relative and the absolute configurations are fixed, as has been illustrated in many chiral molecules so far. For examples, 2(*R*)-bromo-3(*S*)-hydroxybutane refers to the enantiomer (**1**) (Fig. 2.36).

It often happens that *a pure enantiomer is reported with known relative configuration but unknown absolute configuration*. In such cases the *R-S* system of nomenclature is modified as follows (IUPAC Commission 1976) [11]. The atoms are numbered such that the chiral center carrying the highest priority ligand, *e.g.*, C-Br in (**1**) or (**2**) or C-I in (**3**) is given the lowest number (lowest locant). The molecule is written in such a way that the lowest chiral locant gets the *R* configuration. The other chiral centers are then assigned in the usual way, and each descriptor is asterisked (pronounced as *R*-star or *S*-star, indicating thereby that they represent relative configuration). Thus, 2(*R**)-bromo-3(*S**)-hydroxybutane or

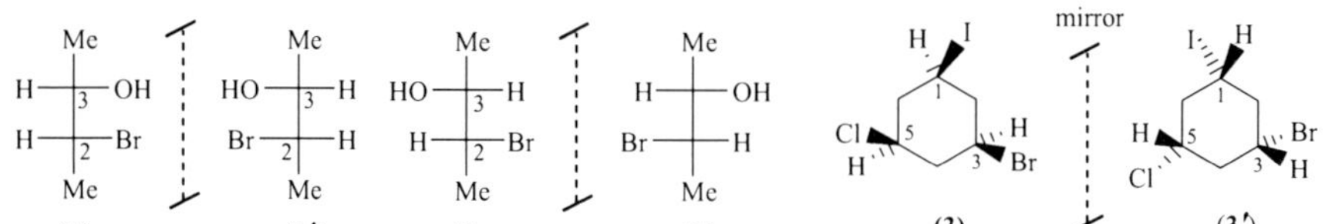

(1) (1′) (2) (2′) (3) (3′)

2(*R)-Bromo-3(*S**)-hydroxybutane** (1) or (1′) (≡ *erythro*-3-Bromo-2-butanol)

2(*R)-Bromo-3(*R**)-hydroxybutane** (2) or (2′) (≡ *threo*-3-Bromo-2-butanol)

1(*R)-Iodo-3(*S**)-bromo-5(*R**)-chlorocyclohexane** represent either (3) or (3′)

Fig. 2.36 *R**-*S** Nomenclature of some compounds of known relative configuration

Tetracovalent chiral compounds with tetrahedral stereocenters. *Examples:*

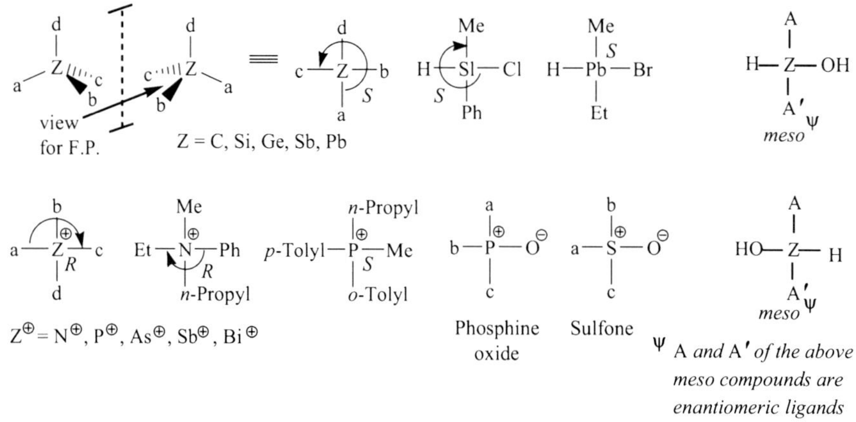

Fig. 2.37 Tetrahedral stereocenters of tetracovalent chiral compounds

simply *erythro*-3-bromo-2-butanol represents either compound (**1**) or its mirror image (**1′**); 2(*R**)-bromo-3(*R**)-hydroxybutane (or threo-3-bromo-2-butanol) represents either (**2**) or (**2′**); and 1(*R**)-Iodo-3-(*S**)-bromo-5(*R**)-chlorocyclohexane represents either (**3**) or (**3′**).

2.6.2.6 Specification of Other Tetracovalent Chiral Atoms

If four different achiral ligands are attached to a tetracovalent atom other than carbon, *e.g.*, Si, Ge, As^+, P^+, N^+, the resulting compounds also possess stereogenecity (*R,S*) as well as chirotopicity; they will have non-superposable mirror images and hence they are optically active. However, such tetracovalent atom carrying two chiral ligands of same structure but of opposite configuration will be still *stereogenic* (*r,s*) but *achirotopic*, making the compound *meso* (Fig. 2.37).

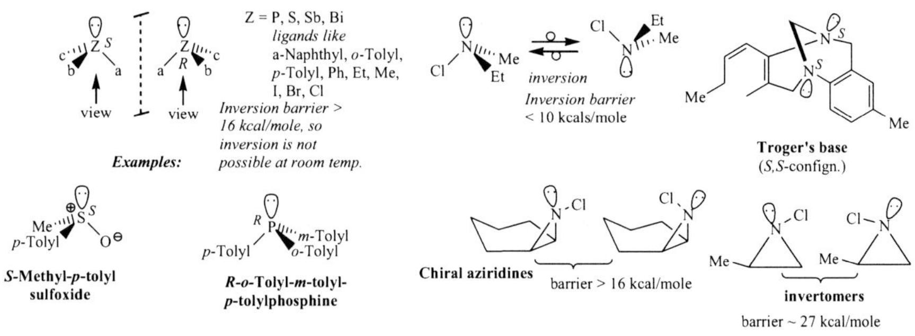

Fig. 2.38 Pyramidal stereocenters of tricovalent chiral compounds

2.6.2.7 Specification of Tricovalent Chiral Compounds (*with pyramidal stereocenter*)

Tricovalent chiral compounds with P, As, Sb, Bi, etc. at the stereocenter are incapable of inversion of the pyramidal stereocenter to form their mirror images (due to their high energy barrier), and hence exhibit optical isomerism. The barriers to nitrogen inversion are usually far too low to permit isolation of the two stereo-isomers. Since the inversion of the sp^3 hybridized nitrogen pyramid involves an sp^2 hybridized transition state, increase of p character of the bonds to N slows down the rate of inversion as in substituted aziridines (internuclear angle $60°-90°$); the rate of inversion is also slowed down by electronegative substituent like Cl. These two factors cooperate in substituted aziridines to increase the energy barrier between the two isomers, which are also called *invertomers* (Fig. 2.38), permitting their isola-tion. In Troger's base (Fig. 2.38) the N atoms are present at the bridgeheads and hence pyramidal inversion is not possible without bond cleavage.

2.6.3 Stereochemistry of Alkenes. E,Z Nomenclature [12–14]

(a) **Stereochemistry.** The C–C σ bond strength is about 83 kcal mol^{-1}, and the strength of the π bond is only 62 kcal mol^{-1} due to its less favorable lateral overlap; addition of these two numbers gives the generally accepted total energy of a $C = C$ double bond as 145 kcal mol^{-1}, much more than the rotational barriers in alkanes (*e.g.*, 3.6 kcal mol^{-1} for C2–C3 bond in butane). During the process of rotation of the Z to the E isomer or *vice versa*, the p orbitals of the two olefinic carbon atoms become orthogonal, with no overlap. Thus, the π-bond is completely broken in the transition state.

The bond length of $C = C$ in unstrained unconjugated ethenes ranges from 1.335 to 1.35 Å (133.5–135 pm), but is extended in conjugated alkenes in which $C = C$ bond is weakened. The bond angle of substituted olefins is not 120° and varies. Electron diffraction showed that H–C–H angle of ethane is 117.8°. In propene the C = C–C angle is 124.3° and the C = C–H angle is 119°. In *cis*-2-butene the C = C–C angle is 125.8°. The Me–C–Me angle in 2-methylpropene, Me–C(Me) = CH_2, is 115.3°. Apparently, R–C–R′ angle in RR′C = C is generally smaller than 120°, while R–C = C angle is larger.

(b) **Nomenclature.** *Cis-trans* isomerism in substituted olefins, earlier called geometrical isomerism, is now regarded as a type of diastereomerism, since *cis* and *trans* isomers are optically inactive (if all olefinic ligands are achiral) stereoisomers and are not enantiomers. Since this isomerism owes its existence to the presence of a π-bond, it is called *π-diastereomerism*, to distinguish it from σ-*diastereomerism*, exhibited by cyclic compounds. This terminology, however, is not commonly used.

For molecules of the type $C_{ab}{=}C_{ab}$ or $C_{ab}{=}C_{ac}$, the terms *cis* and *trans* are unambiguous. But if all four substituents are different, this nomenclature leads to ambiguity. This problem is solved by arranging the pair of ligands at each olefinic carbon in CIP sequence. If the groups of higher priority are on the same side the configuration is *seq-cis*, later replaced by the symbol *Z* (from the German *zusammen* meaning "together"); if they are on the opposite sides, the configuration is *seq-trans*, replaced by the symbol *E* (from the German *entgegen* meaning "opposite" [12]. Examples illustrating application of the rules for assignment of *E* and *Z* are shown in Fig. 2.39.

The *E-Z* Nomenclature is always applicable and unequivocal, including cases where *cis* and *trans* nomenclature becomes ambiguous (Fig. 2.40). The *E-Z* system is useful for unambiguous nomenclature [13, 14] of oximes (such as **F′** and **G**) and of compounds containing non-cumulated (conjugated) double bonds (such as **H** and **I**), and of cumulenes with odd number of cumulated double bonds with two = C_{ab} as terminal groups (such as **J**), as illustrated in Fig. 2.40. The compounds **F′** and **G** are *E* and *Z* isomers rather than *anti* and *syn* isomers (old nomenclature), respectively. In case of compounds **H** and **I**, the *locants must be used in conjunction with the E,Z-descriptors*. In the cumulenes with odd number of double bonds (planar, achiral) (three double bonds in case of **J**), the two ligands in each terminal carbon are in the same plane (cf. olefin) and hence *E-Z* nomenclature is applicable to them also. Thus, the cumulene **J** is a *Z*-isomer.

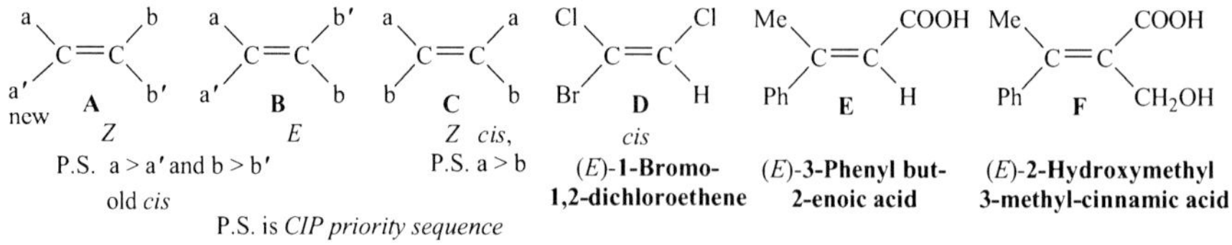

Notes: • *The descriptors (E) and (Z) are always italicized and is placed in parenthesis in front of the name*
• *Structure* **D** *reveals that E and Z do not always correspond to trans and cis respectively. Compound* **E** *and* **F** *have (E) stereochemistry, but they cannot be assigned cis or trans.*

Fig. 2.39 (*Z*)- and (*E*)-alkenes. Unambiguous nomenclature for any substituted olefin

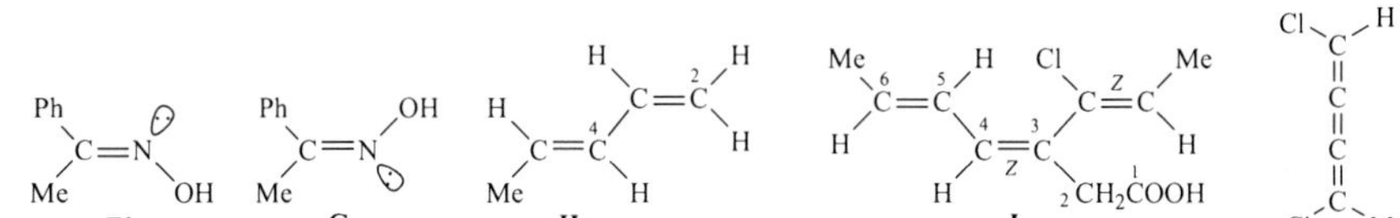

old F′ G H I J

anti (old names) *syn*

new (*E*)-isomer (*Z*)-isomer

of acetophenone oxime

(2*Z*,4*E*)-2,4-**Hexadienoic acid**

3-[*Z*-1-**Chloropropenyl**]-3*Z*, 5*E*-3,5-**heptadienoic acid**

(*Z*)-isomer

Fig. 2.40 Additional examples of *E*-*Z* nomenclature

2.7 Projection (Fischer, Newman, Sawhorse) and Perspective (Flying Wedge and Zigzag) Formulas of Molecules with Two or More Chiral Centers. Working out Stereoisomers

2.7.1 Molecules with Two Unlike (Unsymmetrical) Chiral Centers (AB Type)

Many natural products, *e.g.*, steroids, terpenoids, alkaloids, and carbohydrates contain two or more chiral centers, the stereochemistry of which should be thoroughly understood. An acyclic molecule containing two or more chiral carbons is constitutionally unsymmetrical if the two end groups are nonequivalent, $R(Cab)_n R^1$, where $n \geq 2$. Interconversion of Fischer projection, Newman projection, sawhorses, and flying-wedge formulas of $RCabCabR^1$ is depicted in a tabular form in Fig. 2.41. The conventions followed in writing the sawhorse and flying-wedge formulas will be revealed from their careful inspection.

Designation of Diastereomers. Several systems of designation of AB type diastereomers (Fig. 2.42) are now known of which a few are discussed below:

2.7.1.1 *Erythro* and *Threo* Nomenclature

A compound containing two adjacent nonequivalent chiral centers, $RCabCabR^1$, *i. e.*, of AB type, gives rise to two diastereomeric (±)-pairs, in total 4 stereoisomers (Fig. 2.42), *e.g.*, the aldotetroses, erythrose and threose (Fig. 2.42), and 3-bromo-2-butanol (Fig. 2.43), each one contains two nonequivalent chiral centers.

Erythro isomer: Examining different two-dimensional (Fischer projection, Newman, and sawhorse) and three-dimensional (flying wedge) formulas of an enantiomer of the *erythro* isomer of $R^1CabCabR^2$ in Fig. 2.41, one can define an *erythro* isomer in one of the following alternative ways:

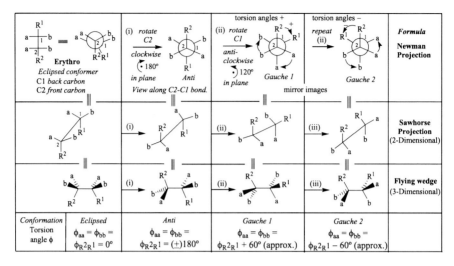

Fig. 2.41 *Interconversion of Fischer, Newman, sawhorse, and flying-wedge formulas of erythro-*
$R^1CabCabR^2$

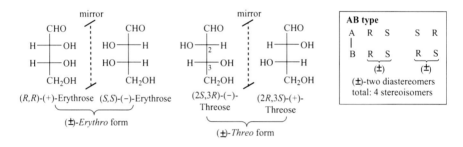

Fig. 2.42 Fischer projection formulas of $(\pm)$-erythrose and $(\pm)$-threose (AB type)

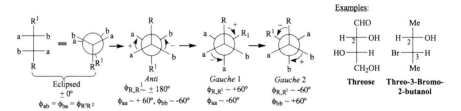

Fig. 2.43 One enantiomer of *threo*-$R^1CabCabR^2$; Fischer and Newman projection formulas;
Conformers. Examples.

(i) "At a glance" nomenclature is possible in the Fischer projection (F.p.) formula
having the main chain vertical in which the *erythro* diastereomer will *have*
both horizontal pairs of matched like (or similar) groups (or ligands), (e.g.,
two OH groups and two H's in erythrose) on the same side, i.e., they are

eclipsed (Torsion angle $\phi \approx \pm 0°$); the other pair of ligands (CHO and CH_2OH in erythrose) will also be eclipsed.

(ii) In the Newman projection, sawhorse, and flying wedge formulas of the eclipsed conformer ($\equiv$ F.p. formula), *at least two pairs of like (or similar) ligands are eclipsed* ($\phi \approx \pm 0°$) (*vide* first column of Fig. 2.41).

(iii) In the *anti* conformer (second column of Fig. 2.41) *at least two pairs of like (or similar) or identical ligands are antiperiplanar* ($\phi \approx \pm 180°$).

(iv) In any of the two gauche conformers (third and fourth columns of Fig. 2.41), *each pair of the two like (or similar) or identical ligands will be gauche* having $\phi \approx +60°$ (third column) or $-60°$ (fourth column).

Threo isomer: The diastereomer of *erythro* RCabCabR[1] is called *threo* isomer. It will have (by default) the following characteristics. Figure 2.43 shows only the Fischer and Newman projection formulas, from which sawhorse and flying formulas can be written (*cf.* Fig. 2.41).

The different formulas of *threo* isomer results by *exchange of any two ligands* (e.g., a and b) *at any one chiral center* (C1 or C2) of the *erythro* isomer in the Fischer, Newman, and sawhorse projection or flying wedge formulas (*cf.* Fig. 2.43).

(v) In the F.p. formula (having the main chain vertical) the *threo* diastereomer *will have both horizontal pairs of unlike (or dissimilar) ligands* (e.g., the OH and H in threose) *on the same side*, i.e., they are eclipsed ($\phi \approx \pm 0°$). Hence, both pairs of like (or similar) ligands will be on the opposite side.

(vi) In the Newman projection formula (sawhorse and flying wedge formulas are not shown) of the eclipsed conformer ($\equiv$ F.p. formula), at least two pairs of unlike (or dissimilar) ligands at the chiral carbons will be eclipsed, having torsion angle $\phi = \pm 0°$.

(vii) In the *anti* conformer of Newman projection formula, two pairs of like (or similar) ligands *will not be antiperiplanar* having $\phi = \pm 180°$.

(viii) In either gauche conformer the *torsion angle ϕ between two pairs of like (or similar) ligands will be of opposite sign.*

The above statements for recognizing the *threo* isomer are illustrated in Fig. 2.43. The *erythro–threo* nomenclature is possible for compounds of the type RCabCacR[1]. Of course, the easiest way to identify the *erythro* and *threo* isomers is to examine their Fischer or Newman projection formulas.

Threo diastereomer: (*compare with the first row of Fig. 2.41*)

The ambiguity in threo–erythro nomenclature arises in cases **A** and **B** (Fig. 2.44), when at least all of R, R[1], x, and y are alkyls or aryls, when matching of the two pairs of ligands at the two chiral centers is not possible; examples are cases **C** and **D**. However, this nomenclature in cases **A** and **B** is still possible if x and y are ligands with hetero atoms such as OR, NR[1]R[2], or halogen, and R and R[1] are alkyl or aryl groups (examples **E, F, G**).

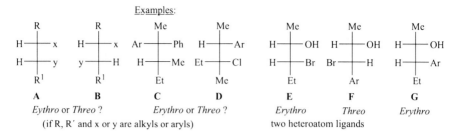

Fig. 2.44 Some cases where ambiguity arises (**a–d**) or does not arise (**e, f, g**)

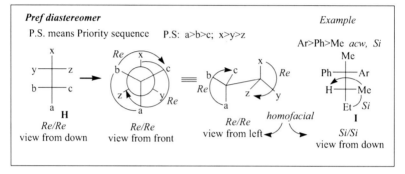

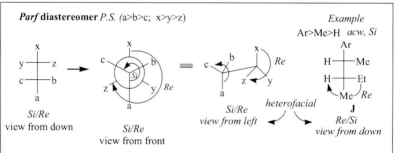

Fig. 2.45 Delineation of *pref* and *parf* diastereomers (one enantiomer of each)

2.7.1.2 *"Pref"* and *"Parf"* Nomenclature

In cases like **A** and **B** (Fig. 2.44), and also in the general case **H** (Fig. 2.45), when only one or no pair of ligands on the two chiral carbons can be matched the ***pref–parf*** nomenclature, developed in 1982 by Carey and Kuehne [15], can be used conveniently. The original version of this nomenclature has been simplified in the following way: Three ligands at each chiral center (case **H**) are to be ordered by CIP priority sequence rules as a > b > c and x > y > z. One does this by viewing the ligands from any one side of the F.p. formula joining the two chiral centers, say, *from bottom/down side*. If the three ligands on each chiral center appear in the clockwise (*Re*), or anticlockwise (*Si*) order, when viewed from any end, the relative configuration may then be specified as *Re/Re* (or *Si/Si*) for one diastereomer and as *Re/Si* (or *Si/Re*) for the other. The former case is denoted by Carey and Kuchne system as **pref** (**pr**iority **ref**lective) and the latter case is denoted by **parf** (**p**riority **a**nti**r**e**f**lective) (Fig. 2.45).

For an AB system usually *erythro* isomer corresponds to the *pref* isomer, but the *erythro–threo* nomenclature has no correspondence with the *pref–parf* nomenclature. For example, compound K (Fig. 2.46) appears to be *erythro* isomer by convention, but *parf* according to *pref–parf* system (Fig. 2.46). In fact, *pref–parf* nomenclature is used only when *threo–erythro* nomenclature is not possible. For *pref–parf* nomenclature of compounds having more than two chiral centers, *e.g.*, (E) and (F), see Fig. 2.49.

An **advantage** of the **pref–parf** system is that it can be used for specifying relative configuration of two noncontiguous chiral centers, assuming that they are directly linked, disregarding the intervening achiral centers (Fig. 2.47).

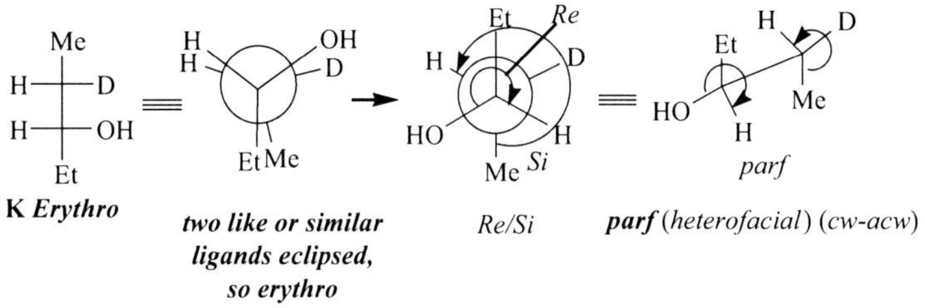

Fig. 2.46 No correspondence between *erythro–threo* and *pref–parf* nomenclatures

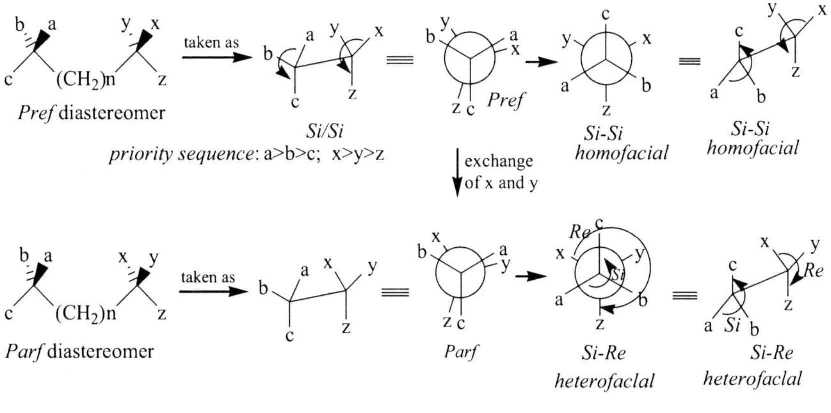

Fig. 2.47 *Pref–parf* specification of two diastereomers having two noncontiguous chiral centers

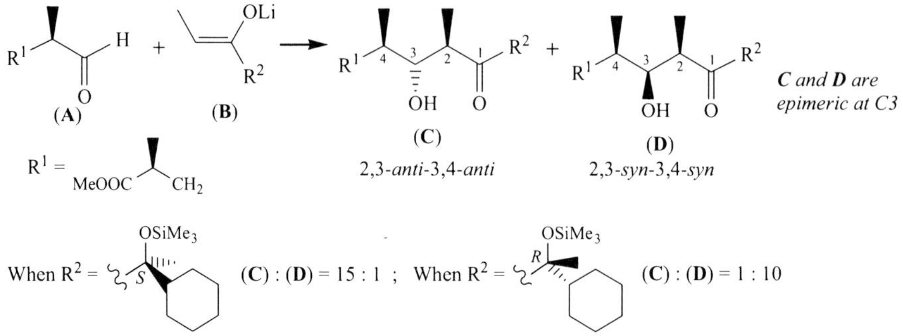

Fig. 2.48 Stereoselective formation of chiral aldols with multiple chiral centers and *syn, anti* designations

2.7.1.3 *Syn* and *Anti* System

This simple system of nomenclature of relative configuration has been introduced by Masamune [16] for aldol type compounds containing multiple chiral centers. Thus, the chiral aldehyde (**A**) was transformed stereoselectively into the diastereomers 3,4-*anti*-aldol (**C**) and into 3,4-*syn*-aldol (**D**) in different ratios using Li-enolates of different ketones (**B**) (Fig. 2.48). The longest carbon chain is written in a zigzag fashion. If two substituents (usually alkyl and hydroxyl) are on the *same side of the carbon chain plane*, the prefix *syn* is used. If they are on the *opposite side*, the prefix *anti* is used, as illustrated in Masamune's stereoselective aldol formation (Fig. 2.48).

2.7.1.4 *Like* (*l*) and *Unlike* (*u*) System

Prelog and Helmchen [10] have proposed a similar system of designation for the relative configuration of acyclic molecules with multiple adjacent chiral centers, in terms of the R,S designation of the adjacent chiral centers. If the adjacent chiral centers of the diastereomer are of like chirality ($R*R*$) or of unlike chirality ($R*S*$), they are termed *l* (like) or *u* (unlike) respectively, starting from the lowest numbered chiral center (locant). Here $R*R*$ and $R*S*$ refer to relative stereochemistry (Sect. 2.6.2.4). This is illustrated by compounds (**E**) and (**F**) containing four and five consecutive chiral centers (Fig. 2.49) which are represented by prefixes *u u u* and *l l l u*, respectively.

2.7.1.5 Brewster's System of Nomenclature

A general stereochemical notation, balanced in its emphasis in geometry (here relative configuration) and topography (here absolute configuration), has been developed in 1986 by Brewster [17] from the *R-S* (*R*S*) system "by use of *ul*

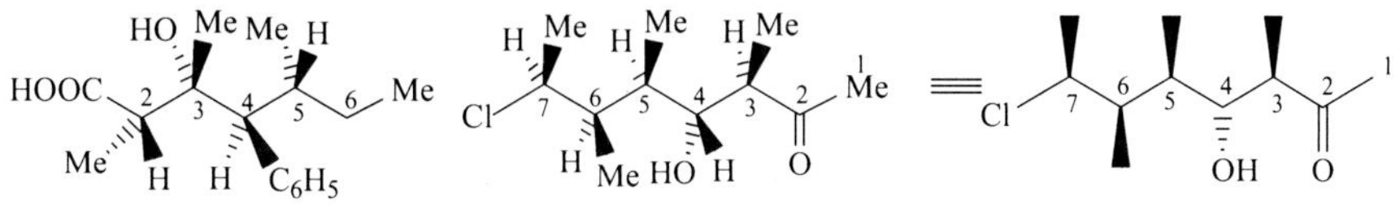

(E)   (F)   (F)

2S, 3R, 4S, 5R   3R, 4R, 5R, 6R, 7S (CIP)

　　u　u　u (Prelog)   　l　l　l　u (Prelog)

Represents (E) or its enantiomer (E′).   Represents (F) or its enantiomer (F′).

S (2l, 3u, 4l, 5u) Represents (E) (Brewster).   R (3l, 4l, 5l, 6l, 7u) Represents (F) (Brewster).

R(2l, 3u, 4l, 5u) Represemts (E′) Brewster).   S (3l, 4l, 5l, 6l, 7u) Represents (F′).

2,3-pref, 3,4-pref, 4,5-parf {Si/Si, Si/Si, Si/Re}   3,4-parf, 4,5-parf, 5,6-parf, 6,7-parf (Re/Si, Si/Re,

　　Represents (E) or (E′).   Re/Si, Si/ Re) Represents (F) or (F′).

Fig. 2.49 Prelog's and Brewster's stereochemical notations for acyclic compounds having multiple chiral centers

notation" of Prelog et al. and "the concept of external referencing." Thus, the absolute configuration of the compounds (E) (Fig. 2.49) may be designated as S (2 l, 3u, 4 l, 5u), where S represents the absolute configuration of the lowest locant (here C2), which is written outside the parenthesis (external reference), while l or u notation (being like or unlike the external reference) for each chiral center is placed inside the parenthesis. The chirality of each chiral center of the enantiomer of (E), designated (E′), is defined by R (2 l, 3u, 4 l, 5u), which means that the nomenclature depicts the chirality as a property of the whole molecule. Thus, topography (absolute configuration) at each center is readily recoverable: RS (2 l, 3u, 4 l, 5u) represents a racemic mixture (E) and (E′). A partially resolved mixture consisting of 80 % (E) and 20 % (E′) is represented by [80S, 20R] (2 l, 3u, 4 l, 5u). Any unknown chirality (say at C3) in compound (E) may be indicated by putting an x as in S (2 l, 3x, 4 l, 5u), which means that the chirality (absolute configuration) of C3 is not known.

2.7.2 Molecules with Two Like (Symmetrical) Chiral Centers (AA Type)

Molecules with two like chiral centers are of AA type and are *constitutionally symmetrical* having two diastereomers: one (±)-pair and one *meso* isomer (Fig. 2.50). It may be noted that if any one ligand on any chiral carbon (C2 or C3) is substituted by a new ligand (different from the existing ones), *meso* isomer becomes *erythro*, and active isomer becomes *threo*. The *anti* forms of (−)-tartaric acid and (+)-tartaric acid, each with **C2** point group, are shown in the figure.

Optical inactivity of meso-tartaric acid. The optical inactivity of the meso-tartaric acid may be explained in terms of its three staggered conformers, *viz.*, the

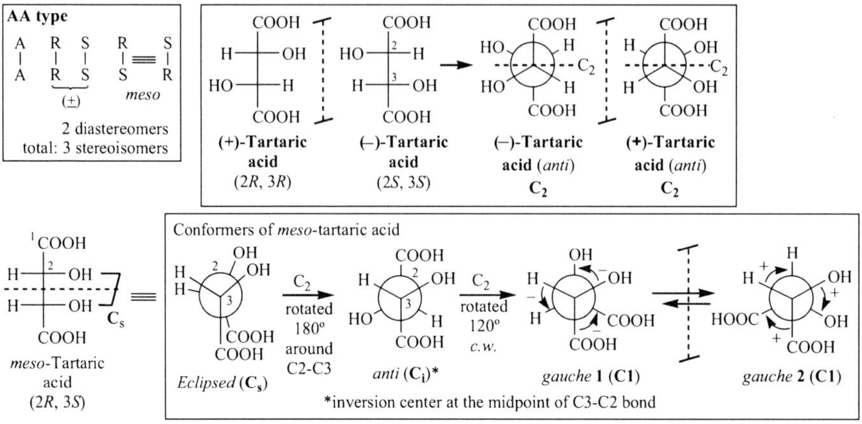

Fig. 2.50 Stereoisomers of tartaric acid (AA type). Optical inactivity of *meso*-tartaric acid

anti conformer and two gauche conformers which constitute over 99 % of the total number of molecules present in a sample of *meso*-tartaric acid. The eclipsed form shown in Fig. 2.50 along with the two other eclipsed forms obtained by rotation of any chiral carbon of the eclipsed form (say C2) around C2–C3 by 120° and 240° (*cw* or *acw*) constitute much less than 1 % of the total number of molecules present. The *anti* conformer with S_2 ($\equiv C_i$ point group) is optically inactive. The gauche 1 and gauche 2 conformers although optically active (each with C_1 point group) are *equally populated* (statistically symmetrical) *enantiomers,* as evident from the same but opposite torsion angle signs between the like ligands in each of them—negative in gauche **1** and positive in gauche **2**. *Thus, the combined effect of all molecules is to make meso-tartaric acid optically inactive.*

The terminology *homochiral* (*RR* or *SS*) for optically active stereoisomers and *heterochiral* (*RS* or *SR*) for *meso* stereoisomers has been adopted [18].

2.7.3 Molecules with Three Unlike Chiral Centers (ABC Type)

An aldopentose, $CHO-(CHOH)_3CH_2OH$, a constitutionally unsymmetrical molecule belonging to ABC type, has three chiral centers and exists as eight (2^n, where n = number of chiral centers) stereoisomers involving four diastereomers, each one having its enantiomer (Fig. 2.51). Of the eight stereoisomers, the one with (2R, 3R, 4R) absolute configuration is shown in its zigzag and Fischer projection formulas in Fig. 2.51.

Stereoisomer	1¦2	3¦4	5¦6	7¦8
C2 A	R¦S	S¦R	R¦S	S¦R
C3 B	R¦S	R¦S	S¦R	S¦R
C4 C	R¦S	R¦S	R¦S	R¦S
	(±)	(±)	(±)	(±)

Examples:

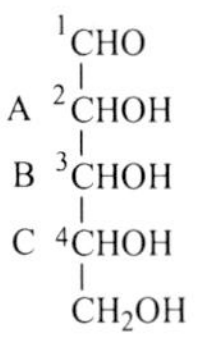

Total stereoisomers = $2^n = 2^3 = 8$ (all of them lack S_1 or S_2 and are optically active)

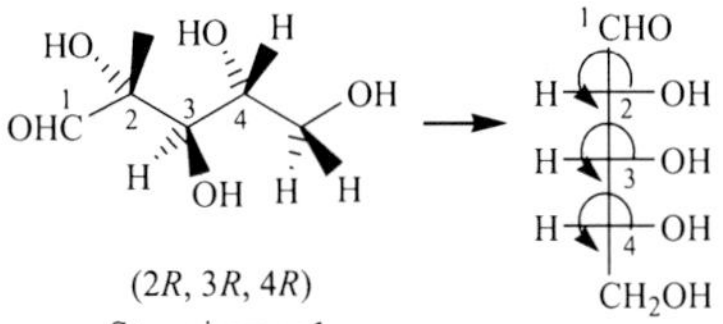

^{1}CHO
A ^{2}CHOH
B ^{3}CHOH
C ^{4}CHOH
CH$_2$OH

Aldopentose

(2R, 3R, 4R)
Stereoisomer 1

^{1}CHO
H—²—OH
H—³—OH
H—⁴—OH
CH$_2$OH

(2R, 3R, 4R)
Aldopentose

Fig. 2.51 Stereoisomers of ABA type with an example

2.7.4 Constitutionally Symmetrical Molecules Having Three Chiral Centers (ABA Type)

An acyclic molecule containing three or more chiral centers is constitutionally symmetrical if the chiral atoms equidistant from the geometrical center of the molecule are identically substituted. The two end groups of such a molecule are necessarily equivalent. Let us take n as the number of chiral centers in such a molecule. Thus, a constitutionally symmetrical molecule, say 2,3,4-trihydroxyglutaric acid, belongs to ABA type (Fig. 2.52) and exists as $2^{n-1} = 2^{3-1} = 4$ stereoisomers. Two of them (stereoisomers **1** and **2**) are enantiomers, and the remaining two, i.e., stereoisomers **3** and **4** are optically inactive *meso* forms. The latter two (**3** and **4**) are diastereomeric with each other and also with the two enantiomers **1** and **2**.

Monomethyl ester (**B**) [half ester of (**A**)] (Fig. 2.52) becomes constitutionally unsymmetrical and belongs to ABC type like an aldopentose (Fig. 2.51), and exists as four diastereomers, each having a (±)-pair, making eight stereoisomers in total (Fig. 2.52).

2.7.5 Stereogenecity and Chirotopicity (Fig. 2.52)

The status of the C2, C3, and C4 centers of the stereoisomers **1**, **2** (enantiomers), and **3** and **4** (Fig. 2.52) will now be discussed in terms of the observations of Mislow and Siegel [19] in 1984. Usually, a chiral center has two distinct attributes or characters:

1. *Stereogenecity* refers to bond connectivity—whether permutation or exchange of a pair of ligands gives a stereoisomer. (*R-S*)-Designation is associated with stereogenecity.
2. *Chirotopicity* is determined by local symmetry, i.e., whether S_n axis is present locally. A chirotopic atom resides in a chiral environment. Thus, all the five

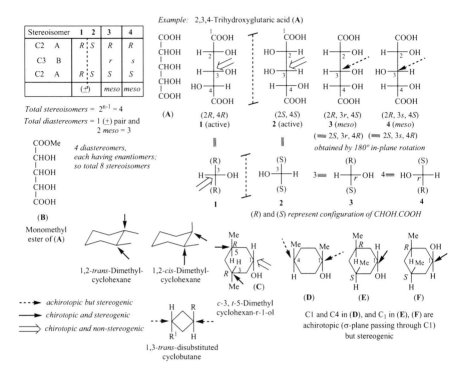

Fig. 2.52 Stereoisomers of compounds of ABA type. Stereogenecity and chirotopicity

atoms in bromofluorochloromethane (BrCl FCH) are chirotopic and belong to C_1 point group, although only the C atom is a stereogen. Three categories of centers are possible:

(i) *Chirotopic and stereogenic.* In most cases, the chiral centers, *e.g.*, C2 and C4 of isomers **1–4**, are *chirotopic* as well as *stereogenic*, i.e., the two attributes overlap. The C1 and C2 of *cis-* and *trans*-1,2-dimethylcy-clohexanes are stereogenic as well as chirotopic.

(ii) *Chirotopic but nonstereogenic.* The C3 centers of enantiomers **1** and **2** are chirotopic since they are devoid of any S_n axis (plane or center of symmetry), and no local symmetry is present. But they are nonstereogenic since two ligands of C3 in each are identical in both structure and chirality (R, R) or (S, S). Consequently, exchange of H and OH, for that matter, any two different ligands, does not give any isomer. Another example is the C1 atom of *c*-3,*t*-5-dimethyl-cyclohexan-*r*-1-ol (**C**) (Fig. 2.52), since the two faces (α-) and (β-) are exchangeable by C_2 operation, and exchange of H and OH does not give any new stereoisomer.

(iii) *Achirotopic but stereogenic* [19]. In the meso isomers **3** and **4**, C3 is achirotopic but stereogenic—thus again showing that stereogenecity and

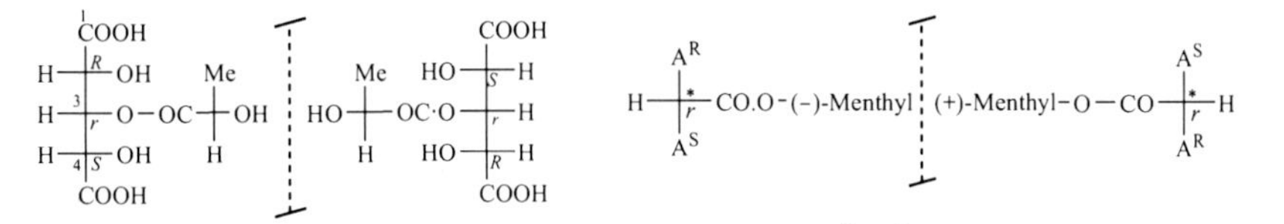

3*r*
3-Hydroxy gr. *esterified by*
R-(−)-lactic acid
C3 is reflection invariant

3*r*
Enantiomers

if −CO.O > A^R > A^S *both enantiomers have the r-configuration of the * asterisked C – and hence reflection invariant*

Fig. 2.53 Examples of reflection invariant carbon, which is *chirotopic* as well as *stereogenic*

chirotopicity are two distinct properties and can be delinked. Exchange of H and OH at C3 of **3** gives the other *meso* isomer **4**. C3 in both **3** and **4** can be given configurational descriptors: *r* to **3** and *s* to **4**, since the priority sequence *R* > *S*. C3 of the two molecules **3** and **4** are invariant to reflection; their configuration is denoted by lower case symbols, *r* and *s*. Even if C3 is made chiral by esterifying OH with *R*-lactic acid, the configurational specification *r* or *s* remains invariant to reflection (Fig. 1.53). In classical stereochemistry such an achirotopic but stereogenic center is called *pseudoasymmetric* and may be designated as C*RS*bc where *R* and *S* represent two enantiomorphous ligands [19].

Some more examples of molecules belonging to the category (iii) are 1,3-disubstituted cyclobutane and the three dimethylcyclohexanols (**D**), (**E**), and (**F**) (Fig. 2.53).

Some examples of compounds with *reflection invariant carbon being chirotopic as well as stereogenic* are depicted in Fig. 2.53.

2.7.6 *Molecules with Four (ABCD Type) or More Unlike Chiral Centers in a Chain*

An aldohexose, OHC−$(CHOH)_4$−CH_2OH, a constitutionally unsymmetrical molecule belonging to ABCD type, has four chiral centers and exists as $2^n = 2^4 = 16$ (where *n* is the number of chiral centers) stereoisomers involving *eight diastereomers*; each one is optically active and has its enantiomer. Figure 2.54 depicts the D-enantiomer of each of the eight pairs of diastereomeric aldohexoses (ABCD type).

Much of the early development of stereochemistry was stimulated by investigations of various sugars. It is, therefore, pertinent to depict in Fig. 2.55 the aldohexoses, only the D-enantiomers (C5 OH on the right side possessing *R*-configuration) of the eight diastereomers. Of these aldohexoses D-(+)-glucose occurs mostly as plant natural product glycosides. D-Gulose and D-idose are levorotatory; all other D-aldohexoses are dextrorotatory.

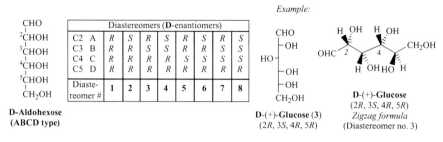

Fig. 2.54 Eight diastereomers of ABCD type (only D-(+)-Glucose shown)

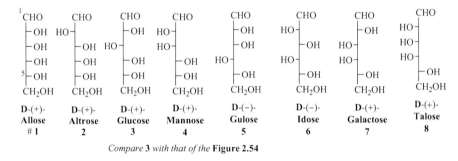

Fig. 2.55 Diastereomeric D-aldohexoses

2.7.7 Constitutionally Symmetrical Molecules with Four or More Like Chiral Centers in a Chain (ABBA, ABCBA, etc. Types)

A constitutionally symmetrical molecule has been defined in Sect. 2.7.4. Such a molecule with four chiral centers (the first and second being similar to the 4th and 3rd chiral centers, respectively) belongs to ABBA type, *e.g.*, tetrahydroxyadipic acids (or hexaric acids), which exist as six diastereomers (four racemic pairs and two *meso* forms) and hence ten stereoisomers (Fig. 2.56).

Stereoisomers of compounds with even number of constitutionally symmetrical chiral centers in a chain. In general, in the series of the type Cabd $(Cab)_{n-2}$ Cabd, where n is the number of chiral centers and n is **even**, there are 2^{n-1} optically active stereoisomers and $2^{(n-2)/2}$ *meso* forms. So, in case of hexaric acid, $n = 4$ and total number of stereoisomers $= 2^{n-1} + 2^{(n-2)/2} = 2^3 + 2^{2/2} = 8 + 2 = 10$. Perhydro-phenanthrene (Sect. 2.15.2) also belongs to ABBA.

Stereoisomers of compounds with odd number of constitutionally symmetrical chiral centers in a chain. In the series of the type Cabd $(Cab)_{n-2}$ Cabd where n is the number of chiral centers and n is **odd**,

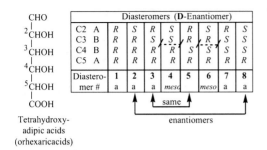

CHO
|
^{2}CHOH
|
^{3}CHOH
|
^{4}CHOH
|
^{5}CHOH
|
COOH

Tetrahydroxy-
adipic acids
(orhexaricacids)

Diasteromers (D-Enantiomer)							
C2 A	R	S	R	S	R	S	R	S
C3 B	R	R	S	S	R	R	S	S
C4 B	R	R	R	R	S	S	S	S
C5 A	R	R	R	R	R	R	R	R
Diastero-	1	2	3	4	5	6	7	8
mer #	a	a	a	meso		meso	a	a

same

enantiomers

a = optically active

Isomer 8 is enantiomeric with isomer 2.
Isomer 5 is same as isomer 3.
So hexaric acid (like perhydrophenanthrene)
has four (±) pairs and two meso forms, i.e., 6
diastereomers, and in total 10 stereoisomers,
as evident from the Table.

Fig. 2.56 Stereoisomers of hexaric acid (ABBA type)

total number of stereoisomers $= 2^{n-1}$, of which number of *meso* forms $= 2^{(n-1)/2}$.

Thus, when $n = 3$, total number of stereoisomers $= 2^{n-1} = 2^2 = 4$, of which number of meso forms $= 2^{(n-1)/2} = 2^{(3-1)/2} = 2$; so number of optically active isomers $= 4-2 = 2$ ($1 \pm$ pair) (see Fig. 2.47).

The next higher homologue with n odd, $HOOC-(CHOH)_5COOH$ (heptaric acid) is of the ABCBA type. Here n = 5, hence total number of stereoisomers $= 2^{5-1} = 2^4 = 16$, of which number of the meso forms $= 2^{(5-1)/2} = 2^2 = 4$. Hence, number of optically active isomers $= 16-4 = 12$. So there will be 6 racemic pairs and 4 meso forms, *i.e.*, 10 diastereomers. All $2^n \div 2 = 2^5 \div 2 = 16$ diastereomers of ABCDE system can be written down following the system in Fig. 2.54. The six racemic and four meso isomers of ABCBA system can also be ascertained from the former system (ABCDE) (*cf.* Fig. 2.56).

2.7.8   Chiral Compounds with Asymmetric Carbon Atoms in Branched Chains

On rare occasions one encounters molecules in which chiral carbon atoms cannot be aligned in one chain. We depict the cases (**1**) and (**2**) in Fig. 2.57. Cases (**2**)–(**5**), where two or more of the chiral ligands attached to the central carbon atom are alike, are more complicated as delineated in Figs. 2.57 and 2.58.

2.8   Chirality and Dimension. One-, Two-, and Three-Dimensional Chiral Simplexes

Some simple concepts of stereochemistry (in the molecular level) may be conveniently revealed on the basis of "*chiral simplex*" (applicable to *1D*, *2D*, or *3D*), introduced by Prelog and Helmchen [10]. The term means the simplest structure

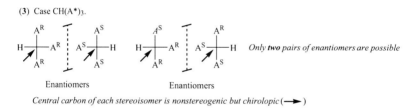

An example of such a meso compound possessing a four-fold alternating axis of symmetry (S_4) and no other element of symmetry, prepared and shown to be optically inactive [18], *is given below*

$A^R = CH_2OCOCH_2R^R$

$A^S = CH_2OCOCH_2R^S$

where R is the (−) or (+)-
menthyloxy group

--▶ *achirotopic and nonstereogenic*

Fig. 2.57 The stereoisomers of cases CA*B*D*E* and CA*₄. A molecule possessing S_4

(3) Case CH(A*)₃.

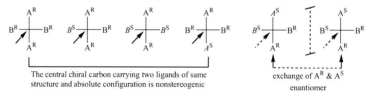

Only ***two*** *pairs of enantiomers are possible*

Central carbon of each stereoisomer is nonstereogenic but chirolopic (──▶)

(4) Case CH(A*)₂B*. *Here* ***four*** *enantiomeric pairs (four diastereomers) and* ***eight*** *stereoisomers are possible as shown below*

──▶ *chirotopic but nonstereogenic*

--▶ *chirotopic and stereogenic*

Each diastereomer has its enantiomer

*Exchange
of A^R & A^S*

One enantiomer each of four enantiomeric pairs

(5) Case C(A*)₂(B*)₂. *Here* ***five*** *enantiomeric pairs and* ***ten*** *stereoisomers are possible as shown below:*

The central chiral carbon carrying two ligands of same
structure and absolute configuration is nonstereogenic

exchange of A^R & A^S
enantiomer

Fig. 2.58 The streoisomers of cases CH(A*)₃, CH(A*)₂B*, and C(A*)₂(B*)₂

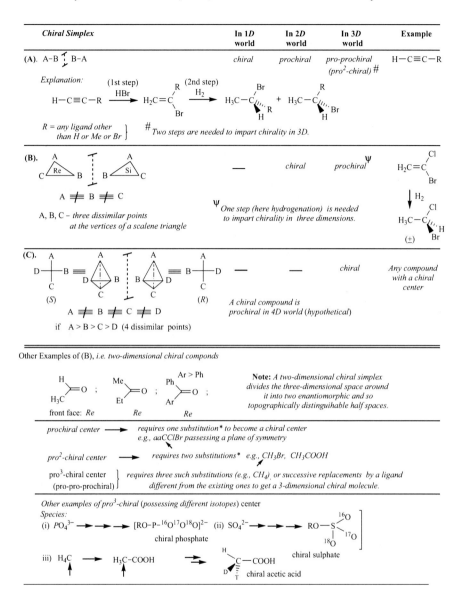

Fig. 2.59 One-, two-, and three-dimensional chiral simplexes. Pro-, pro^2-, and pro^3-chiral molecules

that could be chiral in *1D*, *2D*, or *3D* world, as illustrated in Fig. 2.59 with examples.

A linear molecule, with two dissimilar points, like A–B (case **A**) is a one-dimensional chiral simplex, *i.e.*, chiral in one dimension (*y*-axis) since its mirror image cannot be superposed on it without the help of a second dimension, i.e., by its rotation around the orthogonal x axis. So a linear molecule A–B is prochiral in a two-dimensional world, since one more dimension (*Z* axis) is necessary to make it chiral in three dimensions. The molecule A–B is pro-prochiral in the three-dimensional world since two steps are necessary to convert it to a chiral molecule in our *3D* world.

Likewise, a dissimilarly substituted trigonal tricoordinated carbon (case **B**) is chiral in two dimensions (two-dimensional world) and prochiral in tetrahedral terracoordinated three dimensions.

By the same token, a dissimilarly substituted tetrahedral tetracoordinated carbon is chiral in three dimensions, *e.g.*, any chiral compound with a chiral center. However, one may assume that any chiral compound in three dimensions, which is a three-dimensional chiral simplex, would have been prochiral in an imaginary four-dimensional world, had there been any fourth dimension (*cf.*, time). Some examples of two-dimensional chiral simplex and of pro-, pro^2-, and pro^3-chiral molecules/species are illustrated in Fig. 2.59.

2.9 Prochirality and Prostereoisomerism. Topicity of Ligands and Faces: Homotopicity. Enantiotopicity. Diastereotopicity. Nomenclature [21–25]

2.9.1 Introduction

A nonstereogenic center can be changed to a stereogenic center by replacement of any one of the two *apparently identical ligands* by a different one. Two examples, propionic acid (**1**) or phenylacetic acid (**2**), are given in Fig. 2.60. In each case, the prostereogenic (also prochiral) C2 center is converted to a stereogenic (also chiral) center by replacement of one *homomorphic* ligand (here a hydrogen atom) by a ligand other than those present in the molecule. The term *homomorphic* originates from Greek *homos* meaning "same" and *morphe* meaning "form" [23, 24]. Homomorphic ligands are identical only when separated from the rest of the molecule. Replacement of the other homomorphic ligand by the same new ligand will lead to

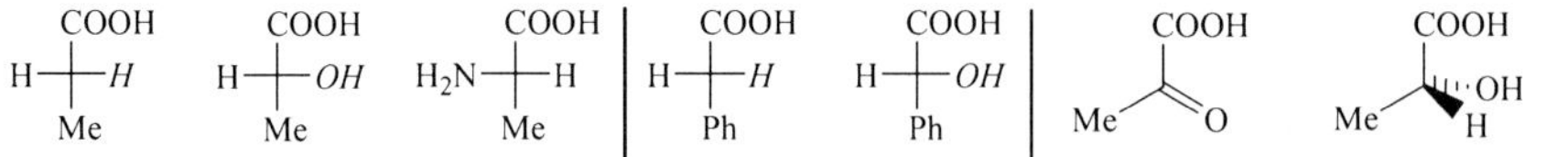

Propionic acid (**1**) (*prochiral*) | *R*-**D**-Lactic acid (*chiral*) | *S*-**L**-Alanine (*chiral*) | Phenyl-acetic acid (**2**) (*prochiral*) | *R*-**D**-Mandelic acid (*chiral*) | Pyruvic acid (**3**) (*prochiral*) | *R*-**D**-Lactic Acid (*R*-**D**-2-Hydroxy-propionic acid)

Fig. 2.60 Prochiral (prostereogenic) [(**1**), (**2**), and (**3**)] and chiral (stereogenic) molecules the corresponding enantiomer. In such cases the topicity (Greek *topos* meaning "place" or "physical surrounding") of the two *homomorphic* ligands (here H's at C2), in other words, their special relation with the rest of the molecule is different; they reside in enantiomeric environments. Such ligands are called *stereoheterotopic*, in this case more specifically *enantiotopic*.

By the same token the two faces (front and rear of the molecular plane) of pyruvic acid (**3**) (Fig. 2.60) are enantiotopic, since the two faces are in enantiomeric environments; thus, addition of hydride (or any nucleophile) to the two faces in turn gives rise to enantiomers, *viz.*, (*R*)- and (*S*)-2-hydroxypropionic acid (*R*- and *S*-lactic acid).

Molecules having stereoheterotopic ligands or faces exhibit *prostereoisomerism* (if they can be reacted upon). Prostereoisomerism is attributed to such a center carrying stereoheterotopic ligands, for example, to C2 of *prostereogenic* and *prochiral* compounds (**1**) and (**2**), or to a center carrying stereoheterotopic faces, *e.g.*, pyruvic acid (**3**).

The principle of stereoselective synthesis (Sect. 2.11.1) is based on the different behavior of stereoheterotopic ligands and faces in the chemical reactions. The concept of stereoheterotopicity has been discussed in 1967 [22], that of prochirality in 1966 [21], and the topic has been reviewed in detail by Eliel in 1982 [25].

Two criteria, *viz.*, (i) substitution or derivatization (in case of ligands) or addition (in case of faces) and (ii) a symmetry criterion are used to determine the topic relationship of homomorphic ligands and faces, as illustrated by several examples of each of homotopic ligands (Fig. 2.61), homotopic faces (Fig. 2.62, enantiotopic ligands (Fig. 2.63), enantotopic faces (Fig. 2.64), diastereotopic ligands (Fig. 2.66), and diastereotopic faces (Fig. 2.67). Any of the two criteria is sufficient to determine the topicity in each case.

2.9.2   Homotopic Ligands

The first example of homotopic ligands is of Za_2b_2 type, belonging to the C_{2v} point group (Fig. 2.61). The homotopic ligands are recognized by either of the two criteria already stated. Two examples of homotopic ligands are illustrated, recognized by both substitution and symmetry criteria, in this Figure. Four more examples of homotopic ligands of simple molecules like chloromethane, ethylene, allene, and acetic acid are also included indicating the homotopicity of the homomorphic ligands by symmetry criteria.

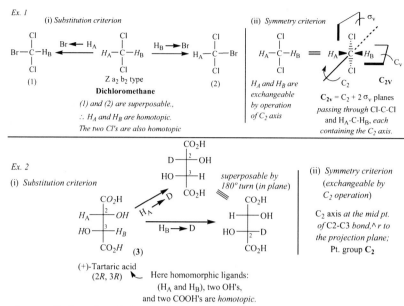

Ex. 1

(i) *Substitution criterion*

(ii) *Symmetry criterion*

$Z\ a_2b_2$ type

Dichloromethane

(1) and (2) are superposable.,

∴ H_A and H_B are homotopic.

The two Cl's are also homotopic

H_A *and* H_B *are*
exchangeable
by operation
of C_2 axis

$C_{2v} = C_2 + 2\,\sigma_v$ *planes*
passing through Cl-C-Cl
and H_A-C-H_B, *each*
containing the C_2 axis.

Ex. 2

(i) *Substitution criterion*

superposable by
180° turn (in plane)

(ii) *Symmetry criterion*
(exchangeable by
C_2 operation)

C_2 axis at the mid pt.
of C2-C3 bond,∧r to
the projection plane;
Pt. group C_2

(+)-Tartaric acid
(2R, 3R)   Here homomorphic ligands:
(H_A and H_B), two OH's,
and two COOH's are *homotopic.*

Comment: (i)   *In a molecue the homomorphic ligands exchangeable by a C_2 axis are homotopic*
(ii) *No chiral reagent or enzyme can differentiate the homotopic ligands*
(iii) *Homotopic ligands are isochronous (possess same chemical shift in NMR).*

Some other examples:

Ex. 3   *Ex. 4*   *Ex. 5*   *Ex. 6*

C_{3v} D_{2h} *Eclipsed* D_2d
but averaged
conformation

$C_3 + 3s_v$ $D_2 = C_2 + 2C_2$ **Acetic acid**   *H's at each terminal*
carbon are homotopic

Symmetry criterion: All H's of the above molecules, exchangeable by operation of C_2/C_3 axis, are homotopic
One can make the same inference by applying substitution criterion in each case.

Fig. 2.61 Homotopic ligands recognized by substitution and symmetry criteria

2.9.3   *Homotopic Faces*

Homotopic faces are recognized by the two criteria as illustrated in Fig. 2.62 with the help of four examples.

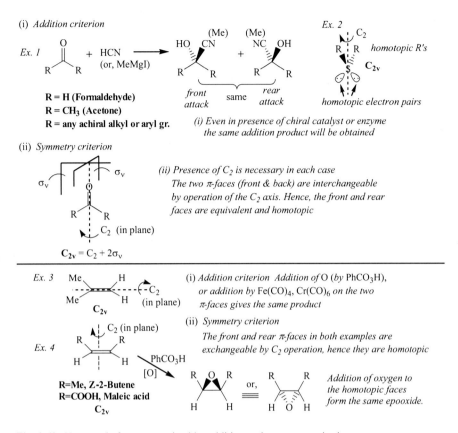

Fig. 2.62 Homotopic faces recognized by addition and symmetry criteria

2.9.4 Enantiotopic Ligands

Enantiotopic ligands are recognized by any one of the two criteria as illustrated in Fig. 2.63 with six examples.

2.9.5 Nomenclature of Geminal Enantiotopic Ligands. Pro-R and Pro-S

In compounds **1–3** of Fig. 2.63, of the two homomorphic ligands $H_A H_B$ or $H_C H_D$ (attached to the same carbon) if replacement of any one, say H_A, by an atom of higher priority, but of lower priority than the other two ligands (in such case by D), gives rise to (S) configuration of the new chiral molecule, H_A is then termed as **pro-S** and is expressed as H_S. H_B in that case, by default, will be termed as **pro-R** and is expressed as H_R. H_A and H_B may be replaced by D to give the (S) and (R) configuration, respectively, of the resulting new chiral molecules. This terminology

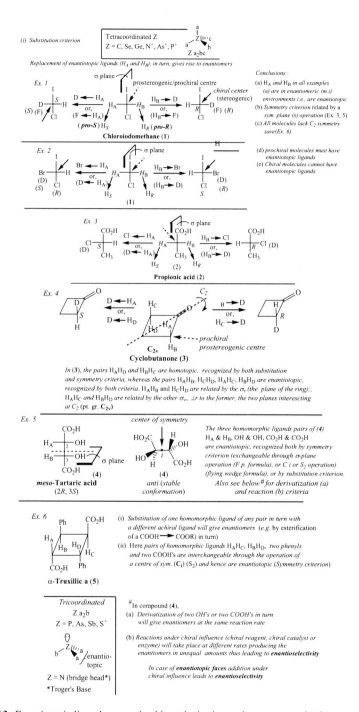

Fig. 2.63 Enantiotopic ligands recognized by substitution and symmetry criteria

is applicable to any enantiotopic homomorphic germinal ligands (attached to the same carbon) [21].

2.9.6 Enantiotopic Faces

Enantiotopic faces are recognized by any one of the two criteria as illustrated in Fig. 2.64 with three examples.

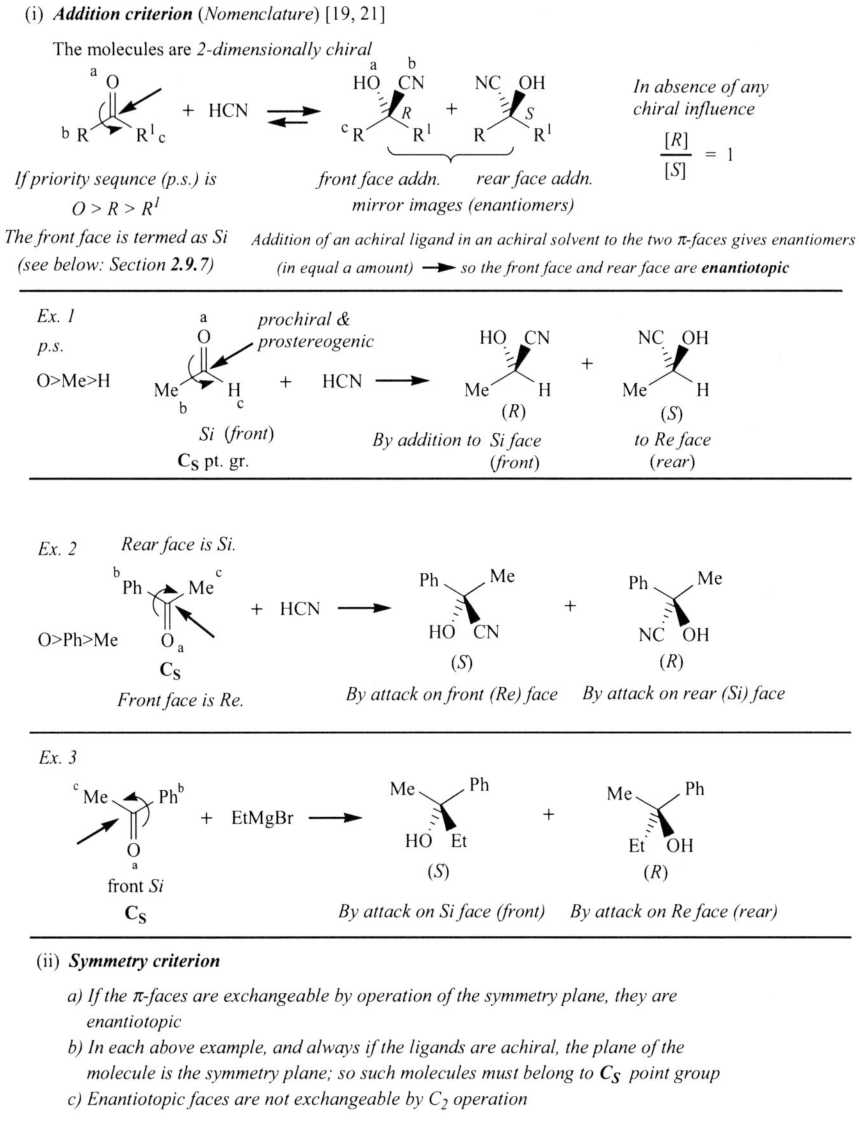

(i) **Addition criterion** (*Nomenclature*) [19, 21]

The molecules are 2-dimensionally chiral

$$\underset{b\,R}{\overset{a\,O}{\bigg|}}\underset{R^1\,c}{\diagdown} + HCN \rightleftharpoons \underset{c\,R}{\overset{HO\,CN}{\diagup}}\underset{R^1}{\diagdown}(R) + \underset{R}{\overset{NC\,OH}{\diagup}}\underset{R^1}{\diagdown}(S)$$

In absence of any chiral influence

$$\frac{[R]}{[S]} = 1$$

If priority sequnce (p.s.) is

$O > R > R^1$

front face addn.   rear face addn.

mirror images (enantiomers)

The front face is termed as Si   Addition of an achiral ligand in an achiral solvent to the two π-faces gives enantiomers
(see below: Section 2.9.7)   (in equal a amount) ⟶ so the front face and rear face are **enantiotopic**

Ex. 1

p.s.

$O>Me>H$

$$\underset{b}{\overset{a\,O}{\bigg|}}\underset{H\,c}{\diagdown}Me \quad \text{prochiral \& prostereogenic} + HCN \longrightarrow \underset{Me}{\overset{HO\,CN}{\diagup}}H \;(R) + \underset{Me}{\overset{NC\,OH}{\diagup}}H \;(S)$$

Si (front)

C_S pt. gr.

By addition to Si face (front)   to Re face (rear)

Ex. 2   Rear face is Si.

$O>Ph>Me$

$$\underset{O\,a}{\overset{b\,Ph\quad Me\,c}{\diagdown\!\diagup}} + HCN \longrightarrow \underset{HO\,CN}{\overset{Ph\quad Me}{\diagdown\!\diagup}}(S) + \underset{NC\,OH}{\overset{Ph\quad Me}{\diagdown\!\diagup}}(R)$$

C_S

Front face is Re.

By attack on front (Re) face   By attack on rear (Si) face

Ex. 3

$$\underset{O\,a}{\overset{c\,Me\quad Ph\,b}{\diagdown\!\diagup}} + EtMgBr \longrightarrow \underset{HO\,Et}{\overset{Me\quad Ph}{\diagdown\!\diagup}}(S) + \underset{Et\,OH}{\overset{Me\quad Ph}{\diagdown\!\diagup}}(R)$$

front Si

C_S

By attack on Si face (front)   By attack on Re face (rear)

(ii) **Symmetry criterion**

a) If the π-faces are exchangeable by operation of the symmetry plane, they are enantiotopic

b) In each above example, and always if the ligands are achiral, the plane of the molecule is the symmetry plane; so such molecules must belong to C_S point group

c) Enantiotopic faces are not exchangeable by C_2 operation

Fig. 2.64 Enantiotopic faces. Addition and symmetry criteria, and nomenclature

2.9.7   Nomenclature of Enantiotopic Faces

The π-faces of a trigonal atom (olefinic/carbonyl carbon) having three different ligands will be enantiotopic and are two-dimensionally chiral (Sect. 2.8).

   If one looks at the plane of a π-face, and finds that the CIP priority sequence of the three dissimilar ligands is clockwise, the face is called *Re*. If the sequence is anticlockwise, the face is called *Si*, these being the first two letters of **Rectus** and **Sinister**, respectively [21]. Examples 1–3 are shown in Fig. 2.64. If one face is *Re*, the other face will be *Si*.

   In case of a double bond containing two specifiable trigonal atoms, each face may be uniquely defined by two symbols, one for each specifiable atom, as illustrated by the compounds in Fig. 2.65. The π-faces of an oxime are also specified in this Figure. The α- and β-faces of each olefin excepting maleic acid being exchangeable only by a σ-plane (*xy* or paper plane) operation and not by any *C₂* operation, each olefin produces enantiomeric epoxides or complexes with the double bond in equal amount in absence of any chiral influence (see also Fig. 2.73).

2.9.8   Diastereotopic Ligands

Substitution criterion. If the sequential replacement of two homomorphic ligands in a molecule by a different achiral ligand gives rise to diastereomeric products, those two ligands are said to be diastereotopic. Such ligands are generally distinct both

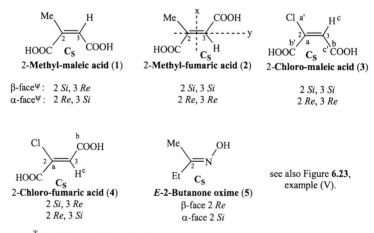

Notes: $^{\Psi}$ *β- and α-faces specify front and rear faces respectively when the structures are written in the xy (paper) plane, as shown. After rotation around in-plane x-axis (vertical) or, around in-plane y-axis (horizontal) the β face becomes α and vice-versa but (Re, Si) specification of a specified face remains unaltered.*

Fig. 2.65 Nomenclature of enantiotopic π-faces of some unsymmetrically substituted olefins

(i) **Substitution criterion**

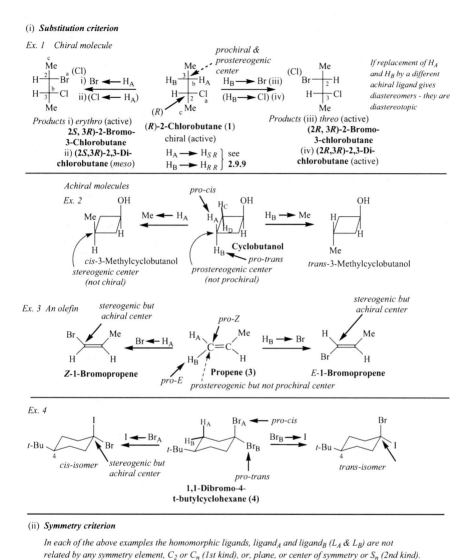

Fig. 2.66 Diastereotopic ligands. Nomenclature

chemically and spectroscopically. They react at unequal rates and their NMR signals will be different (anisochronous).

Symmetry criterion. Diastereotopic ligands must not be related (exchangeable) by a C_n or S_n axis. The molecule (**1**), depicted in Fig. 2.66, is devoid of any symmetry element; the molecules (**2**), (**3**), and (**4**) (Fig. 2.66) although contains a plane of symmetry, its operation does not interchange the ligands.

2.9.9   Nomenclature of Diastereotopic Ligands

A *chiral* molecule containing a pair of homomorphic diastereotopic geminal ligands at a prochiral center must have a chiral element (center, axis, or plane) also. To designate the diastereotopic pair of ligands, one ligand is arbitrarily given preference over the other to treat the prochiral center as a chiral one. Now applying CIP rules if the center becomes (R), then that particular homomorphic ligand is called ***pro-R*** (hypothetical configuration symbol) and is expressed as the ligand subscripted with R, *i.e.*, the ligand is termed $(ligand)_R$. Let us take (R)-2-chlorobutane (**1**) as an example (Fig. 2.66, first row). The hydrogen atoms H_A and H_B at C3 are diastereotopic. If preference is given to H_A over H_B in the sequence rule, the hypothetical configurational symbol for C3 of (**1**) would be S. The configurational symbol of H_A thus becomes H_{SR} after adding to the subscript S a further subscript R, the symbol of the absolute configuration of the chirality present (in this case at C2). By default H_B is H_{RR}. This procedure is applicable for any such homomorphic pair of ligands at a prochiral center, *e.g.*, COOH, COOH or Ph, Ph or Me, Me, etc. of a compound having a chiral element. This type of double indexing system [26] was first introduced in 1982 in a somewhat modified form. It has the advantage that it makes the diastereotopicity immediately obvious.

Ex. 2, Ex. 3, and Ex. 4 of Fig. 2.66 illustrate the nomenclature of different types of *achiral* molecules [*viz.*, cyclic compounds or olefins or cumulenes with odd number of double bonds (having successive planes of π bonds orthogonal to each other)] having diastereotopic homomorphic geminal ligands. The substitution products shown are also achiral since the original plane of symmetry is retained.

The substitution products of homomorphic geminal diastereotopic ligands may become chiral if the original symmetry plane is destroyed, *e.g.*, H_C and H_D at C2 of (**2**) and H_A and H_B at C3 of (**4**) in Fig. 2.66. Thus, C2 of (**2**) and C3 of (**4**) are prostereogenic as well as prochiral centers.

2.9.10   Diastereotopic Faces. Nomenclature

The diastereotopic faces are designated as follows. First, the face is designated as *Re* or *Si* as stated earlier in Sect. 2.9.7. To this symbol the specified absolute configuration of the chiral element present in the molecule is added. Thus, in case of a chiral compound (**1**) (Fig. 2.67), the front face becomes *ReS* by adding *S*, the absolute configuration of the chiral center present, to the *Re* face. Thus, the diastereotopic rear face is designated *SiS*.

Diastereotopic nomenclature in cases of compounds with diastereotopic faces has been illustrated in Fig. 2.67 with examples of an achiral ketone 3-Methylcyclobutanone (**2**), achiral allene with diastereotopic faces 1-chloroallene (**3**), and a chiral sulfide (**4**). In each case the nomenclature of the diastereotopic faces is self-explanatory (*cf.* Ex. 1).

Diastereotopic faces are recognized by application of any one of the usual two criteria as illustrated in Fig. 2.67.

i) **Addition criterion**: *Addition to diastereotopic π-faces gives diastereomers*

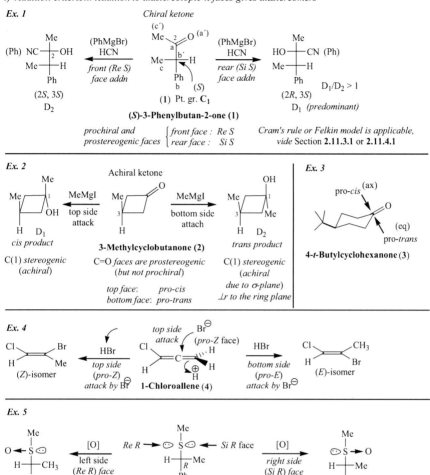

The two diastereomeric sulphoxides are formed in different proportions

(ii) **Symmetry criterion**: *The two π-faces of compounds (1) to (4) are not exchangeable by operation of any C_2 or σ plane or S_n axis, so they are diastereotopic. Same is the case with the two lone pairs of (5), which are also diastereotopic.*

In each case, diastereoselectivity is observed because of the involvement of diastereomeric transition states.

Fig. 2.67 Diastereotopic faces. Nomenclature

2.9.11 Interesting Examples of Topicities of Homomorphic Ligands [27]

In Fig. 2.68 homotopicity, enantiotopicity, and the diastereotopicity of the CH_2OH groups in the *cases 1, 2,* and *3,* respectively, are illustrated by selective oxidations as well as by symmetry criteria.

In *case 1,* like the CH_2OH groups 2-OH & 5-OH, as well as 3-OH & 4-OH, exchangeable by C_2 operation, are homotopic, and hence the homotopic OHs when derivatized would lead to the same derivative, and the corresponding acetates will also be homotopic.

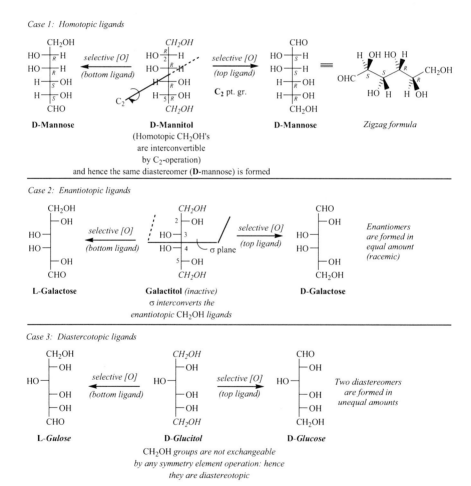

Fig. 2.68 Determination of topicity of CH_2OH groups of a few hexoses by selective oxidation

In *case 2*, like the CH_2OH groups 2-OH & 5-OH, as well as 3-OH & 4-OH are enantiotopic, and hence the corresponding acetates of the enantiotopic OH groups will lead to enantiomers.

In *case 3*, like the CH_2OH groups, any pair of OH groups are diastereotopic, and hence the corresponding acetates of any pair of diastereotopic OH groups will also be diastereomers.

2.9.12   Interrelation of Topicity of Ligands with Isomerism

Topicity of ligands is interrelated with isomerism in general. A classification diagram for topicity is drawn (Fig. 2.69), which may be compared with that drawn for isomerism (Fig. 2.24).

2.9.13   Molecules with Prostereogenic but Proachirotopic Center and Multi-Prochiral Centers

Some interesting examples of topicity and nomenclature are delineated in Fig. 2.70.

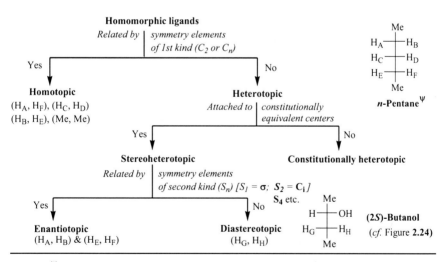

$^\psi$Constitutionally heterotopic ligands

H's at C2 and C4 compared to H's at C3 are constitutionally heterotopic - being bonded to ligating centres of different constitutions

Fig. 2.69 Classification of homomorphic ligands based on topicity

1) *A molecule with prostereogenic but proachirotopic centers* (**1**)

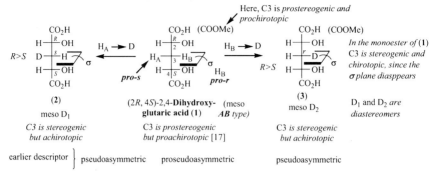

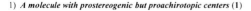

Here, C3 is *prostereogenic and
prochirotopic*

In the monoester of (**1**)
C3 is *stereogenic and
chirotopic, since the
σ plane diasppears*

(**2**)

meso D₁

C3 *is stereogenic
but achirotopic*

(2R, 4S)-2,4-**Dihydroxy-
glutaric acid** (**1**)   (meso
AB type)

C3 *is prostereogenic
but proachirotopic* [17]

(**3**)

meso D₂

C3 *is stereogenic
but achirotopic*

D₁ and D₂ *are
diastereomers*

earlier descriptor } pseudoasymmetric   proseudoasymmetric   pseudoasymmetric

Substitution criteria

• *Symmetry criterion*: H$_A$ & H$_B$ *are not interchangeable by any symmetry operation, so they are diastereotopic*
• *Pairs of homomorphic ligands* (H, H & HO, OH) *at C2 and C4 in* (**1**) *are, however, exchangeable with a σ-plane
  (containing H$_A$-C3-H$_B$) operation, and hence are enantiotopic. Hence replacement of these two H's by D or any other
  substituent, or derivatization of the two OH's will give rise to enantiomers.*

2) *Molecules with one or more than one prochiral centers*

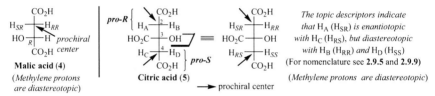

*The topic descriptors indicate
that* H$_A$ (H$_{SR}$) *is enantiotopic
with* H$_C$ (H$_{RS}$), *but diastereotopic
with* H$_B$ (H$_{RR}$) *and* H$_D$ (H$_{SS}$)
(*For nomenclature see* **2.9.5** *and* **2.9.9**)

Malic acid (**4**)

(*Methylene protons
are diastereotopic*)

Citric acid (**5**)

⟶ prochiral center

(*Methylene protons are diastereotopic*)

Note: • *Citric acid contains three prochiral centres, C2, C3, C4, and both enantiotopic and diastereotopic H's*
       H$_A$ *is* **pro-S**, H$_B$ (**pro-R**), H$_C$ (**pro-S**) *and* H$_D$ (**pro-R**).

       • *C3 is a prochiral center since it carries two homomorphic group ligands* –CH$_2$COOH,
         to which topic descriptors, **pro-R** *and* **pro-S** *may be assigned, as usual* (*cf.* **2.9.5**).

Fig. 2.70 Topicity nomenclature of methylene protons in two interesting examples

2.9.14   Topic Relationship of Ligands and Faces

Topic relationship of ligands and faces, so far discussed, are summarized in
Fig. 2.71. This figure has much in common with the table given by Mislow and
Raban in 1967 [22] and by Eliel in 1980 [28].

2.10   Stereoheterotopic Ligands and NMR Spectroscopy

NMR (like IR and UV) is an achiral probe. Nuclei that reside in different environments can be distinguished by NMR spectroscopy showing different chemical
shifts. Such nuclei exhibiting nonequivalent chemical shifts are called
anisochronous; they split. Nuclei showing same chemical shift are called isochronous, which do not split. The *isochrony* or *anisochrony* of nuclei is dependent on

Topic Relationship of Ligands and faces

Topicity	Substition/addition/ reaction criterion (by *achiral groups*)	Symmetry criterion	Behavioral difference
Homotopic	Identical product	Ligands related through C_n and faces by C_2 axis	No difference by any method
Enantiotopic	Enantiomeric products	Ligands related through σ, i or S_n Faces related by σ	Usually not distinguishable Distinguishable, in principle, in chiral media (NMR), by chiral reagents and enzyme
Diastereotopic	Diastereomeric products	Ligands and faces not related by any symmetry element	Distinguishable in principle by all physical and chemical methods

Fig. 2.71 Summary of topic relationship of ligands and faces

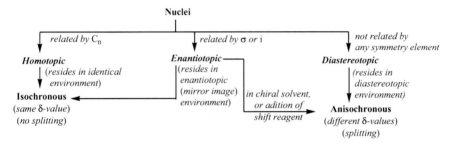

Fig. 2.72 Topicity. NMR chemical shift. Signal multiplicity

their topicity. The relationship between topicity, isochrony/anisochrony, and signal multiplicity are discussed in the form of a chart in Fig. 2.72. Anisochrony is often very small (1 ppm or less); a case, when it is undetectable, is called *accidental isochrony*. Use of higher field or different solvents in ^{1}H NMR or ^{13}C NMR may be helpful in increasing anisochrony.

A very instructive example is illustrated in Fig. 2.73.

2.10.1   Anisochrony Arising out of Diastereotopic Faces

Isochrony arising out of the homotopic faces of the maleic acid (−)-menthyl ester (**1**) and anisochrony arising out of the diastereotopic faces of fumaric acid (−)-menthyl ester (**2**) are much instructive [18] and are illustrated in Fig. 2.73.

The two olefinic protons H_A and H_B of the maleate (**1**), being exchangeable by the C_2 axis (in the molecular plane) operation, are *homotopic*, and hence are *isochronous*, and appear as a 2H singlet. Again, its front and rear π-faces are also interchangeable through the same C_2 axis operation and hence are homotopic. So the complexation by Fe(CO)$_4$ from the homotopic α- or β-face leads to the same complex (*cf*. epoxidation). But H_A and H_B of the Fe(CO)$_4$ complex being *diastereotopic* are *anisochronous* and have different chemical shift values. They

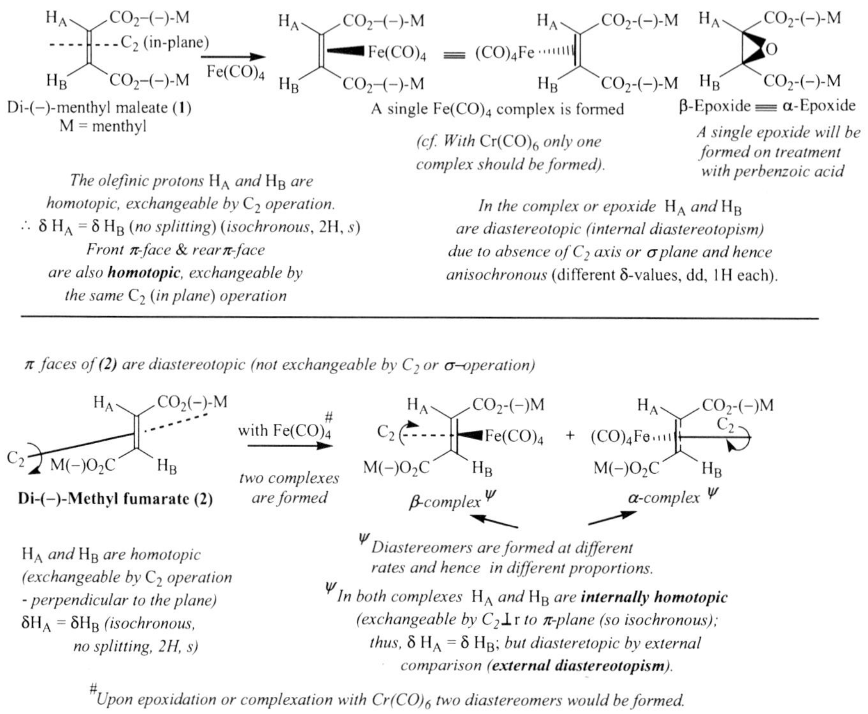

Fig. 2.73 Anisochrony arising out of diastereotopic faces. An instructive example

split and appear as a double doublet (dd) (1H each) in the NMR. Thus, the maleate (**1**) should form, on treatment with perbenzoic acid, a single epoxide of which H_A and H_B being diastereotopic should appear as a double doublet in the NMR.

In case of the fumarate (**2**) also, H_A and H_B, being exchangeable by C_2 (orthogonal to the π-plane) operation, are *homotopic* and appear as a 2H singlet. However, its two π-faces being not exchangeable by any symmetry operation are *diastereotopic*. So with $Fe(CO)_4$ two diastereomeric complexes are formed *at different rates and hence in different proportions*. In each diastereomeric complex, the olefinic protons are *internally homotopic* (exchangeable by a C_2 axis—orthogonal to the π-plane) and so isochronous. However, the two olefinic protons in one diastereomer are diastereotopic with the two olefinic protons of the other diastereomer *by external comparison*, so they are termed **externally diastereotopic**. Consequently, the two olefinic protons of the two diastereomeric complexes show two 2H singlet peaks of different intensities.

The two olefinic esters (**1**) and (**2**) thus may be distinguished by NMR through complexation with faces of different topicities.

Although each of (**1**) and (**2**) show a 2H singlet (because in both compounds the two olefinic protons are homotopic), they may be distinguished by ^1H NMR through complexation (or epoxidation) with homotopic or diastereotopic faces, respectively.

2.11   Asymmetric Synthesis

2.11.1   Introduction. Principles of Stereoselection: Enantioselection. Diastereoselection

2.11.1.1   Lack of Stereoselection

In the absence of chiral influence by way of reactants, reagents, or solvents, a reaction of a starting material having a prochiral center generating a chiral center will produce equal amounts of enantiomers, i.e., a racemic product, since such a reaction proceeds through enantiomeric transition states (TS) of equal energy. Thus, the activation energies being same the reactions take place at the same rate. The principles of the *lack of stereoselection* and of *enantioselection* are illustrated with the help of energy diagrams (Figs. 2.74 and 2.75, respectively).

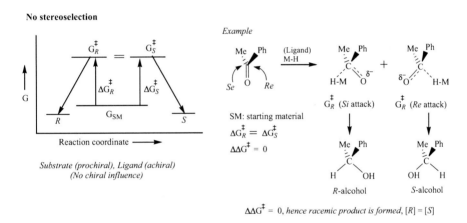

Fig. 2.74 Principle of absence of stereoselection. Energetics

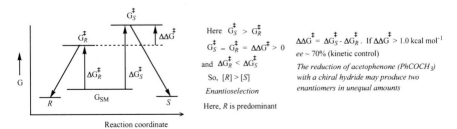

Fig. 2.75 Principle of enantioselection. Energetics

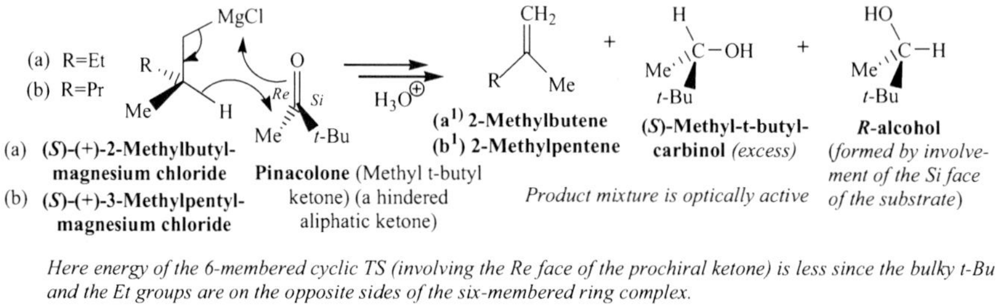

(a) R=Et
(b) R=Pr

(a) (S)-(+)-2-Methylbutyl-
　　magnesium chloride
(b) (S)-(+)-3-Methylpentyl-
　　magnesium chloride

Pinacolone (Methyl t-butyl
ketone) (a hindered
aliphatic ketone)

(a¹) 2-Methylbutene
(b¹) 2-Methylpentene

(S)-Methyl-t-butyl-
carbinol (excess)

Product mixture is optically active

R-alcohol
(formed by involve-
ment of the Si face
of the substrate)

Here energy of the 6-membered cyclic TS (involving the Re face of the prochiral ketone) is less since the bulky t-Bu and the Et groups are on the opposite sides of the six-membered ring complex.

Fig. 2.76 Examples of first enantioselective Grignard reactions

2.11.1.2 Enantioselection

In 1950 Mosher and La Combe reported two papers [29, 30] on the first asymmetric Grignard reactions (Fig. 2.76) which are examples of enantioselection (Fig. 2.76). The existing chiral center (or centers) is said to bring about *asymmetric induction*. Thus, addition of a chiral nucleophile to the two enantiotopic faces of a prochiral carbonyl compound leads to enantioselectivity.

2.11.1.3 Diastereoselection

The concept of diastereoselection was first clearly outlined by Emil Fischer in 1894, based upon the conversion of one sugar to the next higher homologue via the cyanohydrin reaction. Thus, the substrate molecule should possess a prochiral center as well as at least one chiral center, and the reagent may be achiral, as in the reaction of a sugar with HCN. The principle of diastereoselection is illustrated with the help of an energy diagram (Fig. 2.77), which is self-explanatory. The expression for $\Delta G°$ in thermodynamic control and that for $\Delta\Delta G^{\neq}$ in kinetic control are shown. In case *the reaction is reversible*, the product which is formed at a slower rate but at the same time if it is of lower free energy (more stable) will be major, and the reaction is said to be of thermodynamic control. Like Fischer's cyanohydrin reaction, reactions to which Cram's rule and Prelog's rule are applicable are also examples of diastereoselection. Moreover, enantiomers have widely different reactivities especially in biological systems. The enantiomer may have different biological activities, and only a particular enantiomer may have the desired drug effect (Fig. 32.2). It is well known that only L-amino acids can participate in protein synthesis. Thus, diastereoselection is a result of involvement of diastereomeric transition states (with different free energies); consequently, the diastereomer, which is formed irreversibly via less activation energy and hence at a faster rate, is the predominant product of kinetic control.

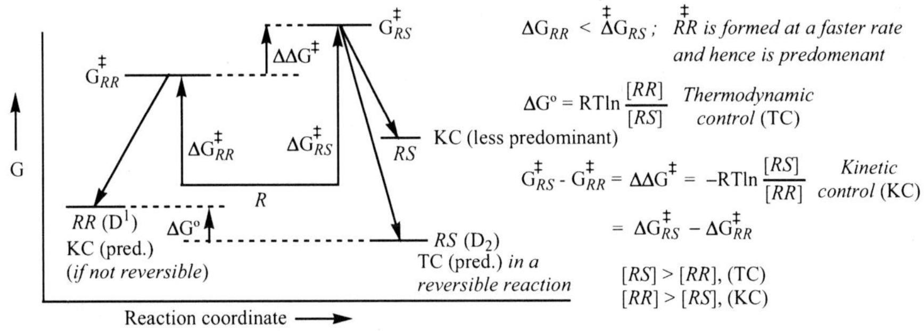

$G^{\ddagger}_{RS}$
$\Delta\Delta G^{\ddagger}$
$G^{\ddagger}_{RR}$
$\Delta G^{\ddagger}_{RR}$
$\Delta G^{\ddagger}_{RS}$
G
RS
KC (less predominant)
R
RR (D1)
ΔG^{o}
KC (pred.)
(if not reversible)
RS (D2)
TC (pred.) in a reversible reaction
Reaction coordinate
$\Delta G^{\ddagger}_{RR} < \Delta G^{\ddagger}_{RS}$; RR is formed at a faster rate and hence is predomenant
$\Delta G^{o} = RT\ln\frac{[RR]}{[RS]}$ Thermodynamic control (TC)
$G^{\ddagger}_{RS} - G^{\ddagger}_{RR} = \Delta\Delta G^{\ddagger} = -RT\ln\frac{[RS]}{[RR]}$ Kinetic control (KC)
$= \Delta G^{\ddagger}_{RS} - \Delta G^{\ddagger}_{RR}$
[RS] > [RR], (TC)
[RR] > [RS], (KC)

If ΔG or $\Delta\Delta G^{\ddagger} \geq 1.0$ kcal/mole, d.e. *would be* $\geq 70\% = \%D^1 - \%D^2$ where D^1 *is the predominant diastereomer*

Fig. 2.77 Principle of diastereoselection. Energetics

Product stereoselectivity

Ex. 1

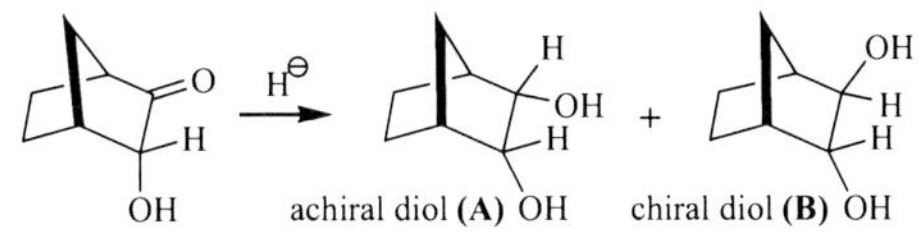

O
H
OH
H
OH
H
achiral diol (A) OH
OH
H
H
chiral diol (B) OH

*By use of an adequately bulky hydride reagent if the achiral diastereomeric diol (**A**) is formed exclusively, still it is an asymmetric synthesis, and the reaction is said to be 100% stereoselective as a result of 100% diastereoselection. This is an example of product stereoselectivity.*

Ex. 2

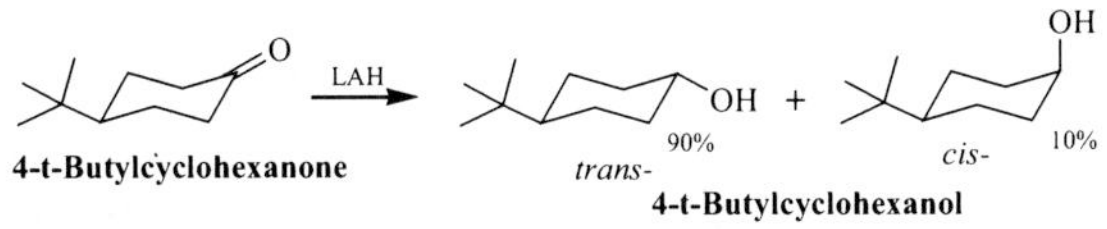

O
LAH
4-t-Butylcyclohexanone
OH +
90%
trans-
OH
cis-
10%
4-t-Butylcyclohexanol

Fig. 2.78 Stereoselective reactions. Product stereoselectivity

2.11.2 Asymmetric Synthesis. Definition. Stereoselective and Stereospecific Reactions. Product/Substrate Stereoselectivity. Regioselectivity

An *asymmetric synthesis* is a reaction in which a *prochiral unit* in a substrate molecule is converted by a reactant into a *chiral unit* in such a manner that the stereoisomeric products are produced in unequal amounts.

The prochiral unit in a substrate molecule must have enantiotopic or diastereotopic ligands or faces. In rare cases, only one stereoisomer is formed with the complete exclusion of the other stereoisomer, and the reaction is termed *100 % stereoselective*. One stereoisomer may be optically inactive also (Ex. 1 of Fig. 2.78). This is an example of *product stereoselectivity*.

Example 2 provides another simple case of *product stereoselectivity* (Fig. 2.78).

When two diastereomers or two enantiomers react at different rates to give two stereoisomers (Fig. 2.79)—the reaction is termed *stereospecific*. Enzymes usually

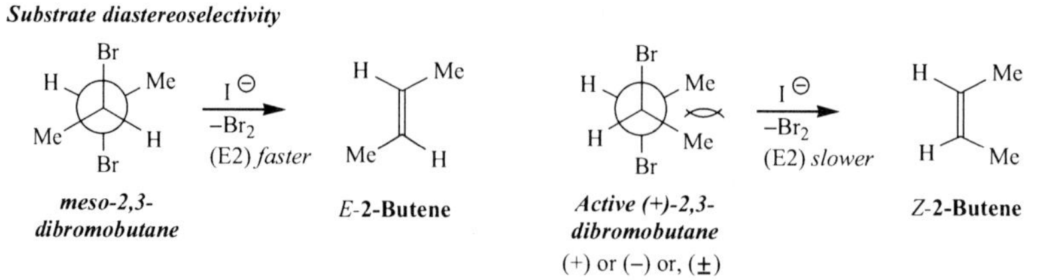

Substrate diastereoselectivity

meso-2,3-dibromobutane → *(E2) faster* → **E-2-Butene**

Active (+)-2,3-dibromobutane
(+) or (−) or, (±) → *(E2) slower* → **Z-2-Butene**

This is a case of substrate selectivity or, better substrate diastereoselectivity. In some particular case one diastereomer may not react at all under the same condition.

Fig. 2.79 Stereospecific reactions. Substrate selectivity

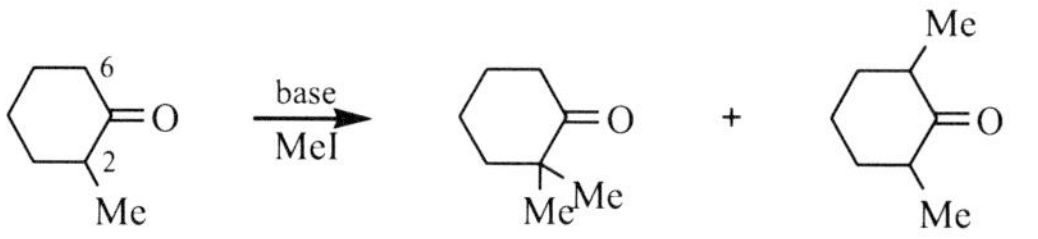

2-Methylcyclohexanone → base, MeI → **2,2-Dimethylcyclohexanone** (mostly) + **2,6-Dimethylcyclohexanone** (a little)

The generated C2 anion is more reactive than that generated at C6, because of the inductive effect of the methyl group.

Fig. 2.80 An example of regioselectivity

react with one enantiomer only, and sometimes with both enantiomers in different ways (Fig. 32.2) and thus show total substrate enantioselectivity.

Regioselectivity. If a substrate is capable of reacting at more than one center (polydent molecules) and reacts at one center with a higher rate than the other centers, this is known as *regioselectivity* (Fig. 2.80).

2.11.2.1   Enantiomeric Excess. Diastereomeric Excess. Optical Purity

In a mixture of a pure enantiomer (*R* or *S*) and a racemate (*RS*), *enantiomeric excess* (*ee*) is the percent excess of the enantiomer over the racemate. Thus, *percent enantioselective excess* (*% ee*) or *percent diastereoselective excess* (*% de*) of *R* is illustrated by the following expression:

$$\% \text{ ee or } \% \text{ de of R} = \frac{[R] - [S]}{[R] + [S]} \times 100 = \% \text{ R} - \% \text{ S} = \% \text{ stereoselectivity of R,}$$

where R and S represent two enantiomers or configurations of the chiral center created in the two diastereomers.

Enantiomeric excess usually corresponds to the older expression, **optical purity,** *op*, $([\alpha]_{\text{obs.}}/[\alpha]_{\text{max}}) \times 100 \%$, *i.e.,* the absolute value of the ratio of the observed specific rotation of a sample made up of two enantiomers, which is otherwise chemically pure, to the corresponding specific rotation of one pure enantiomer, expressed as percentage. The term $[\alpha]_{max}$ signifies the specific rotation of one enantiomerically pure sample. Currently, *ee* or *de* is usually measured by NMR and by chromatographic analysis, and hence the term optical purity is gradually becoming outmoded.

1,2-Addition of Achiral Nucleophiles to Chiral Ketones and Aldehydes (Diastereoselection) (Sects. 2.11.3–2.11.5).

2.11.3 Cram's rule

2.11.3.1 Cram's Open Chain Model

A carbonyl group attached to a chiral center (*e.g.*, RCOCLMS) (L, M, and S represent large, medium sized, and small ligands, respectively) undergoes nucleophilic addition with metal hydride or achiral organometallic reagents (*e.g.*, R^1MgX or R^1M) to produce two diastereomeric products of which one predominates. The relative configuration of the asymmetric center, created in the predominant diastereomer, and its absolute configuration—if the configuration of the chiral carbon already present in the substrate is known—is predicted by *Cram's rule* based on some arbitrary TS models [31–34]. The rule is essentially empirical but has excellent predictive value.

In the open chain model illustrated in Fig. 2.81, the substrate molecule is complexed with the reagent from the right side of the carbonyl π-plane to give (A), from which R^1 is transferred to the trigonal carbon from the right diastereotopic face (from the side of the small ligand S) (route **a**), in preference to the complex (A^1) which transfers R^1 from the left diastereotopic face (route **b**).

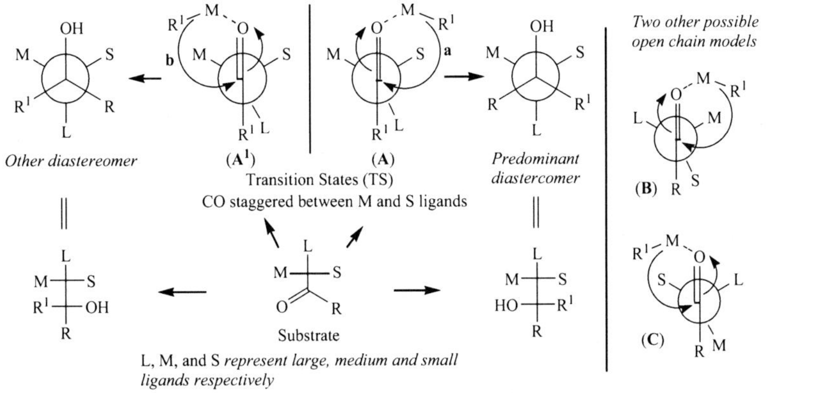

Other diastereomer   (A^1)   (A)   *Predominant diastereomer*

Transition States (TS)
CO staggered between M and S ligands

Substrate

L, M, and S *represent large, medium and small ligands respectively*

*The metallic part gets complexed with O of CO, making it effectively the bulkiest group, better placed between M and S, and thus R^1 is preferentially transferred from the side of S (path **a**) leading to the **predominant diastereomer**.*

Fig. 2.81 Cram's open chain model

Thus, 1,2-asymmetric induction takes place by the chiral center already present. Reagents R^1M: EtLi, MeLi R^1Li, etc.; $LiAlH_4$, $NaBH_4$; R^1MgX: R^1 = Me, Et, Ph, iPr, iPrCH$_2$, etc., X = Br, I.

One other possible open chain TS model (**B**) (Fig. 2.81) with carbonyl group staggered between L and M and the nucleophile a attacking the carbonyl carbon from the side of M (smaller than L) correctly predicts the stereochemical course to give the predominant product, but it involves stronger steric interactions of the bulkiest complexed O (of CO) with gauche L and M groups making the TS energy much higher and hence is not tenable. The third possible TS model (**C**) (Fig. 2.81) with CO staggered between S and L involving the preferential nucleophilic attack on the left diastereotopic face (the side of S) leads to the less predominant product, and hence is eliminated.

Cram's rule is equally applicable to racemic substrate when the products are also racemic.

Limitations of Cram's Open Chain Model

(i) It applies only to kinetically controlled reactions; for Meerwein–Pondorf–Varley (MPV) reaction Cram's model may hold good for a short reaction time.
(ii) It does not apply to catalytic reduction. These two limitations hold good for other models also, to be discussed later.
(iii) It does not apply when the small ligand S is OH, OR, NH$_2$—capable of complexing with the reagent R^1M or R^1MgX. In such cases cyclic model will be applicable.

Some examples of the applications of Cram's rule are shown in Fig. 2.82.

Stereoselectivity through the open chain nonrigid model (Fig. 2.82) is not usually high except for a case in which the difference of bulkiness of the ligands M and S is high.

2.11.3.2   Cram's Chelate or Cyclic Model

If the substrate chiral ketone contains at the α-position an OH, NH$_2$, or OMe group which is capable of coordinating with the reagent, Cram's rule based on a rigid *chelate or cyclic model* predicts the stereochemistry of the predominant product, as delineated in Fig. 2.83. The metallic part of the reagent is doubly coordinated, as shown, to form a 5-membered ring. The nucleophile preferentially approaches the electrophilic carbon from the side of the ligand S. If the chelating group is M (usual case), the cyclic model predicts the same stereochemistry as the open chain model (case 1). If the chelating group is the ligand S or L, the cyclic model predicts the correct stereochemistry of the predominant product, while the open chain model predicts the opposite stereochemistry (case 2). For such substrates the chelate model should always be applied, irrespective of the bulkiness of the chelating ligand.

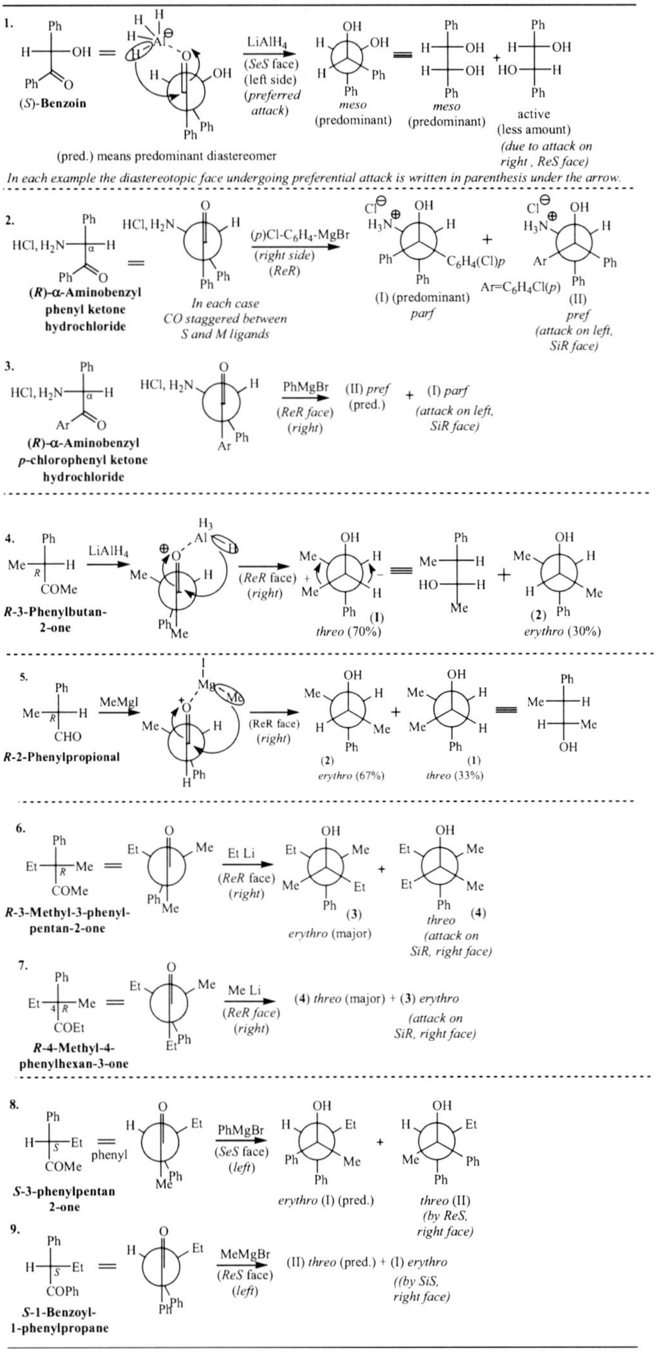

Fig. 2.82 Cram's rule. Open chain model, nine examples

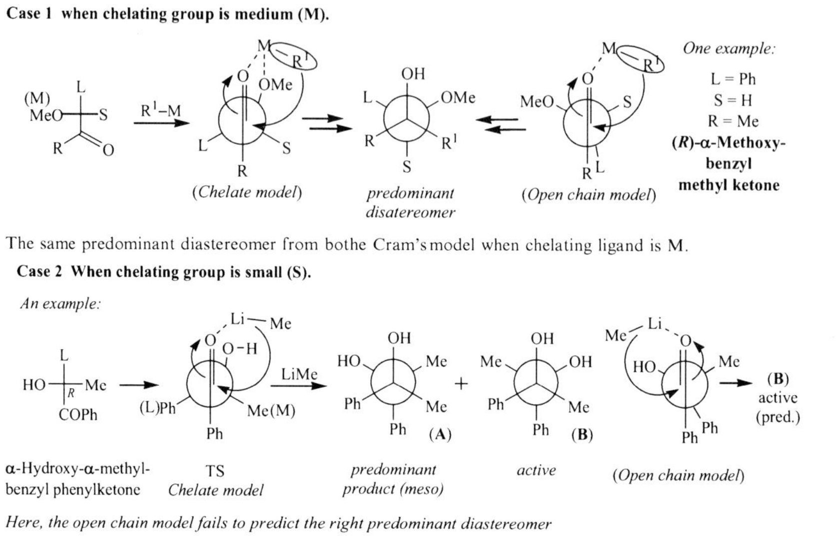

Case 1　when chelating group is medium (M).

The same predominant diastereomer from bothe Cram's model when chelating ligand is M.

Case 2　When chelating group is small (S).

An example:

α-Hydroxy-α-methyl-benzyl phenylketone

Here, the open chain model fails to predict the right predominant diastereomer

Fig. 2.83 Cram's chelate (cyclic) model. Examples

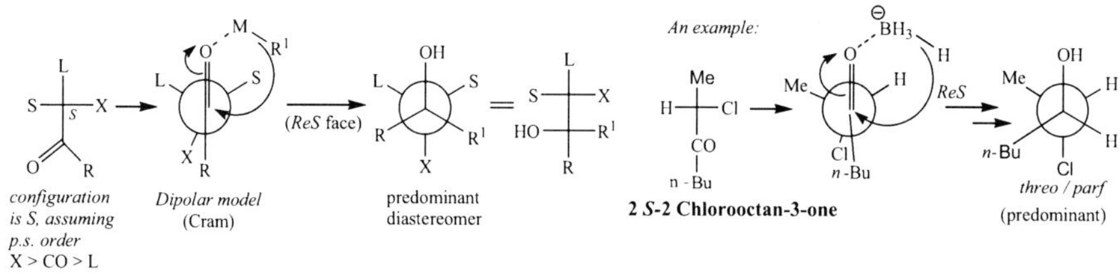

Fig. 2.84 Cram's dipolar model. An example

2.11.3.3　Cram's Dipolar Model

Cram suggested a *dipolar model* for prediction of stereochemistry when the substrate carries a strongly electronegative group, *e.g.*, a halogen atom at C_α position [32, 34]. To minimize the dipole repulsion and to increase the electrophilic character of the carbonyl carbon, the C=O bond and the C–X bond (having opposite dipoles) are placed anti in the dipolar model (Fig. 2.84). The nucleophile adds from the side of the small ligand S, giving the predominant product, as shown. Cram's model correctly predicts the stereochemical course of the reactions but does not always succeed to give a quantitative assessment of the asymmetric induction based on steric interactions. A few alternative models have been suggested [33] of which the Felkin–Anh model [35, 36] has been widely accepted.

2.11.4   Felkin–Anh Models [35, 36]

2.11.4.1   Felkin–Anh Open Chain Model

In this model three reactive transition state (TS) conformations (**A**), (**B**), and (**C**) (Fig. 2.85) are considered in terms of the orbital and the nonbonded steric interactions. Because of the destabilizing four-electron interactions between the HOMO (highest occupied molecular orbital) of the nucleophile and the HOMO of the carbonyl group [(**a**) in Fig. 2.86], the incoming nucleophile in each case approaches the carbonyl at an angle of about 109° with the carbonyl plane, little away from the orthogonal approach, corresponding to the Bürgi–Dunitz trajectory [37–39]. The TS (**A**) encounters minimum steric interaction in addition to the orbital destabilizations (discussed later) and involves mainly the two-electron stabilizing interactions between the HOMO of the nucleophile and the lowest unoccupied molecular orbital (LUMO) of the carbonyl group. Hence, it leads to the predominant diastereomer (**D^1**) by nucleophilic attack on the right side. The TS conformation (**C**) might also lead to the same predominant diastereomer (**D^1**), but its contribution should be *very little* because of its high TS energy due to nonbonded interactions R↔L and L↔R^1 (approaching nucleophile). The TS conformation (**B**) involves the

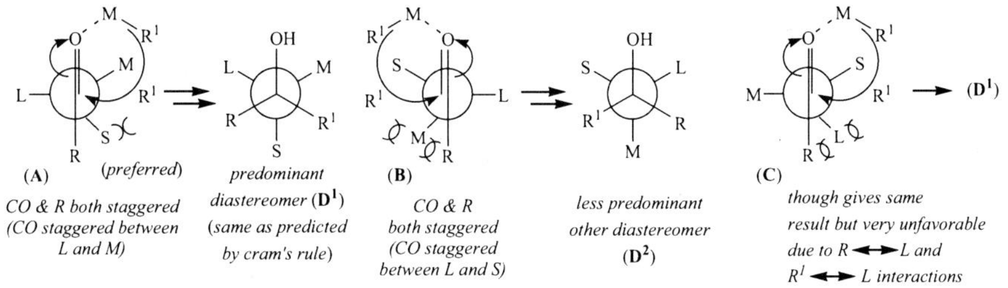

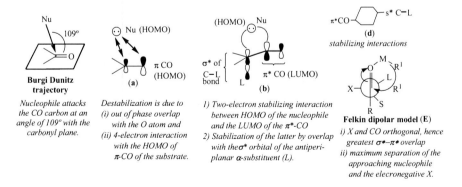

Fig. 2.85 The Felkin–Anh model for the nucleophilic addition to the carbonyl group.

Fig. 2.86 Felkin dipolar model. Orbital interactions in the Felkin–Anh model

nonbonded interactions M↔R and R^1↔M; hence, it possesses higher free energy and leads to the less predominant diastereomer ($\mathbf{D}^2$).

Moreover, the antibonding π* orbital is stabilized by its overlap with the σ*orbital.

Thus, consideration of the *molecular orbitals involved* reveal that the *orthogonal approach* of the nucleophile leads to its destabilization due to (i) out of phase overlap with the oxygen atom and (ii) from the four-electron interaction with the HOMO of the π-CO of the substrate [(**a**) in Fig. 2.86], and hence the approach of the nucleophile invokes the *Bürgi–Dunitz trajectory* to minimize these destabilizing factors. The approach of the nucleophile involves further stabilization if the largest ligand L is placed orthogonal to the C = O double bond, thus providing greatest overlap between the antibonding π* orbital (lowest unoccupied molecular orbital, LUMO) of the carbonyl group and the antibonding orbital σ* of the antiperiplanar α-substituent, in addition to the stabilizing two electron $n-σ*$ interaction with the electron pair of the nucleophile as shown in (**b**).

In fact, any example of the Cram's open chain model (see Fig. 2.83) can be explained in a more quantitative manner by Felkin–Anh's model (**A**) (Fig. 2.85).

2.11.4.2 Felkin–Anh Dipolar Model

Cram's dipolar model for the prediction of the predominant diastereomeric product may also be replaced by a different Felkin–Ahn model (**E**) (Fig. 2.86), based on the tenets already mentioned, with the additional proviso that in the TS the halogen or any strongly electronegative ligand is placed orthogonal to the C=O bond, thus providing greatest overlap between σ* orbital of the C–X bond and the π* orbital of the C=O, as well as allowing maximum separation of the electronegative α-substituent (X) and the negatively charged nucleophilic reagent (Fig. 2.86). Of course, Felkin–Anh dipolar model also leads to the same predominant diastereomer as the Cram's dipolar model (Fig. 2.84).

2.11.5 Prelog's Rule

Prelog's rule, though empirical, allows the prediction of the steric course of an asymmetric synthesis carried out with a chiral α-ketoester (usually a glyoxylate or pyruvate) of a chiral secondary or tertiary alcohol, SMLC-OH (S, M, and L stand for small, medium sized, and large substituents, respectively). The prediction has been confirmed in the majority of the cases.

It is the outcome of generalization of the results of the asymmetric synthesis carried out by McKenzic group in early twentieth century. In McKenzie's first example [40] the phenylglyoxalate ester of (−)-menthol was reduced with aluminum amalgam to give the (−)-mandalate ester of (−)-menthol as the predominant product over the corresponding (+)-mandalate ester (Fig. 2.87).

For application of Prelog's rule [41, 42], by convention, in the transition state the α-ketoester (**1A**) (Fig. 2.88) is so oriented that the two carbonyl groups are

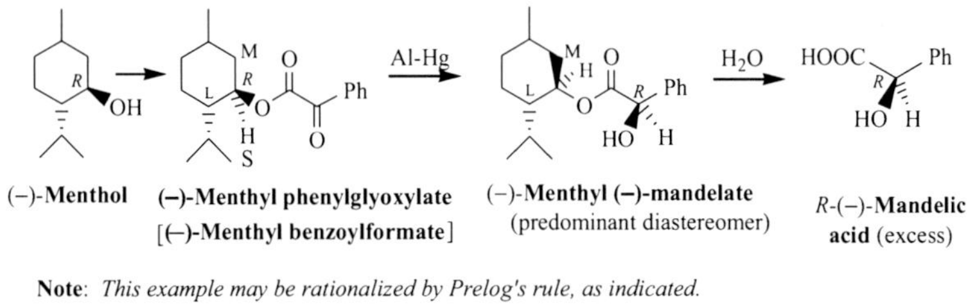

Al-Hg
H2O
HOOC
Ph
(−)-Menthol
(−)-Menthyl phenylglyoxylate
[(−)-Menthyl benzoylformate]
(−)-Menthyl (−)-mandelate
(predominant diastereomer)
R-(−)-Mandelic acid (excess)
Note: This example may be rationalized by Prelog's rule, as indicated.

Fig. 2.87 Asymmetric synthesis of (−)-mandelic acid

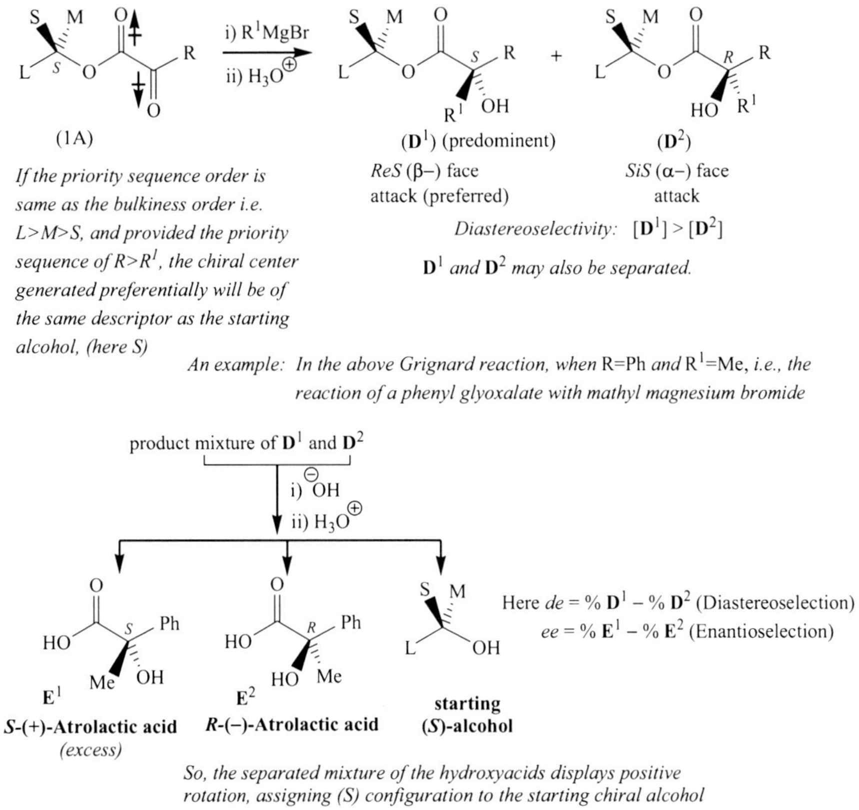

i) R1MgBr
ii) H3O⊕
(1A)
(D1) (predominent)
(D2)
If the priority sequence order is same as the bulkiness order i.e. L>M>S, and provided the priority sequence of R>R1, the chiral center generated preferentially will be of the same descriptor as the starting alcohol, (here S)
ReS (β−) face attack (preferred)
SiS (α−) face attack
Diastereoselectivity: [D1] > [D2]
D1 and D2 may also be separated.
An example: In the above Grignard reaction, when R=Ph and R1=Me, i.e., the reaction of a phenyl glyoxalate with mathyl magnesium bromide
product mixture of D1 and D2
i) OH⊖
ii) H3O⊕
Here de = % D1 − % D2 (Diastereoselection)
ee = % E1 − % E2 (Enantioselection)
E1
S-(+)-Atrolactic acid
(excess)
E2
R-(−)-Atrolactic acid
starting (S)-alcohol
So, the separated mixture of the hydroxyacids displays positive rotation, assigning (S) configuration to the starting chiral alcohol

Fig. 2.88 Prelog's rule. Diastereoselection. Enantioselection

antiperiplanar (to minimize the dipolar repulsion), and the L group occupies the same plane as the two carbonyl groups and the alkyl oxygen bond; thus, the bulky ester carbonyl group (complexed with the reagent) becomes staggered between two smaller groups S and M (Fig. 2.88). In this TS conformation the nucleophile (usually the Grignard reagent R^1MgX) will predominantly approach the ketocarbonyl function from the side of the small group S. This rule thus correlates the configuration of the starting chiral alcohol SMLC-OH with the predominant α-hydroxyester formed, and therefore, of the α-hydroxy acid obtained in excess by hydrolysis along with the starting alcohol. Prelog's rule holds good since such nucleophilic addition is kinetically controlled.

2.11.5.1 Attempted Rationalization of Prelog's Model

The generalization for predicting the stereochemistry of the products from the α-keto-ester asymmetric synthesis was put forward as an empirical model; yet its reasonable success implies some correlation of the model with the conformation of the TS (Fig. 2.89).

It is pertinent to note that trans-coplanar conformation does not represent the most stable ground state; it may represent the most stable TS, because of maximum separation of charge between negative centers is possible in this form. Thus, it represents better distribution of charge density under the influence of the reagent in the TS. X-ray crystal studies revealed that the ϕ between two CO's of (2) (Fig. 2.89) is 104°. This conformation may exist in solution also.

2.11.5.2 More Examples of The Application of Prelog's Rule

The atrolactic acid synthesis and the Prelog's rule have been widely used for the determination (or confirmation) of the absolute configuration of secondary alcohols, e.g., alicyclic alcohols, monoterpene alcohols, sesquiterpene alcohols, triterpene alcohols, and steroid alcohols [43]. Some examples of Prelog's rule are shown in Fig. 2.90. For the asymmetric atrolactic acid synthesis, the chiral secondary alcohol is converted to its phenylglyoxylate ester. The latter is then reacted with MeMgBr to give the α-hydroxy ester in excess predicted by Prelog's rule.

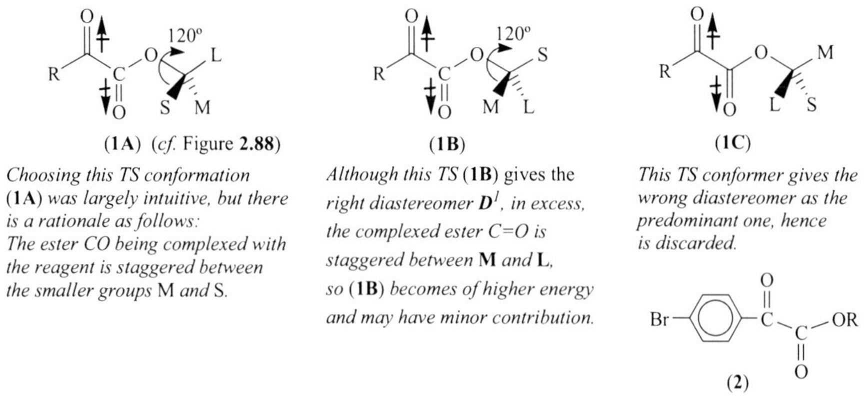

(1A) (cf. Figure 2.88)   (1B)   (1C)

Choosing this TS conformation (1A) was largely intuitive, but there is a rationale as follows: The ester CO being complexed with the reagent is staggered between the smaller groups M and S.

Although this TS (1B) gives the right diastereomer D^I, in excess, the complexed ester C=O is staggered between M and L, so (1B) becomes of higher energy and may have minor contribution.

This TS conformer gives the wrong diastereomer as the predominant one, hence is discarded.

Fig. 2.89 Rationalization of Prelog's rule

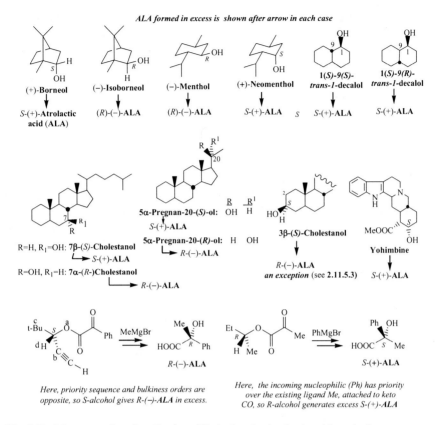

Fig. 2.90 More examples of application of Prelog's rule. Atrolactic acid synthesis

The product mixture upon hydrolysis gives the predicted atrolactic acid in excess (as shown in Fig. 2.88). In case the *priority sequence order is the same as the bulkiness order and (R) (here Ph) has higher priority than the incoming nucleophile (R)^1 (here Me), (R)-alcohol gives the (R)-hydroxy acid (here (R)-atrolactic acid) in excess and the (S)-alcohol gives (S)-hydroxy acid* (here (S)-atrolactic acid). The other two possibilities of getting (R)-hydroxy acid from (S)-alcohol and *vice versa* are illustrated in the last row of Fig. 2.90.

2.11.5.3 Exception to and Anomalies of Prelog's Rule

In atrolactic acid synthesis with (S)-3β-cholestanol (see Fig. 2.90), R-(−)-atrolactic acid is obtained in only 1.7 % enantiomeric excess [43]. Thus, S-alcohol is providing the R-atrolactic acid. This may be taken as an exception to Prelog's rule. Perhaps, here bulkiness order is not the same as the CIP priority sequence. Examination of models reveals that the steric crowding at C2 (with its environment) is greater than that at C4. However, asymmetric induction being so low, no definite conclusion can be drawn from this result.

Atrolactic acid synthesis of the tetramethyl ethers of both $(-)$-epicatechin (R-3-ol) and $(+)$-catechin (S-3-ol) leads to the preferential formation of R-$(-)$-atrolactic acid. These results must be considered as an anomaly rather than an exception to the rule [43] (vide Sect. 14.5.3, Fig. 14.25).

2.11.6   Horeau's Rule

Horeau developed an empirical method [44, 45] for the correlation of configuration of secondary alcohols based on the principle of *kinetic resolution.* During the esterification of an optically active secondary alcohol, say, $(-)$-RR^1CHOH with *an excess of racemic* $(\pm)$-R^2COOH the transition states, $[(+)$-$RR^1CHOH.(+)$-$R^2COOH]^{\#}$ and $[(+)$-$RR^1CHOH.(-)$-$R^2COOH]^{\#}$ are diastereomeric and thus are unequal in enthalpy (*cf.* Fig. 2.77). Hence, the activation energies and the reaction rates of esterification with the enantiomeric acids will be unequal. Thus, the enantiomeric acid involving lower activation energy will be esterified at a faster rate, leaving the other enantiomeric acid in excess being unreacted (kinetic resolution) (Fig. 2.91).

An enantiomerically pure (or enriched) secondary alcohol is treated with an excess of $(\pm)$-α-phenylbutyric anhydride (Fig. 2.91) (or sometimes the corresponding chloride). The residual anhydride is hydrolyzed and the optical rotation of the resultant α-phenylbutyric acid is measured. The Horeau's rule states that (provided, the CIP priority L > M holds good) an alcohol with R-configuration gives an excess of S-$(+)$-phenylbutyric acid and vice versa (Fig. 2.91). Brewster [46] and Horeau [47] have published detailed reviews on this method and its limitations.

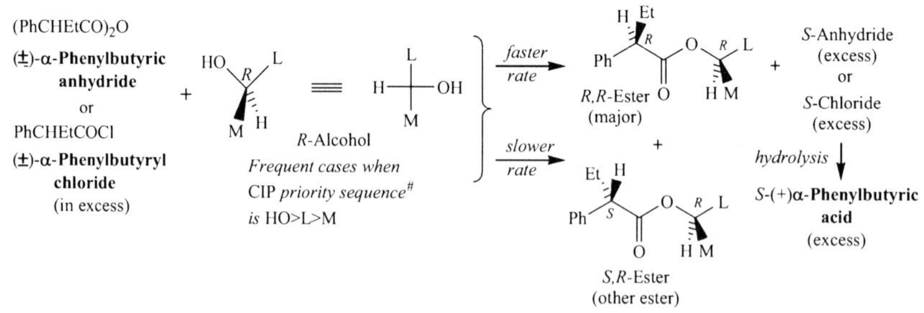

Thus, excess of S-(+)-α-phenylbutyric acid implies R-configuration of the unknown alcohol.
Likewise, excess of R(−)-α-phenylbutyric acid implies S-configuration of the unknown alcohol.

[#]In rare cases when priority sequence is HO>M>L, the conclusion will be reverse, i.e. excess S-acid implies S-alcohol and excess R-acid implies R-alcohol. Examples: M= −C≡CH and L= t-Bu; M= −CH=CH₂, L= −CHMe₂

Fig. 2.91 Horeau's rule illustrated

Horeau's method has the advantage that in case the optically active alcohol is not available, the racemic alcohol itself may be treated with optically active α-phenylbutyric anhydride taken in much excess. The rotation of the unreacted alcohol is measured. Based on kinetic resolution now it may be concluded that if *S*-(+)-phenylbutyric anhydride is used, the unreacted alcohol would have *R* configuration (provided L gets priority over M) and *vice verse* (Fig. 2.91). The sign of rotation of the remaining unreacted alcohol needs to be determined. The rule is strictly empirical, and no rationalization has yet appeared.

The Horeau's procedure can be used [48, 49] for microscale determination if a highly sensitive method (GC, HPLC, MS, or CD) is available to determine the composition of the diastereomeric esters formed.

2.11.7 Sharpless Enantioselective Epoxidation

Sharpless in 1980 discovered a new metal-catalyzed asymmetric epoxidation process [50] which is by far more selective than any previously described methods for such asymmetric transformation. The strained epoxide ring serves as a valuable synthon for many natural products including insect pheromones. One of the most attractive aspects of this new process is the use of its readily available simple components.

In this method a prochiral primary allyl alcohol (mono-, di-, or tri-substituted, R^1, R^2, and R^3 are achiral) is treated with *t*-butylhydroperoxide (TBHP) in the presence of titanium (IV) isopropoxide and optically active diethyl tartrate (DET) (or di-isopropyl tartrate, DIPT) to produce the corresponding epoxide in a highly enantioselective manner, as shown in Fig. 2.92. The system provides consistent enantioselection.

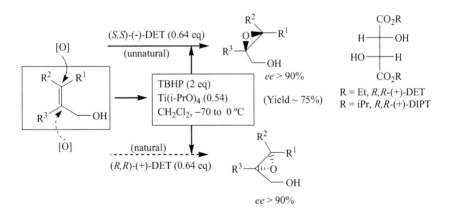

Fig. 2.92 Prediction of the chiral epoxides formed enantioselectively from prochiral allylic alcohols by Sharpless rule

With Sharpless reagent using (+)-DET geraniol undergoes regiospecific epoxidation of the allylic 2,3-double bond to form 2S,3S-epoxygeraniol in 90 % ee in accordance with the Sharpless rule [50]; here attack takes place almost entirely at the 2Re-,3Si (α-face) [cf. ex. (V) of Sect. 6.2.6]. Extensive application of the Sharpless epoxidation method to numerous multistep syntheses demonstrates its utility and reliability [51, 52].

The catalyst is sensitive to preexisting chirality in the substrate. Thus, the epoxidation of racemic secondary allylic alcohols with Sharpless reagent proceeds rapidly with only one enantiomeric alcohol, leaving behind the other slower reacting enantiomer.

2.11.7.1 Kinetic Resolution of Racemate Allyl Alcohols

Resolution of racemate allyl alcohols may be accomplished *by treating with half equivalent of TBHP in the presence of the chiral catalyst like Ti-(i-Pro)₄ complexed with an active DET (or DIPT).*

The epoxidation pattern of *E*-cyclohexylpropenylcarbinol using (+)-Diisopropyltartrate is shown in Fig. 2.93 [53].

Interestingly, for steric reasons in the transition states, the *S*-enantiomer (in Fig. 2.93) reacts much faster than the *R*-enantiomer, thus under the reaction condition permitting the isolation of the unreacted *R*-isomer.

Kinetic resolution of the racemate allylic alcohols (*cf.* Fig. 2.93) has been achieved by Sharpless epoxidation method [54].

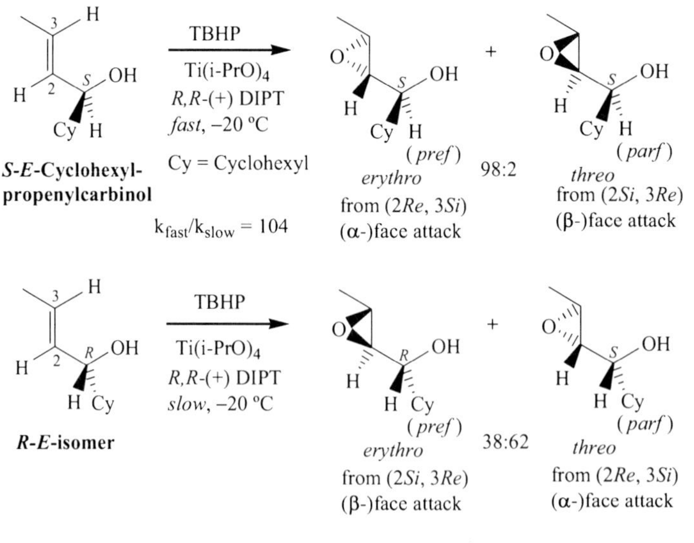

Fig. 2.93 Kinetic resolution of racemate allyl alcohol by Sharpless epoxidation method

2.11.7.2 Mechanism of the Sharpless Reaction

Nature of the titanium alkoxide system has the following properties crucial to the success of the reaction:

(i) Exchange of monodentate alkoxide ligands is rapid in solution.
(ii) Of the four covalent bonds of Ti^{IV} two participate for the divalent chiral auxiliary (tartrate) and one each for TBHP and the allyl alcohol.
(iii) The coordination chemistry of Ti^{IV} (d°) alkoxide system is flexible since this system displays a range of coordination numbers and geometries. This property may be responsible for the catalysts' ability to accommodate sterically widely differing substrates.
(iv) Ti^{IV} alkoxides are weak Lewis acids and thus activate a coordinated alkylperoxo ligand toward nucleophilic attack by olefin of a bound allylic alcohol.

The mechanism of the reaction has been extensively studied by NMR, and numerous crystal structures for the complexes involved have been determined. The active catalyst was initially believed to be the 10-membered structure [55]. Comprehensive analyses [56, 57] favor a mechanism (Fig. 2.94) based on the structure (I) to be converted to (II), by use of R,R-(+)-DET, epoxidizing much predominantly the enantiotopic α-face of the allylic segment (when written as in Fig. 2.92).

Most importantly, the species containing equimolar amount of titanium tetraalkoxide and tartrate is shown to be the most active catalyst in the reaction mixture, mediating the reaction at a much faster rate than titanium tetraalkoxide alone [56].

In (I) due to weak coordination the carbonyl oxygen atoms readily dissociate and recoordinate to the metal centers, providing a means of exchanging the alkoxide ligands for the substrate molecules [R,R,-(+)-DET], TBHP, and allylic alcohol as shown in (II). The lower energy conformation (here oxygen is delivered from the α-face of the double bond of the β-substituted allylic alcohol) of the allylic segment dictates which enantiotopic face is epoxidized. Many allylic alcohols with different substituents have been highly enantioselectively epoxidized following the Sharpless rule.

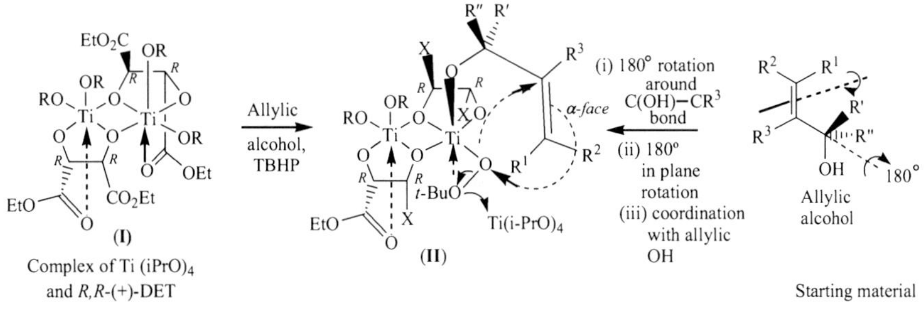

Fig. 2.94 Mechanism of the Sharpless epoxidation of an allylic alcohol by the active catalyst (I) from R,R-(+)-DET

2.12   Conformation of Saturated Six-Membered Ring Compounds

2.12.1   Conformational Aspects of Cyclohexane

2.12.1.1   Geometry of Cyclohexane Chair. Bond Lengths. Bond Angles. Torsion Angles

Baeyer considered that cyclohexane is a strained planar molecule. In 1890 Sachse first pointed out that cyclohexane might be nonplanar puckered chair or boat shaped and unstrained (see D. H. R. Barton, reference 2 of Chapter 11). As evident from X-ray and electron diffraction experiments by Hassel in 1943, cyclohexane exists almost exclusively in its chair conformation. The electron diffraction experiments reveal the C–C and C–H bond lengths, bond angles, the intraannular (C–C–C–C), and the external H–C–C–H torsion angles, which are shown in (1A) and (2A) (Fig. 2.95). The C–C–C– bond angle (111.4°) is more than the regular tetrahedral angle (109°) and little less than the C–C–C bond angle (112.4°) in *n*-propane. The (C–C–C–C) torsion angle (55°) deviates from the optimum of 60° in n-propane, because of the constraint in the ring. Such a decrease in the torsion angle is resisted due to an increase in torsional strain involved. A compromise is reached to

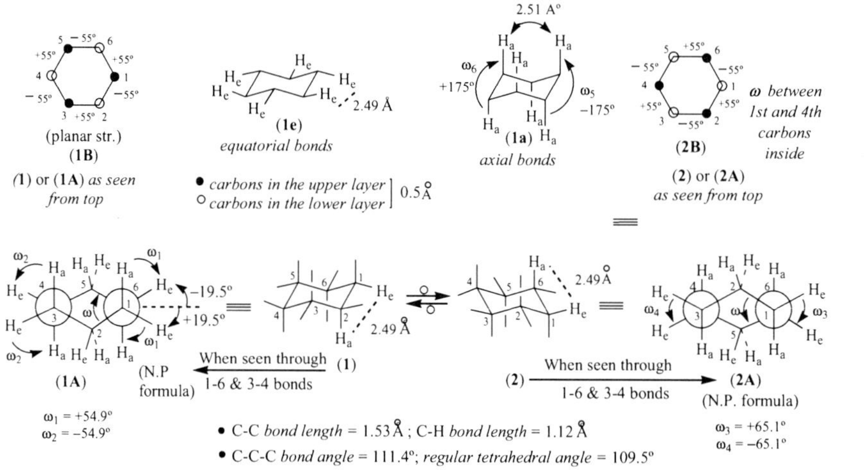

- C-C bond length = 1.53 Å ; C-H bond length = 1.12 Å
- C-C-C bond angle = 111.4°; regular tetrahedral angle = 109.5°

- C-C-C bond angle in propane = 112.4°
- Torsion angle, ω, of C-C-C-C, or ω₁ or ω₂ Hₐ-C-C-Hₑ = +54.9° or −54.9°. Torsion angle ω₃ or ω₄, Hₑ-C-C-Hₑ = +65.1° or −65.1°
- *For determining the sign of ω from chair or boat form (without drawing the N.P formulas) see* **Section 2.15.1.2.**
- *The distance between hydrogen atoms at positions (i) 1a & 2e = 1e & 2a = 1e &2e = 2.49 Å (diverse directions) (ii) 1a & 3a = 2.51 Å (parallel directions). The distance between substituent(s) in case (i) increases (more in 1e, 2e), but in case (ii) remains unaltered if the substituents are same, but increases slightly when the substituents are different.*

Fig. 2.95 Salient features of the geometry of cyclohexane chair and its flipped conformer

minimize the total strain when the bond angle slightly increases and the torsion angle becomes little smaller than that of an open chain molecule like n-propane. Thus, cyclohexane is not an entirely strain-free molecule; some angle strain and torsional strain remain.

2.12.1.2 Equatorial and Axial Bonds

Cyclohexane has two geometrically different sets of H atoms, six are approximately parallel with the vertical C_3 axis, alternatively up and down, as shown in structure (**1a**) (Fig. 2.95) called axial (a) hydrogen atoms; the remaining six H atoms are distributed around the periphery of the ring making alternately $+19.5°$ and $-19.5°$ angles with the equatorial (horizontal) plane of the molecule and are called equatorial (e) as shown in the structure (**1e**) (Fig. 2.95) (see D. H. R. Barton, reference 2 of Chapter 11).

The decrease in the intraannular torsion angle to $54.9°$ brings with it a decrease in the external H–C–C–H torsion angle of *cis* (*ae or ea*) located H atoms from usual $60°$ (in ethane) to $54.9°$ [structure (**1A**)] and that of *trans* (*aa*) located H atoms from usual $180°$ to $174.9°$, and an increase in the corresponding *trans* (*ee*) located H atoms from $60°$ to $65.1°$. In ^{1}H NMR spectrum the vicinal coupling constants for such *cis* and *trans* protons vary according to Karplus relationship (see Sect. 4.2.6).

Since the distances between different pairs of adjacent hydrogens and between 1,3-diaxial hydrogens are 2.49 Å and 2.51 Å, respectively (Fig. 2.95), which are more than twice the van der Waals radius (1.20 Å) of hydrogen, there is no nonbonded interaction in the chair form of cyclohexane.

2.12.1.3 Symmetry of Cyclohexane Conformations

a. *Chair* (**1**). The principal axis of symmetry is the C_3 axis passing vertically through the center of the chair; this is also an S_6 axis. Additionally, the chair conformation possesses three C_2 axes bisecting pairs of opposite sides (str. **3** and **4**) (Fig. 2.96), a center of symmetry, and also three vertical σ planes intersecting at the C_3 axis and passing through the diagonal (opposite) carbon atoms. Each σ_v plane also bisects the angle between two C_2 axes; hence, the σ_v planes are also known as σ_d. Thus, the cyclohexane chair belongs to the point group D_{3d} (Fig. 2.96) and cyclohexane molecule is achiral.

b. *The flexible forms—boat and twist–boat. Symmetry*. The two extremes of the flexible forms of cyclohexane are boat (**5**) and twist–boat (**8**) (Fig. 2.96). These are interconverted by *pseudorotations*, which are low-energy processes, and involve only changes in torsional strain and other nonbonded interactions, but do not involve bond angle variation. This can be verified by molecular model (Dreiding or Fischer) study. In the true boat (**5**) the bowsprit and flagpole hydrogens are only about 1.8 Å apart, whereas the sum of the van der Waals radii of two hydrogen atoms are about 2.4 Å. Thus, there is a significant *bowsprit–flagpole (bs–fp)* interaction.

To minimize the *bs–fp* interaction, the *bs* and *fp* H's are pulled a little apart resulting in the twist–boat in which the eclipsing of the adjacent H's at C2 and C3 and at C5 and C6 are also somewhat alleviated. The important stereochemical features including the symmetry point group of the boat and twist–boat forms are delineated in Fig. 2.96. The symmetry point groups of boat and skew–boat forms of cyclohexane are C_{2v} [$C_2 + 2\sigma_v$ orthogonal)] and D_2 [$C_2 + 2C_2$ (orthogonal)], respectively (Fig. 2.96).

2.12.1.4 Enthalpy (H) or Potential Energy (E) Difference

Here b-ec stands for butane–eclipsing interactions and b–g stands for butane–gauche interactions (see Fig. 2.96).

H_{boat} = two *b-ec* (1,2,3,4 carbons + 4,5,6,1) interactions + 4 *b-g* (The other four successive 4 carbons) = 2 × 4.4 to 6.1) + (4 × 0.9) = (8.8 to 12.2) + 3.6 = 12.4 to 15.8 kcal/mol
    H_{chair} = 6 b–g interactions (6 combinations of consecutive 4 carbons) = 0.9 × 6
= 5.4 kcal/mol
    H_{boat}–H_{chair} = ΔH (or ΔE) = 12.4–15.8) – 5.4 = 7.0 to 10.4 kcal/mol
    Since potential energy of skew–boat is 1.6 kcal/mol less than that of boat, the energy difference between skew–boat and chair form = 7.9 to 10.4—1.6 = 5.4 to 8.8 kcal/mol.

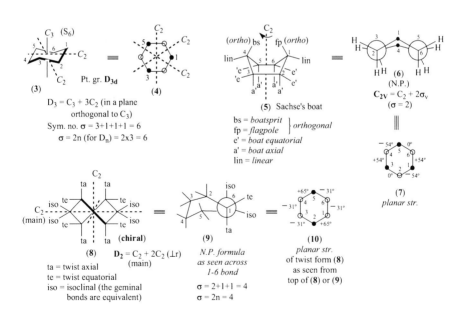

Fig. 2.96 Salient features of the geometry of cyclohexane boat and twist form

The enthalpy difference between flexible (skew–boat) form and chair form is estimated to be 5–6 kcal/mol based on heat of combustions and equilibrium determinations.

So the free energy difference $\Delta G = \Delta H - T\Delta S$ (entropy of flexible form is 5 e.u. greater than that of chair form at 298°K) $= 5{,}560 - 5 \times 298 = 4{,}000$ cals $= 4$ kcal/mol.

Now from the expression $\Delta G = 4{,}000$ cals $= RT\ln K$, one can calculate the equilibrium constant $K = 1{,}000$ (approx.) which means that only one molecule in a thousand will be in the skew–boat form.

2.12.1.5   Cyclohexane Ring Inversion

Cyclohexane chair form undergoes at ambient temperature *ring inversion* or *ring reversal* to another chair form of same energy (degenerate interconversion). Hassel called it *flipping*. Since the diastereomeric six equatorial hydrogens are exchanging with six axial ones in this transformation or *topomerization*, it is a case of *diastereotopomerization*.

Flipping or ring inversion or Hassel interconversion (Fig. 2.97) is accompanied by the following:

(i)   Interchange of planes occur (distance between the two planes is ~0.5 Å).
(ii)   The axial bonds are converted to equatorial bonds and *vice versa*.
(iii)   α-Bonds and β-bonds remain unaltered.

The multistep inversion of cyclohexane chair (**1**) into the equienergetic chair (**1′**) through sequentially the transition state (**TS1**), skew–boat forms, and the transition state (**TS1′**) [enantiomeric with (**TS1**), is depicted in Fig. 2.98]. The interconversion of the skew–boat forms takes place through the intermediate Sachse's boat form, as evident from manipulation of molecular models. A pathway involving chair → envelop/half chair-like TS → intermediate boat form → inverted envelop/half chair TS → inverted chair is also possible. This is called σ-pathway since both the equivalent transition states and the intermediate boat form retain a symmetry plane of the ground state chair form. *But force field calculations make the existence of the boat form of cyclohexane as an intermediate in the cyclohexane inversion itinerary highly unlikely* [58], and hence the σ-pathway is ruled out. The skew–boat (or twist) form is obtained from the so-called *boat form* in the potential energy

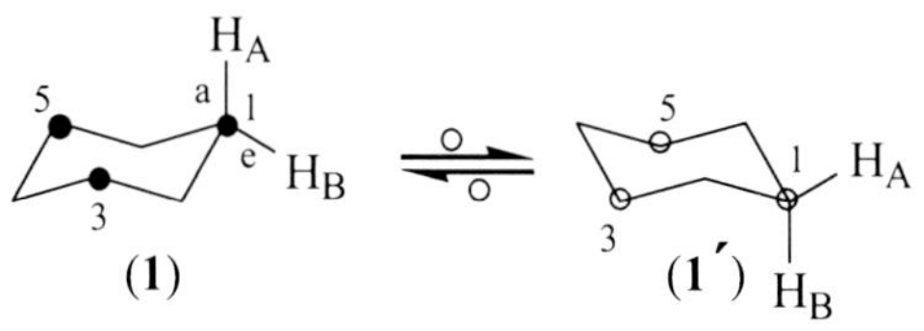

Fig. 2.97 Ring inversion of cyclohexane

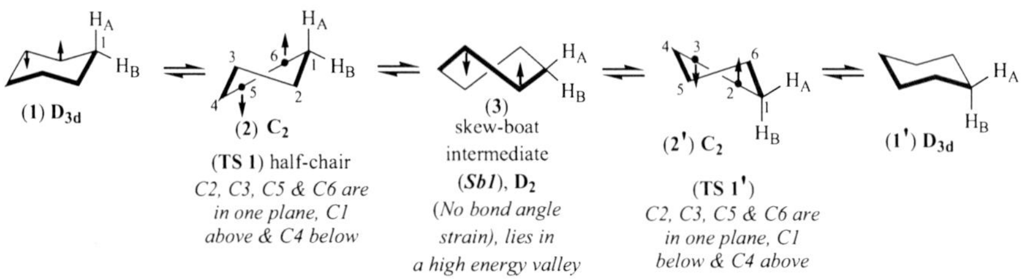

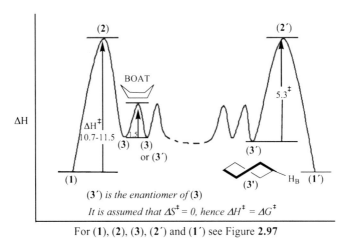

- *Arrows show the direction of suitable twisting involving extensive bond angle deformation with increased torsional strain*
- *The enantiomeric (TS1) and (TS1′) have a C_2 axis (point group C_2), but no σ-plane.*
- *(Sb1) may be reversibly converted to two other skew-boat forms, one identical and the other enantiomeric, by pseudorotation without change of bond angles, but changing only torsion angles.*

Fig. 2.98 Conformational itinerary involved in cyclohexane ring inversion (C_2 pathway)

(3′) is the enantiomer of (3)

It is assumed that $\Delta S^{\ddagger} = 0$, hence $\Delta H^{\ddagger} = \Delta G^{\ddagger}$

For (**1**), (**2**), (**3**), (**2′**) and (**1′**) see Figure **2.97**

Fig. 2.99 Energy profile ($\Delta H^{\neq}$ in kcal/mol) of ring inversion of cyclohexane

diagram by slight deformation to alleviate the *fp* ↔ *bs* interaction and is less stable than the chair conformer by 4.7–6.2 kcal/mol according to various indirect experiments and also by force field calculations. The calculations also suggest that the true boat form is of about 1–1.5 kcal/mol higher energy than that of the twist form, and that the (**TS1**) or (**TS1′**) is of 10.7–11.5 kcal/mol higher energy than the chair form. The boat form is apparently at the energy maximum in the interconversion of the twist (or skew–boat) conformers of which two are mirror images. This pathway (Fig. 2.98) is called C_2 pathway, since the C_2 axis of the ground state chair form is retained along this pathway.

The energy profile diagram of ring inversion of cyclohexane following the C_2 pathway is depicted in Fig. 2.99.

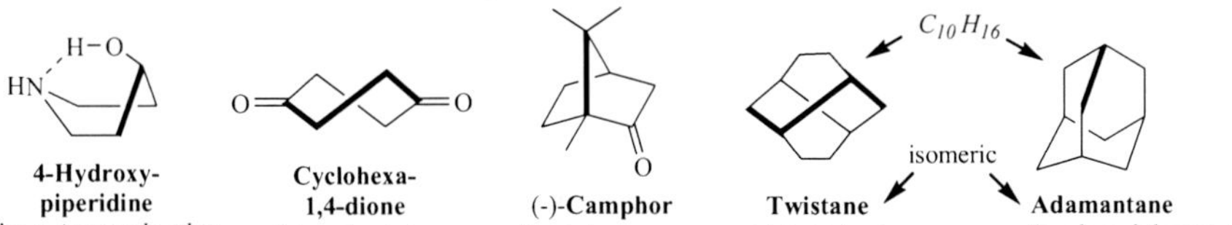

4-Hydroxy-piperidine	Cyclohexa-1,4-dione	(-)-Camphor	Twistane	Adamantane
(due to intramolecular H-bonding)	(skew-boat due to electronic effect)	(cyclohexanone ring is boat)	(A twist boat is locked)	(each cyclohexane ring in chair form)

Fig. 2.100 Some compounds having fixed boat or skew–boat conformation of the cyclohexane ring

2.12.1.6 Stable Boat or Skew–boat Conformers

The chair form of cyclohexane is usually the most stable conformer. However, under certain conditions of structures the boat or the skew–boat forms can be stabilized in preference to the chair form. In case of certain molecules, as shown in Fig. 2.100, the configurational requirement does not allow the chair form of a particular ring. This is also true in cases of *trans-cisoid-trans*-perhydrophenanthrene (Fig. 2.136), *trans-transoid-trans*-perhydroanthracene (Fig. 2.140), the D/E rings in multiflorenol and friedelin, and the E ring in bauerenol and eapacannol (see Fig. 10.31, also see eupacannol in Fig. 2.111).

2.12.2 Monosubstituted Cyclohexanes. Conformational Energy

Monosubstituted cyclohexanes with the more stable equatorial substituent (E) flips into its nonequivalent diastereomeric less stable chair conformations with the substituent axial, as shown for methylcyclohexane in Fig. 2.101. Corresponding Newman projection formulas are also shown. The energy barrier of ring inversion in substituted cyclohexanes remains practically unaffected. The energy profile diagram is similar to that of cyclohexane ring inversion (Fig. 2.99); of course, the two chair conformers have different enthalpies and the rates of forward and backward interconversions are different. For example, in case of methylcyclohexane the equatorial conformer (E) will have two gauche–butane interactions (~1.8 kcal/mol) less enthalpy than the corresponding axial conformer (A), as shown in Fig. 2.101, both in chair formulas and Newman projection (N.P.) formulas.

Conformational Energy The energy (enthalpy or potential energy or free energy) of a given conformational isomer over and above that of the conformational isomer of minimum energy is called conformational energy.

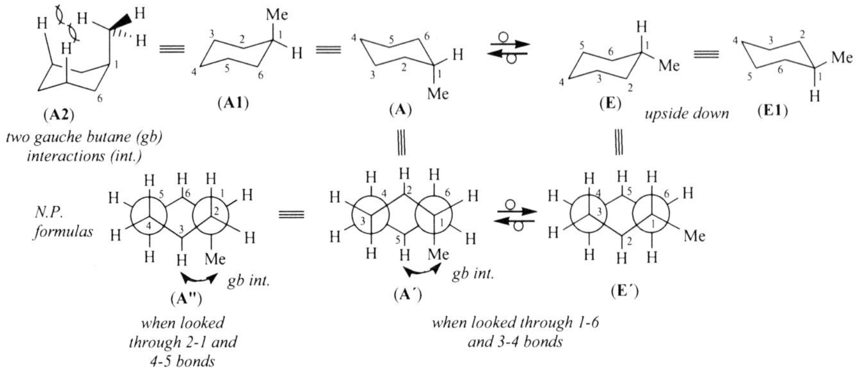

Note: In (**A′**) or (**A″**) (like A2) the distance between axial methyl protons and *syn*axial protons is ~ 1.83 A (~ van der Waals radii of two H's)

Fig. 2.101 Conformers of methylcyclohexane

As already mentioned before, the two diastereomeric chair forms (**A**) and (**E**) (Fig. 2.101) are of unequal free energy and hence their populations are different. Their equilibrium constant K is given by the equation:

$$\Delta G^\circ = -RT\ln K \ldots (\text{equation } \mathbf{a}), \text{where } K = [(\mathbf{E})]/[(\mathbf{A})].$$

The difference of free energy between equatorial (**E**) and axial (**A**) conformers, ΔG°, is negative (the more stable conformer to be written on the right side of the equilibrium). *Thus,* $-\Delta G^\circ$, *which is positive, is known as the conformational free energy of the substituent* (also known as A-value). Usually steric grounds (as explained in Fig. 2.101 in case of methylcyclohexane), in some cases electronic factors may dictate the more stable conformer. In the Newman projection (**E′**) for the equatorial conformer no extra gauche–butane interaction is present, whereas in Newman projections (**A′**) and (**A″**) of the axial conformer one extra *gb* is seen in each; the chair form (**A2**) shown both *gb* interactions. Thus, the enthalpy difference, ΔH, between the two diastereomeric conformers $= 0.8 \times 2 = 1.6$ kcal/mol (approx.). If $\Delta S = 0$, $\Delta G = \Delta H$ (from the equation $\Delta G = \Delta H - TAS$) (equation **b**). This value corresponds to over 90 % population of the e conformer (**E**) (but NMR studies indicate it to be 95 %).

The conformational free energies ($-\Delta G^\circ$ values) of a number of common substituents are listed in Table 2.5.

The Table 2.5 reveals the following interesting facts

(i) In the halogen series the effective bulk of F being least, $-\Delta G^\circ$ value is least, $-\Delta G^\circ$ values for Cl, Br, and I are almost equal due to the increase in their bond lengths resulting in increase in the distance from the *syn*axial hydrogens. Moreover, larger atomic volume causes the increased polarizability of electrons and the ease of deformation to reduce the energy of the axial conformer. In general, the elements of the first two rows in the Periodic Table with relatively short C–X bonds and low polarizability show larger $-\Delta G^\circ$ values

Table 2.5 Conformational free energies $(-\Delta G°)$ in kcal/mol (approx.)[a]

Substituent $-\Delta G°$ (~t, °C)	Substituent $-\Delta G°$ (~t, °C)	Substituent $-\Delta G°$ (~t, °C)	Substituent $-\Delta G°$ (~t, °C)
D 0.006 (25)	CN 0.2 (−79)	OH[b] 0.60 (25)	Me 1.74 (27)
T 0.011 (−88)	NH₂[c] 1.23 (−80)	OH[d] 1.04 (−83)	Et 1.79 (27)
F 0.25 (−86)	NH₂[e] 1.7 (20)	OMe 0.63 (−93)	CH(Me)₂ 2.21 (27)
Cl 0.53 (−80)	NO₂ 1.1 (−80)	OC(Me)₃ 0.75 (36)	C(Me)₃ 4.8 (−120)
Br 0.48 (25)	COOMe 1.2 (25)	OPh 0.65 (−93)	Ph 2.8 (−100)
I 0.47 (−78)	COOH 1.4 (25)	OCOMe 0.68 (25)	CH₂Ph 1.68 (−71)

[a]More detailed tabulations have been compiled in the textbook by Eliel, Wilen, and Doyle [59]
[b]in cyclohexane
[c]in toluene-d₆
[d]in CS₂
[e]in MeOCH₂CH₂CH₂OH

   than the heavier elements with longer bonds to carbon and higher
   polarizability.
(ii) Groups like >NH, -NHMe, OH, etc. which may form H-bonds have different
   $-\Delta G°$ values in protic and aprotic solvents (for OH see ref. [60]).
(iii) Substituents at the second atom (in italics), for example, *O*Me, *O*Et, *O*Ts,
   *O*COMe do not significantly affect the effective bulk. The $\Delta G°$ values of Me,
   Et, *i*Pr do not differ as expected, but it increases sharply for *t*-Bu group having
   greatest effective bulk. The *t*-butyl substituted cyclohexanes have almost
   exclusively equatorial conformation and have been called **anancomeric**,
   meaning *fixed in one conformation* (derived from Gk *unankein* meaning to
   fix by some fate or law).
(iv) The $\Delta G°$ values are approximately additive, and they may be used for di- and
   polysubstituted cyclohexanes.
(v) The additivity of $\Delta G°$ values does not hold good in geminally substituted
   cyclohexanes [61].

2.12.3   1,1-Disubstituted Cyclohexanes

1,1-Disubstituted cyclohexanes having a plane of symmetry (C_S) passing through
C1 and carrying achiral substituents do not show any configurational isomerism
even if the substituents are different. They exist in two interconvertible conformers
when the substituents are different. The energy barrier is usually of the similar order
as that of cyclohexane. For example, in 1-methylcyclohexanol (**1**) the ratio of the
two diastereomeric conformers (**1a**) and (**1e**) (Fig. 2.102) should correspond to the
difference in conformational free energies of the two substituents (1.74 − 1.04
= 0.70 kcal/mol). But actually (**1**) exists as a 70.30 mixture of (**1a**) and (**1e**)
conformers in Me₂SO at 35 °C, corresponding to a free energy difference of
0.48 kcal/mol. Here the bulkier group Me in (**1a**) is predominating but to a much
less amount due to some leveling effect.

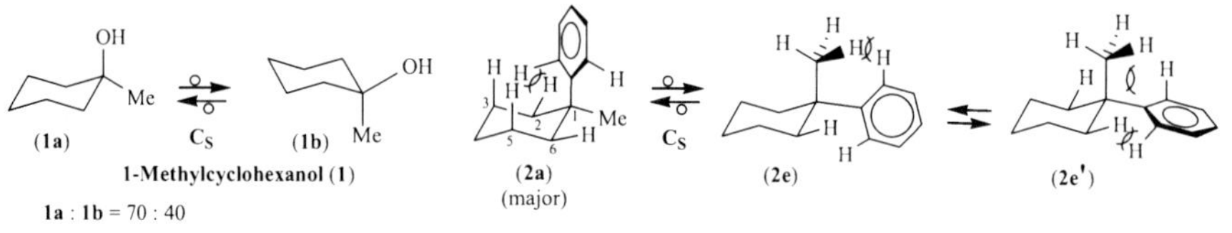

Note: The compound (**2a**) flips into a β-equatorial phenyl and an α-axial methyl conformer which when rotated 180° around a horizontal axis gives the conformer (**2e**) with α-e-Ph and β-a-Me substituents.

Fig. 2.102 Conformations of two geminally substituted cyclohexanes

In case of 1-methyl-1-phenylcyclohexane (**2**) (Fig. 2.102), contrary to expectation, the conformer (**2a**) with axial phenyl and equatorial methyl is preferred over the other conformer (**2e**) with equatorial phenyl and axial methyl by 0.32 kcal/mol in spite of the fact that the difference of their conformational free energy $(2.8 - 1.74 = 1.06$ kcal/mol) would lead to the opposite conclusion. This anomaly may be explained as follows (Fig. 2.102). In the conformer (**2a**) the phenyl ring is so oriented that the ortho H's are away from axial H3 and axial H5 and so the *synaxial* interactions of the phenyl with these H's becomes minimal. However, there may be some interactions between the ortho H's and the adjacent e-H2 and e-H6. On the contrary, the phenyl group in the flipped conformer (**2e**) (see **Note** in the Figure) would interact strongly with the Me H's. Hence, the phenyl group must rotate by 90° (*cw*) about its pivotal bond to give the conformer (**2e′**), in which also strong interactions between the ortho H's and the adjacent *e*-H's exist. Thus, the additivity of $\Delta G°$ values does not hold good in geminally substituted cyclohexanes [59].

2.12.4 Non-geminal Disubstituted Cyclohexanes

Disubstituted cyclohexanes, other than the germinal ones (Sect. 2.12.3), exist in three sets of positional isomers, *viz.*, 1,2-, 1,3-, and 1,4-isomers. Each set has a *cis– trans* pair of diastereomers. Each diastereomer can flip into interconvertible chair conformers. The relative enthalpy of each conformer depends upon the steric interactions present. Depending upon its symmetry property a particular isomer may exhibit enantiomerism. These features in case of all dimethylcyclohexanes are illustrated in Fig. 2.103. Calculation of relative free energy and entropy for each isomer are also shown as **Notes** in the Figure. Some important stereochemical features of dimethylcyclohexanes having *different substituents* are stated below.

cis-1,2-Isomers. The *cis*-1,2-isomer having *different substituents* is resolvable since the flipped conformers are different entities and not enantiomers. The bulkier group exists predominantly in the equatorial orientation.

1,3-Isomers. Both *cis*- and *trans*-isomers with *different substituents* are resolvable, since each of them has C_1 pt gr. and each flips into a different chiral variety.

Planar Structure (Optical activity)	Conformation (symmetry pt. gr.)	Inter-actions	Relative enthalpy ΔH kcal/mole	Kcal/mole ΔH exptl. (ΔG exptl.)	Flips into	Regarding chirality
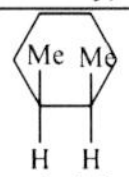 Me Me / H H / cis-1,2 / σ-plane / (meso)	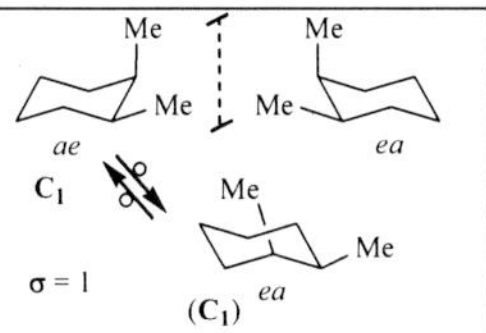 ae, ea, C₁, σ = 1, (C₁)	2 (axial Me) + (1e ↔ 2a Me's) = 3 gb.	2.7 〉 ΔH (calc) 1.8	1.87 (1.66)	A non-superposable mirror image (energy barrier much less than 10 kcal/mole)	Unresolvable (±)-pair
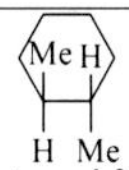 Me H / H Me / trans-1,2 / C₂ pt.gr. / (active)	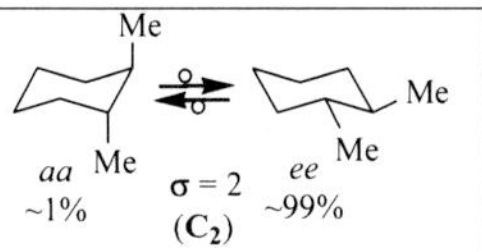 aa Me ~1%, ee Me ~99%, σ = 2, (C₂) (The C₂ passes through the midpoints of 1-2 & 4-5 bonds)	ee 1 gb / aa 4 gb	1.8 〉 0.9 / 3.6		another chiral variety of the same enantiomer (lacks an σ-plane or i)	Resolvable (±)-pair
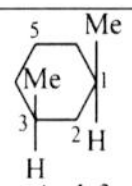 Me / Me / H / cis-1,3 / (σ-plane) / (meso)	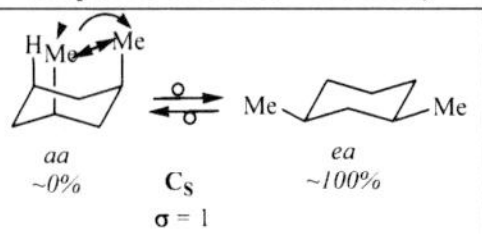 aa ~0%, ea ~100%, Cs, σ = 1 Each variety has a vertical σ-plane passing through C2 and C5 and orthogonal to the puckard plane	aa syn axial MeMe, + 2 gb	5.4 / 0 〉 −1.8 / 1.8	−1.96 (−1.56)*	another achiral variety	True meso
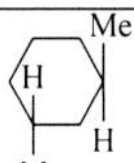 Me / H / H / Me / trans-1,3 / C₂ (active)	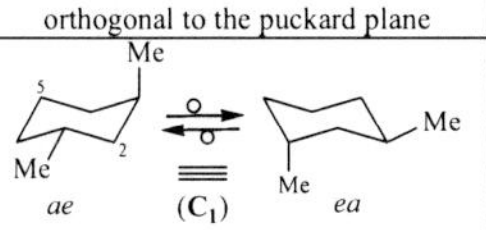 ae, (C₁), ea, σ = 1 (possess only C₁ symmetry)	2 gb for one a-Me	1.8		a superposable (homotopic form)	Resolvable (±)-pair
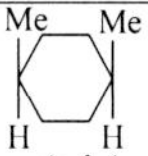 Me Me / H H / cis-1,4 / (meso) / σ-plane through C1 & C4	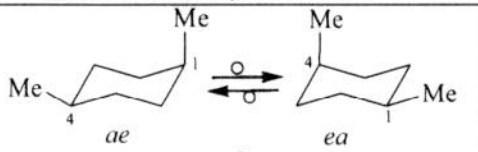 ae, Cs, ea, σ = 1 (A vertical σ-plane passess through C1 and C4)	2 gb for one a-Me	1.8 〉 1.8	1.90 (1.55)*	a superposable (homotopic form)	True meso
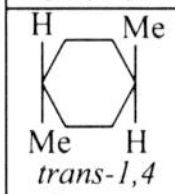 H Me / Me H / trans-1,4 / (meso) / σ-plane through C1 & C4	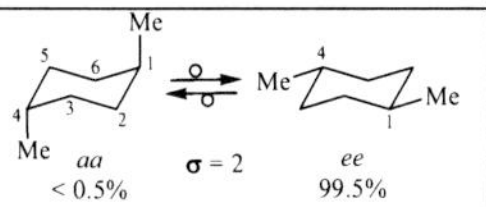 aa <0.5%, ee 99.5%, σ = 2 (C₂h for each): C₂ axis biscet 2-3 and 5-6 bonds; σh orthogonal to C2, passing through C1 and C4)	No gb for ee form / 4 gb for aa form	0 / 3.6		another achiral variety	True meso

Notes: • Symmetry consideration of the planar structures gives the correct inference regarding resolvability.

• G = Free energy; H = Enthalpy (or Potential Energy E); S = Entropy

• ΔG (Relative free energy) = $G_{cis} - G_{trans}$; $\Delta H = H_{cis} - H_{trans}$; $\Delta S = S_{cis} - S_{trans}$

• $\Delta G = \Delta H - T\Delta S$ (ΔG^*exptl. is calculated from experimental ΔH and ΔS data)

• Total relative entropy = S_{calcd}

S_{calcd} (for trans-1,2) = $-R\ln \sigma$ (σ, symmetry no. is 2) + $R\ln$ (entropy of mixing) (due to dl) + S (due to ee, aa equilibrium)

= (−1.38) + 1.38 + 0.11

S_{calcd} (for cis-1,2) = 0 + 1.38 + 0 = 1.38

*Entropy of mixing = $-R (x_1\ln x_1 + x_2\ln x_2)$ where x_1 & x_2 are mole fractions of each component in the mixture

= $-R (\frac{1}{2}\ln\frac{1}{2} + \frac{1}{2}\ln\frac{1}{2})$ = $R\ln 2$ = 1.38 cal/deg. mole

Fig. 2.103 Conformations energies, symmetries, and optical activity of the non-geminal dimethylcyclohexanes

The *cis*-isomer exists in a preferred *ee* conformation and the *trans*-isomer in an *ea* conformation with the bulkier group predominantly in the *e* orientation.

Though the 1,3-*cis* isomer exist almost 100 % in the diequatorial (*ee*) conformation, if necessary it can adopt the diaxial (*aa*) conformation by ring inversion to bring the substituents within reacting distance, *e.g.*, formation of an anhydride from *cis*-1,3-cyclohexanedicarboxylic acid or formation of intramolecular H-bond in *cis*-1,3-cyclohexanediol.

1,4-Isomers. The *cis*- and *trans*-1,4-isomers are always meso because they possess a vertical σ-plane passing through C1 and C4, even if the substituents are different. The *cis*-1,4-isomer exists preferentially in the conformation having the bulkier group in the *e* conformation, whereas the *trans*-1,4-isomer exists almost exclusively in the *ee* conformation.

2.12.4.1 Some Typical Disubstituted Cyclohexanes (Fig. 2.104)

(a) In *trans*-1,2-dihalocyclohexanes the diaxial (*aa*) conformer (which does not exist in dimethyl series) is substantially populated, sometimes as the major conformer (more than the *ee* conformer (**2**)). The percentage of *aa* conformer (**1**) having no dipole repulsion between the halogen atoms increases in the series: Cl < Br < I, since the increased bond length decreases the synaxial interactions. The *ee* conformer (**2**) is also stabilized by solvation with polar solvents. The effects of dipole–dipole repulsion in (**3**) and synaxial interactions in (**4**) result in the 50 % of each isomer in the equilibrium mixture of these two diastereomers of 4-t-butyl-1,2-dibromocyclohexane.

(b) The *cis*- and *trans*-isomers of cyclohexane-1,2-diol show intramolecular H-bonding (little stronger in *cis*). The *ee* conformer (**5e**) of the *trans*-2-halocyclohexanol is stabilized by H-bonding, and the diaxial conformer (**5a**) is free from dipole–dipole repulsion resulting in almost equal population of the two conformers.

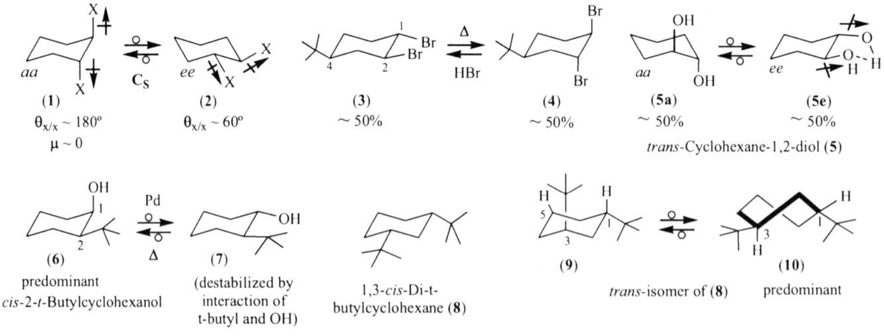

Fig. 2.104 Effect of dipole moment, H-bonding and bulky substituents on conformation in cyclohexane derivatives

(c) Equilibration of *cis*-2-*t*-butylcyclohexanol (*anancomeric* system) with Raney Ni or with Pd upon heating gives a predominant amount of the *cis*-isomer (**6**) over the *trans*-isomer (**7**). The interaction between a methyl group of the t-Bu and OH destabilizes the *trans* isomer. The *cis*-isomer of 1,3-di-*t*-bulylcyclohexane (**8**) exists in the chair conformation with both bulky substituents in equatorial orientation. However, the corresponding *trans* isomer almost exclusively exists in the skew–boat conformation (**10**), since in the chair conformation (**9**) one *t*-butyl group have to occupy axial orientation involving severe interaction with H1 and H5.

2.13 Cyclohexanone

Cyclohexanone exists almost exclusively (99 % at 25 °C) in the chair form and only a small amount (~1 %) in skew–boat forms (Fig. 2.105). The conformational features of cyclohexanone are shown briefly in the Figure.

2.13.1 Torsion Angles, Stability

Torsion angles between pairs of adjacent C–C bonds (Fig. 2.105) show flattening of the ring at the region of the carbonyl group. The chair conformation has only a vertical σ-plane passing through C1 and C4 and belongs to the point group C_S. The skew–boat forms belong to chiral point groups C_1 and C_2. Due to flattening at the site of the CO group, the *e*-H's at C2 and C6 are partially eclipsed with carbonyl oxygen ($\theta = 4.3°$), whereas the corresponding a-H's lean slightly outwards. Cyclohexanone is *slightly destabilized* relative to cyclohexane due to the angle strain and torsional strain. The equilibrium of cyclohexanone cyanohydrin (sp^2 sp^3 equilibrium) lies more toward cyanohydrins side than that of di-*n*-octyl ketone indicating the lower thermodynamic stability of cyclohexanone. Again, cyclohexanone is reduced with $NaBH_4$ at a rate 355 times as fast as di-*n*-hexylketone manifesting the lower kinetic stability of cyclohexanone. The twist–boat forms with C_2 and C_1

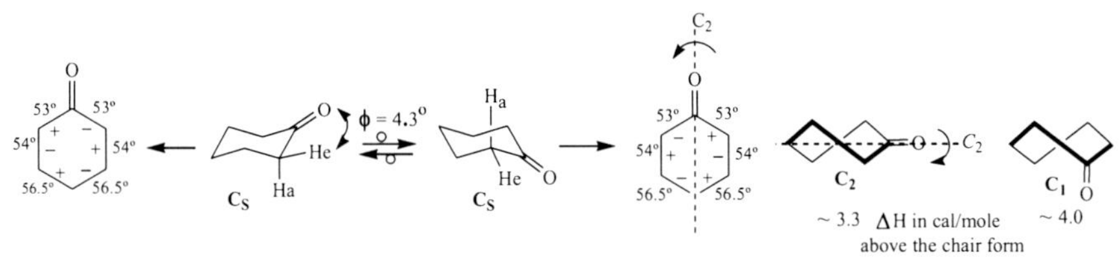

CH$_2$-CH$_2$ bond length = 1.545 Å; CH$_2$-CO bond length = 1.51 Å; C-CO-C bond angle = 116.5°

Fig. 2.105 Conformations, geometry, and torsion angles of cyclohexanone

point groups are having enthalpies (calcd.) 3.3 and 4.0 kcal/mol (approx.), respectively, above that of the chair form (Fig. 2.105).

2.13.2 Ring Inversion

The free energy of activation $\Delta G^{\neq}$ (exptl.) [62] for the ring inversion in cyclohexanone at $-170°$ is ~ 5.0 kcal/mol, much lower than in cyclohexane. This is because of lower torsional barrier around an sp^3–sp^2 C–C bond than an sp^3–sp^3 one. Pseudorotation among the flexible forms is more facile in cyclohexanone than in cyclohexane, and the transition states for C_2 and C_1 skew–boats are not equivalent.

2.13.3 Alkylketone Effects

The alkylketone effect of the substituent $R = (-\Delta G_R)$ (in alkylcyclohexane)—$(-\Delta G'_R)$ (in 2, 3, or 4-alkylcyclohexanone) = the decrease in the conformational free energy of a substituent R in the alkylcyclohexanone with respect to the conformational free energy of the alkyl substituent in the alkyl cyclohexane.

2.13.3.1 2-Alkylketone Effect (Fig. 2.106)

The equatorial (e) substituents at C2 and C6 are nearly eclipsed with respect to CO oxygen and may destabilize the e conformer due to additional steric repulsion—thus decreasing the conformational free energy (ΔG between axial (a) and e) in comparison to that in alkylcyclohexane. Thus, in free energy, the term 2-alkylketone effect was first proposed by Walker in 1955 and Klyne in 1956. The 2-methyl ketone effect was estimated to be 1 kcal/mol, reducing the normal

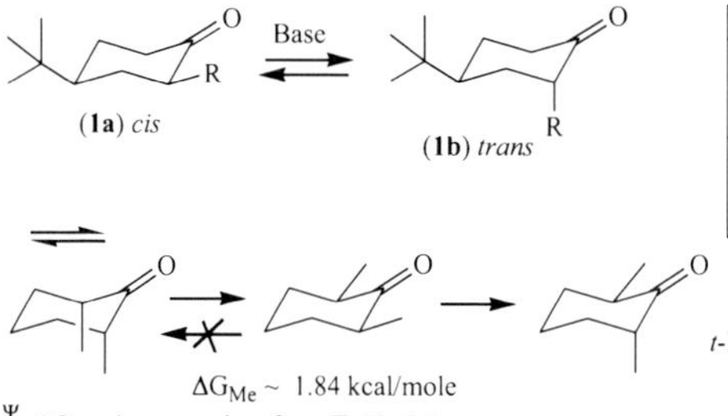

R	$-\Delta G_R$ [Ψ]	$-\Delta G'_R$	2-Alkyl-Ketone effect
Me	1.74	1.84	-0.10
Et	1.79	1.10	0.69
iPr	2.21	0.60	1.61

$\Delta G_{Me} \sim 1.84$ kcal/mole

[Ψ]$-\Delta G_R$ values are taken from **Table 2.5**

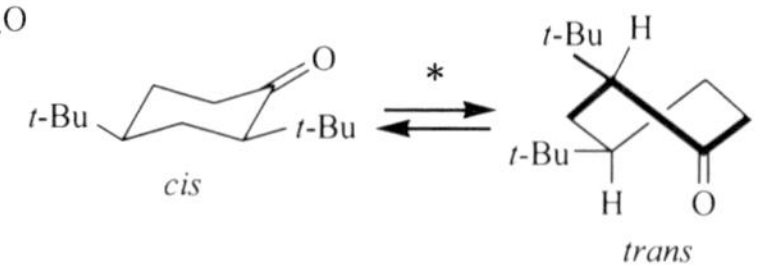

cis-2,4-Di-t-butylcyclohexanone undergoes base catalyzed equilibration to give trans-2,4-di-t-butylcyclohexanone in skew-boat form (and not in chair form, and hence 2-t-butylketone effect cannot be measured).

Fig. 2.106 Measurement of 2-alkylketone effects

difference (ΔG_{Me} ~1.7 kcal/mol) by this amount making the $\Delta G'_{Me}$ (in 2-methyl cyclohexanone) as 0.7 kcal/mol.

But Allinger (1961) demonstrated that 2-methylketone effect is absent. The reason for absence of 2-methylketone effect is that the e-Me is too far (due to greater C–CH$_3$ bond length) from CO oxygen to have any van der Waals repulsion (steric interaction) between them; on the other hand, eclipsing of C–Me and C=O is electronically slightly favorable (*cf.* preferred conformation of propanal is having C=O and C–Me eclipsed). Thus, the 2-methylketone effect appears to be slightly negative, −0.10 kcal/mol. 2-Alkylketone effect can be estimated from the equilibrium data of 2-alkyl-4-*t*-butylcyclohexanones or 2,6-dialkylcyclohexanones (Fig. 2.106). A few alkylketone effects thus obtained are given in the Figure.

2.13.3.2 3-Alkylketone Effect

In 3-alkylcyclohexanone the axial conformer has only one alkyl-hydrogen synaxial interaction (gauche) instead of two such interactions in the axial conformer of alkyl cyclohexane. Thus, there is a decrease in $-\Delta G°$ value [of the alkyl (R) group], which is known as 3-alkylketone effect. In case of 3-methylcyclohexanone (R=Me), this effect is equivalent to one gauche–butane interaction (0.89 kcal or 3.75 kJ/mol). However, in view of the van der Waals repulsion between axial Me and CO oxygen in the axial conformer this value is reduced to a calculated value of 0.6 kcal/mol. Experimental study of the equilibrium between *cis* and *trans* isomers of 3,5-dimethylcyclohexanone (**a** in Fig. 2.107) suggest a value of 0.37 kcal/mol (1.73–1.36 kcal/mol) (taking $\Delta S = O$) for the decrease (3-methylketone effect) in $-\Delta G_{Me}$. The equilibrium data of 2,5-dimethylcyclohexanone (between the *trans*

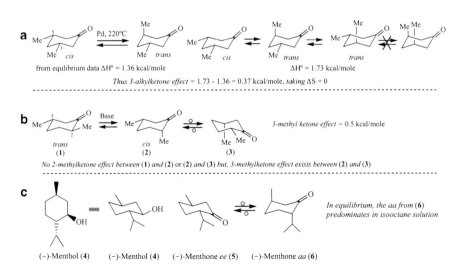

Fig. 2.107 3-Alkylketone effect. Few examples

(**1**) and *cis* (**2**) isomers) (**b** in Fig. 2.107) gives a value of 0.5 kcal/mol [Eliel 1965] which is explained in the Figure.

In (−)-menthone [(**5**) in Fig. 2.107] 2-isopropylketone effect (1.8 kcal/mol) and 3-methylketone effect (0.5 kcal/mol) cooperate with each other in the *aa* conformer (**6**) with respect to the *ee* conformer (**5**)—so much so that in a solvent of low polarity, *e.g.*, isooctane, the diaxial conformer (**6**) predominates, as evident from its CD spectrum. In the polar solvent the *ee* form (**5**) is also stabilized by dipole–dipole interaction with the solvent.

2.13.3.3 4-Alkylketone Effect

4-Alkylketone effect depends upon the relief of strain due to synaxial interactions between an axial 4-alkyl group and the axial H or substituent at C2 and C4, when C1 is converted from the tetrahedral (>CHOH) to the trigonal (>C=0) configuration, since this conversion leads to an outward motion (leaning) of the axial H's or substituents at C2 and C4.

This effect has been invoked to explain (Fig. 2.108).

(a) The greater rate of CrO_3 oxidation of 4,4-dimethylcyclohexanol as compared to that of cyclohexanol (*cf.* triterpene 3β-ol *versus* steroidal 3β-ol)
(b) The larger dissociation constant (K) of 4,4-dimethylcyclohexanone cyanohydrin as compared to that of cyclohexanone cyanohydrin.

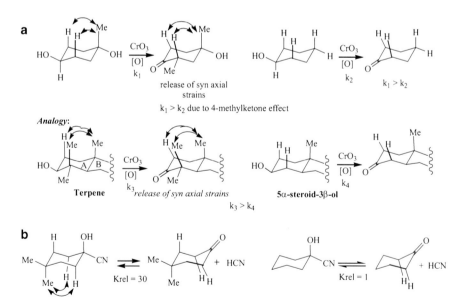

Fig. 2.108 Invocation of 4-alkylketone effect to explain some oxidation rates and relative equilibrium constants

2.13.4   Addition of Nucleophiles to Cyclohexanones. Stereochemical Aspects

The nucleophile can approach the carbonyl carbon of a cyclohexanone derivative from the axial side producing the equatorial alcohol and from the equatorial side producing the axial alcohol.

2.13.4.1   PDC (PSC) and SAC (SSC)

Dauben et al. in 1956 [63] first introduced the concept that in case of unhindered cyclohexanones having no substituent at C3 or C5 position, axial approach of the nucleophile does not encounter any steric hindrance and hence is preferred, producing the more stable equatorial alcohol in major amount (~90 %) (Fig. 2.109), as a result of *product development control* (PDC). This is illustrated with an example of the sodium borohydride reduction of an anancomeric cyclohexanone like 4-*t*-butylcyclohexanone [64] through the transition state (A) (Fig. 208a) which energetically resembles the product. If the attack takes place from the equatorial side, it involves the TS (B) of higher energy due to synaxial interactions with the developing axial OH and hence is formed in minor amount (~10 %).

On the other hand for hindered anancomeric cyclohexanones like *cis*-3-methyl-4-*t*-butylcyclohexanone, the axial approach becomes more hindered due to synaxial interaction with the 3-methyl group in the TS (D), compared to the equatorial approach through TS (C) having much less nonbonded interaction and hence

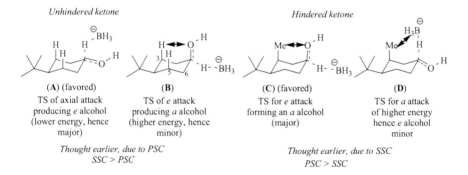

Unhindered ketone   *Hindered ketone*

(A) (favored)	(B)	(C) (favored)	(D)
TS of axial attack producing *e* alcohol (lower energy, hence major)	TS of *e* attack producing *a* alcohol (higher energy, hence minor)	TS for *e* attack forming an *a* alcohol (major)	TS for *a* attack of higher energy hence *e* alcohol minor

Thought earlier, due to PSC
SSC > PSC

Thought earlier, due to SSC
PSC > SSC

Fig. 2.109 Product strain control versus steric strain control

gives rise to the axial alcohol as the predominant product. The predominant formation of the axial alcohol as a result of *steric approach control* (SAC) was first conceptualized by Dauben [63]. The two terms, PDC and SAC, have been replaced later one by Brown [65] by *product stability control* (PSC) and *steric strain control* (SSC) to give more emphasis on the transition state rather than on events prior to that the term PSC refers to a product-like or a late TS along the reaction coordinate, while the term SSC refers to a reactant-like or an early transition state.

2.13.4.2 Observations Against PSC

The role of SSC on the stereochemical course of addition of nucleophiles to cyclohexanones has never been in doubt, but the role of PSC has always been controversial. Some observations against the validity of PSC are as follows:

(i) Usually a higher proportion of an equatorial alcohol is obtained in lithium aluminum hydride reduction than that corresponding to thermodynamic equilibrium, *e.g.*, 90 % versus 80 % in the reduction of 4-*t*-butylcyclohexanone.

(ii) The rate constants of a number of sterically hindered ketones, relative to that of 4-*t*-butylcyclohexanone, support the concept of SSC, but show little effect of PSC [66].

(iii) Reduction of ketones with hydrides are highly exothermic; hence, according to Hammond's postulate the transition states are expected to be reactant-like (*cf.* SSC) and not product-like.

(iv) The deuterium isotope effect (k_H/k_D) in borohydride/borodeuteride reduction of cyclohexanones with varying degrees of steric hindrance [67] is almost constant in spite of widely varying ratios of the equatorial and axial alcohols indicating that the extent of B–H bond breaking at the TS is similar for cyclohexanones having different degrees of steric hindrance—a fact inconsistent with the PSC concept.

While teaching it occurred to the present authors long ago that PSC involves product strain which itself cannot have any role in the TS and activation energy in any reaction of kinetic control, and hence the specific contribution of PSC is unexpected, as is substantiated by the above observations. The same is true for reduction of acyclic carbonyl groups regulated only by steric factors such as Cram's rule and Prelog's rule (Sect. 2.11).

2.13.4.3 Torsional Strain. Role of C2 and C6 Axial Hydrogens

In 1965 it was suggested by Richer [68] that the axial H's at C2 and C6 offer steric resistance to the equatorial approach of a nucleophile (torsional strain) which competes with that offered by the axial H's at C3 and C5 to axial approach (Fig. 2.110). The relative strength of the resistances would, however, depend upon the bulk of the hydride reagent and the exact position of the TS in the reaction coordinate. If C. . . Nu is shorter than 1.6 Å the equatorial approach would be more hindered in case of reduction with hydride reagent.

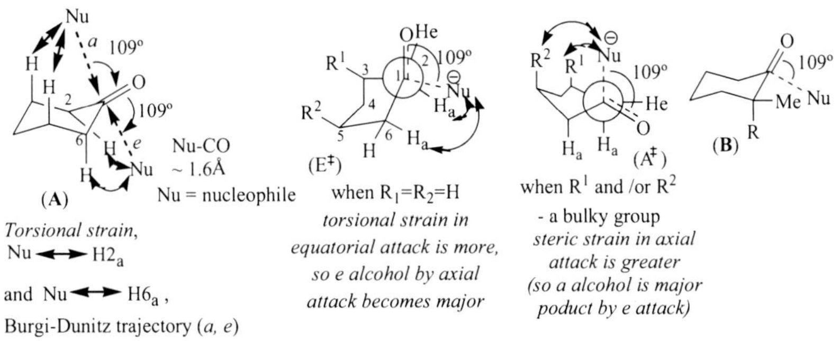

Note: *When GE* ‡ > *GA*‡ → *e Alcohol is major. When GA*‡ > *GE*‡ → *a Alcohol is major*

Fig. 2.110 Felkin transition states for addition of a nucleophile to a cyclohexanone, torsional strain versus steric strain. Formation of equatorial alcohol versus axial alcohol

The most satisfactory and widely accepted alternative to PSC is the concept of *torsional strain* introduced in 1968 by Cherest and Felkin [35]. Two competing factors, namely (1) the steric interaction of the incoming nucleophile with the 3,5-diaxial groups (SSC) in the *axial attack* and (2) the *torsional strain* which arises between the semiformed bond of the nucleophile and the two axial CH bonds at C2 and C6 during the *equatorial approach* (Fig. 2.110) arise, as already stated.

In the absence of any steric hindrance (factor 1) (in case of unhindered ketones), the effect of the torsional strain prevails leading to a preferential axial attack forming the equatorial alcohol. On the contrary, in the presence of one bulky axial substituent at C3 or C5 (or at both), the steric strain control (SSC) (2) prevails reversing the steric course and forming the axial alcohol predominantly. The bulk of the hydride reagent will increase the SSC, more so if C3 and C5 substituents are present and will definitely increase the proportion of axial alcohol. In Fig. 2.110 appropriate transition states are shown for both cases.

This torsional strain effect is not evident in the reduction of 2-methylcyclohexanone (**B**, R=H) and 2,2-dimethylcyclohexanone (**B**, R=Me) indicating that torsional strain is relatively insensitive to the bulk of the group involved (*cf.* rotational barrier of propane is barely larger than that of ethane, Sect. 2.3.4).

The application of Bürgi–Dunitz trajectory [37–39] proposed in 1973, based on molecular orbital interactions (Sect. 2.11.3.2) which states that a nucleophile attacks the carbonyl carbon at an angle of 109° with the carbonyl plane, makes the torsional strain stronger by making it almost like eclipsing strain (Fig. 2.110). Moreover, the torsional strain theory also successfully explains the stereochemistry of the reduction of acyclic ketones (see Sect. 2.11, Felkin models). Both steric strain and torsional strain can be minimized for acyclic substrates.

Several examples of sodium borohydride/lithium aluminum hydride reduction of hindered cyclohexanones are given in Fig. 2.111. The more hindered is the ketone

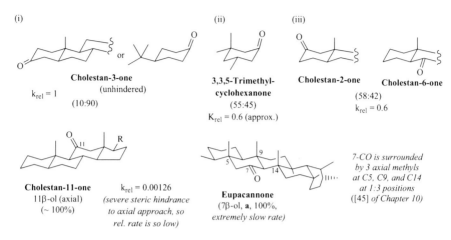

Fig. 2.111   Increased amount of axial alcohol by increased SSC in hindered cyclohexanones

the more equatorial attack leads to the formation of greater percentage of the axial alcohol, and the less is the rate of the reaction.

Hydride reductions of cyclohexanones.

The percentages of *a:e* alcohols are given in parenthesis.

Studies on π-facial selectivity in the addition of nucleophiles to double bonds, especially the carbonyl-based systems have generated more models and hypotheses (~10) (not necessarily mutually exclusive) than any other subject in the field of stereoselective synthesis [69], some of which, *e.g.*, Cram, Prelog, Felkin–Anh, have been discussed earlier.

2.13.5   *Cieplak Hypothesis*

Cieplak in 1981 suggested [69] that the stereochemistry of nucleophilic addition to cyclohexanones is determined by a combination of steric factor, which as usual favors the equatorial approach, and the stereoelectronic factor which involves electron donation from the cyclohexanone C–C and C–H σ bonds into the vacant σ* orbital of the incipient bond formation between the nucleophile and the carbonyl carbon (Fig. 2.112). Since axial C–H bonds (at C2 and C6) next to the carbonyl group are better electron donors than the 2–3 and 5–6 σ bonds, the transition state with axially oriented incipient bond (axial approach) is preferred over the one with the equatorially oriented incipient bond (Fig. 2.112). Thus, the hypothetical model proposed by Cieplak is based on the concept that the carbonyl group undergoes extensive pyramidialization, and that the outcome is primarily a consequence of the aforesaid interactions between the occupied vicinal σ-orbitals by electron donation into the vacant σ*-orbital leading to the stabilization of the transition state.

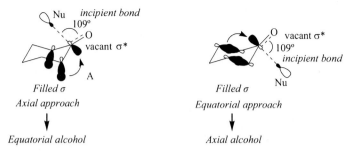

Fig. 2.112 Participating orbitals in the Cieplak model

Fig. 2.113 Hydride reduction of 5-substituted 2-adamantanones

Cheung et al. [70] in 1986 reported that in the nucleophilic addition of NaBH$_4$ to the carbonyl groups of 5-substituted 2-adamantanones, depletion of electron density on the C1–C9 and C3–C4 bonds by halogen or phenyl groups possessing *para* electron-withdrawing substituents favored *syn* approach forming (*E*) (*anti*) alcohols. On the contrary, the electron donating groups like *p*-aminophenyl and *p*-hydroxyphenyl analogs favored the *anti* approach forming the (**Z**) (*syn*) alcohols as major products (Fig. 2.113).

The above observations [70] strongly supported Cieplak view of electronic effects in such asymmetric induction through preferential interaction with newly developing σ* orbital with electron-rich *anti*-periplanar bonds and vice versa.

Cieplak in 1989 [71] observed that the electronegative substitution of the substrate by CF$_3$ at C3 of cyclohexanone or in the reagent, *e.g.*, sodium tri-isopropoxyborohydride reverses the π-facial diastereoselection by increasing the relative proportion of the axial attack leading to the favored formation of the equatorial alcohol. The same type of favored axial attack has been found in a number of widely different reactions including alkyllithium addition and also in cases of peracid epoxidations of methylenecyclohexanes. The findings appear inconsistent with the predictions of Felkin, Klein, Ashby, and Anh models of stereochemistry of reactions in cyclohexane-based systems, but are consistent with Cieplak model [70].

In 1990 Mehta [72] demonstrated that electronic induction by the 2,3-*endo, endo* substituents reverses the π-facial selection (resulting in the reversal of the *E:Z* ratio)

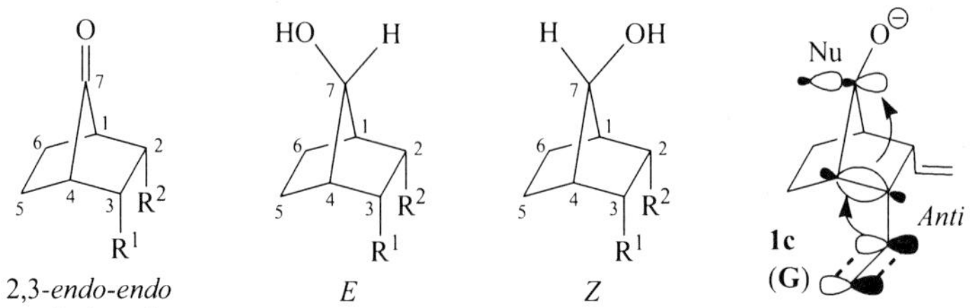

2,3-*endo-endo*
substituted
7-Norbornadiones

1a: $R^1=R^2=$COOMe
1b: $R^1=R^2=$CH$_2$OMe
1c: $R^1=R^2=$vinyl
1d: $R^1=$vinyl, $R^2=$ethyl
1e: $R^1=R^2=$ethyl

E
(from *syn*
face attack)
EWG

Z
(from *anti*
face attack)
EDG

1c
(G)
*Anti face selectivity
in 1c may be
stabilized through
space donation to
σ* orbital in a
perpendicular conformation*

Product Ratios in the Metal Hydride Reductions and Methyllithium Addition

E:Z distribution

Substrate	NaBH$_4$	LiAlH$_4$	Li(t-BuO)$_3$AlH	MeLi
1a	84:16	87:13	77:23	>90:<10
1b	40:60			34:66
1c	36:64	35:65	34:66	27:23
1d	25:75			
1e	20:80	21:79	29:71	17:83

Fig. 2.114 Metal hydride reductions of and methyllithium additions to 2,3-*endo, endo* substituted 7-norbornadiones. *E:Z* ratios of the products support Cieplak hypothesis.

in the nucleophilic addition of hydrides and methyllithium to 7-norbornadione, as shown in a tabular form in Fig. 2.114. The predominant approach of nucleophiles to the *syn* face in **1a** and to the *anti* face in **1e** is fully consistent with the prediction based on the Cieplak's hyperconjugative model, according to which delocalization of σ electrons in the electron-rich antiperiplanar σ bond into the incipient σ* orbital lowers the transition state energy (*cf.* Fig. 2.114). The unexpected *anti*-face selectivity in the cases of **1b** and **1c** (with 2- and 3-substituents supposed to be electron withdrawing groups) (see the Table) may be attributed to *through space donation in a perpendicular conformation* as shown for **1c** in **(G)** in Fig. 2.114. For an authoritative background review and a comprehensive list of references, see [73].

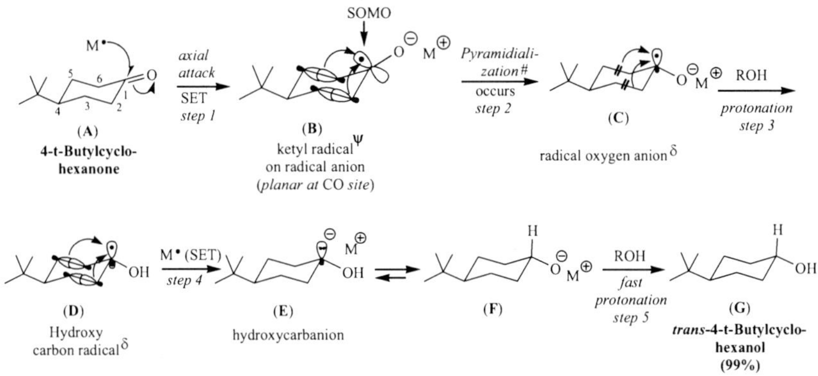

SOMO

(A)
4-t-Butylcyclo-
hexanone

axial
attack
SET
step 1

(B)
ketyl radical [ψ]
on radical anion
(planar at CO site)

Pyramidiali-
zation [#]
occurs
step 2

(C)
radical oxygen anion [δ]

ROH
protonation
step 3

(D)
Hydroxy [δ]
carbon radical

M• (SET)
step 4

(E)
hydroxycarbanion

(F)

ROH
fast
protonation
step 5

(G)
trans-4-t-Butylcyclo-
hexanol
(99%)

[ψ] 70% unpaired electron density is located at C1; and 2-3 and 5-6 s-bonds (HOMO) are favorably disposed to interact with singly occupied p*-orbital (SOMO)

[#] SOMO is extended in the axial direction

[δ] OM in (C) or OH group in (D) is oriented exclusively in the equatorial direction and is antiperiplanar with interacting 2-3 and 5-6 s-bonds

Fig. 2.115 Stepwise mechanism of Birch reduction of an anancomeric saturated cyclohexanone

2.13.6 Highly Stereoselective Reduction of Saturated Cyclohexanones by Dissolving Metals. Birch Reduction

Reduction of saturated anancomeric or rigid unhindered cyclohexanones by LiAlH$_4$ or NaBH$_4$ produces the equatorial alcohols to the extent of 90 % (cf. Sect. 2.13.4).

Reduction of aromatic systems and α,β-unsaturated ketones (1:4 reduction) and of saturated ketones (1:2 reduction) by use of alkali metals (Li, Na or K) dissolved in liquid ammonia in the presence of a proton donor-like alcohol (little amount) is known as **Birch reduction**, which is of versatile synthetic utility [74, 75]. Saturated anancomeric or rigid cyclohexanones are reduced to equatorial cyclohexanols (~99 % diastereoselectivity) by Birch reduction.

The mechanism of such reduction of 4-t-butylcyclohexanone (**A**) proposed by Pradhan [76] is delineated stepwise in Fig. 2.115, based on *Fukui effect* [77], rationalized by Fukui's Frontier Molecular Orbital (FMO) approach. The stereoelectronic effect provides a *kinetic preference of the axial attack of* an electron of the metal (M) on the carbonyl carbon, by a single electron transfer (SET) mechanism leading through a stabilized transition state (TS), to a resonance stabilized ketyl radical; this effect has been termed as Fukui effect. In the Birch reduction 4-t-butylcyclohexanone (**A**) carbonyl carbon, the resonance-stabilized ketyl radical (**B**) with a singly occupied molecular orbital (SOMO) is formed by axial approach of an electron transfer from the metal (SET mechanism). In steps 2–4 the species (**C**), (**D**), and (**E**) are formed successively and the stabilizations of species (**B**) to (**E**) have been indicated in the Figure. The species (**E**) equilibrates to the more stable species (**F**), and either species undergoes protonation very fast to

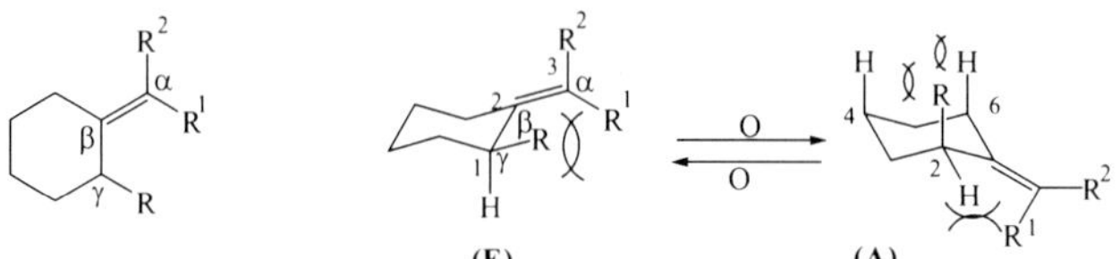

alkalidene cyclohexane
with allylic substituent

(E)
involves $A^{1,3}$ strain
between R and R^1;
when $R = R^1 = Me$,
the distance between
them is ~ 2.45 Å

(A)
involves $A^{1,3}$ strain between
R^1 and H, and synaxial
interaction between R & H4,
and R & H6

When $R = R^1 = Me$ the distance between them = 2.45 Å; enthalpy difference between **(E)** and **(A)** conformers
= $A^{1,3}$ Me/Me interaction – [2(1,3 diaxial Me/H interactions + $A^{(1,3)}$ Me/H interaction.] ≈ 1(1,3-diaxial Me/Me
interaction) –[2(1,3 dixial Me/H interaction. + $A^{(1,3)}$ Me/H interaction] ≈ 3.7 – 3(.9) = 3.7 – 2.7 = 1 kcal/mole.
The axial conformer **(A)** should, therefore, exist to the extent of ~ 85% and the equatorial conformer **(E)** ~ 15% in the
ground state (as obtained from the equation $\Delta G = -RT\ln K$, taking $\Delta S = 0$)

Fig. 2.116 Alkylidenecyclohexane. $A^{1,3}$ Strain

form the final product *trans*-4-t-butylcyclohexanol **(G)** (99 %), giving no chance of
the carbanion **(E)** to undergo inversion at C1.

2.13.7 Alkylidene Cyclohexanes. Allylic$^{(1,3)}$ Strain

The C=O group of cyclohexanone is replaced by $C=CR^1R^2$ to form alkylidenecyclohexanes. The same geometry as in cyclohexanone (Fig. 2.105) is assumed for
alkylidenecyclohexanes (Fig. 2.116). Thus, in a cyclohexane ring containing an
exocyclic double bond and an allylic substituent, the steric interference between the
allylic equatorial substituent at C_γ position and the ethylenic *syn (Z)* substituent
(at C_α position) (Fig. 2.116) is given the trivial designation allylic 1,3- or $A^{(1,3)}$
strain for the sake of semantic simplicity since the groups involved are at the 1 and
3 positions of the allylic system. This is one of the two general stereochemical
theorems introduced by Johnson and Malhotra [78, 79] in 1965; the other one, $A^{(1,2)}$
strain, will be discussed in Sect. 2.14.

In the equilibrium between the equatorial **(E)** and the axial **(A)** conformers, R
and R^1 in **(E)** are nearly eclipsed (the spatial arrangement of $R–C_\gamma–C_\beta=C_\alpha–R^1$ is
almost planar); the dihedral angle θ between $R^1–C_\alpha$ and $C_\gamma–R$ is ~ 4°. *The allylic 1,3
strain ($A^{(1,3)}$ strain) between two substituents/atoms is almost same as the synaxial
interaction between them.* Thus, *in the equatorial conformer **(E)**, even when* R *and*
R^1 *are only moderate in size (medium to large),* the axial conformer **(A)**
predominates.

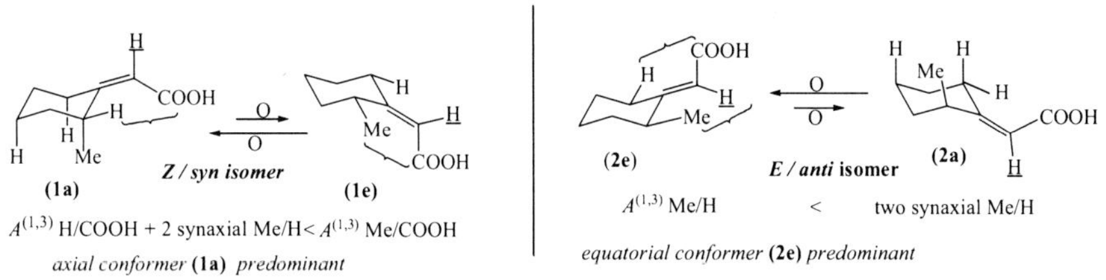

Fig. 2.117 Predominant conformers of the diastereomers of 2-methylcyclohexylidene acetic acid

Some examples illustrating $A^{(1,3)}$ strain follow.

2.13.7.1 Conformational Preference

Of the two geometrical isomers of *2-methylcyclohexylideneacetic acid*, the Z-diastereomer exists predominantly in the Me axial (**1a**) conformer, but the E diastereomer exists predominantly in the Me equatorial (**2e**) conformer, as evident from ^{1}H NMR study. These facts can be explained in terms of the interactions including $A^{(1,3)}$ strain present in the molecules (Fig. 2.117). Likewise, it may be predicted that in case of 3-*methylcyclohexylidenepropionic acid* (underlined H to be replaced by Me in all the structures of Fig. 2.117), both Z- and E-diastereomers exist predominantly in the axial conformer (due to severe $A^{(1,3)}$ Me/Me interaction present in the equatorial conformer of the E isomer).

2.13.7.2 Synthetic Utility [80] of $A^{(1,3)}$ Strain. Stereochemistry of Exoxyclic Enolate Anion Protonation

In the example shown in Fig. 2.116, if R and R′ are small, the equilibrium should lie to the left; if they are large or medium in size it should lie to the right, and the extent of conformation inversion to the axial conformer will depend on the free energy difference between the conformers. In 1955 Zimmerman and coworkers [81, 82] reported the addition of PhMgBr to benzoylcyclohexene to form the *less stable isomer*, cis-1-benzoyl-2-phenylcyclohexane (**4**). It was suggested that 1,4 addition of PhMgBr to benzoylcyclohexene takes place by its equatorial approach at C2, avoiding the steric hindrance in axial approach due to synaxial H4 and H6, to form the bromomagnesium enolate anion (**3**). They thought that C-protonation of the conformer (**3e**) with an acid takes place from the less hindered equatorial direction to give (**4**), since steric interference arising from axial H's at C3 and C5 seriously hindered axial protonation. However, according to $A^{(1,3)}$ strain, (**3**) should exist mainly as the conformer (**3a**) in the ground state, and the protonation at C1 should take place from the axial side in order to have continuous overlap with the π-orbital (stereoelectronic requirement, **SER**).

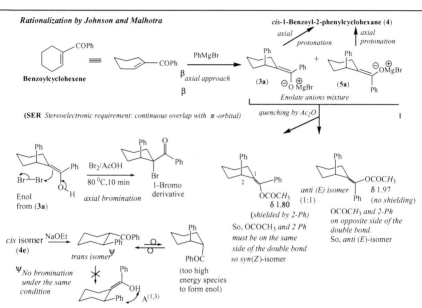

Fig. 2.118 Stereochemistry of protonation of exocyclic enolate anion (avoidance of A$^{(1,3)}$ strain and SER to be satisfied

   Johnson and Malhotra's rationalization [80] has been delineated in Fig. 2.118, invoking the avoidance of A$^{(1,3)}$-strain and axial protonation of the exocyclic enolates to form the less stable *cis*-isomer (**4a**). The easy bromination of the *cis*-isomer and resistance to bromination of the *trans*-isomer under the same condition is also explained by invoking A$^{(1,3)}$ strain during formation of its enol.

2.13.7.3 Another Example of the Use of A$^{(1,3)}$ Strain Concept

Another Example of the Use of A$^{(1,3)}$ Strain Concept will be evident from the axial C-protonation of 1-*aci*-nitro-2-phenyhlcyclohexane using the more stable conformer avoiding severe A$^{(1,3)}$ strain, thus forming the less stable isomer

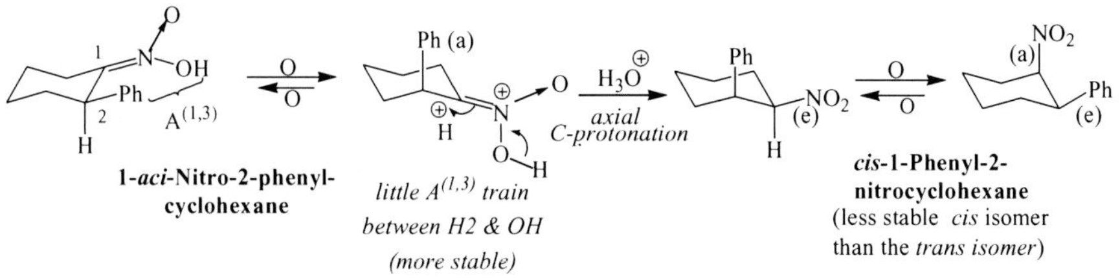

Fig. 2.119 Axial C-protonation of 1-*aci*-nitro-2-phenylcyclohexane. Use of A$^{(1,3)}$ concept

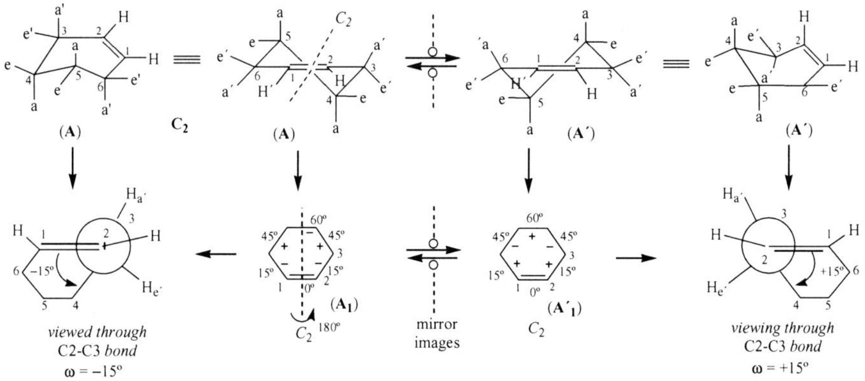

Fig. 2.120 Conformations and ring inversion of cyclohexene. Torsion angles

(Fig. 2.119) (*cf.* Fig. 2.118). Here, also equatorial protonation of the equatorial conformer was postulated by Zimmerman [83].

2.14 Cyclohexene. Conformation. A^{1,2} Strain

2.14.1 Conformation of Cyclohexene. Torsion Angles (Fig. 2.120)

Cyclohexene with two sp^2 carbons possesses a half-chair conformation (**A**) or (**A′**) (Fig. 2.120) confirmed by X-ray crystallographic data of cyclohexene derivatives and also by electron diffraction studies of cyclohexene vapor. The geometry of cyclohexene possesses the following characteristic features:

(i) The C1, C2, C3, and C4 atoms and the two vinylic H atoms are almost in a plane. The C4 and C5 are alternately up and down (or down and up) of this plane.

(ii) Torsion angles shown in the planar projection formulas (**B**) or (**B'**) are supported by theoretical calculations and show considerable flattening of the ring near the double bond.

(iii) The homoallylic carbons C4 and C5 on the opposite sides of the π-plane have almost normal equatorial (e) and axial (a) bonds (perfectly staggered) as in cyclohexane—the axial bonds slightly leaning toward the center.

(iv) The conformation (**A**) or (**A'**) has a C_2 axis in the plane of the double bond, bisecting it and the C4–C5 bond, and belongs to the point group C_2 (chiral).

(v) The conformation (**A**) is convertible to its enantiomer (**A'**) by ring inversion and vice versa, and so they form an inseparable ($\pm$)-pair. They are shown in a different perspective by structures (**A₁**) and (**A₁'**), respectively.

(vi) At the allylic carbons C3 and C6 the axial and equatorial character of the bonds is considerably changed. Vector analysis of cyclohexene by Corey [84] gives the dihedral angle 37° for $H_{e'}$–C6–C1–H (and $H_{e'}$–C3–C2–H) and 83° for $H_{a'}$–C6–C1–H and H_a + C3–C2–H. These are in excellent agreement with the more empirical measurements on Dreiding models. Thus, the substituents at C3 and C6 are imperfectly staggered and are said to occupy pseudoequatorial (e') and pseudoaxial (a') orientations, respectively.

(vii) The interconversion of the two half-chair forms probably goes through a TS with boat-like conformation. The energy barrier ($\Delta G^{\neq}$) determined by ¹H NMR from the coalescence of 1,2,3,3,6,6-d₆-cyclohexene at -164 °C is 5.3 kcal/mol.

(viii) Johnson and Malhotra has proposed allylic 1,2-strain ($A^{(1,2)}$-strain) in 1,6-disubstituted cyclohexene system, the discussion on which follows:

2.14.2   Allylic 1,2-Strain ($A^{(1,2)}$-Strain)

The strain or instability in 1,6-disubstituted 1,2-cyclohexenes (**1**) due to unfavorable interaction between ethylenic substituent and allylic pseudoequatorial substituent of the adjacent C6 is given the trivial designation $A^{(1,2)}$ strain [78, 79]

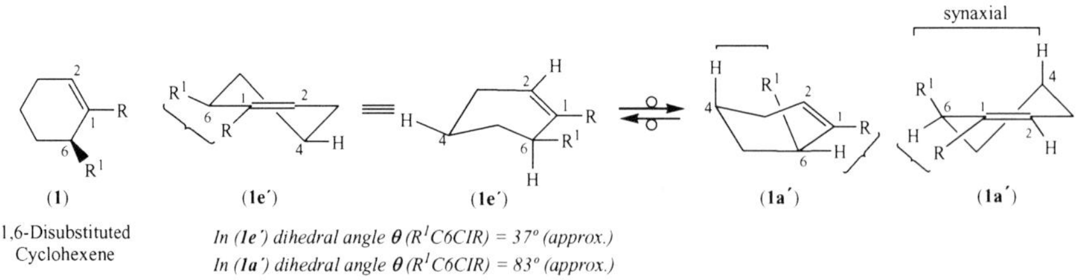

| (1) | (1e') | (1e') | (1a') | (1a') |

1,6-Disubstituted
Cyclohexene

In (1e') dihedral angle θ (R^1C6CIR) = 37° (approx.)
In (1a') dihedral angle θ (R^1C6CIR) = 83° (approx.)

Notes: *The position and intensity of the pseudoallylic t-proton at C6 in NMR gives an estimate of the relative abundance of (1a') and (1e') conformers. The splitting patterns also help the assignment. $A^{(1,2)}$ Interference being weaker than synaxial interaction, will exert a controlling effect only in absence of synaxial interaction [85, 86]*

Fig. 2.121   Ring inverted conformers of 1,6-disubstituted cyclohexene

(Fig. 2.121). The dihedral angles between R^1 and R in the equatorial (**1e**) and axial (**1a**) conformers (Fig. 2.121) are 37° and 83°, respectively.

A$^{(1,2)}$ *strain is not a strong effect* and is manifested only when the groups are moderately large or bulky. However, in 6-methylcyclohexene, the equatorial conformer is more stable, since it does not possess any A$^{(1,2)}$ strain or synaxial strain. A$^{(1,2)}$ strain between two substituents is much weaker than the synaxial interaction between them.

A$^{(1,2)}$ strain operates best when R conjugates with the double bond (*e.g.*, phenyl group). The *ortho*-H of the phenyl group becomes planar with the double bond and comes still much closer to give stronger A$^{(1,2)}$ strain.

Some applications of A $^{(1,2)}$ strain are discussed in the sequel.

2.14.2.1   Conformational Preference

The outstanding examples of A^{1,2} strain in simple 6-disubstituted 1-phenyl (and methyl) cyclohexenes (Fig. 2.122) have been provided by Garbisch [85, 86]. He determined the preferred conformations for C6 substituents in a number of such compounds (Table 2.6) from the widths at half height of the C6 proton magnetic resonance bands [85, 86]. Synaxial interaction between same groups or groups of similar bulk are stronger than the A^{1,2} strain and between them and the former exerts a controlling effect. However, in the absence of synaxial interactions, presence of A^{1,2}-strain makes the particular conformer less predominant. The preferred conformer in each case can be rationalized comparing all the interferences involved in the C6-pseudo axial (a') and C6-pseudo equatorial (e') conformers. The preferred conformations of C$_6$ substituents of 1-phenylcyclohexenes are delineated in Table 2.6.

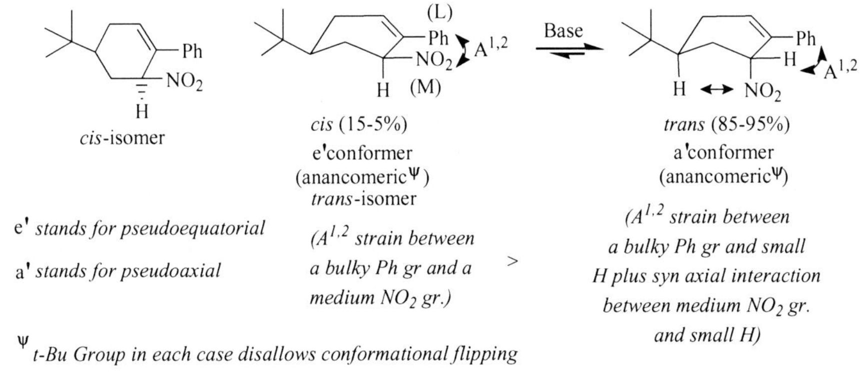

e' *stands for pseudoequatorial*

a' *stands for pseudoaxial*

ψ *t-Bu Group in each case disallows conformational flipping*

Fig. 2.122 Isomerizational preference in base catalyzed equilibration of *cis*- and *trans*-4-t-butyl-6-nitro-1-phenylcyclohexenes

Table 2.6 Preferred conformations of the C6 substituents of 1-Phenyl (or Methyl)-6-substituted or 4,6-substituted cyclohexenes

Preferred conformation of 6R′	1,6 or *trans*-1,4,6-substituted cyclohexenes
Pseudoaxial	R′ = *t*-Bu, Ph, Me₂C(OH) or NO₂ R′ = NO₂ or Br only conformer
Pseudoequatorial	R′ = Me₂C (OH), COMe or, NO₂ *cis*-4-Me-6-NO₂-

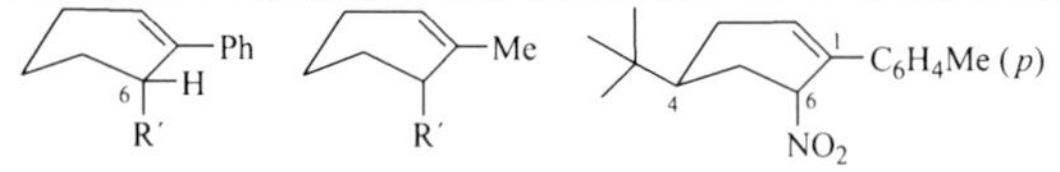

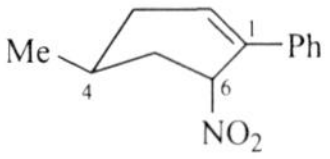

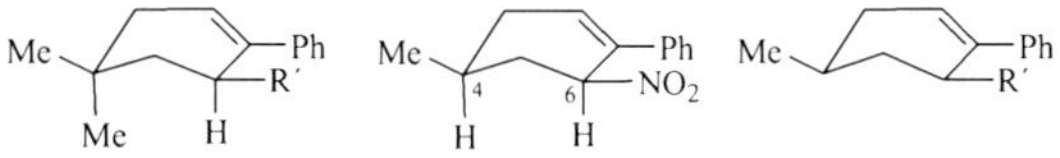

R′ = *t*-Bu, Ph, Me$_2$C(OH) or NO$_2$

R′ = NO$_2$ or Br

—C$_6$H$_4$Me (*p*), NO$_2$, only conformer

Me—...—Ph, NO$_2$

R′ = Me$_2$C (OH), COMe or, NO$_2$

cis-4-Me-6-NO$_2$-

Note. Syn axial R′ ↔ Me interaction in the a′ conf. over-rides A^{1,2} interaction between NO$_2$ and Ph or R′ and Ph groups in the e′-conformer, leading to the preferred e′ conformer.

No Conformational Preference (NCP)

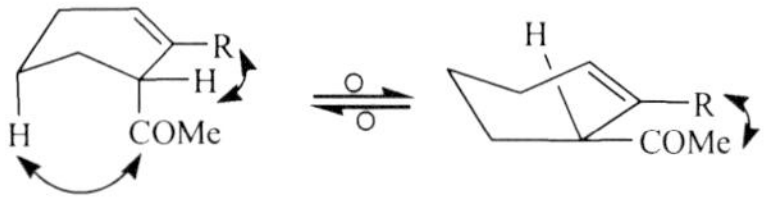

synaxial   R = Me or Ph

2.14.2.2 Isomerizational Preference

One of the best examples is the base catalyzed equilibration of *cis*- and *trans*-4-*t*-butyl-6-nitro-1-phenylcyclohexenes leading to an equilibrium composition of 85–95 % of the *trans* isomer (pseudoaxial nitro group) (Fig. 2.122).

2.14.2.3 Pseudoallylic 1,2-Strain in Enamines

The value of A^{1,2} strain becomes apparent only from its generality. If we replace the ethylenic substituent of 1-alkyl- or 1-aryl-6-substituted cyclohexene by a nitrogen or an oxygen atom, we get an enamine (**2**) or an enol ether. Thus, cyclohexanone can be converted to 1-pyrrolidinyl-6-methylcyclohexene (**2**), which may be regarded as a pseudoallylic system (Fig. 2.123). This reaction is known as Stork reaction [87, 88]. Now, the pseudoallylic 1,2-strain, Me ↔ pyrrolidinyl α-H in (**2e′**), is greater than the sum of the pseudoallylic 1,2-strain, e′H ↔ pyrrolidinyl α-H, and the synaxial interaction between 6-Me and 4-H in (**2a′**). Thus, the pseudoaxial conformer (**2a′**) becomes predominant in the ring inversion of (**2**) (Fig. 2.123). The enamine upon hydrolysis forms pure 2-methylcyclohexanone

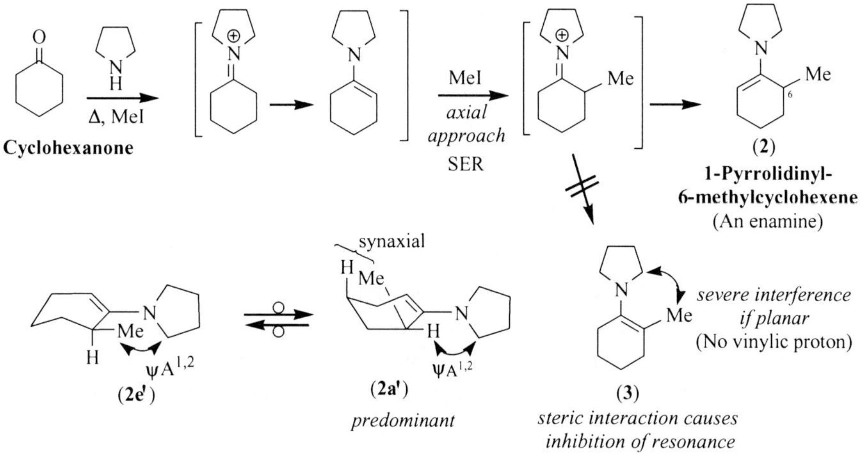

Fig. 2.123 Pseudoallylic 1,2-strain in a 6-methylenamine derivative

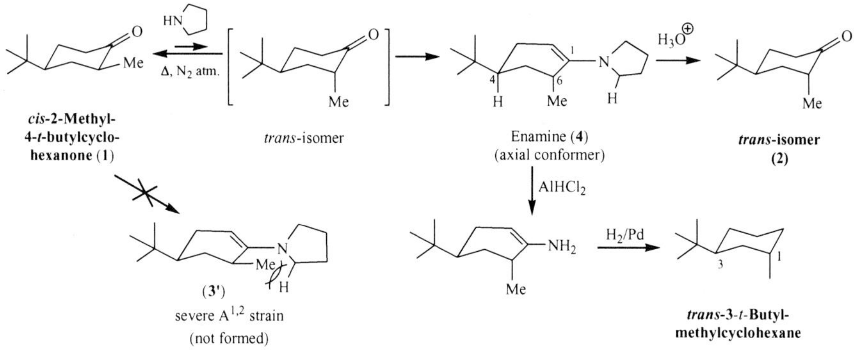

Note. *All compounds including the starting material are racemates, but, as usual, only one enantiomer is written in each case.*

Fig. 2.124 Complete conversion of the *cis*-2-methyl-4-t-butylcyclohexanone into the corresponding less stable *trans* isomer

without any further methylation at C2 or C6—this being the best method of its formation. The absence of any vinylic proton in the NMR establishes the non-formation of (3), which would involve severe interaction between the olefinic methyl and a planar pyrrolidinyl α-H, as shown in (3′) in Fig. 2.124.

Experimental verification of the pseudoaxial conformation of the 6-Me group in (2) is obtained by complete conversion of the more stable *cis* isomer (1) of 2-methyl-4-*t*-butylcyclohexanone to the less stable *trans* isomer (2) [89], which can be explained in terms of the non-formation of the enamine (3′) from the *cis* isomer due to severe $A^{1,2}$ strain (Fig. 2.124). The base (pyrrolidine) catalyzed

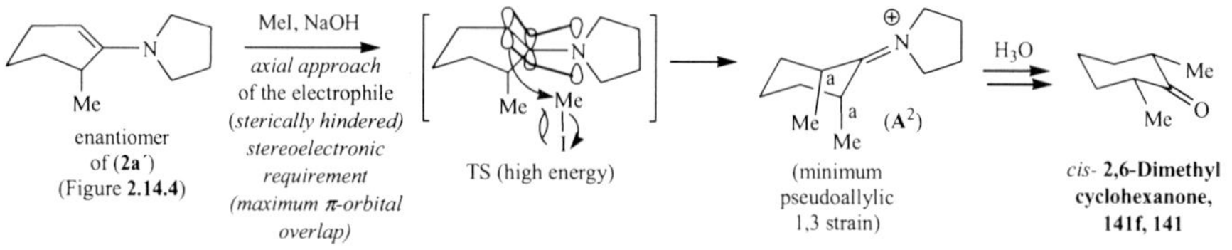

Fig. 2.125 Preparation of *cis*-2,6-Dimethylcyclohexanone (**A**)

equilibration produces the intermediate *trans* isomer (~5 %), which with axial Me is converted to the stable enamine (**4**) involving much weak $A^{1,2}$ strain between He and α-H of pyrrolidine, and the equilibrium is shifted till the whole amount of *cis* isomer is transformed to (**4**). The latter upon hydrolysis gives the *trans* isomer (**2**) quantitatively. The last two steps establish that the acid hydrolysis did not cause any change in orientation (here axial) of the Me group. Thus, the enamine (**4**) must have the Me group axial.

   Another proof of pseudoaxial orientation of 6-substituent of an enamine:

   The 6-methylenamine derivative enantiomer of (**2a′**) of Fig. 2.123 may undergo further methylation (or alkylation), but with difficulty, and the resulting Schiff base (**A²**) (Fig. 2.125) produces upon acid hydrolysis *cis*-2,6-dimethylcyclohexanone (**A**) (without any other accompanying polymethylated cyclohexanone). This is a convenient method for the preparation of (**A**). The sequence of reactions is illustrated in Fig. 2.125.

2.14.2.4   Synthesis of Solenopsin A. Application of $A^{1,2}$ Strain concept [90]

The concept of $A^{1,2}$ strain has been elegantly applied for the synthesis of solenopsin A(**X**), starting from the precursor imine (**B**), as delineated in Fig. 2.126.

2.15   Fused Ring Systems

2.15.1 *Decalins*

Of the fused carbocyclic systems with 6-membered rings like 6–3, 6–4, 6–5, 6–6, and 6–7 systems, the most important fused system is the 6–6 system, [4.4.0] bicyclodecane, trivially named as *decalin* (=decahydronaphthalene).

   This system is very useful for consideration of conformation and reactivity, more quantitatively than other bicyclic systems. The decalin system occurs abundantly in natural products, like diterpenes, triterpenes, steroids, and alkaloids.

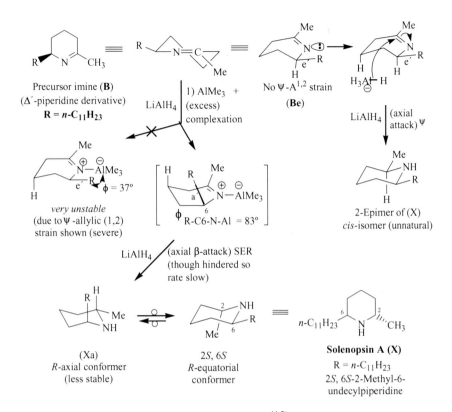

Fig. 2.126 Synthesis of solenopsin A by application of $A^{(1,2)}$ concept

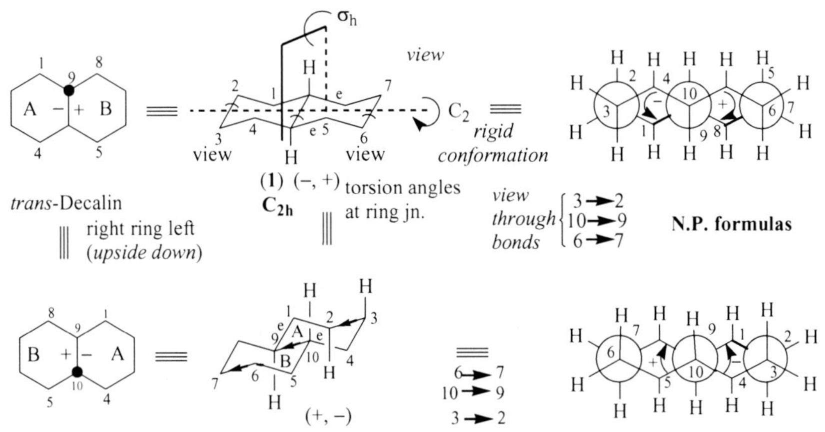

Upon monosubstitution at any position C_2 disappears, and σ also disappears, if the substitution is in a position other than C-9 and C10, and the substituted product is optically active.

Fig. 2.127 Conformation and symmetry point group of *trans*-decalin

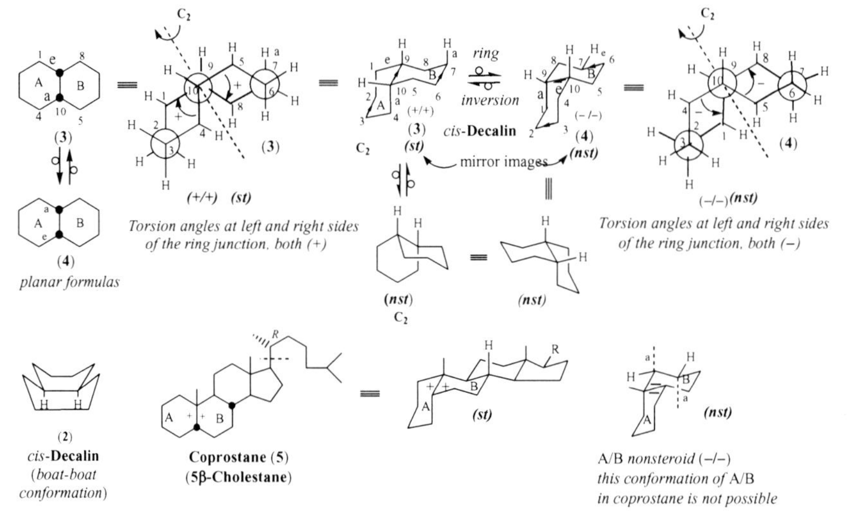

Note. *Monosubstitution product of cis-decelin at any position other than C9 and C10 gives a diastereomer upon inversion and hence is optically active.*

Fig. 2.128 Steroid (st) (+/+) and nonsteroid (nst) (−/−) conformations of *cis*-decalin

2.15.1.1   Brief History

- The existence of *cis* and *trans* decalins with two puckered six- membered rings was first predicted by Mohr in 1918. He also predicted chair/chair conformation (1) (Fig. 2.127) for *trans*-decalin and boat–boat conformation (2) (Fig. 2.128) for *cis*-decalin. On the contrary, two planar six-membered rings, as postulated by Baeyer, can only be fused as cis and not as trans.
- In 1925 Mohr's prediction was confirmed by isolation of *cis*- and *trans*-decalins by Hückel, and *trans* isomer was found to be more stable.
- In 1943 Hassel recognized that the chair form of cyclohexane is at least 5.6 kcal/mol more stable than its boat/skew–boat form, and for the first time predicted the chair–chair conformation for both *trans*-decalin (**1**) (Fig. 2.127) and *cis*-decalin (**3**) (Fig. 2.128).
- This was also substantiated by electron diffraction data in 1946.

The conformations of the two diastereomeric forms, *trans*-decalin and *cis*-decalin, are discussed in the sequel.

2.15.1.2   *trans*-Decalin. Conformation. Torsion Angles. Symmetry

In *trans*-decalin the ring junction H's at C9 and C10 are on the opposite sides. *The sign of any torsion angle involving three consecutive bonds is written against the central bond*. Torsion angles around all C–C bonds in both *trans*-decalin and *cis*-decalin are ~55–56°, the same as in cyclohexane. In *trans*-decalin the two rings (chair) are fused through *e,e* bonds and cannot undergo ring inversion. From the Newman projection formula of *trans*-decalin, obtained by looking through the bonds mentioned in Fig. 2.127, it is quite clear that the torsion angles (ϕ or ω) in rings A and B on the left and right side of the common central bond are (−) and (+), respectively (see also Sect. 2.12.1.1). The signs are reversed if the B ring is on the left side (Figure). The sign of ϕ can be determined directly from the chair/chair conformation of the *trans*-decalin using the following rule. One does not have to draw the Newman projection formula for this purpose, as is shown on the right side of the Figure.

Rule of determining the sign of the torsion angle (ω or ϕ): While moving our eyes in a clockwise direction from the top of a ring (in a slanting manner) in the chair or boat or skew–boat form, the *β-axial* H or substituent in that ring *is preceded by* (+) *sign of the torsion angle* ω or ϕ and *is followed by* (−) *sign*.

Corollary: The α-axial *H* or substituent in that ring is preceded by (−) sign and is followed by (+) sign of the torsion angle ω or ϕ.

Symmetry. The chair–chair conformation of *trans*-decalin (Fig. 2.127) has a center of symmetry at the midpoint of C9–C10 bond and is achiral. Additionally, it has a C_2 axis passing through the midpoints of C2–C3, C9–C10, and C7–C6 bonds, and a σ_h plane passing through the ring junction, and perpendicular to the C_2 axis. Thus, its point group is C_{2h}.

2.15.1.3   *cis*-Decalin. Conformations. Torsion Angles. Symmetry (Fig. 2.128)

In the *cis*-decalin the ring junction H's at C9 and C10 are on the same side of the puckered plane of the molecule, and the two rings are fused through *ea* or *ae* bonds. Thus, ring inversion takes place when the ring fusion bonds *ea* [to a ring, say(A)] (**3**) are converted to ae (**4**). From the Newman projection formulas of *cis*-decalin (Fig. b), it is clear that torsion angle ϕ at the left and right side of the common central C9-C10 bond in rings A and B are (+,+) for left side formula (**3**) and (−,−) for the right side formula (**4**), which are undergoing rapid interconversion. Such a concerted change in torsion angles of a *cis* fused junction is quite feasible. One can easily see this interconversion in its Dreiding or Fischer model.

The signs of the torsion angle ϕ at the ring junction in the rings A and B can be determined directly from either chair–chair conformations (**3**) (+/+) and (**4**) (−/−), by application of the rule, stated above.

The conformation (**3**) of *cis*-decalin is called *steroid* conformation since it resembles the A/B rings of 5β-steroids like coprostane (**5**), whereas the conformation (**4**) which cannot exist in 5β-steroids is called *nonsteroid*.

Symmetry. *cis*-Decalin chair–chair conformation (**3**) or (**4**) is chiral since either has no reflection symmetry (no σ plane or center of symmetry). Each conformer possesses a C_2-axis which passes through the midpoint of C9-C10 bond to which it is orthogonal, and it bisects the dihedral angle between 9-H and 10-H bonds (ϕ 9-H/10-H). So the point group is C_2 (chiral). But it forms an *unresolvable* (±) *pair* because of the rapid interconversion of the two flipped forms: *steroid and nonsteroid (which are enantiomers), at even low temperatures*, the energy barrier between them being much lower than 15 kcal mol^{-1}. Although it is appreciably higher than that in cyclohexane, the barrier to ring inversion $\Delta G^{\neq}$ in *cis*-decalin is *12.3–12.6 kcal mol^{-1}* (51.5–52.7 kJmol^{-1}).

The inversion path involves the following transformations: chair–chair (CC) → chair–twist (CT) → twist–twist (TT) → alternate TT → CT → CC; the alternate TT is reconverted to the CT and CC by a reversal (mirror imaged) of the path by which it was formed.

2.15.1.4   Ring Inversion in *cis*-Decalin

In rigid *trans*-decalin the equatorial and axial protons are distinguishable which appear in ^{1}H NMR as two broadbands due to spin–spin coupling.

cis-Decalin and its derivatives undergo ring inversion (*cf.* cyclohexane)—already discussed. This can be studied by ^{1}H NMR. The equatorial and axial protons are averaged out due to ring inversion and appear as a narrow band at ambient temperature. However, at temperatures lower than the coalescence temperature, they can be split up into two broadbands. The *coalescence temperature* gives a value of 12.76 kcal mol^{-1} at −18° in CS$_2$ for the free energy barrier of ring inversion. 2,2-Difluoro-*cis*-decalin exhibits a barrier of 12.26 kcal mol^{-1} at −30 °C in ^{19}F NMR. The energy barrier of *cis-decalin* is thus considerably higher than that of cyclohexane and *cis*-1,2-dimethylcyclohexane ($\Delta G^t = 10$ kcal/mol). Ring inversion data of many cyclic compounds are available in a monograph by Oki [91]. Some coalescence temperatures and energy barriers are given in Fig. 2.129.

	~ Coalescence temp.	Energy barrier (kcal mol^{-1})
cis-Decalin	≥ −30°C	12.3 – 12.5
Cyclohexane[#]	− 66.7°C	10.1
cis-Hydrindane	− 121°C	6.5

[#]The rate of chair inversion is 105 sec^{-1}.

Fig. 2.129   Some coalescence temperatures

Isomer	Symmetry no. σ	Entropy cal. / deg. mole^{-1}				
		$-R \ln \sigma$	$R \ln 2$ due to $\pm$ forms	Total S	Calcd. ΔS $S_{cis} - S_{trans}$	Exptl. ΔS
cis-Decalin	2	-1.38	1.38	0	$+1.38$	0.55
trans-Decalin	2	-1.38	$-$	-1.38		

Fig. 2.130 Entropy difference in decalins

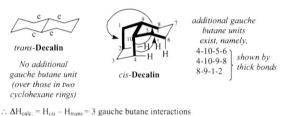

trans-**Decalin**

No additional
gauche butane unit
(over those in two
cyclohexane rings)

cis-**Decalin**

additional gauche
butane units
exist, namely,
4-10-5-6
4-10-9-8   shown by
8-9-1-2   thick bonds

$\therefore \Delta H_{calc.} = H_{cis} - H_{trans} = 3$ gauche butane interactions

$= 3 \times 0.8$ kcal mol$^{-1} = 2.4$ kcal mol^{-1} (= 10.05 k J mol^{-1})

Good agreement with $\Delta H_{exptl.}$ (2.10 to 2.72 kcalmol^{-1}) ($\equiv$ 8.8 to 11.4 k J mol^{-1}) determined by temperature dependence of *cis* ⇄ *trans* equilibrium and also from heat of combustion data.

According to ***Auwers-Skita rule or*** *conformational rule*, *cis*-decalin having higher enthalpy has higher b.p., density, refractive index than those for *trans*-decalin.

cis-Decalin : b.p. 193°C.743 mm; d_4^{20} 0.895; n_D^{20} 1.4811

trans-Decalin   184°C/747 mm;   0.870;   1.4697

Fig. 2.131 Enthalpy and physical constants difference of *cis* and *trans*-decalins

2.15.1.5   Entropy Difference in Decalins

Symmetry number of *cis*- and *trans*-decalin is 2. The difference in entropy of *cis*-decalin results from the fact that the *cis*-isomer is a (±)-pair (although non-resolvable). The total entropy difference is shown in Fig. 2.130.

Thus, ΔS experimental value (0.55 cal K^{-1} mol^{-1} $\equiv$ 2.3 J K^{-1} mol^{-1}) is much less than the calculated value 1.38 cal K^{-1} mol^{-1} $\approx$ 5.8 J K^{-1} mol^{-1}.

This is contrary to the popular belief that *cis*-decalin is more flexible than *trans*-decalin; in that case ΔS would have been more rather than less. Here ring inversion ability possessed by *cis*-decalin should not be confused with the flexibility of the system as in case of *trans*-decalin.

2.15.1.6   Enthalpy and Physical Constants. Auwers-Skita Rule

Difference in enthalpies of *cis*-decalin and *trans*-decalin can be counted in terms of gauche interactions arising out of carbon atoms of two different rings. In *trans*-

decalin each ring is fused through e bonds, and hence no additional gauche unit is introduced. The calculated and experimental results of ΔH (enthalpy difference) are presented in Fig. 2.131.

2.15.1.7   Free Energy Difference in Decalins

The free energy difference has been determined experimentally from the equation $\Delta G = \Delta H - T\Delta S$ as 2.39 kcal mol^{-1}, which is in good agreement with the equilibrium data for 1-decalones [$trans \rightleftarrows cis$ (5–10 %) from $\Delta G = - RT \ln K$]. The comparison is, however, not strictly valid because of 2- and 3-alkylketone effects (*cf.*, Sect. 2.13.3) in 1-decalone.

a   *9-Methyldecalins*

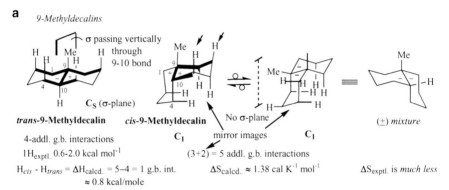

trans-9-Methyldecalin

C_S (σ-plane)

4-addl. g.b. interactions

$1H_{exptl.}$ 0.6-2.0 kcal mol^{-1}

$H_{cis} - H_{trans} = \Delta H_{calcd.} = 5-4 = 1$ g.b. int.
≈ 0.8 kcal/mole

cis-9-Methyldecalin

C_1

mirror images

$(3+2) = 5$ addl. g.b. interactions

$\Delta S_{calcd.} \approx 1.38$ cal K^{-1} mol^{-1}

C_1

No σ-plane

$(\pm)$ *mixture*

$\Delta S_{exptl.}$ is *much less*

$\Delta H_{exptl.} = 0.6$ to 2 kcal mol^{-1} *(determined by temperature dependence of cis-trans equilibrium and heat of combustion data)*

b   *9,10-Dimethyldecalins*

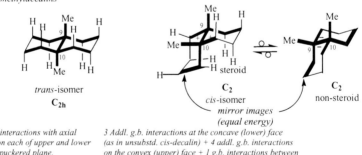

trans-isomer

C_{2h}

cis-isomer

C_2

mirror images
(equal energy)

steroid

C_2

non-steroid

4 Addl. g.b. interactions with axial H's shown on each of upper and lower faces of the puckered plane, Total 8 addl. g.b.

$\therefore \Delta H = H_{cis} - H_{trans}$
≈ 0 *(prediction)*

3 Addl. g.b. interactions at the concave (lower) face (as in unsubstd. cis-decalin) + 4 addl. g.b. interactions on the convex (upper) face + 1 g.b. interactions between Me's at C-9 and C-10 (This interactions is probably equal to the usual g.b. interaction)
Total: 8 g.b.
Equilibrium expt. or interactions heat of combustion data are not available

Ring inversion barrier of cis-9-methyldecalin is ≈ 12.5 kcal mol^{-1} (cf. cis-decalin), and that of cis-9, 10-dimethyldecalin is 14.6 kcal mol^{-1}

Fig. 2.132   Conformations and enthalpy differences of 9-methyldecalins and 9,10-dimethyldecalins

2.15.1.8 Effect of Introduction of Angular Methyl Group/s

The enthalpy difference between *cis*-decalin and *trans*-decalin after introduction of 9-methyl group and 9,10-dimethyl groups are delineated in Fig. 2.132a and b, respectively. The additional gauche–butane interactions between an axial methyl group and synaxial hydrogen atoms are shown in each case. The axial hydrogen atoms at the concave face of *cis*-decalin sustaining three additional gauche–butane interactions are also shown (*cf.* Fig. 2.131).

  trans-Decalins unsubstituted at positions other than bridgeheads are always *achiral*, having a σ-plane vertical to the common 9–10 bond. *cis*-Decalins substituted only at C9 or C10 or at both carbons although chiral exist as unresolvable (±)- or racemic mixtures (*cf. cis*-decalin). However, a *cis*-decalin substituted at any position/s other than bridgeheads will exist as a resolvable (±)-pair, since flipping now converts it into a diastereomer (and not an enantiomer).

2.15.1.9 *cis*-Decalones and *trans*-Decalones

When a carbonyl group is introduced to give 1- or 2-decalones, the two bridgehead carbons, C9 and C10, become chiral, and both the *trans*- and *cis*-decalins exist as resolvable (±)-pair. Thus, *trans*-1-decalone can be resolved into two enantiomers (1) and (1′) (Fig. 2.133), having unequivocal conformations. However, each enantiomer of *cis*-1-decalone (2) or (2′) exists as two nonequivalent conformers (2a) and

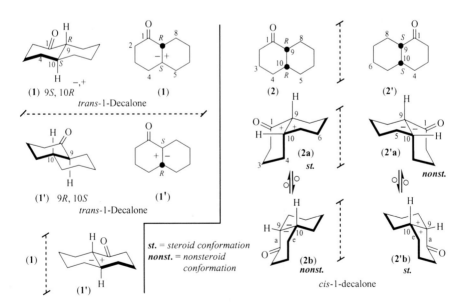

Fig. 2.133 Conformations of *trans*-1-decalones and *cis*-1-decalones

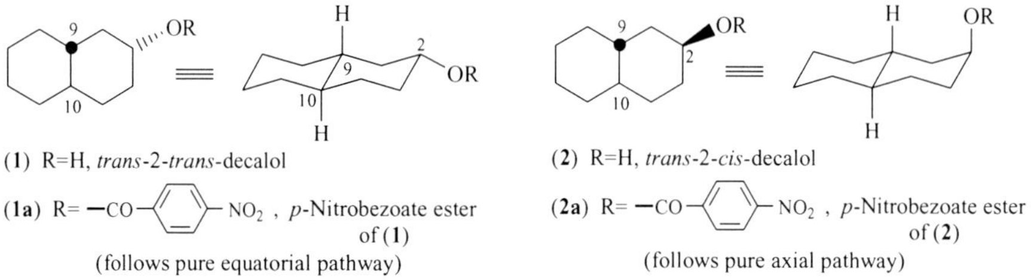

(1) R=H, *trans-2-trans*-decalol

(1a) R= —CO—⟨ ⟩— NO₂ , *p*-Nitrobezoate ester
                              of (1)
         (follows pure equatorial pathway)

(2) R=H, *trans-2-cis*-decalol

(2a) R= —CO—⟨ ⟩— NO₂ , *p*-Nitrobezoate ester
                              of (2)
         (follows pure axial pathway)

Fig. 2.134 Conformational analysis of *trans*-2-decalol diastereomers

(2b) or (2'a) and (2'b), respectively, by ring inversion. Conformer (2a) having steroid conformation flips into (2b) having nonsteroid conformation. Similarly, conformer (2'a), the enantiomer of (2a), having nonsteroid conformation, undergoes ring inversion to form a nonequivalent conformer (2'b) having steroid conformation (Fig. 2.133).

2.15.1.10   *trans*-2-Decalols. Conformational Analysis

Trans-2-Decalol exists as two diastereomers, *viz.*, *trans-2-trans*-decalol (1) and *trans-2-cis*-decalol (2) (Fig. 2.134); they have the hydroxyl group *trans* and *cis*, respectively, to the 9-H atom. Since *trans* decalin possesses unique rigid conformation (ring inversion being not possible), the OH groups in the diastereomers (1) and (2) (Fig. g) must be equatorial and axial, respectively. Thus, the conformational analysis here is easy and gives definite conclusions. Compound (1) is converted to its *p*-nitrobenzoate ester (1a) at a faster rate than compound (2) under the same condition, *because of the steric hindrance* in the latter case. Likewise, the saponification of the ester (1a) follows pure equatorial pathway whereas the reactions of (2) or (2a) follow pure axial pathway involving steric hindrance.

2.15.1.11   *cis*-2-Decalols. Conformational Analysis

Let us now consider the relative reactivity of the *p*-nitrobenzoates of isomeric *cis*-2-dacalols [92]. The configurations of these epimeric decalols were established by Dauben et al. [93, 94]. Because of the flexibility (due to ring inversion) of *cis*-decalols, conformational analysis does not give definite conclusions regarding stereochemistry of the molecules. The relative rates of saponification of the diastereomers are delineated in Fig. 2.135 in terms of the equatorial 1e and axial 1a conformers arising due to ring inversion of 1 and in terms of the equatorial 2e and axial 2a conformers arising out of the ring inversion of 2 and by the application of Winstein–Holness equation (i) (Fig. 2.135); here k_e and k_a denote the specific rate

Relative rates of saponification of the diastereomers

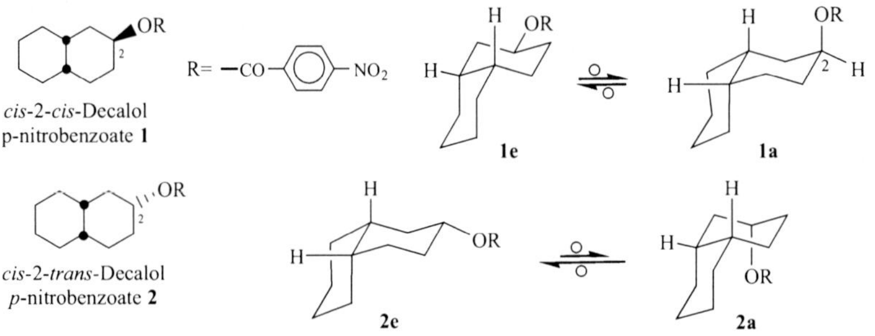

cis-2-cis-Decalol
p-nitrobenzoate **1**

$R= -CO-\langle \rangle -NO_2$

1e ⇌ **1a**

cis-2-trans-Decalol
p-nitrobenzoate **2**

2e ⇌ **2a**

Winstein-Holness equation: overall rate $k = k_e N_e + k_a N_a$

In case of compound **1**

$$k_1 = k_{1e}N_{1e} \quad + \quad k_{1a}N_{1a}$$

both greater, | both smaller,
e pathway | but not negligible;
| a pathway

∴ The ester **1** undergoes saponification
by both e and a pathways

$G_{1e} \approx G_{2e} < G_{1a} << G_{2a}$
∴ $N_{1e} > N_{1a};\ N_{2e} >> N_{2a}$
Again, $\Delta G^{\neq}_{1e} < \Delta G^{\neq}_{1a}$ ∴ $k_{1e} > k_{1a}$
and $\Delta G^{\neq}_{1e} < \Delta G^{\neq}_{2a}$
(cases of steric hindrance)
∴ $k_{2e} >> k_{2a}$

In case of compound **2**

$$k_2 = k_{2e} N_{2e} \quad + \quad k_{2a} N_{2a} \approx k_{2e} k_{2a} \text{ (approx.)}$$

both much | both negligibly smaller, and the product
greater | is even more negligible

Note: The equatorial pathway for
the saponification of an ester
is faster than the axial pathway
(case of steric hindrance).

∴ Compound **2** undergo saponification by equatorial pathway.

∴ k_2 should be greater than k_1.

Experimentally it has been found [94] that the ester **2** is saponified 1.5 times more rapidly than that of **1**.

Fig. 2.135 Conformational analysis of *cis*-2-decalol *p*-nitrobenzoate diastereomers

constants of the equatorial and axial conformers, and Ne and Na denote the mole fractions of the equatorial and axial conformers, respectively.

2.15.2 *Perhydrophenanthrenes (PHP's). Stability. Point Groups. Optical Activity*

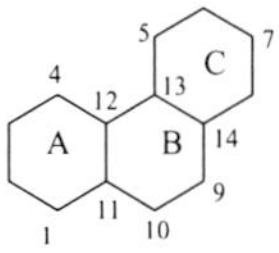

PHP

Three cyclohexane rings when fused successively to an angular arrangement gives rise to perhydrophenanthrene (PHP). It constitutes the ABC rings of terpenoids and steroids. It contains two equivalent pairs of chiral centers—the four bridge-head carbons, C11 and C12 and C13 and C14. Thus, it constitutes an **ABBA** system and can have four pairs of enantiomers and two meso forms. (cf. CH_2OH-$(CHOH)_4$-CH_2OH); total *10 stereoisomers* and *six diastereomers* (Fig. 2.54).

In the nomenclature of the diastereomers, the prefixes *cis* and *trans*, here abbreviated as **c** and **t**, define the stereochemistry of the ring junctions A/B or B/C, whereas the other terms *cisoid* and *transoid*, abbreviated as **ci** and **tr**, define the steric relationship of the nearest bridgehead atoms (which also denotes the orientation of the terminal rings with respect to each other).

Planar structure (A) (**Point group**) (*chirality*)	*Conformation* (B) e equatorial a axial	*No. of gauche* butane interactions (C)	Relative $\Delta\Delta H^0$ calcd.		$\Delta\Delta G^0$ exptl. [91] (F)	*σ plane in* planar projection formula† (G)
			kcal mol^{-1} [89] (D)	Molecular mechanics [90] (E)		
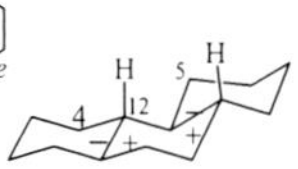 (1) **t-tr-t** *four e ring fusion* *bonds (rfb)* (*chiral*) (**C₁**) 11α, 12β, 13α, 14β R S S R		1 (C4-C12-C13-C5)	0	0	0	*Absent* *so resolvable*
		Note: A dot at a ring juncture indicates a forward or upward hydrogen, whereas the absence of a dot indicates a backword or downward hydrogen. In case of (±) pairs only one enantiomer with H12 β- is shown.				
(2) **c-tr-t** *three e rfb* (*chiral*) (**C₁**) S S S R		1+3 (for cis-decalin)	2.4	2.44	2.25	*Absent,* *so resolvable*
		†*the chirality or otherwise can be judged by looking for a vertical σ-plane passing through the 9-10 and 12-13 bonds in the planar structure*				
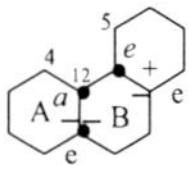 (3) **c-ci-t** *three e r f b* (*chiral*) (**C₁**)	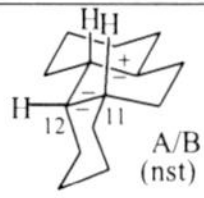	1+3 (for cis-decalin)	2.4	2.57	2.66	*Absent* *so resolvable*

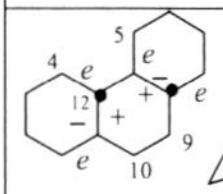

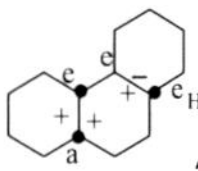

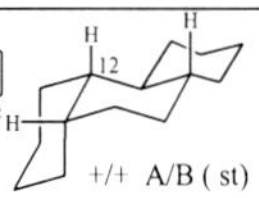

(A)	(B)		(C)	(D)	(E)	(F)	(G)
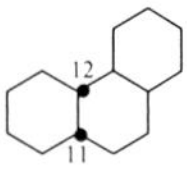 (4) **c-tr-c** (**C₁**) *chiral* *two e rfb)*	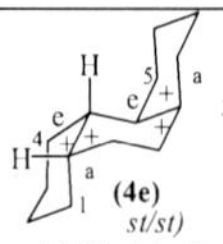 (4e) *st/st)* 1,4,diaxial *rf* bonds. A/B B/C 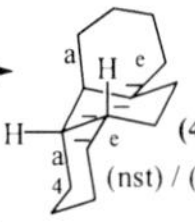 +/+ +/+	(4a) *(nst) / (nst)* 1,2-diaxial *rf* bonds –/–, –/–	1+6(two cis-decalins) ↕ 6(two cis-decalins)	4.8 (4e) (4,5 di eq) ↕ 4.0 (4a)	4.01	4.6	*Absent, so* resolvable; *flips into a diastereomer*

Fig. 2.136 (continued)

	Conformation / interaction	C	D	E	Notes
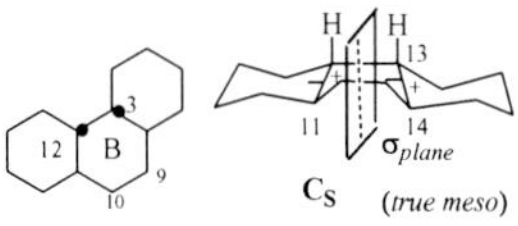 C_s (true meso) **(5) t-ci-t** **Achiral** **(4be rfb)** In ring B the sign at 11-12 bond is + " " 13-14 " is −, (should have been + in chair form) Impossible in chair conformation; so B ring is boat.	B-ring true boat [89]	7.2	7.03	8.98	*Present in planar as well as conformational structure.* *So true meso*
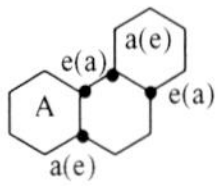 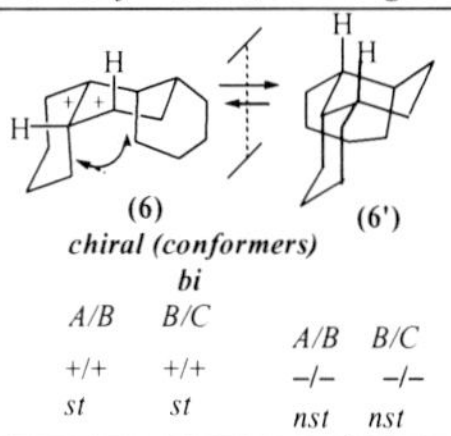 **(6) c-ci-c bi** unresolvable **(2e, 2a, rfb)** **(6)** chiral (conformers) **bi** **(6')** A/B B/C A/B B/C +/+ +/+ −/− −/− st st nst nst	$1+6+$ syn axial $CH_2 \leftrightarrow CH_2$ interaction  Me↔Me	$4.8+$ >3.6 ¥ = >8.4 total $(5.2-1.6$ $= 3.6)$ for Me↔Me	9.01	7.43	*present in planar str, so unresolvable;. actually unresolvable (±) pair formed by flipping.(cf. cis-decalin).* *So not true meso*

Notes : Approx. agreements between H^0(calcd.) and G^0(exptl.) energies are satisfactory in view of the differences, though small, in entropy of mixing and entropy of symmetry between isomers.

• For the highly strained isomers (5)and (6) the calculated. energies are the reverse of the experimental.

• ¥ This is between 1, 3 synaxial ring fusion bonds; being rigid it would be more than synaxial Me□ Me interaction.

Fig. 2.136 Perhydrophenanthrene (PHP)diastereomers: conformations, torsion angles, energies, resolvabilities

The six perhydrophenanthrene diastereomers (**1**), (**2**), (**3**), (**4**), (**5**), and (**6**) are included in Fig. 2.136 in order of increasing enthalpy, *i.e.,* decreasing stability. This basic ring system is of wide occurrence in nature, and the relative stabilities of the diastereomers are of interest. In this figure the calculated relative enthalpies $(\Delta\Delta H^0)$ based on nonbonded interactions (**C**) [95], and on molecular mechanics [force field calculations (**D**)] [96], and the experimental relative free energies $(\Delta\Delta G^0)$ (**E**) [97] are expressed in kcal/mol (approx.)

Conformations of the enantiomers, (marked with superscript 1) of the chiral perhydrophenanthrene diastereomers (**1**), (**2**), (**3**), and (**4**) are shown in Fig. 2.137.

It should be noted that the torsion angles of the central ring at 11–12 and 13–14 bonds at the ring junctions are the same in (**1**) to (**4**) and (**6**), as expected for chair conformation. But in the isomer *t-ci-t* (**5**) these signs are opposite which suggest a non-chair conformation for the central ring. The same conclusion is also drawn from the orientation (*a,e*) of the ring fusion bonds. If the ring B were a chair, the ring fusion bonds would be *e, e, a, a* with respect to B, and it is known that the *a, a* fusion in the *trans*-decalin moiety is sterically not possible. Because of the rigidity of the *t-ci-t* (**5**) diastereomer, the central ring assumes a *true boat* form (instead of a flexible boat).

The predicted order of stability of the PHP's has been confirmed by equilibration of the 9-ketones in the cases shown in Fig. 2.138.

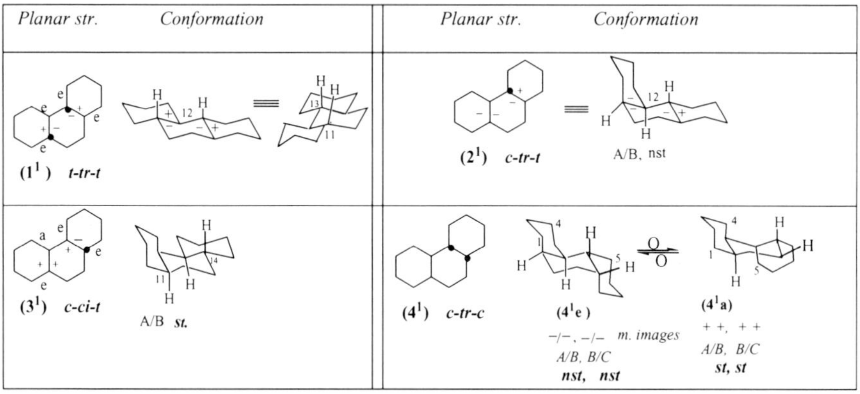

Fig. 2.137 Conformations of the enantiomers of the active perhydrophenanthrene diastereomers of Fig. 2.136

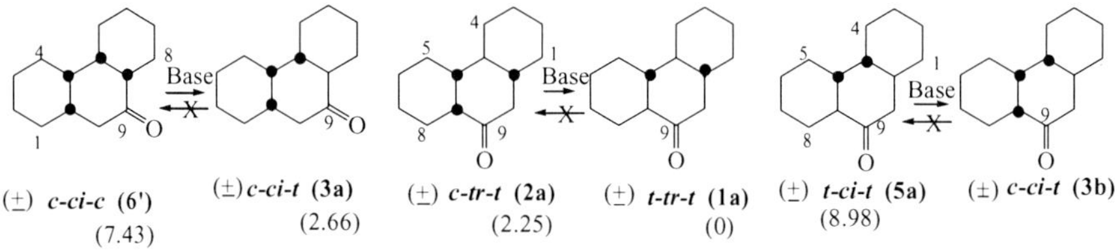

Approx. rel. enthalpy (H) in kcal/mole (within parenthesis)

Fig. 2.138 Confirmation of the predicted stability of perhydrophenanthrene-9-ones

2.15.2.1   Stereochemistry of Some Perhydrophenanthrones and All Perhydrodiphenic Acids (PHDPA's)

An elegant piece of work by Linstead and coworkers [98–104] presents a challenging exercise in stereochemical reasoning.

Several perhydrophenanthrones were correlated with PHDPA's in the following way:

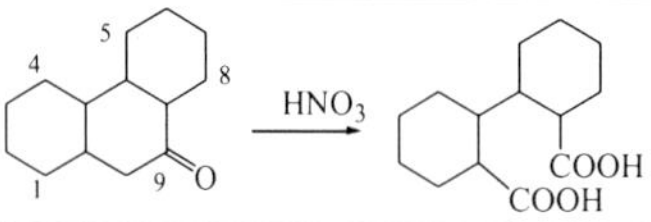

Epimerization in this reaction must be guarded if same configuration is to be assumed

Syntheses of PHDPA's and the process of their configurational assignments from the study of catalytic hydrogenation of diphenic acid giving *cis* isomers (**A**), (**B**), and (**C**) of PHDP's are summarized in Fig. 2.139. The hydrogenated diastereomers were subjected to preferential epimerization of the monomethyl esters and

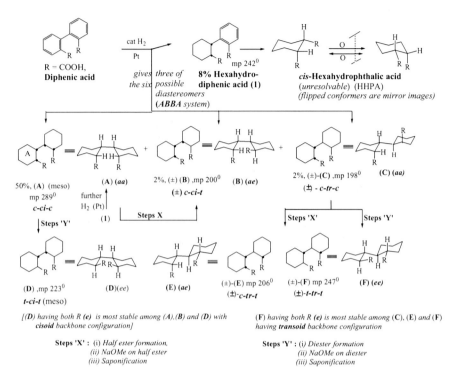

Fig. 2.139 Synthesis and configurational assignments of the perhydrodiphenic acids

dimethyl esters with sodium methoxide giving ultimately the most stable diastereomer (**D**) among the *cisoid* isomers, and the most stable diastereomer (**F**) among the *transoid* isomers (Fig. 2.139). The following important points should be noted:

- Formation of *cis*-hexahydrophthalic acid from hexahydrodiphenic acid (**9**) and the conversion of the latter to (**A**) by further hydrogenation indicates that ring A in (**A**) is *cis* 1,2- disubstituted.
- Epimerization takes place at the α-carbon to the ester and not at the backbone carbons. Thus, (**A**), (**B**), and (**D**) belong to the same backbone configuration; this is also valid for (**C**), (**E**), and (**F**).
- Stoermer–Steinbach principle was applied to unresolvable (**A**) and (**D**) as follows: Both of them were converted to the monomethyl ester (dissymmetry introduced), which could be resolved. An active monoester was converted to the inactive diester (plane of symmetry again introduced).
- Thus, **PHDPA's**, (**A**) and (**D**), were established to be meso/unresolvable. Hence, (**A**), (**B**), and (**D**) must belong to cisoid backbone configuration.
- The PHDPA's (**C**) and (**F**) were resolved, so (**C**), (**E**), and (**F**) must belong to the *transoid* backbone configuration.

- Stability order in each series easily determines the relative configuration of the compounds in both *cisoid* and *transoid* series.
- For designation of PHDPAs the two COOH groups must be on the same side of the backbone C–C bond. (*cf.* the nomenclature of the perhydrophenanthrenes).

2.15.3　Perhydroanthracenes: Relative Stability. Torsion Angles. Point Group. Optical Activity

In perhydroanthracene (**PHA**) diastereomers (Fig. 2.140) the rings are fused in a linear arrangement. In the nomenclature of the diastereomers, the prefixes *c* and *t* define the stereochemistry of the ring junctions A/B and B/C, whereas the other terms *cisoid* and *transoid*, abbreviated as *ci* and *tr*, define the steric relationship of the nearest bridgehead atoms. Here, all the four chiral centers are equivalent, corresponding to **AAAA** system and there are five diastereomers—two of them are enantiomeric pairs and the other three are *meso*. The diastereomers (A), (B), (C), (D), and (E) in Fig. 2.140 are in order of increasing enthalpy and hence decreasing stability.

The relative stability of the isomers was originally predicted by Johnson [95]. Their planar structures, conformations, the relative enthalpies (calculated by molecular mechanics and determined experimentally [105]), and resolvability are outlined in Fig. 2.140. In contrast to the least stable PHP diastereomer, here the ***t-ci-t*** isomer (**A**) is the most stable isomer having no extra gauche interaction among the atoms of different rings. It possesses a plane of symmetry and hence is meso. The ***c-ci-t*** isomer (**B**) may also be called ***c-tr-t*** isomer, if one looks at the bridgehead atoms in the clockwise or anticlockwise direction, respectively. The ***c-tr-c*** isomer (**C**) flips into an identical conformer and possesses a center of symmetry and hence is *meso*. The instability of the isomer (**D**) is due to the twist–boat conformation of the central ring. The ***c-ci-c*** isomer (**E**) possesses 1,3-*syn*diaxial methylene interaction and two *cis*-decalin interactions and is the least stable one. Interestingly, an alternate conformer of the isomer (**E**) with the central ring as a twist–boat (**E^2**) is of only slightly higher energy than the chair (**E^1**) and contributes to the extent of 13 % at 271°C. The signs of torsion angles of the two junctions at 11–12 and 13–14 bonds in the central ring are consistent with the chair form in each isomer excepting (**D**). In case of the isomer (**D**) the same sign at the ring junctions 11–12 and 13–14 in the central ring in (**D$_2$**) and in (**D$_1$**) also indicates its boat or better twist–boat conformation.

Planar str. Conformation	No. of gauche int (s) (kcal/mole), relative to **A**[a]	ΔH⁰ kcal/mole		Sym. pt. gr. (Sym.no σ)	Flipping	Resolvability (opt. active)
		Calcd.[b]	Exptl.[c]			
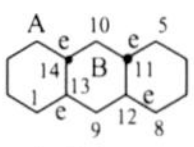 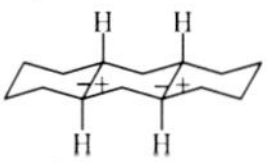 *t-ci-t* (**A**) four e rfb (meso) rfb : ring fusion bonds	0 (0)	0	0	**C₂ₕ** (2)	not possible	no (no) meso
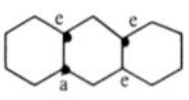 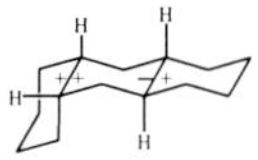 (**B**) *c-ci-t or c-tr-t* (three e rfb) (±)(chiral)	one cis-decalin, 3 (2.7)	2.62	2.76	**C₁** (1)	not possible	yes (yes)
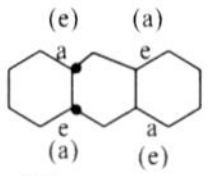 (**C**) *c-tr-c* (two e rfb) (meso) 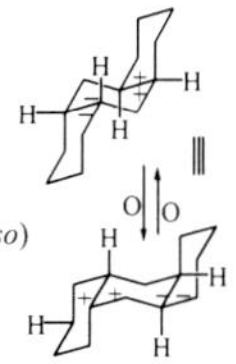	two cis-decalins, 6 (5.4)	5.56	5.58	**Cᵢ** (both forms) (1)	gives an identical form	no (no) meso
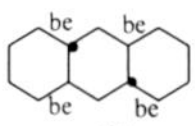 (**D**) *t-tr-t* (four boat e rfb) (±) (chiral) **D₁** 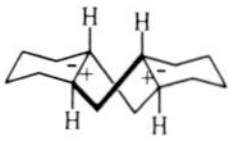	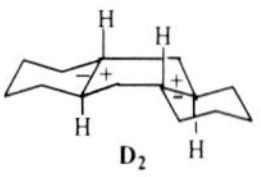 skew boat (5.6)	5.86	4.15	**C₂** (2)	not possible	yes (yes)
  (**E**) *c-ci-c* (two e rfb) (meso)	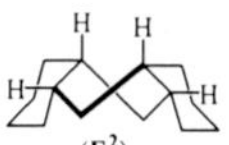 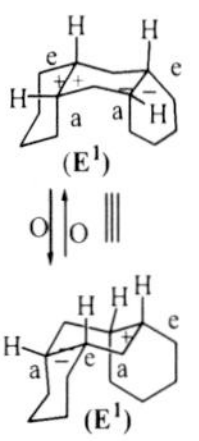 two cis-decalins + syn-diaxial int. (5.4+3.6 = 9.0) (**E₂**) is of slightly higher energy than (**E¹**)	8.13	8.74	**Cₛ** (1)	gives an identical form	no (no) meso

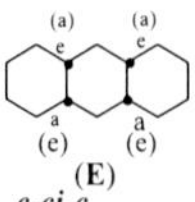

Notes: [a] ΔH⁰ calculated based on the number of gauche interactions [89]
[b] ΔH⁰ calculated using molecular mechanics (force field method) [98]
[c] ΔH⁰ determined experimentally from the composition of an equilibrium mixture of the hydrocarbons as a function of temperature [98].
[d] Both **c-ci-c** isomers of PHP (**5**) and PHA (**E**) are the least stable diastereomers of PHP and PHA respectively.

Fig. 2.140 Conformation, stability, and optical activity of perhydroanthracenes

2.16   Stereoisomerism: Axial Chirality, (*R*,*S*) Notations

Different aspects of stereoisomerism in organic compounds, having chiral centers and pseudoasymmetric centers acting as stereogenic units, have been discussed in the preceding sections. Other elements of chirality, *viz.*, axis, plane, and helicity, also act as stereogenic units. Appropriately substituted allenes, spiranes, alkylidenecycloalkanes, biaryls, and adamantoids behave as stereogenic units, due to the presence of chiral axis. Likewise, appropriately substituted *trans*-cycloalkenes, cyclophanes, and their analogs display stereoisomerism due to the presence of a chiral plane. A helix is non-superposable on its mirror image; it possesses an inherent chirality, a stereogenic unit known as helicity.

The configurational nomenclature of these stereogenic units will be illustrated briefly with examples in this and the following sections.

2.16.1   Stereochemistry of Allenes. Configurational Nomenclature

Of the three carbons of allene C1 and C3 are sp^2 hybridized and C2 is sp hybridized. The orbital picture of allene (**1**) is shown in Fig. 2.141. The shaded *p* orbitals as well as the unshaded p orbitals overlap with each other separately to form orthogonal π-bonds placing the ligands at C1 in a plane orthogonal to that of the ligands at C3. The structure (**1**) can be projected to a Newman-like projection formula (**1a**) to *R*-configuration, as shown, viewed from the left side with the front ligands in a vertical plane and the rear ligands in a horizontal plane (Fig. 2.141). The (*R*,*S*) nomenclature is independent of the direction of viewing, and the same specification will follow when viewed from the right side. The allenes of the general formula $C_{ab} = C = C_{ab}$ possess a C_2 axis, but no σ-plane, and belong to the point group C_2. If the three or four of the ligands are different as in $C_{ab} = C = C_{ac}$ or $C_{ab} = C = C_{cd}$, the C_2 axis disappears, and the molecules are totally asymmetric and possess C_1 point group. This is true for all axially chiral molecules.

For configurational nomenclature of allenes and other axially chiral molecules, the standard subrule 0 (Sect. 2.6.2), which states that "the near groups precede the far groups," is considered first, ahead of other subrules. *R,S* nomenclature in cases of some specific examples of a few optically active allenes and their enantiomers are illustrated in Fig. 2.142.

The molecule is viewed from any end of the chiral axis and Newman-like projections are drawn; the groups near the viewer are numbered 1 and 2, whereas the groups at the far end are numbered 3 and 4, following the priority sequence rule. The order 1 → 2 → 3, clockwise or anticlockwise, gives the configuration as *R* or *S*, respectively. This type of nomenclature is applicable for other types of compounds with axial chirality, to be discussed in the sequel. Interchange of the two geminal groups at any end in these molecules leads to the enantiomer.

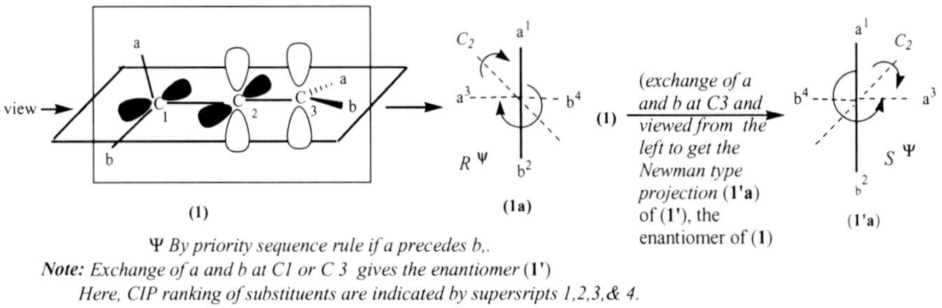

Ψ *By priority sequence rule if a precedes b,.*

Note: *Exchange of a and b at C1 or C3 gives the enantiomer (1')*
Here, CIP ranking of substituents are indicated by supersripts 1,2,3,& 4.

Fig. 2.141   Orbital picture of an allene, its Newman projection, its enantiomer, and their (R,S) nomenclatures

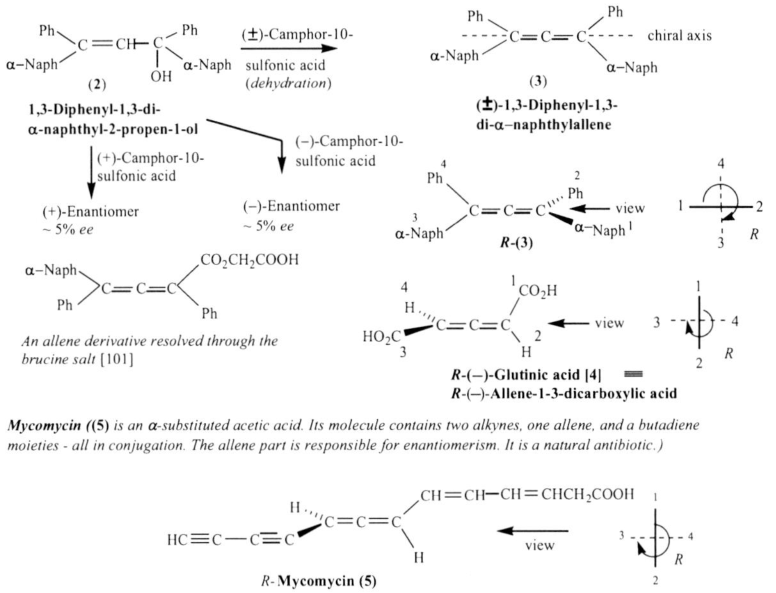

Mycomycin ((5) is an α-substituted acetic acid. Its molecule contains two alkynes, one allene, and a butadiene moieties - all in conjugation. The allene part is responsible for enantiomerism. It is a natural antibiotic.)

[#] **Nomenclature:** *One has to look along the chiral axis of the biaryl (or any axially chiral) molecule and project on a plane orthogonal to the chiral axis; now, after assigning the priority sequence 1 (or a) and 2 (or b) to the near ligands shown by a solid line, and 3 (or c) and 4 (or d) to the far ligands shown by a dotted line the (R,S) specification can be made as shown in the examples given above.*

Fig. 2.142   A few optically active allenes and their (R,S) specification

Maitland and Mills [106, 107] prepared the first optically active allene (**3**), 60 years after Van't Hoff's prediction. 1,3-Diphenyl-1,3-di-α-naphthyl-2-propen-1-ol (Fig. 2.142) was separately dehydrated with (±)-, (+)-, and (−)-camphor-sulfonic acid to give (±)-, (+)-(~5 % ee), (−)-(~5 % ee), respectively. The absolute configuration of the (+)- or (−)-enantiomer was, however, not determined.

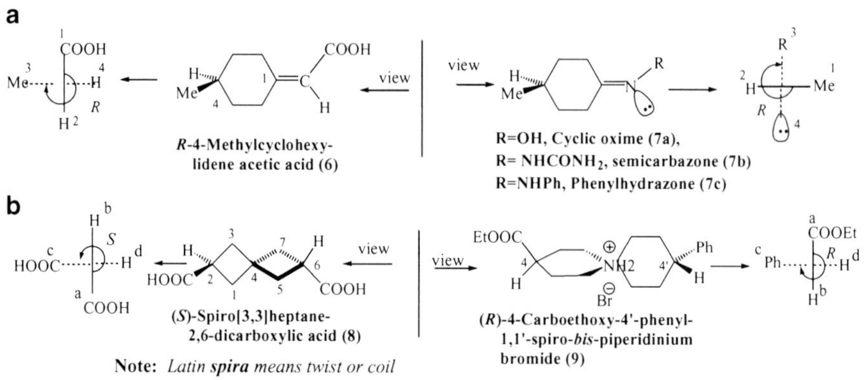

a

R-4-Methylcyclohexy-
lidene acetic acid (6)

R=OH, Cyclic oxime (7a),
R= NHCONH₂, semicarbazone (7b)
R=NHPh, Phenylhydrazone (7c)

b

(S)-Spiro[3,3]heptane-
2,6-dicarboxylic acid (8)

(R)-4-Carboethoxy-4'-phenyl-
1,1'-spiro-*bis*-piperidinium
bromide (9)

Note: Latin **spira** means twist or coil

Fig. 2.143 Configurational nomenclatures of some known resolved axially chiral compounds (**a**) derivatives of an alkalydenecyclohexane and 4-carboxycyclohexylidene imine, (**b**) two spiranes

Naturally occurring (−)-glutinic acid and the antibiotic mycomycin [108] are also examples of optically active allenes (Fig. 2.142) [109].

For the (R,S) specification of other types of axially chiral molecules the same procedure [#] is followed (see Figs. 2.143, 2.145, 2.146, 2.147, 2.151, and 2.152).

2.16.2   Chiral Spiranes and Analogs. Configurational Nomenclature

Spiranes, alkylidenecycloalkanes, adamantanes, and catenanes with appropriate substituents can be dissymmetric like allenes. The name "*spirane*," derived from the Latin *spira* meaning twist or coil, implies that spiranes are nonplanar. One of the double bonds of an allene upon replacement by a ring, such as (**6**) (Fig. 2.143), gives rise to alkyledenecycloalkanes, which are also called hemispiranes. When both the double bonds of an allene are replaced by two rings, a *spirane*, e.g., (**8**), is generated. Here, also the two terminal disymmetrically substituted methylene planes are orthogonal to each other. The compounds having pairs of nonequivalent geminal substituents at both ends will exhibit enantiomerism, and (R,S) specification is achieved in such compounds as in case of allenes (Fig. 2.143). In all such compounds if one terminal methylene is substituted by same ligands, and the other terminal methylene carries different ligands, the molecule becomes prochiral—carrying a prochiral axis. When one of the same ligands at one terminal is replaced by a different ligand, the molecule becomes chiral, and the prochiral axis in the precursor molecule becomes chiral axis in the generated chiral molecule.

Spiro[4, 4]nonane-1,6-dione (**10**) [110], *a spirane having a chiral axis as well as a chiral center,* is shown in Fig. 2.144. Configurational nomenclature of (**10**) and its

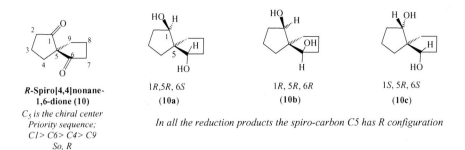

R-Spiro[4,4]nonane-
1,6-dione (10)
C₅ is the chiral center
Priority sequence;
C1> C6> C4> C9
So, R

1R,5R, 6S
(10a)

1R, 5R, 6R
(10b)

1S, 5R, 6S
(10c)

In all the reduction products the spiro-carbon C5 has R configuration

Fig. 2.144   Spiranes with one chiral center (**10**) and three chiral centers (**10a, 10b, 10c**).

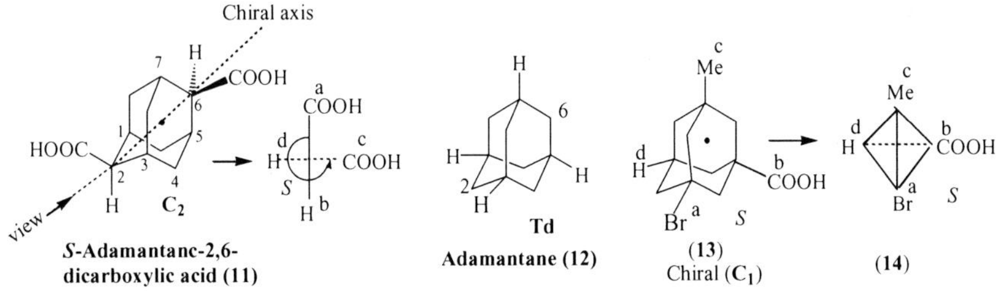

Chiral axis

S-Adamantane-2,6-
dicarboxylic acid (11)

Adamantane (12)

(13)
Chiral (**C₁**)

(14)

Fig. 2.145   Adamantane, chiral adamantanoids, and specification of chirality

three distereomeric reduction products (**10a**), (**10b**), and (**10c**) are shown in Fig. 2.144. These compounds exhibit both central chirality and axial chirality; in such cases *central chirality has precedence for configurational nomenclature.*

2.16.3   Chiral Adamantoids. Configurational Nomenclature

Adamantane-2,6-dicarboxylic acid (**11**) (Fig. 2.145) is an axially chiral compound. In it C2 and C6 methylenes are dissymmetrically substituted and exist in orthogonal planes. The adamantoid (**11**) is shown to have *S* configuration. The imaginary chiral axis in adamantoids passes through the substituted terminal carbon atoms C2 and C6 and the geometrical center of the ring system.

It may be mentioned here that if the four bridgehead carbon atoms of adamantane (**12**) bear four different ligands as in (**13**), the molecule becomes chiral. Adamantane [111] is a highly symmetrical molecule having **T_d** symmetry like CH₄. On the other hand, the adamantane derivative (**13**), carrying four different chiral centers at the four bridgeheads, becomes completely asymmetric with the point group **C₁**, and exists only as a (±)-racemate pair. The four different ligands in (**13**) form a tetrahedral arrangement like (**14**), and the chirality of the molecule may be

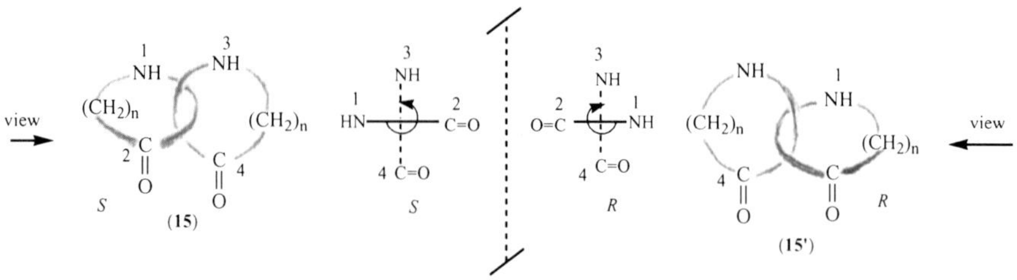

Fig. 2.146 Configurational nomenclatures of a chiral catenane and its enantiomer

associated with a center, represented by a dot, in the unoccupied space of the adamantane skeleton (Fig. 2.145). Like a centrodissymmetric compound exchange of any pair of ligands abcd in (**13**), or of the pair ab at C2 or C6 in (**11**), will lead to reversal of chirality and will give the enantiomer.

2.16.4   Chiral Catenanes. Configurational Nomenclature

The name catenane (from Latin *catena* meaning chain) was coined by Wasserman [112] to molecules of a type of unusual topology containing two or more *dissimilar* intertwined (or knotted) rings. The two rings are *to be held* with their planes perpendicular to each other as shown in (**15**) (Fig. 2.146) with regard to the arrangement of the four distinguishable groups in the chains. Configurational nomenclature may be given from similar projection formulas. Thus, the catenane (**15**) and its enantiomer (**15'**) are determined to possess *S* and *R* configurations, respectively.

2.16.5   Biphenyl Derivatives and Atropisomerism

2.16.5.1   Introduction

In biphenyls an sp^2–sp^2 pivotal single bond joins the two phenyl rings. The distance between two same side ortho Hs in unsubstituted biphenyl (**1**) is 2.90 Å (approx.) > van der Waals radii of two H's, 2×1.2 Å = 2.24 Å. Hence, the rotation about the pivotal bond is not hindered by steric factor (Fig. 2.147). A dissymmetrically substituted (at C2 and C6) phenyl group of such a biphenyl lacking a vertical plane of symmetry is two-dimensionally chiral. A planar combination of two such groups in opposite ways would lead to cisoid [(**2**), C_{2v}] and *transoid* [(**4**), C_4, C_{2h}] conformers.

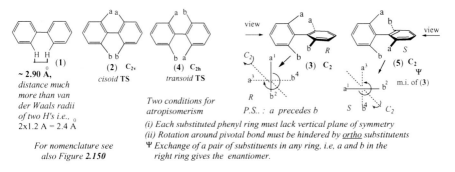

Fig. 2.147 Principle of atropisomerism. *R,S* nomenclature

On the contrary, a nonplanar combination of such groups would lead to two enantiomers **(3)** and **(5)**, each belonging to C_2 point group. When the energy barrier exceeds 19–20 kcal/mol (80–85 kJ mol^{-1}), the enantiomers are separable at room temperature. This type of enantiomerism was first discovered by Christie and Kenner in 1922 in the case of 6,6'-dinitro-2,2'-diphenic acid [113] (Fig. 2.149). Richard Kuhn later in 1933 called it *atropisomerism* and such enantiomers as *atropisomers* (from Greek *a* meaning "not" and *tropos* meaning "turn"), since such molecules do not turn around the molecular axis (due to steric hindrance).

2.16.5.2 Energy Profile Diagram

Energy Profile Diagram (approx.) is shown in Fig. 2.148 for a 360° displacement of torsion angle (ω) about the pivotal bond or showing the interplanar angle in *ortho* substituted biphenyls. The following points are to be noted:

(i) Nonplanarity of the two aromatic planes is caused by the steric demands of the *ortho* substituents. This is opposed by π electron overlap of the two rings; maximum stabilization occurs when the rings are coplanar. Even the biphenyl itself is nonplanar in the ground state (the inter-ring torsion angle being 44° in the vapor phase, but in the crystal phase the rings are coplanar because of packing forces [114].

(ii) The planar diastereomeric conformations **(2)** and **(4)** represent the energy maxima due to severe steric interference of the ortho substituents on either side; the former [*cisoid* TS, **(2)**] having similar groups on the same side has higher energy than the *transoid* TS **(4)** having dissimilar groups on the same side.

(iii) Recemization of the two enantiomers **(3)** and **(5)** takes place with greater ease through the transoid conformation **(4)** involving less activation energy than that involving the cisoid conformation **(2)**.

(iv) The bulkier the *ortho* substituents are, the higher is the energy barrier between the enantiomers.

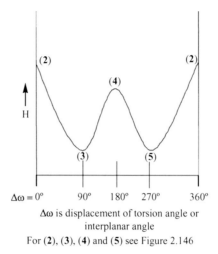

Δω is displacement of torsion angle or
interplanar angle
For (2), (3), (4) and (5) see Figure 2.146

Fig. 2.148 Energy profile diagram of *ortho* substituted biphenyl

(v) Due to complete absence of resonance stabilization at $\omega = 90°$ or $270°$, the decrease of resonance stabilization may be more than the decrease in steric repulsion (which may be even absent) at these positions and a situation may be that of a double minima when ω is around $90°$ and $270°$, and of a small maximum at ω $90°$ and $270°$.

(vi) Excess resonance energy of a biphenyl over two benzene rings is a function of the angle of twist (ω) between the phenyl groups. At even as large twists as $45°^{\tilde{}}$ there is still 50 % resonance energy.

2.16.5.3 Examples of Atropisomerism

Sufficient bulky groups must be present in *ortho* positions of biphenyls to destabilize the planar conformations due to steric repulsion and to generate atropisomers. They may be *ortho ortho* disubstituted, trisubstituted, tetrasubstituted, and even monosubstituted. Examples are given in Fig. 2.149.

2.16.5.4 Orders of Steric Hindrance and of Buttressing Effect

These are depicted in Fig. 2.150.

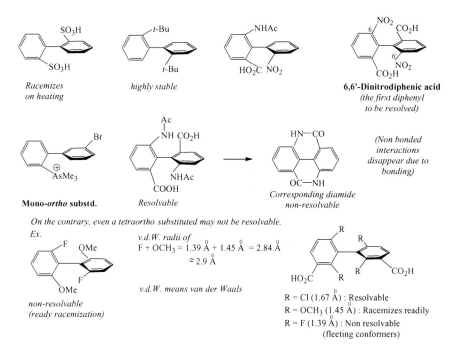

Racemizes on heating

highly stable

6,6'-Dinitrodiphenic acid
(the first diphenyl to be resolved)

Mono-*ortho* substd.

Resolvable

Corresponding diamide non-resolvable

(Non bonded interactions disappear due to bonding)

On the contrary, even a tetraortho substituted may not be resolvable.
Ex.

v.d.W. radii of
F + OCH$_3$ = 1.39 Å + 1.45 Å = 2.84 Å
≈ 2.9 Å

v.d.W. means van der Waals

non-resolvable (ready racemization)

R = Cl (1.67 Å) : Resolvable
R = OCH$_3$ (1.45 Å) : Racemizes readily
R = F (1.39 Å) : Non resolvable
(fleeting conformers)

Fig. 2.149 Resolvable (stable) atropisomers

Order of Steric hindrance : Br > Me > Cl > NO$_2$ > COOH > OMe > F > H > D
roughly corresponds to v. d. W. radii of atoms or groups

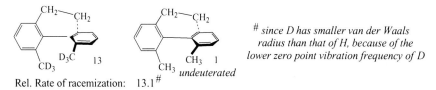

Rel. Rate of racemization: 13.1$^{\#}$

 # since D has smaller van der Waals radius than that of H, because of the lower zero point vibration frequency of D

***Buttressing Effect:**

Rate of racemization of (**A**) is much lower than that of (**B**) due to the buttressing effect of the bulky NO$_2$ group in (**A**) adjacent to the OMe group.

**Order:* NO$_2$ > Br > Cl > Me (different from bulk order)

Fig. 2.150 Order of steric hindrance and buttressing effect on configurational stability of biphenyls

2.16.5.5   Configurational Nomenclature of Chiral Biphenyls (R,S or aR, aS)

Selection rules (new convention*, 1966 [5]) for axial chirality has been introduced in case of atropisomers.

Subrule 0: Proximity rule: Groups about the near end of the chiral (molecular) axis precede over groups about the far end (same as in other axially chiral molecules).

a*Ordering of groups: Only the four atoms C2, C6, C2′, and C6′ which correspond to the four vertices of the *elongated tetrahedron* and which contribute more to the chirality properties of the molecule are considered for sequencing as shown in Fig. 2.151.

c. **P, M-Nomenclature.** Alternatively, molecules with chiral axes may be viewed as helicenes (since they resemble the helicenes to be discussed in the sequel). Their configuration may be designated as *P* or *M*. If the turn from the higher priority front ligand "a" to the higher priority rear ligand "c' is clockwise, the configuration is *P*, and if anticlockwise it is *M* (see Fig. 2.151). In general, *aR* corresponds with *M* and *aS* with *P*. The compound **A** possesses *aS* or *P* configuration and **B** possesses *aR* or *M* configuration according to the new convention introduced in 1966 [5]. According to the old convention published in 1956 [8], the nomenclature is the opposite in these cases.

2.16.5.6   Some Interesting Examples of Axially Chiral Molecules Exhibiting Atropisomerism

A few examples are illustrated in Fig. 2.152. Molecules of the type in which one planar ring is replaced by a dissymmetrically substituted trigonal atom (two-dimensionally chiral) may exhibit atropisomerism if sufficient steric hindrance exists around the pivotal bond. A substituted stilbene (**A**) has been resolved. In the substituted naphthylamine (**B**), the peri nitro group gives rise to restricted rotation around the pivotal bond (chiral axis). Their (R,S) nomenclatures are also illustrated (Fig. 2.152).

Again, of the two oximes of 1-acetyl-2-hydroxynaphthalene-3-carboxylic acid, the *E*-isomer (**C**) (Fig. 2.152) is not resolvable (since the restriction around the pivotal aryl-carbon bond is not sufficient), whereas the *Z*-isomer (**D**) is resolvable because of restricted rotation around the pivotal bond.

Natural products possessing axial chirality have been reported, *e.g.*, natural dimeric coumarins, desertorin A, B, and C and triumbelletin (Fig. 13.71) and colchicine alkaloids (Fig. 24.1). Additionally, many asymmetric catalysts also possess axial chirality. Configurational nomenclature has been specified for each such compound (*vide* Figs. 6.14, 6.37, 6.38, 7.66, 7.67, 7.69, 13.56, 13.71, 17.10, 18.9, and 24.1).

a Ordering of groups

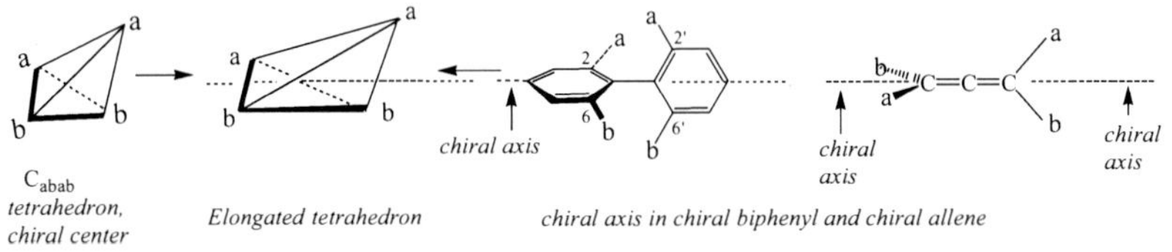

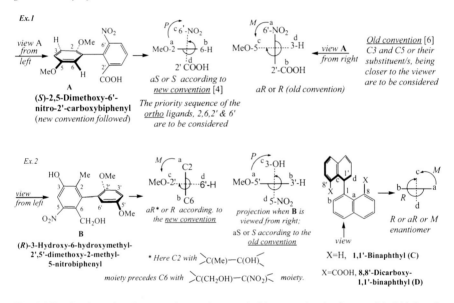

C_{abab}
tetrahedron,   *Elongated tetrahedron*   *chiral axis in chiral biphenyl and chiral allene*
chiral center

Ordering starts from the first off axis chiral biphenyl atoms (near the observer) i.e., 2nd and 6th atoms of the front ring.
- The priority sequence is fixed by the four ortho carbons after due complementation for quadriligancy proceeding along the branch of higher priority.
- molecule viewed from either end leads to the same configurational descriptor. *R* or *S*.

b. Nomenclature : *One has to look along the chiral axis of the biaryl molecule and project on a plane orthogonal to the chiral axis; now, after assigning the priority sequence 'a' and 'b' to the near ligands shown by a solid line, and 'c' and 'd' to the far ligands shown by a dotted line the (R,S) specification can be made as shown in the following examples. In many cases the nomenclature remains same according to new and old conventions. Here we are citing two examples where nomenclature changes according to the convention. Viewing from left or right gives the same specification.*

Fig. **2.151** Configurational nomenclatures (new and old conventions) of some chiral biphenyls and binaphthyls

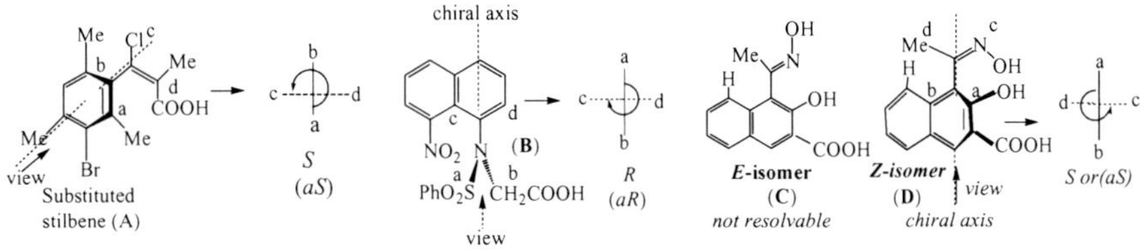

Fig. **2.152** Atropisomerism in acyclic analogs of biphenyl and in oximes

2.17   Planar Chirality

2.17.1   Introduction

A molecule possessing a chiral plane exhibits planar chirality. A chiral plane is not definable as easily as a chiral center or axis. A chiral plane contains as many of the atoms of the molecule as possible. Chirality is due to the fact that at least one ligand (usually more) is not in the chiral plane.

Moelcules with planar chirality include *ansa* compounds, *paracyclophanes*, *metacyclophanes* (Fig. 2.153), and a few *trans-cycloalkenes*. In the enantiomers the methylene chain is on either side of the aryl ring or C=C bond. The interconversion between the enantiomers is prevented by the inability of the alicyclic ring, being too small, to swing from one side to the other of the aryl or olefinic plane.

2.17.2   The (R,S) Specification of Planar Chirality

The specification of these compounds (Fig. 2.153) is done by application of the following *selection rule*.

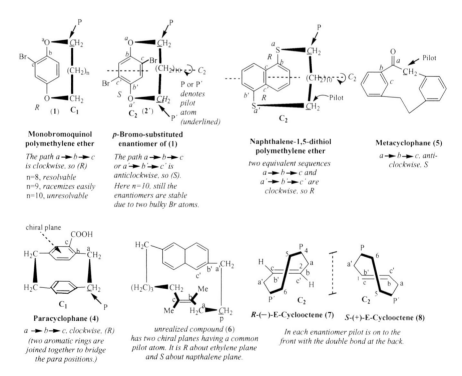

Monobromoquinol polymethylene ether

The path a→b→c is clockwise, so (R)
n=8, resolvable
n=9, racemizes easily
n=10, unresolvable

p-Bromo-substituted enantiomer of (1)

The path a→b→c or a'→b'→c' is anticlockwise, so (S).
Here n=10, still the enantiomers are stable due to two bulky Br atoms.

Naphthalene-1,5-dithiol polymethylene ether

two equivalent sequences
a→b→c and
a'→b'→c' are
clockwise, so R

Metacyclophane (5)

a→b→c, anti-clockwise, S

Paracyclophane (4)

a→b→c, clockwise, (R)
(two aromatic rings are joined together to bridge the para positions.)

unrealized compound (6)

has two chiral planes having a common pilot atom. It is R about ethylene plane and S about napthalene plane.

R-(−)-E-Cyclooctene (7)   S-(+)-E-Cyclooctene (8)

In each enantiomer pilot is on to the front with the double bond at the back.

Note: (1), (2) and (3) are *ansa* compounds; in Latin *ansa* means handle.

Fig. 2.153 Some compounds of planar chirality and their (R,S) specification

Selection rule [8]: The most preferred atom directly bound to atoms in the chiral plane is selected as the *pilot atom* (spectator point). It is the first out-of-plane atom linked to the sequence-preferred end of the chiral plane. The *sequencing starts with the first in-plane atom directly bound to the pilot atom* (the underlined C) and going along the in-plane sequence (marked as a, b, and c) involving the more preferred atom at each branch. The order in which a, b, and c appear when seen from the pilot atom, specifies the absolute configuration, *i.e.*, *R* for clockwise and *S* for anticlockwise order (see Fig. 2.153).

The compounds of planar chirality: (**1**), (**2**), and (**3**) are *ansa* compounds, (**4**) and (**5**) are cyclophanes, (**6**) is an imaginary compound containing two chiral planes having a common pilot atom. Cyclooctene is the smallest ring which can accommodate a *trans* double bond. Two enantiomeric configurations (**7**) and (**8**) are possible.

The interconversion of the two enantiomers (**7**) and (**8**) of *trans* or *E*-cyclooctene requires the swinging of the tetramethylene chain over and below the plane of the double bond (chiral plane) which is opposed by angle strain (ring strain). The two enantiomers have been resolved by Cope and coworkers [115, 116] who determined the absolute configuration of the enantiomers [117]; the (−)-enantiomer has been shown to be *R* and the (+)-enantiomer *S*. The molecule possesses C_2 point group, having only the C_2 axis passing through the center of the double bond and bisecting the 5–6 bond. In the higher homologues increase in the mobility of the polymethylene chain decreases the rotational barrier. Thus, *trans*-cyclononene exists in enantiomeric forms only at −80 °C, and *trans*-cyclodecene is an extremely mobile system.

2.18   Helicity and *P,M*-Designation

A helix is inherently chiral and may be considered as displaying axial chirality—its axis serving as the chiral axis. However, it is more convenient to discuss such type of chirality as helicity. A helix possessing a C_2 axis orthogonal to the chiral axis is called *palindromic*, and it looks same from either end. If a helix moves from one end to the other in a clockwise direction, it is designated *P* (plus) and if it moves in an anticlockwise direction, it is designated *M* (minus) (Fig. 2.154). It is known that a polypeptide chain derived from natural L-amino acids often coil to form an α-helix with *P* helicity, but it is not palindromic since its two ends are nonequivalent and hence it lacks a C_2 axis. Its point group is C_1. The nucleic acids present in DNA or RNA possess the skeleton of a double *P*-helix, preserved by strong intramolecular H-bondings [118].

Molecular overcrowding [119] is exemplified by hexahelicine (**A**) [120] and 4,5-disubstituted phenanthrenes as in (**B**) [121], (**C**) [122], and (**D**) and (**E**) [123] (Fig. 2.154).

The ring structures, due to overcrowding, assume helicity since the two terminal rings, as in (**A**), or the substituents at 4 and 5 positions of phenanthrene, as in (**B**) to (**E**), are large enough to prevent their existence in the same plane as that of the aromatic rings, thus giving rise to this type of optical isomerism, due to molecular asymmetry [118] arising out of helicity. With respect to some parts of the central

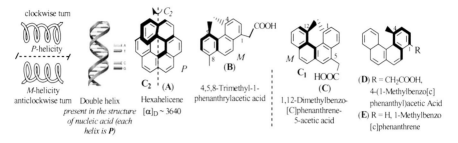

Fig. 2.154 Helical structures. Helicity due to molecular overcrowding

rings, the ring planes on the upper right side, as written in the Figure, are gradually bent up, and the ring planes on the upper left side are gradually bent down by small angles, ($10°–14°$) in case of P helicity, as have been evident from X-ray diffraction analysis [118]. In case of M-helicity the ring planes will bend just in opposite direction. Thus, in each case two helical enantiomers exist. The synthetic racemates (**A**) to (**E**) have been resolved. The P, M nomenclature and the point groups of these molecules are depicted in Fig. 2.154.

Hexahelicene having very high rotation, $[\alpha]_D$ ~3,640, racemizes slowly at its mp 266 °C. Higher helicenes readily racemize at their mps (about 200°) [124]. The absolute configuration, however, has not been assigned. As shown earlier in Sect. 2.16.5.5c, chiral biphenyls may also be viewed as helicenes and designated as P or M (see Fig. 2.151). For the relationship between the sign of the CE and absolute configuration of helicenes, see Sect. 2.19.6.2.

2.19   Chiroptical Properties. Optical Rotation. ORD, CD [125–127]

2.19.1   Origin of Optical Rotation. Circular Birefringence, Its Effect

Optical rotation and optical isomerism have been briefly discussed in Sect. 2.2.8. Now we would dwell on the origin of optical rotation. The electric field associated with the light wave in ordinary radiation oscillates in all directions perpendicular to the direction of propagation along the z-axis. Such radiation is isotropic or unpolarized. If the radiation is filtered to remove all oscillations other than in one direction, say in the xz plane, then the light becomes linearly polarized, since the projection in xy plane is linear, and the light becomes anisotropic.

When a plane-polarized (better called linearly polarized or LP) monochromatic light wave passes through a dissymmetric medium (nonracemic sample of a chiral substance), the plane of polarization rotates giving an optical rotation—a chiroptical property exhibited by nearly all chiral molecules. Two enantiomers exhibit optical rotations, equal in magnitude, but opposite in sign.

The electrical field vector (E) of an LP wave oscillates in the xy plane along the direction of propagation (z-axis). The LP wave is the resultant of two chiral components—a right circularly polarized (RCP) ray and a left circularly polarized (LCP) ray, whose projections on the xy plane are circles (Fig. 2.155).

In a symmetric or isotropic medium, the two components RCP and LCP travel at the same velocity. The resultant of the two rays in phase constitutes the linearly polarized light.

RCP ray + LCP ray = LP ray.

These rays make diastereomeric relationship with the two enantiomers and so interact differently. When the linearly polarized light is passed through a nonracemic sample of a chiral or dissymmetric compound (or medium), the velocities of LCP and RCP rays become different; the two circularly polarized rays will have different refractive indices, $viz.$, n_L and n_R, being the refractive indices of left and right circularly polarized rays; this is known as $circular$ $birefringence$. If $n_L > n_R$, $V_R > V_L$ since $n = V_O/V$, V_O and V are velocities in vacuum and the medium, respectively; thus, the plane of polarization or the LP wave will rotate toward right, i.e., the observed rotation α will be positive. The compound is said to be dextrorotatory. Thus, if $n_R > n_L$, $V_L > V_R$ and hence the observed rotation α will be levorotatory.

The observed rotation changes continuously as the LP proceeds (α varies directly with length l). Optical rotation due to circular birefringence for 1 cm path length is given by the expression $\alpha = (n_L-n_R).$ π/λ radian $= \Delta n$ $180/\lambda$ degree. The specific rotation at a wavelength λ may be given by the expression $[\alpha]_\lambda$ in degrees $= \alpha \times 10/c$ for 1 dm path length, c = concentration in g/ml. The optical rotation of a chiral compound is usually reported as the specific rotation at sodium D line (589 Å), $i.e.$, as $[\alpha]_D$. The molar rotation at a wavelength λ, $[M]_\lambda$ or $[\Phi]_\lambda = ([\alpha]_\lambda M)/100$ (cf. Sect. 2.2.9).

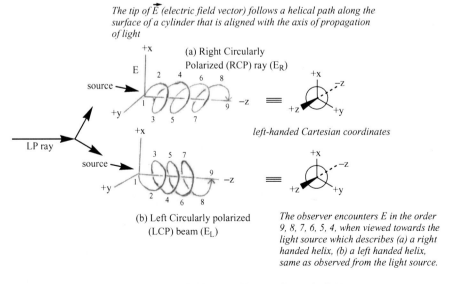

The tip of $\vec{E}$ (electric field vector) follows a helical path along the surface of a cylinder that is aligned with the axis of propagation of light

(a) Right Circularly Polarized (RCP) ray (E_R)

left-handed Cartesian coordinates

(b) Left Circularly polarized (LCP) beam (E_L)

The observer encounters E in the order 9, 8, 7, 6, 5, 4, when viewed towards the light source which describes (a) a right handed helix, (b) a left handed helix, same as observed from the light source.

Fig. 2.155 LP wave is the resultant of RCP and LCP rays (E electric field vector)

2.19.2   *Optical Rotatory Dispersion. Plain Curve*

The specific rotation $[\alpha]$ of a chiral compound depends upon the wavelength of the monochromatic light wave. The measurement of specific rotation as a function of wavelength is called optical rotatory dispersion (ORD). Specific rotations of chiral compounds are generally reported for sodium D-line (589 nm wavelength), which is quite far from UV region. Consequently, the values of specific rotations are much lower than those recorded near the UV absorption maximum; if $[\alpha]_D$ is too low to be detected, and it is not possible to ascertain whether such a compound is dextrorotatory or levorotatory, the enantiomer may be mistaken as a racemic variety. In old literature the $[\alpha]_D = \pm 0$ has been assigned to such chiral compounds whose rotation at 589 nm could not be measured.

Specific rotations of such compounds having very low $[\alpha]_D$ values or any chiral compound having no UV absorption if measured at shorter wavelengths undergo many fold increment as one approaches the absorption maximum below 210 nm (due to $\sigma \rightarrow \sigma^*$ transition). Such an ORD curve is called a plain curve (Fig. 2.156). A plain curve results when measurement is done at wavelengths away from the absorption maximum (λ_{max}), and only circular birefringence is operative. By measuring a plain curve one can assay in a precise way the enantiomeric purity of a natural product or a bioorganic compound having no UV absorption. A chiral compound, if racemic, remains inactive throughout the measureable wavelength.

A plain curve for certain molecules may or may not cross the zero rotation axis. Thus, the *ortho* isomer of α-(iodophenoxy)-propionic acid shows levorotation at the D-line whereas the *meta* and *para* isomers show $[\alpha]_D$ positive, although all three isomers have the same absolute configuration (Fig. 2.156) [125]. Each of the three position isomers shows plain positive curve; the observed positive rotation increases with decrease of wavelength. In case of negative plain curve the observed negative rotation increases with decrease of wavelength.

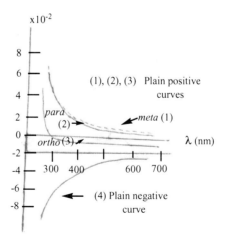

Fig. 2.156   ORD curves α-(*m*-, *p*-, and *o*-iodophenoxy)-propionic acids (*from C. Djerassi "Optical Rotatory Dispersion"*, **1960**, *McGraw-Hill Book Company*). Plain positive and negative curves

2.19.3 Circular Birefringence and Circular Dichroism. Cotton Effect

It has already been mentioned that when the LP rays of different wavelengths pass through a medium having a λ_{max} in the region studied, in addition to circular birefringence, the intensity of emergent light diminishes due to absorption. The two circularly polarized light rays, RCP and LCP, are absorbed to different extents; this phenomenon is termed as circular dichroism. Both these phenomena are wavelength dependent. Absorptions of RCP and LCP rays are maximum at the absorption maximum (λ_{max}). The refractive index increases with the decrease of the wavelength but rapidly falls to a minimum in the region of absorption showing abnormal dispersion. At the absorption maximum there is no effect on refractive index ($n = 1$ for both RCP and LCP rays). Figure 2.157 shows the variation of Δn ($= n_L - n_R$), refractive indices dispersion (I) of LCP and RCP, and of absorbance (A) (II) over a range of wavelength (including the absorption maximum, λ_{max}); Fig. 2.158 shows ORD curve with positive Cotton effect (III) and a positive CD spectrum (IV).

If an optically active compound absorbs in the UV or visible region, the two components RCP and LCP of the LP will be absorbed to different extents (anisotropic absorption), and this phenomenon of differential absorption is called *circular dichroism*, usually abbreviated as CD. The differential absorption is caused by electronic (along with vibrational) transition of energy states associated with a chiroptic chromophore in a chiral molecule. Let A_L and A_R represent the absorbances of LCP and RCP rays, respectively, then in case of CD $A_L \neq A_R$. According to Lambert–Beer's law $A_L - A_R = \Delta A = \Delta \varepsilon c l$, where $\Delta \varepsilon$ represents $\varepsilon_L - \varepsilon_R$, ε_L and ε_R are the molar absorption coefficients for LCP and RCP rays, respectively, c is the

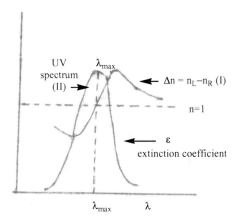

Fig. 2.157 Variation of Δn with λ (curve I). Variation of extinction coefficient (ε) with λ (UV spectrum, curve II)

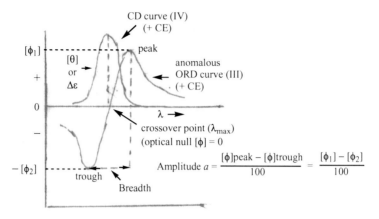

Fig. 2.158 Anomalous ORD curve with a positive CE and molar rotation $[\varphi]$ as the ordinate (curve III) and a positive CD spectrum with molar ellipticity $[\theta]$ as the ordinate (curve IV)

concentration in moles per liter (mol L^{-1}), and l is the path length per cm. The term $\Delta\varepsilon$ is called *differential dichroic absorption*. If $A_L > A_R$ its electric field vector E_L will be smaller than E_R; the tip of the resultant vectors will trace a flattened helix whose projection on xy plane is an ellipse. Such a ray is called elliptically polarized (EP), characterized by a major axis and a minor axis. The elliptical polarization may be right handed or left handed. The major axis of the ellipse traces the angle of rotation α due to also the unequal velocity of RCP and LCP, and the ellipticity ψ is defined by $\tan\psi = b/a$, where b and a are the minor axis and major axis, respectively, of the ellipse. For small difference of A_L and A_R ψ for 1 cm is given by the simplified equation $\psi = {}^1\!/_4\,(A_L - A_R)$. In general, the major axis $a = (A_L + A_R)$ of this ellipse is much greater than the minor axis $b = (A_L - A_R)$ and may be treated as a plane polarized light for the purpose of measurement. Elliptically polarized light is the most general form of polarized light. In special cases like for linear polarized light, the eccentricity of the ellipse, $(a - b)/a$ becomes 1 since $b = 0$, and for circularly polarized light eccentricity becomes 0 since $a = b$.

Like in case of rotation, a *specific ellipticity* $[\psi]$ and a *molar ellipticity* $[\theta]$ may be defined as shown below:

$[\psi] = \psi/c\,l$ in 10^{-1} deg cm^{2}g^{-1} and $[\theta] = [\psi]\,M/100$ in 10 deg cm^2 mol^{-1} (*cf.* Fig. 2.13),

where the symbols c, l, and M denote concentration in g/ml, length of the cell in dm, and molecular weight respectively, as in cases of specific rotation and molar rotation (Sect. 2.2.9), (Fig. 2.13).

Using a similar procedure as in optical rotation (Sect. 2.19.2), the specific ellipticity of the chiral medium may be defined for 1 dm path length and concentration c in g/ml as follows:

$[\psi]_\lambda^T$ in degrees $= (\psi.1{,}800)/c\pi$.

Ellipticities and *molar ellipticities* depend upon the temperature, wavelength, and concentration of the sample which should always be specified.

The combination of circular birefringence and circular dichroism gives rise to an important chiroptical phenomenon, namely the Cotton effect. In contrast to the plain curve, an anomalous ORD curve exhibits a peak (maximum) and a trough (minimum). This anomaly is called Cotton effect. The anomalous ORD arises from the superposition of two anomalies, *viz.*, the anomalies of n_L and n_R giving rise to the anomaly of Δn (Fig. 2.157). The Cotton effect is called positive when the rotation first increases with the decrease of wavelength and negative when the rotation magnitude decreases with decrease of wavelength. In other words in an anomalous ORD curve the Cotton effect is said to be positive if the peak is at a higher wavelength than the trough (Fig. 2.158). Conversely, the Cotton effect is termed negative if the trough is at a higher wavelength than the peak. The optical null (crossover point) closely corresponds to the ε_{max} of the UV spectrum in the absence of superposition of two or more close lying electronic transitions.

The *molar amplitude* of the ORD curve is expressed by the expression $a = ([\phi]_1 + [\phi]_2)/100$, where $[\phi]_1$ and $[\phi]_2$ are the absolute values of the molar rotations at the first and second extrema (peak and trough, respectively, in Fig. 2.158 displaying positive Cotton effect). The difference of the wavelengths of the two extrema is called the *breadth* (Fig. 2.158) of the ORD curve.

The chiroptical phenomenon CD is usually observed in the nontransparent regions of the spectrum due to conjugated chromophores of strong absorption with high extinction values. In such regions circular birefringence is not operative.

In the CD spectrum showing positive Cotton effect, only a peak appears near the λ_{max} of the chromophore in the molecule (Fig. 2.158) and conversely, a trough appears in the CD spectrum showing negative Cotton effect. The CD spectra are inherently simpler to interpret. In fact CD has been essentially replacing ORD as the main chiroptical technique in the study of chiral compounds [128].

Classification of chromophores. By ORD and CD measurements, the chromophores may be classified into two broad categories,

1. Chromophores that are *inherently achiral* such as the functional groups and the corresponding λ_{max} in nm in the parenthesis, as follows:

   Ketone (280–300), α,β-unsaturated ketone (330–360, 230–260), carboxylic acid (215–220), α,β-unsaturated acid (~250), ester (215–220), lactone (215–235), conjugated diene (~270), substituted phenyl (250–280), sulfoxides (~210), nitro compound (~270), amides and lactam (220–235).

   Each of the above chromophores when considered separated from the rest of the molecule contains at least one mirror plane and hence are inherently symmetrical but exhibit CE due to asymmetric perturbation exerted by chiral surrounding or by the molecular skeleton itself, *e.g.*, a carbonyl group (of local C_{2v} symmetry) in a steroid molecule.

2. Chromohores that are *inherently chiral*. In compounds of this category the chirality is built into the chromophore. Examples are helicenes and chiral compounds having skeletons like biaryls, allenes, spiro compounds, cycloalkenes, twisted 1,3-dienes, etc.

Applications of ORD and CD Curves with Cotton Effects *Some empirical rules.*
Many chiral natural products are ketones or contain groups which can be converted
into ketones (*e.g.,* secondary OH). The carbonyl $n \rightarrow \pi^*$ transition absorbs at 280–
300 nm, relatively free from the interference of other chromophores. Moreover, the
extinction coefficient is low ($\varepsilon \approx 20$–80 near 290 nm) and light is transmitted even at
λ_{max} leading to easy measurement of ORD and CD. Several empirical rules have
been formulated theoretically to correlate the sign of the Cotton effect with its
chiral environment. Several such rules are discussed in the sequel.

2.19.4 The Axial Haloketone Rule and Its Applications

The axial haloketone rule [125, 129] as a special case of the more general octant
rule (*vide infra*). The rule is applicable to an axial α-halocyclohexanone moiety (the
halogen is Cl, Br or I, but not F). One should view along the O=C bond in the
direction of the ring with the carbonyl carbon placed at the "head" of the chair
(or boat). If the axial α-halogen (X) appears at the right of the line of view
(Fig. 2.159), the compound will show a strong positive CE; if it appears at the
left, a strong negative CE will be observed. Presence of axial α-halogen also causes
(i) a bathochromic shift of the CE or of the first extremum at higher wavelength
(like the λ_{max} in the UV spectrum), (ii) an increase in the amplitude of the CE, and
(iii) may cause the inversion of the sign of the CE with respect to the parent
cyclohexanone, depending upon the configuration of the chiral atom bearing the
halogen atom. The CE sign of the parent compound with known absolute configuration and conformation without any halogen atom can be ascertained by use of
the octant rule or by actual experiment.

   The sign of the CE depends on the (1) *structure*, (2) *conformation*, and (3) *configuration* of the halocyclohexanone in the neighborhood of the carbonyl group. If
any two of these factors and the sign of the CE are known, the third factor will be
determined. In case, these three factors are known, the sign of the CE can be
predicted. Axial fluorine α-substituents have been shown to have opposite effects
when compared with other halogen atoms at axial α-position.

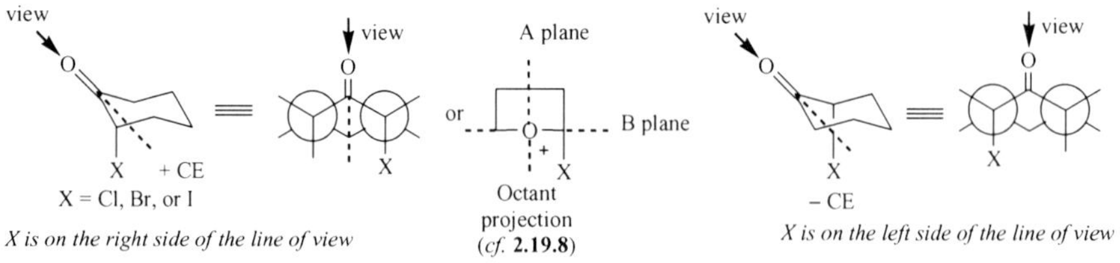

Fig. 2.159 Axial haloketone rule

Based on studies on steroidal ketones, it was revealed that equatorial α-halogen (or acetoxy) substituents on either side of the CO group have little effect on the ORD curve; the sign of the CE at ~300 nm remains unchanged relative to that found for the unsubstituted parent ketone. In cases where the α-halogen is equatorial, no significant bathochromic shift or increase in the amplitude of the CE is observed. This is also evident from the octant projection diagram (*vide infra*) as the equatorial α-substituents fall in the *yz* (B) plane and have no contribution to the sign of the CE.

The axial haloketone rule has been later expanded to include SR, SO₂R, NR₂, and other substituents [130].

2.19.4.1   Position of the Halogen Substituent

Two examples of kinetically controlled bromination of nonracemic samples of chiral compounds (**A**) and (**B**) are provided in Fig. 2.160; in each case the constitution and conformation of the parent ketone are known. The bromination product of the compound (**A**), the α-bromo derivative, exhibited negative CE establishing that axial bromination took place at C5. In case of compound (**B**) negative CE of the axial β-bromination took place at C11 [125] (p 123).

2.19.4.2   Absolute Configuration by Comparison Method

ORD and CD curves showing CEs are extremely useful for determining the absolute configuration by a comparison method: a dispersion curve of a compound of unknown configuration is compared with some compounds of similar structure of known absolute configuration. Some examples have been cited in the sequel. Here, one example is given in Fig. 2.161. The absolute configuration of (−)-*cis*-1-decalone (Ia) and *trans*-1-decalone (IIa) were established by comparing their ORD curves with those of their 10-methyl analogs (Ib) and (IIb) of known absolute configuration [131]. The flipped conformer (steroid form), if present, must have minor contribution in case of **1b** and hence of **1a**.

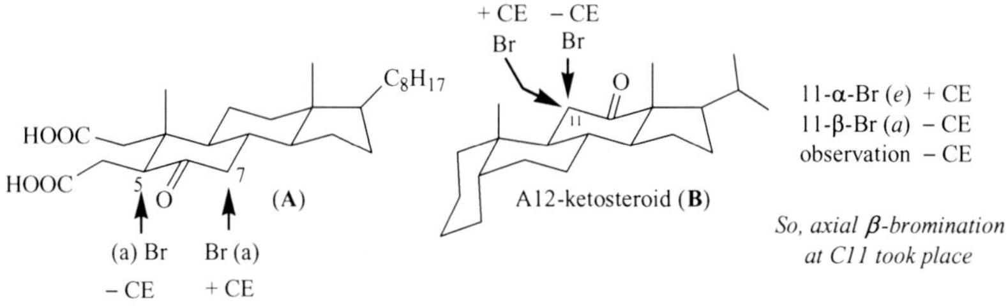

Note: *Axial orientation of bromine in the compounds established by UV and/or IR.*

Fig. 2.160   Position of bromine substituent in (**A**) and (**B**) by use of axial haloketone rule

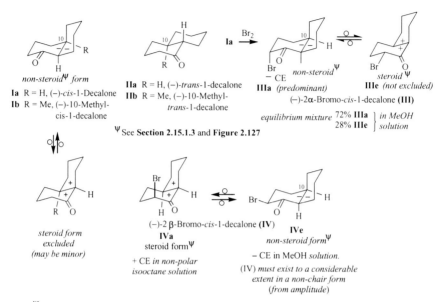

$^\Psi$See **Section 2.15.1.3** and **Figure 2.127**

Note: $^\Psi$*The steroid and non-steroid forms seem to be designated [123] in the opposite way by oversight*

Fig. 2.161 Absolute configurations of **Ia**, **IIa** by comparison method and of **IIIa**, **IVa** from axial haloketone rule, and solvent effect

2.19.4.3 Absolute Configuration by Axial Haloketone Rule. Conformational Mobility

Kinetically controlled bromination of $(-)$-*cis*-1-decalone (**1a**) afforded in pure form all six possible monobromo-*cis*- and *trans*-1-decalones whose ORD curves were fully consistent with the assigned structures [131]. In Fig. 2.161 only the nonsteroid form **IIIa** of the 2α-bromo (axial) derivative and the steroid form (**IVa**) of the 2β-bromo (axial) derivative and their CE by application of axial haloketone rule are shown. It should be noted here that in case of **IVa**, the carbonyl carbon being not at the head or upper layer of carbons, but at the lower layer of carbons of the cyclohexanone ring, and the axial halogen being present at the left of the line of view through O=C (**IVa**) exhibits + CE. The structure **IVa** if turned upside down, the α-axial bromine will be at the right side of the line of view, consistent with the positive sign of the CE.

Measurement of the ORD curves of $(-)$-2α-bromo-*cis*-1-decalone (**III**) in solvents of different polarities demonstrated the existence in methanol solution of a 70–30 equilibrium between the nonsteroid (**IIIa**) and steroid (**IIIb**) conformations of the ketone (**III**) (Fig. 2.162). The existence of the $(-)$-2β-Bromo-*cis*-1-decalone predominantly as the axial conformer (**IVa**) in nonpolar isooctane solution showing + CE and as the equatorial conformer (**IVe**) in methanol solution to a

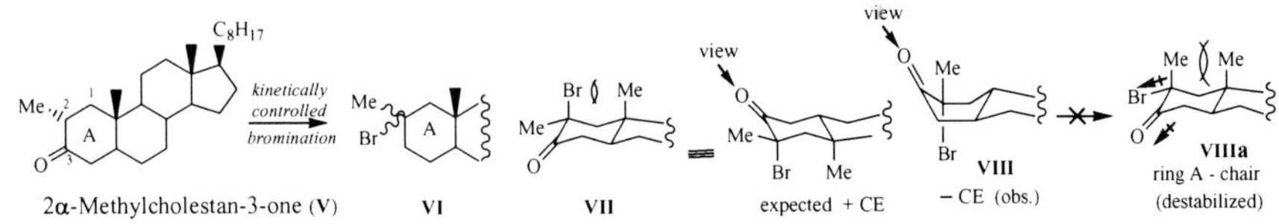

Note: *In VII the steroid rings including ring A has been inverted to bring carbonyl carbon at the head of the chair.*

Fig. 2.162   Demonstration of boat form by application of axial haloketone rule

considerable extent showing –CE have also been demonstrated. The low amplitude of the ORD of **IV** in methanol solution (compared to that of **III** in methanol) also indicates its existence to a considerable extent in a non-chair form.

2.19.4.4   Boat Form of Ring A of a Steroid Bromoketone

The axial haloketone rule is also applicable to cyclohexanones existing in a boat conformation, the rule being same as in the case of chair form.

Kinetically controlled bromination of 2α-methylcholestan-3-one (**V**) or its enol acetate leads to 2-bromo-2-methylcholestan-3-one (**VI**), whose spectral properties [λ_{max} (CHCl$_3$) 5.84 μ, λ_{max} 313 nm] indicate an axial bromine atom [132] (Fig. 2.162). The normally expected 2β-bromoketone in the chair form **VII** should exhibit a strong + CE, contrary to the observed—CE. The only structure which conforms to the—CE is the structure **VIII**, a diastereomer of **VII** with ring A in the boat conformation. Bromination takes place by an axial-like electrophilic attack by Br$^+$ from the α-face through the skew–boat transition state avoiding the syn-axial interaction with 10β-methyl, to form **VIII**. The latter structure does not flip to the chair form **VIIIa** with equatorial Br which is destabilized by 1,3-dimethyl syn-axial interaction and also by the unfavorable dipole–dipole interaction between the equatorial C–Br bond and the carbonyl group.

2.19.5   The Octant Rule and Its Applications

The octant rule is an empirical generalization relating the sign of the Cotton effect of the carbonyl chromophore of saturated cyclohexanones at around 290 nm with the configuration of the chiral centers present in the neighborhood of the chromophore. The octant rule was first formulated by Mislow et al. [133] and Moffitt et al. [134] in 1961. Consider a carbonyl group (say of a cyclohexanone having a dissymmetric part) lying along the *z*-axis, with its attached groups C2 and C6 lying in plane *yz* (plane B), of a left-handed Cartesian coordinate system of two other mutually perpendicular planes *xy* (plane C) and *xz* (plane A), the midpoint of the CO group being at the origin (Fig. 2.163). These are nodal and symmetry planes of

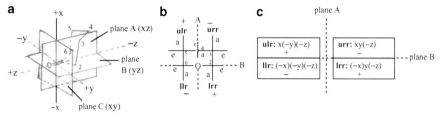

Note: *The projections of A plane and B plane on xy plane are x-axis and y-axis respectively.*

Fig. 2.163 (**a**) Stereoprojection of the cyclohexanone ring (chair form) in the octant projection diagram (opd) (**b**) Projection of cyclohexanone bonds on the *xy* (or C) plane. View facing the CO oxygen with signs of rear octants (**c**) Sign of contribution of a perturber in rear (or back) octants

the orbitals involved in the relatively weak $n \rightarrow \pi^*$ transition of the carbonyl absorption. Thus, it is expected that the effect of an atom at any point P (x, y, z) in inducing asymmetry in the electronic process associated with the CO absorption (270–310 nm) will be equal in magnitude and opposite in sign to that induced by the same atom situated at the reflection of P through any of the planes A, B, and C and will vanish identically when P lies in any one of these planes (the product xyz being zero) (Fig. 2.163). The effect is characteristic of the perturbing atom and is additive of the other atoms.

The coordinate system divides space around the CO group into eight octants or sectors. The principal premise of the very simple octant rule is that the contribution to the sign of the $n \rightarrow \pi^*$ Cotton effect which a given atom at a point P (x, y, z) makes to anomalous rotary dispersion will be determined by the simple product xyz of its coordinates [(C) in Fig. 2.163].

For example: contribution (to the Cotton effect) of an atom located in the lower right rear (**lrr**) octant (Fig. 2.163), whose coordinates are $-x, +y, -z$, is positive in a left-handed coordinate system. The atoms located in the mirror image (enantiomorphic) lower left rear (**llr**) octant would induce contribution of negative Cotton effect.

2.19.5.1 Determination of the Preferred Conformation

Example 1. Application of the octant rule allows one to ascertain the preferred conformation of a flexible molecule, if the absolute configuration is known. Let us take the example of (+)-3-methylcyclohexanone, known to have R configuration (Fig. 2.164). It can exist in an equatorial (**1e**) or axial (**1a**) conformer. Octant rule projections for equatorial and axial conformers show that they would display a positive CE and a negative CE, respectively. Since it actually exhibits a positive CE, the equatorial conformer predominates, consistent with the principle of conformational analysis. Conversely, if the preferred conformation (equatorial) had been established independently (say by NMR), the absolute configuration would have been determined as R (Fig. 2.164).

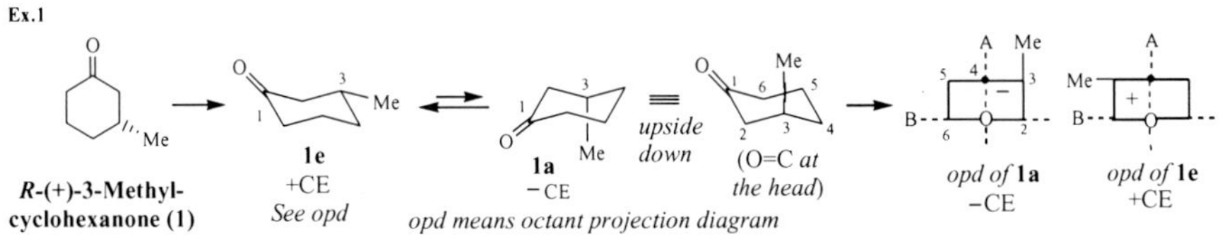

Ex.1

Fig. 2.164 Application of the octant rule to *R*-(+)-2-Methylcyclohexanone

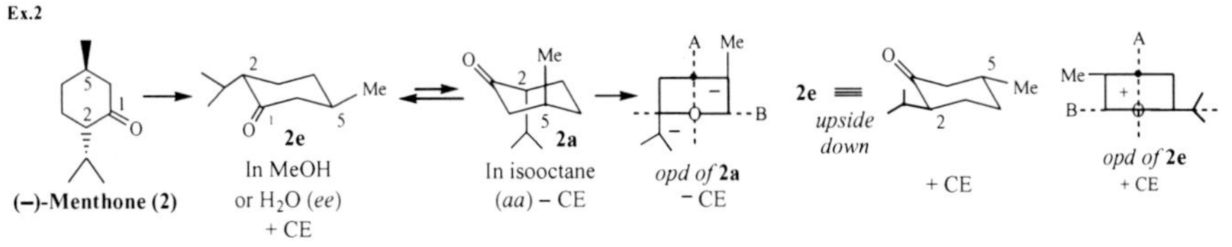

Ex.2

Fig. 2.165 Changes of conformation and CE sign of (−)-menthone with change of solvent polarity in CD spectra

Example 2. *Conformational mobility of menthone:* Each conformer of a flexible molecule (−)-menthone like **1** will have its own ORD or CD curve. Change of the conformer population caused by a change of solvent polarity or temperature leads to the change of its ORD or CD curve. (−)-Menthone (**2**) (Fig. 2.165) is a typical example having two conformers **2e** and **2a** in equilibrium. In a solvent of high polarity like water, (−)-menthone exhibits a positive CE (only a peak) in its CD curve due to the predominance of the diequatorial conformer **2e** (probably solvent effect). In methanol (a solvent of moderate polarity) two Cotton effects due to the two conformers appear, one strong positive peak below 300 nm and the other weak negative peak above 300 nm. In isooctane, a nonpolar solvent, a strongly positive CE peak appears at about 300 nm. In solvents of intermediate polarity both the CE's are exhibited in different proportions [135]. Such CD curves having two maxima of opposite signs are called *bisignate* [136]. The diaxial conformer **2a** is stabilized by the synergistic operation of a 2-alkylketone effect in **2a** and 3-alkylketone effect in **2e**, but the function of the polarity is not clear.

That conformational mobility is a function of solvent polarity can be demonstrated by the use of the axial haloketone rule in the following example. Bisignate curves are observed in case of *trans*-2-chloro-5-methyl cyclohexanone (**3**) (Fig. 2.166) obtained from chlorination of *R*-(+)-3-methylcyclohexanone (axial haloketone rule, Sect. 2.19.4.3, Fig. 2.161). The spectral properties of the chloroketone (**3**) indicate that in octane (of much lower polarity) solution it has axial halogen, and the CE in octane is negative establishing its *trans* configuration [137]. The ORD curve in methanol (of much higher polarity) is found to exhibit a positive CE indicating that the conformational equilibrium of the *trans* isomer has changed from the diaxial to the diequatorial form, presumably because in methanol of high dielectric constant the dipole repulsion between carbonyl and adjacent equatorial chlorine is not as serious as in octane of much lower dielectric constant.

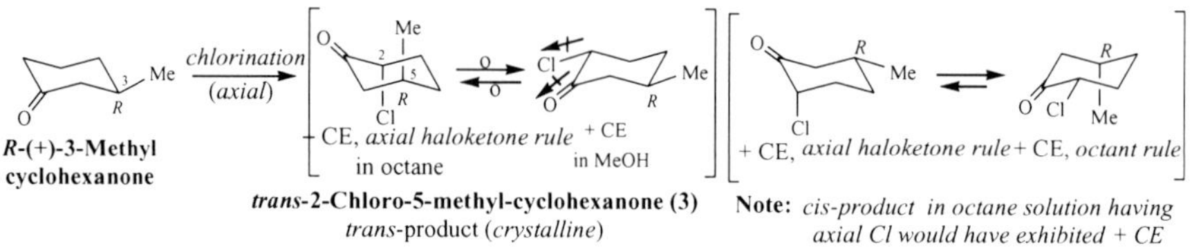
R-(+)-3-Methyl cyclohexanone
chlorination (axial)
+ CE, axial haloketone rule in octane
+ CE in MeOH
trans-2-Chloro-5-methyl-cyclohexanone (3)
trans-product (crystalline)
+ CE, axial haloketone rule + CE, octant rule
Note: cis-product in octane solution having axial Cl would have exhibited + CE

Fig. 2.166 Conformational mobility of *trans*-2-chloro-5-methylcyclohexanone in octane and methanol

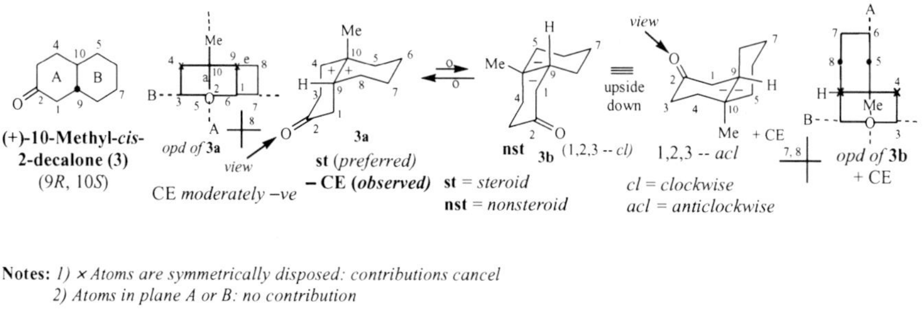
(+)-10-Methyl-cis-2-decalone (3)
(9R, 10S)
opd of 3a
view
CE moderately −ve
3a
st (preferred)
− CE (observed)
st = steroid
nst = nonsteroid
nst 3b (1,2,3 -- cl)
upside down
1,2,3 -- acl
cl = clockwise
acl = anticlockwise
+ CE
opd of 3b
+ CE
Notes: 1) × Atoms are symmetrically disposed: contributions cancel
2) Atoms in plane A or B: no contribution

Fig. 2.167 Preferred steroid conformation of (3) by application of octant rule

Example 3. ***Preferred conformation of cis-decalone derivatives*** can be determined by application of octant rule. (−)-*cis*-10-Methyl-2-decalone (**3**) may exist in steroid (**3a**) and/or nonsteroid (**3b**) conformation/s (Fig. 2.167). By the application of the octant rule, one finds that the steroid form (**3a**) should show a negative CE while the nonsteroid form (**3b**) a positive CE. RD study shows that the compound **3** exhibits a negative CE and hence should have the preferred steroid conformation. Here also the absolute configuration of **3** being known, the preferred conformation can be deduced. The ORD curve of **3** resembles that of coprostan-3-one having rings A/B in rigid steroid conformation (*vide* **2.19.5.3**); for further detailed study see [138].

2.19.5.2 Determination of Absolute Configuration of *trans*-Decalones

A *trans*-decalone of known structure possesses a unique rigid conformation; its absolute configuration can be determined from the sign of the Cotton effect it exhibits in its ORD spectrum, by application of the octant rule. Thus, the absolute configuration of (+)-*trans*-10-methyl-2-decalone can be determined as **4** by its ORD study which shows a positive CE. The mirror image configuration (**4′**) would exhibit a negative CE (Fig. 2.168); for further details see [138].

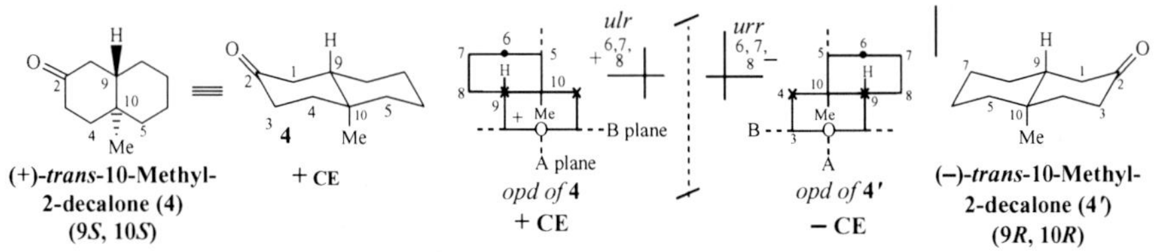

Fig. 2.168   Absolute configuration of (+)- and (−)-*trans*-10-methyl-2-decalone by application of the octant rule

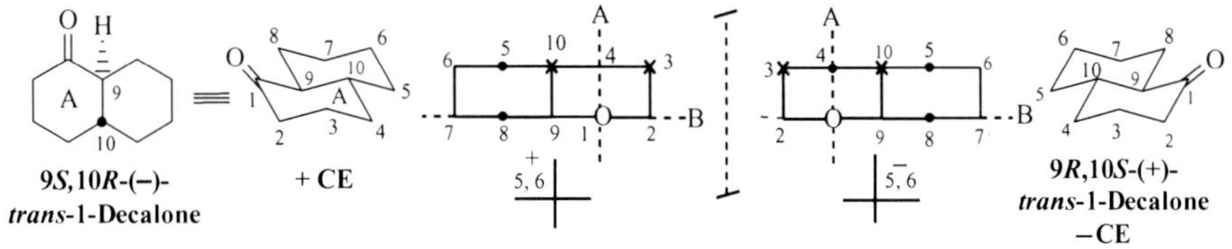

Fig. 2.169   CE signs of *trans*-1-decalone enantiomers by application of the octant rule

Prediction of the CE sign by application of the octant rule. If the structure and absolute configuration of a rigid molecule having a fixed conformation are known, the sign of the CE to be exhibited by it can be predicted by application of the octant rule. Conversely, the absolute configuration of such a molecule with known structure, conformation (fixed), and sign of the CE (by experiment) can be deduced by application of the octant rule. One more example is given in Fig. 2.169. For further details see [138].

2.19.5.3   Tricyclic Ketones: Perhydrophenanthrenones and Perhydroanthracenones

Only a few *trans-transoid-trans* (**t-tr-t**) perhydrophenanthrenones and *trans-cisoid-trans* (**t-ci-t**) perhydroanthracenones have been studied by CD spectroscopy [138, 139] (see Sects. 2.15.2 and 2.15.3 for nomenclature). Their octant projection diagrams (*opd*) are shown in Fig. 2.170. Steroid numbering has been used for convenience of comparison with corresponding steroid ketones (instead of the usual numbering). All compounds were found to obey the octant rule. The observed Cotton effect sign is same as predicted by octant rule in each case. In *opd* the carbon atoms in a front octant become evident from a watchful examination of Dreiding models.

No other diastereomeric ketones with different ring junction stereochemistry (*cf.* Sects. 2.15.2 and 2.15.3 have yet been investigated by CD; and *this is a potentially rich source for future research.*

2.19.5.4 Tetracyclic Ketones: Steroids

The perhydrophenanthrene skeleton with *t-tr-t* and *c-tr-t* stereochemistry is found in most steroids, *viz.*, cholestane and coprostane derivatives, respectively. Innumerable ORD and CD spectra recorded on such steroid ketones during late 1950s and 1960s formed the basis of the octant rule [125, 133, 134, 140, 141]. Rotatory dispersions of innumerable available steroid ketones and triterpene ketones have been compiled by Crabbe [142] and Kirk et al. [143–145].

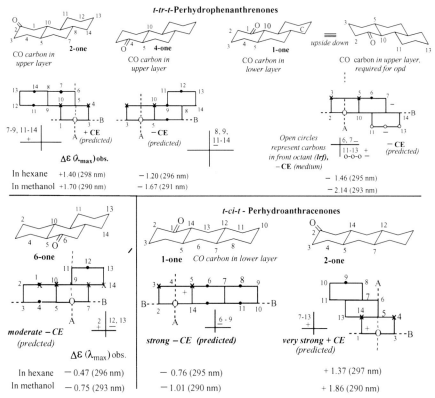

Notes: (i) *The skeletal carbons which are in mirror image positions with respect to A or B plane cancel.*
(ii) *The sign and magnitude of the amplitude of the nπ* * *CD Cotton effect depend upon the contribution of the uncancelled perturbing atoms and their distances from the carbonyl chromophore.*
(iii) *The skeletal carbons which lie on octant planes have no contribution to CE*
(iv) *The skeletal carbons close to the C=O group have strong contributions.*
(v) *Open circles represent carbons in front octant.*

Fig. 2.170 CE signs from octant projection diagrams of perhydrophenanthrenones and Perhydroanthracenones. [Adapted with permission from David A. Lightner and Jerome E. Gurst, *Organic Conformational Analysis and Stereochemistry from Circular Dichroism Spectroscopy*, Copyright © 2000 by John Wiley & Sons, Fig. 8–2, p. 234.]

Observed $n \rightarrow \pi^*$ Cotton effects of some steroid ketones from ORD and CD spectroscopy, which were consistent with the predicted Cotton effects based on octant projection diagrams (*opds*), are shown in Fig. 2.171 [146]).

The octant rule correctly predicts the signs as well as the magnitudes of cholestan-1-one (very weakly negative CE), -2-one (very strongly positive CE), -3-one (strongly positive CE), -12-one (moderately strong positive CE), and of coprostane-1-one (very strongly negative CE), -2-one (moderately strong negative CE), -6-one (very strongly negative CE), and −7-one (moderately strong positive

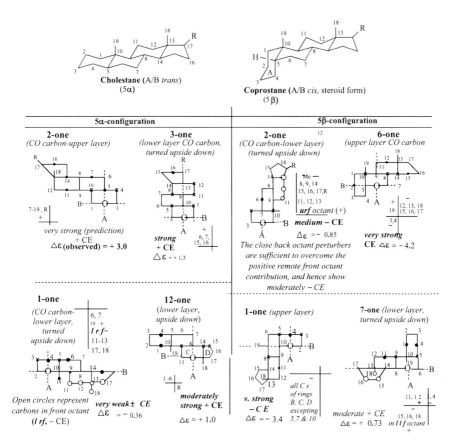

Notes: • *The skeletal carbons which are in mirror image positions with respect to A or B planes cancel.*
•   *The skeletal carbons which lie on octant planes have no contribution to CE.*
•   *The skeletal carbons close to the C=O group have strong contributions (adjacent axial position has strongest contribution)*
•   *The skeletal carbons remote from CO group are weak contributors.*
•   *Open circles represent carbons in front octant.*

Fig. 2.171 Octant projection diagrams and observed/predicted signs of the CE of some cholestanones and coprostanones. [Adapted with permission from David A. Lightner and Jerome E. Gurst, *Organic Conformational Analysis and Stereochemistry from Circular Dichroism Spectroscopy*, Copyright © 2000 by John Wiley & Sons, Table 8.3, pp. 237–238.]

CE), based on their octet projection diagrams (*opds*), as delineated in Fig. 2.171 [146]. The carbons in a front octant can presumably be ascertained by carefully projecting a Dreiding model of the molecule in a plane (xy) orthogonal to the line of view (z-axis) through O=C bond. The ketosteroids showing front octant contributions may complicate the projection as well as prediction of the CE sign. The 15, 16, and 17 ketosteroids, with 5α- and 5β- configurations, which have also been studied, make the projections complicated, but prediction of CE sign has still been correct excepting in case of coprostan-17-one [146]. In making accurate predictions, magnitudes of the individual contributions of the atoms, which depend on their positions as well as distances from the chromophore, should also be estimated from their *opds*.

2.19.6   Helicity Rule or Chirality Rule

In Sect. 2.19.3 we have mentioned that Cotton effects have been studied in two types of chromophores, inherently achiral and inherently chiral. Molecules consisting of both these types of chromophores have also been treated with success. Inherently chiral chromophores like skewed dienes, unsaturated ketones biaryl atropisomers, and helicenes exhibit much stronger CE than the inherently achiral chromophores present in the molecules. It has been derived theoretically and demonstrated experimentally that the direction of the Cotton effect depends upon the sense of helicity of the inherently chiral chromophores producing a positive effect. Inherently chiral chromophores follow the *helicity rule* or *chirality rule* which states that *P*-helicity (a right-handed helix) or positive chirality gives rise to positive optical rotation and positive CE, and *M*-helicity or negative chirality is correlated with negative optical rotation and negative CE (also see Sect. 2.19.63, last but 2 sentences). Some examples illustrating the helicity rule are shown in Fig. 2.172.

2.19.6.1   Conjugated Dienes and Enones: Steroids

R-α-Phellandrene (**1**) (Fig. 2.172) shows negative CE at room temperature and above. However, on reducing the temperature the magnitude of CE decreases, becomes zero, and down to $-160\,°C$ becomes positive [147]. The greater stability of the pseudoaxial conformer (**1a**) at room temperature may be due to higher entropy caused by the free rotation of the pseudoaxial isopropyl group relative to that of the pseudoequatorial conformer (**1e**). This isopropyl group in the latter may be facing some restriction to rotate freely. This thermodynamically preferred conformer is present at low temperature [147, 148].

The contribution of a chiral twisted diene or enone is so high that it usually overrides the contributions of other dissymmetric parts or the chiral center/s present. The helicity rule is applied to the lowest energy $\pi \rightarrow \pi^*$ transition in dienes (around 230–260 nm). The ORD curves of a number of s-*cis* (cisoid)

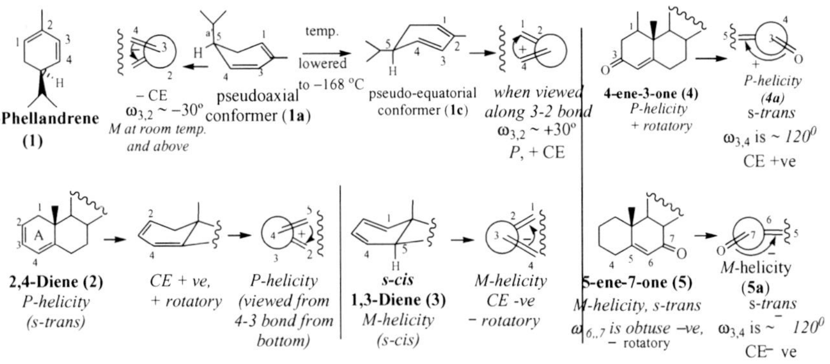

Fig. 2.172 Helicity rule applied to α-phellandrene and steroidal conjugated dienes and enones

2,4-dienes, e.g., (2) and 1,3-dienes, e.g., (3) (Fig. 2.172) derived from steroids show a strong positive CE and a strong negative CE, suggesting P-helicity and M-helicity respectively, the two double bonds being at acute torsion angles (+ω or −ω < 60°). The helicities may be confirmed from the acute torsion angles of the diene portions (Fig. 2.172), as evident from their Dreiding models.

The helicity rule has also been applied to α,β-unsaturated ketones (4) and (5) (Fig. 2.172), derived from steroids. Such enones are in *transoid* or *s-trans* conformation. Cholest-4-en-3-one (4) (in hexane) shows a positive CE at ≈250 nm whereas choles-5-en-7-one (5) shows a negatine CE (at ≈250 nm), confirming *P-helicity* and M-helicity, respectively [149]. The helicities can be confirmed by the obtuse positive and negative torsion angles (ω) between C=C and C=O bonds, respectively, from a careful examination by Dreiding models.

The homoconjugated enone (C=C.CH$_2$.C=O) rule has been shown to be compatible with the ketone octant rule, with C=C considered as a substituent, dominating the sign of the CE (generalized octant rule) [150].

2.19.6.2 Biaryl Atropisomers and Helicenes

According to helicity rule the biaryl atropisomers and helicenes having P-configuration are expected to show positive rotation and positive CE in their CD spectra, while those with M-configuration negative rotation and negative CE. Thus, the biphenyl derivative **B** (new convention) and 1,1′-binaphthyl (**C**) and 8,8′-dicarboxy-1,1′-binaphthyl (**D**) (see Fig. 2.151) possessing M-configuration are expected to exhibit negative CE and to be levorotatory, while the biphenyl derivative **A** (see Fig. 2.151) of P-configuration (new convention) to exhibit positive CE and be dextrorotatory.

Likewise, the helicenes (**A**), (**D**), and (**E**) of P-helicity (see Fig. 2.154) are expected to show positive CE and be dextrorotatory, while the helicenes (**B**) and (**C**) of M-helicity (see Fig. 2.154) are expected to show negative CE and be levorotatory. Helicenes are characterized by high optical rotation.

The absolute configuration of the helicenes can be predicted from chiroptical data only after careful assignment of the transition. In cases of helical molecules during excitation, electron movement takes place in a helical manner. For example, hexahelicene, wherein the electron transition moment direction is identical for either enantiomer, the magnetic moment direction is reversed according to whether the helix is *P* (right handed) or *M* (left handed). The parallel magnetic transition moments for *P*-hexahelicene generate a positive CE, whereas the antiparallel magnetic transition moments for *M*-hexahelicene lead to a negative CE which refers to the longest wavelength of high intensity at about 325 nm, presumably corresponding to the π–π* transition between HOMO and LUMO [126, 151, 152, p. 1012] and [153, p 69].

Thus, the absolute configuration of chiral biphenyls, biaryls, helicenes, and substituted helicenes, determined by CD studies from their observed CE signs, should be confirmed by X-ray or some possible unambiguous method.

Theoretical treatment by Mason et al. [153, 154] gives a more precise correlation of the CD spectra and absolute configuration of certain biaryls; the CD spectra were shown to be a function of the dihedral angle between the two aryl ring planes. 1,1'-Binaphthyl was unambiguously assigned to possess the critical interplanar dihedral angle of 100–110° [153].

2.19.6.3   Correlation of Optical Rotation with Ligand Polarizability: Brewster's rule

Attempts have been made by many scientists to provide a theoretical basis of optical activity during three decades since 1930. It is known that optical rotation at sodium D-line (589 nm) originates from the long wavelength tail of one or more UV Cotton effect/s. Optical rotation at different wavelengths arises because chiral compounds are circularly birefringent. The refractive index is related to the polarizability of ligands (atoms or groups) attached to the asymmetric center/s. Polarizability causes the sensitivity of the ligands to deformation by electric fields and to their relative positions.

In 1959 Brewster formulated a useful model of optical activity of open chain compounds [155] and of saturated cyclic compounds [156]. He suggested that a center of optical activity can usefully be described as a screw pattern of electron polarizability. Brewster approached to calculate the sign and magnitude of optical rotation based on two independent components: (i) contribution to the rotation by the difference in polarizability of atoms or groups attached to asymmetric atoms known as *atomic asymmetry* (or local chirality) and (ii) contribution from conformational dissymmetry (or chiral conformations). Both components lead to chiral screw pattern of polarizability and contribute to rotation in flexible molecules.

Atomic asymmetry refers to a chiral model (Fig. 2.173) Cabcd, where the ligands a-d are atoms or small groups having average cylindrical or conical symmetry. Such a compound is dextrorotatory, if the polarizability order is a > b > c > d, and the order a > b > c is clockwise, placing "d" at the rear of the line of view [cf. *R,S* nomenclatures, see Sect. 2.6.2.2, (i) and (ii) with corollary (1), priority sequence order being replaced by polarizability order]; the compound will be levorotatory if the polarizability order is anticlockwise (Fig. 2.173). It is difficult to measure the magnitude of rotation. The contribution of the atomic asymmetry component is small especially when the polarizabilities of two attached atoms or groups do not differ much. The polarizability of an atom or group is derived from the atomic refraction [157]. The order of polarizabilities of several common atoms or group, derived from the atomic refraction [157], is given in Fig. 2.173. The atomic refractions of the attached atoms in $C \equiv C$ and $C = C$ are taken as half the value of the group refraction.

In case of CN, C_6H_5, and CO_2H a complicated share of the group refraction is assigned to the attached carbon. Polarizability is affected by the nature of the attachment atoms, e.g., NH_2 and OH, when they are α- to a phenyl group, are to be ranked just after the chlorine atom, ahead of groups whose attachment atom is carbon. A conformational dissymmetry contribution to the rotation takes place by intramolecular hydrogen bonding, e.g., lactic acid (Fig. 2.173) or some other type of intramolecular interaction.

If the aforesaid two types of contribution to rotation predict the same sense of rotation, the configuration may be determined accurately. However, if the two components contribute opposite senses of rotation the model may lead to ambiguous results. The usual low contribution of atomic asymmetry to optical rotation is significantly increased, when one or more ligands absorb in UV region, e.g., phenylmethylcarbinol or mandelic acid (Fig. 2.173).

Brewster [155, 156, 158, 159] has developed an empirical approach to the contribution of *conformational dissymmetry* to the sign and approximate magnitude

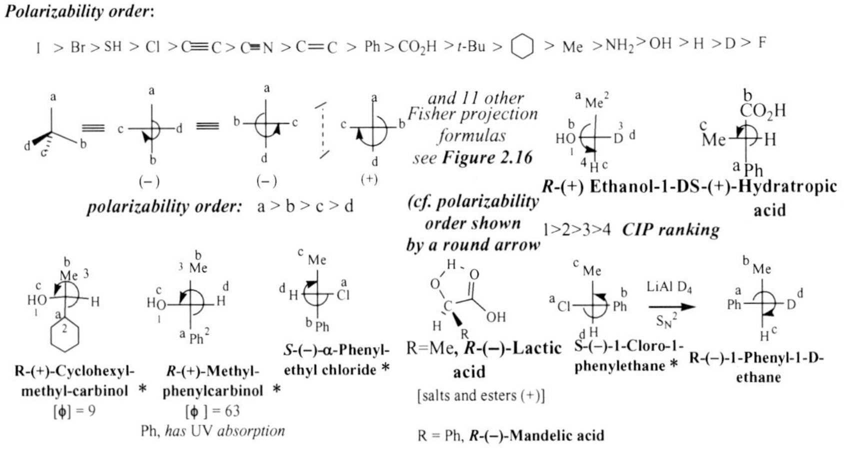

Polarizability order:

$$I > Br > SH > Cl > C\equiv C > C\equiv N > C=C > Ph > CO_2H > t\text{-Bu} > \bigcirc > Me > NH_2 > OH > H > D > F$$

Note: *The asterisked compounds possess the same sense of polarizability order and the priority sequence order.*

Fig. 2.173 Prediction of the sign of optical rotation in compounds with one chiral center and known absolute configuration and exhibiting atomic asymmetry (Brewster's rule)

of optical rotation which is beyond the scope of the present treatise. Based on major theoretical models of optical activity Brewster concluded [156] that as a general rule a system in which electrons are constrained to right-handed helical path will give a positive Cotton effect and will be dextrorotatory at long wavelength (*cf.* helicity rule, Fig. 2.172).

Atomic asymmetry component has been illustrated by several examples in Fig. 2.173. The absolute configuration of monochiral compounds of known structure can also be determined from the sign of optical rotation.

2.19.6.4 Absolute Configuration of Chiral Allenes: Lowe's Rule

During 1959–1965 the absolute configurations of eight chiral allenes have been determined by either the conversion of an optically active molecule of known absolute configuration into an allene or by converting a dissymmetric allene into a molecule of known absolute configuration by stereochemically unambiguous reaction [160]. Three such allenes (**1**), (**2**), and (**3**) are shown in Fig. 2.174.

Brewster's idea [155] of describing a chiral center as an asymmetric screw pattern of polarizability has been utilized by Lowe to put forward an empirical rule [160] for the absolute configuration of a chiral allene from the sign of its optical rotation at sodium D-line. Lowe's rule can be stated in a clarified general way as follows:

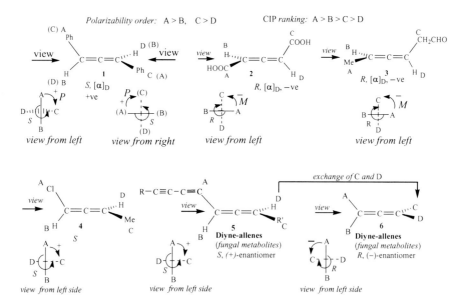

Fig. 2.174 Absolute configuration of allenes of known sign of $[\alpha]_D$ by Lowe's rule

The allene is viewed (as in case of nomenclature of all axially dissymmetric compounds, see Sect. 2.16) along its chiral (molecular) axis to get the ligands A and B (A being more polarizable than B) of the front carbon attached either at the vertical axis or horizontal axis and the ligands C and D (C being more polarizable than D) of the back carbon (attached at the axis orthogonal to that of the front carbon). If the ligand C appears at a clockwise direction with respect to the ligand A, the handedness of the screw pattern of polarizability is also clockwise, the enantiomer should be dextrorotatory at the sodium D-line; if A to C makes an anticlockwise turn, the enantiomer is levorotatory. Levorotation thus results from an anticlockwise screw pattern of polarizability. The rule, *which holds well if any of the ligands A–D in an allene does not exhibit conformational dissymmetry*, is illustrated for allenes **1–6** in Fig. 2.174.

Most of the naturally occurring allenes are fungal metabolites. All the dissymmetric fungal allenes contain a rigid diyne-allene system and are represented by the general formulas **5** and **6**. For those metabolites where the group R does not introduce conformational dissymmetry, Lowe's rule predicts that the (+)-enantiomers have the *S*-configuration **5,** and the (−)-enantiomers have the *R*-configuration **6.**

It seems that Lowe's rule may be extended to other axially dissymmetric compounds having similar geometry of the ligands of the end carbon; thus, the absolute configurations of a spiran (with both the double bonds of allene replaced by rings) and of an alkylidenecycloalkane (with one double bond of allene replaced by a ring) should be predictable correctly from their observed sign of $[\alpha]_D$.

2.19.6.5   The Exciton Chirality Method or The Dibenzoate Chirality Rule

As we know, in the UV-visible absorption, an electron moves from an occupied molecular orbital to a higher energy unoccupied molecular orbital, *viz.,* in $n \rightarrow \pi^*$ or $\pi \rightarrow \pi^*$ transition. The movement of an electron causes a dipole or polarization charge known as *electric dipole transition moment.* When two chromophores are in spatial proximity and are so oriented that a chiral array results, and when one chromophore is excited in the CD spectrum, the chromophores may interact with each other to generate distinctive Cotton effect (CE) couplets, known as *exciton coupling*, and the CD spectrum appears in a typical *bisignate* form. This allows the determination of the configuration of the *chiral array*. The term exciton was coined by Davydov in 1962 in connection with his theory of molecular excitons [161]. The energy difference between the common excited states is called *Davydov splitting*, which gives rise to $\Delta\lambda$, the wavelength difference between the two maxima of opposite signs (see Fig. 2.175b).

Harada and Nakanishi in their *benzoate sector rule* [162] utilized the Cotton effect at 225 nm ($\Delta\varepsilon \approx 3.5$) due to π–π^* intramolecular charge transfer band (230 nm, ε 14,000) of the benzoate chromophore. They extended this rule to *the dibenzoate chirality rule* [163], which correlates the chirality of glycols with signs of the intense and split (bisignate) $\pi-\pi^*$ Cotton effects of their dibenzoate derivatives. They later proposed to call this rule as *the exciton chirality method* [164]. The following characteristic features of the exciton chirality method are to be noted:

1. In a typical two-chromophore system (bisaromatic ester) a pair of CD bands (maxima) appears, one band at a longer and the other with opposite sign at a shorter wavelength relative to the absorption wavelength of the monomeric chromophore: thus the CD is bisignate.
2. Davydov splitting causes the appearance of the second CE at a shorter wavelength, the sign of which is opposite of the first CE. Together they resemble an ORD curve.
3. The electric dipole transition moment of each aromatic ester group is oriented collinearly with its long axis (Fig. 2.175a).
4. The sign of the CE at the longer wavelength corresponds to the helicity rule, discussed earlier. The right-handed screw pattern or the positive exciton chirality of the chiral array gives a positive CE, and the left handed screw pattern or negative chirality gives a negative CE.
5. The conformations of the bisbenzoates or bisaromatic esters of vicinal or non-vicinal diols if known, the *chiral array* between the individual electric transition moments (μ) aligned approximately parallel to the alcoholic C–O bonds of the ester groups (Fig. 2.175) will also be known. The dihedral angle (ω) between the two transition moments therefore corresponds to that between the two C–O bonds.
6. For glycol benzoate chromophores the amplitude of the exciton coupling is maximal when the dihedral angle ω between the electric dipole transition moments is about 70° and is zero when $\omega \approx 0°$ or 180° [165].
7. CE couplet should be away from other strong CE bands. Substitution in the benzoate ring especially by a p-dimethylamino- or a p-methoxycinnamate group displaces the exciton couplet to longer wavelengths $\approx$311 nm. The p-dimethylaminocinnamate group that absorbs at even longer wavelengths ($\approx$361 nm) may also be used.
8. Chromophores with more intense electric dipole transition moments (like a porphyrin chromophore) may be used in order to magnify the CE intensities in long range interactions (*cf.* 5α-steroids **8–10** of Fig. 2.175) [166].

2.19.6.6   Absolute Configuration of the 5α-Steroid Diols by Exciton Chirality Method

The benzoate chromophore shows two $\pi \rightarrow \pi^*$ intramolecular charge transfer transitions in the UV region: 280 (ε 1,000) and 230 nm (ε 14,000). These absorption bands have the transition moments along the short axis and the long axis of the benzoate chromophore, respectively. The CD spectrum of 2α,3β-dibenzoyloxy-5α-cholestane (**1**) (Fig. 2.175) like that of glycol dibenzoate gives rise to two strong CEs of similar amplitude but of opposite signs around 233 nm (first CE) and 219 nm (second CE) [163]. The resulting splitting indicates that the two CEs are mainly due to a dipole–dipole interaction between the electric transition moments of the intramolecular charge transfer band of the two benzoate chromophores, and that the CEs are separated from each other by $\Delta\lambda = 15$ nm due to *Davidov splitting* [161]. Compound (1) has a negative chirality of the two OH groups, and in accordance with the *exciton chirality rule* the sign of the first CE is also negative ($\Delta\varepsilon_{234} = -13.9$; $\Delta\varepsilon_{219} = 14.6$) (Fig. 2.175) [163, 164]. As expected, since 2β,3-β-dibenzoyloxy-5α-cholestane (2) possesses positive chirality of the two C–OH bonds, the sign of the first CE is positive [164] (Fig. 2.175).

Most 5α-cholestane-diols, whether with vicinal hydroxyls and distant hydroxyls, derivatized as *p*-dimethylaminobenzoates [compounds (3) to (10)] [166] give bisignate CEs in their CD, originating from exciton coupling and consistent with the exciton chirality rule (Fig. 2.175). In all the examples there is excellent correlation between the observed CEs in the CDs of steroids of known absolute configuration and the predicted positive or negative exciton chirality. The magnitude of the exciton CD Cotton effects decreases with increase in distance between the chromophores. Even in steroid-3,15-diol or 3,17-diol in which the chromophores are, respectively, more than 13 Å or 16 Å apart, quite appreciable $\Delta\varepsilon$ values of expected CEs are observed [166].

2.19.6.7 Absolute Configuration of *trans*-Cyclohexane-1,2-diol Enantiomers

The proximity, orientation, and the nature of the chromophores are important in exciton chirality rule. Thus, the CD spectra of bis-*p*-dimethylaminobenzoates of 5α-cholestan-2α,3β-diol and 1R,2R-*trans*-cyclohexanediol (11) (Fig. 2.175), both having the same absolute configuration, and vicinal diequatorial orientation are essentially identical with some difference in $\Delta\varepsilon$ values. The CD spectrum of the 2α,3β-dibenzoate derivative (1) has same type of negative exciton chirality (EC) with different $\Delta\varepsilon$ values (but with almost identical $\Delta\varepsilon$ values also of the bis-*p*-dimethylaminobenzoate esters in both cases). The diester of 1S,2S-*trans*-cyclohexanediol (12) possesses positive EC and shows bisignate CE with the first CE positive and the $\Delta\varepsilon$ values equal in magnitude but opposite in sign of (11), as expected, the nature of the EC of (12) being same as that of the 3β,4α diester (5) (Fig. 2.175), showing some difference in $\Delta\varepsilon$ values.

2.19.6.8 Prediction of the First CE Signs of Vicinal and Non-vicinal Dihydroxy-5α-Steroid Diesters

The sign of the first CE exhibited by suitable aromatic diesters of the following 5α-steroids containing two vicinal or non-vicinal OH groups of known absolute configuration (different from those mentioned in Fig. 2.175) *can be predicted* from the sign of the exciton chirality (EC) of each, as is evident from careful

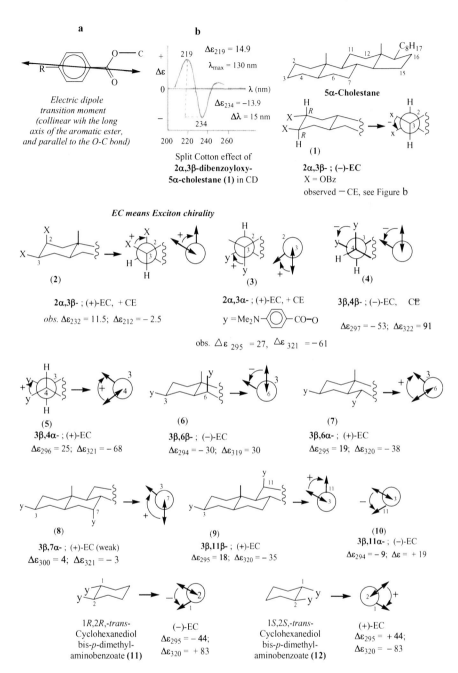

Fig. 2.175 Predicted and observed exciton chirality of bisbenzoyloxychotestane or bis-*p*-dimethylamino-benzoyloxycholestane derivatives and also of (1*R*,2*R*)- and (1*S*,2*S*)-*trans*-cyclohexanediol. [Adapted with permission from David A. Lightner and Jerome E. Gurst, *Organic Conformational Analysis and Stereochemistry from Circular Dichroism Spectroscopy*, Copyright © 2000 by John Wiley & Sons, the data (Δε values) of compounds (**3**) to (**10**)from the Table 14.3 (p. 436) and of compounds (**11**) and (**12**) from the Table 14.2 (p.433).]

1β, 2α (ee), (+ EC), (+)	6α, 7α (ea), (+ EC), (+)	1α, 11α (ae), (+ EC), (+)
1α, 2α (ae), (− EC), (−)	6α, 7β (ee), (− EC), (−)	1β, 11α (ee), (− EC), (−)
1β, 2β (ea), (− EC), (−)	7β, 8β (ea), (− EC), (−)	4α, 7α (ea), (+ EC), (+)
3α, 4α (ae), (− EC), (−)	11α, 12α (ea), (− EC), (−)	4α, 7β (ee), (− EC), (−)
4α, 5α (ea), (+ EC), (+)	11α, 12β (ee), (+ EC), (+)	7β, 14α (ea), (+ EC), (+)
5α, 6α (ae), (− EC), (−)	11β, 12β (ae), (− EC), (−)	9α, 11α (ae), (+ EC), (+)

Fig. 2.176 Predicted sign of the first CE of possible vicinal and non-vicinal 5α-steroiddiol bisaromatic esters from the sign of the observed EC (by proper examination of their Dreiding models)

examinations of their Dreiding models. The *a* or *e* orientations of the ester groups, the EC sign and *the predicted sign* of the first CE to be observed in the CD spectrum are shown in successive parentheses for each 5α-cholestanediol diester with the locations and α/β orientations of the ester groups in Fig. 2.176.

The *bichromophoric exciton chirality method* utilizing two different types of exciton chromophores together, which have been selectively introduced at two different types of hydroxyls, has been applied for determining the relative and absolute configurations in acyclic 1,2,3-triols [167].

For more details of CD and other applications of CD, see [166] and [168].

References

1. C. Winther, *Z. Phys. Chem.*, **1907**, *60*, 621.
2. E. Fischer, *Ber*, **1891**, *24*, 2683.
3. D. H. R. Barton, *The Principles of Conformational Analysis*, An Inaugural Lecture delivered at Berbeck College, London, 11 th February, **1954**, 10 pages.
4. W. Klyne and V. Prelog, Conformational Terminology, *Experientia*, **1960**, *16*, 521.
5. R. S. Cahn, Sir Christopher, K. Ingold and V. Prelog, Specification of Molecular Chirality, *Angew. Chem. Int. Edn. Engl.*, **1966**, *5*, 385–415. For *Conformational chirality* see p.406.
6. M. F. Rosanoff, On Fischer's Classification of Stereo-Isomers. *J. Am. Chem. Soc.*, **1906**, *28*, *114–121*.
7. W. Klyne, Nomenclature of Some Stereoisomeric Compounds..Specification of Configurations, *Chem. Ind..*(London), **1951**, 1022–1025.
8. R. S. Cahn, C. K. Ingold and V. Prelog, The Specification of Asymmetric Configuration in Organic Chemistry, *Experientia,* **1956,** *12*, 81–94.
9. R. S. Cahn and C. K. Ingold, Specification of Configuration about Quadrivalent Asymmetric Atoms, *J. Chem. Soc.* (London), **1951**, 612–622
10. Vladimir Prelog and Günter Helmchen, Basic Principles of the CIP-system and Proposals for a Revision, *Angew. Chem. Int. Edn. Engl.*, **1982**, *21*, 567–583.
11. IUPAC Commission on Nomenclature of Organic Chemistry, "*Rules for the Nomenclature of Organic Chemistry: Section E, Stereochemistry*", *Pure & Appl. Chem.*, Pergamon Press, Oxford, **1976**, *45*, 11–30; pertinent page 22.
12. James E. Blackwood, Casimir L. Gladys, Kurt L. Loening, Anthony E. Petrarca, and James E. Rush, Unambiguous Specification of Stereoisomerism about a double bond, *J. Am. Chem. Soc.,* **1968**, *90*, 509–510.
13. IUPAC Tentative Rules for the Nomenclature of Chemistry Section E. Fundamental Stereochemistry, *J. Org..Chem.,* **1970**, *35*, 2849–2867.

14. L. C. Cross and W Klyne, Rules for the Nomenclature of Organic Chemistry, *Pure Appl. Chem.*, **1976**, *45*, 11.

15. Francis A. Carey and Martin E. Kuehne, Beyond Erythro and Theo. A Proposal for Specifying Relative Configuration in Molecules with Multiple Chiral Centers, *J. Org. Chem.*, **1982**, *47*, 3811–3815.

16. Satoru Masumune, Sk. Asrof Ali, David L.Snitman, and David S. Garvey, Highly Stereoselective Aldol Condensation Using an Enantioselective Chiral Enolate, *Angew. Chem. Int. Ed. Engl.*, **1980**, *19*, 557–558.

17. James H. Brewster, On the Distinction of Diastereomers in the Cahn-Ingold Prelog (RS) Note, *J. Org. Chem.* **1986**, *51*, 4751–4753.

18. V. Schurig, Internal vs. External Diastereotopism in Homochiral *cis* vs. *trans* Olefin Metal π–Complexes, *Tetrahedron Lett.*, **1977**, 3977–3980.

19. Kurt Mislow and Jay Siegel, Stereoisomerism and Local Chirality, *J. Am. Chem. Soc.*, **1984**, *106*, 3319–3328.

20. G. E. McCasland, Robert Horvat, and Max R. Roth, Substances whose Optical Inactivity can be Attributed only to Fourfold Alternating Axial Molecular Symmetry. III. A Second Example (*meso*-Pentaerythritol Tetramenthyloxyacetate. *J. Am.Chem.Soc.*, **1959**, *81*, 2399–2402.

21. Kenneth R. Hanson, Applications of the Sequence Rule. I. Naming the Paired Ligands g,g at a Tetrahedral Atom Xggij. II. Naming the Two Faces of a Trigonal Atom Yghi, *J. Am. Chem. Soc.*, **1966**, *88*, .2731-2742.

22. Kurt Mislow and M. Raban in *Topics in Stereochemistry,* Vol.1, Eds. N. L. Allinger and E. L. Eliel, **1967**, Wiley, New York.

23. A. Hirschmann and Kenneth R. Hanson, Elements of Stereoisomerism and Prostereoisomerism, *J. Org. Chem.*, **1971**, *36*, 3293–3306.

24. A. Hirschmann and Kenneth R. Hanson, Prochiral and Pseudoasymmetric Centers: Implications of Recent Definitions *Tetrahedron*, **1974**, *30*, 3649–3656.

25. Ernest L. Eliel, *Prostereoisomerism (Prochirality) Top. Curr. Chem.* **1982**, *105*, 1–76.

26. J. Retey and J. A. Robinson, *Stereospecificity in Organic Chemistry and Enzymology*, **1982**, Verlag Chemie, Weinheim

27. Abhik Ghosh, A Quiz on Stereochemistry, *J. Chem. Educ.*, **1987**, *64*, 1015.

28. Ernest L. Eliel, Stereochemical Non-equivalence of Ligands and Faces (Heterotopicity), *J. Chem. Educ.*, **1980**, *57*, 52–55.

29. Harry S. Mosher and Edward La Combe, Asymmetric Reductions. I. The Action of (+)-2-Methylbutylmagnesium Chloride on Methyl t-Butyl Ketone, *J. Am. Chem. Soc.*, **1950**, *72*, 3994–3999.

30. Harry S. Mosher and Edward La Combe, Asymmetric Reductions. II. The Action of (+)-3-Methylpentylmagnesium Chloride on Methyl *t*-Butyl Ketone, *J. Am. Chem. Soc.*, **1950**, 72, 4991–4994.

31. Donald J. Gram and Fathy Ahmed Abel Elhafez, Studies in Stereochemistry. X. The Rule of Steric Control of Asymmetric Induction" in the Synthesis of Acyclic Systems, *J. Am. Chem. Soc.*, **1952**, 74, 5828–5835.

32. Donald J. Gram and Donald R.Wilson Studies in Stereochemistry, XXXII. Models for 1,2-Asymmetric Induction, *J..Am. Chem. Soc.*, **1963**, *85*, 1245–1249.

33. Ernest L. Eliel in "*Asymmetric Synthesis*, Vol.1, **1983**, Ed. J. D. Morrison, Academic Press, New York.

34. D. R. Boyd and M. A. McKervey, Asymmetric Synthesis, *Quart. Rev.*, **1968**, *22*, 95–122.

35. Marc Chérest and Hugh Felkin, Torsional Strain Involving Partial Bonds. The Steric Course of the Reaction Between Allyl Magnesium Bromide and 4-t-Butylcyclohexanone, *Tetrahedron Lett.*, **1968**; 2205–2208.

36. N. T. Anh, Regio- and Stereo-Selectivities in Some Nucleophilic Reactions, *Top. Curr. Chem.*, **1980**, *88*, 145–162.

37. H. B. Bürgi, J. D. Dunitz, and Eli Shefter, Geometrical Reaction Coordinates. II. Nucleophilic, Addition to Carbonyl Group, *J. Am. Chem. Soc.*, **1973**, *95*, 5065–5067.

38. H. B. Bürgi, J. D. Dunitz, J. M. Lehn, and G. Wipff, Stereochemistry of Reaction Paths at Carbonyl Centers, *Tetrahedron*, **1974**, *30*, 1563–1572.

39. H. B. Bürgi, J. M. Lehn, and G. Wipff, An *ab initio* Study of Nucleophilic Addition to a Carbonyl Group, *J. Am. Chem. Soc.*, **1974**, *96*, 1956–1957.

40. Alexander McKenzie, Studies in Asymmetric Synthesis. I. Reduction of Menthyl Benzoylformate. II. Action of Magnesium Alkyl Haloids on Menthyl Benzoylformate, *J. Chem. Soc*, **1904**, *85*, 1249–1262.

41. Vladimir Prelog, Untersuchungen über asymmetriche Synthesen 1. Über den sterischen Verlauf der Reaction von α-Keto-säure-estern optisch aktiver Alcohol emit Grignard'schen Verbindungen, *Helv. Chim. Acta*, **1953**, *36*, 308–319.

42. Vladimir Prelog, *Bull. Soc. Chim.* France, **1956**, 987.

43. J. C. Fiaud, Prelog's Methods in *Stereochemistry, Fundamentals and Methods*, Ed. H. B. Kagan, Vol.3, "Determination of Configurations by Chemical Methods", George Thieme Publishers, Stuttgart, **1977**, 19–49, (Chapter 2 Pertinent pages 20–24; and pertinent references cited.

44. Alain Horeau, Principe et Applications d'une Nouvelle Methode de Determination des Configurations dite "par Dedoublement Partiel" *Tetrahedron Lett.,* **1961**, 506–512.

45. Alain Horeau, Determination des Configurations par "Dedoublement Partiel-II Precisions Complements" *Tetrahedron Lett.,* **1962**, 965–969.

46. J. H. Brewster, "Assignment of Stereochemical Configuration by Chemical Methods" in *Elucidation of Organic Structures by Physical and Chemical Methods,* Ed. K. W. Bentley and G. W. Kirby, Vol. IV, Part III, Wiley-Interscience, New York, **1972**, pp. 1–249.

47. A. Horeau, "Determination of the Configuration of Secondary Alcohols by Partial Resolution" in *Stereochemistry, Fundamentals and Methods*, Ed. H. B. Kagan, Vol.3, Determination of Configurations by Chemical Methods, George Thieme Publishers, Stuttgart, **1977**, Chapter 3, pp.51-94.

48. R. Weidmann and A. Horeau, Determination des Configurations d'Alcohols Secondaire par "Dedoublement Partiel" IX(I) – Semimicromethode non Polarimetrique, *Tetrahedron Lett.*, **1973**, 2979–2982.

49. Alain Horeau and Andrée Nouaille, Micromethode de Determination de la Configuration des Alcohols Secondaires par Dedoublement Cinetique Emploi de la Spectrographie de Masse, *Tetrahedron Lett.,* **1990**, *31*, 2707–2710.

50. Tsutomu Katsuki and K. Barry Sharpless, The First Practical Method for Asymmetric Epoxidation, *J. Am. Chem. Soc.*, **1980**, *102*, 5974–5976.

51. Bryant E. Rossiter, Tsutomu Katsuki, and K. Barry Sharpless, Asymmetric Epoxidation Provides Shortest Routes to Four Chiral Epoxy Alcohols which are Key Intermediates in Syntheses of Methymycin, Erythromycin, Leukotriene C-1, and Disparlure, *J. Am. Chem. Soc.*, **1981**, *103*, 464–465.

52. A. Pfenninger, Asymmetric Epoxidation of Allylic Alcohols – The Sharpless Epoxidation, *Synthesis*, **1986**, 89–116.

53. Victor S. Martin, Scott S. Woodford, Tsutomu Katsuki, Yashiro Yamada, Masonari Ikeda, and K. Barry Sharpless, Kinetic Resolution of Racemic Allylic Alcohols by Enantioselective Epoxidation, A Route to Substances of Absolute Enantiomeric Purity, *J. Am. Chem. Soc.*, **1981**, *103*, 6237–6240.

54. Yun Gao, Robert M. Hanson, Janice M. Khender, Soo Y. Ko, Hiroko Masamune, and K. Barry Sharpless, Catalytic Asymmetric Epoxidation and Kinetic Resolution. Modified Procedures Including in Situ Derivatization, *J. Am. Chem. Soc.*, **1987**, *109*, 5765–5778.

55. K. Barry Sharpless, Scott S. Woodard, and H. G. Finn, On the Mechanism of Titanium-Tartrate Catalyzed Asymmetric Epoxidation, *Pure & Appl..Chem.*, **1983**, *55*, 1821–1836.

56. Scott S. Woodard, M. G. Finn, and K. Barry Sharpless Mechanism of Asymmetric Epoxidation. 1. Kinetics, Most importantly, the species containing equimolar amounts of Ti and

tartrate is shown to be the most active catalyst in the reaction mixture, mediating reaction at much faster rates than a titanium tetraalkoxide alone, *J. Am. Chem. Soc.*, **1991**, *113*, 106–113.

57. M. G. Finn and K. Barry Sharpless, Mechanism of Asymmetric Epoxidation. 2. Catalyst Structure, *J. Am. Chem. Soc.*, **1991**, *113*, 113–126.

58. G. M. Kellie and F. G. Riddell, Non-Chair Conformation of Six-Membered Rings, *Top. Stereochem.*, **1974**, *8*, 225–269.

59. Ernest L. Eliel, Samuel H. Wilen, and Michael P. Doyle, *Basic Organic Stereochemistry*, **2001,** John Wiley, New York, pp 688, pertinent pages 443–445.

60. Ernest L. Eliel and Siegfried H. Schroeter, Conformational Analysis. IX. Equilibrations with Raney Nickel. The Conformational Energy of Hydroxyl Group as a Function of Solvent, *J. Am. Chem. Soc.*, **1965**. *87*, 5031–5038.

61. N. L. Allinger, S. J. Angyal and G. M. Morrison, *Conformational Analysis*, **1965**, Wiley, New York.

62. F. A. L. Anet, Gwindolyn N. Chmurny and Jostein Krane, Ring Inversion in Cyclohexanone, *J. Am. Chem. Soc.*, **1973,** *95*, 4423–4424.

63. William G. Dauben, Gerhard J. Fonken, and Donald S. Noyce, The Stereochemistry of Hydride Reductions, *J. Am. Chem. Soc.,* **1956**, *78*, 2579–2582.

64. Donald C. Wigfield, Stereochemistry and Mechanism of Ketone Reductions by Hydride Reagents, *Tetrahedron*, **1979**, *35*, 449–462.

65. Herbert C. Brown and Harold R. Deck, Selective Reductions. VIII, The Stereochemistry of Reduction of Cyclic and Bicyclic Ketones by the Alkoxy Substituted Lithium Aluminium Hydrides, *J. Am. Chem. Soc.,* **1965**, *87*, 5620–5625.

66. E. L. Eliel and Y. Senda, Reduction with Metal Hydrides – XIX Competitive Reduction of Alkylcyclohexanones : Regarding the Cecepts of Steric Approach Control and Product Development Control, *Tetrahedron*, **1970**, *26*, 2411–2428

67. Donald C.Wigfield and David J. Phelps, Transition State Analysis: Evidence against Product Development Control in the Sodium Borohydride Reduction of Ketones, *Chem. Commun.*, **1970**, 1152–1153.

68. Jean-Claude Richer, On the Stereochemistry of the Reduction of Cyclic Ketones with Lithium Tri-*t*-butoxyaluminium Hydride, *J. Org. Chem.*, **1965**, *30*, 324–325

69. Andrzej Stanisiaw Cieplak, Stereochemistry of Nucleophilic Addition to Cyclohexanone. The Importance of Two-Electron Stabilizing Interaction, *J. Am. Chem. Soc.,* **1981**, *103*, 4540–4552.

70. C. K. Cheung, L. T. Tseng, M. –H. Lin, S. Srivastava, and W. J. le Noble, Geometric Equivalents of Enantiomers in Studies of the Stereochemical Course of Substitution at Carbon. Electronic Effects in Nucleophilic Addition to Carbonyl Groups and to Carbocations. Virtual Proof of the Existence of σ Participation by Unstrained Carbon-Carbon Bonds. *J. Am. Chem. Soc.,* **1986**, *108*, 1598–1605.

71. Andrzej S. Cieplak, Bradley D. Trait, and Carl R. Johnson, Reversal of π-Facial Diastereoselection upon Electronegative Substitution of the Substrate and the Reagent, *J. Am. Chem. Soc.*, **1989**, 111, 8447–8462.

72. Goverdhan Mehta and Faiz Ahmed Khan, Electronic Control of π-Facial Selectivities in Nucleophilic Addition to 7-Norbornanones, *J. Am. Chem. Soc.*, **1990**, 112, 6140–6142.

73. Goverdhan Mehta and Jayaraman Chandrasekhar, Electronic Control of Facial Selection, *Chem. Rev.*, **1999**, *99*, 1438–1467, and references cited.

74. A. J. Birch and G. S. R. Subba Rao, Reductions by Metal-Ammonia Solutions and Related Reagents in *Advances in Organic Chemistry : Methods and Results*, Vol. 8, Ed. E. C. Taylor, Wiley, New York, **1972**, 1–65.

75. A. J. Birch, The Birch Reduction in Organic Synthesis, *Pure Appl. Chem.* **1996**, *68*, 553–556 and references cited.

76. Suresh K. Pradhan, Mechanism and Stereochemistry of Alkali Metal Reductions of Cyclic Saturated and Unsaturated Ketones in Protic Solvents, *Tetrahedron*, **1986**, 6351–6388 (Tetrahedron Report No.R212).

77. K. Fukui in *Theory of Orientation and Stereoselection*, **1975**, Springer-Verlag, Berlin.

78. Francis Johnson and Sudarshan K. Malhotra, Steric Interference in Allylic and Pseudo-Allylic Systems. 1. Two Stereochemical Theorems, *J. Am. Chem. Soc.,* **1965**, *87*, 5492–5493.

79. Francis Johnson, Allylic Strain in Six-Membered Rings, *Chem. Rev.*, **1968**, *68*, 375–413.

80. Francis Johnson and Sundarshan K. Malhotra, Steric Interference in Allylic and Pseudo-Allylic Systems. II. Stereochemistry of Exocyclic Enolate Anion Protonation, *J. Am. Chem. Soc.*, **1965**, *87*, 5493–5494. Zimmerman & coworkers' work interpreted.

81. Howard E. Zimmerman, The Stereochemistry of Ketonization Reaction of Enol, *J. Org. Chem.*, **1955**, *20*, 549–557.

82. Howard E. Zimmerman and Theodore W. Cutshall, The Stereochemistry of Ketonization. X. Enols from α-Haloacids X., *J. Am. Chem. Soc.*, **1959**, *81*, 4305–4308. and earlier papers.

83. Howard E. Zimmerman and Thomas E. Nevins, The Stereochemistry of Ketonization. V. - aci-Nitro Tautomerism, *J. Am. Chem. Soc.* **1957**, *79*, 6559–6561.

84. Elias J. Corey and Richard A. Sneen, Calculation of Molecular Geometry by Vector Analysis. Application to Six-membered Alicyclic Rings, *J. Am. Chem. Soc.*, **1955**, *77*, 2505–2509.

85. Edgar W. Garbisch, Jr.,, Conformations. 1. Synthesis, Proton Magnetic Resonance Spectra, and Ultraviolet Spectra of Substituted 1-Phenylcyclohexenes, *J. Org. Chem.*, **1962**, *27*, 4243–4249.

86. Edgar W. Garbisch, Jr. Conformations. II. Proton Magnetic Resonance. Determination of C-6 Substituent Conformations in 6-Substituted 1-Phenylcyclohexenes, *J. Org. Chem.*, **1962**, *27*, 4249–4253.

87. Gilbert Stork, Ross Terrell, and Jacob Szmuszkovicz, A New Synthesis of 2-Alkyl and 2-Acyl Ketones, *J. Am. Chem. Soc.* **1954**, *76*, 2029–2030.

88. Gilbert Stork, A. Brizzolara, H. Landesman, J. Szmuszkovicz, and R. Terrell, The Enamine Alkylation and Acylation of Carbonyl Compounds, *J. Am. Chem. Soc.*, **1963**, *85*, 207–222, and references cited.

89. Francis Johnson and Alan Whithead, The Stereochemistry of 2-Substituted Cyclohexanone Enamines and the Corresponding Schiff's Bases, *Tetrahedron Lett.*, **1964**, 3825–3829

90. Keiji Marnoka, Tohru Miyazaki, Mamoru Ando, Yasushi Matsumura, Soichi Sakane, Kazunobu Hattori, and Hisashi Yamamoto, Organoaluminium – Promoted Beckmann Rearrangement of Oxime Sulfonates, *J. Am. Chem. Soc.*, **1983**, *105*, 2831–2843.

91. M. Oki, *Applications of Dynamic NMR Spectroscopy to Organic Chemistry*, Vol. 4, **1985**, VCH, Deerfield Beach, Florida.

92. William G. Dauben and Kenneth S. Pitzer, Conformational Analyhsis, in *Steric Effects in Organic Chemistry*, Ed. Melvin S. Newman, John Wiley, New York, **1956,** Chapter 1, pp 1–9 pertinent page 50.

93. William G. Dauben and Earl Hoerger, The *cis*-Decahydro-2-naphthoic Acid and their Relationship to the *cis-2*-Decalols and *cis-2*-Decalylamines, *J. Am. Chem. Soc.*, **1951**, *73*, 1504–1508.

94. W. Hückel, *Ber*, **1934**, *67*, 129.

95. W.S. Johnson, Energy Relationships of Fused Ring Systems, *J. Am. Chem. Soc.,* **1953**, *75*, 1498–1500.

96. Norman L. Allinger, Berner J. Gordon, Irene J. Tyminski, and Michael T. Wuesthoff, Conformational Analysis. LXX. Perhydrophenanthrenes, *J. Org. Chem.*, **1971**, *36*, 739–745.

97. Helmet Hőnig and Norman L. Allinger, The Stereoisomers of perhydrophenanthrene, *J. Org. Chem.* **1985**, *50*, 4630–4632.

98. R. P. Linstead, W. E. Doering, Selby B. Davis, Philip Levine and Richard W. Whetstone, The Stereochemistry of Catalytic Hydrogenation. I. The Stereochemistry of the Hydrogenation of Aromatic Rings, *J. Am. Chem. Soc.*, **1942**, *64*, 1985–1991.

99. R. P. Linstead and W. E. Doering, The Stereochemistry of Catalytic Hydrogenation. II. The Preparation of the Six Inactive Perhydrodiphenic Acids, *J.Am. Chem. Soc.,* **1942**, *64*, 1991–2003.

100. R. P. Linstead and W. E. Doering, The Stereochemistry of Catalytic Hydrogenation. III. Optically Active Perhydrodiphenic Acids. A Proof of the Configuration of the Backbone, *J. Am. Chem. Soc.*, **1942**, *64*, 2003–2006.

101. R. P. Linstead and Selby B. Davis, The Stereochemistry of Catalytic Hydrogenation. IV. Hexahydrodiphenic Acids, *J. Am.Chem. Soc.*, **1942**, *64*, 2006–2009.

102. R. P. Linstead, Selby B. Davis and Richard R. Whetstone, The Stereochemistry of Catalytic Hydrogenation. V. The assignment of *cis*- and *trans*-Configurations, *J. Am. Chem. Soc.*, **1942**, *64*, 2009–2014.

103. Selby B. Davis and R. P. Linstead, The Stereochemistry of Catalytic Hydrogenation. Part IX. Confirmatory Evidence of the Configurations of the Perhydrodiphenic Acids, *J. Chem. Soc.* **1950**, 1425–1427.

104. R. P. Linstead and Richard R. Whetstone, The Stereochemistry of Catalytic Hydrogenation. Part X. The Hydrogenation of *trans*-as-Octahydrophenanthrene Derivatives.A New Form of the Perhydrophenanthrene Ring as a Model for the Inversion of the 6-Ketosteroids, *J. Chem.. Soc.*, **1950**, 1428–1432.

105. Norman L. Allinger, and Michael T. Wuesthoff, Conformational Analysis. LXXII. The Perhydroanthracenes. An Equilibration Study, *J. Org. Chem.*, **1971**, *36*, 2051–2053.

106. Peter Maitland and William H. Mills, Experimental Demonstration of the Allene Asymmetry, *Nature*, **1935**, *135*, 994–994.

107. Peter Maitland and William H. Mills, Resolution of Allene Hydrocarbon into Optical Antipodes, *J. Chem. Soc.*, **1936**, 987–988.

108. Walter D. Celmer and I. A. Solomon, Mycomycin. III. The Structure of Mycomycin, an Antibiotic Containing Allene, Diacetylene and *cis,trans*- Diene Groupings, *J. Am. Chem. Soc.*, **1953**, *75*, 1372–1376.

109. E. P. Kohler, J. T. Walker, and M. Tishler, The Resolution of an Allenic Compound, *J. Am. Chem. Soc.*, **1935**, *57*, 1743–1745.

110. H. Gerlach, Über die Chiralität der enantiomeren Spiro[4,4]nonan-1,6-dione, *Helv. Chim. Acta*, **1968**, *51*, 1587–1593.

111. H. W. Whitlock Jr. and M. W. Siefken, Tricyclo[4.4.0.0^{3,8}] decane to Adamantane Rearrangement, *J. Am. Chem. Soc.*, **1968**, *90*, 4929–4939

112. Edel Wasserman, The Preparation of Interlocking Rings: A Catenane, *J. Am. Chem. Soc.*, 1960, *82*, 4433–4434.

113. George Hallatt Christie and James Kenner, The Molecular Configurations of Polynuclear Aromatic Compounds. Part I. The Resolution of r-6:6′-Dinitro and 4:6:4′:6′-Tetranitro-diphenic Acids into Optically Active Components, *J. Chem. Soc.*, **1922**, *121*, 614–620.

114. L. L. Ingraham, Steric Effects on Certain Physical Properties, in *Steric Effects in Organic Chemistry*, M. S. Newman, Ed.,Wiley, New York, **1956**, pp 479–522, pertinent page 483.

115. Arthur C. Cope, C. R. Ganellin, and H. W. Johnson, Resolution of *trans*-Cycloöctene; Confirmation of the Asymmetry of *cis-trans*-1,5-Cycloöctadiene, *J. Am. Chem. Soc.*, **1962**, *84*, 3191–3192.

116. Arthur C. Cope, C. R. Ganellin, H. W. Johnson, T. V. Van Auken, and Hans J. S. Winkler, Molecular Asymmetry of Olefins. I. Resolution of *trans*-Cyclooctene, *J. Am. Chem. Soc.*, **1963**, *85*, 3276–3279.

117. Arthur C. Cope and Anil S. Mehta, Molecular Asymmetry of Olefins. II. The Absolute Configuration of *trans*-Cyclooctene, *J. Am. Chem. Soc.*, **1964**, *86*, 5626–5630.

118. M. Goodman, A. S. Vardini, N. S. Choi, and Y. Matsuda, Polypeptide Stereochemistry, *Top. Stereochem.*, **1970**, *5*, 69–166.

119. Melvin S. Newman, 'Molecular Complexes and Molecular Asymmetry' in *Steric Effects in Organic Chemistry*, Ed. Melvin S. Newman, John Wiley, New York, **1956**, Chapter 10, 471–478.

120. Melvin S. Newman and Daniel Lednicer, The Synthesis and Resolution of Hexahelicene, *J. Am. Chem. Soc.*, **1956**, *78*, 4765–4770.

121. Melvin S. Newman and Allen S. Hussey, The Synthesis and Optical Resolution of 4,5,8-Trimethyl-1-phenanthrylic Acid, *J. Am. Chem. Soc.*, **1947**, *69*, 3023–3027.

122. Melvin S. Newman and Richard M. Wise, The Synthesis and Resolution of 1,12-Dimethylbenzo[c]phenanthrene-5-acetic acid, *J. Am. Chem. Soc.*, **1956**, *78*, 450–454

123. Melvin S. Newman and William B. Wheatley, Optical Activity of the 4,5-Phenanthrene Type: 4-(1-Methylbenzo[c]phenanthryl)-Acetic Acid and 1-Methylbenzo[c]phenanthrene,, *J. Am. Chem. Soc.*, **1948**, *70*, 1913–1916.

124. R. H. Martin and Marie-Jeanne Marchant, Thermal Racemization of [6], [7], [8], [9] Helicenes, *Tetrahedron Lett.*, **1972**, 3707–3708.

125. Carl Djerassi, *Optical Rotatory Dispersion: Applications to Organic Chemistry*, McGraw-Hill, New York, **1960**.

126. Ernest L.Eliel, Samuel H. Wilen, and Michael P.Doyle, *Basic Organic Stereochemistry*, John Wiley, New York, **2001**, Chapter 12, pp.534-607, and relevant references.

127. D. Nasipuri, *Stereochemistry of Organic Compounds*, (*Principles and Applications*), New Age International (P) Limited, Second Edition, **1994**, Chapter 15, pp.478-508, and relevant references.

128. P. M. Scopes, Applications of the Chiroptical Techniques to the Study of Natural Products. *Fortschr. Chem. org. Naturst.*, **1975**, *32*, 167–265.

129. Carl Djerassi and W. Klyne, Optical Rotatory Dispersion Studies. X. Determination of Absolute Configuration of α-Halocyclohexanone, *J. Am. Chem. Soc.*, **1957**, *79*, 1506–1507.

130. Carl Djerassi, Jeanne Osiecki, Rosemarie Riniker, and Bernhard Riniker, Optical Rotatory Dispersion Studies. XV. α-Haloketones (Part 2), *J. Am. Chem. Soc.*, **1958**, *80*, 1216–1225.

131. Carl Djerassi and J. Staunton, Optical Rotatory Dispersion Studies. XL1. α-Haloketones (Part 9). Bromination of Optically Active *cis*-1-Decalone. *J. Am. Chem. Soc.*, **1961**, *83*, 736–743.

132. Carl Djerassi, Neville Finch, and Rolf Mauli, Optical Rotatory Dispersion Studies. XXX. Demonstration of Boat Form in a 3-Keto Steroid, *J. Am. Chem. Soc.*, **1959**, *81*, 4997–4998.

133. Kurt Mislow, M. A. Glass, Albert Moscowitz, and Carl Djerassi, A Generalization of the Octant Rule, *J. Am. Chem. Soc.*, **1961**, *83*, 2771–2772.

134. William Moffitt, R. B. Woodward, A. Moscowitz, W. Klyne, and Carl Djerassi, Structure and the Optical Rotatory Dispersion of Saturated Ketones, *J. Am. Chem. Soc.*, **1961**, *83*, 4013–4018.

135. Keith M. Wellman, P. H. A. Laur, W. S. Briggs, Albert Moscowitz and Carl Djerassi, Optical Rotatory Dispersion Studies. XCIX. Superposed Multiple Cotton Effects of Saturated Ketones and Their Significance in the Circular Dichroism Measurement of (−)-Menthone, *J. Am. Chem. Soc.*, **1965**, *87*, 66–72.

136. W. Klyne and D. N., Kirk, The Signs of Chiral Properties: Suggested Terms "Consignate" and "Dissignate", *Tetrahedron Letters*, **1973**, 1483–1486. In this paper the authors have suggested the use of right handed Cartesian coordinates (See Octant rule).

137. Carl Djerassi, L. E. Geller, and E. J. Eisenbraun, Optical Rotatory Dispersion Studies. XXVI. A-Haloketones (Part 4). Demonstration of Conformational Mobility in α-Halocyclohexanones, *J. Org. Chem.*, **1960**, *25*, 1-6. See also Ref. 131.

138. David A. Lightner and Jerome E. Gurst, *Organic Conformational Analysis and Stereochemistry from Circular Dichroism Spectroscopy*, Wiley-VCH New York, **2000**; Sect.7.2, Decalones, pp 189–207; Tricyclic Ketones. Perhydrophenanthrones. Perhydroanthracones, pp 233–234.

139. Benito Alcaides and Franco Fernandez, Polycyclic Analogues of *trans*-Decalones. Part 6. Synthesis, Optical Resolution and Circular Dichroism of *trans-transoid-trans* Perhydrophenanthrene-1-one and *trans*-transoid-*trans*-Perhydrophenanthrene-2-one, *J. Chem. Soc.*, Perkin Trans I, **1983**, 1665–1671.

140. Carl Djerassi and W. Klyne, Optical Rotatory Dispersion: Application of the Octant Rule to Some Structural and Stereochemical Problems, *J. Chem. Soc.*, **1962**, 4929–4950.

141. Carl Djerassi and W. Klyne, Optical Rotatory Dispersion: Further Application of the Octant Rule of Structural and Stereochemical Problems, *J. Chem. Soc.*, **1963**, 2390–2402.

142. P. Crabbé, *Optical Rotatory Dispersion and Circular Dichroism in Organic Chemistry*, Holden-Day, San Francisco, **1965**.

143. D. N. Kirk and W. Klyne, Optical Rotatory Dispersion and Circular Dichroism. Part LXXXII, An Empirical Analysis of the Circular Dichroism of Decalones and Their Analogues, *J. Chem. Soc., Perkin Trans.* **1**, **1974**, 1076–1103.

144. David N. Kirk, Chiroptical Studies, Part 98. The Short-wavelength Circular Dichroism of Ketones in 2,2,2-Trifluoroethanol, *J. Chem. Soc.* Perkin I, **1980**, 1810–1819.

145. D. N. Kirk, The Chiroptical Properties of Carbonyl Compounds, *Tetrahedron*, **1986**, *42*, 777–818 (Tetrahedron Report Number 196) and relevant references cited.

146. David A. Lightner and Jerome E. Gurst, *Organic Conformational Analysis and Stereochemistry from Circular Dichroism Spectroscopy*, Wiley-VCH New York, **2000**, Tetracyclic Ketones. Steroids, pp 234-239.

147. W. Burghatahler, Herman Ziffer, and the Ulrich Weiss, The Configuration of Levopimaric Acid and α-Phellandrene : *J. Am. Chem. Soc.*, **1961**, *83*, 4660–4661

148. Günther Snatzke, E. Sz. Kovats, and Günther Ohloff, Circulardichroismus von α-Phellandren, *Tetrahedron Lett.* **1966**, 4551–4555.

149. G. Snatzke and F.Snatzke in '*Fundamental Aspects and Recent Developments in ORD and CD*', Eds. F. Ciardelli and P. Salvadori, Heydon, London, **1973.**

150. Albert Moscowitz, Kurt Mislow, M.A.W. Glass, and Carl Djerassi, Optical Rotatory Dispersion Associated with Dissymmetric Non-conjugated Chromophores. An Extension of the Octant Rule, *J. Am. Chem. Soc.*, **1962**, *84*, 1945–1955.

151. Ernest L. Eliel, *Stereochemistry of Carbon Compounds*, McGraw Hill, New York, **1962**.

152. Ernest L. Eliel and Samuel H. Wilen, *Stereochemistry of Organic Compounds*, John Wiley, New York, **1994**, Chapter 13, pp 991–1071 and relavant references.

153. S. F. Mason, in *Molecular Optical Activity and Chiral Discrimination*, Cambridge University Press, Cambridge, **1982**, p. 69.

154. S. F. Mason, R. H. Seal, and D.R. Roberts, Optical Activity in the Biaryl Series, *Tetrahedron*, **1974**, *30*, 1671–1682.

155. James H. Brewster, A Useful Model of Optical Activity. 1. Open Chain Compounds, *J. Am. Chem. Soc.*, **1959**, *81*, 5475–5483.

156. James H. Brewster, The Optical Activity of Saturated Cyclic Compounds, *J. Am. Chem. Soc.*, **1959**, *81*, 5483–5493.

157. A. I. Vogel, Physical Properties and Chemical Constitution. Part XXII. Miscellaneous Compounds. Investigation of the So- called Co-ordinate or Dative Link in Esters of Oxy-acids and in Nitroparaffins by Molecular Refractivity Determinations. Atomic, Structural, and Group Parachors and Refractivities, *J. Chem. Soc.*, **1948**, 1833–1855.

158. J. H. Brewster, Some Applications of the Conformational Dissymmetry Rule, *Tetrahedron*, **1961**, *13*, 106–122.

159. James H. Brewster, Helix Models of Optical Activity, *Top. Stereochem*, **1967**, *2*, 1–72.

160. G. Lowe, The Absolute Configuration of Allenes, Chem. Commun. **1965**, 411–413.

161. A. S. Davydov, *Theory of Molecular Excitons*, **1962**, M. Kasha and M. Oppenheimer, Jr. (translators), McGraw-Hill, New York.

162. Nobuyuki Harada, Motoaki Ohashi, and Koji Nakanishi, The Benzoate Sector Rule, a Method for Determining the Absolute Configuration of Cyclic Secondary Alcohols, *J. Am. Chem. Soc.*, **1968**, *90*, 7349–7351.

163. Nobuyuki Harada and Koji Nakanishi, Determining the Chiralities of Optically Active Glycols, *J. Am. Chem. Soc.*, **1969**, *91*, 3989–3991.

164. Nobuyuki Harada and Koji Nakanishi, The Exciton Chirality Method and Its Application to Configurational and Conformational Studies of Natural Products, *Acc. Chem. Res.*, **1972**, *5*, 257–263.

165. N. Harada and K. Nakanishi, *Circular Dichroic Spectroscopy. Exciton Coupling in Organic Stereochemistry*, University Science Books, **1983**, Mill Valley, CA.

166. David A. Lightner and Jerome E. Gurst, *Organic Conformational Analysis and Stereochemistry from Circular Dichroism Spectroscopy*, Wiley-VCH New York, **2000**; 423–456, Chapter 14 entitled Exciton Coupling and Exciton Chirality.

167. William T. Wiesler, and Koji Nakanishi, A Simple Spectroscopic Method for Assigning Relative and Absolute Configuration in Acyclic 1,2,3-Triol, *J. Am. Chem. Soc.*, **1969**, *111*, 3446–3447.
168. Ernest L. Eliel and Samuel H. Wilen, *Stereochemistry of Organic Compounds*, John Wiley, New York, **1994**, Chapter 13, pp 991–1071 and relevant references.

Further Reading

Ernest L. Eliel, *The Stereochemistry of Carbon Compounds*, **1962**, McGraw-Hill, New York.

Ernest L. Eliel in *Asymmetric Synthesis*, Vol.1, 1983, Ed. J. D. Morrison, Academic Press, New York.

Ernest L. Eliel, Samuel H. Wilen, and Lewis N. Mander, *Stereochemistry of Organic Compounds*, **1994**, John Wiley, New York (pp 1267).

D. Nasipuri, Stereochemistry of Organic Compounds (Principles and Applications), **1991**, New Age Intrernational (P) Limited, Second Edition, **1994** (pp 564).

Ernest L. Eliel, Samuel H. Wilen, and Michael P. Doyle, *Basic Organic Stereochemistry*, **2001**, John Wiley, New York (pp 688).

Ernest L. Eliel, N. L. Allinger, S. J. Angyal and G. A. Morrison, *Conformational Analysis*, **1965**, Wiley, New York; reprinted. Reprinted **(1981)** by American Chemical Society, Washington, DC.

P. Deslongchamps, *Stereoelectyronic Effects in Organic Chemistry* **1983**, Ed. J. E. Baldwin, Pergamon Press, New York.

Euseb io Juaristi, *Introduction to Stereochemistry and Conformational Analysis*, **1991**, John Wiley, New York.

David A. Lightner and Jerome E. Gurst, *Organic Conformational Analysis and Stereochemistry from Circular Diehroism Spectroscopy,* WILEY-VCH, **2000**, PP 487.

Henri Kagan *Organic Stereochemistry*, Edward Arnold (Publishers) Ltd., **1979**, pp.66.

M. Nógrádi, Stereochemistry : Basic Concepts and Applications, Akademiai Kiadó, Budapest, **1981**; currently Pergamon Press, Oxford,

Kurt Mislow, *Introduction to Stereochemistry*, W. A. Benjamin, New York, **1965**, pp 193.

Subrata Sen Gupta, *Basic Stereochemistry of Organic Compounds*, Fourth Edition, **2011**, M/S Subrata Sengupta (pp 629), special feature: solved problems and exercises in each chapter.

Yoshiharu Izumi and Akira Tai, *Stereodifferentiating Reactions* (the Nature of Asymmetric Reactions), **1977.**

J. Retey and J. A. Robinson, *Stereospecificity in Organic Chemistry and Enzymology*, **1982**, Verlag Chemie, Weinheim.

A. Bassindale, *The Third Dimension in Organic Chemistry*, **1984**, Wiley, New York.

Steric Effects in Organic Chemistry, Ed. Melvin S. Newman, John Wiley, New York, **1956**.

David A. Lightner and Jerome E. Gurst, *Organic Conformational Analysis and Stereochemistry from Circular Dichroism Spectroscopy,* Wiley-VCH New York, **2000**, pp 487.

Dipak K. Mandal, The R/S System : A New and Simple Approach to Determining Ligand Priority and a Unified Method for the Assignment and Correlation of Stereogenic Center Configuration in Diverse Stereoformulas, *J. Chem..Educ.* ;**2000**, 77, 866–869.

IUPAC Commission on Nomenclature of Organic Chemistry, *"Rules for the Nomenclature of Organic Chemistry: Section E, Stereochemistry"*, *Pure & Appl. Chem.*, Pergamon Press, Oxford, **1976**, *45*, 11–30.

IUPAC Tentative Rules for the Nomenclature of Organic Chemistry, Section E. Fundamental Stereochemistry, *J. Org. Chem.* **1970**, *35*, 2849–2867.

Karl-Heinz Hellwick and Carsten D. Siebert, *Stereochemistry Workbook; 191 Problems and Solutions*, translated by Allan D. Dunn, Springer-Verlag, Berlin, Heidelberg, **2006**, pp198.

Chapter 3
Important Biological Events Occurring in Plants

3.1 Photosynthesis

"Photosynthesis is an integrated system in which light harvesting, photo-induced charge separation, and catalysis combine to carry out two thermodynamically demanding processes, the oxidation of water and reduction of carbon dioxide."
  - A part of the comment that appeared on the cover of the *Journal of Organic Chemistry*, **2006**, *71, issue no. 14,* (July 7, 2006).

3.1.1 Light Reaction: Formation of NADPH, ATP, and O_2

"Photosynthesis is undoubtedly the most important single metabolic innovation in the history of life on the planet" [1]. It is looked upon as the mother of all biological events of life since essentially all free energy needed for biological systems originates from solar energy *via photosynthesis,* the most successful of all natural energy storage processes. Thus, all forms of life evolved to exploit oxygen for their well-being depend for their energy, directly or indirectly on this phenomenon, which truly belongs to an interdisciplinary field involving radiation physics, solid state physics, chemistry, enzymology, physiology, and ecology. The present discussion is mainly based on chemistry. Photosynthesis is a green plant process, the means of converting light into chemical energy, taking place into organelles called chloroplasts. In this process chlorophyll, the green coloring matter of plants utilizes the solar energy in converting CO_2 and H_2O into carbohydrates, and oxygen and some water are evolved. Bacteria with bacteriochlorophyll can also achieve photosynthesis—bacteria do not evolve oxygen like green plants [2]. Since oxygen is a by-product of the green plant photosynthesis, the latter is sometimes called *oxygenic photosynthesis,* and it is the major source of oxygen in the atmosphere. The chemistry of the green plant or green algae photosynthesis will be discussed briefly. The plant photosynthesis utilizes the solar light. The bacterial photosynthesis utilizes the light in the infrared region. However, photons of far-infrared region

S.K. Talapatra and B. Talapatra, *Chemistry of Plant Natural Products,*
DOI 10.1007/978-3-642-45410-3_3, © Springer-Verlag Berlin Heidelberg 2015

have too low energies to be utilized for any photochemical reaction. The splitting regions (approximately) of white sunlight, as seen after diffraction through prism or in rainbows (VIBGYOR), are shown in Fig. 3.1.

When white sunlight falls on **chlorophyll**, it absorbs light waves of all the wave lengths except at the region ~480–550 nm; the region is attributed to green color of the VIBGYOR. Because of the nonabsorption and reflection of light of the green region, chlorophyll is green in color and hence the leaves are green. Richard Willstätter (1872–1942) (Appendix A, A30) won the 1915 Nobel Prize for elaborating the structure of chlorophyll. There are two major types of chlorophyll: chlorophyll-*a* (**Chl-a**) and chlorophyll-*b* (**Chl-b**) having the same structural scaffold, differing only in the ring B in having a –CHO group in **Chl-b** in place of a CH_3 group in **Chl-a** (Fig. 3.2). This small change in the substituent causes a difference in energy absorption. The basic structure is named porphyrin possessing a tetrapyrrole skeleton, which contains an 18-membered ring with nine conjugated double bonds. The basic skeleton is thus aromatic and has its two canonical (resonance) forms contributing equally to their resonance hybrid. In Fig. 3.2, one canonical form is shown. However, like aromatic compounds either canonical form is used in the literature to represent the basic skeleton of the compounds shown in Fig. 3.2. Mg^{2+} is present at the center and is covalently bonded with two N atoms, and coordinately bonded to the other two N atoms of the porphyrin. **Chl-a** and **Chl-b**, thus called

Visible light (400-700 nm)

Color	Violet		Blue		Green		Yellow		Orange		Red	
λ in nm	400		450		500		550		600		650		700

Fig. 3.1 The electromagnetic radiation in the visible spectrum

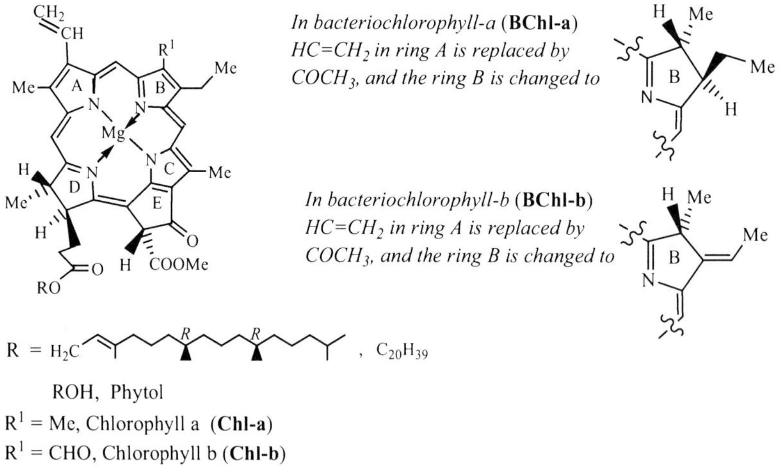

In bacteriochlorophyll-a (**BChl-a**)
$HC=CH_2$ *in ring A is replaced by*
$COCH_3$, *and the ring B is changed to*

In bacteriochlorophyll-b (**BChl-b**)
$HC=CH_2$ *in ring A is replaced by*
$COCH_3$, *and the ring B is changed to*

R = H_2C ———————— , $C_{20}H_{39}$

ROH, Phytol
R^1 = Me, Chlorophyll a (**Chl-a**)
R^1 = CHO, Chlorophyll b (**Chl-b**)

Fig. 3.2 Stereostructures of **Chl-a**, **Chl-b**, **BChl-a** and **BChl-b**

magnesium porphyrins, are present in plants in the ratio 3:1. This ratio as well as their structures has remained the same during evolution, a very amazing feat of Nature. **Chl-a** does not absorb light in a wide range of the visible spectrum. This nonabsorbing region is known as the "green window". The first absorption maxima of **Chl-b** and **Chl-a** are around 460 nm and 430 nm, respectively, and their second maxima are around 650 nm and 700 nm, respectively. **Chl-b** can transfer light energy very efficiently to **Chl-a**. Thus, the absorption gap of **Chl-a** green window is narrowed by the light absorption of **Chl-b**. Hence, the latter increases the efficiency of plants for utilizing sunlight energy.

The antenna protein containing many chromophores absorbs light. **Chl-a** is the central pigment of the plant photosynthesis. The trapped electromagnetic radiation of the sun causes the flow of energy within the photochemical apparatus containing specialized reaction centers, allowing the trapped energy to participate in a series of biochemical reactions. Because of the presence of a number of conjugated double bonds in the chlorophyll skeleton, much less amount of energy (photon of red light) is required to cause excitation of electrons to the lower energy first singlet state (half-life 4×10^{-9} s). The second singlet state requires higher energy (photon of blue light) and is of short half-life (10^{-12} s) to effect chemical reactions. Excited chlorophyll as such directly cannot transfer the energy to the right location. It initiates electron-transfer chain through electron acceptors like *quinones, cytochromes,* etc. resulting in a charge separation. In effect, the first singlet state splits water molecule into H and OH radicals. The OH radicals yield some oxygen and some water, while H reduces the oxidized coenzyme **NADP$^\oplus$** to **NADPH** (catalyzed by **ferredoxin** NADP$^\oplus$ reductase, a **flavoprotein** with an **FAD** prosthetic group). The excess energy is stored as an energy-rich **ATP**, acting as a temporary source of energy. Its terminal γ-phosphoanhydride bond is hydrolyzed to yield **ADP** with the release of energy to be used in the biochemical processes, which are especially energetically unfavorable. ATP is rebuilt from ADP and Pi with the adequate input of energy. ATP is known as 'energy currency' in the cell. This forms the part of *light reaction* (Fig. 3.3).

The reduced form of the coenzyme NADPH, a two electron donor, thus formed, reduces CO_2 to 3-phosphoglycerate. Inorganic carbon is converted into the first organic molecule of the **carbon cycle (Calvin cycle)** to be discussed in the sequel.

The overall green plant photosynthesis reaction is represented by the century-old oversimplified basic expression (3.1) in which we find two thermodynamically demanding processes, oxidation of water and reduction of CO_2 in the combined form.

$$CO_2 + H_2O + h\nu \rightarrow C\,(H_2O) + O_2 \qquad (3.1)$$

Priestley in 1770s first showed by performing the following experiment that oxygen is liberated during photosynthesis. He put a mouse under each of two bell jars, one of which contained a plant while the other none. He found that the mouse in the bell jar containing the plant lived much longer than the other mouse. From this observation "...he correctly concluded that the plant, through interaction with

Fig. 3.3 A simplified
representation of
photosynthesis in green
plants

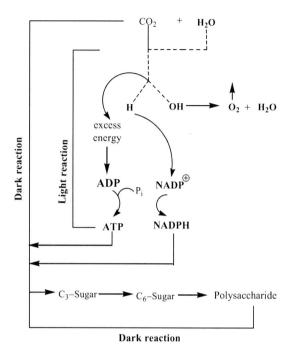

Fig. 3.3 A simplified representation of photosynthesis in green plants

light, was modifying the air by producing a new substance. He set out to prove what this substance was and thereby discovered oxygen" [2]—a gas so convivial to our very existence. Melvin Calvin (NL, 1961) observed that plants/green algae upon irradiation with light in presence of $^{14}CO_2$ produced labeled 3-phosphoglycerate, and he concluded that this C_3-sugar was the intermediate in the fixation of CO_2 to saccharides. The unique enzymatic machinery present in the chloroplasts of green plants catalyzes the conversion of CO_2 into simple organic compounds. This process is called *CO_2 fixation* or *carbon fixation*. The reactions involved make up a cyclic pathway constantly regenerating the key intermediates. The pathway was elucidated by Melvin Calvin in early 1950s and is often called the **Calvin cycle**. These simple products of photosynthesis are converted in plants into complex biomolecules like sugars, polysaccharides, and their metabolites.

Formation of NADPH and ATP takes place by the action of light and is called *light reaction*. In the *dark reaction*, the NADPH and ATP drive the reduction of CO_2 to more useful organic compounds. A very simplified representation of photosynthesis is shown in Fig. 3.3.

To make a glucose molecule, six molecules of CO_2 are needed as shown in the following equation:

$$6\ CO_2 + 12\ H_2O + h\nu \xrightarrow{\text{chlorophyll}} C_6H_{12}O_6 + 6\ O_2 \uparrow + 6\ H_2O$$

and the overall balanced equation for Calvin cycle may be represented as:

$$6CO_2 + 18ATP + 12NADPH + 12H_2O \rightarrow C_6H_{12}O_6 + 18ADP + 18Pi$$
$$+ 12NADP^+ + 6H_2O.$$

Interestingly, water appears on both sides of the equation. It is due to the fact that OH obtained from H_2O gives oxygen as well as some water molecules.

3.1.2 Dark Reaction (Calvin Cycle): Formation of 3-, 4-, 5-, 6-, and 7-Carbon Sugars

Detailed studies by Melvin Calvin (NL 1961) and his collaborators revealed that photosynthetic CO_2 fixation proceeds by a cyclic process which has been named **Calvin Cycle** (Fig. 3.4). Since reduction occurs and pentoses are formed in the cycle, the latter is also termed as **reductive pentose phosphate (RPP) pathway**. The Calvin cycle consists of *three stages* (Fig. 3.4) to be explained in succession.

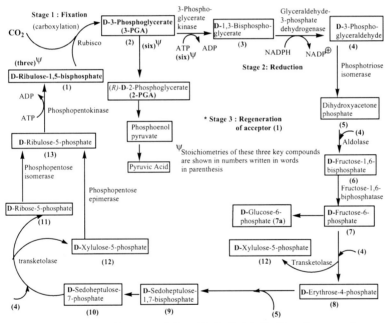

* Stage 3 **Regeneration of acceptor (1) from (4) via (5), (6), (7), (8) and so on, as shown.**

(*For mechanisms of all conversions see* **Figures 3.5 to 3.10**)

D,L-Nomenclature: *When the Fischer projection formula of the molecule is written placing (i) the more oxidized end of the chain at the top, and (ii) the longest carbon chain vertically, the enantiomer with the OH group at the lowermost (highest numbered) chiral center [e.g.,C2 in (2), (3) and (4), C3 in (8), C4 in (1) and (12), C5 in (6), (7) and (7a), and C6 in (9) and (10)]* **on the right-hand side is given D-configuration** - *the corresponding enantiomer of each sugar is designated* **L.**

Fig. 3.4 Calvin cycle or carbon cycle: Formation of 3-, 4-, 5-, 6- and 7-carbon sugars in three stages

Stage 1: The Fixation of CO$_2$ by Carboxylation of a C$_5$ Sugar to Form D-3-Phosphoglycerate (2) In the fixation of CO$_2$, the C$_5$ sugar D-ribulose-1,5-bisphosphate (for D,L-nomenclature *see* Fig. 3.4 and Sect. 2.6.1) acts as the acceptor on which carboxylation takes place in presence of an enzyme ribulose-1,5-bisphosphate carboxylase-oxygenase, commonly known as **rubisco/Rubisco** [3], *the most abundant protein in plant, and for that matter on earth. With its 16 subunits, it is one of the largest enzymes in nature. It is a lazy enzyme* and its catalytic rate is 3 s^{-1} only, *i.e.*, it fixes only three molecules of CO$_2$ per second and causes its rich concentration in *chloroplast* [4]. During the process, an unstable C$_6$ compound is formed which suffers rapid hydrolysis to yield two molecules of D-3-phosphoglycerate. Incorporation of CO$_2$ into 3-phosphoglycerate has been verified through the use of ^{14}CO$_2$. Rubisco needs bound divalent Mg^{+2} for its activation. Prior to carboxylation, D-ribulose-1,5-bisphosphate undergoes enolization via a complex formation with the enzyme active site, and Mg^{+2}. The enediol form couples with CO$_2$ when a new carbon–carbon bond is generated. The adduct being unstable undergoes hydrolysis to yield two molecules of D-3-phosphoglycerate (Fig. 3.5).

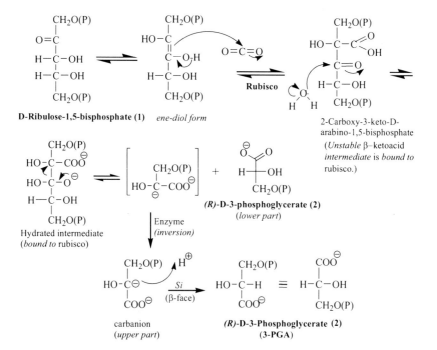

Fig. 3.5 Stage 1 of the Calvin cycle: *Carboxylation*

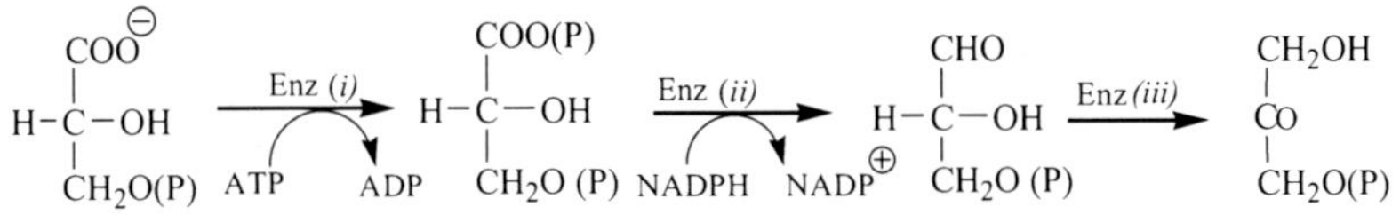

$$
\begin{array}{ccccc}
\overset{\ominus}{\text{COO}} & & \text{COO(P)} & & \text{CHO} & & \text{CH}_2\text{OH} \\
| & \xrightarrow{\text{Enz }(i)} & | & \xrightarrow{\text{Enz }(ii)} & | & \xrightarrow{\text{Enz }(iii)} & | \\
\text{H}-\text{C}-\text{OH} & & \text{H}-\text{C}-\text{OH} & & \text{H}-\text{C}-\text{OH} & & \text{CO} \\
| & & | & & | & & | \\
\text{CH}_2\text{O(P)} & \text{ATP} \quad \text{ADP} & \text{CH}_2\text{O (P)} & \text{NADPH} \quad \overset{\oplus}{\text{NADP}} & \text{CH}_2\text{O (P)} & & \text{CH}_2\text{O(P)}
\end{array}
$$

(R)-D-3-Phosphoglycerate   *(R)*-D-1,3-Bisphospho-   **(R)-D-3-Phospho-**   **Dihydroxyacetone**
(3-PGA) (**2**)   glycerate (**3**)   **glyceraldehyde (4)**   **phosphate (5)**

Compound numberings correspond to those in **Figure 3.4**

Enzymes : *(i)* 3-Phophoglycerate kinase
          (ii) Glyceraldehyde 3-phosphate dehydrogenase
          (iii) Triose phosphate isomerase

Fig. 3.6 Stage 2 of the Calvin cycle: Formation of D-3-phosphoglyceraldehyde (**4**) and dihydroxyacetone phosphate (**5**)

Stage 2: Formation of 3-Phosphoglyceraldehyde (4) and Dihydroxyacetone phosphate (5), the More Versatile Biosynthetic Precursors, by Reduction of 3-Phosphoglycerate (2) Conversion of D-3-phosphoglycerate (**2**) to D-3-phosphoglyceraldehyde (**4**) takes place in two steps (Fig. 3.6). In the first step, 3-phosphoglycerate kinase catalyzes the transfer of phosphate from ATP to D-3-phosphoglycerate (**2**), yielding D-1,3-bisphosphoglycerate (**3**). In the second step, (**3**) is reduced by NADPH, being catalyzed by glyceraldehydes-3-phosphate dehydrogenase to form D-3-phosphoglyceraldehyde (**4**), which isomerizes to (**5**) in presence of triose phosphate isomerase.

Stage 3: The Regeneration of the CO$_2$ Acceptor D-ribulose-1,5-bisphosphate (1) The third stage of CO$_2$ fixation involves the remaining set of reactions of the Calvin cycle. The products of stage 2, *viz.*, the triose phosphates (**4**) and (**5**) pass through a series of successive transformations to six-, four-, five-, and seven-carbon sugars, eventually leading to the regeneration of the starting material, D-ribulose-1,5-bisphosphate (**1**) in presence of a specific enzyme in each stereospecific step as shown in Figs. 3.7–3.9. Thus, a continuous flow of CO$_2$ into carbohydrates is maintained in the Calvin cycle. The H$_S$ of (**5**) attacks on the *Si* (or α-) face of the carbonyl group of (**4**), being catalyzed by aldolase, to produce (**6**). In the next step, the transfer of CH$_2$OH–CO– of (**7**) to the *Re* face of –CHO of (**4**) is effected by the transketolase enzyme whose prosthetic group, thiamine pyrophosphate (TPP), acts as the carrier. *Regeneration of D-ribulose-1,5-bisphosphate* (**1**) *and its formation from* (**4**) *and* (**5**) *via* (**6**) *to* (**13**) *in succession* are outlined and explained stepwise in Fig. 3.9.

The mechanism of the catalytic function of TPP is shown in Fig. 3.8.

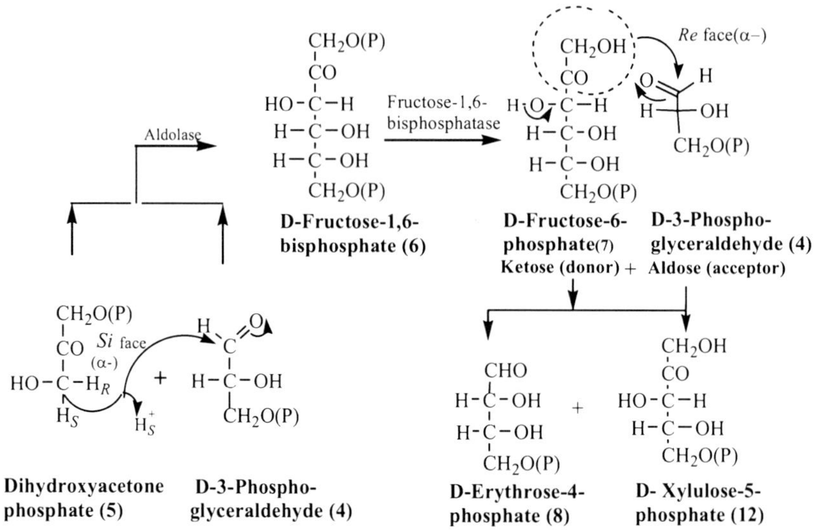

Compound numberings correspond to those in *Figure 3.4*

Fig. 3.7 Stage 3 of the Calvin cycle (1st part): Formation of D-erythrose-4-phosphate (**8**) and D-xylulose-5-phosphate (**12**)

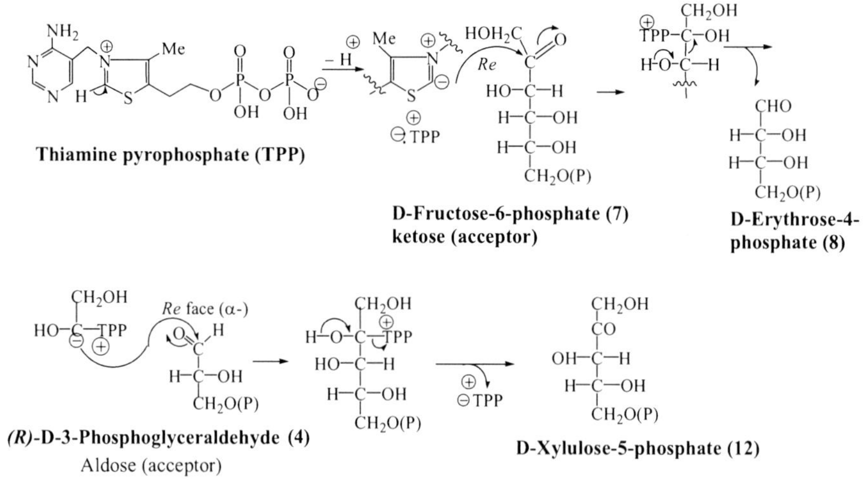

Compound numberings correspond to those in **Figure 3.4**.

Fig. 3.8 Function of TPP in the conversion of (**7**) to (**8**) and (**12**)

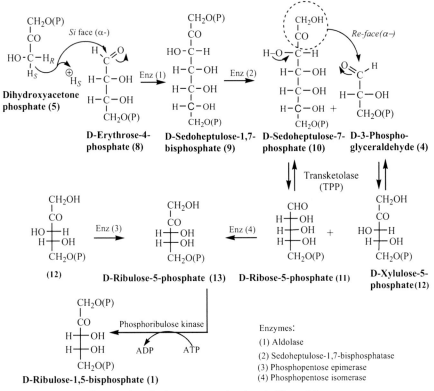

Compound numberings correspond to those in **Figure 3.4**

Fig. 3.9 Stage 3 of the Calvin cycle (remaining part): Regeneration of D-ribulose-1,5-bisphosphate (**1**)

3.1.2.1 Some Comments and Implications Regarding Calvin Cycle Molecules

(i) The reactions can be summarized as follows:

$$C_3\,(5) + C_3\,(4) \xrightarrow{\text{Aldolase}} C_6\,(6) \rightarrow C_6\,(7)$$

$$C_6\,(7) + C_3\,(4) \xrightarrow{\text{Ketolase}} C_4\,(8) + C_5\,(12)$$

$$C_4\,(8) + C_3\,(5) \xrightarrow{\text{Aldolase}} C_7\,(9) \rightarrow C_7\,(10)$$

$$C_7\,(10) + C_3\,(4) \xrightarrow{\text{Ketolase}} C_5\,(4) + C_5\,(12)$$

Adding,

$$5\,C_3 \longrightarrow 3\,C_5$$

$$C_5\,(13) \longrightarrow C_5\,(1)$$

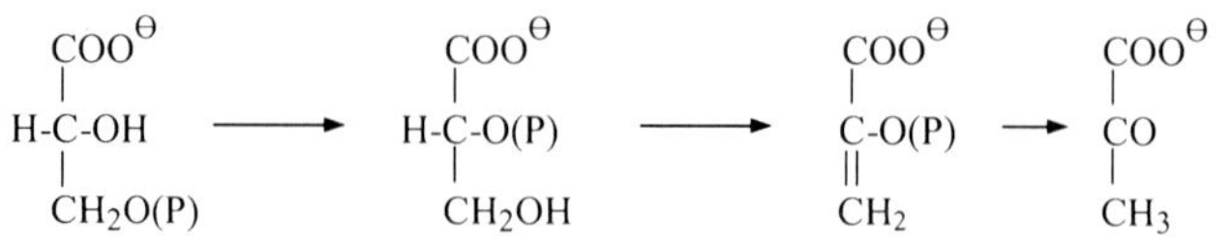

$$
\begin{array}{cccc}
\overset{\ominus}{COO} & \overset{\ominus}{COO} & \overset{\ominus}{COO} & \overset{\ominus}{COO} \\
| & | & | & | \\
\text{H-C-OH} \longrightarrow & \text{H-C-O(P)} \longrightarrow & \text{C-O(P)} \rightarrow & \text{CO} \\
| & | & || & | \\
\text{CH}_2\text{O(P)} & \text{CH}_2\text{OH} & \text{CH}_2 & \text{CH}_3
\end{array}
$$

| 3-Phosphoglycerate | 2-Phosphoglycerate | Phosphoenol | Pyruvate |
| (3-PGA) (2) | (2-PGA) | pyruvate | |

Fig. 3.10 Stepwise conversion of 3-PGA to pyruvate

(ii) Each of these C_3, C_4, C_5, and C_6 molecules plays an important role in the biosynthesis of natural products.

(iii) (R)-D-3-Phosphoglyceraldehyde, (**4**), generated in the 2nd stage of the Calvin cycle, serves as the mother molecule for the other sugars (4-, 5-, 6-, and 7-carbon sugars) including the CO_2 acceptor molecule D-ribulose-1,5-bisphosphate (**1**), preserving the stereochemical integrity of the chiral center of (**4**), which becomes the lowermost chiral carbon. Hence, according to IUPAC nomenclature, all the sugars are named D-sugars (Chap. 2). Apparently, in the plant cells, no enzyme is available which can epimerize the designated chiral carbon of the sugars producing an L-diastereomer, although epimerization at other centers in presence of specific epimerases is known.

(iv) Most of the D-glyceraldehyde-3-phosphate (**4**) is recycled to D-ribulose-1,5-bisphosphate (**1**). The remaining (**4**) may be used immediately as a source of energy, converted to sucrose for transport *via* its precursor D-fructose-6-phosphate (**7**), or stored as *starch* for future use. The part of dihydroxyacetone phosphate (**5**), which is not used in Calvin cycle, leaves the chloroplast and can be degraded via glycolysis to provide energy.

(v) 3-Phosphoglycerate (**2**) (3PGA) gives rise to pyruvate at the primary metabolic level as shown in Fig. 3.10.

(vi) 3PGA may also be formed by the breakdown of D-glucose *via* D-fructose 1,6-diphosphate (**6**)

(vii) Pyruvic acid is a very important key compound. It forms acetyl coenzyme A; the latter enters into the formation of fatty acids, polyketides (Chap. 14), and aromatic compounds and also terpenoids and steroids *via* mevalonic acid pathway (Chaps. 5, 5, and 11).

3.1.3   *C₄-Plant Photosynthesis, C₃- and C₄-Plants*

As discussed above, a C_3-acid, phosphoglyceric acid (PGA) is the primary fixation product in the **Calvin cycle** of the carbon fixation process. PGA is formed mostly in the chloroplast present in the leaf *mesophyll* tissues involving *rubisco* having both carboxylase and oxygenase catalytic activities. Because of the latter activity of the enzyme which is less suppressed by comparatively low CO_2 intake from air (CO_2 concentration being much less than that of O_2), a considerable part of fixed carbon is oxidized and lost as CO_2—a phenomenon known as *photorespiration* resulting into a significant loss of energy. Thus this process is inefficient.

Subsequent to the discovery of Calvin cycle, Hugo Kortschak obtained an unexpected result while studying the photosynthesis with $^{14}CO_2$ in sugarcane (at the Sugarcane Research Institute, Hawaii), in which C_4-acids, viz., oxaloacetate and malate appeared as the primary fixation products. Kortschak did not publish this observation in any journal. Ten years later, a Russian scientist Yuri Karpilov made a similar observation in the maize plant. This observation challenged the universality of the Calvin cycle. However, Hatch and Slack [5, 6] solved this riddle. It was then generalized that in some plants, prior to the Calvin cycle, CO_2 is prefixed as some C_4 acids like *oxaloacetate* and *malate*, as the primary fixation products. It was suggested that such fixation takes place in leaf cells with Kranz (meaning garland/wreath)-type anatomy (a few exceptions) having a wreath type arrangement of cells, with the *inner bundle sheath cells* being surrounded by *mesophyll tissue cells.*

Initial fixation to D-*malate* takes place in the outer *mesophyll* tissues, which lacks rubisco. Atmospheric CO_2 is converted to HCO_3^- catalyzed by *carbonic anhydrase*, which is absent in C_3-plants. The bicarbonate anion reacts with *phosphoenol pyruvate* (PEP), being catalyzed by PEP carboxylase to form oxaloacetate, which is then enzymatically reduced to malate (Fig. 3.11). The metabolic product malate is transferred to *bundle sheath* cell zone through a mechanism involving *plasmo-desmata* (intercellular connections), since the bundle sheath cell wall is impermeable to the fixation products. D-Malate forms pyruvate and releases CO_2 to *rubisco* as its substrate to enter into Calvin cycle (Fig. 3.11). The pyruvate moves to mesophyll tissue cells, gets phosphorylated to form PEP, which is reused in the oxaloacetate formation, and the cycle is repeated.

In some plants (e.g., millet and forage plants) glutamate aspartate transaminase concentration being higher, oxaloacetate is transaminated to aspartate (in presence of glutamate), which moves to bundle sheath cells. There it is reconverted to oxaloacetate that in turn enzymatically releases CO_2 and forms PEP (Fig. 3.11). Initially, the concentration of oxaloacetate is not sufficiently high to allow its participation in the metabolic flux to diffuse from mesophyll tissue cells to bundle sheath cells. Hence, it moves via malate or aspartate. This pathway is known as Hatch and Slack pathway.

Thus, two major pathways, C_3-acid pathway (Calvin cycle and RPP cycle) and C_4-dicarboxylic acid pathway (Hatch and Slack pathway and C_4-pathway), are followed in photosynthesis, and according to the primary fixation products the plants are classified as C_3-*plants* (most plant species belong to this class) and C_4-*plants* (mostly pasture plants, crops, forage plants, wild weeds, desert plants, etc.). In C_3-plants *rubisco* is the primary enzyme while in C_4-plants phosphoenol pyruvate carboxylase (*PEP carboxylase*) is the key enzyme.

In C_4-plants, there is no rubisco in the mesophyll tissue cells, but some rubisco is present exclusively in bundle sheath cells. C_4-Plants need less rubisco, less nitrogen, and also less water for growth. So in warm weather, C_4-plants grow advantageously. However, C_4-plants are affected by chilly weather.

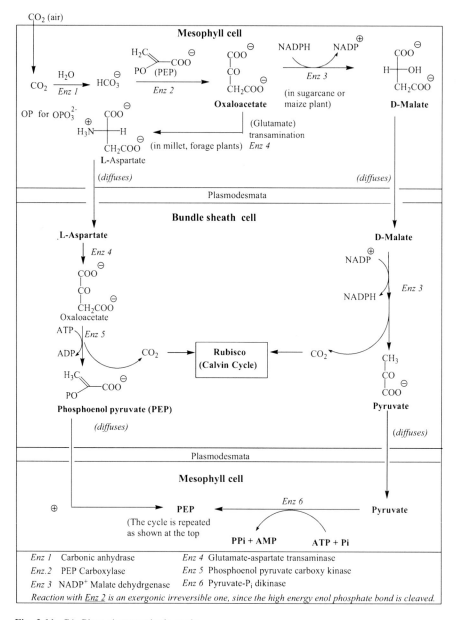

Fig. 3.11 C4- Plant photosynthetic cycle

3.1.3.1   Identification of C_3 and C_4 Metabolism Products by Mass Spectrometry

The natural isotopes of carbon are ^{12}C (98.89 %) and ^{13}C (1.11 %). Because of the kinetic isotope effect rubisco reacts faster with $^{12}CO_2$ than with $^{13}CO_2$ in C_3 plants, resulting in the lowering of $^{13}C/^{12}C$ ratio in the products of C_3 photosynthesis than this ratio in the atmospheric CO_2. In C_4-metabolism, the enzymatically prefixed CO_2 with less kinetic isotope effect reacts with rubisco in bundle sheath cells and thus such lowering of the $^{13}C/^{12}C$ isotope ratio in the photosynthesis products is less. Hence by studying the $^{13}C/^{12}C$ ratio in the product mass spectrometrically, it is possible to predict the metabolism pathway followed to form the product. For example, sugar obtained from two different sources, e.g., sugar beet (C_3-plant) as well as sugarcane (C_4-plant) can be related to its source by its isotope ratio analysis. In C_4 Pathway, photorespiration is almost absent due to high concentration of CO_2 released by the prefixed C_4 products, and the suppression of the enzymatic oxygenase activity of rubisco due to absence of O_2.

3.1.3.2   Crassulacean Acid Metabolism (CAM)

The plants like all cacti and many desert plants growing in arid zones, the succulent ornamental plant *Kalanchoe* and plants growing in tropical rain forests, including half of the orchids manage the extreme shortage of water by keeping their leaf stomata closed during daytime to avoid loss of water, and reduce the area of their leaves. They open their leaf stomata in the night to allow the entry of CO_2 and to mask evaporation. CO_2 is prefixed as C_4-acids and stored in the vacuole until next morning when they are degraded to release CO_2 to rubisco in the chloroplasts; the pyruvate formed is subsequently phosphorylated by pyruvate phosphate kinase. The major difference in this case is the time lag between CO_2 fixation (storage) and its release. The mechanism has been first studied in Crassulaceae plants, and CO_2 is stored as an acid and hence the name crassulacean acid metabolism (*CAM*). Pineapples and the agave sisal, yielding natural fibers are important *CAM plants*.

3.2   Biological Oxidation: Reduction (NADPH $\rightleftharpoons$ NADP$^+$)

As mentioned earlier (Chap. 1), there are enzymes which are coenzyme or cofactor dependent in their functions. Enzyme promoted oxidation–reduction in the biological systems uses several coenzymes. The function of one such coenzyme, nicotinamide adenine dinucleotide phosphate [in the reduced (NADPH) (**15**) and oxidized (NADP$^\oplus$) (**14**) forms] (Fig. 3.12) will be discussed briefly. They act in a large number of oxidation–reduction reactions. Sometimes, NADPH has been compared to the laboratory reagent NaBH$_4$ by describing it as Nature's NaBH$_4$. However, the

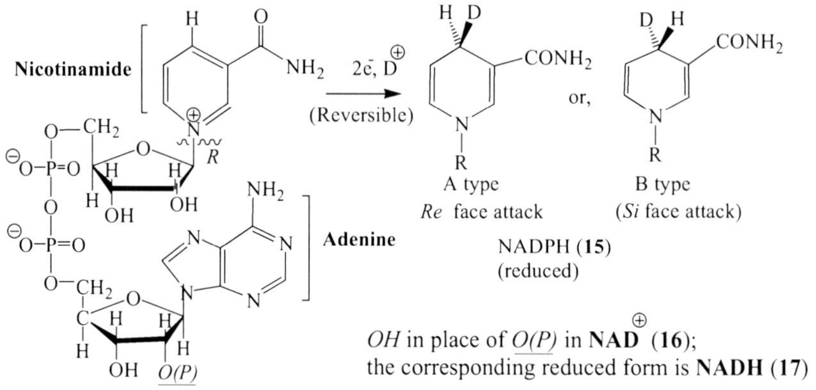

Nicotinamide adenine dinucleotide phosphate (NADP⊕)(14) ;
the corresponding reduced form is **NADPH (15)**

Fig. 3.12 Structures of NADP⊕, NADPH, NAD⊕ and NADH

latter being achiral and incapable of being reversed like a redox system, such a comparison is hardly reasonable. However, Hantzch 1,4-dihydropyridine has been widely used as a biomimetic reducing agent (cf. NADPH) (Chap. 13).

The coenzyme is tightly bound to the enzyme by multiple noncovalent interactions in a dissociable fashion. Many oxidation–reduction involving enzymes carry out electron transfer through the agency of NADPH/NADP$^+$. The nicotinamide part of the coenzyme is responsible for the biological oxidation–reduction reaction and constitutes an example of a biological redox system; its $CONH_2$ part ensures its redox potential; and the R part controls the binding of the coenzyme to the enzyme. The word **'redox'** is the abbreviated form of '**red**uction–**ox**idation'.

The term oxidation finds its genesis in terms of addition of oxygen (or removal of hydrogen) and reduction consists of removal of oxygen (or addition of hydrogen). They can better be explained in terms of electron gain (reduction) and electron loss (oxidation) during the process.

The nicotinamide end of the coenzyme participates in the oxidation–reduction of a substrate molecule by way of a hydride removal and acceptance, at the same time the concerned moiety of the coenzyme undergoing reduction and oxidation (Fig. 3.13). Enzymatic reactions are stereospecific; of the two diastereotopic hydrogen atoms ($H_R = H_A$ and $H_S = H_B$, see Chap. 2) of NADPH (C4 being prochiral), H_R with its bonded electron pair (hydride) is transferred to the substrate to effect its reduction, NADPH itself being converted to its oxidized form NADP$^+$. In the reverse process, i.e., during oxidation of the substrate, the same hydrogen (H_A) with its bonded electron pair is delivered to the *Re* face of NADP$^+$ to convert it to NADPH, and H_A now becomes H_R. Some enzymes use the ***pro-S*** (or H_S) in a similar fashion.

The stereospecificity of several oxidation–reduction processes caused by biological redox system with respect to the substrate and the coenzyme has been

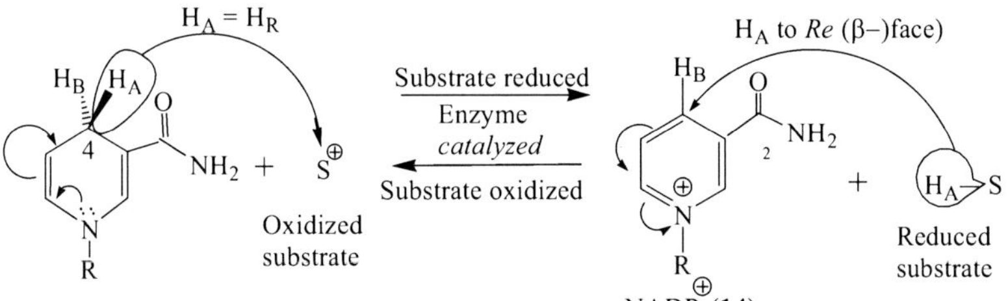

$H_A = H_R$

H_B H_A O

NH$_2$ +

$S^{\oplus}$

Oxidized
substrate

N
|
R

NADPH (15)
1,4-Dihydronicotinamide moiety
Here, $H_A = H_R$; $H_B = H_S$

Substrate reduced
→
Enzyme
catalyzed
←
Substrate oxidized

R as in **Figure 3.12**

H_A to *Re* (β−)face)

H_B O

NH$_2$

N
|
R
$\oplus$

NADP$^+$ (14)
Nicotinamide moiety

+

H_A S

Reduced
substrate

Fig. 3.13 Stereospecific delivery and acceptance of hydride by NADPH and NADP + respectively.

proved by labeling the key hydrogen by deuterium/tritium. The stereochemical information is delivered at the site of attack depending on the *diastereofacial differentiation*, as illustrated in Fig. 3.13.

Incubation of the alcohol dehydrogenase with CH_3CD_2OH in presence of the cofactor NAD$^+$ (**16**) generates the reduced cofactor (**17a**) (Fig. 3.14). The latter is reoxidized by CH_3CHO to NAD$^+$ (**16**), the former being reduced to 1*R*-deuteroethanol, demonstrating that the D atom transferred to the cofactor at C4 at its *Re* face is stereospecifically removed in the reverse reaction to attack the *Re* face of CH_3CHO, forming the chiral 1*R*-deuteroethanol. In the reverse reaction (1*R*)-deuteroethanol when incubated with NAD$^+$ (**16**) in presence of the enzyme regenerates only 4*R*-^{2}H-NADH (**17a**) with complete transfer of D to its *Re* face. Evidently, the rigorous stereochemical control of enzyme catalyzed reactions is brought about by the precise positioning of the substrate and the coenzyme at the enzyme active site.

The interconversion NAD$^+$ (16) $\rightleftharpoons$ NADH (17a) and their phosphorylated analogs NADP$^+$ $\rightleftharpoons$ NADPH can be monitored in vitro by UV spectroscopy. Both the oxidized forms show one absorption band around 260 nm, whereas reduction to NADH or NADPH produces a new broad absorption band with a maximum of 340 nm. Thus, the production of NADH or NADPH during an enzyme-catalyzed oxidation of a substrate can be conveniently followed by observing the appearance of the λ_{max} at 340 nm. We cite here another example of diastereospecific enzymatic reduction by NADH. It occurs during *souring of milk*.

Glycolysis, a part of *fermentation*, is a ten-step pathway converting one molecule of glucose to two molecules of pyruvate. A lactic acid bacterium uses a direct route to effect an enantiospecific reduction of pyruvate to *S*-L-lactate with NADH in presence of the enzyme lactate dehydrogenase (Fig. 3.15). Here, an electron pair from NADH attacks the *Re* (front) face of pyruvate. *Souring of milk* involves this endergonic reaction.

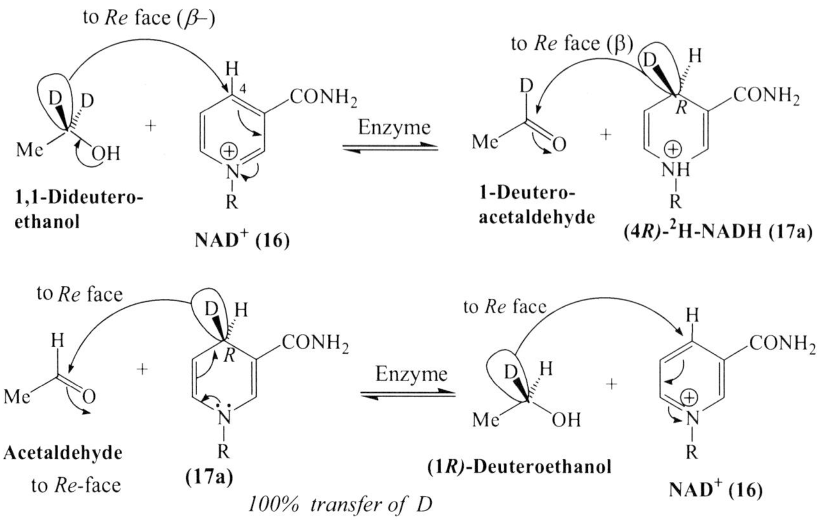

Fig. 3.14 Stereospecificity of alcohol dehydrogenase

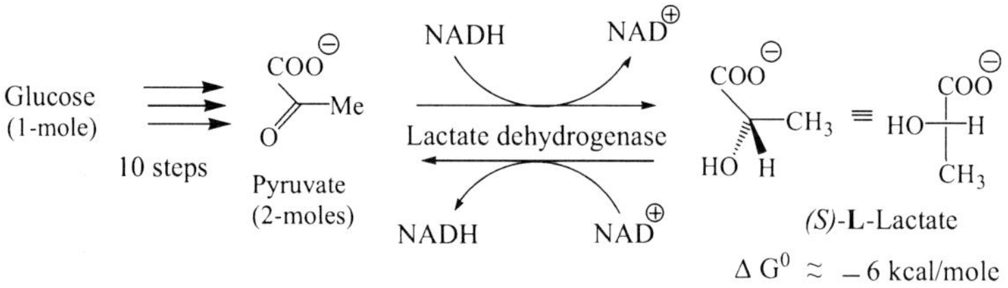

Fig. 3.15 Enantiospecific reduction of pyruvate to L-lactate (during souring of milk)

3.2.1   Flavin Coenzymes

Two coenzymes, flavin adenine dinucleotide (FAD) and flavin mononucleotide (FMN), are derived from vitamin B_2 or riboflavin. FMN is riboflavin phosphate (Fig. 3.16). Functionally both types are equivalent. Some enzymes use FAD while some others use FMN. Highly conjugated tricyclic isoalloxazine ring system, the functional part of both coenzymes, serves as a two-electron acceptor and is responsible for its strong redox character toward diverse classes of compounds. Molecules possessing such a ring system are called flavins. Flavin coenzymes are used by enzymes called flavoproteins (flavoenzymes). The latter bind FMN or FAD in most cases noncovalently but tightly so that the coenzymes can be reversibly dissociated. The flavins undergo two-electron oxidation and reduction reactions. They have a stable species, a semiquinone free radical produced by one-electron reduction that can be detected spectrophotometrically. The oxidized flavins are bright yellow;

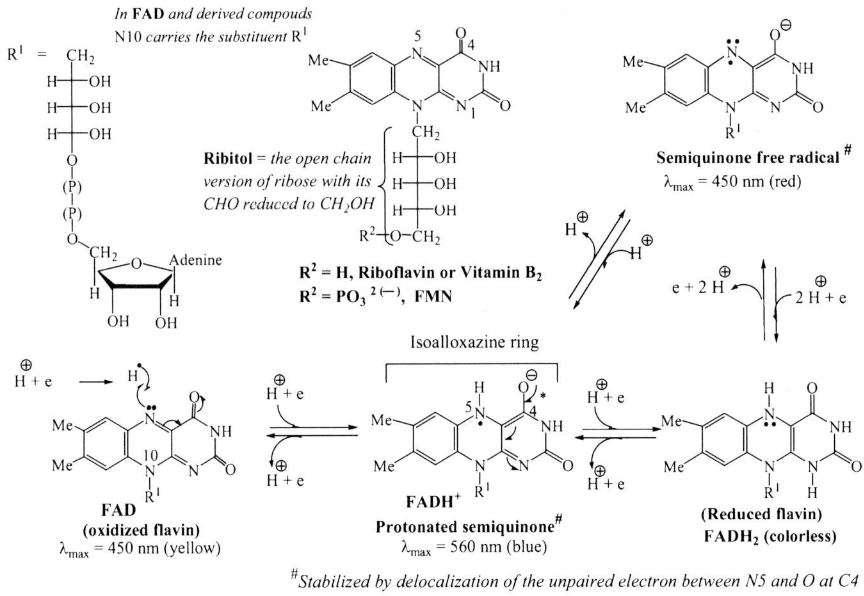

Fig. 3.16 Flavin coenzymes and the mechanism of their redox reactions

protonated semiquinone is blue, the semiquinone free radical is red, but the fully reduced flavins are colorless. The flavins have catalytic versatility more than nicotinamide coenzymes, since they can interact with two-electron or one-electron donor–acceptor pairs. Mechanism of stepwise reduction of FAD to FADH$_2$ via FADH$^+$ (protonated semiquinone) is shown in Fig. 3.16.

3.2.2   Combined Use of NADPH and FAD

There are cases when the substrate molecule is not directly reduced by NADPH (or NADH). Transfer of hydride from NADPH (or NADH) takes place indirectly. Two electrons are delivered from NADPH to a second enzyme cofactor, e.g., FAD in two steps to the N5 in the isoalloxazine ring, accompanied by accepting one proton from the medium in each step to form FADH$_2$. The latter delivers two electrons to the substrate to reduce it, itself being oxidized to FAD (Fig. 3.17).

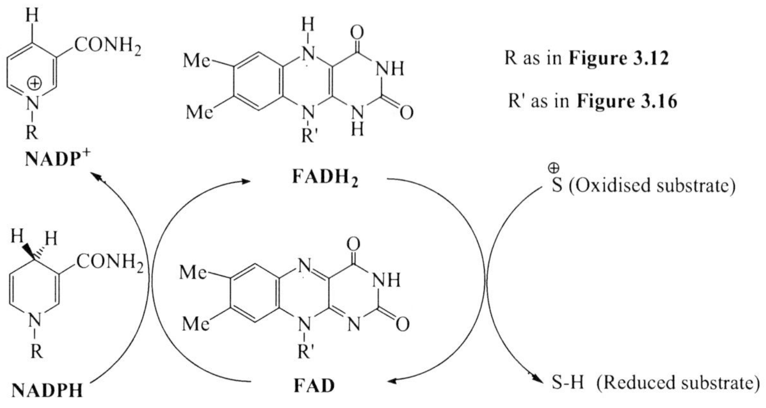

Fig. 3.17 Reduction by NADPH/FAD through $FADH_2$ in presence of enzyme

3.3 Phosphorylation (ATP→ADP) and Regeneration (ADP→ATP)

3.3.1 Function of ATP: Its Conversion to ADP

Adenosine triphosphate (**18**) (Fig. 3.18), abbreviated as ATP, is one of the major cofactors in many metabolic biosynthetic pathways. It acts as the phosphate donor to the appropriate substrates and is biologically a potential source of metabolic energy. It is considered to be the key molecule for capturing and storing chemical energy. It moves through the cells providing energy for all processes of the cells [1]. It contains one ester and two anhydride phosphoryl groups (β- and γ-), and the $5'CH_2$ group where nucleophilic attack could take place. Its γ-phosphate group is transferred to various acceptors to activate them for their subsequent participation in various biological events. In unusual cases, attack on C5′ of the adenosyl part takes place, e.g., the formation of *S*-adenosyl methionine (SAM or AdoMet) (see Sect. 3.6).

The function of ATP in various biochemical reactions has been illustrated, whenever necessary (e.g., formation of mevalonic acid and γ,γ-dimethylallyl pyrophosphate, see Chap. 5). It has been observed that during pyrophosphorylation of alcoholic OH group (e.g., mevalonic acid to mevalonic acid pyrophosphate, Chap. 5, Fig. 5.3), two molecules of ATP are needed. Each ATP molecule donates its γ-phosphate group to the substrate and is converted to ADP. Presumably, the incapability of ATP for donating two phosphate groups in succession (ATP→ADP→AMP) is prevented by the spatial incompatibility of ADP to work as coenzyme with phosphorylating enzyme for such a reaction. The ATP can be hydrolyzed to adenosine diphosphate (ADP) (**19**) by shedding off the γ-phosphate.

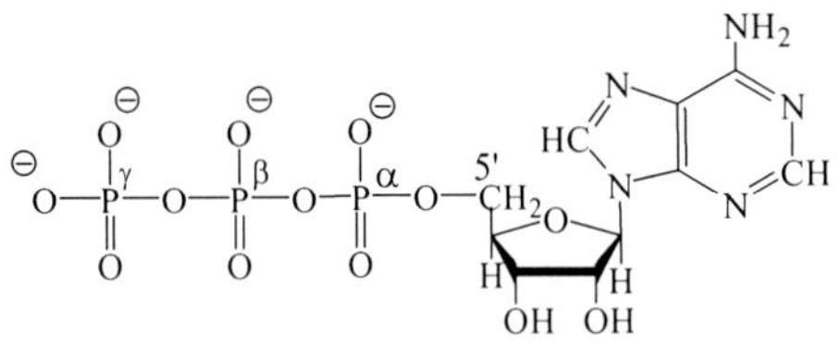

5$'$CH$_2$ when attached to –O–(P)(P)
gives adenosine diphosphate **(ADP) (19)**

5$'$CH$_2$ when attached to –O–(P)
gives adenosine monophosphate **(AMP) (20)**

(P) stands for

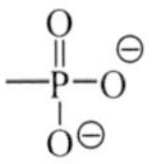

Adenosine triphospate
ATP (18)

Fig. 3.18 Structures of **ATP(18)**, **ADP(19)** and **AMP(20)**

$$\text{ATP} + \text{H}_2\text{O} \rightleftharpoons \text{ADP} + \text{Phosphate (P}_i\text{)}; \Delta G^\circ = -30.5\,\text{kJ/mol} \ (\textit{exergonic reaction})$$

The negative standard free energy ΔG^0 is a measure of the ability of its phosphoryl group transfer potential. However, it has been shown that standard free energy of the hydrolysis of ATP depends on the water concentration. In the absence of water, the above reaction is directed towards ATP formation, and much energy is not needed. The reaction site of ATP formation in the enzyme can exclude water molecules and may proceed without the consumption of energy. Actually, energy requirement is necessary for certain conformational change in the catalytic sites of the enzyme and finally to release the ATP from the binding site.

3.3.2   Conversion of ADP to ATP

Adenosine diphosphate (ADP) can be reconverted to ATP by uniting with inorganic phosphate and with an adequate input of energy. In the leaf cells, during photosynthesis ADP is converted to ATP (photophosphorylation) (Fig. 3.3) by solar energy, and also in almost all cells by aerobic oxidation (oxidative phosphorylation) using chemical energy.

3.3.3   Formation of Proteins from Amino Acids

ATP favors thermodynamically unfavorable reactions like peptide bond formation. Calculations show that polymerization of amino acids to protein in the absence of ATP should be *endergonic, i.e.,* ΔG becomes positive.

Amino acid $\rightarrow\rightarrow\rightarrow$ Protein; here ΔG_1 is positive (*endergonic reaction*)

However, ATP activates the amino acid by cellular reaction involving ATP breakdown to adenosine monophosphate (**20**) (AMP), giving more negative ΔG

(−45.6 kJ/mol). The sum total free energy change for the formation of protein becomes negative. Since $\Delta G_2 << \Delta G_1$ (the magnitude of the negative ΔG_2 being much more than the positive ΔG_1), the overall reaction becomes thermodynamically favorable.

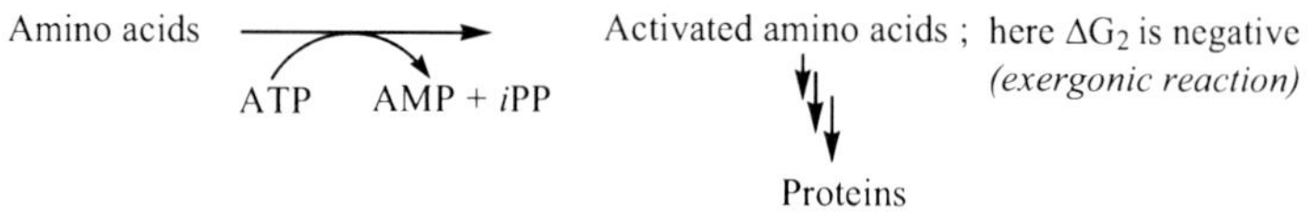

Amino acids $\xrightarrow{\hspace{2cm}}$ Activated amino acids ; here ΔG_2 is negative

ATP   AMP + iPP   *(exergonic reaction)*

Proteins

3.3.4 Biosynthesis of Starch with the Help of ATP

Similarly in the biosynthesis of starch, polymerization of glucose takes place through the participation of ATP (Fig. 3.19). Different plants perceive a variety of environmental and endogenous influences, and as a consequence various chemical components including starch, differ in the composition. Amylose, one form of starch contains repetitive units of glucose united *via* α-1,4- disaccharide linkage. α-D-Glucose-1-phosphate reacts with ATP reversibly to give ADP-glucose. The pyrophosphate formed is hydrolyzed by the enzyme *pyrophosphatase* and in this way the formation of ADP-glucose becomes irreversible. The glucose activated by

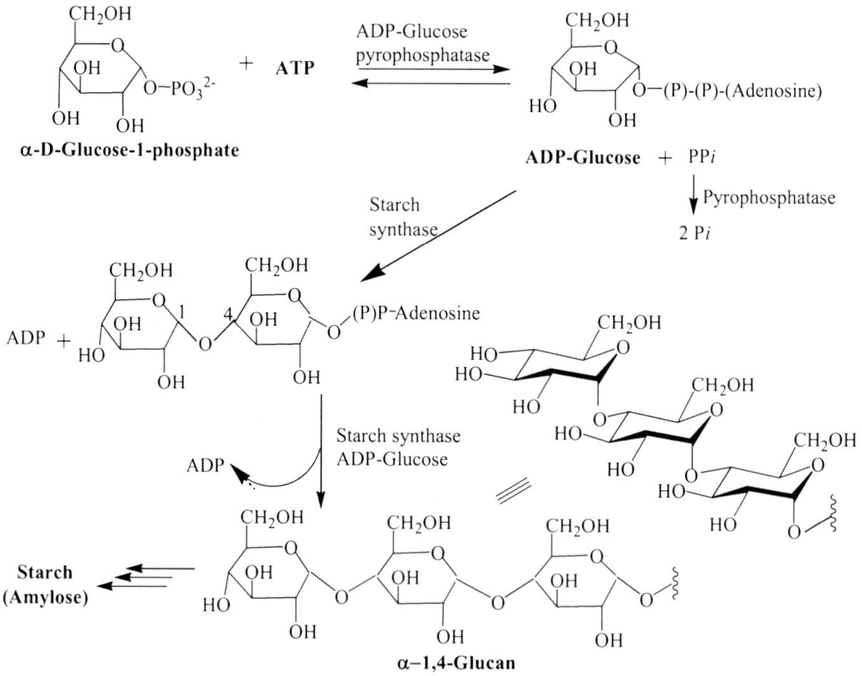

Fig. 3.19 Biosynthesis of starch (amylose) from -D-glucose

ADP is transferred by starch synthase from ADP-glucose to the OH group in the 4-position of the terminal glucose residue in the polysaccharide chain in the formative stage of starch. Similar deposition of glucose residue (α-1,4-disaccharide linkage) continues, catalyzed by starch synthase to form starch.

3.4 Acetyl Coenzyme A

The discovery of the cofactor, acetyl coenzyme A (**21**) (Fig. 3.20), an essential biochemical reagent for many diverse biological reactions, led to a major advance in the field of biochemistry. Structurally, it has three units: β-mercaptoethylamine, pantothenate (vitamin B$_5$), and 3'-phosphoadenosine-5'-diphosphate units, as shown in its structure (**21**).

Breslow provided evidence (IR, NMR) for the detection of stable thiazolium zwitterions of model thiazolium compounds by deuterium exchange studies in D$_2$O [7]. He thus suggested a mechanism involving such a zwitterion for the thiamine pyrophosphate (TPP$^+$) (Fig. 3.21) catalyzed in vivo biochemical reactions, e.g., decarboxylation of α-keto acids like pyruvic acid and benzoyl formic acid.

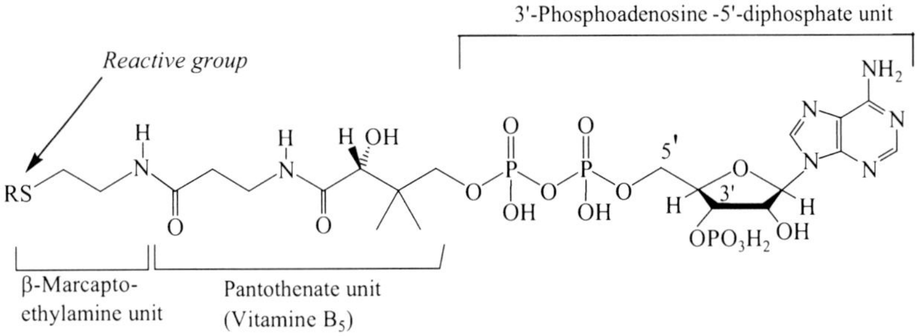

(21) Acetylcoenzyme A, R = COMe, (Acetyl SCoA)
(22) Coenzyme A, R=H (HSCoA)

Fig. 3.20 Structures of acetyl coenzyme A and coenzyme A

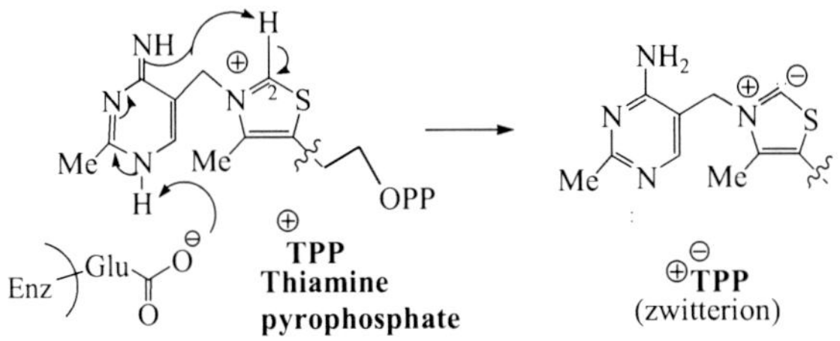

Fig. 3.21 Generation of TPP carbanion

Many other acids are also known to be activated as the corresponding acyl coenzyme A, e.g., malonyl coenzyme A, glutaryl coenzyme A, benzoyl coenzyme A, etc. The acyl coenzymes are responsible for many biological reactions.

3.4.1   Formation of Acetyl Coenzyme A from Pyruvic Acid

The stepwise mechanism of the formation of acetyl coenzyme A (21) by TPP$^+$ catalyzed decarboxylation of pyruvic acid is shown in two parts:

(a) **Generation of TPP carbanion.** It has been shown from NMR and IR studies that the pyrimidine ring of TPP participates in the generation of the carbanion at C2 of the thiazole ring. It is initiated by the abstraction of the proton from the pyrimidine tautomer by the glutamate carboxyl group of the enzyme as shown in Fig. 3.21.

(b) **Participation of carbanion of TPP ($-$:$^+$TPP) in the decarboxylation of pyruvic acid.** The thiazole carbanion participates in the nucleophilic attack on the ketocarbonyl of pyruvic acid to form the adduct (Fig. 3.22). The latter then suffers decarboxylation in the presence of the enzyme pyruvate decarboxylase

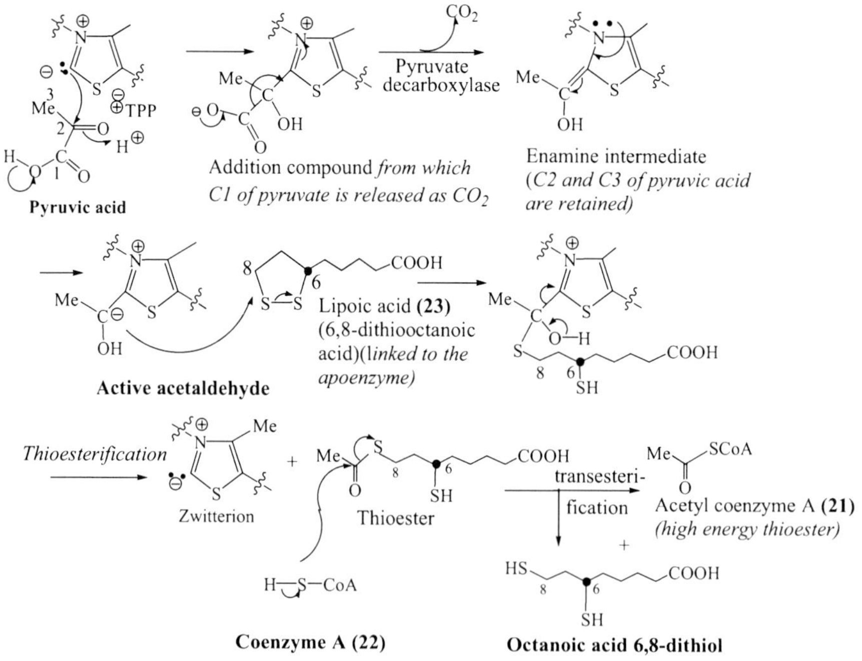

Fig. 3.22 Formation of acetyl coenzyme A from pyruvic acid

$$ATP \quad + \quad CH_3COO^{\ominus} \quad \longrightarrow \quad CH_3COAMP \quad + \; iPP$$

$$CH_3COAMP \; + \; HSCoA \quad \longrightarrow \quad AMP \quad + \quad CH_3COSCoA$$

$$i\,PP \quad \longrightarrow \quad 2\,i\,P$$

Adding, $ATP + CH_3COO^{\ominus} + HSCoA \quad \underset{\text{synthetase}}{\overset{\text{Acetyl coenzyme}}{\rightleftharpoons}} \quad AMP + CH_3COSCoA + 2iP$

(22)   (21)

Fig. 3.23 Formation of acetyl coenzyme A (**21**) from coenzyme A (**22**)

to form an **active acetaldehyde** via an enamine intermediate. The former attacks the disulfide bond of the lipoic acid residue linked to the apoenzyme. The thioester thus formed by transesterification yields acetyl coenzyme A and octanoic acid 6,8-dithiol.

3.4.2 Formation of Acetyl Coenzyme A from Coenzyme A

Acetyl coenzyme A (**21**) is also formed by acetylation of coenzyme A (**22**) by acetate; the reaction is catalyzed by acetate thiokinase as shown in Fig. 3.23. The pyrophosphate (*i*PP) breaks down to inorganic phosphate (*i*P).

3.4.3 Functions of Acetyl Coenzyme

Acetyl coenzyme A (**21**) performs the pivotal role as the starter in the biosynthesis of molecules of different skeletal patterns. It is capable of transferring a C_2-unit, like CH_3CO or CH_2CO moiety to appropriate substrates. It can undergo oligomerization via malonyl coenzyme A to give molecules with active ketomethylene groups, and it finds ways to enter into the *biosynthesis of polyketides, aromatic phenolic compounds and fatty acids* (Chap. 14). It is the key molecule of the several metabolic pathways like *mevalonic acid pathway* (Chap. 5) leading to many terpenoids and steroids and the citric acid cycle.

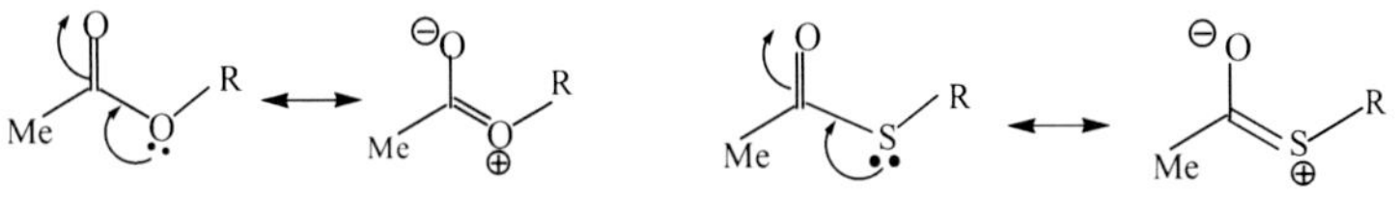

more favorable   *less favorable*

The reactivity of the thioester part of acetyl coenzyme A is due to two factors: (i) the thioester group is less stable than the ester group since the electron delocalization is more prominent in the latter; smaller volume of oxygen atom and its closer orbital energy than that of sulfur helps the overlapping of the lone pair more efficiently with the carbon orbitals, and hence the double bond character of C–S bond is less compared to the C–O bond in their respective esters. (ii) This makes the C–S bond weaker in addition to the more polarizability of the sulfur atom than oxygen; the thioester group is thus a better leaving group than the ester.

3.4.4 Enzymatic Conversion of Choline to Acetylcholine by Acetyl Coenzyme A

Acetylcholine (**ACh**) (**25**) and norepinephrine are important neurotransmitters which use different pharmacological receptors to mediate their end-organ responses. Acetylcholine is formed from choline (**24**) (Fig. 3.24) by the action of acetyl coenzyme A bound to the surface of the enzyme choline acetyltransferase (**ChAT**). The imidazole moiety of the enzyme promotes the removal of the alcoholic proton of choline (**24**) which is also bound to **ChAT**, generating a more nucleophilic center in choline, thus facilitates the acetyl group transfer and produces acetylcholine (**ACh**) (**25**) and coenzyme A (**22**).

High energy transfer potential. Acetyl coenzyme A is a small water-soluble high energy metabolite. It suffers hydrolysis, and the free energy change of the reaction is largely negative:

$$\text{Acetyl CoA} + \text{H}_2\text{O} \rightleftharpoons \text{Acetate} + \text{CoA} + \text{H}^+ \Delta G^\circ = -7.5 \text{ kcal mol}^{-1}$$

The reaction is thermodynamically favorable. The acetyl coenzyme A has a high acetyl transfer potential, the reaction being exergonic.

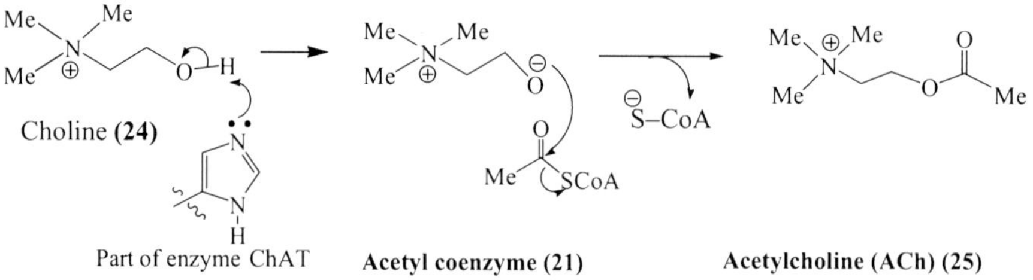

Fig. 3.24 Enzymatic conversion of choline to acetylcholine

3.5 Transamination, Isomerization, and Decarboxylation

Transamination is a very important biological event, a multistep reaction, catalyzed by enzymes called aminotransferases or transaminases. In a transamination reaction, the α-amino group of an L-amino acid$_1$ is donated to the α-carbon atom of an acceptor like an α-keto acid$_2$, *in a stereospecific manner*, leaving behind an α-keto acid, the corresponding α-keto acid analog of the amino acid$_1$. The overall reaction and a specific example are shown in Fig. 3.25. In the transamination reactions, the removal of the α-amino groups constitutes the first step in the catabolism of most of the amino acids. There is no net deamination in such reactions, e.g., α-ketoglutarate becomes aminated, while the α-amino acid is deaminated. The transamination reaction of different L-amino acids with α-ketoglutarate produces L-glutamate. The latter enters either into biosynthetic pathways or into a sequence of reactions producing nitrogenous waste products for excretion.

The aminotransferases are specific for a particular L-amino acid that donates the amino group and are named accordingly. The reactions catalyzed by the transaminases are completely reversible, having an equilibrium constant of nearly 1.0 ($\Delta G^0 \approx 0$ kcal/mol).

The transamination is mediated by the most versatile of all coenzymes (or cofactors), pyridoxal-5'-phosphate (PLP), the vitamin form of which is the corresponding alcohol, pyridoxine (vitamin B$_6$). The overall established mechanism of the PLP catalysis is delineated in two distinct halves in Figs. 3.26 and 3.27.

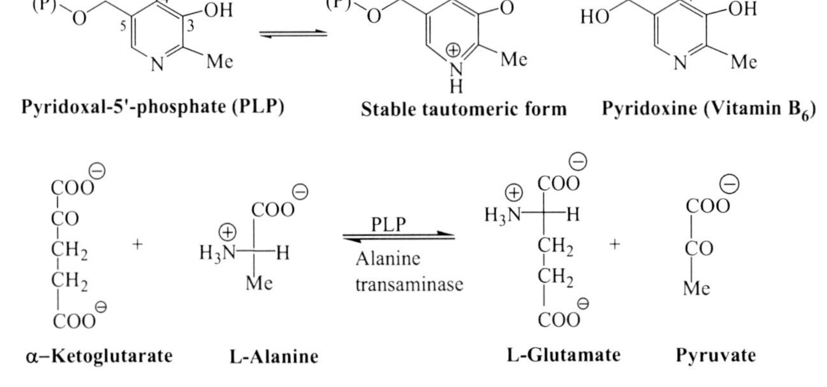

Fig. 3.25 Transamination (Aminotransferase reaction) :

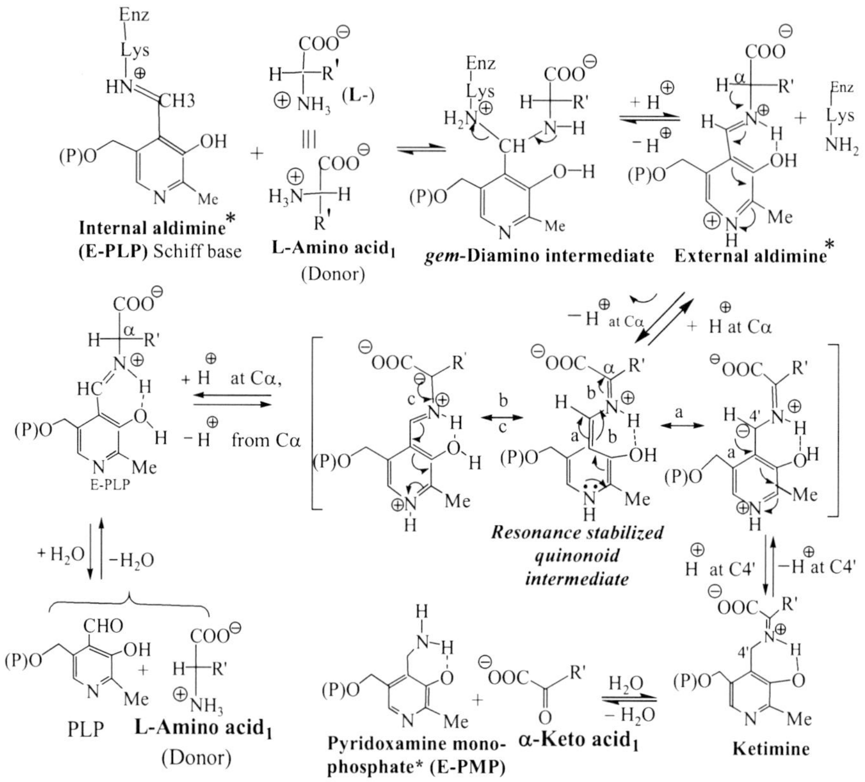

* *Tightly bound with the enzyme by multiple noncovalent interactions*

Fig. 3.26 Mechanism of the first half of the PLP catalyzed transamination reaction

The cofactor PLP is initially bound through multiple noncovalent interactions, and also covalently with the active site of the PLP-dependent enzyme through an imine (Schiff base) linkage, and is commonly called *internal aldimine* or E–PLP, by reaction of 4-CHO of the PLP with the ε-amino group of a lysine residue of the enzyme. The E-PLP is then attacked by the α-amino group of the donor L-amino acid (substrate) to form a *gem*-diamine intermediate that rapidly collapses to a new Schiff base of the coenzyme–substrate complex. The new aldimine thus formed is referred to as the '*external aldimine*'. Loss of proton at the α-carbon (of the amino acid), aided by protonation of the pyridine nitrogen (acting as electron sink), gives rise to a resonance stabilized quinonoid intermediate. This species can be reprotonated through addition of H^+ at C4′. Hydrolysis, of the resulting ketimine gives pyridoxamine phosphate (PMP) and a new keto acid (as in the first half of the transamination reaction). Reprotonation of the quinonoid intermediate may also take place at the amino acid α–C to regenerate the catalyst PLP and the original amino acid (Fig. 3.26), through the reformation of E-PMP.

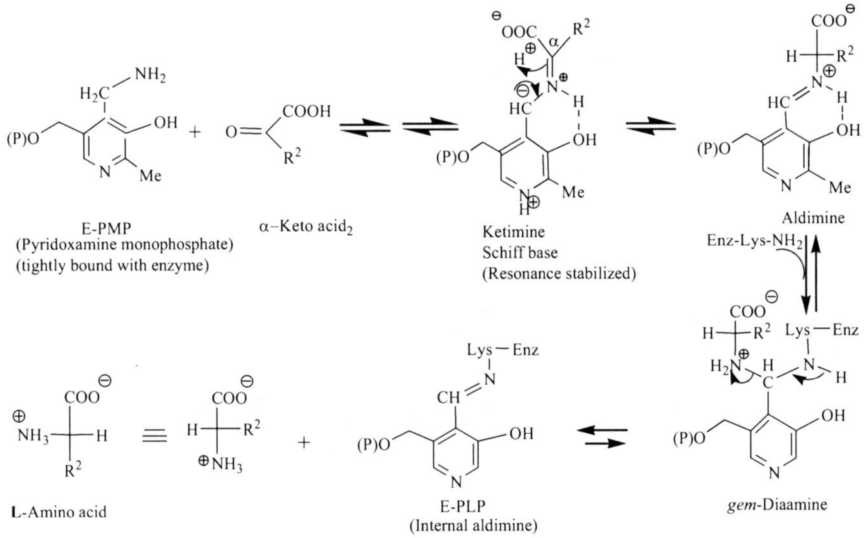

Fig. 3.27 Mechanism of the second half of the PLP catalyzed transamination reaction

Thus the first half of the *transamination reaction* constitutes the following:

$$\alpha\text{-Amino acid}_1 + \text{E-PLP} \rightleftharpoons \alpha\text{-Keto acid}_1 + \text{E-PMP}$$

The second half: $$\alpha\text{-Keto acid}_2 + \text{E-PMP} \rightleftharpoons \alpha\text{-Amino acid}_2 + \text{E-PLP}$$

Adding, $$\alpha\text{-Amino acid}_1 + \alpha\text{-Keto acid}_2 \rightleftharpoons \alpha\text{-Amino acid}_2 + \alpha\text{-keto acid}_1$$

Fig. 3.28 Transamination reaction

The second half of the transamination process, catalyzed by the specific aminotransferase is completed in a reversal of the preceding reaction pathway (Fig. 3.27). The second α-keto acid which replaces the released one, condenses with the generated PMP (tightly bound with the enzyme and known as E-PMP) to form the corresponding ketimine Schiff base, deprotonation at C4′—the pyridine ring acting as the electron sink, leading to the formation of the resonance stabilized Cα carbanion, followed by protonation to form an aldimine. The latter gives rise to a different L-amino acid, and regenerates E-PLP, through the intermediacy of a *gem*-diamine (Fig. 3.27). The E-PMP generated in the first half of the transamination is thus recycled to the E-PLP formed in the second half (Fig. 3.28).

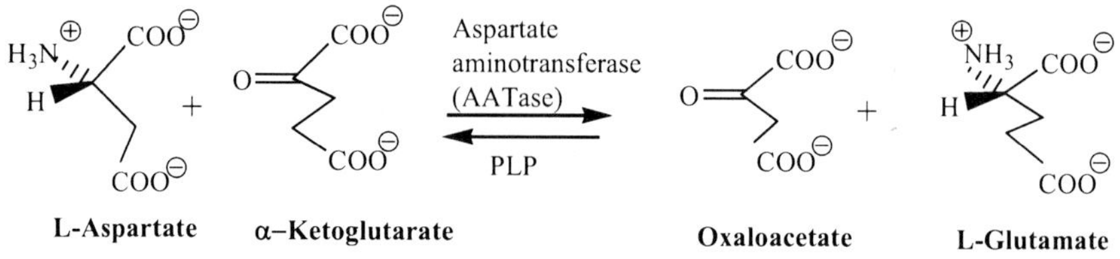

L-Aspartate   α–Ketoglutarate   **Oxaloacetate   L-Glutamate**

E-PLP + L-Aspartate   ⇌   E-PMP + Oxaloacetate
E-PMP + α–Ketoglutarate   ⇌   E-PLP + L-Glutamate

Fig. 3.29 The most studied transamination reaction

3.5.1 Transamination by Aspartate Aminotransferase

Aspartate aminotransferase (**AATase**) is *the most thoroughly studied* of all the PLP-dependent enzymes involved in amino acid metabolism. The reversible interconversion of L-aspartate and α-ketoglutarate to oxaloacetate and L-glutamate is catalyzed by AATase (Fig. 3.29). The enzyme operates through a mechanism already discussed with two distinct half-reactions constituting a full catalytic cycle as illustrated in Fig. 3.29.

3.5.2 Some Interesting Concepts of The PLP-Catalyzed Transamination Reactions

3.5.2.1 Racemization and Decarboxylation

The external aldimine is a common intermediate in all PLP-dependent amino acid reactions. The pyridinium ring in the Schiff base (external aldimine) (Fig. 3.26), being electron deficient, can act as an electron sink and can trigger the breaking of all the four bonds to Cα, *i.e.,* Cα–H or Cα–COOH or Cα–R or Cα–N, being labilized by PLP-dependent enzymes. Abstraction of the Cα proton by an enzymatic base gives rise to the resonance stabilized Cα carbanion which may also undergo inversion and pick up a proton from the stereochemically opposite face to form the Cα-epimeric aldimine. The latter on hydrolysis gives rise to the corresponding D-amino acid (Fig. 3.30). The racemases thus provide D-amino acids for bacterial cell wall synthesis.

Breaking of the Cα–COOH bond effects decarboxylation and eventually leads to the formation of the amine RCH_2NH_2 as outlined in Fig. 3.30. This type of reaction is physiologically very important for mammals. Decarboxylation of glutamic acid produces γ-aminobutyric acid (GABA), a major inhibitory neurotransmitter. Dopamine, the immediate precursor of the hormone epinephrine, is

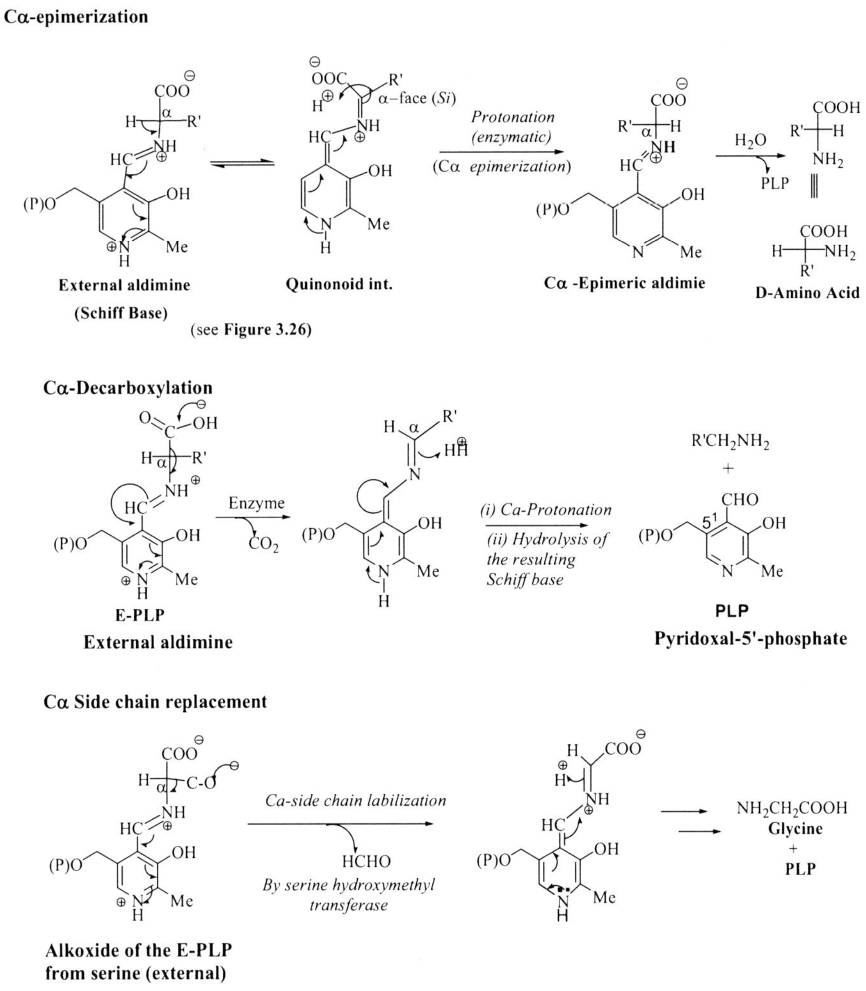

Cα-epimerization

External aldimine   Quinonoid int.   Cα -Epimeric aldimie   D-Amino Acid

(Schiff Base)

(see **Figure 3.26**)

Cα-Decarboxylation

E-PLP   PLP

External aldimine   Pyridoxal-5'-phosphate

Cα Side chain replacement

Alkoxide of the E-PLP
from serine (external)

Fig. 3.30 PLP-dependent amino acid Ca epimerization (racemization), decarboxylation and Ca replacement reactions via external aldimine)

formed by the decarboxylation of dihydroxyphenylalanine (DOPA). Likewise, histamine and serotonin, which have important complex biological functions, are formed by the decarboxylation of histidine and 5-hydroxytryptophan, respectively.

3.5.2.2 Cα Side Chain Replacement

An example of such a reaction is that catalyzed by serine hydroxymethyl transferase (Fig. 3.30). Here, the replacement of the hydroxymethyl group by a proton is initiated by the formation of the serine alkoxide ion which collapses to formaldehyde, and the Cα carbanion eventually gives glycine.

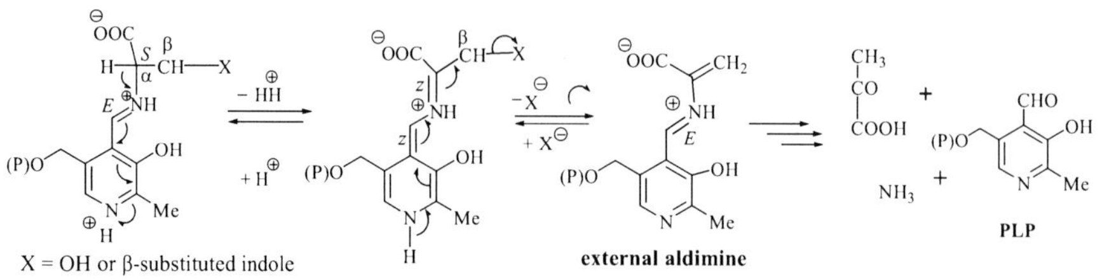

X = OH or β-substituted indole   external aldimine

Fig. 3.31 β-Elimination reactions catalyzed by PLP-dependent enzymes

3.5.2.3 PLP-Catalyzed Reaction at β-Carbon Atom of Amino Acids

The catalytic power of PLP is observed in reactions involving groups on the β-carbon (Fig. 3.31). Specific enzyme catalyzed β-elimination takes place when a good leaving group is present. Serine dehydratase or tryptophanase is an example of an enzyme when serine or tryptophane is used. The Cα−H bond is just broken to give a delocalized carbanion, which subsequently expels the leaving group (here H_2O; indole when tryptophan is used) to form the external aldimine. The latter is hydrolyzed to the free enamine which decomposes to pyruvic acid, ammonia, and PLP.

3.5.2.4 Stereochemical Concepts of the Pyridoxal Phosphate (PLP) Catalyzed Reactions

Much work has been done in this area and the salient interesting points are summarized below:

Reaction specificity imposed by the enzyme protein upon the system is achieved by control of the conformation around Cα−N bond of the substrate cofactor complex. The bond to be broken must be orthogonal to the plane of the conjugated π system. In this conformation, the breaking σ-bond achieves maximal orbital overlap with the π system, resulting in a substantial rate enhancement. Thus, the pyridine ring, the C4′ amino nitrogen and Cα of the coenzyme–substrate complex must lie in a plane for resonance stabilization. Hence, the three conformers (A), (B), and (C) (Fig. 3.32) represent the orientations of the complex for the enzyme-catalyzed cleavage of the Cα−H bond, the Cα−COOH bond, and the Cα−Cβ bond, respectively.

The reactions of PLP enzymes take place on one face of the planar PLP–substrate complex—the *exposed* or *solvent* face is the *Si face* at C4′ of the cofactor—the other face being covered by the protein. A chemically intuitive concept is that the enzyme binds the relatively rigid PLP cofactor at the points: pyridine N, and the phosphate, and the third point is probably the single distal carboxyl group on the substrate, resulting in a particular conformation by the Cα−N bond. It has been

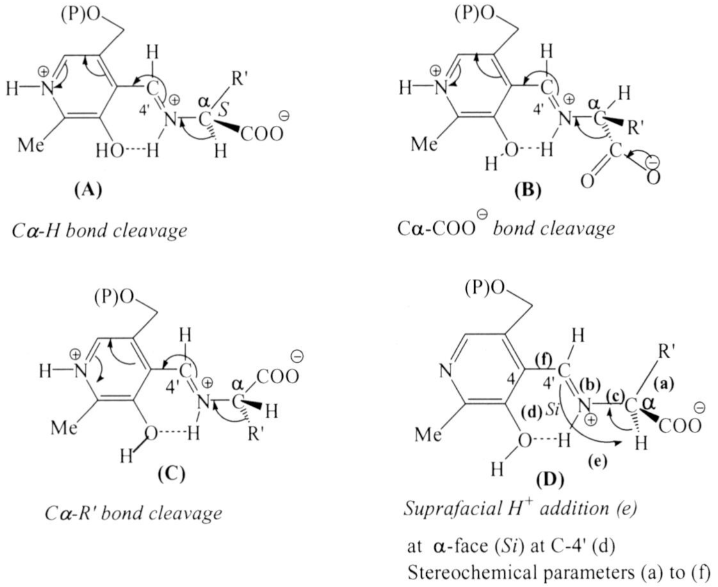

(A)

Cα-H bond cleavage

(B)

Cα-COO$^{\ominus}$ bond cleavage

(C)

Cα-R' bond cleavage

(D)

Suprafacial H^{+} addition (e)

at α-face (Si) at C-4' (d)
Stereochemical parameters (a) to (f)

Fig. 3.32 Optimal conformations about the Cα-N bond for cleavage of the Cα-H(A), Cα-COO$^{\ominus}$(B), or Cα-R'(C) bonds of L-amino acid, and suprafacial H^{+} addition. Stereochemical parameters of enzymatic transformation (**D**)

shown that group interchanges take place *in a retention* mode and proton transfer with *suprafacial geometry*.

The stereochemistry of the whole transamination reaction, catalyzed by the given specific enzymes and the cofactor PLP, thus involves the following six parameters [(**D**), Fig. 3.32] established through experiments of many workers.

(a) **Configuration at Cα of the substrate**: Only the L-enantiomer (and not D-) of an α-amino acid reacts.

(b) **Configuration of the C4' = N double bond** must be E (or *trans*): a Z (or *cis*) double bond will not be coplanar with the pyridine ring due to steric interference by the adjacent ring substituents (*ortho*).

(c) **Conformation around the Cα–N bond**. As discussed earlier, the bond to the Cα to be broken must be perpendicular to the plane of the conjugated π system [as shown in conformers (A), (B), and (C) in Fig. 3.32].

(d) **Site of proton addition or deprotonation at C4'** has been shown to be on the *Si* face or α-face of the structure (D) of the coenzyme–substrate or the coenzyme–intermediate complex, as revealed from studies with many specific transaminases.

(e) **Mode of prototropic shift from Cα of the substrate to C4' of PLP** has been demonstrated to be *suprafacial* (not *antarafacial*) by experiments involving internal transfer of tritium and deuterium. Internal proton transfer in aldimine

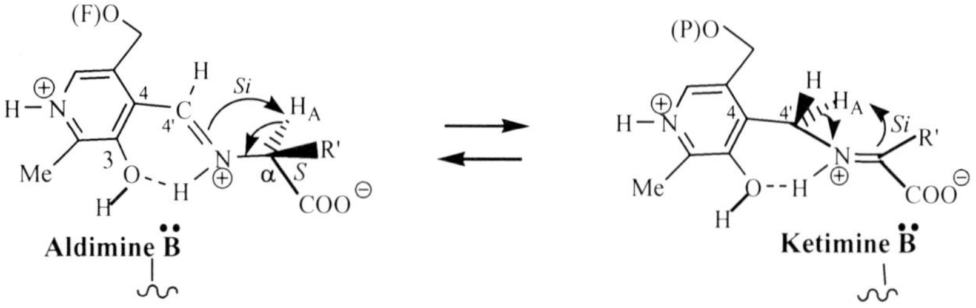

Note: *Protonation of pyridine N makes the α-H_A / 4' H_A more acidic.*

Fig. 3.33 Aldimine ketimine tautomerization in the transamination of an L-amino acid

ketimine tautomerization (Fig. 3.33) strongly suggests that the deprotonation and protonation are mediated by the simple base which necessitates *suprafacial* nature of the process. Thus, for such reaction the conformation around Cα−N single bond must be such that Cα−H is exposed on the side of the complex corresponding to the *Si face* at C4', *i.e.,* in an L-amino acid–coenzyme complex the carboxyl group and C4' would be *trans* to each other (Fig.3.33).

(f) *Conformation around C4–C4' bond*. Model studies and calculations show that in the absence of the enzyme, pyridoxal Schiff bases prefer a *cisoid* conformation (C4' = N imine bond is on the same side of C4–C4' bond as the C3–OH). Of course, during catalytic process a reorientation of the cofactor takes place. X-ray diffraction studies revealed that the cofactor-lysine Schiff base is in the *cisoid* conformation with the C4' = N roughly coplanar with the pyridine ring, and the *Si* face against a β-sheet of the protein. Moreover, the *cisoid* form of all aldimines and ketimines are expected to be more stabilized than the transoid form, by H-bonding between oxygen of OH and 4'-NH, as shown in the conformers (A) to (D).

3.6   Addition of C_1-Unit with AdoMet (SAM)

3.6.1   *Methylation*

In the biosynthetic process of C_1-unit addition, the enzymatic transfer of a methyl group takes place from L-**methionine**. The latter, prior to its participation in methyl transfer, is activated by its conversion to *S*-adenosylmethionine (**SAM or AdoMet**) by an unusual S_N2 reaction with ATP by the action of methionine adenosyl transferase, in which the nucleophilic sulfur atom of methionine attacks the C5' on the ribose unit expelling the inorganic triphosphate (*i*PPP) rather than attacking any phosphorus atom. Thus, an unstable *sulfonium ion* is formed having a

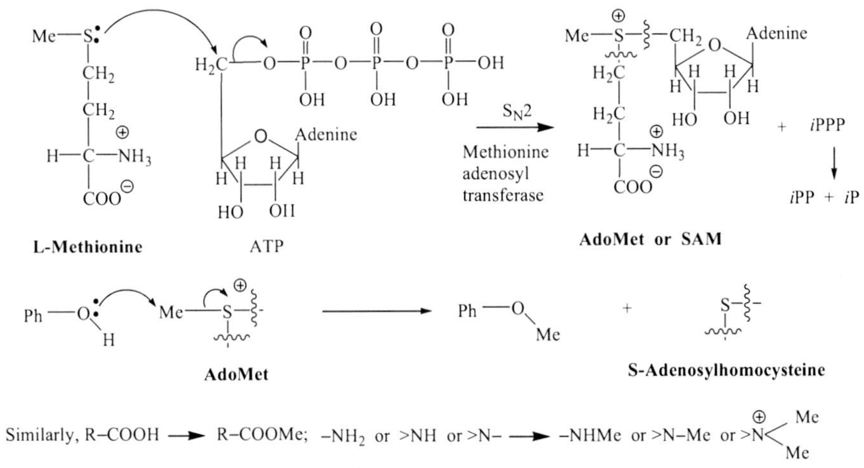

Fig. 3.34 Biogenetic O-methylation or N-methylation by the strong electrophile
S-adenosylmethionine (SAM or AdoMet)

high thermodynamic tendency to transfer its methyl group and gets neutralized. The
expelled triphosphate is enzymatically hydrolyzed to pyrophosphate (*i*PP) and
orthophosphate (*i*P) (Fig. 3.34). SAM is a potent strong methylating agent and it
undergoes facile attack by oxygen or nitrogen nucleophiles.

3.6.2 Formation of Methylenedioxy Bridge and Its Reductive Opening [8, 9]

The methylenedioxy moiety is formed via regiospecific monomethylation of an
ortho-dihydroxybenzene system, followed by the removal of a hydride (*cf.* anti
elimination of e$^-$ and H$^•$) to form a carbocation, stabilized by resonance by a
nonbonding lone pair of the adjacent oxygen (see Fig. 3.35). The *ortho* OH then
attacks the carbocation center to form a methylenedioxy bridge, a common functional moiety in many natural products.

Studies with labeled AdoMet showed the stereochemical implication of the
transformation of a labeled methoxy into a labeled methylenedioxy group, followed
by reductive opening of the latter to a labeled methoxy function, as illustrated in the
biosynthesis of *protoberberine type* isoquinoline alkaloids by enzymes from plant
cell cultures, *e.g.,* conversion of 2-O-demethyljatrorrhizine (1) to labeled
jatrorrhizine (4) via (2) and labeled *berberine* (3) (Fig. 3.35).

The reaction sequence involves the transfer of a chiral methyl group from SAM
to oxygen of a protoberberine alkaloid (1) to form (2) with inversion of configuration. The latter is then fed into callus cultures of *Berberis koetineana*, which

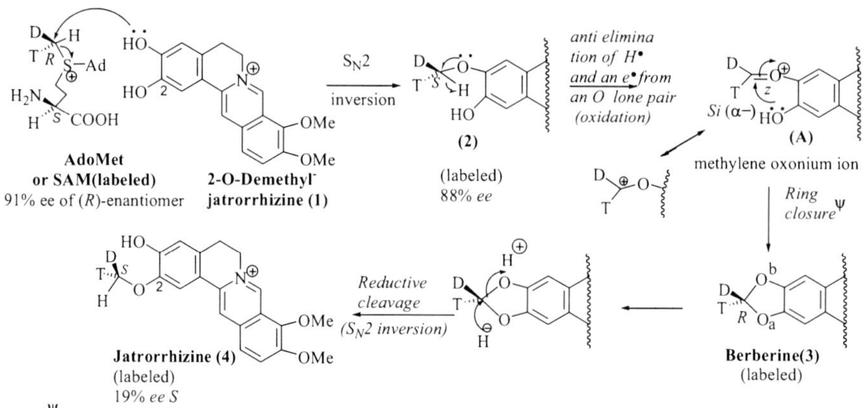

$^{\Psi}$Ring closure takes place by attack of O atom at the Si face of ion (A) with overall retention of configuration of D and T; however, due to low energy barrier to rotation of ion (A) significant amount of racemization occurs.

Fig. 3.35 Formation of the methylenedioxy bridge of berberine and its subsequent reductive opening to form the hydroxy/methoxy functions of jatrorrhizine with stereochemical mechanism.

converted it via berberine (3) into jatrorrhizine (4). By degradation, it was found that the chiral methyl groups in the protoberberine (2) and jatrorrhizine were both having (S) configuration.

It follows that probably the oxidative closure to form the methylenedioxy bridge occurs with retention of configuration, and the labeled methylenedioxy bridge reductively opens up with inversion of configuration. Thus, the migration of the chiral methyl from the original oxygen to the adjacent (vicinal) oxygen is formally equivalent to inversion. The methyleneoxonium ion (A) has a significantly lower energy barrier to rotation, and can to a significant extent undergo configurational isomerization prior to attack of oxygen on the methylene carbon, accounting for partial racemization.

3.6.3   N-Methylation and Formation of a Methylene Bridge Between Nitrogen and Carbon [10]

The conversion of >NH to –NCH$_3$ by transfer of the chiral methyl from labeled Ado-Met to the nitrogen proceeds clearly with complete inversion of configuration. The subsequent transformation of the chiral N-methyl group into a chiral methylene bridge between nitrogen and carbon takes place in presence of the specific enzyme. This sequence of reactions has been demonstrated to lead to labeled scoulerine, a *tetrahydroprotoberberine* type alkaloid, as shown in Fig. 3.36.

From the tritium NMR analysis of scoulerine formed, it is apparent that the replacement of a hydrogen of the chiral N-methyl group by the aromatic ring carbon

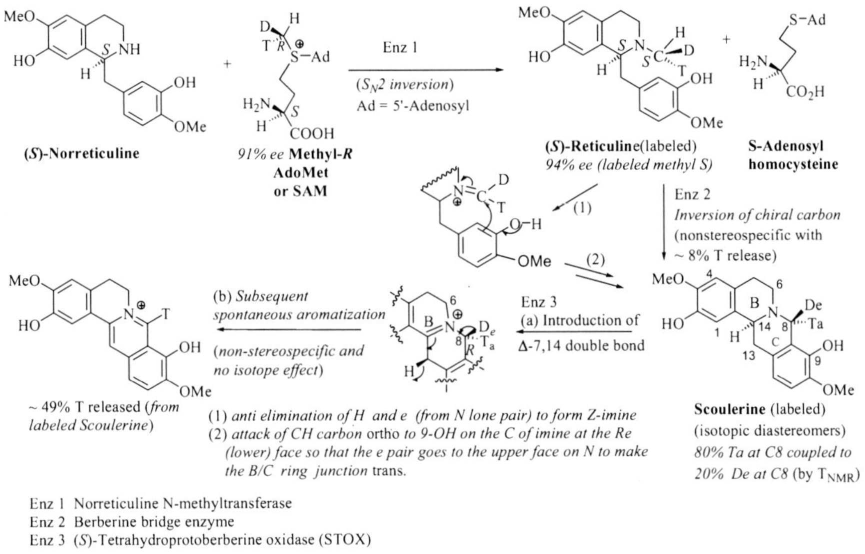

Fig. 3.36 Enzymatic reaction sequence in the N-methylation of norreticuline, and generation of the berberine bridge followed by its further oxidation

has occurred *with inversion*, and that the hydrogen abstraction has an isotope effect k_H/k_D 4. It is observed that in the berberine bridge the enzyme releases about 8 % T from (S)-Me group of (S)-**reticuline**, consistent with the observed isotope effect of 4. Subsequent addition of the enzyme STOX leads to 49 % T release, nearly half of the remaining T from the two *isotopically diastereomeric* substrates indicating *non-stereospecific nature of the reaction* lacking any isotope effect. Thus it is postulated that STOX only catalyzes the introduction of a $\Delta^{7,14}$ double bond, and subsequent aromatization occurs spontaneously (Fig. 3.36).

3.7 C- and O-Alkylation

3.7.1 C and O-Alkylation of Phenols

Both C- and O-alkylation of phenols are quite prevalent in natural products. The alkyl groups mostly consist of methyl and prenyl groups. SAM (AdoMet) serves as the methylating agent (Sect. 3.6), and various prenyl groups as their pyrophosphates act as the prenylating agents. Figure 3.37 depicts different types of probable C- and O-prenylation and cyclization of polyphenolic compounds like resorcinol.

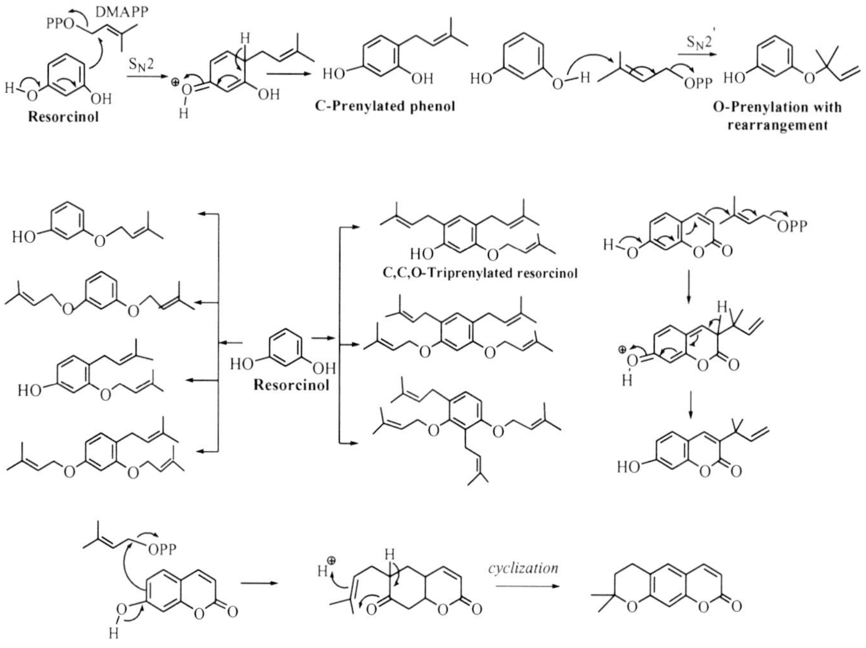

Fig. 3.37 Different types of probable C- and O-prenylations of polyphenols like resorcinol, and cyclization

The formation of C_5, C_{10}, C_{15}, and C_{20} prenyl pyrophosphates has been discussed in detail in Chap. 5. Of all the prenyls, C_5 and its modified forms occur most prevalently. Since the leaving groups of the natural alkylating agents do not vary, their influence on the proportion of the C- and O-alkylation, as observed in the laboratory reactions with different leaving groups, does not occur. Further, the specificity of the catalytic enzymes also plays an important role in the *regiospecificity* of alkylation, as is observed in coumarins, xanthones, flavones, etc. Coumarins provide best examples of such specificity. Figure 3.38 displays some examples of regiospecific methylation, C- and/or O-prenylation/s, prenyl chain modification of *resorcinol, umbelliferone*, and *6-hydroxyumbelliferone* to produce natural coumarins [11]. Sometimes prenyl groups are locked in a cyclic moiety (*e.g., pyranocoumarin* or *furanocoumarin* derivatives). In coumarins, perhaps the largest number of biogenetic modifications of the simple isoprene unit occurs, and also quite often they appear in a cyclic form [8]. C-Alkylation takes place at the available *ortho-* position of an activating group like OH before it may undergo O-alkylation [11].

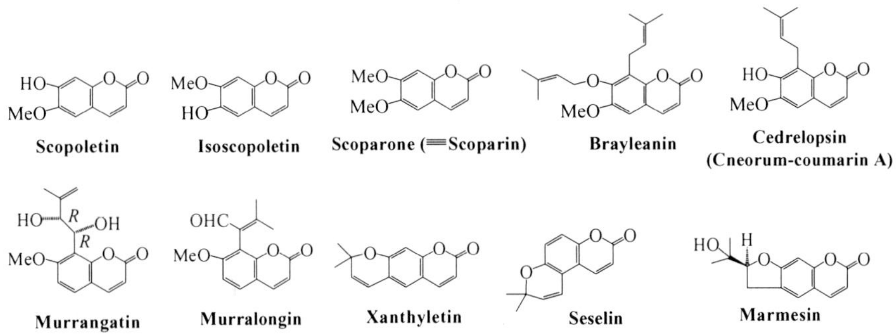

Fig. 3.38 Regiospecific methylation, C- and O- prenylation, prenyl chain modification, cyclization etc. to form some natural coumarins

3.7.2   C-Methylation and Modification of Cycloartenol Side Chain to Form Phytosterols [12, 13]

All phytosterols except cholesterol contain extra carbon atom/s at C24 of the side chain compared to its precursor cycloartenol. Cholesterol occurs mainly in the animal organisms and less abundantly in plant cells. Cycloartenol is metabolized in the plant cells to yield phytosterols including cholesterol (Chap. 11). The presence of a double bond (Δ^{24}) in the side chain of the precursor plays a vital role as an obligatory π-bond nucleophile towards the electrophilic SAM (AdoMet). The electrophilic addition of the methyl group takes place from the Si (α-) face of the double bond at C24. Through labeling experiments it has been shown that the resulting C25 carbocation **A** is quenched by delivery of hydride either from NADPH at the Si (α-) face of the cation to form **campesterol**, or from the adjacent C23 methylene at the appropriate Re face, followed by loss of H$^+$ from C22 to form **ergosterol** having (24R) and E stereochemistry of the 22–23 double bond Fig. 3.39). The newly formed sp^3 chiral carbon always appears in one epimeric form—a characteristic of enzymatic reaction. The carbocation **A** may also be quenched to a double bond by loss of H$^+$ from the methyl group, followed by methylation by SAM to form the cation **B**, which upon delivery of hydride by NADPH at the Si (α-) face of the cation forms **sitosterol**. Alternatively, the cation **B** may be quenched by a hydride shifted from C23 giving rise to the C23 cation, followed by loss of the appropriate proton to form **stigmasterol** having (24S) and E stereochemistry of the 22–23 double bond. Thus, in cases of ethyl group (C$_2$) containing sterols (sitosterol and stigmasterol) two methylations by SAM take place (Fig. 3.39).

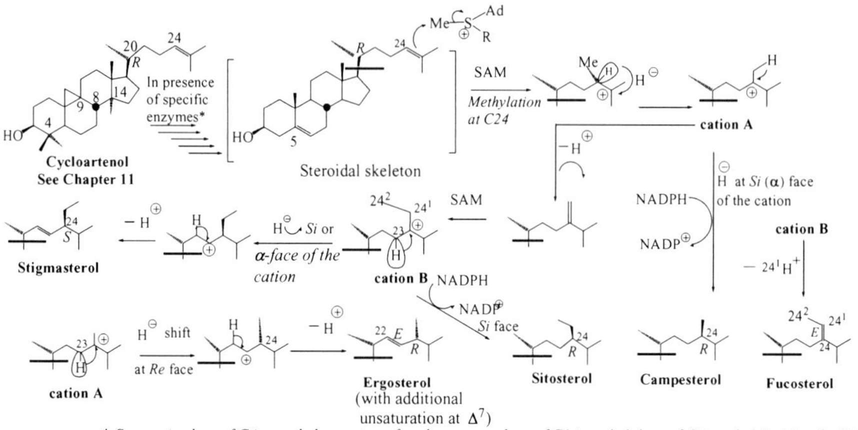

* *Successive loss of C4α-methyl, opening of cyclopropane, loss of C14-methyl, loss of C4-methyl (originally β-) and formation of 5,6 double bond.*

Each step in this Figure is catalyzed by a specific enzyme present in plant cells.

Fig. 3.39 Side chain methylation and modifications to form different phytosterols

3.8 Other Important Biological Events

Many reactions/rearrangements like Wagner–Meerwein rearrangement, Baeyer–Villiger oxidation, Michael addition, Markovnikov addition, less likely anti-Markovnikov addition, diradical coupling reactions, etc., and rarely 3,3-sigmatropic rearrangement do occur in plant cells, being catalyzed by suitable enzymes during the biosynthesis of natural products with various skeletal patterns. These have been discussed in the biosyntheses of pertinent natural products in different chapters.

References

1. Bill Bryson, *A Short History of Nearly Everything*, Black Swan, **2004**, p.362.
2. G. Feher, Identification and Characterization of the Primary Donor in Bacterial Photosynthesis: A Chronological Account of an EPR/ENDOR Investigation (The Bruker Lecture), *J. Chem. Soc. Perkin Trans.* 2, **1992**, 1861-1874.
3. M. B. Bishop and C. B. Bishop, Photosynthesis and Carbon Dioxide Fixation, *J. Chem. Educ.*, **1987**, *64*, 302-305.
4. R. Sterner and B. Höker, Catalytic Versatility, Stability and Evolution of the $(\beta\alpha)_8$-Barrel Enzyme Fold, *Chem. Rev.*, **2005**, *105*, 4038-4055.
5. M. D. Hatch and C. R. Slack, Photosynthesis by Sugarcane Leaves. A New Carboxylation Reaction and the Pathway of Sugar Formation, *Biochem. J.*, **1966**, *101*, 103-111.
6. M. D. Hatch, C_4-Photosynthesis. An Unlikely Process Full of Surprises, *Plant Cell Physiology*, **1992**, *33*, 332-342.

7. Ronald Breslow, On the Mechanism of Thiamine Action. IV. Evidence from Studies on Model Systems, *J Am Chem Soc,* **1958**, *80*, 3719-3726.

8. Heinz G. Floss, Thomas Frenzel, David R. Houck, Lai-Duien Yuen, Pei Zhou, Lynne D. Zydowsky, and John M. Beale, Stereochemistry of One-carbon Metabolism in Aerobes and Anaerobes, in *Molecular Mechanisms in Bioorganic Processes,* Ed. C. Bleasdale and B. T. Goldberg, Royal Society of Chemistry, **1990**, pertinent pages 31-37, and relevant references cited.

9. Motomasa Kobayashi, Thomas Frenzel, Jonathan P. Lee, Meinhart H. Zenk, and Heinz G. Floss, Stereochemical Fate of O-Methyl Groups in the Biosynthesis of Protoberberine Alkaloids, *J. Am. Chem. Soc.,* **1987**, *109*, 6184-6185.

10. Thomas Frenzel, John M. Beale, Motomasa Kobayashi, Meinhart H. Zenk, and Heinz G. Floss, Stereochemistry of Enzymatic Formation of the Berberine Bridge in Protoberberine Alkaloids, *J. Am. Chem. Soc.,* **1988**, *110*, 7878-7880.

11. R. D. H. Murray, Naturally Occurring Plant Coumarins, *Fortschr. Chem. Org. Naturstoffe*, **1997**, *72*; **1991**, *58*; **1991**, *35*; **1997**, *72* **2002**, *83*

12. Geoffrey D. Brown, The Biosynthesis of Steroids and Terpenoids, *Nat. Prod. Rep.,* **1998**, 653-696.

13. H. H. Rees and T. W. Goodwin, Biosynthesis of Terpenes, Steroids and Carotenoids in *Biosynthesis*, Volume 1, *Specialist Periodical Reports*, The Chemical Society, London, **1972**, pp. 59-118, pertinent pp. 93-99.

Further Reading

P. Suppan, *Principles of Photochemistry* (Monograph for Teachers), The Chemical Society, London, **1973**, pp. 59-62.

T.A. Geissman and D.H.G. Crout, *Organic Chemistry of Secondary Plant Metabolism,* Freeman, Cooper & Company, San Francisco, **1969**.

A. Cox and T.J. Kemp, *Introductory Photochemistry*, McGraw-Hills Book Company (UK) Limited, England, **1971**, pertinent pages 166-171.

Jeremy M. Berg, John L. Tymoczko and Lubert Stryer, *Biochemistry,* 6th Edition, Chapter 20, W.H. Freeman & Company, New York, **2007**.

Hans-Walter Heldt, *Plant Biochemistry*, Academic Press (An Imprint of Elsevier), **2005** (3rd edn.), pp.45-66, 165-192, 213-242.

Albert L. Lehninger, David L. Nelson and Michael M. Cox, *Principles of Biochemistry*, 2nd Edn., Worth Publishers, New York, **1992**.

Christopher K. Mathews, K.E. van Holde and Kevin G. Ahern, *Biochemistry*, 3rd Edn., An Imprint of Addition Wesley Longman Inc., San Francisco, New York, **2000**.

Robert Ruffolo Jr., Physiology and Biochemistry of the Peripheral Nervous System, Chapter 8 in *Human Pharmacology* (Molecular to Clinical), Editors, Theodora M. Brody, Joseph Larner, Kenneth P. Minneman and Harold C. Neu, Mosby, St. Louis, 2nd Edn., **1995**.

Chapter 4
Natural Products Chemistry: A General Treatment

"The plant kingdom is a virtual goldmine of new chemical compounds waiting to be discovered." [1]

"Isolation of secondary plant metabolites constitutes a chemical research project with significant potential for future benefit to mankind." [1]

4.1 Introduction. Isolation

The most important part of the natural products chemistry work is the isolation of the constituents in the pure state from their source/s. Since natural products possess different structural patterns and properties the procedures for their isolation need a lot of manipulations and modifications over the usual general procedures (Figs. 4.1, 4.2, 4.3, 4.4, and 4.5) given as models. Further, with the advent of new isolation techniques and refinements of the old ones, it is now possible to isolate most of the secondary metabolites including the trace ones present in the crude extract. This was almost an impossible task till the mid-twentieth century when only the major components could be isolated. The earlier laborious procedures have now been reduced to some time-saving efficient procedures. The crude extracts of the plant materials should first be fractionated adopting some suitable procedures, and the different fractions may then be subjected to easily available classical separation techniques, such as column chromatography (CC) (Sect. 4.1.6), flash chromatography (FC) (Sect. 4.1.7), and thin-layer chromatography (Sect. 4.1.8), as needed. These techniques have been discussed in some detail, based on the experience of the authors in sixties to eighties. These details will be handy and useful, especially to those who are working in different corners of the world with profuse plant resources, but lacking the expensive materials/tools needed. The fractions remaining after isolation of the major constituents may then be subjected for isolation of minor or/inseparable constituents to sophisticated expensive separation techniques like HPLC, MPLC, GC-MS, etc., in laboratories where these are

S.K. Talapatra and B. Talapatra, *Chemistry of Plant Natural Products*,
DOI 10.1007/978-3-642-45410-3_4, © Springer-Verlag Berlin Heidelberg 2015

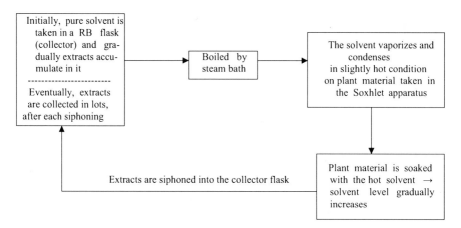

Fig. 4.1 Recirculating system of solvent extraction in a Soxhlet apparatus

available. The high-resolution NMR spectroscopy if available in any organic chemistry laboratory may preferably be used for monitoring the constituents (natural or synthetic) starting from their crude states.

4.1.1 Herbarium Specimen. Voucher Specimen

Prior to the commencement of the isolation work, a *voucher specimen* of the plant under chemical investigation, preferably with flowers, the latter being a better marker, should be preserved after establishing its identity by comparing it with an authentic *herbarium specimen. Herbarium* is a collection of specimens of dried and pressed plant materials (usually leaves and flowers), properly mounted on sheets arranged systematically. Each sheet carries the botanical name and family of the plant and place, season, and date of its collection. The plate should record particularly the characters which may be lost after drying and should be housed in a botanical museum in such a way that it can be preserved for centuries. Such specimen sheets are used for authentic identification of plants. The factors affecting the ecology of plants, e.g., the place or location and the time of collection, and the maturity of the plant under investigation (if known) should be recorded in the voucher specimen. These factors should be mentioned while publishing the work. The maturity of the plant is usually not recorded in the literature.

4.1.2   *Ecological Influence on Plant Constituents. Plant Names. Plant Parts*

Ecological factors mentioned above are important because the chemical constituents of plants (natural products) are quite often the function of these parameters. The chemical constituents of the same plant grown at different places and collected in different seasons might vary in kinds and yieldwise. The maturity of plants, especially in cases of higher plants, usually causes variations in natural products contents: some unreported compounds may be present, and some isolates obtained earlier may be missing. That the geographical parameters quite often play an important role in the formation of the secondary metabolites is evident from an example of a plant, namely, *Vinca rosea* (syn. *Catharanthus roseus*). The plants grown in India and Madagascar and other places vary remarkably in their alkaloidal content [2]. To cite an example: dramatic changes due to seasonal variations have been observed in the monoterpene constituents of the oil from both *Rosmarinus officinalis and Salvia officinalis* [3, 4].

Plant names   The botanical name of a plant has two parts. It starts with the genus (e.g., *Vinca*—conventionally the name of the genus always starts with a capital letter), followed by the name of the species (e.g., *rosea* which always starts with a small letter), and they appear in italics in print (*Vinca rosea*). The name is underlined singly when written by hand (e.g., <u>Vinca rosea</u>). Sometimes two or more different names appear for a particular plant, identified independently by different botanists. Even after establishing their identity, these names are retained in the literature as synonymous (syn.) (e.g., *Vinca rosea* Linn., syn. *Lochnera rosea* (Linn.) Reichb. f., syn. *Catharanthus roseus* (Linn.) G. Don [3, 5]), for convenience of researchers. This nomenclature known as "Binomial nomenclature" introduced by a Swedish Botanist Carolus Linnaeus (Carl von Linne', 1707–1778) first appeared in Linnaeus' "Species Plantarum" in 1753 and then in a book *Systema Naturae* written in Latin language. This nomenclature is applicable to both plants and animals. Initially Latin names were introduced. Later Greek names have also been used. By this practical method a plant could be quickly placed in a named category. After the botanical name of the plant usually the name/s or short name/s or initial/s of the identifier/s appear/s. The name of the family to which the plant belongs should also appear in parentheses, not italicized, and starts with a capital letter (e.g., *Vinca rosea* Linn.) (Apocynaceae)—the suffix "ae" generally appears at the end of the name of the family. Some examples are cited in Table 4.1.

Plant parts   The part/s of the plant (leaves, flowers, stems, stem-bark, roots, trunk bark, root-bark, and heart wood (in cases of trees); fruits, exudates, seeds), with which the chemical work is done, should be recorded since the chemical components of different parts of the same plant, especially a tree, vary in kinds and concentrations. In cases of small plants like herbs, creepers, and sometimes even small shrubs, the whole plant materials are conveniently used for chemical work.

Table 4.1 Names of some plants and their families [5]

Name of the plant	Family
Conium maculatum Linn.	Umbelliferae
Jatropha curcas Linn.	Euphorbiaceae
Lawsonia alba Lam	Lythraceae
Alstonia scholaris R.Br.	Apocynaceae
Stephania glabra (Roxb.) Miers	Menispermaceae
Cinchona calisaya Wedd.	Rubiaceae
Aegle marmelos Correâ	Rutaceae

Part/s of the plant, one has worked with, should always be mentioned for precise information and reproducible data.

4.1.3   Literature Survey. Phytochemicals. Chemotaxonomic Significance

One essential element of any research is the search of relevant literature. Thorough literature survey on the chemical studies of the plant/s under investigation is a must. In cases of chemically virgin plants, work on other species of the same genus and sometimes on other genera of the same family should be searched to know the structures of their chemical constituents. This will be of immense help in the manipulation of the isolation process (Figs. 4.2, 4.3, 4.4, and 4.5) and also for easy identification of the known compounds. Literature survey, especially the biosynthetic knowledge of the compounds already reported from the plant species of the same or different genera of the family, helps largely for the structural elucidation of new compounds: the plant species of the same genus are likely to elaborate compounds with same or biogenetically related skeletal patterns on genetic grounds. Chemical studies on innumerable plants have shown compatibility between taxonomical classifications and the phytochemicals produced in the same genus or allied genera within the family. A few such observations are cited in the sequel.

Rutaceae plants are well known for elaborating coumarin derivatives (Sect. 13. 2), which are absent in Apocynaceae plants. Again indole alkaloids occur abundantly in a number of genera of Apocynaceae plants, while Rutaceae plants do not produce them. Likewise, many plant families may be correlated with the types of compounds they produce—and a chemotaxonomic relationship is established. In some cases the taxonomical classification (genus or even family) has been revised based on the class of the phytoconstituents.

With the help of new analytical and instrumental techniques such as CC, preparative TLC and HPLC, GC-MS, ^{1}H and ^{13}C NMR (one and two dimensional), EIMS, HRMS, and FABMS spectra, etc., a large number of individual compounds can be isolated from a plant and can be characterized unambiguously in a relatively

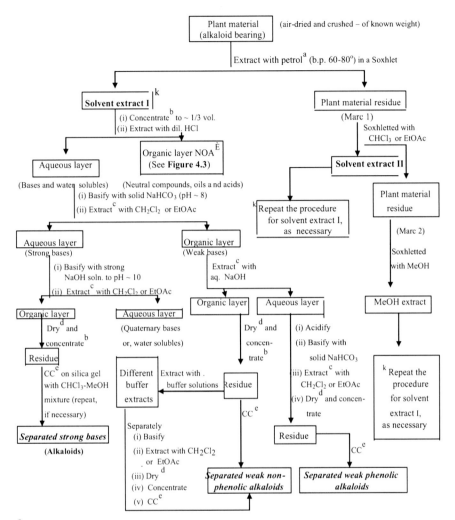

[a] Usually CHCl$_3$ and EtOAc extracts and not petrol extract may respond to Dragendorff's reagent. In case of nonalkaloid bearing plants of some families, or alkaloids not extractable by petrol, subsequent procedures for alkaloids are omitted.

[b] Concentrate always in a flash evaporator under reduced pressure at ~ 40 °C.

[c] Extract >2 instalments with ~ 1/3 volume of the phase being extracted.

[d] Organic extracts should be dried over anhydrous Na$_2$SO$_4$ or MgSO$_4$ to get the residue for CC and/or TLC.

[e] CC (column chromatography), followed by preparative TLC and/or HPLC/MPLC, as required.

Fig. 4.2 A general scheme for separation of strong bases, weak non-phenolic and phenolic bases and the quaternary basic fraction from petrol, CHCl$_3$/EtOAc, and MeOH extracts of plant material

short period of time. The chemical knowledge gained not only helps the taxonomists but also stimulates interest of the chemists and biochemists involved in biosynthetic work. The occurrence of a given compound or its congeners in two related species often gives evidence of common steps in biosynthesis.

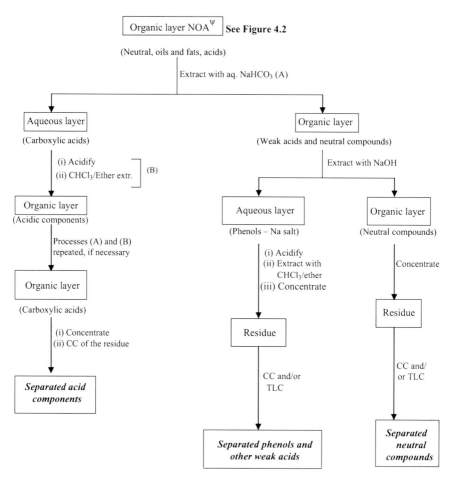

Fig. 4.3 A general scheme for separation of *neutral compounds, acids, phenols,* and other *weak acids* from petrol and CHCl₃ or EtOAc extracts of plant material

Extensive studies on the chemotaxonomic significance of various types of natural products have been carried out. Only a few of such studies will be briefly mentioned. Part 18 of the series, "Phytochemistry and Chemotaxonomy of the Convolvulaceae," concerns the occurrence and distribution of 74 tropane (including 4 new) and 13 biogenetically related pyrrolidine (including nicotine) alkaloids in 18 *Merremia* species (Convolvulaceae) [6] of tropical occurrence. The extensive GC-MS study with aerial parts as well as with roots led to the isolation and structure elucidation of four novel 3-acyloxytropanes. Each species, *excepting two*, included in this study is capable of producing simple tropanes, and thus the presence of such

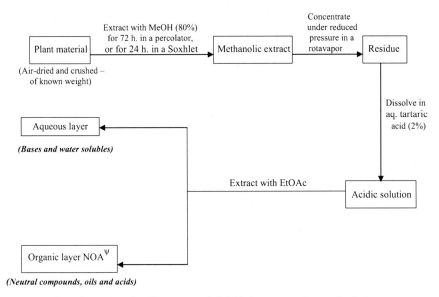

Fig. 4.4 Direct extraction of plant materials by MeOH (80 %) and isolation of alkaloids, neutral compounds, and acids

metabolites is a common trait in this genus. Those *two species* are segregated from the genus and transferred to a novel one, *Xenostegia*. This unique observation in lacking tropanes thus supports this transfer chemotaxonomically.

The chemotaxonomic significance of the isolated compounds from the water-soluble part of extracts of two *Digitalis* species and some other related species of Plantaginaceae family (N.O. Digitalideae) has been discussed [7]. Recent extensive chemosystematic investigations of the family Scrophulariaceae have led to significant changes in its limitation. Many former members of the family have been assigned to a largely expanded Plantaginaceae. Within this family the chemotaxonomy of the genus *Plantago* has been recently reviewed [7]. Thus, enormous amount of accurate chemical work is to be done using modern techniques for the analysis of all possible isolated constituents of different species of the same genus or related genera for getting significant chemotaxonomic information and for revealing the many closely interwoven biosynthetic routes. Hence the use of the chemical constituents of plants as an aid to their classification is now a familiar concept. However, a few fortuitously isolated substances cannot always give much phylogenic information. In the end, the presence or absence of the enzyme systems involved and the alterations in their substrate specifications and catalytic activities will provide the most illuminating information.

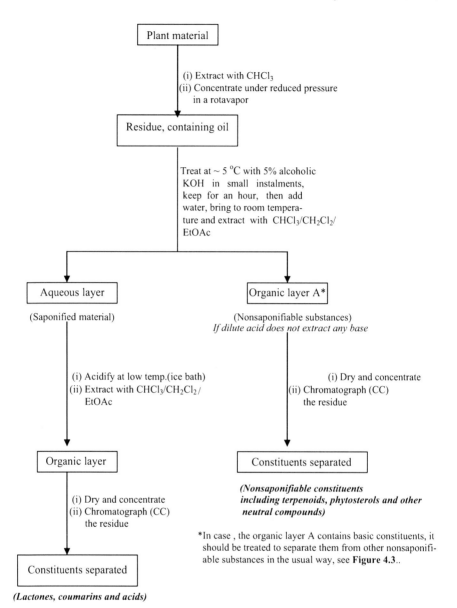

The following text appears within the figure:

Plant material

(i) Extract with CHCl₃
(ii) Concentrate under reduced pressure
    in a rotavapor

Residue, containing oil

Treat at ~ 5 °C with 5% alcoholic
KOH in small instalments,
keep for an hour, then add
water, bring to room temperature and extract with CHCl₃/CH₂Cl₂/
EtOAc

Aqueous layer

(Saponified material)

Organic layer A*

(Nonsaponifiable substances)
If dilute acid does not extract any base

(i) Acidify at low temp.(ice bath)
(ii) Extract with CHCl₃/CH₂Cl₂/
    EtOAc

(i) Dry and concentrate
(ii) Chromatograph (CC)
    the residue

Organic layer

(i) Dry and concentrate
(ii) Chromatograph (CC)
    the residue

Constituents separated

*(Nonsaponifiable constituents
including terpenoids, phytosterols and other
neutral compounds)*

*In case , the organic layer A contains basic constituents, it
should be treated to separate them from other nonsaponifiable substances in the usual way, see **Figure 4.3**..

Constituents separated

(Lactones, coumarins and acids)

Fig. 4.5 Separation of saponifiable constituents of plant extracts from nonsaponifiable ones

4.1.4 *Isolation of Plant Constituents: Solvent Extraction. Buffer Extraction. Thimble Extraction. Steam Distillation*

The plant materials are usually sun-dried, crushed to powder in a grinding machine, and weighed. The weight is necessary for calculating the percentage yield of the constituents of the plant materials. A few general methods of isolation will be outlined (Figs. 4.1, 4.2, 4.3, and 4.4) briefly in the sequel. *These are just the skeletal procedures for the specified types of compounds.* For better results one can make a lot of modifications, designed by experience, and by trial and error methods. Though the present generation of chemists is fortunate to have more sophisticated and time-saving equipments for the effective separation of the natural products from their sources, the old methods responsible for the present developments are worth inclusion in the current books on natural products of plant origin in order to get an idea of the principles involved. They may be useful for simple separation of the major components.

Extraction in a Soxhlet apparatus The dried and powdered plant material is extracted in a suitable Soxhlet apparatus [8, 9] successively with organic solvents of increasing polarity, e.g., petrol (b.p. 60–80 °C), chloroform or ethyl acetate, and methanol (or ethanol) to effect some sort of separation during the initial extraction process. The Soxhlet apparatus is a device for repetitive use of nearly the same volume of solvent to extract the plant materials through boiling, condensing, soaking the plant materials, and siphoning the solvent extract. Somewhat hot extraction (due to the vapor of the solvent) takes place. The Soxhlet apparatus, made of glass, of different sizes (5 l, 3 l, 2 l, 1 l, 500 ml, 250 ml, 100 ml, 50 ml) are available. It is a recirculating system and its operation may be represented as shown in Fig. 4.1.

Extraction in a percolator The dried and powdered plant materials (known weight), if available in kgs, can also be extracted in a metal percolator (with a lid at the top and a tap at the bottom) at room temperature, by keeping the plant materials soaked thoroughly with a less volatile polar solvent (e.g., rectified spirit/ methanol, etc.) for a week or so and then taking out the extract completely by opening the tap at the bottom. The extraction is repeated. Fresh (undried) plant materials (usually leaves) are sometimes used for extraction at room temperature, especially for light and heat sensitive constituents. However, in case of fresh leaves, the extracted chlorophyll interferes with the separation procedure. If the plant material is available in less quantity (0.1–0.5 kg), suitably designed small metal/ glass percolatorsand volatile solvents like chloroform or ethyl acetate may be used.

General schemes for the isolation of weak bases, strong bases, and quaternary basic fractions (Fig. 4.2) and for the isolation of acids, phenols, and other weak acids and neutral compounds (Fig. 4.3) from solvent extracts of plant materials are presented. At every stage one should monitor the organic layer by micro-TLC (*vide infra*) and proceed accordingly. The schemes may be simplified if constituents of basic or acidic nature are absent. *These procedures may be followed for work-up of chemical reaction mixtures also, as needed.*

Buffer extraction Complex mixture of alkaloids (as evident from Dragendorff test[1] and micro-TLC), present in the basic fraction of the plant extract, can be successfully separated by extraction with buffer solutions of different pHs, taking advantage of the different basicities of the alkaloids present, as indicated in the lower part of the Scheme in Fig. 4.2. For this purpose, a countercurrent distribution apparatus may be conveniently used.

The following procedure (Fig. 4.4) may be conveniently used for extraction and isolation of alkaloids from different parts (leaves, twigs, stems, stem-bark, trunk-bark, roots, or root-bark) of plants bearing alkaloids (positive Dragendorff test).

Lactones and coumarins can be separated from the nonsaponifiable components by saponification. This procedure is generally followed to isolate the lactones or coumarins from plant oils (Fig. 4.5).

Components of different extracts Petrol (b.p. 60–80 °C) or hexane generally extracts less polar monoterpenoids, sesquiterpenoids, di- and triterpene hydrocarbons, carotenes, fats, and waxes. Chloroform usually extracts sesquiterpene lactones, diterpenoids, less polar triterpenoids, sterols, oxygenated flavonoids, coumarins, alkaloids, etc. Ethyl acetate extracts more oxygenated di- and triterpenoids, sterols, and alkaloids. Methanol (90 %) or ethanol (90 %) extracts highly oxygenated (hence polar) di- and triterpenoids, flavonoids, steroidal glycosides, and alkaloids, etc. Finally, water extracts highly polar substances like glycosides, free sugars, and amino acids, etc. Solvents may be arranged in order of increasing polarity and "solvent power" toward polar functional groups as stated in the sequel, under '*Eluting Solvents*'.

Thimble methods *In a more convenient procedure* the dried and powdered plant material (weighed) is extracted with ethyl acetate and/or 90 % aqueous ethanol/methanol. The crude residue (may be solid, resinous, or tarry), obtained by concentration/evaporation of the extracts in a rotavapor under reduced pressure at a temperature <40 °C, is thoroughly mixed with a requisite amount of celite. The mixture is taken in a suitable thimble (made from good thick filter paper sheet by rolling, stapling, and closing one end), which is then put into a Soxhlet apparatus of appropriate size, and successively reextracted with petrol (b.p. 60–80 °C), or hexane, chloroform, and ethyl acetate, and finally with 90 % methanol to effect broad fractionation. Different solvent extracts are collected and concentrated. The residues obtained are processed to separate the acidic, basic, and neutral fractions in the usual way; each fraction is then chromatographed to isolate the pure constituents.

Various essential oils containing monoterpenes, sesquiterpenes, and some other volatile compounds as components are obtained from plant materials by water

---

[1] A small crystal of potassium bismuth iodide when added to an acidic solution of an alkaloid taken in a watch glass, generally a yellow to orange precipitate is obtained.

distillation, by steam distillation, and sometimes by a combination of the two, when the components are codistilled.

Direct steam distillation   Plant materials, cut into small pieces (generally flower petals), are heated in water for some time in a distilling flask and then distilled. Here the steam is produced in situ. As the steam is removed with volatile compounds, water is added dropwise from the separatory funnel fitted with the flask. The collected distillate contains essential oils and other volatile compounds. During efficient distillation for a longer period, sometimes the esters, if any, may get hydrolyzed.

Live steam distillation   Crushed plant materials are arranged on grids of a vessel from the bottom of which steam as such or under pressure is allowed to pass through the grids. The steam comes in contact with the plant material and carries with it the volatile components, which are collected in the distillate. Geraniol, citral, menthol, caryophyllene, etc., are obtained by this method. Here steam is used as one of the immiscible phases and the mixture of immiscible liquids boils at a constant temperature lower than the boiling points of the pure constituents present in the mixture and of water.

Quite often grids containing the plant materials are preheated by steam coils and then are allowed to come in contact with steam directly. Steam carrying the essential oils is collected in the distillate. In the coil, the steam is generally allowed to pass under low pressure.

Another efficient method is the *supercritical fluid extraction method* in which liquid carbon dioxide is used for the extraction of the essential oils.

4.1.5   Chromatography: Different Techniques [10–14]

Chromatography is the most powerful tool for the separation of two or more components present in a mixture. It is most widely used in chemical laboratories as well as in chemical industries for effecting isolation and purification. Since its invention, which is a great boon to chemical sciences, its full potential has been explored and utilized in basic research, industry, and medical sciences with great success. It is a physical method of separation effected by the distribution of the components between two phases—one of which, being in the form of a porous bed or bulk liquid layer or film, is immobile (*stationary phase*), while the other one is a fluid (mobile phase) that percolates over or through the stationary phase on which the sample is applied.

Various chromatographic techniques are in use. The common ones are conveniently classified based on the nature of the phases employed for the separation (Fig. 4.6) [10, 14]. Two distinct phases are required to set up the distribution resulting from repeated adsorption/desorption events during the movement of the sample components along the stationary phase in the direction of the flow of the mobile phase. Thus, gas–gas chromatography does not exist and liquid–liquid chromatography is restricted to immiscible solvents. It is interesting to understand the mechanism of the different chromatographic techniques (Fig. 4.6).

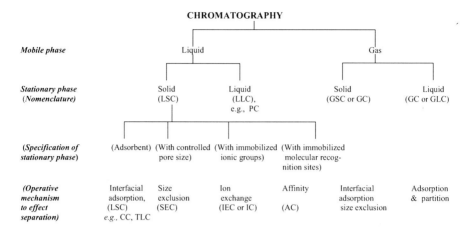

Fig. 4.6 Classification of the common chromatographic techniques

If the mobile phase is a gas, the stationary phase may be a solid or a liquid, and the separation techniques are called *gas–solid chromatography* (**GSC**) and *gas–liquid chromatography* (**GLC**), respectively. The simple term **GC** implies both these techniques, but it usually means GLC (unless otherwise specified), since this is the most commonly used technique. Here separation is caused by differences in interfacial adsorption and gas–liquid partitioning. In **GSC** the retention mechanism depends upon interfacial adsorption. If a solid of controlled pore size, such as zeolite, is used as the stationary phase, the size-exclusion mechanism operates.

If the mobile phase is a liquid, the stationary phase can be any one of the following:

(i) A solid (*liquid–solid chromatography,* LSC) with interfacial adsorption as the dominant distribution process.

(ii) A solid of controlled pore size (*size-exclusion chromatography,* **SEC**); here, the ratio of the solute size to the dimensions of the stationary phase pore sizes determines the distribution constant.

(iii) A solid with immobilized ionic groups coming in contact with the solutes in the mobile phase having electrostatic interactions operative as the dominant distribution process (*ion-exchange chromatography*, **IEC**, or *ion chromatography*, **IC**);

(iv) A solid possessing immobilized molecular recognition sites when in contact with the solute in the mobile phase (*affinity chromatography*, **AC**); here, the dominant distribution process is the three-dimensional specificity of the molecular recognition between the receptor sites and the solute (a technique more applicable in biotechnology)

(v) a porous solid coated with a thin film of immiscible *liquid* (*liquid–liquid chromatography*, **LLC**); here, the dominant distribution process is partitioning.

We will briefly discuss only a few chromatographic techniques which find extensive use in the separation and purification of small molecules like secondary metabolites or organic reaction products from their mixtures. Details of operations, operational data, and the instrumental components may be found in a number of books [10–14].

*The genesis of the term "**chromatography**"* appears to be interesting [15]. The first compounds separated by Tswett (Appendix A, A-27) using his technique were plant pigments, which appeared as colored bands on the column. He used the term *chromatography* in his first paper on the subject in 1906. Tswett coined the term for this technique from the Greek words *"chroma"* (color) and *"graphein"* (to write), as if, the results were *written* in *color* on the column. Some speculated that in Russian *"Tswett"* means color; perhaps Tswett named the technique after his own name (Tswett's writing ≅ '*chromatography*'). However, the term *chromatography* was known throughout the nineteenth century in connection with artists' material, especially artists' colors and pigments. George Field was the foremost *color man* of the nineteenth century; he published his treatise on *chromatography* in 1835. "It is interesting that a term, originally only used by artists, is now almost exclusively in the province of scientists" [15].

4.1.6   Column Chromatography

Adsorption Column Chromatography (LSC) It has been used most extensively for separation of simple to complex mixtures of natural products from different fractions (Figs. 4.2, 4.3, 4.4, and 4.5) or of reaction products and also for purification of individual compounds. *It has profound influence on the progress of natural products research and hence the parameters affecting the separation and some other important aspects will be treated somewhat in detail.* The method depends upon reversible adsorption of the substances in different degrees at a solid surface (adsorbent) and also upon their solubility in the solvents or solvent mixtures being used. Hence the substances are displaced from the adsorbent by the eluting solvent of varying increasing polarities at different rates, leading to their separation.

Adsorbents The most commonly used *adsorbents* are silica gel, SiO_2, xH_2O (used for polar to less polar substances), and alumina, Al_2O_3, xH_2O, obtainable in different activities (used for a range of nonpolar or less polar substances). More restricted adsorbents include charcoal (for sugars, amino acids), sucrose (for chlorophyll), calcium hydroxide (for carotenoids), etc.

Deactivation of adsorbents Anhydrous alumina or silica gel is highly activated. They can be *deactivated* by adding water to different extents. Water binds very tightly to the adsorbent taking up sites on the particles which could otherwise be used for equilibration.

Anhydrous neutral alumina (grade I) (purchased or obtained by heating alumina at 200 °C for 3 h with occasional stirring) is deactivated by adding small amounts of water: 3 % (grade II), 6 % (grade III), 10 % (grade IV), and 15 % (grade V). Deactivated alumina can be prepared by adding the requisite amount of water to a clean big beaker, swirling it to distribute the water over the inner walls, and adding the adsorbent immediately, while continuing the swirling motion. The adsorbent should then be transferred to a distilling flask of a flash evaporator and rotated and blended for about an hour.

Capillary tube standardization process This can be used for determining the activity of alumina or silica gel produced. A m.p. capillary tube *is filled* to the top with each grade of adsorbent and a drop of benzene is put on the adsorbent at the open end. The closed end of the tube is broken and the other end (wet) is immersed in a 0.5 % solution of *p*-phenylazoaniline in benzene for a moment. The tube is then immersed in a shallow layer of benzene taken in a small jar/beaker. Migration of the solvent by capillary action is allowed to a point near the top of the capillary column. The tube is then removed and the R_f value of the dye band for each activity grade is measured. The R_f value stands for "**R**atio to front" and is defined by the following fraction, expressed as a decimal. The R_f value is characteristic of a compound under some specific conditions (solvent, adsorbent, and thickness of the layer)

$$R_f = \frac{\text{the distance traveled by the spot/dye-band}}{\text{the distance traveled by the solvent front}}$$

Both distances are measured from the center of the applied spot/band. The distance traveled by the dye is measured up to the center of the band/spot. The standardization process can also be done using microscopic TLC slides (*vide infra*).

R_f values for the dye on alumina are *approximately* as follows: grade I 0.0; grade II 0.13; grade III 0.25; grade IV 0.45; and grade V 0.55.

Silica gel may be regarded as the most versatile of all adsorbents and can be used with all solvents. *Activity grade I silica gel* (anhydrous) can ordinarily be prepared by heating it at 150–160 °C with occasional stirring for 3–4 h. Deactivated silica gel grades II–V are made by adding water to a concentration of 10, 12, 15, and 20 %, respectively, following the same way as described in case of deactivated alumina. The capillary tube standardization process, already described for alumina, or microscopic TLC slide standardization (*vide infra*), can be used for determining the activity of silica gel. Activity grade I has an R_f value of 0.0 and grade III has an R_f 0.65 (approx.). Usually adsorbents of lower activity, grade II or III, are used to avoid too strong adsorption or any possible rearrangement.

For column chromatography the ***particle size*** of alumina is generally of 60–120 mesh and that of silica gel is 40–100 mesh. These are commercially available from different manufacturing firms. The **mesh number system** *gives a measure of the number of openings per linear inch in a screen.* This can easily be determined, as screens are made from wires of standard diameters; however, opening sizes can vary slightly due to wear and distortion. Some mesh numbers and their equivalent opening sizes in mm are given below (reproduced from website):

Mesh no.	Opening (mm)	Mesh no.	Opening (mm)	Mesh no.	Opening (mm)
20	0.841	80	0.177	200	0.074
42	0.354	100	0.149	250	0.063
60	0.250	150	0.105	400	0.037

Some solid **adsorbents** can be arranged as follows *in order of increasing strength of binding interactions,* as given below, toward polar compounds or solutes called **eluates** or **elutants**:

Silica gel, florisil, alumina, activated charcoal.

Eluting solvents   The common solvents used are called **eluents**. They are arranged *in order of increasing polarity and "solvent power" toward polar functional groups* as follows:

Petrol (b.p. 40–60 °C or 60–80 °C), hexane, cyclohexane, carbon tetrachloride, benzene, chloroform, methylene chloride, diethyl ether, ethyl acetate, acetone, ethanol, methanol, water, acetic acid.

The elution sequence (fastest to slowest) ***of elutants*** of different kinds are approximately as follows:

Hydrocarbons, olefins, ethers, aromatics, ketones, aldehydes, esters, alcohols, amines, acids, strong bases.

Hydrocarbons will be eluted with a nonpolar solvent (like hexane or petrol) whereas acids or strong bases will be eluted with a polar solvent. Of course, the molecular weights and the number and kind of functional groups of the natural or reaction products also play a vital role in this regard. Generally one starts elution with a nonpolar solvent to **elute** (bring down) relatively nonpolar compounds from the column and then gradually increases the solvent polarity to **elute** the compounds of increasingly greater polarity. The polarity of the eluent can be increased gradually by adding suitable greater percentages of a more polar solvent. The eluent to be used can be determined by running micro-TLC (*vide infra*).

Principle of separation. Interactions and distribution equilibrium   Intermolecular forces of varying strength cause organic molecules to bind to the adsorbents. Strengths of different types of interactions in the decreasing order are as follows [12]:

Salt formation > coordination > H-bonding > dipole–dipole interaction > van der Waals forces.

The first three types are direct interactions and are typical of polar organic compounds. Salt formation may take place between acids and alumina or between bases and silica gel. Coordination interaction may occur between bases and Al of alumina. Dipole–dipole interactions may occur between polar molecules and alumina or silica. Nonpolar compounds weakly bind to the adsorbent using only van der Waals forces. Of course, compounds of extremely high molecular weights, even if nonpolar, may bind strongly by such multiple weak forces. In general, the more polar the functional group, the stronger will be the bond to the adsorbent.

Solubility also plays a significant role: nonpolar solvents dissolve the nonpolar compounds best, whereas polar solvents dissolve polar compounds more effectively than nonpolar solvents. Thus any compound with some carbonyl function, adsorbed on an adsorbent, may not be eluted by hexane, but may be completely eluted by methylene chloride, chloroform, or ethyl acetate. Because of the competition for the solutes a distribution equilibrium is set up for each solute between the adsorbent and the solvent. Thus many such equilibria in series are set up as a consequence of the solutes in the solution coming in contact with the adsorbent down the column. These ultimately result in the difference in the partitioning of each solute from the other, leading to their differential elution by different eluents, and thus achieve separation.

Columns Cylindrical tubes of various types, usually made of glass, are used for column chromatography. They are packed with the adsorbents and clamped *exactly vertically*. The upper end is open and the lower end is fitted with a stopcock (preferably made of teflon) for controlled exit device (like a burette). Alternatively, a piece of flexible polyethylene (inert in most solvents) tubing is often attached to the bottom of the column and a screw clamp is used to stop or regulate the flow. Rubber or tygon (which dissolves in many solvents) should not be used. For large-scale chromatographic separation it is often convenient to use a solvent reservoir made by fusing the top of the column to the bottom part of a round-bottomed flask. The eluting solvent may also be added from a long stem capped separatory funnel, adequately opened and well inserted into the column of the eluent, filling the space above the adsorbent column to make the column *self-filling*. The available volume at the top of the column must be large enough to prevent any overflow.

Column preparation For efficient separation of the mixtures the column should be prepared properly. The column is partially filled with a nonpolar solvent like hexane or petrol (b.p. 40–80 °C), or the solvent mixture to be first used during chromatography. A loose plug of glass wool (or cotton) is pushed down softly into the bottom of the column with a long glass rod, and all entrapped air is forced out as bubbles. Clean, white sand is poured into the column to form a thin layer. The surface of the layer is leveled by tapping the column. Then the column is best prepared by either wet-packed (or slurry) or dry-packed method.

The slurry or wet pack method The slurry is prepared in a container by adding the adsorbent, a little at a time, to a quantity of the solvent (not in the reverse order). Enough adsorbent is added to the solvent, with swirling, to form a thick but flowing slurry. The swirling is continued to make it homogeneous and free of entrapped air bubbles. The column is now half filled with the solvent and the stopcock is opened to drain the solvent slowly into a large beaker. The slurry is then poured in portions through a funnel into the top of the draining column. The column is then tapped gently and constantly on the side with a plastic or wooden rod or a piece of thick pressure tubing until all the material has settled. The process is continued until the desired column height is obtained. The collected solvent in a conical flask should be recycled through the column to make it firmly packed. *The column should never be allowed to run dry during packing or subsequent elution.*

Dry pack method The column is filled with a less polar solvent, which is allowed to drain slowly. *The dry adsorbent* is added from a beaker through a funnel, a little at a time, while the column is tapped constantly, as in the previous method. Addition is continued to a desired height of the column. An evenly packed column will thus be obtained. Solvent should be recycled a few times through the column before starting elution.

Alternatively, the column may be packed dry without any solvent. The adsorbent is introduced in small quantities, tapping constantly, as described before. After packing, the solvent to be used initially is allowed to percolate down the column. *This method may lead to uneven packing, cracking, and some entrapped air bubbles and hence is not recommended,* especially when the solvent, having an exothermic heat of solvation with the adsorbent (silica gel or alumina), is used.

Column size and adsorbent quantity The amount of adsorbent should usually be 20–50 times (by weight) the amount of material to be separated by chromatography. For separation of complex mixtures, or for separation of the components (having close R_f values) of a simple mixture the adsorbent–solute mixture may be raised up to 200:1. Monitoring the crude mixture by micro-TLC may give an idea about the suitable ratio to be used. Moreover, the column should have the height:diameter about 10:1. Difficultly separable compounds may require larger columns and more adsorbent. A smaller column and less amount of adsorbent may be enough for easily separable compounds.

Sample application The sample containing the mixture of solutes is dissolved in a minimum volume of the solvent to be used first for elution, if it is quite soluble. If not, it is dissolved in minimum amount of a polar solvent (like chloroform), since the mixture should form a *narrow band* on the top of the column, *ideal* for optimum separation of the components. The concentrated solution of not very thick consistency (which might choke the column head) or the neat liquid (if the sample mixture is liquid) is added carefully with a small pipette or a long dropper touching the inside of the column close to the surface of the adsorbent, and slowly draining it, not disturbing the surface. The small layer of liquid or solution is then drained into the column until the top surface just begins to dry. Some eluting solvent (taken in a test tube) is carefully added with the pipette or dropper. The small layer of solvent is now drained into the column until it touches the upper surface. Another small layer of the solvent is added and the process is repeated till the sample is expected to be adsorbed. The level surface of the adsorbent may then be protected by carefully filling some solvent, covering the top with a circular filter paper and sprinkling clean white sand, which settles down and forms a small protective layer on the top of the adsorbent. Separation is often better if the sample is allowed to stand for a short time on the column before elution, allowing a true equilibrium to be established. Elution is then started.

Alternatively, to a solution of the sample mixture in a minimum amount of a polar solvent like chloroform or methanol taken in an evaporating dish, a small quantity of the adsorbent is added The solvent is evaporated, first by stirring with a spatula under a fan and then in a vacuum desiccator. The dry powder thus obtained is added carefully to the wet column having a big layer of the solvent at the top. Elution is then started after putting a small protective layer on the surface in the manner described above.

Elution   The starting eluting solvent or eluent should be determined by running a micro-TLC (*vide infra*) of the mixture. The minimum polar solvent or solvent mixture showing a spot of $R_f \sim 0.3$ should be used first. Usually the starting eluent is hexane or petrol (b.p. 40–60 °C). If this eluent shows more than one spot having close R_f values, the adsorbent:mixture ratio should be increased. The polarity of the eluent can be increased gradually by adding to petrol or hexane successively greater percentages of a more polar solvent, benzene/ethyl acetate/chloroform (e.g., 10 %, 15 %, 20 %, 30 %, 50 %, 75 %), or the more polar solvent alone, or by adding successively greater percentages of methanol to ethyl acetate/chloroform (e.g., 1 %, 2 %, 5 %, 10 %, 20 %, etc.), as needed. The transition from one solvent to another should not be too rapid in most eluent changes, especially when they differ greatly in their heats of solvation in binding to the adsorbent, since the generated heat may crack the column. *Automatic gradient elution technique* may be conveniently introduced. Most organic compounds can be separated on silica gel or alumina using hexane–benzene mixtures followed by mixtures of benzene and chloroform, ethyl acetate, or methylene chloride, and then by pure chloroform, ethyl acetate, or methylene chloride. Increasing percentage of methanol in chloroform, ethyl acetate, or methylene chloride may elute more polar compounds.

Rate of flow   The effectiveness of a separation depends upon the *rate of flow* of the solvent through the column. It should be optimum—not too rapid; then the solutes will not have sufficient time to equilibrate with the adsorbent as they pass down the column. If the rate of flow is too low, or stopped for a period, the solute band may diffuse in all directions. In any of these two cases the separation will be poor. Usually, *the optimum rate of flow will be approximately 5–50 drops of effluent per minute, depending upon the diameter of the column.* The running of the column should not be stopped or set aside overnight, or for any prolonged period of time to avoid diffusion of the bands. In this respect FC, to be dealt with next, is much better. The bands, if not colored, may be seen under UV lamp (if they fluoresce), and may be collected separately as they elute from the column. A common method of monitoring separation of a colorless compound is to collect fractions of constant volume in preweighed flasks, to find out the weights of the residues after complete evaporation under *vacuo*, and to plot the weights against fraction numbers. The volume of each fraction collected (e.g., 10 ml, 25 ml, 50 ml, 100 ml, 250 ml, or 500 ml) depends on the size of the column and the ease of separation. Fractions may be collected continuously by setting the column stem directed to a tube of an *automatic fraction collector* (carrying many tubes in a concentric circular fashion) and adjusting for collecting the eluate fractions of some fixed volume, each.

Fractions showing the same spot/s in micro-TLC slides, when developed in the same solvent/solvent mixture, are combined for further separation, if necessary, and purification. *Visualization of the spots is routinely done using microscopic slides, to be described later, for monitoring the collected fractions.*

Chromatogram A chromatographic experiment gives the information contained in the *chromatogram*. The latter consists of a plot of concentration or weight profile of the sample components as a function of the flow of the mobile phase (fraction numbers) or as a function of time. The developed and spotted slides or plates (in TLC) are also usually termed chromatograms.

4.1.7 Flash Chromatography [16–18]

FC, first introduced [16] in 1978, provides a very powerful and rapid technique for separating compounds with similar polarities. It is mainly applied for (i) laboratory scale purification of mixture of organic compounds, (ii) isolation of target compounds from different fractions of the extracts of natural products, and (iii) separation of geometric isomers and diastereomers. It is also used for simplification of mixtures prior to preparative high-pressure liquid chromatography (HPLC), when needed. For low molecular weight neutral organic compounds relatively fine silica powder is used, with a smaller particle size range (e.g., May and Baker Sorbsil C 60, 40–60 μm, i.e., ≈600–400 mesh or Merck C 60, 40–63 μm, type 9385). Such finely powdered alumina is also used occasionally. The finely powdered silica (or alumina) gives better surface contact leading to more effective adsorption than ordinary column chromatography (CC). The latter operating under gravity and atmospheric pressure is slow and leads to band dispersion reducing the resolution. In flash chromatography modest pressure of less than 2 atm., with some sort of pressure-release valve, is applied from the top of the column to accelerate solvent flow resulting in fast and increased resolution between bands with reduced tailing. Labile compounds, susceptible to degradation, or rearrangement in the chromatographic system, can be isolated in higher purity because of the shorter contact time. *Every separation is different, and like any chromatographic technique one gains expertise by experience.*

Column preparation Glass columns of suitable length (because of the limited operating pressure) usually 25 cm, sometimes 10–15 cm for small amount of the mixture, are used. Originally long columns were used, but these are difficult to load. A set of four or five columns ranging in diameter from about 5 mm to 50 mm should be on hand, since a ratio of ≈20:1 (silica:mixture) should be sufficient; for more difficult separation of compounds ranging by ~1 % in their R_f values in TLC, this ratio may be increased up to 100:1. The more spots the mixture shows in TLC, the greater proportion of silica is to be used for separation of each component.

The column, containing at its base a circular glass frit of suitable size, or a small glass wool or cotton wool plug and a layer of pure sand, is partially filled with the adsorbent using the slurry-packing technique. At first, the column is partially filled with a small volume of a weak solvent. A dilute suspension of the adsorbent in the same solvent is then added slowly, the excess solvent being drained out slowly, putting small pressure on the adsorbent bed periodically.

Sample application   Usually a solution of the sample in a *minimum volume* of the weak solvent like hexane or petrol (b.p. 40–60 °C or 60–80 °C) is added to the column and forced into the adsorbent bed to form a narrow sample zone. For samples of low solubility in weak less polar solvents a solution of the sample in a minimum volume of a strong more polar solvent like $CHCl_3$ is added to a small amount of the adsorbent (1–2 g of the adsorbent per g of the sample) taken in a standard joint r-b flask to make a slurry. The solvent is then stripped completely in a rotavapor under reduced pressure and finally under high vacuum to give a free-flowing powder. The latter is then added to the solvent on top of the column. Finally, a thin layer of pure (acid-washed) sand, glass wool, or cotton wool is added to the top of the column so that the column bed is not disturbed during addition of the eluent. *(It should be ensured in both column and flash chromatography that the column should always be under solvent during column packing, sample application, and the elution sequence; and the top of the column should never be disturbed, and never be allowed to run dry.)*

Elution and fractionation   Eluting solvent or solvent mixture is chosen by use of TLC of the sample mixture employing the same adsorbent. If several spots/zones appear in the TLC (after developing in the iodine chamber), the solvent strength should be adjusted so that the central zone or the zone of interest has an R_f value of ~0.3. As for example, ethyl acetate–petrol mixture may be started with. The impurities will be usually very polar having R_f ~0.1, or very nonpolar (fatty material) moving with the solvent front, which may be ignored. The individual components may be effectively separated by using stepwise gradient elution, the optimum mobile phase velocity being ~5 cm/min. Well-packed columns are expected to provide about 5–20 theoretical plates per cm of bed height. For samples containing components of wide polarity one can start with hexane or petrol (b.p. 40–60 °C or 60–80 °C) and add to it increasing volumes of strong polar solvent such as ethyl acetate, methylene chloride, or acetone, and finally with increasing proportions of methanol. The number of fractions to be collected at each step is determined by monitoring the composition of each fraction. The fractions are subsequently combined, based on similarity of their composition as monitored by TLC or GLC. *TLC on microscopic slides is mostly used since it is quick and inexpensive compared to various detectors like UV, visible, or IR spectrum, or refractive index which may be online, continuous, and adequate for this purpose.*

Since compressed air or gas is used for pressurizing the flash columns, some sort of pressure-release valve should be incorporated so that unsafe pressure buildup cannot occur.

When flash chromatography lacks the resolving power needed to separate the components, it can be repeated to get the desired separation by use of the expertise gained through experience. The fractions containing the components of interest, isolated through flash chromatography, if needed, should be subjected to high-resolution techniques, such as *medium-* or *high-pressure liquid chromatography* (MPLC or HPLC) for isolation of the pure components.

Miniature flash chromatography For rapid isolation of pure compounds (having similar polarity, as evident from TLC) from their mixture (~10–50 mg), or for rapid purification of crude sample (10–50 mg) obtained as some reaction product or as some natural product isolation fraction, a *miniature form of flash chromatography* may be conveniently used.

For this purpose a Pasteur pipette or a long pyrex glass dropper may be employed as a column. The pipette/dropper containing a cotton wool plug at the bottom is half to three-quarter filled with silica gel, the quantity being decided as stated before. A weak solvent like hexane or petrol is added to the top of the thin column at its inner wall from a pipette or dropper. The column may also be prepared by adding the solvent system (eluent) (petrol–ethyl acetate mixture) showing the $R_f \sim 0.3$ of the main component or components in TLC. When the eluent appears at the bottom running through the column under gravity, pressure is applied using a pipette teat to force the eluent to pass through at a faster rate. Two column volumes of the eluent is passed through the adsorbent. The sample is then applied to the top in the usual way, preferably by the slurry method, already described. A small layer of glass wool or pure sand is put at the top to protect the adsorbent from any disturbance. Pressure is again applied using the pipette teat. Here, of course, the pressure applied is not constant and the eluent does not pass through at a constant rate, which apparently does not affect the separation or purification.

4.1.8   Thin-Layer Chromatography [11–13]

This technique is being extensively used since mid-1960s for identification of mixtures of two or more organic compounds and for separation on a small scale. This technique is comparatively inexpensive and much less time-consuming. It involves interfacial adsorption/partition between liquid (mobile phase) and solid (stationary phase), like CC. This technique may also be termed **LSC** (Fig. 4.6). In this case the mobile phase is allowed to ascend a thin layer of the adsorbent (stationary phase), coated on the surface of a glass plate or microscopic slide (or a standard aluminium sheet which can be purchased) and dried—called *thin-layer plate* or *thin-layer slide*.

The adsorbents usually used for TLC are silica gel G (silicic acid) or alumina G (aluminium oxide). "G" stands for gypsum, $CaSO_4$, $\frac{1}{2}H_2O$ (better known as plaster of Paris), and is present in the adsorbent in ~10–13 % by weight, as a binder. In the presence of water or moisture it sets in a rigid mass, $CaSO_4$, $2H_2O$, and binds the adsorbent to the backing support (plate or slide). The particle size of the adsorbent is small (200–300 mesh), compared to that (usually 60–120 mesh) used for CC. Silica gel is normally used for separation of neutral and acidic compounds, whereas alumina is frequently used for separation of bases. TLC is useful as an adjunct to column chromatography.

Microscopic slides Small TLC plates are made from microscopic slides (~7.5 cm × 2.5 cm). They are extremely useful for (i) identifying the components in a mixture, (ii) determining the appropriate solvent for a CC separation, (iii) monitoring the eluate fractions of a CC and also the course of any reaction, and (iv) ascertaining the identity of two compounds or purity of a compound.

The microscopic slides should always be handled by the top edge to avoid fingerprints on the slide surface—before coating, which may not allow uniform coating, and also to avoid disrupting after coating. A pair of new or clean microscopic slides is carefully held together by external air pressure, generated by pressing them together with a micro drop of water (from a fine dropper). They may be held with a strong forceps or by hand with the thumb and the forefinger at the top sides. The pair is smoothly dipped into a slurry of the adsorbent, already prepared in a wide-mouthed screw-cap bottle, ~6.5 cm—leaving ~1 cm at the top of the slides uncoated. They are then withdrawn slowly and steadily. The slurry should be shaken immediately before dipping. *The dipping operation should be done within 2 s.* After withdrawing the pair of slides the cap should be replaced on the bottle, and the slides should be held for a few minutes for evaporation of most of the solvent. The coated slides may then be separated and kept on their backs (uncoated side) for complete drying. Several pairs of slides, *thus evenly coated and dried*, are kept ready for use.

Slurry preparation The slurry is prepared in a 4 oz or 125 ml (approx.) wide-mouthed screw-cap bottle. The silica gel G should be added to methylene chloride or chloroform, used as solvent, with constant swirling. The reverse addition may cause formation of lumps. Amounts of the adsorbent and solvent (usually 1 g of silica gel is required per 3 ml of the solvent) should be adjusted so that the depth of the slurry in the bottle becomes ~7 cm. After addition the cap should be tightly secured on the bottle which is then shaken well by a swirling motion to ensure thorough mixing and stored for future use. Evaporation loss may require addition of requisite amount of solvent from time to time. Each time just before using the slurry should be shaken vigorously by a swirling motion.

Thin-layer chromatography plates These are larger than microscopic slides and are usually of the size approximately 20 cm × 5 cm, 20 cm × 10 cm, or 20 cm × 15 cm. The TLC plates must be coated by an applicator (a commercial spreading device) with a uniform layer of the adsorbent, usually of 0.25 mm thickness for analytical purposes and of up to 2.0 mm thickness for preparative purposes. With the Stahl–Desage applicator, model S 11, the thickness of the adsorbent layer may be varied from 0.25 mm to 2 mm by proper adjustment. With such applicator the glass plates of uniform thickness must be placed beside one another. The more expensive models available now have designed plate holders leveling all plates, and thickness to ~±0.01 mm is guaranteed. Instructions for preparation of slurry and use of the applicator are given by the manufacturers and are found in the literature. The coated plates of silica gel G or alumina are allowed to stand for 30 min, activated at 110 °C for a minimum of 1 h, and stored in a dry box or desiccator until used. The layers have activity II–III.

Application of the sample   A solution of the sample in a volatile solvent (chloroform, methylene chloride, or acetone), as less polar as possible, is applied on the plate or slide by means of a capillary micropipette (prepared from a drawn m.p. capillary tube) by briefly touching it to the adsorbent surface at about 1.5 cm from the edge. This is popularly known as *spotting*. The spot should be as small as possible. The spotting may be done several times, if needed, after allowing the solvent to evaporate. Up to three spots, suitably apart, may be applied to a microscopic slide.

Development chamber   For development of microscopic slides a coupling jar or a wide-mouthed screw-cap bottle should be used. The inside of the bottle should be lined with a filter paper cut to such a length as to keep a vertical opening of ~2.5 cm for observation. The filter paper should be thoroughly moistened with the chosen developing solvent. Thus the chamber is kept saturated with the solvent vapor leading to the increase of the development speed. The level of the solvent in the bottle is adjusted to a depth of 0.5 cm; the bottle is capped and can be used when needed. Special tanks, used to develop large TLC plates, are available in the market.

Choice of developing solvent   Hexane or petrol (b.p. 40–60 °C) with varying proportions of benzene will be a good solvent for least polar compounds like hydrocarbons. For compounds containing a wide variety of functional groups benzene with varying proportions of methylene chloride (or chloroform) will be suitable. For more polar compounds methylene chloride or ethyl acetate with varying proportions of methanol gives satisfactory separation.

A suitable solvent is rapidly determined in the following way. Several sample spots are applied on a coated microscopic slide at a minimum distance of 1 cm from each other. A different solvent or solvent mixture (already available on the rack) is gently touched to the centre of each spot by a microcapillary, as already stated. The solvent front and the compound will advance outward *in concentric circles around each spot.* The solvent front should be marked with a pencil by dotted lines since it will disappear when the adsorbent gets dried. *From the appearance of the rings the suitability of the solvent or solvent mixture may be judged.* The compound rings, if colorless, will be seen under UV light or will stain in iodine chamber (*vide infra*).

Development of the chromatogram   After application of the spots and selection of the solvent the slide is carefully placed in the chamber containing the solvent for development so that the coated portion does not touch the filter paper liner and the solvent level in the chamber is 5–8 mm below the applied spots. The cap is then replaced; the solvent rapidly rises up the slide by capillary action. One should watch carefully and the slide should be removed from the bottle when the solvent has traveled to within 5 mm of the top end of the adsorbent layer. The position of the solvent front is immediately marked gently with a pencil/needle as a dotted line. After the slide has dried, the spots (after visualization) are outlined on the slide with a pencil. Visualization methods are discussed briefly in the sequel.

The R_f for each spot (compound) is measured as described earlier. It varies from zero (if the compound does not move) to unity (if the compound moves with the solvent front) and will be a decimal fraction (when the compound moves). It serves for qualitative identification under a given set of conditions (vide *supra*). Two compounds may have, by coincidence, same R_f value under a given condition, just as two compounds may have the same m.p. In such a case, different solvent systems should be tried, and the solvent system showing different R_f values may effect their separation by CC using same adsorbent and similar solvent system. The relative R_f values of the compounds, obtained in microscopic slides under some specified condition, may still be published since these give some idea about their relative mobility and may be helpful for separation/identification.

Visualization Separation by a TLC experiment can be followed visually for colored compounds. Many organic compounds are, however, colorless. The separated components must be made visible by some method or reagent, called visualization method or reagent, which may be *nondestructive* or *destructive*.

Nondestructive methods

(i) *Exposure to iodine*. Most organic compounds (except saturated hydrocarbons and alkyl halides) react with iodine to form unstable yellow or brown complexes. The developed and dried TLC microscopic slide is placed in a 4-oz wide-mouthed screw-cap glass bottle containing some crystals of iodine. The capped bottle is mildly warmed on a steam bath when the former is filled with iodine vapor, and the spots begin to appear. After the spots become quite intense, the slide is removed from the bottle, and the spots are immediately outlined by a needle/pencil. On keeping for some time, the spots in most cases fade, as the iodine sublimes off leaving the compounds unchanged.
Documentation. Microscopic slide chromatograms can best be recorded in a lab. notebook in the following way. A cellulose tape is pressed onto the visualized layer. The tape and the portion of the layer, which adheres to it, are then removed. An additional layer of tape is placed, covering the layer. The resulting "tape sandwich" is then placed in a lab. notebook. Alternatively, one can draw a picture of the chromatogram in a lab. notebook.

(ii) *UV method*. The most common method of visualization is by a UV lamp, when compounds often look like bright spots. Some types of compounds fluoresce under UV and become very bright. The bright spots under UV should be marked with a needle/pencil.

Destructive methods Several chemical methods are available as stated below for plates using adsorbent with binder:

(i)   The slides or plates can be sprayed with conc. H_2SO_4 and heated to 100–110 °C in an oven for a few minutes. Organic compounds get destroyed and charred and appear as black spots.

(ii) There are chemical methods, specific only for particular functional groups, permanently altering the compounds through reactions. Preparation and use of the spray reagents, e.g., H_2SO_4–$Na_2Cr_2O_7$, H_2SO_4–$K_2Cr_2O_7$, $HClO_4$ (organic compounds appear as black spots), $SbCl_3$–$CHCl_3$ (steroids, steroid glycosides, etc. → various colors), 2,4-DNPH (for aldehydes and ketones → yellow to red spots), bromocresol green (carboxylic acids → yellow to green spots), Dragendorff's reagent (alkaloids and organic bases → yellow to orange), ferric chloride (phenols → various colors), ninhydrin (amino acids, amino sugars → blue), etc., are available [11].

Some more chromatographic methods which are quite involving, but useful, especially HPLC and MPLC, and also paper chromatography (PC), gel chromatography, reverse phase chromatography (RPC), etc., are briefly discussed in the sequel, only to give an elementary idea about the techniques and their principles. For detailed information, the relevant chapters of the books [10–14, 17] may be consulted.

The recent versions of the apparatus/equipments used for the different chromatographic techniques, as discussed in this chapter, including their operating manuals, etc., are available with their manufacturers. They are also available in advanced organic chemistry laboratories (research institutes and universities) of many countries.

Preparative TLC This technique was being used extensively for separating small quantities of compounds (up to 1 g) of similar polarity until late 1980s. However, it has been largely superseded by the advent of flash chromatography and preparative HPLC and MPLC.

Large plates (20 cm × 20 cm, 20 cm × 15 cm, or 20 cm × 10 cm) are coated with silica gel G or alumina by an applicator with a thickness of 1–2 mm, as required—depending upon amount of the material. The sample to be separated is applied in a thin line on one side of the dried and activated layer, about 1 cm from the layered bottom of the plate. A number of TLC plates, as needed, may be used. The plates are then developed as usual by dipping in a suitable chamber containing about 0.5 cm of the desired eluting solvent to resolve the mixture into bands. The bands are visualized by a nondestructive method already described. The adsorbents in those bands are separately scraped from the glass plates and the adsorbed material is extracted with a polar solvent. Filtration removes the adsorbent and evaporation of the solvent from the filtrate gives the separated components admixed with some adsorbent. Further purification of the latter by repeated preparative TLC and crystallization from a suitable medium often becomes necessary to get the components in the pure state.

Impregnated silica gel as an adsorbent Silica gel impregnated with different inorganic salts like silver nitrate [19] and mercuric acetate or with an organic compound like trinitrobenzene has been found to be selective; it shows discriminating adsorption behavior toward closely related compounds with special reference to olefinic compounds. In the latter cases, this selective behavior depends on the configuration, steric location, and the number as well as conjugation of the double bonds present in the concerned molecules. An adduct is formed with each of the constituent olefins endowed with different absorptivity and hence could be eluted from the column separately. From the adduct the olefin could be isolated by steam distillation, when it is steam volatile, or by treatment with ammonia. Silica gel is treated [19] with silver nitrate solution in a rotavapor when silver nitrate solution gets uniformly smeared on silica gel surface, thus making the silica gel ready for the chromatographic column. TLC plates or slides can also be drawn with a slurry of silver nitrate impregnated silica gel of TLC grade.

4.1.9   Paper Chromatography

Though apparently PC appears to involve solid (paper)–liquid (mobile) phases, in practice it involves liquid–liquid phases and is mechanistically termed as LLC. The paper is made up of high-quality cellulose which absorbs water from the atmosphere and retains it in the pores. The absorbed water thus serves as the stationary phase rather than the cellulose (solid) itself. The sample is applied at one end and the mobile phase is allowed to act either by the ascending way or by the descending way in a chamber specific for this purpose. The compounds of the sample get partitioned between the two liquid phases and a series of distribution equilibria is set up; the latter being different for different components for a particular pair of phases, the separation could be achieved.

Two-dimensional PC run in an orthogonal direction of the first development by the same ascending or descending method makes better resolution of the components.

Mostly, ionic compounds or polar compounds like sugars could be effectively separated. Spots are developed on the paper by spraying different reagents and the choice depends on the nature of the compounds separated. The mobile phases used are polar and in most cases water is present as a cosolvent.

4.1.10   Gas Chromatography

In this technique the components of the mixture to be separated are partitioned between the moving gas phase and the stationary liquid phase. Here the column is usually made up of copper or stainless steel or glass tubings of various diameter and length. A high boiling, less volatile liquid (low vapor pressure) or a low melting solid is supported generally on uniformly crushed firebricks (common supporting material) in the following way. The former is taken in a low boiling volatile liquid (CH_2Cl_2, $CHCl_3$) and mixed with firebrick particles using a rotary evaporator. When the solvent is removed, firebricks are uniformly coated with the liquid or the low melting solid acting as the stationary phase. With these uniformly coated particles, the tubing is packed carefully and uniformly. The column is then coiled and placed in a gas chromatograph oven. One end of the column is used for gas entrance and the other end as outlet. The latter is connected to a detector. The sample is injected in the form of either liquid or solution. The injected material is vaporized and the components will travel through the column at different rates due to difference in their partition coefficients; the components are collected at different times and get detected by the detection device. *Retention time* is constant for each component for a particular column system. Retention time of a compound is defined as the time between the sample injection and the appearance of the maximum peak shown by the concerned compound, as pointed out by the deflection of the pen. Outcoming components in the gaseous state are collected in a cooled trap; the coolant may be liquid nitrogen (b.p. $-196\ °C$), liquid helium (b.p. $-260\ °C$), dry ice-acetone ($-65\ °C$), etc.

The area of the gas chromatographic peak is proportional to the amount of the sample that appeared. Nonpolar, nonionic, water-insoluble hydrocarbons and multifunctional compounds could be separated by this method. Other equivalent terms for this process are *vapor phase chromatography* and *gas–liquid partition chromatography*.

4.1.11 High-Performance/Pressure Liquid Chromatography

It is one of the most successful and routinely used chromatographic techniques. Unlike GC, its application is not limited by the sample volatility or thermal stability. It is capable of separating a wide variety of compounds that include macromolecules, natural products, ionic compounds, polymeric compounds, and also labile and multifunctional compounds. It is the most versatile tool of modern days for both qualitative and quantitative analyses. It allows many types of inter-actions operative within the sample, mobile phase, and stationary phase, which in turn result in effective separation. At one end of the column sample injection device is present, and at the other end a detector is attached. Compounds are mostly detected from their refractive indices. The packing material and the length of the column are chosen carefully. Sometimes separation is more effective at high temperature which decreases the viscosity of the mobile phase, increases the sample solubility in the mobile phase, and causes effective mass transfer; the net effect is a better resolution. Generally the internal diameter of the tube is 4–5 mm and the packing material particle size is 3–5 μm.

At the beginning the sample material stays at the top of the column. With the beginning of elution, the strength of the mobile phase (eluent) gradually increases and the migration of the components starts; the components are collected in the liquid mobile phase. Solvents with low viscosity are always preferred for HPLC. *High-grade dry solvents should be used.*

Several modes of HPLC other than the partition chromatography or interfacial adsorption chromatography, *viz.,* size-exclusion chromatography, ion-exchange chromatography, etc., are also in use today.

HPLC and GC coupled with mass spectrometry (MS) deliver more information about the compound. Sometimes MS acts as the detector for collecting fractions (having same M^+) for further study—especially for bioactive fractions.

Various solvent extraction methods and different chromatographic techniques, e.g., capillary electrophoresis, micellar electrokinetic capillary chromatography, centrifugal partition chromatography, and two-dimensional PC, have been employed for the separation and isolation of specific coumarins [20].

4.1.12   Medium Pressure Liquid Chromatography

MPLC, in principle, is quite close to flash chromatography and HPLC, but is more efficient than flash chromatography in separating the constituents from their mixture. It is capable of handling larger quantity and is economical since the column could be reused several times. However, the column should be cleaned before reuse from the residual leftover from the previous use. The most vital part of MPLC is its pump to be operated at 100 psi with a controlled flow-rate system and a provision for not building up high pressure during operation. Columns are made up of teflon or good quality glass and are coiled. The total length of the column may be tailored according to the need. The columns may be packed ordinarily with flash chromatography-quality silica (40–60 μm) and for difficult separations with 15–20 μm silica. Other packing materials include alumina, ion-exchange resin, and sephadex. The chromatogram is connected to a fraction collector and the fractions may be identified by a refractive index or a UV detector. The flow rate depends on the diameter of the column tube. The sample is injected with a good syringe at a reduced flow rate to avoid the building up of the pressure in the system. The columns can be purchased from the manufacturers. *Solvents used must be of high grade.*

4.1.13   Reverse Phase Chromatography

The name itself suggests the order of elution of the components. The most polar compound is eluted first, and the other components are eluted in the order of decreasing polarities. Accordingly hydrocarbons are eluted last. The packing material of the column is generally hydrophobic bonded material, usually with an octadecyl (C_{18}) or octyl (C_8) functional group and a polar mobile phase, often partially or fully aqueous. Polar compounds dissolve in the polar mobile phase first and will move faster than the nonpolar components. The solvents like methanol, acetonitrile, etc., which are available in high-grade purity, are used along with water, as the mobile phase.

4.1.14   Gel Permeation Chromatography

In this chromatographic technique the stationary phase commonly uses cross-linked polymer having sieve-like structure. In this method the molecules get separated according to their shapes and sizes, and the dominant distribution process is SEC. When a mixture is allowed to pass through such a column used for SEC, small molecules stay in the pores, while the big molecules, which the pores fail to hold, move and elute faster than the small molecules. Sephadex (cross-linked dextrans of microbial origin) is commonly used for this purpose. It has an ability to absorb water, and as a consequence, the material swells and expands, and holes are created in the matrix. Equivalent terms of the process are *gel filtration, gel chromatography, or molecular sieve chromatography.* Biomolecules are mostly separated by this method.

4.1.15   Bioassay-Guided Investigation

From time immemorial the plant kingdom remains a major source of various remedial measures against physical ailments of man. Today, search for bioactive molecules from plants has been one of the major thrusts of natural products chemistry research. The objective is to isolate bioactive molecules having fair to profound medicinal values and to discover lead molecules directed toward the synthesis of comparatively simple molecules with improved medicinal properties and economic viability. In this endeavor, during fractionations of crude plant extracts, each fraction should be monitored by ^{1}H NMR and GC-MS. Fractions showing the presence of good signals and peaks for organic compounds are collected, and their biological activities are studied. Fractions showing promising bioactivity are collected in amounts, monitored each time during collection by HPLC and/or GC-MS. Both structural work as well as different biological activities in detail on the compounds isolated from the bioactive fractions should be carried out to reach the goal of discovering a drug.

4.1.16   Homogeneity and Physical Constants of the Isolated Compounds

Homogeneity of an isolated compound is the primary requirement for its characterization. Assessment of homogeneity basically needs the application of various chromatographic techniques; the following applicable characterizations are to be observed for this purpose:

(i) Appearance of a single spot on TLC plate/slide in different solvent systems
(ii) Appearance of a single molecular ion peak in its mass spectrum
(iii) In case of a polar compound a single spot on paper chromatogram
(iv) A single peak in GLC/HPLC
(v) A sharp/consistent m.p. in case of a crystallizable solid
(vi) In case of a chiral compound, optical purity may be established if there is no significant change of $[\alpha]_D$, after repeated purification
(vii) For a liquid, refractive index measurement, till there is no significant change after repeated purification by GLC or HPLC.

4.2   Structural Elucidation

4.2.1   General Approach

After the establishment of the homogeneity of the isolate, some of its physical properties, *viz.,* m.p./b.p., molecular formula, solubility in common solvents, and optical rotation if any, are studied, followed by its structural elucidation. In fact, structural elucidation is a prerequisite for any further research in natural products chemistry.

Optical rotation The chirality (mirror image nonsuperposibility) of a natural product is exhibited by its ability to rotate the plane of polarized light. The observed sense of rotation of a chiral natural product—if clockwise—identifies it as the dextrorotatory (+) enantiomer, or if anticlockwise—as the levorotatory (−) enantiomer. Usually optically pure natural products would reveal the sense and magnitude of rotation of one enantiomer only. *When the absolute configuration is determined, it is mandatory to tag it with the sign of rotation, i.e., it must be mentioned whether it is the absolute configuration of the (+) or (−) enantiomer. The magnitude of rotation depends upon the solvent and concentration. Sometimes even the sign of rotation also changes with different solvents, as well as with the concentration in the same solvent, e.g., nicotine* [21]. For a detailed discussion of optical rotation, see Sects. 2.2.8 and 2.2.9.

The various strategies developed for structural elucidation during the nineteenth and twentieth centuries will now be discussed through brief illustrations. We will find, how the tedious and time-consuming degradative methods and various chemical reactions, once practiced with great success although taking many years, have been gradually replaced by IR, UV, and NMR spectroscopic and mass spectrometric analyses [22–66]. The spectral analyses have manifold advantages, as stated below, over the earlier degradative and other classical methods.

(i) It requires a very small amount of the material and much less time.

(ii) It is nondestructive in nature (except mass spectrometry which needs less than a mg), and the sample may be recovered and reused, if necessary.

(iii) Presence of some chromophore/s and some functional group/s can be detected immediately from the UV and IR spectra of the concerned compound.

(iv) Spectral data especially of ^1H [37–46] and ^{13}C [47–53] are available in the literature for compounds with a wide range of structural patterns. These data are extremely useful for structure deductions of new compounds from their spectral data [24–31, 37–55, 58] comparison within much less time, and a very quick identification is possible for known compounds. Thus such time-saving data give either the full structure or a part structure.

(v) Stereochemical and conformational information can be derived by judicial analysis of ^1H and ^{13}C NMR spectral data (chemical shifts and coupling constants, NOE, and various two-dimensional correlation studies). A number of excellent books [24–31, 37–45, 47–51, 53] on these very important topics have appeared from time to time. For an introduction to 2D NMR spectroscopy references [56, 57] and for details some references quoted under "Further References" may be consulted.

(vi) The mass spectral data [59–66] furnish the correct molecular weight and molecular formula as well as can predict the structural framework on the basis of the fragmentation pattern of the molecule.

(vii) ORD and CD studies [67–74] are capable of yielding information on the absolute stereochemistry of the chiral molecule (vide Chap. 2).

(viii) X-ray crystallographic analysis [75–77] reveals the structure, stereochemistry, and conformation of the chiral molecule and provides clues to the interpretation of NMR spectral data of complex organic molecules. In fact, in such cases X-ray data may complement the NMR data and vice versa.

All these methods collectively may not be necessary to establish the structure. Judicial application of the methods as demanded by the complexity of the molecule should be made. There are remarkable proliferations in the NMR and mass spectral analyses, which are now in extensive use. Carsten Reinhardt in his book [78] states that "Physical instruments threatened to destroy the methodological autonomy of chemistry [now] engineers and physicists appeared on the scene, attempting to displace chemistry by electronics". However, the reviewer of the book, Jerome A. Berson of Yale University commented, "We welcomed with enthusiasm the new power placed at our disposal by instrumental advances" [78].

In this connection we are predisposed to quote Woodward [79], "While it is undeniable that organic chemistry will be deprived of one special and highly satisfying kind of opportunity for the exercise of intellectual *élan* and experimental skill when the tradition of purely chemical structure elucidation declines, it is true too that not infrequent dross of such investigation will also be shed; nor is there any reason to suppose that the challenge for the hand and the intellect must be less, or fruits less tantalizing, when chemistry *begins* at the advanced vantage point of an established structure."

And to remain competitive in the field of chemical research one has to take advantage of the powerful instrumental analysis. However, some of the classical chemical methodologies, which served earlier as the basic part of chemical research work, especially in the structural studies of natural products, though not much in vogue now, have been discussed in most of the chapters (Chaps. 6–29) in fair detail for compounds on which they were usefully applied.

A general approach for the structural determination of natural products (or any new compound) is now outlined wherein classical chemistry and instrumental analysis play the partnership roles.

(1) *Molecular formula determination can be achieved and confirmed by*

    (a) elemental analysis giving the empirical formula and molecular weight determination
    (b) high-resolution mass spectrometry (MS) giving the correct molecular formula
    (c) ^{13}C NMR (number of carbon atoms) and ^{1}H NMR (number of protons) and identification of the functional groups.

Nowadays the procedure (a), especially the molecular weight determination, requiring time and material is not generally practised. However, it is helpful and convincing to have the elemental analysis data while reporting a new compound. Earlier, before the advent of NMR and MS, much emphasis used to be given on the elemental analyses. Incidentally, it may be mentioned as a relevant example [80] that the Editor of the *J. Am. Chem. Soc.* commented on the elegance of the authors' methodology, but did not accept a paper of Konrad Bloch on the characterization of 14-norlanosterol, an intermediate in cholesterol biosynthesis, for not providing the m.p. and the elemental composition, thus failing to satisfy editorial requirements. Of course, the *Journal of Biological Chemistry* accepted the manuscript as such [80]. The elemental analysis gives the empirical formula; the actual molecular formula = n (empirical formula), where n is an integer like 1, 2, or 3, etc. For an unknown compound the molecular formula may be determined by the analysis of the compound and its suitable derivatives or best by its high-resolution mass spectrometry.

(2) *Detection* of the *chromophore/s* and *functional group/s present* in the molecule *by* UV and IR *spectral measurements*
(3) Number of unsaturation/s and ring/s (if present) together being called unsaturation number (UN)
(4) Classical degradative methods to determine the basic skeleton and the presence of chromophores and functional groups
(5) ^{1}H and ^{13}C NMR spectral analysis.
(6) Mass spectral (MS) analysis.
(7) ORD/CD studies.
(8) X-ray crystallographic analysis

4.2.2 *Unsaturation Number. Degradative Methods. Derivatization*

During prespectroscopic days (before NMR and MS) in the nineteenth century to mid-twentieth century, structural determinations were based on degradative methods and chemical reactions of some functional groups and/or of the structural framework. Some of them are now briefly described.

Determination of unsaturation and the number of rings (Unsaturation Number) Double bond and triple bond, though almost inert to H_2 alone, undergo facile catalytic hydrogenation. One mole of H_2 consumption corresponds to one ethylenic double bond (C=C), while one ethynic triple bond (C≡C) consumes two moles of H_2. In practice, the volume of hydrogen consumed is converted into moles; the number of double bond/s and/or triple bond/s present in the molecule can thus be determined. This is illustrated by a simple example involving hydrogenation of geraniol to give its tetrahydro derivative (Fig. 4.7). Thus, the parent molecule must contain *two* double bonds.

In cases of *sterically hindered, tetrasubstituted, and trisubstituted double bonds* which resist catalytic hydrogenation, peracid titration is done. In the conversion of a double bond to an epoxide, one mole equivalent of peracid is consumed, and the number of double bonds can be determined from the number of mole equivalent/s of peracid consumed.

In cases of homocyclic *ring compounds* the number of rings can easily be obtained as follows. The molecule is formally converted into the corresponding saturated hydrocarbon; its hydrogen content is to be compared with the number of hydrogen atoms present in the corresponding saturated open chain hydrocarbon having the formula C_nH_{2n+2}, where n represents the number of carbon atoms present in the parent compound. If the compound contains double bond/s, the number of hydrogen atoms required to saturate the double bond/s should be deducted from the molecular formula of the saturated open chain hydrocarbon. For example, the saturated hydrocarbon caryophyllane, $C_{15}H_{28}$, obtained by catalytic hydrogenation of caryophyllene, $C_{15}H_{24}$, contains four hydrogen atoms more than the parent compound (Fig. 4.8); hence caryophyllene contains two double bonds. Compared to the corresponding theoretical open chain saturated hydrocarbon, $C_{15}H_{32}$, caryophyllane contains four hydrogen atoms less, which are accountable for being bicyclic.

A general equation determining the number of double bonds and/or triple bond/s and rings, known as unsaturation number (**UN**), present in any organic compound having C, H, O (or any dicovalent atom like S), N (or any tricovalent atom like P or B), and X (halogen), is given below.

For a compound $C_nH_mO_pN_qX_r$ the UN $= [(2n + 2) - (m - q + r)]/2$ (4.1)

The same equation holds good for other bicovalent atoms like S, Se, to be ignored like O, and also for other tricovalent atoms like P, B, As, to be treated within q like N.

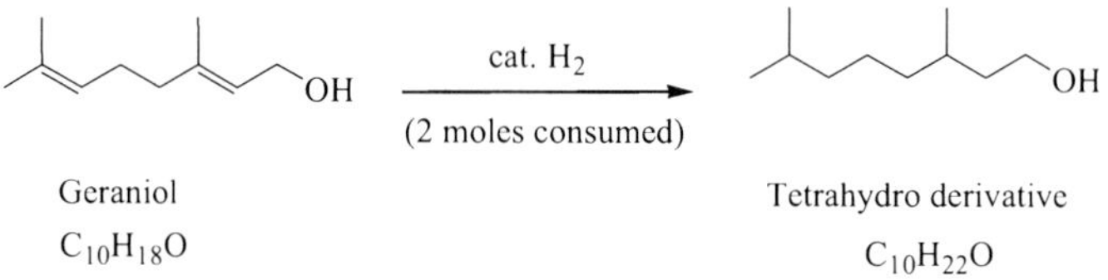

Geraniol

$C_{10}H_{18}O$

Tetrahydro derivative

$C_{10}H_{22}O$

Fig. 4.7 Hydrogenation of geraniol

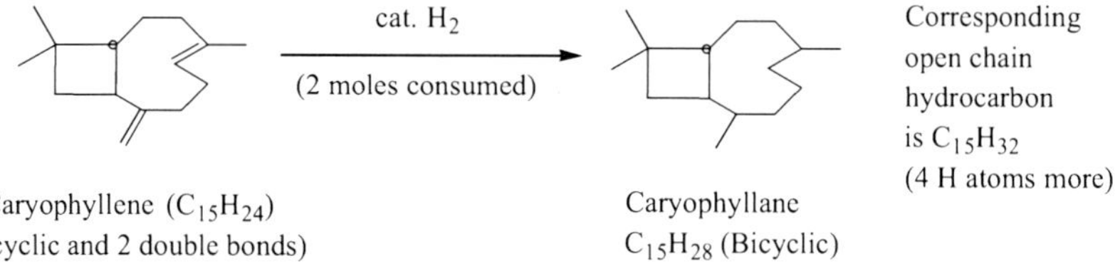

Caryophyllene ($C_{15}H_{24}$)
(Bicyclic and 2 double bonds)

Caryophyllane
$C_{15}H_{28}$ (Bicyclic)

Corresponding
open chain
hydrocarbon
is $C_{15}H_{32}$
(4 H atoms more)

Fig. 4.8 Hydrogenation of caryophyllene

The above equation is derived from the consideration of the following points:

(i) The known molecular formula is compared with the theoretical formula expected for a completely saturated open chain molecule C_nH_{2n+2}.

(ii) Each O atom present as C=O, or S atom present as C=S, will be taken as a double bond.

(iii) Each O atom present as –OH or –OR, standing in place of –H or –R, is ignored.

(iv) Each S atom present as –SH or –SR, standing in place of –H or –R, is ignored.

(v) Each tricovalent atom (N, P, B, etc.) brings one additional H atom. (The simplest example is the H count of CH_3NH_2 compared to CH_4.) >C=NH should be regarded as one double bond and –C≡N as two double bonds, like C≡C.

(vi) Each halogen atom replaces one H atom, so needs addition of one H atom.

(vii) Each unit of unsaturation or each ring removes two H atoms. Hence the UN (unsaturation number) equals half of the shortfall of H atoms from $2n + 2$— the numerator of the general equation stated above, after taking care of the above points (ii) to (vi), whichever are necessary, to deduce the UN for any particular molecule.

A few examples (see Fig. 4.9 for the structures of the compounds) are given below to illustrate the validity of (4.1). Of course, for finding out the UN of an unknown compound one must know the correct molecular formula.

Example 1
For 5-chlorocyclohex-2-ene-1-one (**1**), C_6H_7OCl, *or its enantiomer or any position isomer* $n = 6$, $m = 7$, $r = 1$. Applying (4.1), UN $= [(6 \times 2 + 2) - (7 + 1)]/$

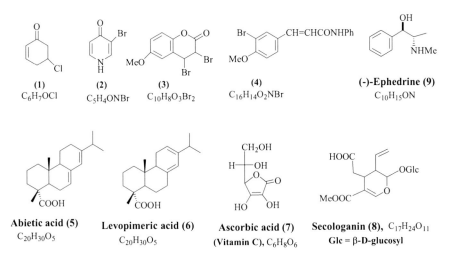

Fig. 4.9   Structures of compounds (**1**)–(**9**)

$2 = (14 - 8)/2 = 3$, which constitutes two double bonds (including one C=O) and one ring.

Example 2

For 3-bromo-1,4-dihydropyridine-4-one (**2**), C_5H_4 ONBr, or its any possible position isomer $n = 5$, $m = 4$, $q = 1$, $r = 1$, so UN $= [(5 \times 2 + 2) - (4 - 1 + 1)]/2 = (12 - 4)/2 = 4$, which constitutes three double bonds (including one C=O) and one ring.

Example 3

For 7-methoxy-3,4-dihydro-3,4-dibromocoumarin (**3**), $C_{10}H_8O_3Br_2$, or its any possible position isomer UN $= [(2 \times 10 + 2) - (8 + 2)]/2 = 6$ (four double bonds and two rings).

Example 4

For the benzanilide of 3-bromo-4-methoxycinnamicacid (**4**), $C_{16}H_{14}O_2NBr$, or its any possible position isomer   UN $= [(2 \times 16 + 2) - (14 - 1 + 1)]/2 = (34 - 14)/2 = 10$, constituted of two rings and eight double bonds (including one C=O).

Example 5

For abietic acid (**5**), $C_{20}H_{30}O_2$, or its any possible position isomer [e.g., compound (**6**)] UN $= [(2 \times 20 + 2) - 30]/2 = 6$ [three double bonds (including one C=O) and three rings].

Example 6
For levopimaric acid (**6**), UN = same as abietic acid

Example 7
For ascorbic acid (vitamin C) (**7**), $C_6H_8O_6$ UN = $[(2 \times 6 + 2) - 8]/2 = 3$ [two double bonds (including one C=O) and one ring]

Example 8
For secologanin (**8**), $C_{17}H_{24}O_{11}$, so UN = $(36 - 24)/2 = 6$, of which four are double bonds (including 2C=O's) and two are rings.

Example 9
For (−)-ephedrine (**9**), $C_{10}H_{15}ON$, UN = $[22 - (15 - 1)]/2 = 4$ (three double bonds and one ring)

Degradative methods A major tool in the past has been the structural elucidation by degradative methods. The unknown compounds are chemically degraded, the degraded products are identified, and the structure of the compound is constructed on the basis of its degradation products. This involves lots of chemistry, bench work, and intuition, and unlike the picture puzzles guide no definite structure is available prior to its construction. The structural elucidations on the basis of such degradation in cases of complex alkaloids morphine (Chap. 23) and quinine (Chap. 25) are indeed stunning achievements in the absence of NMR and MS spectral facilities in those days. Of several degradations, a few still survive and the *Hofmann degradation* is the most informative one [coniine, conhydrine, and ψ-conhydrine (Chap. 17) and morphine (Chap. 23)]. *Alkali fusion method*, a favorite reaction of the past, used to give sumptuous structural information of the compound tried for this reaction.

Dehydrogenation with S/Se/Pd, especially of perhydropolynuclear natural products like terpenoids, steroids, alkaloids, etc., furnishes information about their structural skeletons. Such a study was first done by Vesterberg in 1903; he obtained retene from abietic acid on its dehydrogenation with sulfur. However, Ruzicka developed it to be a major method of structure elucidation. The identification of the aromatic compounds produced suggests the number of rings and their orientation, i.e., the major connectivity of the skeletal carbons in the molecule. Some examples are cited in Fig. 4.10. Diels first introduced the use of Se for dehydrogenation of polycyclic compounds (Chap. 31).

Derivatization From the IR and NMR spectral analyses the presence of some functional groups is indicated. Their presence could be confirmed chemically by derivatization (Fig. 4.11). The spectral studies of the derivatives furnish further structural information on the parent compound.

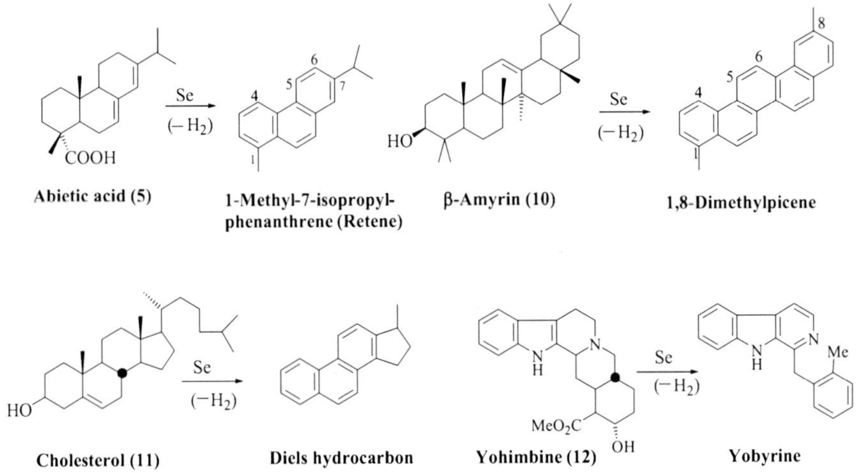

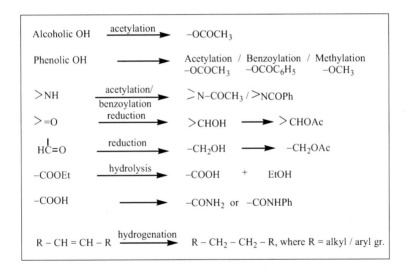

Fig. 4.10 Se dehydrogenation products of some natural products

Alcoholic OH	acetylation →	$-OCOCH_3$		
Phenolic OH	→	Acetylation / Benzoylation / Methylation $-OCOCH_3$ $-OCOC_6H_5$ $-OCH_3$		
>NH	acetylation/ benzoylation →	$>N-COCH_3$ / >NCOPh		
>=O	reduction →	>CHOH	→	>CHOAc
HC=O	reduction →	$-CH_2OH$	→	$-CH_2OAc$
−COOEt	hydrolysis →	−COOH	+	EtOH
−COOH	→	$-CONH_2$ or −CONHPh		
R − CH = CH − R	hydrogenation →	R − CH_2 − CH_2 − R, where R = alkyl / aryl gr.		

Fig. 4.11 Derivatization of some common functional groups

4.2.3 Spectral Analysis. General Discussion

The general regions of the spectrum of the electromagnetic radiations are shown in Table 4.2, ***although there are no sharp boundaries between regions***. The spectral measurements (except mass spectrometry) are based on the ability of molecules to absorb radiant light at different specific wavelengths (having different

Table 4.2 Regions of the electromagnetic radiation spectrum

	Gamma (γ) Rays and Cosmic Rays	X-Rays	UV (electronic transition)	Visible — Blue	Visible — Red	IR (bond vibration) Near Far	Microwave (rotational motion)	Television	Radio
Unit									
λ cm	10^{-10}	10^{-8}	10^{-6}	4×10^{-5}	7×10^{-5}	10^{-2}	10^{2}	10^{4}	
λ μm	10^{-6}	10^{-4}	10^{-2}	4×10^{-1}	7×10^{-1}	10^{2}	10^{6}	10^{8}	
λ nm	10^{-3}	10^{-1}	10	4×10^{2}	7×10^{2}	10^{5}	10^{9}	10^{11}	

	γ-Rays	X-Rays	UV	Visible	IR	Microwave	Television	Radio
Wave number cm^{-1}	10^{10}	10^{8}	10^{6}	2.5×10^{4}	1.4×10^{4}	10^{2}	10^{-2}	10^{-4}
Frequency Hz	3×10^{20}	3×10^{18}	3×10^{16}	7.5×10^{14}	4.3×10^{14}	3×10^{12}	3×10^{8}	3×10^{6}
Energy kcal / mole	2.9×10^{6}	2.9×10^{4}	2.9×10^{2}	72	40	2.9×10^{-1}	2.9×10^{-5}	2.9×10^{-7}

- 1 μm = 10^{-4} cm, μ means "one millionth part of." Thus the unit of length 1 μm (micrometer) = 10^{-6} meter [(or 10,000 angstroms (Å), 1 Å = 10^{-8} cm or, 10^{-10} m] (μ is always used as prefix). However, the unit micron (μ) is often used synonymously with the unit micrometer (μm) by many chemists, especially in case of IR.
- 1 nm = 10^{-9} m or, 10^{-7} cm (n abbreviated for nano, meaning 10^{-9}, is also used as prefix).

- Wave number is defined as the number of waves per cm (i.e., $1/\lambda$) and has the dimension cm^{-1}. **Thus, wave no. = 1 cm / λ (in μ) = 10^{4} μ / λ (in μ); hence λ (in μ) = 10^{4} / wave no. (cm^{-1}).**
- Frequency, ν, refers to the number of waves passing through a point per second. Thus ν = c/λ .

- $E = h\nu$ E = Energy associated with the electromagnetic radiation at different wave lengths, and is directly proportional to ν
  $\nu = c/\lambda$ (frequency) and is inversely proportional to wave length λ; the symbol h is the Planck's constant (= 1.57 x 10^{-34} cal/ sec)
- $E = hc/\lambda$ The symbol c is the velocity of light (= 3 x 10^{10} cm / sec). The symbols h and c are natural constants. Thus, the lower is the wave length λ, the higher is the associated energy. Energy per mole = Nhc/λ, where N = Avogadro's number, 6.02 x 10^{23} mole^{-1}.

energies) (Table 4.2), which in turn causes different types of excitations. These excitations include electronic transition, bond deformation, rotational and vibrational excitations, and nuclear spin inversion. The excitations are caused by different amounts of energy the molecules absorb and hence are observed at different wavelengths of the electromagnetic spectrum. The energy being quantized, the absorption of energy should be such that they correspond to the energy gap between the two states involved in the transition. The molecule rotates about the axis between its constituent atoms and also vibrates. The quantized energy required for these modes of excitements being less than that required for electron transition, they remain operative and interact with electron transition making the absorption lines of the latter broader as we find in cases of UV spectrum. Ultraviolet (UV) and infrared (IR) spectral data of the compound are initially measured to gain some information about some **chromophores** and **groups**. The bonds the nucleus is held with influence the nucleus from which electrons are excited from the ground state. The atoms and the groups carrying such atoms affecting such absorptions are known as **chromophores**. Several books [22–46] are available dealing with the spectroscopic analysis of organic molecules in fair details.

4.2.4 Ultraviolet Spectroscopy

Ultraviolet spectroscopy was developed first among the various types of spectrometry. But with the advent of IR spectroscopy, especially of NMR—the most sophisticated tool available to the chemists—the usefulness of UV spectroscopy has become relatively less. Yet it is being routinely used to further confirm the presence of different chromophores.

As mentioned, the molecules selectively absorb electromagnetic radiations, sometimes strongly and sometimes weakly. The regions of absorptions are known as bands. UV-active molecules absorb within the span of UV range (UV absorption region 200–400 nm) between X-ray and visible light. The longest wavelength is little shorter than that of the violet light. The bands together represent the UV spectrum of the molecule in which absorbance is plotted against wavelength.

Let I be the intensity of light passing through a sample in solution and I_0 be the intensity of the incident light before it passes through the sample; the ratio I/I_0 is called *transmittance* and is usually expressed as a percentage [$\%T = 100(I/I_0)$]. The *absorbance A*, also called *optical density* (in older literature), is based on the transmittance and, at a specified wavelength λ, is given by $A_\lambda = \log(I_0/I)$.

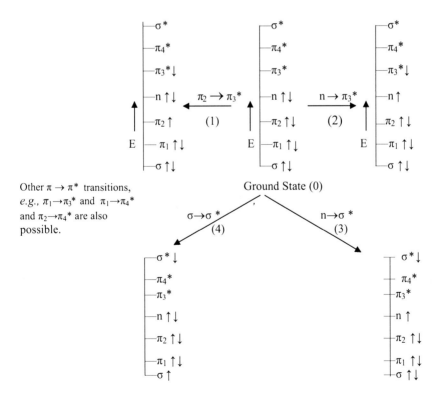

Fig. 4.12 Molecular orbitals involved in electronic transitions (visible, UV, and far UV regions)

According to *Lambert–Beer law*, also known as *Beer–Lambert law*, $A = \varepsilon cl$ for a given wavelength, where ε is the *molar absorptivity*, formerly known as *molar extinction coefficient*, c is the molar concentration of the solute, and l is the length (cm) of the sample cell. So absorbance is linear with c and l and is directly proportional to them. Thus $\varepsilon = A$ at unit length and concentration. Absorptivity is controlled by the nature of the absorbing system and the probability of the electronic transition. Values above 10^4 (up to 10^6) and below 10^3 are called *high-intensity* and *low-intensity* absorptions, respectively. The group of atoms producing an absorption in UV or visible region is called a *chromophore*.

The spectrum involves three major types of ordinarily recognizable electronic transitions involving the valence electrons (Fig. 4.12) from a bonding/nonbonding orbital to an antibonding orbital. The σ–σ* transition requires much energy and is generally observed in the far ultraviolet region below 200 nm.

4.2.4.1   Different Types of Electronic Transitions

Π–π* Transition (*type 1*) is caused by the transition of an electron from a bonding orbital π to an antibonding orbital π*. In a compound containing an isolated double bond, this transition requires much energy and generally occurs at a wavelength below 200 nm; this strong absorption is called *end absorption* indicating the probability of the presence of a double bond. However, if the double bond is present in conjugation, the energy gap becomes less (Fig. 4.12), and absorption takes place at a higher wavelength, which is known as *bathochromic shift (red shift)* of the absorption band. Depending on the region of absorption the extent of conjugation in the molecule could be predicted. Substitutions on the olefinic double bond also cause such bathochromic shifts (shift from lower to higher wavelength). Examples: unconjugated to conjugated acyclic or cyclic olefins. Conjugated double bonds give a strong absorption: s-*cis* (253 nm) and s-*trans* (213 nm) in cases of 1,3-butadiene and substituted cyclic or acyclic 1,3-dienes. Sometimes *hypsochromic shift (blue shift)* (higher to lower wavelength) may also take place. An increase in intensity is termed *hyperchromic effect* and a decrease in intensity—*hypochromic effect*. Substituents which may not give rise to UV absorptions themselves but increase the intensity of absorption of the chromophore are called *auxochromes*; examples are methyl, halogen, hydroxyl, alkoxy, amino groups, etc.

n–π* Transition (*type 2*) involves the excitation of electrons from nonbonding atomic orbital n to antibonding orbital π*. Compounds containing –C≡N, –N=N–, >=O, –N=O, or >C=S group exhibit n–π* transition. However, the absorption is weak when the double/triple bond is isolated.

n–σ* Transition (*type 3*) takes place when an electron is excited from a nonbonding n to an antibonding σ* orbital. It requires higher energy and hence takes place at a shorter wavelength, ~200 nm. Single bonded heteroatom containing groups, e.g., C–S (acyclic or cyclic), $R–NH_2$, C–Cl, R–X, R–COOH, $R–CONH_2$, $RCOOR_1$, etc., participate in such transition, which are detectable only in vacuum.

σ–σ* Transition (*type 4*) can take place only in case of hydrocarbons (cyclic or acyclic). The σ–σ* transition from a bonding to an antibonding orbital requires much higher energy and hence takes place at a much shorter wavelength in the *far UV region below 200 nm* and may not be detected even in vacuum.

The molecular orbitals of a molecule representing different electronic transitions in visible, UV, and far UV regions are qualitatively shown in Fig. 4.12. The UV spectrum is useful in finding the presence or absence of group/s capable of *type 1* and *2* transitions. The chromophores like substituted conjugated dienes, conjugated enones, etc., often occurring in natural products, are detectable by UV spectra. The frequency or the wavelength of a particular absorption depends on the electronic transition of the molecular orbitals involved. The value of the extinction coefficient ε is increased if the transition is symmetry allowed (e.g., $\pi \rightarrow \pi^*$ transition of conjugated dienes) and is significantly decreased if it is not symmetry allowed (e.g., $n \rightarrow \pi^*$ transition of unconjugated ketones).

The light absorption maxima positions of dienes were first systematized by Woodward in 1941, based on an extensive study of the UV spectra of terpenoids, steroids, and other molecules. These have been modified by Fieser and by Scott based on the absorptions of a very large number of dienes and trienes. The modified rules [32] are given in Table 4.3. These rules are much helpful in identifying the conjugated diene systems and their environment, present in natural or synthetic molecules. The experimental results (observed λ_{max} values) are in good agreement with those derived from the empirical rules. The light absorption maxima positions of the dienes were first systematized by Woodward in 1941, based on an extensive study of the UV spectra of terpenoids and steroids and other molecules. These have been modified by Fieser and Scott based on the absorption of a very large number of dienes and trienes. These rules are much helpful in identifying the conjugated diene systems and their environment, present in the natural or synthetic molecules.

Table 4.3 Woodward–Fieser–Scott empirical rules for conjugated polyene absorptions

	Parent diene:	homoannular (cisoid or *s-cis*)	253 nm
		heteroannular (transoid or *s-trans*)	214 nm
Increments for			
	a) each alkyl substituent or ring residue		5 nm
	b) each double bond of exocyclic nature		5 nm
	c) each double bond extending the conjugation		30 nm
Polar groups:	d) O-Alkyl		6 nm
	e) Cl/Br		5 nm
	f) O-Acyl		0 nm
	g) S-Alkyl		30 nm
	h) NAlkyl$_2$		60 nm
	Solvent correction (compare with **Table 4.4**)		0 nm$^\Psi$
		Calc. λ_{max} (EtOH) =	Total

$^\Psi$The polyenes being much less polar are insensitive to solvent changes presumably because of the absence of appreciable solvent-solute interaction.

Table 4.4 Calculation of UV absorption maxima (in EtOH) of some conjugated polyenes

Abietic acid (5)		Levopimeric acid (6)		(7)	
Heteroannular diene	214 nm	Homoannular diene	253 nm	Homoannular diene	253 nm
1 Alkyl substituent	5 nm	1 Alkyl substituent	5 nm	2 Double bonds	
3 Ring residues	15 nm	3 Ring residues	15 nm	extending conjugation	60 nm
1 Exocyclic double bond	5 nm	1 Exocyclic double bond	5 nm	5 Ring residues	25 nm
				3 Exocyclic double bonds	15 nm
				1 OAc substituent	0 nm
Calc.	239 nm	Calc.	278 nm	Calc.	353 nm
Obs.	241 nm	Obs.	275 nm	Obs.	355 nm

Examples illustrating the calculation of the UV absorption maxima (in nm) of some polyenes are given in Table 4.4.

Thus a conjugated chromophore strongly absorbs and the band undergoes bathochromic shift to different extents depending on the geometry of the conjugated system and on the substituent/s or ring residues present on the chromophore.

Similarly, Woodward first devised empirical rules for calculating the expected position of the absorption maxima of conjugated ketones and aldehydes from the known absorption maxima of a number of steroids and terpenoids containing such chromophores. Like the diene rules these rules were also modified by Fieser and by Scott and are outlined in Table 4.5. They are similar to but not as much reliable as those for conjugated polyenes. They are still helpful for predicting the absorption maxima of conjugated enones and were useful to determine the correct structure among several alternatives in many cases during the pre-NMR spectroscopic days. *For further details proper books [22–32] on spectroscopy dealing with UV spectra of the organic molecules may be consulted.*

Solvent correction. Some changes in the UV spectra of conjugated enones are observed when the solvent is changed. Decrease in the polarity of the solvent causes decrease in the λ_{max} value (hypsochromic shift) since the energy necessary for the electronic transition is less in the more polar solvent due to the stabilization of the charge separated excited state by increased interaction with a more polar solvent (Table 4.5).

Tables 4.4 and 4.6 allow prediction of the UV absorption maxima, which are found to be in excellent agreement with the experimentally observed ones.

Empirical rules for predicting the absorptions of αβ-unsaturated acids and esters and for the principal band of substituted benzene derivatives have been outlined in tabular forms in the book by Scott [32]. For absorptions of these and many other types of chromophores one can consult other books [24–31].

The two additional absorption maxima in cases of compounds (9) and (10) are due to the parts of the long chromophoric systems (Table 4.6).

Table 4.5 Woodward–Fieser–Scott empirical rules for conjugated ketone and aldehyde absorptions[*]

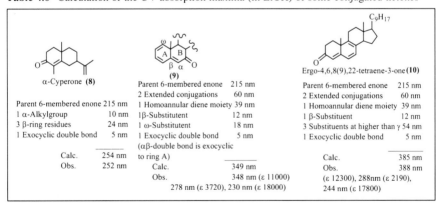

	nm
Parent αβ-unsaturated six-membered ring or acyclic ketone	215
Parent αβ-unsaturated five-membered ring ketone	202
Parent αβ-unsaturated aldehyde	207
Increments for	
a) extended conjugation by one double bond	30
b) homoannular diene moiety	39
c) each alkyl group or ring residue	10 (α), 12 (β), 18 (γ, and higher)
d) OH	35 (α), 30 (β), 50 (δ)
e) OAc	6 (α, β or δ)
f) OMe	35 (α), 30 (β), 17 (γ), 31 (δ)
g) Cl	15 (α), 12 (β)
h) Br	25 (α), 30 (β)
i) NR$_2$	95 (β)
j) S-Alkyl	85 (β)
k) each double bond of exocyclic nature	5

Calc. λ_{max} (EtOH) = Total: _____

Solvent correction:

Water	+ 8 nm	Chloroform	– 1 nm	Ether	– 7 nm
Methanol	0 nm	Dioxan	– 5nm	Hexane	– 11 nm

[*] *ε Values are usually above 10,000 and increase with the length of the conjugated system.*

Table 4.6 Calculation of the UV absorption maxima (in EtOH) of some conjugated ketones

α-Cyperone (8)	(9)	Ergo-4,6,8(9),22-tetraene-3-one (10)
Parent 6-membered enone 215 nm	Parent 6-membered enone 215 nm	Parent 6-membered enone 215 nm
1 α-Alkylgroup 10 nm	2 Extended conjugations 60 nm	2 Extended conjugations 60 nm
3 β-ring residues 24 nm	1 Homoannular diene moiety 39 nm	1 Homoannular diene moiety 39 nm
1 Exocyclic double bond 5 nm	1β-Substitutent 12 nm	1 β-Substituent 12 nm
	1 ω-Substitutent 18 nm	3 Substituents at higher than γ 54 nm
	1 Exocyclic double bond 5 nm	1 Exocyclic double bond 5 nm
	(αβ-double bond is exocyclic	
Calc. 254 nm	to ring A)	Calc. 385 nm
Obs. 252 nm	Calc. _____ 349 nm	Obs. 388 nm
	Obs. 348 nm (ε 11000)	(ε 12300), 288nm (ε 2190),
	278 nm (ε 3720), 230 nm (ε 18000)	244 nm (ε 17800)

Measurement The UV spectra are measured in dilute solutions prepared in spectral grade solvents having no absorption in the UV range; aldehyde-free ethanol serves as a good solvent. The dilution is further manipulated on the basis of extinction coefficient (ε), since absorption is measured during the study, and in practice the relative absorbance of light by the solution is measured, while comparing it to the solvent taken in an identical UV cell serving as the control. An idea about the molar extinction coefficient ε of an absorption maximum, having the unit cm^{-3} mol^{-1} (but not usually used), is an important guide for diluting the solution properly for the measurement of the UV or visible spectrum.

4.2.5 Infrared Spectroscopy

Although in Table 4.2 the IR region of the electromagnetic radiation spectrum is shown to be from 0.7 to 100 μm (14,000– 100 cm^{-1}), it is conventionally divided into the three regions: the near-, mid-, and far-infrared, named for their relation to the visible spectrum. The *mid-infrared region* from approximately **2.5 to 25 μm (4,000– 400 cm^{-1})**, *most commonly referred to as the IR part*, is used to study the *fundamental vibrations* and the associated *rotational–vibrational* structures. The higher energy *near-IR region*, approximately **0.7–2.5 μm (14,000–4,000 cm^{-1})**, can excite *harmonic or overtone vibrations*. The *far-IR region*, **25–1,000 μm (400–10 cm^{-1})** approx., lying adjacent to the microwave region, has low energy and may be used for *rotational spectroscopy*. These classifications are not strict divisions and not based on molecular or electromagnetic properties.

Each bond in a molecule has a vibrational frequency. If IR radiation is passed through a sample, an energy exchange takes place when a bond in a molecule has a vibrational frequency identical to the frequency of the incident radiation leading to its absorption. After absorption, the molecule vibrates at increased amplitude. A molecule can have many vibrational frequencies due to numerous independent vibration modes. For example the molecule having XY_2 moiety (Y=H or any other atom) may undergo various types of vibrations which involve (i) *bond length changes* or **stretching** [(a) **symmetric**—when both bonds simultaneously become longer/shorter, and (b) **asymmetric**— when the bonds alternately become longer/ shorter], and (ii) *bond angle changes* or *bending/deformation* [(a) **in-plane movements—scissoring** when bond angle increases or decreases, and **rocking** when bonds move in the same direction, and (b) **out-of-plane movements— wagging** when both bonds simultaneously move toward/away from the viewer, and **twisting**—when each bond alternately makes such movements]. These vibrations have different energies and hence cause IR absorptions at different frequencies. Such an analysis can be extended to a structural unit of any type. In general, asymmetric stretching vibrations occur at higher frequencies, and stretching vibrations occur at higher frequencies than bending vibrations: the order of decreasing frequencies is *scissoring > wagging ~ twisting > rocking* which is also the expected decreasing order of energy. The frequencies of vibrations depend on the masses of its atoms, strength of the connecting bonds, and the geometry of the molecule. Hence each pure compound is characterized by its own unique IR absorption spectrum containing a great number of peaks. Because of the presence of identical groups, sometimes certain parts of the IR spectra of two different compounds appear to be similar. However, the region ~**1,350–900 cm^{-1}** (**7.4–11 μm**) containing large number of peaks is difficult to interpret fully. The peaks in this region are characteristic of the particular compound and can be used, just as a *fingerprint* used for a human. This region is known as the *fingerprint region*. When two compounds obtained from two different sources are suspected to be identical with each other, the superimposable IR spectra serve as the sure evidence of their identity. Superimposibility of the IR spectrum (peak for peak along with relative absorption) of a natural or synthetic compound with that of an authentic sample establishes the identity of the former with certainty.

IR-inactive molecules If the resultant dipole moment of the bonds of a molecule does not change as a function of time when subjected to IR radiation, the transfer of energy cannot take place, and the IR radiations cannot be absorbed, and the molecule becomes **IR inactive**. Such molecules will have symmetrical stretching vibrations, e.g., H_2, CH_4, O_2, N_2, C_6H_6, symmetric, and pseudosymmetric alkenes and alkynes, e.g., $CH_2{=}CH_2$, $MeCH{=}CHMe$, $(Me)_2C{=}C(Me)_2$, and $MeCH{=}CHEt$, $CH{\equiv}CH$, $MeC{\equiv}CMe$, and $MeC{\equiv}CEt$, etc.

The IR spectra exhibit a number of absorption bands associated with various structural units of organic compounds. We examine the IR spectrum of a compound to search for the presence of –N–H, –O–H, C–H, >C=O, C≡C, C≡N, –COOH, –CONH-, C=N, C=C, N=O, C–C, C–N, C–O, C–Cl, etc. vibrations; the approximate regions of their stretching vibrations are given in Table 4.7. Bending, twisting, and other types of bond vibrations have been omitted for clarity [23]. One should record the broad visual patterns of this table quite firmly in mind. From the base value of any particular stretching vibration one can find out the kind of absorption from the pertinent correlation charts of various group frequencies in a number of the available books [22–31, 33–36]. An overwhelming number of compounds possessing these groups variously associated with other bonds or groups have been studied and the positions of the bands have been generalized. These bands are useful in identifying the presence of such group/s in the compounds under investigation. Here, we will briefly talk about O–H and C=O groups often encountered.

Wave numbers are expressed as reciprocal of the wavelength expressed in cm. A higher wave number corresponds to higher energy. Wavelength values are encountered more in older literature. Wavelengths (μ or μm) are converted to wave numbers by using the following relationship.

$$\text{wave no.} = 1 \text{ cm}/\lambda \text{ (in μ)} = 10^4 \text{ μ}/\lambda \text{ (in μ); hence } \lambda \text{ (in μ)}$$

$$= 10^4/\text{wave no.}\left(\text{cm}^{-1}\right) \text{ or frequency}$$

Hydroxyl groups The –OH group may be alcoholic, phenolic, or carboxylic. Sometimes –OH is present in the free state when the peak appears to be sharp. In case of hydrogen bonding, the bond appears to be broad and shifted (bathochromic) from the normal frequency of O–H stretching. The IR spectra of intermolecularly hydrogen-bonded compounds are concentration dependent. In

Table 4.7 The approximate regions of the absorptions (stretching vibrations only) of various common types of bonds

⟵ Frequency (cm⁻¹) (approx)

4000	2500	2000	1800	1650	1550	1350	900	700
O–H (free) (sh) 3650-3600 H–bonded OH (br) 3400-2200 C-H N-H (sh)	C≡C C≡N (~ 2250) (sh)	No Characteristic bands	C=O For details see **Figure 4.13**	C=N C=C	N=O	N=O	C–C C–N C–O Fingerprint region †	C–Cl

| 2.5 | 4.0 | 5.0 | 5.6 | 6.1 | 6.5 | 7.4 | 11.1 | 14.3 |

Wavelength (μ) (approx.) ⟶
br broad   sh sharp

(μ)(cm⁻¹) = 10000   †Characteristic patterns for each compound

very dilute solution the hydrogen bonding is destroyed. However, in cases of intramolecularly H-bonded compounds, their IR spectra are independent of concentration. In cases of carboxylic acids, the hydrogen bonding is so strong that they remain in the dimeric form in the pure liquid. The entire range of bands due to hydroxyl group lies between 3,650 and 2,200 cm⁻¹. The free hydroxyl group appears at 3,650–3,600 cm⁻¹ (sharp), hydrogen-bonded OH (3,400–3,200 cm⁻¹), and very strongly bonded hydroxyl group, as in a carboxylic acid (**11**), appears at 3,400–2,200 cm⁻¹ as a very broad band. A cyanide (C≡N) group appears at ~2,250 cm⁻¹ as a very sharp band.

$$R-C \underset{O-H ---O}{\overset{O---H-O}{\big<}} {\big>} C-R$$

(11)

Carbonyl groups They may be present as free carbonyl, conjugated carbonyl, carboxylic carbonyl, ester carbonyl, cyclic carbonyl (4,5,6,7 ... membered), conjugated cyclic carbonyls and lactones of different ring sizes, amide carbonyl, etc., appear within the normal range of carbonyl ~1,830–1,640 cm⁻¹, and exhibit peaks at specific frequencies. General frequency ranges of the vibrational modes of various types of carbonyl groups are given in Fig. 4.13. In case of ring compounds, ring strain plays an important role. Additionally, the presence of various other groups could also be detected [26–32]. If the IR spectrum of a natural or synthetic product shows one or more than one such peaks, the compound is likely to contain such a group/s.

Sample preparation The samples for IR spectral studies may be prepared *in one of the following ways.* The choice depends on the nature of the compound.

(i) *As a KBr pellet.* The compound/sample, if solid, is finely powdered with a non-absorbing metal halide, usually KBr (IR-grade), and compressed under

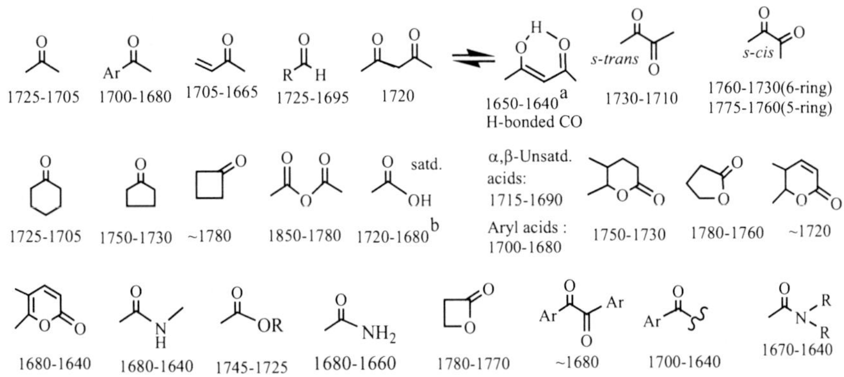

[a]Hydrogen bonded carbonyl band moves to a lower frequency. α,β-unsaturated CO 1715-1690, Aryl CO 1700-1680 cm^{-1}.
[b]A lower frequency CO (1720-1680) along with a broad H-bonded OH (3400-3200) suggest the presence of a COOH group in the molecule.

Fig. 4.13 The absorption maxima (cm^{-1}) of C=O groups (vibrational modes) of different types and in different environments

high pressure into a thin transparent film of KBr carrying the compound (a matrix). The resulting pellet is inserted in a holder and is placed in the spectrometer. No absorption is added to the spectrum due to KBr.

(ii) ***In the form of a mull.*** The compound/sample, if solid, is finely ground in an agate mortar (using a pestle) with liquid paraffin, a high molecular weight hydrocarbon (usually Nujol), and the dispersion (mull) is pressed between two IR transparent salt plates. The peaks due to Nujol are sometimes superimposed. Strong peaks due to Nujol appearing at 2,924 (sp^3 C–H stretching), 1,462 (CH_2 bending), 1,377 (CH_3 bending) cm^{-1} (approx.), and a weak peak at 760 cm^{-1} (long chain band) are to be ignored. Thus the main disadvantage of using this method is to obscure the bands of the analyzed compound, if present in these regions. However, if the Nujol bands are stronger than the sample peaks appearing in those regions, more sample and less Nujol should be used.

(iii) ***As a pure liquid.*** A drop of liquid completely free from water is placed between a pair of finely polished NaCl or KBr plates (*salt plates*) and squeezed carefully when a thin film of the liquid is formed in between. The thin plate is then placed in the spectrophotometer. An IR transparent cell of appropriate length may also be used for this purpose. Since no solvent is used, the spectrum thus obtained is referred to as a ***neat*** spectrum.

(iv) ***As a solution.*** A solution of the sample, if solid or liquid, is made in a suitable solvent (CCl_4—most commonly used, $CHCl_3$, or CS_2, etc.) and its spectrum is taken in a suitable IR transparent cell. Some regions of the spectrum are obscured by the absorption bands of the solvents: (CCl_4, 850–700 cm^{-1}; $CHCl_3$, 1,250–1,175 and below 850 cm^{-1}; CS_2, 2,200–2,100 and 1,600–1,400 cm^{-1}); these are to be taken into account while interpreting the spectra.

Calibration of the spectrum The IR spectrophotometer should be calibrated for accuracy of the bands appearing in a spectrum by comparison with the bands of a standard. Usually a thin film of polystyrene is used for this purpose. Polystyrene shows a number of sharp absorption bands of which the band at 1,601 cm^{-1} (6.246 μm) or 1,944 cm^{-1} (5.144 μm) is recorded on the chart paper for calibration, after running the spectrum of the compound being measured.

Fourier Transform Infrared (FTIR) Spectrometers The design of the optical pathway of such spectrometers produces a pattern called an interferogram. Its wavelike pattern contains all the frequencies which make up the IR spectrum. The FTIR instrument acquires the interferogram in less than a second, and thus it can collect dozens of interferogram of the sample within a minute and accumulate them in the memory of a computer. The sum of the interferograms is subjected to a mathematical operation known as *Fourier Transform (FT)* which separates the accumulated individual frequencies, and a spectrum with a better signal-to-noise ratio can be plotted. Thus FTIR instrument has a greater speed and a greater sensitivity than a traditional dispersion instrument and needs a much less amount (<0.1 mg) of the material. Moreover, the computer software automatically subtracts the spectrum first taken of the background [containing IR-active gases like CO_2 and H_2O vapor (O_2 and N_2 are not IR-active)] and yields the spectrum of the compound only, being analyzed.

4.2.6 ^1H and ^{13}C NMR Spectroscopy [54–58]

We will now concentrate on Nuclear Magnetic Resonance (NMR) spectral analysis, the most rewarding of all spectroscopic measurements. It is capable of revealing information about the number of magnetically distinct atoms present in the compound under investigation and about its immediate as well as distal chemical environments. Because of remarkable evolution of instrumentation NMR has become the most sophisticated useful tool to organic chemists. Various types of 2D-NMR and correlation diagrams indicate/confirm the complete structure and stereochemistry of the compounds in most cases. A discussion of these tools developed during the last few decades is beyond the scope of this chapter, and the reader is referred to the available excellent books [22, 23, 27, 29, 38–44]. A simple basic treatment and application of some pertinent data will be briefly presented. There are excellent books [37–52] which may lead to thorough understanding of the theoretical basis of the nuclear resonance phenomenon and adequate interpretation of results.

All nuclei carry a charge which spins around nuclear axis. Circulation of nuclear charge (of nuclei with spin number not zero) generates magnetic dipole and angular momentum along the axis of spin. The nuclei of all atoms may be characterized by a nuclear spin quantum number *I*, which may have values *equal to zero or multiples of* ½. The spin number I is related to the mass number and atomic number as follows:

Mass number	Atomic number	Spin number I (multiples of ½)
(i) Odd	Even or odd	1/2, 3/2, 5/2, ...
(ii) Even	Odd	1, 2, 3, ...
(iii) Even	Even	0 (no nuclear spin)

The nuclei having atomic mass and atomic number both even, e.g., $^{12}C_6$, $^{16}O_8$, have zero spin ($I = 0$); such nuclei cannot exhibit nuclear magnetic resonance and are termed "*NMR silent.*" However, many elements have at least (i) one isotope having nuclear spin of magnetic quantum number $I = 1/2$ (e.g., $^1H_1$, $^{13}C_6$, $^{15}N_7$, $^{19}F_9$, $^{29}Si_{14}$, $^{31}P_{15}$) and spherical charge distribution, or (ii) one isotope having nuclear spin $I = 1$ (e.g., $^2H_1$ or D, $^{14}N_7$) and nonspherical charge distribution; such isotopic nuclei are observable by NMR. *In an external static (constant) magnetic field denoted by B_0,* the microscopic magnetic moments align themselves relative to the field in a discrete number of orientations because of quantized energy states involved. *For a spin number I the number of possible orientations with spin states of different energies is given by $2I + 1$.* Thus for a proton there are two possible orientations: a low-energy orientation aligned with the applied static field and denoted by $+1/2$ (α-state, parallel) and a high-energy orientation opposed to the applied field and denoted by $-1/2$ (β-state, antiparallel). These are two energy levels, and in accordance with Boltzman distribution, there is a slight excess of proton population N_α at lower energy over proton population N_β at higher energy. The nuclei possessing angular momentum behave like a bar magnet and experience a precessional motion at right angle to the direction of the static magnetic field. The energy difference $\Delta E = h\gamma \, B_0/2\pi$, where h is Planck's constant, and γ is the *magnetogyric ratio* (also known as *gyromagnetic ratio*) which is a measure of the strength of the nuclear magnet and is constant for a given nucleus having spin. Thus the energy difference is proportional to B_0.

The transition from one orientation to another orientation occurs when the frequency of absorption or emission of electromagnetic radiation is equal to the precessional frequency of the nuclei. NMR occurs in the radiofrequency region ($\lambda > 10^4$ cm) of the electromagnetic spectrum (see Table 4.2). The transition from the lower energy level to the higher one is induced by introducing another magnetic field at right angle to the applied external magnetic field. For the aforesaid transition the proton has to be irradiated with an electromagnetic radiation of frequency ν so that $\Delta E = h\nu$. Thus $\nu = \gamma \, B_0/2\pi$ and this will vary with the structural environment.

Chemical shift Proton environment is affected by the orbital electrons creating a small magnetic field. The net effect is the *decrease in the magnetic field (H) felt by the nucleus* ($H_{at\ nucleus} = H_{ext.} - H_{shielding}$), where $H_{ext.} = B_0$. Thus the proton is shielded. For a slight variation in the structural environment of a proton, variation of radiation frequency is needed to cause the resonance of the proton. This variation is known as **chemical shift**, which thus represents the separation of the resonances of a type of nuclei in a different chemical environment with respect to a standard reference.

The most used *standard reference* is tetramethylsilane (TMS), all the methyl protons of which are more shielded compared to the protons of most of the organic compounds. Further, it gives a strong singlet peak of 12 equivalent protons at a very low concentration and the compound is chemically inert. The scale starts from this peak and is taken as zero (δ), so that positive values appear for less shielded protons. An alternative unit τ has been used to express chemical shift. It holds a relationship with δ as follows: $\tau = 10 - \delta$. Thus, a proton with less magnetic shielding will have a smaller value of τ (or a larger value of δ) and vice versa.

The NMR spectrum, like UV and IR spectra, represents a plot of energy of absorbance versus frequency of electromagnetic radiation. *The chemical shifts are dimensionless and are expressed for the required accuracy, as parts per million (ppm), and the displacement is measured from the reference compound TMS (Fig. 4.14), added in very minute amount in the solution of the sample prepared in a thin-walled specially made glass tube, called an NMR tube.*

This is the reason that the chemical shifts of the protons of a particular compound when measured in different instruments with different magnetic field (external) strength will feel the same fraction of external magnetic field and hence the chemical shifts are independent of the external applied field, but is very much dependent on the electron density around the nucleus under study and the associated atoms with which it is bonded. The decrease in the electron density decreases the shielding effect and the resonance takes place in a lower field. The situation is reversed with increase of electron density around the nucleus. The relationship between δ (chemical shift), τ, and R_f frequency of the ^{1}H NMR instrument is depicted in Fig. 4.14.

A proton which shows observed shift 120 Hz in a 60 MHz instrument will show 1,000 Hz shift in a 500 MHz instrument, and the chemical shift δ will be 2 ppm in each case (Fig. 4.14). Since the difference of the two energy levels depends upon the magnetic field felt by the nucleus, the frequency for transition will depend on the strength of the magnetic field. Chemical shift values of protons of different environments are available from standard books.

Integration The intensity of the absorption peak/s appearing in the ^{1}H NMR spectrum is directly proportional to the number of nuclei responsible for the peak/s. The area is electronically measured in a separate operation with the NMR machine. In cases of ^{1}H NMR, integration is easy and reliable. Integration for each resonance is recorded as vertical lines and the relative height represents the number of protons. In current machines the digital number appearing at the base line under the peak dictates the relative intensity of the peak. In case of ^{13}C NMR spectra, the integration of the signals depends on the relaxation rate of the nucleus concerned and its scalar and dipolar coupling constants. Very often these factors are poorly understood— therefore, the integration of the ^{13}C NMR signal is very difficult to interpret.

Magnetic field strength →   TMS

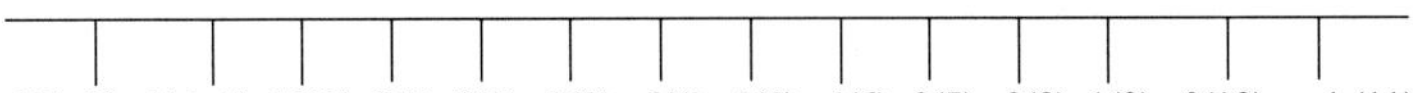

12(−2)	11(−1)	10(0)	9(1)	8(2)	7(3)	6(4)	5(5)	4(6)	3(7)	2(8)	1(9)	0(10)	−1 (11)	δ (τ)
720	660	600	540	480	420	360	300	240	180	120	60	0	−60 Hz	(60 MHz)
1200	1100	1000	900	800	700	600	500	400	300	200	100	0	−100 Hz	(100 MHz)
3600	3300	3000	2700	2400	2100	1800	1500	1200	900	600	300	0	−300 Hz	(300 MHz)

NMR Instrument

60 MH$_3$ (δ = 1 ppm = 60 Hz)

100 MH$_3$ (δ = 1 ppm = 100 Hz)

300 MH$_3$ (δ = 1 ppm = 300 Hz)

δ = 1 ppm is conventionally expressed as 1 δ

δ in parts per million (ppm) = observed shift in Hz x 10^6/ rf oscillator frequency of the instrument used in MHz

= 120 Hz x 10^6 / 60 MHz = 2 = 8 τ (in a 60 MHz instrument)

= 1000 Hz x 10^6 / 500 MHz = 2 = 8 τ (in a 500 MHz instrument), since τ = 10 − δ

Note: The factor 10^6 is included to give convenient numbers, so that δ is dimensionless and is thus expressed in parts per million (ppm). It is thus independent of oscillator frequency.

Fig. 4.14 Relationship between chemical shift in Hz, δ, τ, and R_f frequency of the NMR instrument

Splitting Patterns and Coupling Constants: The splitting patterns and the coupling constants (***J-coupling or scalar coupling—a special case of spin–spin coupling***) between NMR-active nuclei give some of the most useful information for structural elucidation and sometimes reveal the stereochemical features of the molecule. The nuclear spin–spin interaction is responsible for the origin of splitting of the resonance lines (more than one line). A nucleus with spin quantum number I (>0) is equally separated into the spin states represented by ($2I + 1$) when placed in a magnetic field. By spin angular momentum polarization, the effects of the individual spin states are transferred to the adjacent nuclei (coupling nucleus). Chemically equivalent nuclei either by symmetry (homotopic or enantiotopic) (Chap. 2) or by averaging due to free rotation are *isochronous* showing same chemical shift and appear as a singlet. However, some nuclei become isochronous accidentally.

The coupling arises from the interaction of different spin states of anisochronous nuclei through the chemical bonds of a molecule and results in the splitting of NMR signals. These splitting patterns can be complex or simple and, likewise, can be straightforwardly interpretable or deceptive. The coupling provides detailed insight into the connectivity of atoms in a molecule. Coupling to n equivalent (spin ½) nuclei, three bonds apart, splits the signal into a $n + 1$ *multiplet* with intensity ratios following ***Pascal's triangle*** as given below. The intensity ratio in case of ^{1}H NMR is also same as the coefficients of the terms in the expansion of $(a + b)^n$. For example, the intensity ratio of the septet, when $n = 6$, may be derived from the expression $(a + b)^6 = a^6 + 6a^5 b + 15a^4 b^2 + 20a^3 b^3 + 15a^2 b^4 + 6ab^5 + b^6$. The intensities of the peaks of such a multiplet are thus symmetric about the midpoint of the band.

Pascal's triangle

Multiplicity	Intensity ratio
Singlet (s)	1
Doublet (d)	1:1
Triplet (t)	1:2:1
Quartet (q)	1:3:3:1
Quintet	1:4:6:4:1
Sextet	1:5:10:10:5:1
Septet	1:6:15:20:15:6:1

A simple example is given. In the proton spectrum for ethanol described above, the CH_3 group is split into a *triplet* with an intensity ratio of 1:2:1 by the two neighboring CH_2 protons. Similarly, the CH_2 is split into a *quartet* with an intensity ratio of 1:3:3:1 by the three neighboring CH_3 protons. In principle, the two CH_2 protons would also be split again into a *doublet* to form a *doublet of quartets* by the hydroxyl proton, but intermolecular exchange of the acidic hydroxyl proton often results in a loss of coupling information. Coupling to additional spins will lead to additional splittings of each component of the multiplet, e.g., coupling to two different spin ½ nuclei with significantly different coupling constants will lead to a *doublet of doublets* (abbreviation: dd). No coupling occurs between nuclei that are chemically equivalent (that is, have the same chemical shift). Couplings between nuclei that are distant (usually more than three bonds apart for protons in flexible molecules) are usually too small to cause observable splittings. *Long-range* couplings over more than three bonds are often observed in cyclic and aromatic compounds, leading to more complex splitting patterns.

Chemical shift (and the integration for protons) combined with multiplicity and coupling constant/s gives information not only about the chemical environment of the nuclei, but also the number of *neighboring* NMR-active nuclei within the molecule. Coupling constants (J-values in Hz) of interacting anisochronous protons of different environments and hence of different chemical shifts commonly encountered in natural and synthetic organic compounds are shown in Fig. 4.15. One can differentiate and characterize the different dimethoxycoumarins on the basis of the chemical shifts and the coupling constants of the aromatic and olefinic protons. The ^3J-value suggests the strength of interaction between the nuclei concerned. Martin Karplus first found out that the coupling constant (J) between protons attached to vicinal carbons depends on the dihedral angle α between them, and he developed an expression known as **Karplus equation** as follows:

$$^3J_{HH} = A + B \cos \alpha + C \cos 2\alpha$$
$$A = 7 \quad B = -1 \quad \text{and} \quad C = 5$$

These values of the constants A, B, and C are accepted as those that predict the J-values best. The curve plotting ^3J-values obtained by application of Karplus

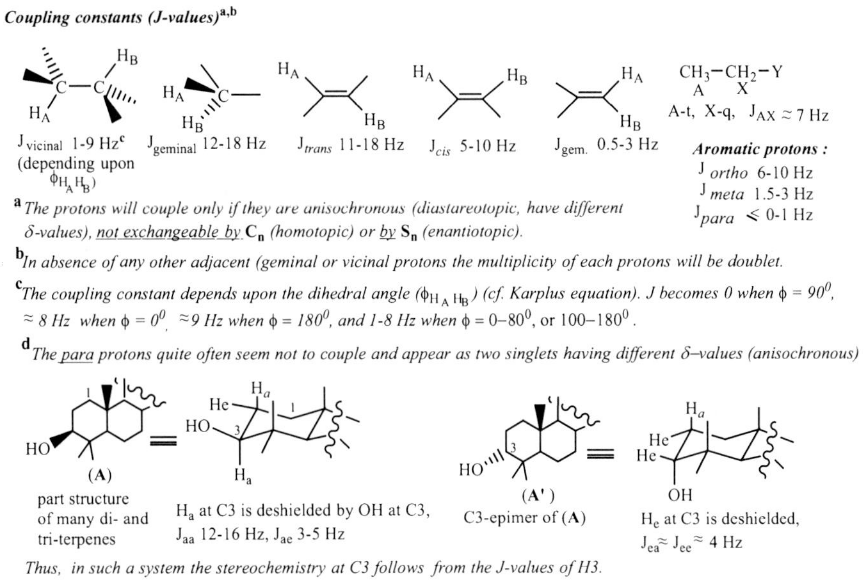

Coupling constants (J-values)[a,b]

$J_{vicinal}$ 1-9 Hz[c] $J_{geminal}$ 12-18 Hz J_{trans} 11-18 Hz J_{cis} 5-10 Hz $J_{gem.}$ 0.5-3 Hz   CH$_3$-CH$_2$-Y
(depending upon   A   X
$\phi_{H_A H_B}$)   A-t, X-q, $J_{AX} \approx$ 7 Hz

[a] *The protons will couple only if they are anisochronous (diastereotopic, have different*   *Aromatic protons :*
δ-values), not exchangeable by **C$_n$** *(homotopic) or by* **S$_n$** *(enantiotopic).* J_{ortho} 6-10 Hz
                                                                                     J_{meta} 1.5-3 Hz
[b]*In absence of any other adjacent (geminal or vicinal protons the multiplicity of each protons will be doublet.* $J_{para} \leqslant$ 0-1 Hz

[c]*The coupling constant depends upon the dihedral angle ($\phi_{H_A H_B}$) (cf. Karplus equation). J becomes 0 when ϕ = 90⁰,*
≈ 8 Hz when ϕ = 0⁰, ≈9 Hz when ϕ = 180⁰, and 1-8 Hz when ϕ = 0–80⁰, or 100–180⁰ .

[d]*The para protons quite often seem not to couple and appear as two singlets having different δ–values (anisochronous)*

part structure   H$_a$ at C3 is deshielded by OH at C3,   (A')   OH
of many di- and J_{aa} 12-16 Hz, J_{ae} 3-5 Hz   C3-epimer of (A)   H$_e$ at C3 is deshielded,
tri-terpenes $J_{ea} \approx J_{ee} \approx$ 4 Hz

Thus, in such a system the stereochemistry at C3 follows from the J-values of H3.

Fig. 4.15 The coupling constants (*J*-values) of interacting nonequivalent protons—common in natural and synthetic products

equation against α (different dihedral angles) is called the **Karplus curve**. But actual experimental data display a wide range of variation for different types of compounds based on the minimum energy conformations, if any, and other factors influencing the interactions. The *J*-value becomes **zero** when α is ~90°, since the side-to-side overlap of the two *perpendicular* C–H bond sp^3 orbitals (and hence their interaction) is at a minimum, whereas *J*-value becomes maximum (8.4–12.2) or (9.6–13) when α is ~0° or ~180°, respectively, since the side-to-side overlap of the two *parallel* C–H bond orbitals (and hence their interaction) is maximum. Overlap of the front lobes occurs at $\alpha = 0°$, whereas that of the back lobes and front lobes of the sp^3 orbitals occurs at $\alpha = 180°$ [23].

In more complex spectra with multiple peaks at similar chemical shifts or in spectra of nuclei other than hydrogen, coupling is often the only way to distinguish different nuclei.

Second-order (or strong) coupling The above description assumes that the coupling constant is small in comparison with the difference in NMR frequencies between the nonequivalent spins. If the shift separation decreases (or the coupling strength increases), the multiplet intensity patterns are first distorted, and then become more complex and less easily analyzed (especially if more than two spins are involved). Intensification of some peaks in a multiplet is achieved at the expense of the remainder, which sometimes almost disappear in the background noise, although the integrated area under the peaks remains constant. In most high-field NMR, however, the distortions are usually modest and the characteristic distortions (*roofing*) can in fact help to identify related peaks. Second-order effects decrease as the frequency difference between multiplets increases, so that high-field (i.e., high-frequency) NMR spectra display less distortion than lower frequency spectra. Early spectra at 60 MHz were more prone to distortion than spectra from later machines typically operating at frequencies at 200 MHz, 300 MHz, 400 MHz, or above.

Magnetic nonequivalence More subtle effects can occur if chemically equivalent spins (i.e., nuclei related by symmetry and so having the same NMR frequency) have different coupling relationships to external spins. Spins that are chemically equivalent but are distinguishable (based on their coupling relationships) are termed magnetically nonequivalent. For example, the four H sites of *1,2-dichlorobenzene* divide into two chemically equivalent pairs (H3,H6) and (H4,H5) by symmetry (C_2 operation), but each pair is magnetically nonequivalent, since an individual member of one of the pairs has different couplings to the spins making up the other pair; thus here, $J_{3,4} = J_{ortho}$, $J_{3,5} = J_{meta}$, whereas, $J_{6,4} = J_{meta}$, $J_{6,5} = J_{ortho}$. Another example is 4-nitrochlorobenzene; here the pairs (H2,H6) and (H3,H5) are chemically equivalent (homotopic), but magnetically nonequivalent. Magnetic nonequivalence can lead to highly complex spectra which can only be analyzed by computational modeling. Such effects are more common in NMR spectra of aromatic and other non-flexible systems, while conformational averaging about C–C bonds in flexible molecules tends to equalize the couplings between protons on adjacent carbons, reducing problems with magnetic nonequivalence.

  Coupling to any spin ½ nuclei such as ^{31}P or ^{19}F works in this fashion (although the magnitudes of the coupling constants are very different). But the splitting patterns differ from those described above for nuclei with spin greater than ½ because the *spin quantum number* has more than two possible values. For instance, coupling to deuterium (a spin 1 nucleus) splits the signal into a 1:1:1 *triplet* because the spin 1 has three spin states. Similarly, a spin 3/2 nucleus splits a signal into a 1:1:1:1 *quartet* and so on.

Correlation spectroscopy It is one of several types of *two-dimensional nuclear magnetic resonance (2D NMR) spectroscopy*. This type of NMR experiment is best known by its *acronym*, COSY. Other types of 2D NMR include Nuclear Overhauser effect spectroscopy (NOESY), J-spectroscopy, exchange spectroscopy (EXSY), total correlation spectroscopy (TOCSY), and heteronuclear correlation experiments, such as HSQC, HMQC, and HMBC. Two-dimensional NMR spectra provide more information about a molecule than one-dimensional NMR spectra and are especially useful in determining the structure of a molecule, particularly for molecules that are too complicated to work with using one-dimensional NMR. The first two-dimensional experiment, COSY, was proposed in 1971 and was later implemented in 1976 [44]. Limitation of space does not allow us to discuss even briefly different types of 2D NMR experiments, for which the readers may consult refs. [43, 44, 47–49, 52, 53, 57, etc.].

Sample preparation Samples are generally prepared in $CDCl_3$ taken in an NMR tube. Depending on the solubility. Sometimes d_6-DMSO, C_6D_6, CF_3COOD,

Table 4.8 Corresponding approximate field strengths of ^{1}H NMR and ^{13}C NMR spectra

^{1}H	^{13}C	^{1}H	^{13}C	^{1}H	^{13}C
60 MHz	15 MHz	200 MHz	50 MHz	400 MHz	100 MHz
100 MHz	25 MHz	300 MHz	75 MHz	500 MHz	125 MHz

Table 4.9 Common ranges for ^{13}C NMR chemical shifts (δ-values in ppm[a])[b]

sp^3 Carbon (when associated only with H or C)	4–32	Methylenedioxy (–O–CH$_2$–O)	90–100
sp^3 Methyl (–CH$_3$)	4–24	Olefinic carbon (>C=C<)	100–145
sp^3 Methylene (–CH$_2$–)	14–50	Acetylenic carbon (–C≡C–)	105–130/150
sp^3 Methine (>CH–)	22–54	Aromatic carbon (e.g., ⬡)	110–150 (105–130)
sp^3 quaternary C (>C<)	30–56	Amidecarbonyl (–NH–CO–)	165–175
Aromatic methyl (Ar–CH$_3$)	12–35	Acid carbonyl (–COOH)	160–175
Olefinic methyl (=CH–CH$_3$)	15–35	Aromatic Aldehyde (Ar–CHO)	170–200
Aromatic methoxy (Ar–OCH$_3$)	55–62	Aldehyde carbonyl (–CHO)	185–200
Oxymethylene (–O–CH$_2$–)	60–65	Aromatic ketone (Ar–CO–)	185–210
Epoxy carbon (>C–C<)	50–80	Aliphatic ketone carbonyl (–CO–)	200–220
Oxymethine (O–CH<)	65–75	Nitrile carbon (–CN)	112–126

[a]For substitution effects and nature of substituents the δ-values are different from the normal ones
[b]In SFORD spectra –CH$_3$ (primary), –CH$_2$ (secondary), >CH– (tertiary) and >C< (quaternary) carbons appear as quartet (q), triplet (t), doublet (d), and singlet (s), respectively

CD$_3$CN, etc., solvents are also used as the solvent. A minute drop of TMS is added acting as internal reference.

^{13}C NMR spectroscopy It is a boon to the natural product chemistry. The natural abundance of ^{13}C is low (~1.1 %); hence compared to the ^{1}H NMR spectrum the ^{13}C NMR spectrum requires more material. The gyro magnetic ratio of ^{13}C being ¼ of that of proton, the field strength for ^{1}H NMR spectrum will be approximately four times that for ^{13}C NMR spectrum, as shown in Table 4.8.

The ^{13}C resonance is much weaker (~6,000 times) than proton and is ordinarily difficult to observe for mainly two reasons: (i) the natural isotope ^{13}C abundance is low (~1.1 %) and (ii) its gyromagnetic ratio is ¼ of that of proton; hence it resonates at a lower frequency than that of proton. For a given magnetic strength, when the protons are observed at 300 MHz, the carbon will be observed at 75 MHz. Like proton ^{13}C also couples with the neighboring nuclei. Thus, in the single frequency off-resonance decoupled (SFORD) spectrum, CH$_3$ carbon appears as a quartet, CH$_2$ carbon as a triplet, CH carbon as a doublet, and a quaternary carbon as a singlet. A noise decoupled (ND) ^{13}C NMR spectrum results when the whole range of ^{1}H NMR spectrum is decoupled, and each carbon of any molecule appears as a singlet. For example, a C$_{30}$ molecule (e.g., a triterpene) usually shows 30 singlet peaks in its ND ^{13}C NMR spectrum. For identification of protonated (primary, secondary, and tertiary) and nonprotonated (quaternary) carbons APT and DEPT spectra are routinely measured. Some common ranges for ^{13}C NMR chemical shifts of different carbons are given in Table 4.9.

Scopes of the discussions on the various types of NMR spectroscopic analysis being much less, the readers are advised to consult some helpful books (names given at the end) for such spectral analysis. The Chemical Shifts **(for both H and C) for commonly used solvents are given in tabular form I Appendix C**.

4.2.7   Mass Spectral Analysis [59–66]

The mass spectrometric analysis in principle differs from the electromagnetic spectral analysis. The molecule does not absorb any part of the electromagnetic radiation frequency; rather it involves the formation of the molecular ion and its subsequent fragmentation to other ions. These ions are separated according to their mass to charge ratios (m/z) when the ion beam enters a magnetic field. The mass spectrum is a record of the relative abundance of the ions and is measured relative to the strongest ion peak termed as **base peak**. The highest m/z peak is the molecular ion peak in EI mass spectrum. There are various ways of ionization. The most used one is the bombardment by energized electron to the molecule in the gaseous phase. Ionization of the compound takes place when bombarded with electron emitted from the hot filament, accelerated by an electric field, and having kinetic energy higher or equal to the ionization potential of the compound. The rate of ionization is extremely fast, and being faster than the bond vibration, the molecular ion thus formed will have the same molecular configuration of the parent molecule and is known as molecular ion peak. Electron with 50–70 eV is used for electron-impact mass spectrometric studies. The molecular ion peak then undergoes fragmentation by breaking of the susceptible bonds to yield ions. On the way to fragmentation some neutral molecules are eliminated.

If the electron energy is raised above the ionization potential of the compound, the molecular ion will undergo fragmentation. When the molecular ion is very unstable it undergoes instant fragmentation prior to the recording of its molecular ion peak. In such cases special techniques are applied to obtain the molecular ion peak and sometimes to limit the fragmentation to fewer ions. Because of the presence of isotopes in high proportion in some atoms the isotopic peaks also appear in the spectrum along with the true molecular ion peak. Depending on the percentage of isotope, the relative intensity of the isotopic peak (e.g., in case of bromo or chloro compound) is observed.

A few important ***methods of ionization*** are now described briefly.

1. ***Electron Impact (EI)***. The most commonly used method is the EI ionization. In this method the molecular ion formation (M^+) (by loss of an electron) and the subsequent structurally significant fragmentations are initiated by bombarding the molecules: in the gaseous phase with energized electrons (50–70 eV) at a low pressure in the ionization chamber. The method is applicable to molecules

capable of vaporization and not prone to thermal decomposition prior to the recording of the molecular ion.

2. **Chemical Ionization (CI).** In this method the ionization of the sample is effected by gas phase ion or molecule interactions. Different reagent ions can be used, and the molecular ion remains associated with the reagent. When the reagent ions are NH_4^+, H_3^+, CH_5^+, the peaks are $(M+NH_4)^+$, $(M+H)^+$ etc. However, whatever may be the reagent ion the protonated molecular ion peak is produced in relatively high abundance. Ions produced are of low energy and the fragmentations are limited. Because of the low-energy ions produced in the process, the method could be utilized in identifying isomers including stereoisomers which is not possible with EI mass spectral study. Sometimes negative ion CI are also studied, but only relatively used in natural product chemistry but more applicable in environmental chemistry.

3. **Field Desorption (FD).** In cases of compounds which fail to vaporize or suffer decomposition during vaporization FDMS is applicable. A solution of the material is applied between two plates or emitters and evaporated to grow small whiskers and are placed in the strong electrostatic field (10^7–10^8 ν cm^{-1}) potential difference is maintained when a molecule loses an electron to form positively charges ions. Mostly molecular ion or protonated molecular ion is formed depending on the nature of the molecule.

4. **Fast Atom Bombardment Mass Spectrometry (FABMS).** In this method ions are produced by bombarding the compound under investigation with a beam of fast atoms. Nowadays solution in a liquid matrix-coated probe surface is used which allows interaction between fast atom beam with fresh solute molecules. Glycerol has been found to a useful liquid matrix since it is less volatile and its solvent capacity for polar compound is appreciative.

Tandem Mass Spectrometric System This technique has been found to be useful in identifying followed by collection of the fractions of interest *when coupled with chromatographic separation techniques* like GC/MS and LC/MS. This technique serves as the guide for the collection of biologically active fractions previously identified by bioassay. It can also act as the detector especially in GC/MS.

It is possible to measure precisely the molecular weight with high-resolution mass spectrometer, suggesting the probable elemental composition of the molecule.

The Major Modes of Fragmentation In cases of polyheteroatom containing natural products, the predictability of fragmentation pattern is low. However, the fragmentation patterns of the major classes of natural products have been rationalized and advantageously utilized in settling the structures of many new compounds. Mass fragmentations of some natural products have been rationalized in relevant chapters.

4.2.8 Electrospray Ionization Mass Spectrometry

Electrospray Ionization Mass Spectrometry (ESI-MS), less commonly Electrospray Mass Spectrometry (ES-MS), is a technique used in mass spectrometry to produce ions. It is especially useful in producing ions from macromolecules without fragmenting them and can be analyzed intact. The development of ESI for the analysis of biological macromolecules [81] was rewarded with the award of Nobel Prize in Chemistry to John Bennet Fenn in 2002. One of the original instruments used by Fenn is on display at the Chemical Heritage Foundation in Philadelphia, Pennsylvania.

In an ESI-MS the molecules to be analyzed pass into the mass spectrometer in a fine spray, typically following chromatographic separation, whether traditional liquid chromatography, HPLC, or nano-LC. As the spray emerges from the ESI source, the molecules are ionized by the nozzle's eclectically charged tip. As the mist travels and evaporates, electrostatic repulsion between like charged ions ultimately forces the molecules apart which may then be analyzed by a wide variety of mass analyzers.

Small organic molecules or pharmaceutical drugs (MW 200–500) in most instances ionize quite well to form quasimolecular ions created by the addition of proton and denoted by $(M+H^+)$. Quasimolecular ions created by the addition of another cation such as sodium ion, $[M+Na]^+$, or by the removal of proton, $[M-H]^-$, may also be observed. Thus interpretation becomes fairly straightforward. The mass spectrometers use so-called soft ionization technique that enables the ionization of the large biological molecules such as proteins.

ESI-MS has become an increasingly important technique in the chemical laboratory for structural study and also for quantitative measurement of metabolites in a complex biological system.

Liquid Chromatography-Mass Spectrometry (LC-MS) Electronspray ionization is the ion source of choice to couple liquid chromatography with MS. The analysis can be performed on line, by feeding the liquid eluting from the LC column directly to an electronspray, or off line, by collecting fractions to be analyzed later in a classical nanoelectronspray-mass spectrometry set up.

4.2.9 X-Ray Crystallography: Relative and Absolute Configuration. Conformation

The basic concepts of X-ray crystallographic analysis for full structural information of compounds have been dealt in fair details in various books [75–77].

X-ray crystallographic analysis requires single crystals characterized by parallelepiped unit cells which occur in a repetitive manner in the lattice. Connecting identical regular arbitrary points in the microscopic structure of the crystals forms the lattice. For structural determination excellent results are obtained with a good single crystal, which is roughly equidimensional with edges. X-ray of suitable wavelength is allowed to fall on the suitably mounted single crystal. The intensity of the diffracted beam is dependent on the distances between the atoms and independent of the orientations. Thus enantiomers will have the same X-ray diagram. The phase changes due to scattering of incident radiation for atoms like C, H, N, and O are nearly same and X-ray diffraction is centrosymmetric. However, if a heavy atom like Br or any atom having mass ~30 more than the above atoms could be introduced into the molecule by chemical means, the work will be easier as these atoms increase the scattering power of the crystal. The phase change increases with atomic mass and the refraction pattern is no longer centrosymmetric. The heavy atom in the unit cell is located by diffraction pattern, which serves as the reference, and the structure factors are calculated from atomic coordinates, electron density map, and the phase number scattering amplitudes. In cases of chiral compounds both relative and absolute configuration can be determined. In the latter case X-ray wavelength should be close to the absorption of wavelength of the inner shell electrons of one of the atoms in the crystal. If the relative stereochemistry of all the chiral centers of a compound is known by chemical methods and if the absolute configuration of its one chiral center is determined by chemical correlation with a compound of known absolute configuration (by X-ray), the absolute configurations of all other chiral centers of the former follow. Before the advent of X-ray crystallography, the absolute configuration of some natural products having known relative stereochemistry was determined by application of Prelog's method (Chap. 2) on the secondary alcoholic function. The conformation of the molecule may also be known from X-ray crystallographic studies.

4.2.10 ORD and CD: Absolute Stereochemistry. Conformation

Optical Rotatory Dispersion (ORD) and Circular Dichroism (CD) studies become useful for determining the absolute stereochemistry of rigid molecules and the preferred conformation of flexible molecules, when applicable. Some rules with examples have been briefly discussed in Sect. 2.19.

4.2.11 Synthesis. Retrosynthesis. Green Chemistry. Atom Economy

Synthesis is an integral part of organic chemistry. In the field of natural product chemistry, synthesis confirms the structure assigned to a natural product on the basis of spectral and chemical methods and also establishes the stereochemistry of a chiral natural molecule by carrying out its asymmetric synthesis.

For designing a synthetic plan for a natural product and for its subsequent implementation one needs to have knowledge of reactions and their scopes and limitations. The target molecule (TM) may be formally fragmented or reasonably chemically disconnected to arrive at the much simpler molecular entities or ionic components (synthons/synthon equivalents) from which the original bonds or connections, as present in the TM, could be chemically recreated in steps leading finally to its formation [82, 83]. To achieve this, one has to proceed in a backward way and hence the name *retrosynthetic approach* for this organizational chart. The path for such search is shown by hollow arrow ==>. Sometimes the target molecule may be disconnected in more than one way leading to different synthesis of a single natural product. It is also true that even though several possibilities are there, in reality, all of them may not be chemically viable.

Synthesis offers a challenge to the organic chemists, provides an intellectual exercise, and causes the development of new methodologies. Three excellent volumes in the Series "Classics in Total Synthesis" [84–86] have appeared in 1996, 2003, and 2011, respectively. The first two volumes contained much-acclaimed accounts of the total synthesis of some 60 complex natural molecules. The third volume features 42 most impressive total syntheses of 25 challenging natural products of the most recent era (i.e., 2003–2010). Each synthesis in these volumes features important strategies and tools employed and explains the key steps of the synthetic pathway. Each synthesis represents a small world of chemistry, logic, and thoughts of great masters. *Retrosynthetic approaches* on most natural product syntheses have been briefly discussed. In course of the syntheses new concepts and methodologies have generated which are generalized to enrich the chemistry in general and the area of application in particular. On the whole, such total syntheses contribute to the overall development of organic chemistry. Corey wrote in the *Foreword* of the Classics III [86], "In aggregate, Classics I–III elegantly and effectively chronicle monumental scientific achievements across a broad front." The reader will find the contents of all the volumes "equally enjoyable, instructive, and inspirational." A critical review [87] contains the structures and the colorful highlights of >75 selected natural products synthesized in the lab of Nicolaou since 1980, and the structures of >100 natural products synthesized in other labs since 2000, including many complex ones and their discussions, are a very useful archive of this field. The other reviews on *Trilogy* [88] and on the structure of the molecule and the art of its synthesis [89], appearing in 2012 and 2013, respectively, are also highly instructive and demands thorough reading. Two informative and instructive books entitled "Classics in Stereoselective Synthesis" [90] and "Elements of Synthesis Planning" [91] have appeared in 2009.

From extensive literature survey one can have the advantage of suitable methodologies, specially the current ones to be applicable or to be attempted for the synthesis of TM. In cases of complex molecules, researcher's insight and intuition play a vital role in the sequential organization of the synthetic planning. Only one enantiomer of optically active natural products is biosynthesized by Nature, with very rare exceptions. Further, it has been observed that the enantiomers of an asymmetric compound, natural or synthetic, possess different biological activities (see Chap. 32). So the synthesis of the natural or synthetic compound with desired biological activity is generally aimed at. *Firstly*, the (±) racemic product is synthesized, followed by resolution through the formation of diastereomeric salts or esters to get the desired enantiomer. The overall yield of the desired enantiomer is, however, always less than 50 %. The rest being the undesired enantiomer is wasted. *Secondly,* by asymmetric synthesis one enantiomer/diastereomer is synthesized in a satisfactory excess of the other (enantiomeric/diastereomeric excess, ee/de) in a particular step and in a good yield. *Thirdly,* one can start with a chiral synthon for the synthesis of the desired molecules having multichiral centers. Suitable carbohydrates and amino acids may serve as good chiral synthons. In the second/third case diastereomers may be obtained in one or more step/s on the way, and the desired diastereomer is needed to be separated by suitable chromatographic technique/s (see Sect. 4.1). Many syntheses including stereoselective syntheses of natural products of different skeletal patterns have been discussed in various chapters.

Though the total synthesis of a natural product offers a great challenge to the organic chemists, its commercial use is not economically viable, especially in cases of complicated molecules. However, in some cases semisynthesis becomes effective, as in the case of (−)-taxol (Sect. 8.4.6.1).

Green Chemistry. Atom Economy [92–94]. Basically, three important interdependent aspects are to be considered in the planning of an efficient synthesis:

(a) Environmentally friendly experimental condition (green chemistry)
(b) Atom economy for each step of a multistep synthesis
(c) Economic viability, especially in cases of large/industrial scale synthesis and the overall yield.

Environmentally friendly experimental conditions refer to the cleaner conditions in which nontoxic and nonhazardous chemicals (synthons, reagents, solvents) should be used, and the waste products are biodegradable and either noninjurious or less injurious to the environment. On the whole, the pollution of the environment should not be caused, or the balance of the ecosystem should not be disturbed by such synthesis, especially in case of an industrial scale synthesis which includes reactions, conversions, etc., in large scales.

Chemical studies performed under such benign experimental conditions are referred to as green chemistry, a term coined only in early nineties. A number of journals, e.g., *Green Chemistry*, have been launched and **dedicated to green chemistry**.

In shifting from traditional chemistry to green chemistry the atom economy [92, 93] and the improvement of yield are the two vital points to be taken care of. The atom economy or atom efficiency [94] for any reaction is given by the following expression:

$$\text{Atom economy} = \frac{\text{Molecular weight of the desired product}}{\text{Sum of the mol. wts. of all products in the stoichiometric equation of the reaction}}$$

Thus efficient synthetic methods include reactions that are both selective and economical in atom count so that maximum number of atoms of reactants appears in the products. Methods that involve combining two or more building blocks with any other reactant, which serves as a catalyst, constitute the highest degree of atom economy [94]. *Enzymes are the most benign and efficient catalysts known, and enzyme-catalyzed reactions have the highest atom economy.*

Firstly, enzymes are chemo-, regio-, diastero-, and enantioselective in their behavior. The step of blocking the undesirably located similar group/s followed by their deblocking is not necessary for enzyme-catalyzed reactions. Such steps are thus eliminated leading to the improvement of the yield. *Transition metal-catalyzed methods* being selective and economical are now being increasingly used for synthetic efficiency. Barry M. Trost has published two elegant review articles [92, 93] on atom economy—a search for synthetic efficiency, and a challenge for organic synthesis using homogeneous catalysis—citing many examples.

Secondly, in the enzyme-catalyzed reactions the first chiral center is generated in a highly enantiomeric excess; enzyme-catalyzed generation of more chiral centers in subsequent steps give rise to highly diastereomeric excess (~100 %).

On the other hand, *an achiral synthesis starting with a symmetric/achiral substrate and using achiral reagents leads to racemic products in each step from the very first one, generating a chiral center, and in subsequent steps generating racemate diastereomers without much desired diastereoselectivity.* In the final step, the desired enantiomeric product may be obtained only after resolution, reducing its yield by 50 %. In chiral syntheses also, starting from a suitable chiral synthon the problem of the diastereoselectivity of the desired product still remains. At present, attempts are usually made to employ proper chiral reagents or catalysts to obviate the formation of undesired diastereomers, thus increasing the diastereoselectivity. Of course, attempts should always be made to use the undesired diastereomers in one way or other.

Thirdly, in green chemistry the **waste** should be minimum. The **waste** is defined as the entire chemical component/s produced in the process other than the desired product, and the ***E factor*** gives the waste per kg product [94]. Thus, the more the waste, the less is the atom economy. The waste is minimized in fine chemicals manufacture by widespread substitution of classical organic syntheses using stoichiometric amounts of inorganic reagents with cleaner catalytic alternatives. The E factors of chemical processes increase dramatically on going from bulk to pharmaceuticals and fine chemicals mainly due to the use of stoichiometric methods. The atom-efficient catalytic processes have been illustrated by Sheldon [94] with industrially relevant examples.

The Role of Catalysis A primary cause of *waste* generation is the use of stoichiometric inorganic reagents [94]. For example, stoichiometric reductions with metals (Na, Mg, Fe, Zn) and metal hydrides ($LiAlH_4$, $NaBH_4$) and oxidations with permanganate or chromium (VI) reagents are rampant for manufacture of fine chemicals. Replacing stoichiometric methodologies with cleaner catalytic alternatives may solve this. The atom economies of the stoichiometric oxidation of a secondary alcohol to the corresponding ketone with chromium trioxide in sulfuric acid and its catalytic oxidation with oxygen may be compared, e.g., in case of

Stoichiometric method:

$$3 \, PhCH(OH)CH_3 \; + \; 2 \, CrO_3 \; + \; 3 \, H_2SO_4 \quad \longrightarrow \quad 3 \, PhCOCH_3 \; + \; Cr_2(SO_4)_3 \; + \; 6 \, H_2O$$

Atom economy $= \; 3 \times 120 \, / \, (3 \times 120 + 392 + 6 \times 18) = \quad 360 \, / \, 860 = 42 \, \% \, (approx.)$

Catalytic method:

$$PhCH(OH)CH_3 \; + \; \frac{1}{2} \, O_2 \quad \xrightarrow{\text{catalyst}} \quad PhCOCH_3 \; + \; H_2O$$

Atom economy $= 120 \, / \, (120 + 18) = 120 \, / \, 138 \; = \; 87\% \, (approx.)$

Fig. 4.16   Atom economy of stoichiometric oxidation versus catalytic oxidation

oxidation of α-phenylethanol to acetophenone, as shown in Fig. 4.16 [94]. Naturally, atom economy or atom efficiency dictates the adoption of the catalytic method, and a suitable catalyst is to be employed for this purpose.

Biocatalysis Biocatalysis has the potential to deliver *greener* chemical syntheses. Some of these opportunities and outstanding challenges are presented in a review article by J. M. Woodley [95].

4.2.12   Biosynthetic Compatibility of the Proposed Structure

The structure is expected to be biosynthetically compatible. The formation should be conceived following the discipline of the asymmetric biosynthesis in the cell. The precursors are to be selected retroasymmetric-biosynthetically to fit preferably into the identified biosynthetic pathways. In fact, structures of many natural products have been proposed, and in some cases corrected, based on biosynthetic compatibility. In quite a number of cases the proposed structures have been confirmed by X-ray studies and unambiguous syntheses.

4.2.13   Conclusions

In summary we can conclude that

- **UV** spectra show the presence of chromophore/s.
- **IR** spectra reveal the type of functional group/s present.
- **NMR** spectra provide information about the magnetically distinct atoms (mostly protons and carbons in organic molecules). Carbon connectivity studies by various correlation NMR spectroscopy give us the structural pattern of the molecule under investigation. The appropriate correlation diagrams are quite often equivalent to X-ray analysis, being capable of delivering complete structural information.

- **X-ray** analysis provides complete structural information of the molecules especially carrying heavy atom (absolute stereochemistry).
- **Optical Rotatory Dispersion (ORD)** and Circular **Dichroism (CD)** studies,
  when applicable, furnish absolute stereochemistry of the molecule.
- **Chemical Correlation** with a compound of known structure and absolute
  stereochemistry, if possible, reveals the absolute stereochemistry of the chiral
  natural molecule.
- **Synthesis** gives definitive support to the structure and stereochemistry.

All the above analyses may not be necessary for establishing the structure of a
new natural product. Selection of the above analytical tools depends on the nature
as well as complexity of the molecule.

4.2.14   Naming of Natural Products

To put a name to a new natural product is essential for its identity. The associated
properties and other chemical details of the compound can be obtained from the
published literature on the compound, or from the laboratory of its isolation—using
its name. Hence, once the structure of a new natural product is established and its
various physical constants (e.g., mp/bp, specific rotation, refractive index, etc.) are
measured, it should be named. Earlier, prior to their structural assignment they used
to be named for the sake of convenience. In most cases the structures of natural
products being complex in nature, the IUPAC nomenclature in terms of homocyclic
and heterocyclic systems present in many natural products poses difficulty to the
readers for getting the name translated into structure and vice versa. Artemisinin
(qinghaosu) (**1**), an important antimalarial drug, was isolated from *Artemisia annua*
(fam. Compositae) [for structures (**1**)–(**41**) see Fig. 4.17] (*the structure numbers of
the natural products mentioned here refer to the present subsection only*). The name
qinghaosu is derived from a Chinese drug *qinghao*, well known for its medicinal
use, especially against fever, dating from 168 BC. According to the IUPAC
nomenclature it is octahydro-3,6,9-trimethyl-3,12-epoxy-12H-pyrano[4,3-j]-1,2-
benzodioxepin-10(3H)-one.

Echinulin, a fungal metabolite of *Aspergillus echinulatus,* later isolated from the
peels of the green fruits named *Parwal* (Hindi) and *Potol (*Bengali) of a higher plant
Trichosanthes dioica (Cucurbitaceae), is established to have structure (**2**). Its IUPAC
nomenclature is 3S,6S,3-[{2-(1,1-dimethylallyl)-5,7-bis(3,3-dimethylally1H-inda-3-
yl)}methyl]-6-methyl-2,5-piperazinedione. Longifolene, a sesquiterpene from *Pinus
longifolia* (Pinaceae), possesses structure (**3**). Its IUPAC nomenclature is decahydro-
4,8,8-trimethyl-9(15)ene-1,4-methanoazulene. To avoid the complexity of the IUPAC
nomenclatures trivial names are always used.

It may be mentioned that another way of numbering the skeleton is the biogenetic numbering, e.g., in cases of sesquiterpenes, farnesol numbering system (**5**) is

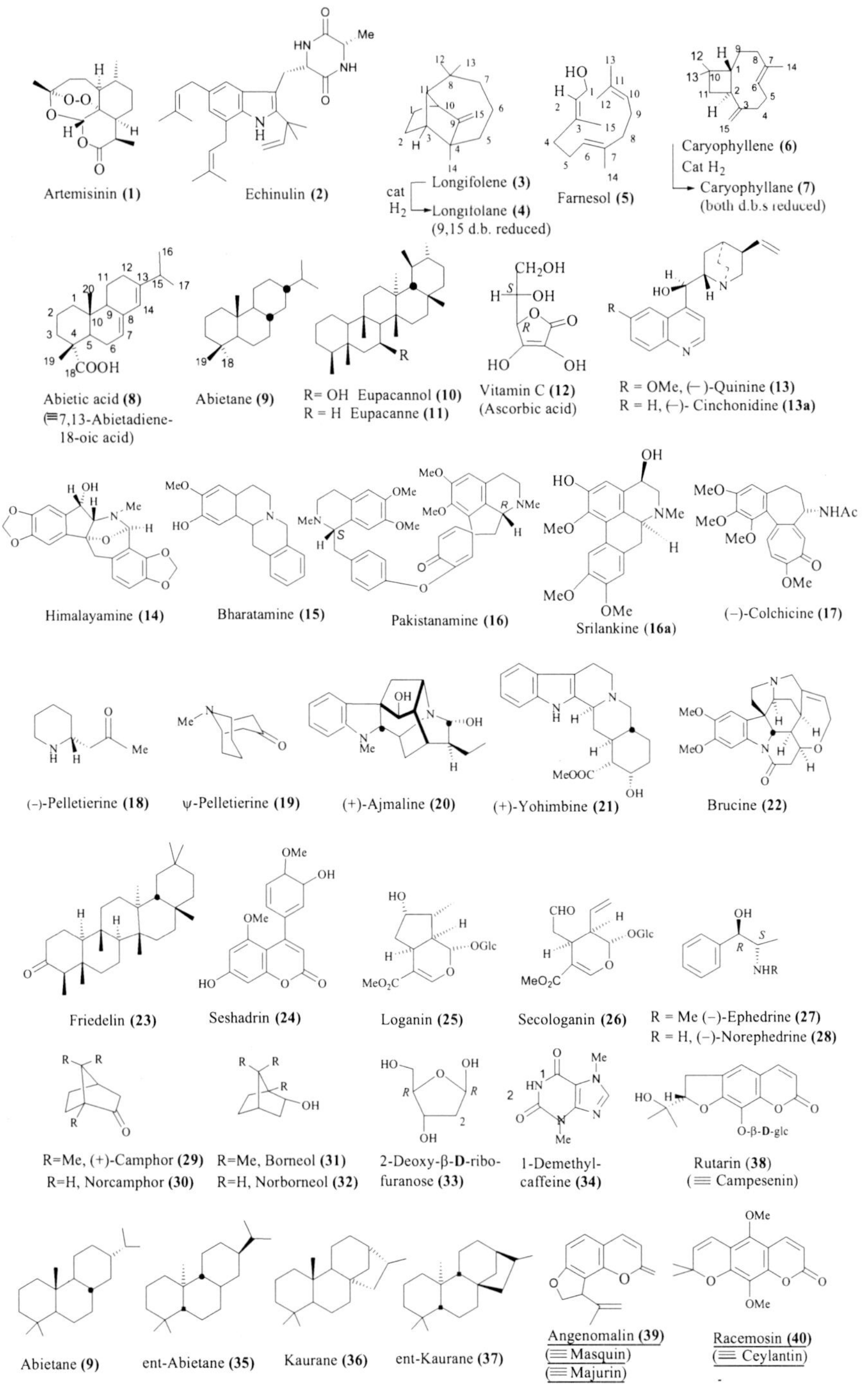

Artemisinin (1)

Echinulin (2)

Longifolene (3)

cat H₂ → Longifolane (4) (9,15 d.b. reduced)

Farnesol (5)

Caryophyllene (6)
Cat H₂
→ Caryophyllane (7)
(both d.b.s reduced)

Abietic acid (8)
(≡7,13-Abietadiene-18-oic acid)

Abietane (9)

R= OH Eupacannol (10)
R = H Eupacanne (11)

Vitamin C (12)
(Ascorbic acid)

R = OMe, (−)-Quinine (13)
R = H, (−)- Cinchonidine (13a)

Himalayamine (14)

Bharatamine (15)

Pakistanamine (16)

Srilankine (16a)

(−)-Colchicine (17)

(−)-Pelletierine (18)

ψ-Pelletierine (19)

(+)-Ajmaline (20)

(+)-Yohimbine (21)

Brucine (22)

Friedelin (23)

Seshadrin (24)

Loganin (25)

Secologanin (26)

R = Me (−)-Ephedrine (27)
R = H, (−)-Norephedrine (28)

R=Me, (+)-Camphor (29)
R=H, Norcamphor (30)

R=Me, Borneol (31)
R=H, Norborneol (32)

2-Deoxy-β-D-ribo-furanose (33)

1-Demethyl-caffeine (34)

Rutarin (38)
(≡ Campesenin)

Abietane (9)

ent-Abietane (35)

Kaurane (36)

ent-Kaurane (37)

Angenomalin (39)
(≡ Masquin)
(≡ Majurin)

Racemosin (40)
(≡ Ceylantin)

Fig. 4.17 Structures of the natural products (1)–(40) of plant origin

used. Here the folding of farnesol corresponds to the longifolene skeleton (**4**) having the same numbering system as in (**5**).

For convenience, complex natural products, especially terpenoids and steroids, are named based on the trivial names of the corresponding basic saturated hydrocarbons. Likewise, sometimes the complex alkaloids are named on the basis of the trivial names of the corresponding basic heterocycles. The individual compounds having the same basic stereostructural skeleton can be given IUPAC names based on the latter, mentioning the substitution locations. Caryophyllene, a sesquiterpene from mentha oil, possesses structure (**6**); the corresponding saturated hydrocarbon has been named caryophyllane (**7**). In terms of this basic hydrocarbon caryophyllene will have the IUPAC nomenclature, caryophylla-3(15),6-diene. Similarly longifolene will be longifola-9(15)-ene. Another example is abietic acid (**8**); the corresponding saturated hydrocarbon skeleton is named abietane (**9**). Hence abietic acid may be named abieta-7,13-diene-18-oic acid. A pentacyclic triterpene eupacannol (**10**), possessing a new skeleton, is named eupacann-7β-ol and the corresponding ketone is named eupacann-7-one, based on the corresponding saturated hydrocarbon eupacannane (**11**). The IUPAC nomenclatures of the compounds discussed in different chapters (Chaps. 6–15) have been mentioned along with their trivial names, wherever possible.

Thus, to avoid complications, the natural products are named mostly after the sources from which they are first isolated, as has been the practice since the early days of research on natural product chemistry. The new compounds of plant origin are named after the botanical names of the plants, barring few exceptions to be mentioned later. Some letters of the name/s of the genus and/or species are used in various pronounceable combinations to name them. In some cases the plant constituents have been named based on their biological activity. Several examples (Table 4.10) will illustrate these conventions.

Vitamins are named by letters as vitamins A, B, C, D, E, K, etc.; perhaps this nomenclature followed the chronology of their discovery, since the structures were not known at that time. They are not structurally related. The name vitamin was coined by C. Funk in 1913. In Latin *vita* means life. The compounds were erroneously thought to be amines and hence the name *vitamin*. Some vitamins belong to B, D, and K groups having numeral subscripts, e.g., B_1, B_2, B_6, B_{12}, D_1, D_2, D_3, K, K_1, K_2, K_3, etc. These compounds are essential for our well-being with disease-healing and health-giving properties, when taken in optimum amounts in the form of foods or medicines.

Quinine was named from the cutoff part of *Cinchona* species whose local name (South Africa) is *quina quina*. The name was given by Pelletier and Caventau. The suffix *ine* is always added in naming the alkaloids with few exceptions (e.g., camptothecin, taxol when considered as alkaloid). However, in German spelling the last letter *e* is dropped.

Sometimes compounds are named after the place/country of collection of the plant material, e.g., himalayamine from *Meconopsis villosa*, the plant being collected from the alpine region of Himalayas, bharatamine, isolated from an Indian (Bharat) plant, and pakistanine, isolated from a plant of Pakistan.

Table 4.10 Plant constituents derived from their sources[a]

Source (plant)	Name	Class of compound
Abies grandis	Abietic acid	Diterpene acid
Aegle mormelos Correâ	Marmesin, aegelenine	Coumarin, alkaloidal amide
Ailanthus malabaricus	Malabaricol	Triterpene
Anhalonium lewinii	Mescalin[b]	Alkaloid (protoalkaloid)
Artemesia annua	Artemisin	Sesquiterpene peroxide
Atropa belladonna	Atropine	Alkaloid
Capsicum annuum	Capsaicin	Alkaloidal amide
Chloroxylon swietenia	Xylotenin	Coumarin
Citrus fruits and plants	Ascorbic acid[c]	Sugar
Cinchona species	Cinchonine, quinine	Alkaloids
Coffea arabica	Caffeine	Xanthine derivative
Dendrobium gibsonii	Dengibsin, dengibsinin	Fluorenones
Dendrobium nobilie	Denbinobin	1,4-Phenanthraquinone
Ephedra sinica	Ephedrine	Alkaloid
Eupatorium cannabinum	Eupacacannol	Triterpene
Gelonium multiflorum	Multiflorinol, gelomulides A-K[d]	Triterpene, diterpene lactones
Gibberella fujikuroi	Gibberellins[e]	19-Diterpene acids with skeletal oxygenation sites
Ginkgo biloba	Ginkgolides A-C[d]	Diterpene lactones
	Bilobanone, Bilobalone	Sesquiterpenes
Nicotiana tobaccum	Nicotine	Alkaloid
Papaver somniferum	Papaverine, morphine[c]	Alkaloids
Rauwolfia serpentina	Reserpine, serpentine	Alkaloids
Seseli sibiricum	Sesebiricol	Coumarin
Taxus brevifolia	Taxol	Diterpene

[a]Literature references of most of the natural products may be found in the relevant chapters

[b]The flowering heads of this plant used to be imported to Europe under the name "*mescal buttons*"; hence the name

[c]*Compounds are named on the basis of their biological activity*: hexuronic acid (vitamin C) was trivially named ascorbic acid (**11**) [Greek *a* (without) and Latin *scorbutus* (scurvy)], since its deficiency causes a disease called scurvy, first observed in sailors, whose diet during their long voyage was found to be devoid of vitamin C. The name was given by Albert Szent-Gyorgyi (NL 1937). Its nomenclature is threo-hex-2-enoic acid-γ-lactone (threo-hexulosono-1,4-lactone-2,3-enediol). Morphine was named after the Greek God of dream and son of God of sleep *Morpheus*, because of its profound sedative and pain killing properties. The name was given by Gay-Lussac

[d]When more than one compound with the same but differently substituted skeletal pattern are obtained from the same plant, or from different species of the same genus, they may be named as A, B, C, preceded by a general name (based on the genus) given to the congeners possessing that skeletal pattern

[e]Gibberellins are named as Gibberellic acid (GA) with suffixes 1, 2, 3, 4, e.g., GA_1, GA_2, GA_3, GA_4...

Some compounds have been named after some famous persons who are related directly or indirectly with the work. The alkaloids pelletierine and ψ-pelletierine were named after Pelletier, regarded as the father of alkaloids. The alkaloid nicotine as well as the genus *Nicotiana* of the plant was named after a French diplomat Jean Nicot de Villemain who introduced tobacco in France. S. Siddiqui during 1930s isolated nine alkaloids from *Rauwolfia serpentine*; he named the main alkaloid "ajamaline" after the name of Hakim Ajmal Khan who had used *R. serpentina* for treatment of mental ailments for nearly two decades. Again, the alkaloid brucine (**12**), isolated from *Strychnos nuxvomica* and many other *Strychnos* species, derived its name from the plant source *Brucea antidysenterica from which it was first isolated.* Incidentally it may be mentioned that the genus of the plant was named after the explorer James Bruce (1730–1794), and the species, probably, was named in view of the plant's physiological property.

In 1899 Constantin Istrati and A. Ostrogovitch of the University of Bucharest isolated a compound from cork and named it fridelin after the famous nineteenth century French chemist, Charles Friedel. The compound later turned out to be a pentacyclic triterpene and has been isolated from many plants. A neoflavone derivative has been named seshadrin after the name of an Indian natural product chemist T. R. Seshadri. These ways of naming the compounds reflect the personal fascination of the researchers to remember the place, honor the man, and also highlight their biological properties and do not fall under the normal purview of the convention.

None of the methods of naming, discussed so far, reflects the nature/structure of the compounds. However, the suffix "ine" after the name points to its being an alkaloid (in German spelling e is dropped), the suffixes "ol" and "one" point to alcohol and carbonyl compounds respectively and the suffix "lide" points to terpene (sesqui-, di-) lactones. The following prefixes before the name of the parent compounds bear some significance.

"Seco" means the presence of a cleaved ring, e.g., loganine (13) and secologanin (14). The prefix "nor" indicates the removal of one or more carbon atoms from the skeletal framework or from a group containing a heteroatom, e.g., ephedrine (15) and norephedrine (16). The lack of two, three, or four, etc., skeletal atoms of the parent compound is indicated by the numerical prefixes "dinor," "trinor" or "tetranor," etc. In certain cases "nor" is used to denote lack of even three skeletal atoms with respect to the parent compounds, e.g., camphor (17) and norcamphor (18) and borneol (19) and norborneol (20). Tetranortriterpenoids (C_{26} compounds) form an important class of natural products. From the name it is clear that they have four carbon atoms less, compared to the skeletal carbon content of triterpenes (C_{30}). The prefix "des" means one unit less. Compounds containing $-OCH_3$ (either a natural product or a reaction product) when demethylated results in a desmethyl compound. The prefix "de" also represents absence of some group or atoms, e.g., deoxy, deactoxy, deacetyl, etc., compounds like 2-deoxy-D-ribofuranose (**21**) and 1-demethylcaffeine (**22**). The prefix "*ent-*" (standing for *enantio-*) indicates stereochemical inversions at all chiral centers compared to the parent compound, e.g., abietane (**23**) and *ent*-abietane (**24**) and kaurane (**25**) and *ent*-kaurane (**26**).

Literature survey shows that a single compound isolated from different sources, sometimes, has been given two or more names. This happens due to its simultaneous independent publications, or may be due to the absence of knowledge of the previous work by the later workers. A few examples are given: the coumarin (**27**) named rutarin (isolated from *Ruta graveoleus*, 1967) and campesinin (from *Seseli campestre*, 1970); the coumarin (**28**) named angenomalin (from *Angelica anomala*, 1967), masquin (from *Pimpene cola*, 1967), and majurin (from *Amni majus*, 1971); and the coumarin (**29**) named racemosin (from *Atalantia racemosa*, 1978) and ceylantin (from *Atalantia ceylanica*, 1984). However, nowadays in view of the electronic library facilities available such a probability may be avoided.

References

1. Denise M. Hosler and Michael A. Mikita, Ethnobotany: The Chemist's Source for the Identification of Useful Natural Products, *J. Chem. Educ.*, **1987**, *64*, 328-332.
2. N. R. Farnsworth, The Pharmacognosy of the Periwinkles: Vinca and Catharanthus, *Lloydia*, **1961**, *24*, 105-138, (pertinent pp. 114-116) and the references cited.
3. N. R. Farnsworth, The Pharmacognosy of the Periwinkles: Vinca and Catharanthus, *Lloydia*, **1961**, *24*, p. 106.
4. A. R. Battersby, D. G. Laing and R. Ramagi, Biosynthesis Part XIX. Concerning the Biosynthesis of (−)-Camphor and (−)-Borneol in *Salvia officinalis, J. Chem. Soc.* Parkin 1, **1972**, 2743-2748.
5. R. N. Chopra, S. L. Nayer and I. C. Chopra, *Glossary of Indian Medicinal Plants*, CSIR, New Delhi, **1956**.
6. Kristina Jenett-Siems, Robert Weigl, Anke Böhm, Petra Mann, Britta Tofern-Reblin, Sonja C. Ott, Azar Ghomian, Maki Kaloga, Karsten Siems, Ludger Witte, Monika Hilker, Frank Müller and Eckart Eich, Chemotaxonomy of the pantropical genus *Merremia* (Convolvulaceae) based on the distribution of tropane alkaloids, *Phytochemistry*, **2005**, *66*, 1448-1464.
7. Rika Mladenova Taskova, Charlotte Heid Gotfredsen, Soren Rosendal Jensen, Chemotaxonomic markers in Digitaldeae (Plantaginaceae), *Phytochemistry*, **2005**, *66*, 1440-1447; and references cited.
8. M. D. Luque de Castro and L E. Garcia Auso, 'Soxhlet Extraction' in *Encyclopedia of Separation Science*, (Chief Ed., Ian D. Wilson), Vol. 6, Academic Press, San Diego, **2000**, pp. 2701-2709.
9. Willium B. Jensen, The Origin of the Soxhlet Extractor, *J. Chem. Educ.*, **2007**, *84*, 1913-1914.
10. C. F. Poole, 'Chromatography' in *Encyclopedia of Separation Science*, (Chief Ed., Ian D. Wilson), Vol. 1, Academic Press, San Diego, **2000**, pp. 40-64.
11. James M. Bobbitt, Arthur E. Schwarting and Ray J. Gritter, *Introduction to Chromatography*, Van Nostrand Reinhold Company, New York, **1968**.
12. Donald L. Pavia, Gary M. Lampman and George S. Kriz, Jr., Introduction to Organic Laboratory Techniques (a contemporary approach), (3rd Ed.), Saunders College Publishing, Philadelphia, **1988**, pp. 593-640.
13. Hobert H. Willard, Lynne L. Merritt, Jr., John A. Dean, Frank A. Settle, Jr., *Instrumental Methods of Analysis*, 7th Edition, Wardsworth Publishing Company, USA, First Indian Edition, CBS Publishers, New Delhi, 1986, pp. 513-654.
14. Satinder Ahuja, 'Chromatography and Separation Science' in *Separation Science and Technology*, Vol. 4, (Ed. Satinder Ahuja), Imprint from Elsevier Science, Amsterdam, **2003**.

15. Michael H. Abraham, 100 years of chromatography – or is it 171? *J. Chromatogr. A*, **2004**, *1061*, 113-114.
16. W. C. Still, M. Kahn and A. Mitra, Rapid chromatographic technique for preparative separations with moderate resolution, *J. Org. Chem.*, **1978**, *43*, 2923-2926.
17. C. F. Poole, 'Flash Chromatography' in *Encyclopedia of Separation Science*, (Chief Ed., Ian D. Wilson), Vol. 6, Academic Press, San Diego, **2000**, pp. 2808-2813.
18. Claeson F. Tuchinda and V. Reutrakul, Some empirical aspects on the practical use of flash chromatography and medium pressure liquid chromatography for the isolation of biologically active compounds from plants, *J. Sci. Soc.,* Thailand, **1993**, *19*, 73-86.
19. A. S. Gupta and Sukh Dev, Chromatography of Organic Compounds I. Thin Layer Chromatography of Olefins, *J. Chromatography*, **1963**, *12*, 189-195.
20. D. H. Murray, Naturally Occurring Plant Coumarins, *Fortschr. Chem. org. Naturstoffe*, **1997**, **72**, 2-119; pertinent page 3, and the references cited therein.
21. E. Eliel, S. H. Wilen and L. N. Mander, *Stereochemistry of Organic Compounds*, John Wiley & Sons, Inc., New York, **1994**, pp. 1076-77, other examples, pp. 1078-79.
22. Donald L. Pavia, Gary M. Lampman and George S. Kriz, *Introduction to Spectroscopy (A Guide for Students of Organic Chemistry)*, 2nd ed., Harcourt Brace College Publishers, Philadelphia, New York.
23. Donald L. Pavia, Gary M. Lampman, George S. Kriz,, and James R. Vyvyan, *Spectroscopy*, Cengage Learning, **2007**, Seventh Indian Reprint, **2011**, p. 17.
24. Daniel J. Pasto and Carl R. Johnson, *Organic Structure Determination*, Prentice-Hall Inc., Canada, 1969.
25. John R. Dyer, *Applications of Absorption Spectroscopy of Organic Compounds*, Prentice-Hall, Englewood Cliff, NJ, 1965; Prentice-Hall of India, New Delhi, **1989**.
26. Joseph B. Lambert, Herbert F. Shurvell, Lawrence Verbit, R. Graham Cooks, and George H. Stout, *Organic Structural Analysis*, Macmillan Publishing Co. Inc., New York, **1976**.
27. Phillip Crews, Jaime Rodri'guez and Marcel Jaspars, *Organic Structure Analysis*, Oxford University Press, New York, Oxford, **1998**.
28. Dudley H. Williams and Ian Fleming, *Spectroscopic Methods in Organic Chemistry*, 4th ed., McGraw-Hill, London, New York, **1988**; Tata McGraw-Hill, 3rd reprint, New Delhi, **1993**.
29. Robert M. Silverstein and Francis X. Webster, *Spectrometric Identification of Organic Compounds*, 6th ed., Wiley, New York, **1998**.
30. Clifford J. Creswell, Olaf A. Runquist, and Malcolm M. Campbell, *Spectral Analysis of Organic Compounds – An Introductory Programmed Text*, 2nd ed., Longman, **1972**.
31. L. M. Harwood and T. D. W. Claridge, *Introduction to Organic Spectroscopy*, Oxford University Press, Oxford, **1977**.
32. A. I. Scott, *Interpretation of the Ultraviolet Spectra of Natural Products*, Pergamon Press, New York, **1964**.
33. Koji Nakanishi and P. H. Solomon, *Infrared Absorption Spectroscopy*, 2nd ed., Holden-Day, San Francisco, **1977**.
34. N. P. G. Roeges, *Guide to Interpretation of Infrared Spectra of Organic Structures*, Wiley, New York, **1984**.
35. C. N. R. Rao, *Chemical Applications of Infrared Spectroscopy*, Academic Press, New York, **1963**.
36. P. Griffiths, *Fourier Transform Infrared Spectrometry*, 2nd ed., Wiley, New York, **1986**.
37. L. M. Jackman and S. Sternhall, *Applications of Nuclear Magnetic Resonance Spectroscopy in Organic Chemistry*, 2nd ed., Pergamon Press, London, Oxford, 1969; 1st ed. authored by L. M. Jackman, **1959**.
38. E. Breitmaier, *Structure Elucidation by NMR in Organic Chemistry. A Practical Guide*, Wiley, New York, **1993**.
39. J. K. M. Sanders and B. K. Hunter, *Modern NMR Spectroscopy – A Guide for Chemists*, 2nd ed., Oxford University Press, Oxford, **1993**.
40. Atta-ur-Rahman, *Nuclear Magnetic Resonance*, Springer–Verlag, New York, **1986**.

41. H. Günther, *NMR Spectroscopy,* 2nd ed., Wiley, New York, Chichester, 1995.
42. R. Abraham, J. Fisher, and P. Loflus, *Introduction to NMR Spectroscopy*, Wiley, New York, Chichester, **1988**.
43. H. Friebolin, *Basic One- and Two-dimensional NMR Spectroscopy, 3rd ed.,* VCH Publishers, Weinheim, **1988**.
44. G. E. Martin and A. S. Zecker, *Two-Dimensional NMR Methods for Establishing Molecular Connectivity*, VCH Publishers, Inc, New York, **1988**.
45. D. Shaw, *FourierTransform NMR Spectroscopy,* Elsevier, Amsterdam, **1996**.
46. Warren Steck and M. Mazurek, Identification of Natural Coumarins by NMR Spectroscopy, *Lloydia,* **1972**, *55*, 418-439.
47. Eberhard Breitmaier and Wolfgang Voelter, *Carbon-13 NMR Spectroscopy* (Third Edition), VCH, New York, **1987**.
48. G. C. Levy, R. L. Lichter, and G. L. Nelson, *Carbon-13 Nuclear Magnetic Resonance for Organic Chemists,* 2nd ed., Wiley, New York, **1986**.
49. H. O. Kalinowsky, S. Berger, and S. Braun, *Carbon-13 NMR Spectroscopy,* Wiley, New York, **1988**.
50. F. W. Wehrli and T. Wirthlin, *Interpretation of Carbon-13 NMR Spectra,* Heyden, London, 1976.
51. F. W. Wehrli, The Use of Carbon-13-Nuclear Magnetic Resonance Spectroscopy in Natural Product Chemistry, *Fortschr. Chem. org. Naturstoffe,* **1977**, *36*, 1-229.
52. Helmut Duddeck and Manfred Kaiser, ^{13}C NMR Spectroscopy of Coumarin Derivatives, *Organic Magnetic Resonance*, **1982**, *20*, 55-68.
53. Atta-ur-Rahman and Vigar Uddin Ahmad, *^{13}C NMR of Natural Products*, Vol. I (Monoterpene and Sesquiterpene), Vol. II (Diterpene), Plenum Press, New York, London, **1992**.
54. Sunil K. Talapatra, Gita Das and Bani Talapatra, Stereostructures and Molecular Conformations of Six Diterpene Lactones from *Gelonium multiflorum, Phytochemistry,* **1989**, *28*, 1181-1185.
55. Bani Talapatra, Gita Das, Asoke K. Das, Kallolmay Biswas, and Sunil K. Talapatra, Stereostructures and Molecular Conformations of Four Additional Diterpene Lactones from *Gelonium multiflorum, Phytochemistry,* **1998**, *37*, 1353-1359.
56. D. E. Alsono and S. E. Warren, NMR Analysis of Unknowns: An Introduction to 2D NMR Spectroscopy, *J. Chem. Educ.,* **2005**, *82*, 1385-1386.
57. L. T. Alty, Monoterpene Unknown Identified Using IR, ^{1}H NMR, ^{13}C NMR. DEPT, COSY and HETCOR, *J. Chem. Educ.,* **2005**, *82*, 1387-1389.
58. N. M. Glagovich and T. D. Shine, Organic Spectroscopy Laboratory: Utilizing IR and NMR in the identification of an unknown substance, *J. Chem. Educ.,* **2005**, *82*, 1382-1384.
59. Stephen R. Shrader, *Introduction to Mass Spectroscopy*, Allyn and Bacon Inc., Boston, **1971**.
60. Herbert Budzikiewicz, Carl Djerassi and Dudley H. Williams, *Interpretation of Mass Spectra of Organic Compounds*, Holden-Day, San Francisco, **1964**.
61. *Idem, Structure Elucidation of Natural Products by Mass Spectrometry,* Vols. I and II, Holden-Day, San Francisco, **1964**.
62. *Idem, Mass Spectrometry of Organic Compounds*, Holden-Day, San Francisco, **1967**.
63. K. Biemann, *Mass Spectrometry: Organic Chemical Applications,* McGraw-Hill, New York, **1962**.
64. F. W. McLafferty and F. Turecek, *Interpretation of Mass Spectra,* 4th ed., University Science Books, Mill Valley, Calif., **1993**.
65. Q. N. Porter and J. Baldas, *Mass Spectrometry of Heterocyclic Compounds*, Wiley-Interscience, New York, **1971**.
66. H. Budzikiewiez, J. M. Wilson, and C. Djerassi, Mass Spectrometry in Structural and Stereochemical Problems XXXII, Pentacyclic Triterpenes, *J. Amer. Chem. Soc.,* **1963**, *85*, 3688-3699.
67. C. Djerassi, *Optical Rotatory Dispersion*, McGraw-Hill, New York, **1962**.

68. P. Crabbe, *Optical Rotatory Dispersion and Circular Dichroism in Organic Chemistry*, Holden-Day, London, **1965**.

69. P. Crabbe and W. Klyne, Optical Rotatory Dispersion and Circular Dichroism of Aromatic Compounds: A General Survey, *Tetrahedron,* **1967**, *23*, 3449-3503.

70. G. Snatzke, *Optical Rotatory Dispersion and Circular Dichroism in Organic Chemistry*, Heydon, London, **1967**.

71. G. Snatzke, Circular Dichroism and Optical Rotatory Dispersion – Principles and Applications to the Investigation of the Stereochemistry of Natural Products, *Angew. Chem. Int. Edn.,* **1968**, **7**, 14-25.

72. G. Snatzke and F. Snatzke in *Fundamental Aspects and Recent Developments In Optical Rotatory Dispersion and Circular Dichroism*, eds. F. Ciardelli and P. Salvadori, Heydon, London, **1973**.

73. Sunil K. Talapatra, Bimala Karmacharya, Shambhu C. De, and Bani Talapatra, (−)-Regiolone, An α-Tetralone from *Juglans regia* Linn. Structure, Stereochemistry and Conformation, *Phytochemistry,* **1988**, *27*, 3929-3932.

74. Bani Talapatra, Avijit Porel, Kallolmay Biswas and Sunil K. Talapatra, Absolute Configuration of Goniodiol, Goniodiol Monoacetate and other Related Dihydropyrones from Synthetic, Circular Dichroism and X-ray Crystallographic Evidence, *J. Indian Chem. Soc.,* **1997**, *74*, 896-903.

75. J. C. P. Schwarz (Ed.), *Physical Methods in Organic Chemistry*, Oliver & Boyd, London, **1965**, pertinent pages 276-305.

76. E. L. Eliel, S. H. Wilson and M.P. Doyle, *Basic Organic Stereochemistry*, John Wiley & Sons, New York, **2001**. pertinent pages 17-19, 75-77, 85-86.

77. R. Parthasarathi, The Determination of Relative and Absolute Configuration of Organic Molecules by X-ray Diffraction Methods in *Stereochemistry: Fundamentals and Methods, Vol. 1, Determination of Configuration by Spectrometric Methods,* ed. Henry B. Kagan, Georg Thieme Publishers, Stuttgart, **1977**, pp 181-234.

78. Carsten Reinhardt, *Shifting and Rearranging*, Science History Publications, Sagamore Beach, 2006; Book Review (Jerom A. Berson), *Angew Chem. Int Ed.,* **2007**, *46*, 4818-4819.

79. R. B. Woodward, M. P. Cava, W. D. Ollis, A. Hunger, H. U. Daeniker, and K. Schenker, The Total Synthesis of Strychnine, *Tetrahedron,* **1963**, *19*, 247-288; pertinent page 248.

80. Konrad Bloch, Sterol Molecule: Structure, Biosynthesis, and Function, *Steroids,* **1992**, *57*, 378-383; pertinent page 381.

81. John Bennett Fenn, M. Mann, C. K. Meng, S. F. Wong, and C. M. Whitehouse, Electrospray Ionization for Mass Spectrometry of Large Molecules, *Science,* **1989**, *246*, 4926.

82. Stuart Warren, *Organic Synthesis: The Disconnection Approach*, John Wiley, 1982.

83. E. J. Corey and X.-M. Cheng, *The Logic of Chemical Synthesis*, Wiley-VCH, **1995**.

84. K. C. Nicolaou and E. J. Sorensen, *Classics in Total Synthesis – Targets, Strategies, Methods,* Wiley-VCH, **1996**, pp 798.

85. K. C. Nicolaou and S. A. Snyder, *Classics in Total Synthesis II – More Targets, Strategies, Methods,* Wiley-VCH, **2003**, pp 636.

86. K. C. Nicolaou and J. C. Chen, *Classics in Total Synthesis III – Further Targets, Strategies, Methods,* Wiley-VCH, **2011**, pp 746.

87. K. C. Nicolaou, Christopher R. H. Hale, Christian Nilewski, and Herakidia A. Ioannidou, Constructing Molecular Complexity and Diversity. Total Synthesis of Natural Products of Biological and Medicinal Importance, *Chem. Soc. Rev.,* **2012**, *41*, 5185-5238.

88. K. C. Nicolaou, Christopher R. H. Hale, and Christian Nilewski, A Total Synthesis Trilogy: Calicheamicin γ_1, Taxol[R], and Brevetoxin A, *Chem. Rec.,* **2012**, *12*, 407-441.

89. K. C. Nicolaou, The Emergence of the Structure of the Molecule and the Art of Its Synthesis, *Angew. Chem. Int. Ed.,* **2013**, *52*, 131-146.

90. E. M. Carreira and L. Kvaerno, *Classics in Stereoselective Synthesis,* Wiley-VCH, **2009**.

91. R. W. Hoffmann, *Elements of Synthesis Planning*, Springer–Verlag, Berlin, Heidelberg, **2009**.

92. Barry M. Trost, The Atom Economy – A Search for Synthetic Efficiency, *Science*, **1991**, *254*, 1471-1477.
93. Barry M. Trost, Atom Economy – A Challenge for Organic Synthesis: Homogeneous Catalysts Leads the Way, *Angew. Chem. Int. Ed. Engl.*, **1995**, *34*, 259-281.
94. Roger A. Sheldon, Atom Efficiency and Catalysis in Organic Synthesis, *Pure Appl. Chem.*, **2000**, *72*, 1233-1246.
95. John M. Woodley, New Opportunities for Biocatalysis: Making Pharmaceutical Processes Greener, *Trends in Biotechnology*, **2008**, *26*, 321-327.

Further Reading

Satinder Ahuja, 'Chromatography and Separation Science' in *Separation Science and Technology, Vol. 4*, (Ed. Satinder Ahuja), Imprint from Elsevier Science, Amsterdam, **2003**.

W. D. Conway, E. L. Bachert, A. M. Sarlo and C. W. Chan, Comparison of countercurrent chromatography with flash chromatography, *Journal of Liquid Chromatography & Related Technologies*, **1998**, *21*, 53-63.

A. E. Derome, *Modern NMR Techniques for Chemistry Research,* Penguin, Oxford, **1987**.

Timothy D. W. Claridge, *High-Resolution NMR Techniques in Organic Chemistry*, 1st ed., Pergamon, Amsterdam, **1999**.

H. Frebolin, *Basic One- and Two- Dimensional Spectroscopy,* 2nd ed. VCH, New York, **1993**.

D. Wittakar, *Interpretation of Organic Spectra*, R.S.C. (Royal Society of Chemistry), London, **2000**.

James R. Hanson, The Development of the Strategies for Terpenoid Structure Determination, *Nat. Prod. Rep.*, **2001**, *18*, 607-617.

Addison Ault and Gerald O. Dudek, *An Introduction to Proton Nuclear Magnetic Resonance Spectroscopy*, Holden-Day, San Fransisco, London, **1976**.

Donald A. McQuarrie, Proton Magnetic Resonance Spectroscopy, *J. Chem. Educ.*, **1988**, *65*, 426-433.

L. J. Bellamy, *The Infrared Spectra of Complex Molecules*, 3rd ed., Chapman and Hall, London, Halsted-Wiley, New York, **1975**.

Atta-ur-Rahman, and M. I. Chaudhary, *Solving Problems with NMR,* Academic Press, New York, **1996**.

Stefan Berger and Dieter Sicker, *Classics in Spectroscopy*, Wiley-VCH, **2009**.

Catalogs and Workbooks

G. Socrates, *IR Characteristic Group Frequencies,* John Wiley, New York, **1994**.

F. F. Bentley, L.D. Smithson, and A.L. Rock, *Infrared Spectra and Characteristic Frequencies ~700–300 cm^{-1}, A Collection of Spectra, Interpretation and Bibliography,* Wiley-Interscience, New York, **1988**.

D. Dolphin and A. E. Wick, *Tabulation of Infrared Spectral Data,* Wiley, New York, **1977**.

E. Pretsch, T. Clere, J. Siebl, and W. Simon, *Tables of Spectral Data for Structure Determination of Organic Compounds,* Springer, English Edition, Berlin, **1983**.

Infrared Band Handbook, 2nd rev. ed., IFL/Plenum Press, Two Volumes, **1970**.

L. F. Johnson and W. C. Jankowski, *Carbon-13 NMR Spectra, a Collection of Assigned, Coded and Indexed Spectra,* Wiley, New York, **1972**.

J. G. Graselli and W. M. Ritchey, eds., *Atlas of Spectral Data and Physical Constants,* CRC Press, Cleveland, Ohio, **1975**.

P. L. Fuchs and C. A. Bunnel, *Carbon-13 NMR Based Organic Spectral Problems, Wiley, New York,* **1972**.

E. Breitmaier, G. Hass and W. Voelter, *Atlas of C-13 NMR Data,* Vols. 1-3, Heyden, Philadelphia, 1979 (3017 compounds).

Jeremy K. M. Sanders, Ewin C. Constable, and Brian K. Hunter, *Modern NMR Spectroscopy; A Workbook of Chemical Problems,* 2nd ed., Oxford University Press, Oxford, **1993**.

H. Dudeck and W. Dietrich, *Structural Elucidation by Modern NMR, A Workbook, 2nd ed., Springer-Verlag,* **1992**.

C. J. Pouchert and J. Behnke, *Aldrich Library of ^{13}C and ^{1}H FT-NMR Spectra, 300 MHz,* Aldrich Chemical Co., Milwaukee, WI, **1993**.

Varian Associates, *High Resolution NMR Spectra Catalogue,* 60 MHz, Varian Associates, Palo Alto, Vol. 1, 1962; Vol. 2, **1963**.

Chapter 5
Biosynthesis of Terpenoids: The Oldest Natural Products

5.1 Biochemical History

Terpenoids constitute the largest and structurally vastly diverse class of natural products with spectacular abundance (~30,000 known compounds) and have an extremely rich and long biochemical history. They are the oldest natural products known, as they are found to occur in fossils and sediments of different ages [1–4]. Terpenoids in various forms play an essential role in the biomembrane reinforcement [3].

5.1.1 Terpenoids as the Precursor of Cholesterol

Cholesterol is biosynthesized from lanosterol in animals and fungi and from cycloartenol in plants (vide Sect. 11.2.5). It is an important component of biomembrane (eukaryotic phospholipid membrane), and it participates in the reinforcement of membrane architecture by forming hydrogen bonds with its OH group and the head group of the phospholipids [3]. Both lanosterol and cycloartenol are formed from squalene 2,3S-oxide (vide Sect. 11.2.5).

5.1.2 Terpenoid Derived Diagenetic Entities [1–4]

An understanding of geochronology along with organic geochemical knowledge would certainly reveal many mysteries of evolution. The occurrence of terpenoid-derived diagenetic entities in the informative fossils, and ancient and recent sediments serve as the pointer to estimate to some extent the degree of maturation of the sediment, and to reconstruct the paleo-environment. It may be mentioned that the diterpenoid phytane (C_{20}-hydrocarbon) having the same carbon skeleton of phytol

S.K. Talapatra and B. Talapatra, *Chemistry of Plant Natural Products*,
DOI 10.1007/978-3-642-45410-3_5, © Springer-Verlag Berlin Heidelberg 2015

moiety of chlorophyll, and supposed to be generated from chlorophyll in a very early stage of degradation, has been found to be present in some fossils. Many other diagenetic entities from terpenoids belonging to higher terrestrial plants, in which they are biosynthesized, have been isolated from fossils and sediments [1–4] (vide Sects. 8.3.9 and 10.3). Their isolation and identification are the indicators of the presence of photosynthetic organisms during that contemporary time. Fossil records suggest that the precursor organisms in many cases were plants [3].

5.1.3   Ruzicka's Isoprene Hypothesis [5–7]

The carbon contents of terpenoids show that the skeletal carbon numbers are multiples of 5, *i.e.*, 5*x*, where $x = 1, 2, 3, 4, 5, 6$, etc.; hence, C_5 is the building block of terpenoids. Their known molecular structures, when formally dissected, appear to have a synchronized combination of C–C(C)–C–C units. Of the several C_5 potential candidates like isovaleric acid, dimethylacrylic acid, and isoprene, the last one has been found to be the obvious choice, as was initially surmised by Otto Wallach (NL 1910). Ruzicka (NL 1939) meaningfully hypothesized [5–7] the compositional regularity involving the isoprene unit in the structures of terpenoids. This hypothesis, termed as biogenetic *isoprene rule* [5, 6], is inspirational; it serves as a very useful guide and forms the basis for the biosynthetic and structural speculations of terpenoids.

5.1.4   Discovery of Isopentenyl Pyrophosphate (IPP): The Biological Isoprene Unit

Isoprene itself has been found to be biologically inactive. Search for biological equivalent of isoprene led to the isolation and identification of isopentenyl pyrophosphate (IPP)—"a long sought biological *isoprene unit*" [8] by Lynen (NL 1964) when he was studying the conversion of (*R*)-(+)-mevalonic acid into squalene [9]. Bloch (NL 1964) also made similar observations independently [10]. The identification of the true biologically active isoprene, IPP, paved the way for detailed biosynthetic studies of terpenoids at the enzymological level. The skeletal carbons of (*R*)-(+)-MVA pyrophosphate, the biogenetic precursor of IPP in the MVA pathway, have been obtained from three molecules of acetyl coenzyme A (Sect. 5.2.1).

5.1.5   Concept of Biogenesis and Biosynthesis

Prior to the discussion of the biosynthetic pathways, the terms *biosynthesis* and *biogenesis* need clarification. Plants synthesize natural products within the cells—the process is called *biosynthesis*, because the syntheses are carried out in the biosystem. But when we formulate the biosynthetic pathway of a structurally known natural product from the assumed precursor/s—simple primary metabolite/s (equivalent to synthon/s in a laboratory synthesis), following the mechanistic discipline and logic of organic chemistry, we call this assumed pathway as the *biogenetic pathway* of formation of that natural product or of the class it belongs to. *Biogenesis* may also be looked upon as the *retrobiosynthetic process*, the precursor being selected retrobiosynthetically, just as we select the synthon retrosynthetically for the laboratory synthesis of an organic compound of known structure. However, in case of retrobiosynthesis, the selection of the precursor is to be restricted within the repository of the known simple metabolites or some metabolites derived from them.

The assumed precursor is selectively labeled by an isotope or isotopes, and feeding experiments are done with it. If the isotope/s gets/get incorporated in the product/s at the expected location/s, it will be established that the synthesis has taken place in the biosystem (plant, in vivo) following the designed pathway. This *experimentally supported biogenesis* is also called *biosynthesis*.

However, unexpected incorporation at a location or locations in some cases compels the investigators to rationalize the biosynthesis following a different pathway, as is the case of non-MVA pathway (Sect. 5.3) for terpene biosynthesis, discussed later.

In summary, it may be said that all natural products of plant origin are biosynthesized in plant cells, and that a hypothetical biosynthetic pathway for the formation of a natural product or its class is referred to as *biogenesis*. When the hypothetical pathway is backed by experiment, it is also known as *biosynthesis*.

5.2   Mevalonic Acid Pathway

5.2.1   Acetyl Coenzyme A to Isopentenyl Pyrophosphate (IPP): Stereochemical Implications

Acetyl coenzyme A ($CH_3COSCoA$) is thought to be involved in its initial nucleophilic addition to biotin bound CO_2 to yield malonyl coenzyme A, generating a better activated nucleophilic methylene group, flanked between two electron withdrawing groups (Fig. 5.1). Malonyl coenzyme A then reacts with a second molecule of acetyl coenzyme A in a Claisen condensation fashion, followed by

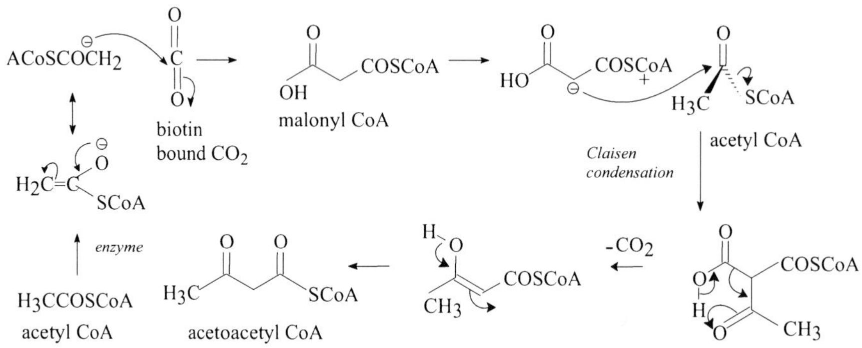

Fig. 5.1 Biosynthesis of acetoacetyl CoA *via* malonyl CoA

decarboxylation to yield acetoacetyl coenzyme A. However, it has been shown that malonyl coenzyme A, though involved in fatty acid and polyketide biosynthesis, is not accepted in the mevalonic acid pathway.

5.2.2   Bioformation of (R)-(+)-Mevalonic Acid

The *mevalonic acid pathway* uses three molecules of acetyl coenzyme A in two steps (Fig. 5.2). The first two molecules undergo Claisen condensation, mediated by acetoacetyl coenzyme A thiolase (AACT), to yield acetoacetyl coenzyme A. The third molecule of acetyl coenzyme A, prior to its participation in aldol condensation with acetoacetyl coenzyme A, gets attached to the catalyzing enzyme to form an acetylated enzyme [11, 12]. The latter then reacts with acetoacetyl coenzyme A to yield a chiral molecule. The enzyme-bound thioester group is then hydrolyzed to yield (S)-β-hydroxy-β-methylglutaryl coenzyme A (HMG-CoA). The thioester group of HMG-CoA is reduced to a hydroxymethyl group in two reductive steps ($-COSCoA \rightarrow -CHO \rightarrow -CH_2OH$) using two molecules of NADPH to yield (R)-(+)-mevalonic acid, [via (R)-mevalonic acid], being catalyzed by HMG-CoA reductase.[1]

The name mevalonic acid is derived from its systematic name β,-δ-dihydroxy-β-methylvaleric acid. It was discovered serendipitously by Karl Folkers and his associates. "The discovery of 3R-mevalonic acid was most

---

[1] This reaction is irreversible and is a committed key step in the biosynthesis of cholesterol (*vide* Sect. 11.2.5) in eukaryotes. Thus cholesterol formation can be regulated by synthesizing drugs which can act as the inhibitor in this enzymic reaction. Incidentally, it may be mentioned that in 2001 Bayer's cerivastatin, one of the cholesterol reducing statins that inhibits 3-hydroxy-3-methylglutaryl coenzyme A reductase involved in cholesterol synthesis, is withdrawn because of the increasing reports of its side effects.

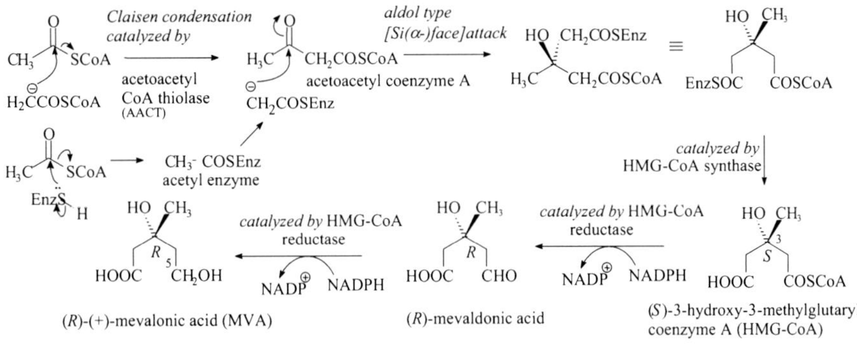

Fig. 5.2 Biosynthesis of (R)-(+)-mevalonic acid

important of all: ...and it is not utilized for any metabolic process other than that of isoprenoid biosynthesis" [13].

5.2.3 Conversion of (R)-(+)-MVA to IPP

The conversion of mevalonic acid to IPP requires three molecules of ATP, though the direct evidence of the involvement of the third molecule of ATP is not available from the intermediates formed on the way. Two molecules are involved in pyrophosphorylation of C_5–OH of mevalonic acid to yield mevalonic acid 5-pyrophosphate (MVAPP) in two steps, releasing two molecules of ADP (Fig. 5.3). It is assumed that during decarboxylative dehydration of MVAPP, the latter undergoes phosphorylation at the 3-OH during its elimination. This conjecture has been proved by labeling 3-OH oxygen by ^{18}O (i.e., using 3-^{18}OH), and the eliminated inorganic phosphate (Pi) is found to contain ^{18}O. Such phosphorylation may be realized by its interaction with γ-phosphate of the third ATP molecule (Fig. 5.3). The double bond of IPP is formed in a concerted elimination process and not by dehydration followed by decarboxylation. This type of enzymatic reactions is not observed elsewhere in enzyme chemistry. Here, we find the conversion of a C_6 compound to a C_5 one through the destruction of the chiral center. Further, it has been found that (R)-(+)-mevalonic acid is the obligatory precursor in this conversion. If the system is fed with (S)-(−)-mevalonic acid, it will not be accepted by the enzyme mevalonate kinase. Again, if (RS)-(±)-mevalonic acid is fed, the system will accept only the (R)-(+) variety while the (S)-(−) variety will remain unattended and will be eliminated as such from the system, being metabolically inert. Thus, this enzyme acts as "an efficient stereochemical filter." Cornforth (NL 1975) commented [14–16], "This is fortunate, since optical resolution of the synthetic acid is difficult: labeled mevalonic acids are usually racemic."

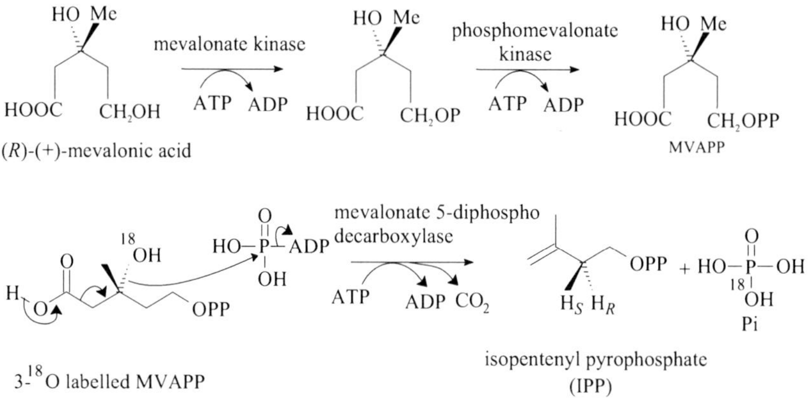

Fig. 5.3 Conversion of (R)-(+)-mevalonic acid to isopentenyl pyrophosphate (IPP)

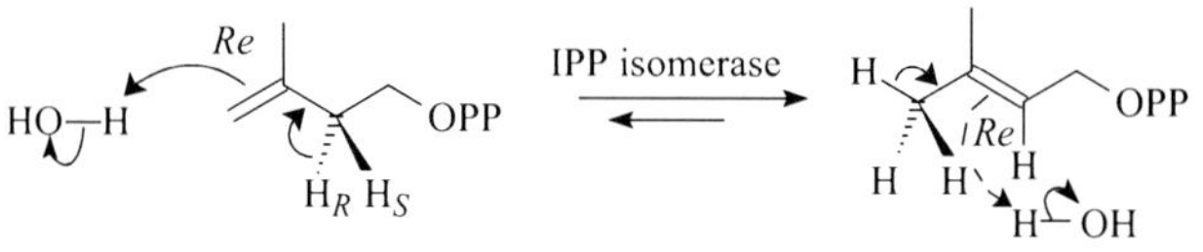

for *Re*, H_R and H_S see **Chapter 2**

(**Sections 2.9.5 and 2.9.6**)

Fig. 5.4 Isomerization of IPP to DMAPP

5.2.4 Isomerization of IPP to γ,γ-Dimethylallyl Pyrophosphate (DMAPP): Stereochemical Implications

The isomerase for converting IPP to DMAPP and the prenyl transferase that catalyze the head to tail condensation between IPP and DMAPP have been discovered by Lynen et al. [14] in 1959. The above isomerization is represented in Fig. 5.4. The equilibrium is inclined more towards γ,γ-dimethylallyl pyrophosphate (DMAPP), its double bond being more substituted. The stereochemistry of this isomerization will be defined by analyzing the following points: (a) which of the two prochiral allylic hydrogen atoms (H_R or H_S) (vide Sect. 2.7) will be eliminated to form a new double bond and (b) to which *enantiotopic face* the hydrogen will be accepted to saturate the existing double bond leading to the formation of a methyl group. In this study, isotope labeling of the key prochiral hydrogen by deuterium or tritium serves as the sure guide towards the stereochemical aspect of the process [13–21].

5.2.5 *Formation of Chiral Acetic Acid (7) [13, 17, 19] from 2-T-MVA Pyrophosphate (1)*

The aforesaid types of enzymatic experiments were elegantly designed and executed by Cornforth [13, 17, 19]. (2R,3R)-(2-T)-Mevalonic acid pyrophosphate (**1**) is subjected to similar decarboxylative dehydration in presence of the specific enzyme catalyst and ATP (cf. Fig. 5.3) to form the Z-isomer of T-labeled IPP (**2**) (Fig. 5.5). The stereochemistry of the decarboxylative dehydration step involving the elimination of the antiperiplanar OH [reacting with ATP to give the labeled IPP (**2**), ADP, and inorganic phosphate] and COOH groups is also shown. The labeled IPP (**2**) then undergoes isomerization in D_2O in the presence of IPP isomerase (cf. Fig. 5.4), which catalyzes the antarafacial 1,3-allylic rearrangement (consistent with the stereoelectronic control) to give the labeled DMAPP (**3**).

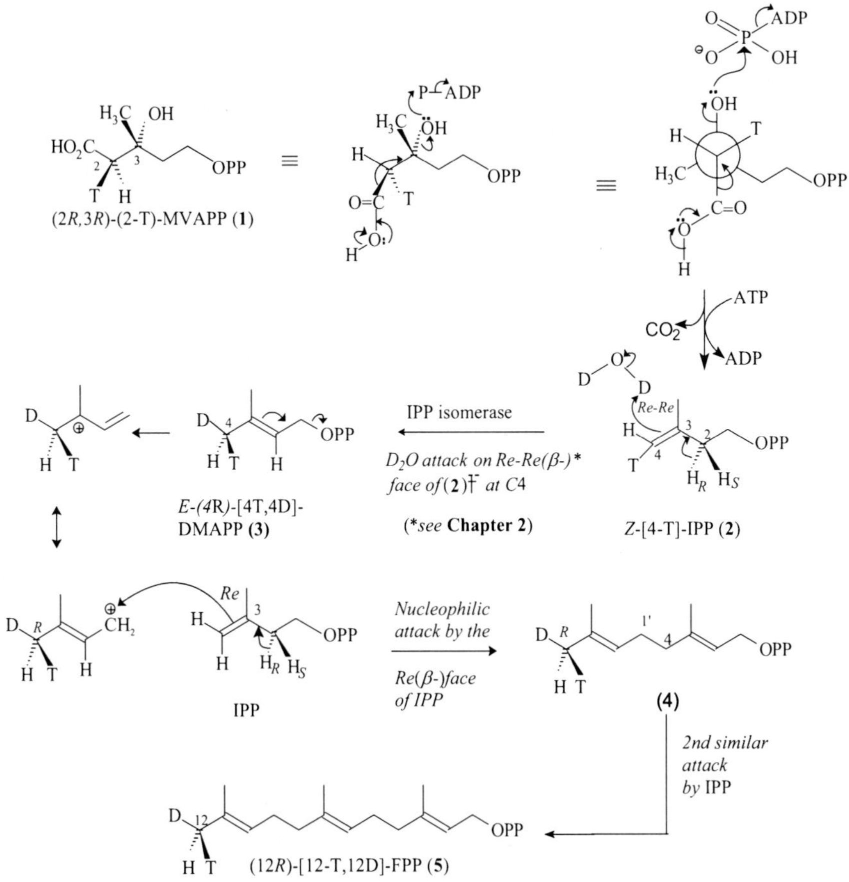

Fig. 5.5 Steric course of the reactions forming labeled IPP (**2**), DMAPP (**3**), and FPP (**5**)

Cornforth unambiguously established by elegant experiments (stated in the sequel) that the electrophilic addition of D^+ takes place at C_4 of (3) (Fig. 5.5) on the enantiotopic *Re–Re* (front) face (vide Sect. 2.7) [the priority sequence of ligands at C_3 being $C_2 > C_4 > CH_3$ (cf. Sect. 2.6.2.3) and that of ligands at C_2 being $C_3 > T > H$ of compound (1)] of the 3–4 double bond in a concerted manner with the loss of H_R from C_2. The geometry of the double bond in (3) becomes E, the generated doubly labeled (by T and D) methyl group (with 4R configuration) being *anti* to the CH_2OPP group. This provides an example of elucidation of prochirality in a biochemical reaction; here, loss of H_R from C_2 and concerted addition of D from the *Re–Re* face of (2) generates the doubly labeled (R)-methyl of DMAPP (3).

5.2.5.1 Formation of Isotopically Substituted Chiral Farnesyl Pyrophosphate (FPP) (5) and Chiral Acetic Acid (7)

The successful implementation of this plan required overcoming the reversibility of the isomerization responsible for the loss of stereochemical integrity of the chiral methyl group. This problem is solved by using a soluble enzyme fraction (containing a suitable prenyltransferase) from pig liver to remove DMAPP (3) by converting it, as soon as it is formed, into the chiral farnesyl pyrophosphate (FPP) (5) containing the labeled methyl group (at its head) by nucleophilic attacks by two molecules of IPP on C_1 of (3) and (4) successively. The corresponding alcohol, farnesol, obtained by enzymatic hydrolysis of (5) with alkaline phosphatase (Fig. 5.6), is subjected to ozonolysis to convert the labeled terminal isopropylidene group to labeled acetone (6). The latter undergoes iodoform reaction with $KI–I_2$ to afford chiral (R)-acetic acid (7) without giving an opportunity for proton exchange between the methyl group of (6) and the solvent. The R-configuration of (7) is established by a configurational correlation with a synthetic sample obtained by Cornforth [13, 17, 19] using a wholly chemical sequence and resolution procedure. Arigoni and Retey [18] used a part-enzymatic synthesis of the labeled (R)- and (S)-acetic acid, which appeared in *Nature* as the paper next to that of Cornforth. A few years later Arigoni [20] developed an interesting chemical synthesis of both (R) and (S) forms of chiral acetic acid. It is also observed that the configuration of the labeled acetic acid mainly generated via E-[4-T] IPP (the geometrical isomer of (2),

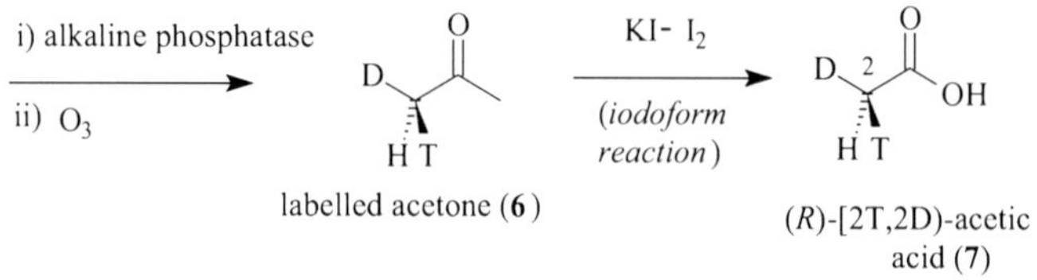

Fig. 5.6 Conversion of labeled farnesyl pyrophosphate (5) to (R)-[2T,2D]-acetic acid (7)

is S. It is thus established that the deuterium must have been added to the Re–Re (front) face of the double bond in Z-[4-T]-IPP (2).

It should be noted that chiral methyl compounds having no other dissymmetric part would exhibit very small optical rotation because of very little polarizability difference of H, D, and T. Moreover, even if all molecules of such a compound possess same chirality, they are usually mixed with majority of achiral methyl molecules. Hence, optical rotation measurement is clearly not practical.

5.2.5.2   The Absolute Configuration of [HDT]-Acetic Acid

The absolute configuration (R or S) of [HDT]-acetic acid produced is determined by bioassay methods using stereochemistry of known enzymatic reactions, *e.g.*, malate synthase reactions. [13, 17–21] In this method, the chiral acetic acid (as acetyl coenzyme A) and glyoxylate are condensed irreversibly on the enzyme malate synthase when they produce malates as shown in Fig. 5.7. This enzyme reaction is known to proceed with inversion of configuration. Tritium is displaced at the least rate, and deuterium is displaced three or four times less easily than protium when H, ^{2}H, and ^{3}H are present in the same methyl group (chiral methyl). The ratio of the abundance of the ^{2}H and ^{1}H in the products is equal to their kinetic isotope effects. Consequently, the products malates of this reaction with (R)-acetyl conenzyme A contains two tritiated species in unequal amounts, the major one retaining ^{2}H and the minor one retaining ^{1}H. Likewise, from (S)-acetyl coenzyme A two other tritiated malate species are formed (Fig. 5.7). The enzyme-catalyzed C–C bond formation takes place by attack on the Si-(α-) face of the glyoxylate to form 2S-malate in both cases (Fig. 5.7, last column).

Treatment of the malates with the enzyme fumarase affect the anti elimination of the elements of water. It is obvious from the figure that the major fumaric acid from the (R)-enantiomer will contain ^{3}H of the chiral methyl while the minor one is devoid of ^{3}H. In the case of the (S)-enantiomer, major fumaric acid will contain ^{2}H and the minor one carry the ^{3}H (Fig. 5.7).Thus, from the isotope analysis of the fumaric acids, the chirality of the unknown chiral acetic acid could be ascertained [13].

It is also found that catalyzed by the specific enzyme, ($3R,4R$)-[4-D]-MVA (8) is converted[2] through labeled IPP (9) and labeled DMAPP (10) into [2,6,10-D_3]-farnesyl pyrophosphate (11) which is enzymatically hydrolyzed to [2,6,10-D_3]-farnesol (12) (Fig. 5.8). The deuterium content of the isolated farnesol, as determined by mass spectroscopy, is consistent with the retention of the deuterium in (8)

---

[2] Actually ($3RS$, $4R$)-[4-D]-MVA was used since only $3R$ (and not $3S$) isomer got incorporated in the biosynthesis, as mentioned earlier.

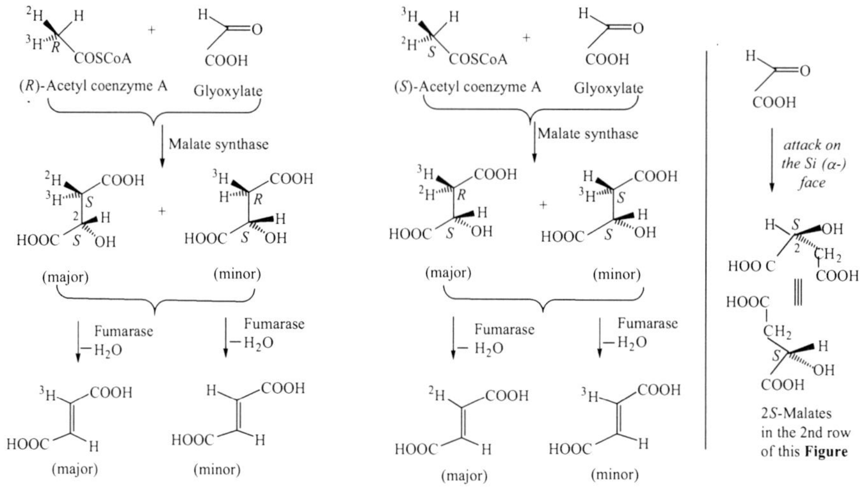

Fig. 5.7   A system for discriminating between the two enantiomeric chiral acetic acid [13]

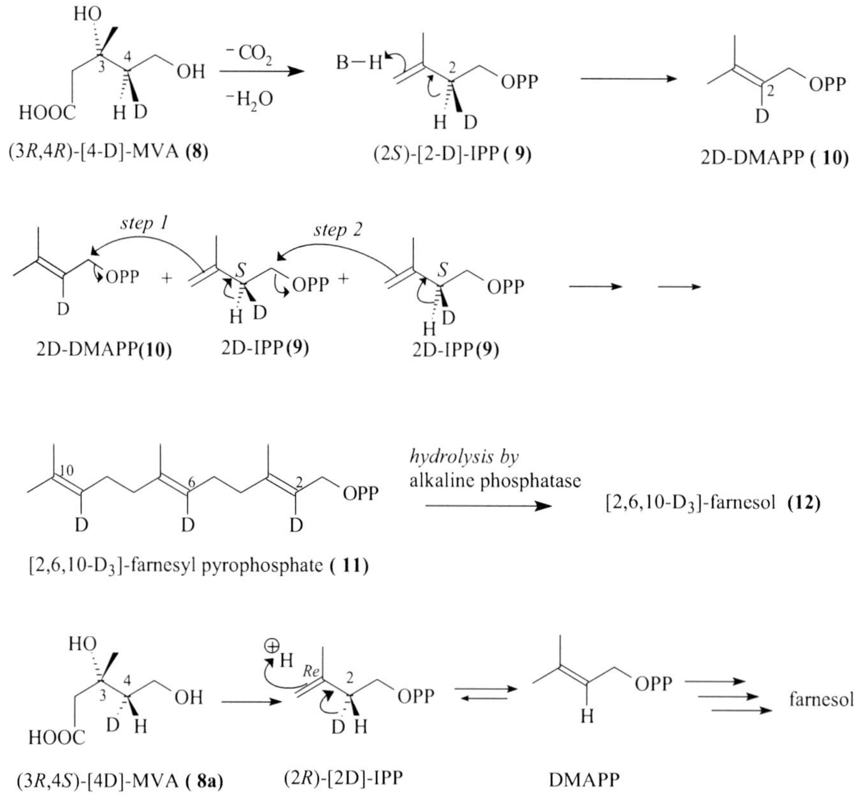

Fig. 5.8   Conversion of (3R,4R)-[4-D]-MVA (8) into [2,6,10-D₃]-farnesol (12) and of (3R,4S)-[4D]-MVA (8a) into farnesol (Cornforth's concept)

after decarboxylative dehydration and isomerization to give [2D]-DMAPP (**10**), proving that during isomerization, 2-H (α-) is lost. As expected (*3R*, *4S*)-[4-D]-MVA (**8a**) loses virtually all of its deuterium during conversion into farnesol through the same sequence of reactions,[3] confirming that 4α-D is lost from the rear face during the isomerization (Fig. 5.8).

5.3 Non-Mevalonoid (Rohmer) Pathway {1-Deoxy-D-Xylulose-5-Phosphate (DXP or DOXP): Mevalonate Independent [22–25]

The (*R*)-(+)-mevalonic acid, thought to be the obligatory precursor in the entire biogenetic pathway of terpenoids, remained unchallenged since the discovery (1950s) of this route in the plants. However, comparatively recently (starting from late 1980s) *poor incorporation* of the labeled acetate/mevalonate in the expected manner was observed during the biosynthetic studies of various terpenoids, especially in a number of monoterpenes like *geraniol* and diterpenes like *taxol*, *ginkolides*, etc. The labeling patterns were inconsistent. This inconsistency questioned the compatibility and the universal acceptability of the mevalonic acid pathway as the cornerstone for the biosynthesis of terpenoids. The authors working on geraniol and other monoterpenes [26] and taxol and taxane skeleton [27] thought for an alternative pathway discovered by Rohmer [22, 23]. He discovered that the C_5 framework of the isoprenoid gets its carbon atoms out of the condensation of a C_2 unit, derived from pyruvic acid, on the carbonyl of a triose, D-glyceraldehyde-3-phosphate. Two major groups [11, 22–25] worked on the non-mevalonate pathway for the formation of isoprenoids.

5.3.1 *Formation of DXP from Pyruvic Acid and D-Glyceraldehyde-3-Phosphate*

The newly discovered route involves 1-deoxy-D-xylulose-5-phosphate (DXP or DOXP) in the early stage of IPP formation. This C_5-sugar is generated (Fig. 5.9) by a transketolase catalyzed addition of C_2 unit ($CH_3CO–$), derived from pyruvate decarboxylation (vide Fig. 3.22), to D-glyceraldehyde-3-phosphate. The DXP thus formed is transformed into 2-C-methyl-D-erythritol-4-phosphate (MEP) by skeletal rearrangement as stated in Sect. 5.3.2.

---

[3] Here the (*3RS*, *4S*) compounds may be used giving farnesol, devoid of deuterium.

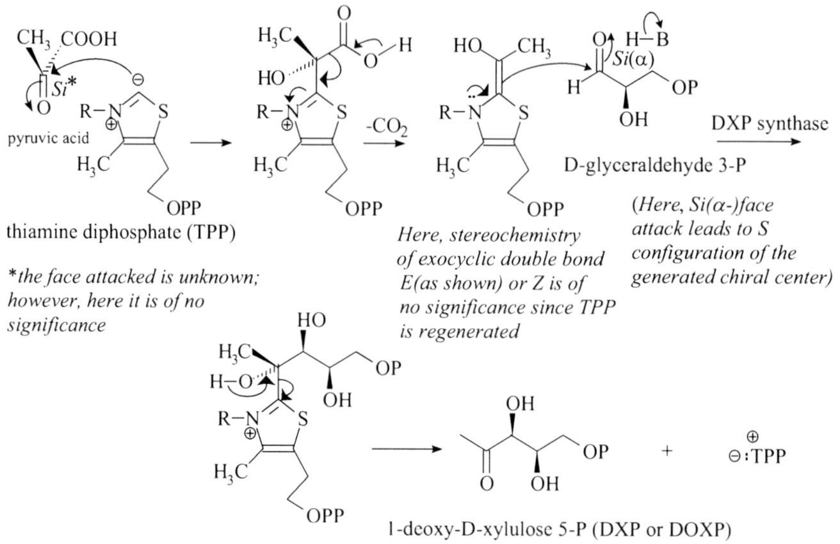

Fig. 5.9 Formation of 1-deoxy-D-xylulose 5-P (DXP or DOXP) (*non-mevalonoid pathway*)

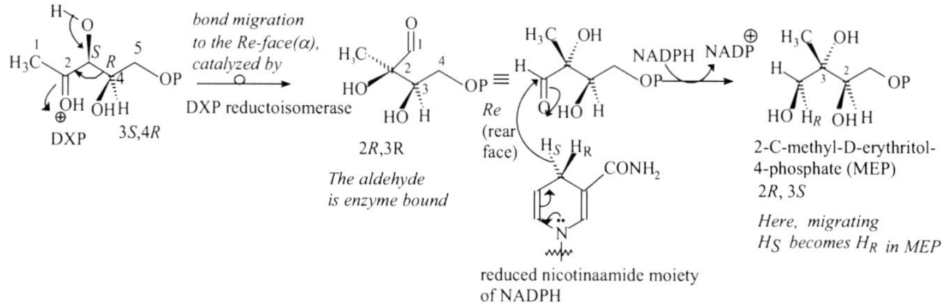

Fig. 5.10 Conversion of DXP to MEP

5.3.2   Conversion of DXP to 2-C-Methyl-D-Erythritol-4-Phosphate (MEP or ME4P)

The following transposition reaction is catalyzed by DXP-reductoisomerase. It resembles the pinacol–pinacolone type rearrangement. The aldehyde remains bound to the enzyme and uses NADPH for its reduction to 2-C-methyl-D-erythritol-4-phosphate (MEP) (Fig. 5.10). It is inconsequential to know which of the two diastereotopic hydrogens from the reduced nicotinamide moiety of NADPH effects the reduction, since the –CHO group is reduced to –CH$_2$OH group and no chiral center is generated (Fig. 5.10). However, through selectively labeled NADPH it has been demonstrated that H$_R$ is delivered to the *Re* (α-) face of the aldehyde and occupies *pro-R* position in the product MEP as shown.

5.3.3 Conversion of MEP to IPP Via a Cyclic Diphosphate: Its Ring Opening, Followed by Repeated Reduction and Dehydration [11]

MEP is converted into 2-C-methyl-D-erythritol-2,4-cyclodiphosphate (14) through the intermediacy of 4-diphosphocytidyl-2-methyl-D-erythritol (12) and its 2-phosphate (13) as shown in Fig. 5.11. The cyclic diphosphate then opens up at C_2 and undergoes loss of the hydroxymethine proton at C_3 to form the 3-keto-4-diphosphate (16) via its enol (15). At this point, it is pertinent to mention that C_3–H

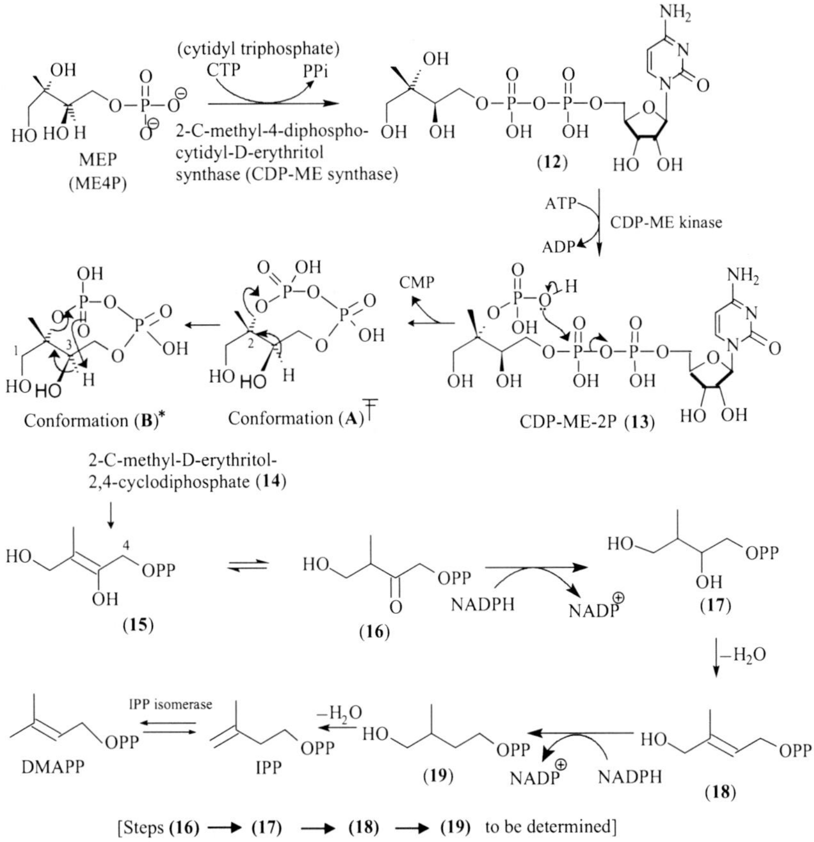

[Steps (16) ⟶ (17) ⟶ (18) ⟶ (19) to be determined]

Ŧ Since in this conformation 3-H and 2-O bonds are cis (syn-clinal), it is not expected to undergo E-2 elimination involving these bonds, as shown.

*Cyclic cis-elimination can take place in this conformation having 3-H and 2-O bonds in cis orientation (syn-clinal) and O of P=O and 3-H in proximity, as shown.

Fig. 5.11 The non-mevalonate/DXP pathway of isoprenoid biosynthesis starting from MEP

as well as C_2–O bonds are *syn*-clinal and hence E_2-elimination, as shown in conformation (**A**) [11], may not be stereoelectronically compatible.

However, *we think that it may undergo ring opening through the participation of an enzyme, or by cis-elimination through a 6-membered cyclic transition state as shown in conformation* (**B**). The next biological events are reduction and dehydration [11] to form the linear diphosphate (**18**). The latter upon repetition of the same biological events yields IPP, which is in equilibrium with DMAPP in presence of IPP isomerase.

An alternative hypothetical mechanism [11] for the conversion of (14) **to** (18)
It has been shown that all the hydrogen atoms of the C–H bonds present in the cyclic diphosphate (**14**) are preserved in the IPP and DMAPP molecules. This observation suggests [11] that (**14**) is converted to the linear diphosphate (**18**) through a series of reactions (Fig. 5.12) by direct involvement of the enzyme (this possibility has already been mentioned), and not through a proton loss at C_3 as shown in conformations (**A**) and (**B**) (Fig. 5.11). In the first step, the cofactor NADP$^+$ of the enzyme abstracts the hydride from the methylene of the –CH_2OH group of (**14**) and is converted to NADPH. The resulting aldehyde makes its adjacent C_2 more electrophilic due to its inductive effect. Nucleophilic attack on the quaternary $C_2{}^*$ by the thiol ($-SH$) group or thiolate ($-S^-$) anion of the enzyme resident cysteine displaces the diphosphate moiety and thus opens up the ring. The terminal aldehyde provides the necessary electron sink and enolizes through the cleavage of the newly formed S–C bond with the concomitant formation of a disulphide bond. The resulting enolate anion triggers the β-elimination of the water molecule. In the last step, the NADPH delivers back the same hydride to C_1 that it got in the first step and thus restores its original oxidation state, leading to the formation of (**18**) (Fig. 5.12). Reductive cleavage of the disulphide bond is necessary to regenerate the cysteine moiety of the enzyme for multiple turnovers.

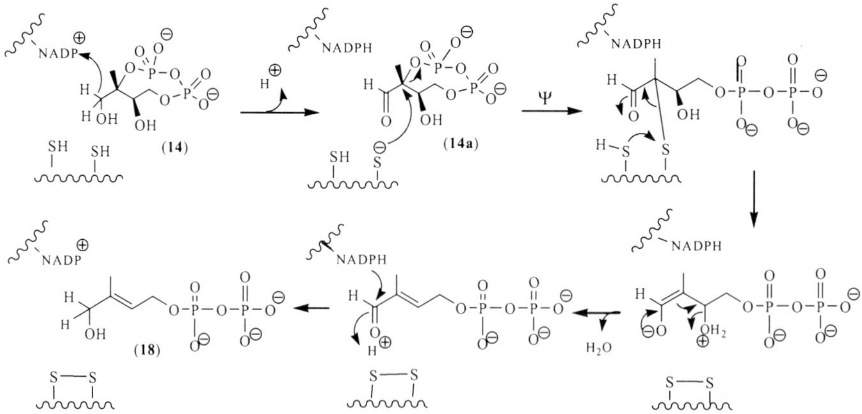

Fig. 5.12 Hypothetical mechanism for the conversion of (**14**) to (**18**) in the DXP pathway

However, it appears to us that such an S_N2 attack on a crowded quaternary carbon with a bulky nucleophile is less likely, and an S_N1 opening of the cyclic diphosphate (14a) with the release of the pyrophosphate anion at C_2 even after complexation with Mg^{2+} is also less likely because of the unstable nature of the resulting carbocation due to its attachment to the electron-deficient C_1 (see step ψ in Fig. 5.12).

An alternative hypothetical mechanism has also been suggested [11] that involves a radical process in each step and constitutes the same sequence of reactions as shown in Fig. 5.12. In this pathway, the hydrogen radical abstractor thiolyl radical $(-S^{\bullet})$ and the hydrogen radical donor thiol $(-SH)$ group of the enzyme have been used instead of the hydride acceptor $NADP^+$ and the hydride donor NADPH. Thus, here bond formation and bond breaking in each step are initiated by a radical and are homolytic.

It has been shown [25, 26] that the IPP and DMAPP molecules are also independently produced in the DOXP pathway (Fig. 5.13) since the depletion of the gene specifying the IPP isomerase does not affect the non-mevalonate pathway biosynthesis. Thus, DMAPP and IPP may be formed independently from (18), also in absence of IPP isomerase via the resonance stabilized allylic carbocation (18a) (Fig. 5.13). Arigoni and others [24] have shown that in the conversion of DXP to IPP and DMAPP *isp*-proteins *viz.*, *isp* C, *isp* D, *isp* E, *isp* F, *isp* G, and *isp* H— which are specified by the corresponding genes—are involved. They have carried out these experiments in vitro also. We thus find a dual origin of IPP. The discovery of the DXP pathway is a major shift in our understanding of terpenoid biosynthesis. The mevalonate-independent pathway is operative in the plastids (chloroplasts).

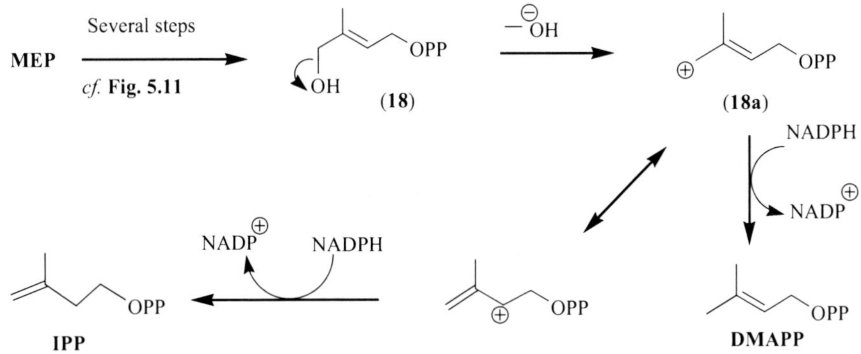

Fig. 5.13 Independent routes for the conversion of MEP to DMAPP and IPP in absence of IPP isomerase

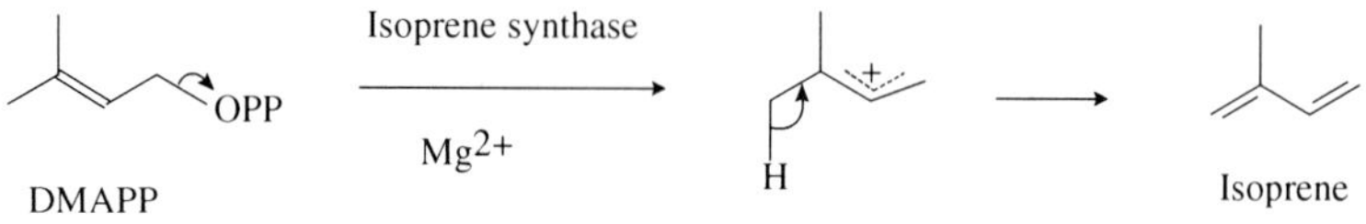

Fig. 5.14 Biosynthesis of isoprene from DMAPP

5.3.4 Emission of Isoprene from Some Plants [28]

Incidentally, it may be mentioned that leaves of some plants, such as oak, poplar, spruce, etc., are capable of producing the *hemiterpene* isoprene (b.p. 33°), which escapes as gas into the air during the day at a temperature $\geq$ 30°. As a consequence, certain percentages of photosynthetically fixed carbon in those plants are emitted as isoprene. It is synthesized in plastids from the non-mevalonoid derived DMAPP via the action of *isoprene synthase* which catalyzes the elimination of the pyrophosphate [11, 28] in presence of Mg^{++} (Fig. 5.14). The global isoprene emission is estimated to be about as high as the global methane emission. There are indications that low amounts of isoprene provide a stabilizing effect against high temperature damage of photosynthetic membrane. In fact, the blue haze sometimes observed over forests at high temperatures is due to emission and accumulation of isoprene from trees.

Incidentally, it may be mentioned that C.G. Williams discovered isoprene in 1860 by dry distillation of rubber and named it without any explanation. Though isoprene does not occur in the free state in nature, several natural C_5 compounds, called *hemiterpenes*, e.g., isopentenyl pyrophosphate (IPP), γ,γ-dimethylallyl pyrophosphate (DMAPP), isoamyl alcohol (3-methylbutan-1-ol), isovaleric acid (3-methylbutanoic acid), tiglic acid (*E*-2-methylbut-2-enoic acid), angelic acid (*Z*-2-methylbut-2-enoic acid), and β-furoic acid (furan-3-carboxylic acid) are known to have the isoprene skeleton.

5.4 Dual Origin of IPP: Labeling Patterns of IPP Derived from Labeled Glucose by Two Different Routes

$^{13}C_1$-D-Glucose has been chosen to show these two distinct pathways (mevalonate and mevalonate-independent DXP pathways) for the terpenoid biosynthesis, since the former forms metabolites in two different routes [29] to yield IPP in which the distribution of the ^{13}C label indicates the path followed (Fig. 5.15).

The MVA pathway is operative in the cytoplasm and is responsible for the biosynthesis of sesquiterpenes, triterpenes, and sterols, whereas DXP pathway is reported to take place in the chloroplasts requiring photosynthetic machinery and is responsible for the formation of various groups of monoterpenes and diterpenes. Such compartmentation (*i.e.*, different subcellular locations) is based on incorporation of various labeled compounds in feeding experiments. However, exchanges

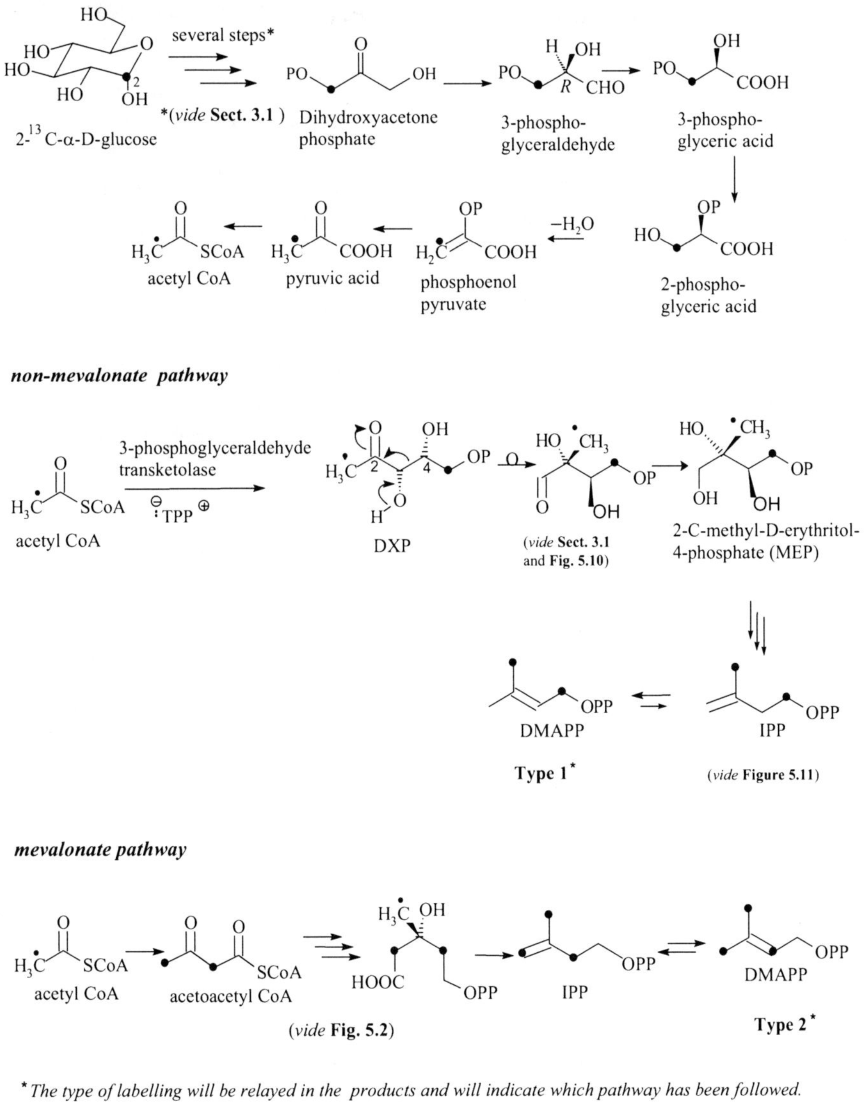

non-mevalonate pathway

mevalonate pathway

The type of labelling will be relayed in the products and will indicate which pathway has been followed.

Fig. 5.15   Different labeling patterns of IPP and DMAPP obtained (feeding 2-^{13}C-α-D-glucose) in non-mevalonate (Type 1) and mevalonate (Type-2) pathways

[11, 25, 29] of common intermediates such as IPP, GPP, and FPP might occur between the two compartments as described by the "Cross-Talk Theory" [29]. Perhaps, prior to the discovery of the non-mevalonate pathway, this was the reason for the incorrect conclusion that the mevalonate pathway was the universal pathway for terpene biosynthesis [11].

5.5 Chain Elongation in Terpenes (Prenyl Transfer) [30]

Chain elongation from C_5 to C_{10}, C_{15} and beyond is an important biosynthetic phenomenon. The prenyl chain elongation is catalyzed by a set of enzymes called '*prenyltransferase*'. Sixteen such different prenyl transferases with different catalytic functions, *i.e.*, chain length specifications, have been identified [30] till 1998. Chain elongation process in terpenes is summarized in a general way in Fig. 5.16.

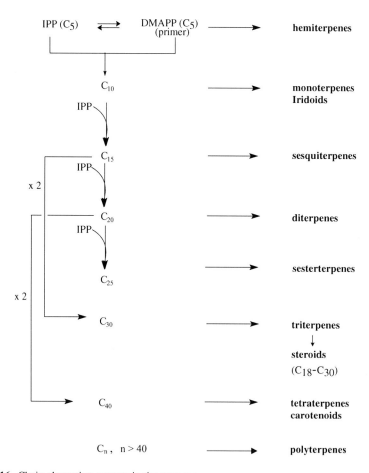

Fig. 5.16 Chain elongation process in the terpenes

5.5.1 Cornforth's Concept and Its Modification by Poulter and Rilling

The mechanism of chain elongation has been extensively studied by Cornforth [8, 14–16] and subsequently by Poulter and Rilling [8, 31]. According to Cornforth, the double bond of IPP is activated by the attack of a nucleophile from the appropriate enzyme XE to trigger its attack on the C_1 of DMAPP from which the pyrophosphate group is eliminated. This is followed by the loss of H_R from the original IPP part and release of XE. The whole process is suprafacial and kinetically S_N2' type (Fig. 5.17). The S_N2' type process demands inversion at C_1 of DMAPP moiety which is observed according to this proposition. However, a time lag between the bond breaking and bond formation was evident later on. This observation is not mechanistically compatible with S_N2' type kinetics. This unexpected experimental result prompted a reinvestigation of the existing mechanism for chain elongation.

Poulter and Rilling [8, 31] modified Cornforth's concept of chain elongation. According to them, the rate determining step in the enzymatic prenyl transfer involves the ionization of the allylic pyrophosphate substrate (S_N1 type) followed by a nucleophilic attack of IPP with the stereospecific removal of its H_R from the allylic methylene in tandem to form a C–C and a C=C bonds (Fig. 5.18). The S_N1 type metal catalyzed generation of dimethylallyl carbocation (DMA$^+$) and its electrophilic addition to IPP with concomitant stereospecific elimination of H_R can take care of the time lag between the bond breaking and bond formation. However, S_N1 type kinetics expects a randomization at C_1 of the added DMA $^{(+)}$, and loss of its stereochemical integrity, and thus apparently cannot explain the observed inversion at C_1 of the DMA$^{(+)}$ carbocation. The expected randomization is prevented by the rotational barrier (~28 kcal/mole) around C_1–C_2 bond of the allylic carbocation [8, 31]. In the absence of definitive idea about the topology of the enzyme–substrate conformation, it has been assumed that the inversion is imposed by the compatible prenyltransferase enzyme itself which holds the DMA$^{(+)}$ carbocation and the proximate cosubstrate IPP in such cooperative conformations that the electrophilic addition to C_4 at the *Re*-face (at C_3) of IPP takes place on the face of the DMA$^{(+)}$ opposite to that from which the OPP departs

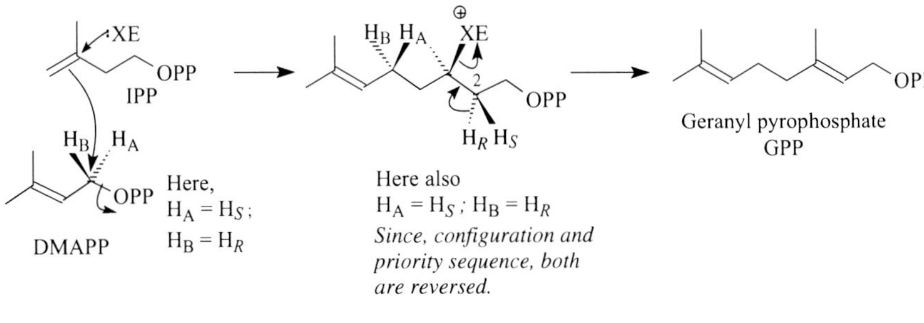

Fig. 5.17 Cornforth's concept of chain elongation (Formation of GPP)

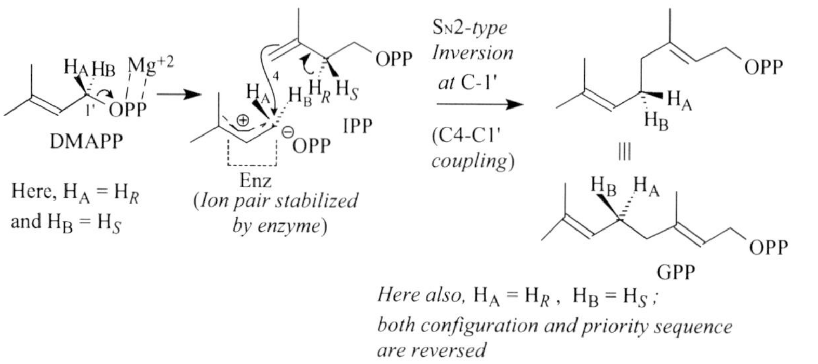

Fig. 5.18 Modification of chain elongation pathway. Formation of GPP (Rilling and Poulter)

(Fig. 5.18). So, basically the process remains suprafacial, as originally suggested by Cornforth, 2-H_R and 1'-OPP being eliminated from the same face.

5.5.2   Formation of C_{10}, C_{15}, C_{20}, C_{25}, C_{30}, and C_{40} Linear Terpenoids and Natural Rubber

The general paradigm associated with the elongation of isoprenoid chain (C_{10}, C_{15}, C_{20}, and C_{25}) involves repetition of the above-mentioned processes: (1) ionization of the allyl pyrophosphate, (2) addition of IPP to the carbocation thus formed at the face opposite to that from which 1'-OPP is eliminated, and (3) stereospecific elimination of 2-H_R from the IPP during its each addition. In summary, ionization, condensation, and elimination take place stereospecifically for each C_5-addition, i.e., 1'–4 coupling reaction [31] leading to the linear terpenoids (see Fig. 5.19).

Here, it is assumed that the atoms C_1' (of DMAPP serving as the electrophilic center) and C_4 (of IPP) are adjacent in the enzyme–substrate complex. Moreover, the formation of the E-C_2, C_3 double bond in IPP part requires that the dihedral angle (ω) between H_R–C_2 and C_3=C_4 bonds should ideally approach 180° (antiperiplanar), pushing the double bond to the β-face (Fig. 5.18). Since OPP binds with Mg^{+2}, it is assumed that the metal is assisting in the catalytic step. The enzyme can ionize the allylic substrate without anchimeric assistance by IPP. The electrophilic attack at the allylic carbocation takes place to the Re (or, β-) face of the C_3, C_4 double bond of IPP, and H_R from the α-face is lost. Hence, the allylic substrate must be located on the Re face of the trigonal center C_3. Molecular orbital calculations and ^{13}C NMR data give definitive evidence for the formation of C_{10}-allyl cation during FPP biosynthesis [32].

Rubber Biosynthesis [33] Natural rubber is produced in a simple extension of the ubiquitous isoprenoid pathway from IPP and allylic-PPs, namely DMAPP, GPP, and FPP. The allylic-PPs are synthesized by soluble enzymes in the same cytosol

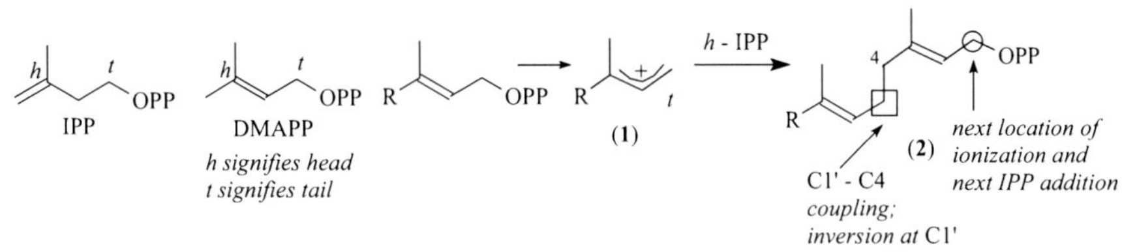

(C5) - *h* - *t* DMAPP	+	(C5) - *h* - *t* IPP	R = CH3	$\xrightarrow{t-h}$	C10	$\longrightarrow$	geranyl PP (GPP)
(C10) - *h* - *t* GPP	+	(C5) - *h* - *t* IPP	R = C5 unit	$\xrightarrow{t-h}$	C15	$\longrightarrow$	farnesyl PP (FPP)
(C15) - *h* - *t* FPP	+	(C5) - *h* - *t* IPP	R = C10 unit	$\xrightarrow{t-h}$	C20	$\longrightarrow$	geranylgeranyl PP (GGPP)
(C20) - *h* - *t* GGP	+	(C5) - *h* - *t* IPP	R = C15 unit	$\xrightarrow{t-h}$	C25	$\longrightarrow$	farnesylgeranyl PP (FGPP)
(C15) - *h* - *t* FPP	+	*t* - *h* - (C15) FPP	C15 + C15	$\xrightarrow{t-t}$	C30	$\longrightarrow$	presqualene PP (PSPP)
							$\downarrow$
							squalene
(C20) - *h* - *t*	+	*t* - *h* - (C20)	C20 + C20	$\xrightarrow{t-t}$	C40	$\longrightarrow$	carotenoids

Fig. 5.19 Mode of attachments in chain elongation leading to C10, C15, C20, C30, and C40 compounds

compartment as the rubber transferase enzyme in the rubber producing evolution-arily divergent plant species (mainly *Hevea brasiliensis*) and could be used by the enzyme. The polymerization (chain elongation) of IPP, probably derived from the mevalonate pathway (Sect. 5.2), is brought about by rubber transferase present in *cytosol*. However, IPP derived from non-mevalonate DOXP pathway (Sect. 5.3) may diffuse from the *plastids* and join the *cytosolic* pool for the rubber biosynthesis [33].

5.5.3   Ruzicka's Nomenclature: Terpene or Terpenoid, Head and Tail Parts of Acyclic Terpene Pyrophosphates (Diphosphates)

At this stage, it is necessary to explain the terminologies 'terpene' and 'terpenoid'. The term terpene and for that matter, different classes of terpenes (mono-, di-, tri-, etc.) are considered to be synonymous with terpenoids—monoterpenoids, diterpenoids, triterpenoids, etc. in the literature.

Ruzicka held the opinion [5–7] of retaining the term terpene and the classes as monoterpenes, diterpenes, etc. According to him, **'oid'** has been introduced in terpene compounds by analogy with **'steroids'** "which denotes a group of compounds having an irregularly varying number of carbon atoms. In the terpenic field, the expression terpenoid should be reserved by analogy for compounds in which the number of carbon atoms varies irregularly, in contrast to the terpenes proper, where the number of carbon atoms is always a multiple of five" [5–7]—a unifying feature. However, the term 'terpenoid' has deeply percolated through the veins of terpene chemistry literature, and it is difficult to avoid the **'oid'** nomenclature. Hence, both the terminologies *'terpenoid'* and *'terpene'* have been used in this book.

Further, in defining the **'head'** and **'tail'** parts of the acyclic terpene pyrophosphates (C_5, C_{10}, C_{15}, etc.) (Fig. 5.19), some confusion has been observed in the literature. Most of the earlier literatures label the terminal carbon atom carrying the pyrophosphate group as the **tail** part, while the isopropenyl/isopropylene group as the **head** part of the molecule. *In this book, this widely used convention of Ruzicka will be followed.* However, others [34–36] are also justified in the sense that *the functional group (OPP) containing carbon (C_1) should be recognized as the* **head** part of the molecule.

Here, two other points should be noted. (i) The importance of pyrophosphate group (OPP) lies in its ability to form a complex with Mg^{+2} or Mn^{+2}; this complexation confers better leaving group ability to OPP, a situation needed for generating a carbocation for chain elongation and cyclization. (ii) In the literature, some authors have designated OPP as 'diphosphate', which seems to be proper, instead of 'pyrophosphate'. The term 'pyrophosphate' has been used in the earlier literatures and is also in use in the current literature along with the 'diphosphate'; we have mostly used the old term pyrophosphate, and sometimes the other term 'diphosphate'.

5.5.4 Formation of Squalene (C_{30}) via Presqualene Pyrophosphate (Involving Cyclopropane/Cyclobutane Ring Opening)

The chain elongation to C_{30} takes place by a tail to tail union of two FPP molecules (Fig. 5.18), and not by an IPP addition to a C_{25} terpene residue. As discussed in the sequel, squalene ($C_{30}H_{50}$) is biosynthesized in a multistep pathway (Fig. 5.20). As expected, a stable intermediate, presqualene pyrophosphate [37–41] (PSPP) is formed in the key step. PSPP was first isolated [38] by Rilling in 1966. Presqualene alcohol was synthesized in 1971 by Rilling et al. [40] and two other groups [42, 43].

Squalene (**6**) is a highly branched (tentacled with methyls) symmetrical hydrocarbon having a C_2-axis. It results from an asymmetric process, for which a number of mechanisms has been proposed and debated. A simple mechanism for the biosynthesis of squalene from farnesyl PP has been proposed on the basis of a π-complex (between farnesyl carbocation, $F^{(+)}$ (**1**) and FPP) (*The numbering of the structures in this figure is independent of the earlier figures*) theory and supported

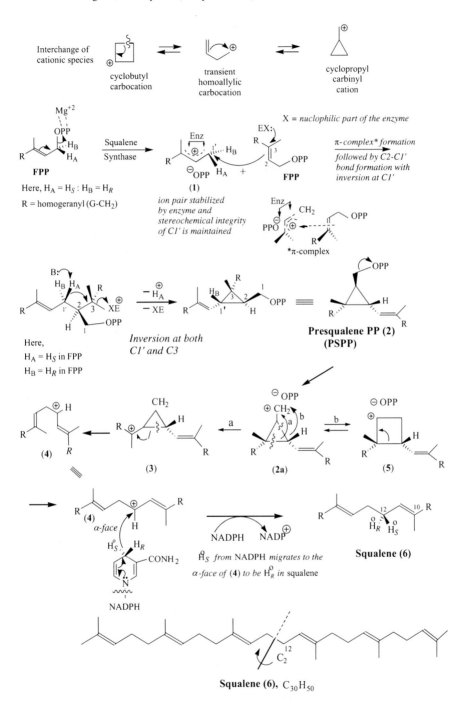

Fig. 5.20 Biosynthesis of squalene from FPP *via* presqualene PP **(2)**

by theoretical calculations [37]. The mechanism, which transpires from these calculations, is based on the ionization of one of the two participating FPP molecules *in tail to tail condensation*. The pyrophosphate group leaves with the help of Mg^{+2}; the generated resonance stabilized farnesyl cation (1) is then trapped by the enzyme squalene synthase, which can function only in presence of Mg^{+2}. It is unlikely that the free carbocation $F^{(+)}$, in the absolute sense, participates in this reaction. In all probability, it remains as an ion pair (1) with the counter ion $^{(-)}OPP$, the ion pair being stabilized by an appropriate enzyme. The trapping of the farnesyl carbocation by enzyme is followed by an S_N2 type interaction with the second molecule of FPP to form a π-complex, an outcome of theoretical calculations [37]. Real S_N2 type reactions are rare in biological systems. Rapid stereospecific deprotonation $\{-H_A(-H_R)\}$ gives PSPP (2), the structure and stereochemistry of which has been confirmed [38–43] by degradation and an unambiguous synthesis of its alcohol. Since one of the allylic methylene protons is stereospecifically eliminated, the deprotonation should be a fast process, being brought about by a base of appropriate strength. Literature provides such examples of cyclopropane ring formation [35]. Of the eight possible stereoisomers (four diastereomers and their optical antipodes) of PSPP (2) possessing three unlike chiral centers, only one with absolute stereochemistry defined as (2) is formed (as shown mechanistically in Fig. 5.20) and is found to be biologically active. Incorporation of various mixtures of isomers of (2) into yeast homogenates containing NADPH results in the efficient conversion of the active isomer (2) to squalene (6) through the steps delineated below.

It has been suggested [38–41, 44] that PSPP (2) suffers ionization (S_N1 type) to form the primary cyclopropylcarbinyl carbocation (2a) which rearranges to yield the cyclopropylmethyl tertiary carbocation (3) by the bond migration shown by arrow *a* in preference to the secondary cyclobutyl carbocation (5) [35] by the bond migration shown by arrow *b*. The rearrangement of (2a) to form the carbocation (3) is expected both from the theory of ring closing kinetics and formation of the more stable carbocation. The cyclopropane ring opening of (3), as shown, forms the more stable squalene allylic carbocation (4), which might also have been produced from the cyclobutyl carbocation (5), if at all formed, by migration of the bond shown. The conversion (2) to (4) via (2a) and (3) is catalyzed by squalene synthase. That the tertiary carbocation (2a) and not the cyclobutyl carbocation (5) is involved in this conversion has been proved by using a nitrogen analogue of cyclopropyl-carbinyl cation as an inhibitor to squalene formation [37].

However, in view of the cation exchange as shown in the equilibriums (top of the Fig. 5.20), the opening of the cyclopropane ring prior to its conversion to squalene may also be considered to proceed via cyclobutyl carbocation (5) [35, 44]. Use of chemically and structurally equivalent inhibitors may show whether cyclobutyl carbocation is involved in the ring-opening path.

As has been delineated in Fig. 5.20, Cornforth has defined *the following steric course of the bioreactions leading to the formation of squalene from two molecules of FPP* by selective labeling experiments [14–16].

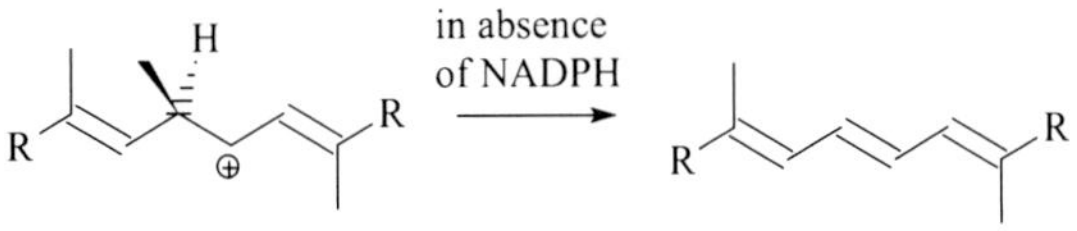

(4) $R = C_{11}H_{19}$

vide **Figure 5.17**

Fig. 5.21 Dimer of FPP with an additional double bond

(i) The enantiotopic hydrogen H_S from C_1 of one FPP molecule is lost and is eventually replaced by a hydrogen from NADPH (the last step).

(ii) The diastereotopic hydrogen H_S at C_4 of NADPH is delivered to the carbocation (4) at C_{12} and it occupies the *pro-R* position in squalene (6), *i.e.*, it becomes H_R in squalene.

It is further concluded that the delivery of the hydride from NADPH takes place at the final step, and in absence of NADPH, the squalene cation is quenched through the loss of adjacent proton to yield a dimer, 12,13-dehydrosqualene, $C_{30}H_{48}$ with one extra double bond (Fig. 5.21).

5.5.5   Different Phases of Terpenoid Biosynthesis

All terpenoids are oligomeric/polymeric products of IPP and DMAPP in which DMAPP serves as the primer.

Terpenoid biosynthesis may be sequentially divided into the following four phases, brought about by specific enzymes present in the plant cells:

(i) The origin of the isoprene units, IPP and DMAPP (mevalonoid and non-mevalonoid pathways).

(ii) Stepwise linear polymerization of the isoprene units to form the acyclic polyprenyl precursors like geranyl PP, farnesyl PP, geranylgeranyl PP, and farnesylgeranyl PP; however, squalene (via presqualene) and C_{40} terpenoids are formed by the dimerization of C_{15} (FPP) and C_{20} (GGPP) linear terpenoids, respectively.

(iii) Proper folding, cyclization, rearrangements of these polyprenyl precursors, and also sometimes shedding from and addition to the carbon skeletons to form an array of innumerable terpenoids of diverse skeleton patterns.

(iv) In most cases, functionalization (e.g., various regiospecific and/or stereospecific hydroxylations, oxidations, reductions, etc.) on the skeletal frames in presence of the specific enzymes leading to varied natural terpenoids.

In the present chapter, the first two phases have been discussed in depth. The other two phases will be discussed in the respective chapters for different classes of terpenoids, as mentioned in the Contents. We have discussed the phases (iii) and (iv) in the biogenesis/biosynthesis of monoterpenoids (Chap. 6) sesquiterpenoids (Chap. 7), diterpenoids (Chap. 8), sesterterpenoids (Chap. 9), triterpenoids (Chap. 10), steroids (Chap. 11), and carotenoids (Chap. 12), starting from the corresponding acyclic precursors included in this chapter.

References

1. P. A. Schenck and J. W. de Leeuw, Molecular Organic Geochemistry, in *Handb. Environ. Chem.* (Part B, Nat. Environ. Biogeochem. Cycles), (Ed.: Otto Hutzinger), Springer, Berlin, **1982**, pp. 111–129.
2. J. R. Maxwell, C. T. Pillinger and G. Eglinton, Organic Geochemistry, *Quart. Rev.,* **1971**, 25, 571–629
3. G. Ourisson and Y. Nakatani, The Terpenoid Theory of the Origin of Cellular Life: the Evolution of Terpenoids to Cholesterol, *Chemistry & Biology,* **1994**, *1*, 11–22.
4. I. R. Hills, E. V.Whitehead, and D. E. Anders, J. J. Cummins and W. E. Robinson, An Optically Active Triterpane, Gammacerane in Green River, Colorado, Oil Shale Bitumen, *Chem. Commun.* (Chem. Soc., London), **1966**, 752–754.
5. L. Ruzicka, The Isoprene Rule and the Biogenesis of Terpenic Compounds, *Experientia,* **1953**, *9*, 357–367
6. A. Eschenmoser, L. Ruzicka, O. Jeger, and D. Arigoni, Zur Kenntnis der Triterpene Eine stereochemische Interpretation der biogenetischen Isoprenregal bei den Triterpenen, *Helv. Chim. Acta,* **1955**, 38, 1890–1904.
7. Albert Eschenmoser and Duillio Arigoni, Revisited after 50 years: 'The Stereochemical Interpretation of the Biogenetic 'Isoprene Rule for the Triterpenes', *Helv. Chim. Acta,* **2005**, 88, 3011–3050.
8. D. E. Cane, The Stereochemistry of Allylic Pyrophosphate Metabolism, *Tetrahedron* (Report No. 82), **1980**, *36*, 1109–1159; and the references cited therein.
9. F. Lynen, H. Eggerer, U. Henning and I. Kessel, Farnesyl-pyrophosphat und 3-Methyl-butenyl-Δ_3-pyrophosphat, Die biologischen Vorstufen des Squalens. Zur Biosynthese der Terpene, III, *Angew. Chem.,* **1958**, 70, 738–742.
10. S. Chaykin, J. Law, A. H. Phillips, T. T. Tehen and K. Bloch, Phosphorylated Intermediates in the Synthesis of Squalene, *Proc. Natl. Acad. Sci. (USA),* **1958**, *44*, 998–1004.
11. Paul M. Dewick, The Biosynthesis of C_5-C_{25} Terpenoid Compounds, *Nat. Prod. Rep.,* **2002**, *19*, 181–222; pertinent page 182, 184, 185 and the references cited therein.
12. Felix Rohdich, Stephan Hecht, Adelbart Bacher and Wolfgang Eisenreich, Deoxyxylulose Phosphate Pathway of Isoprenoid Biosynthesis. Discovery and Function of ispDEFGH Genes and Their Cognate Enzymes, Pure Appl. Chem., 2003, 75, 393–405; and the references cited.
13. J. W. Cornforth, Enzymes and Stereochemistry, *Tetrahedron,* **1974**, *30*, 1515–1524., pertinent page 1520.
14. B. W. Agranoff, H. Eggerer, U. Henning, and F. Lynen, Isopentenyl Pyrophosphate Isomerase, *J Am. Chem. Soc.,* **1959**, *81*, 1254–1255
15. J. W. Cornforth, Terpenoid Biosynthesis, *Chem. Brit.,* **1968**, *4*, 102–106
16. J.W. Cornforth, The Logic of Working with Enzymes, *Chem. Soc. Revs.,* **1973**, *2*, 1–20.
17. J. W. Cornforth, J.W. Redmond, H. Eggerer, W. Buckel and Christine Gutschew, Asymmetric Methyl Groups and the Mechanism of Malate Synthase, *Nature,* **1969**, *221*, 1212–1213
18. J. Luthy, J. Retey and D. Arigoni, Synthesis of Chiral Methyl Group, *Nature,* **1969**, *221*, 1213–1214
19. J. W. Cornforth, The Chiral Methyl Group - Its Biochemical Significance, *Chem. Brit.,* **1970**, *6*, 431-436.

20. C. A. Townsend, T. Scholl and D. Arigoni, A New Synthesis of Chiral Acetic Acid, *J. Chem. Soc. Chem. Comm.*, **1975**, 921–922

21. Alan Fersht, *Enzyme Structure and Mechanism*, W.H. Freeman and Company, New York, 2nd ed., **1985**, pertinent pages 232–235.

22. Michel Rohmer, M'hamed Knani, Pascale Simonin, Bertrand Sutter and Hermann Sahm, Isoprenoid Biosynthesis in Bacteria a Novel Pathway for the Early Steps Leading to Isopentenyl Diphosphate, *Biochem. J.*, **1993**, *295*, 517–524.

23. Michel Rohmer, Myrium Seemann, Silke Horbach, Stephanie Bringer-Meyer and Hermann Sahm, Glyceraldehyde 3-Phosphate and Pyruvate as Precursors of Isoprene Units in an Alternative Non-Mevalonate Pathway for Terpenoid Biosynthesis, *J. Am. Chem. Soc.*, **1996**, *118*, 2564–2566.

24. Petra Adam, Stephan Hecht, Wolfgang Eisenreich, Johannes Kaiser, Tobias Gräwert, Duilio Arigoni, Adelbert Bacher and Felix Rohdich, Biosynthesis of Terpenes: Studies on 1-Hydroxy-2-methyl-2-(E)-butenyl-4-diphosphate Reductase, *Proc. Natl. Acad. Sci.*, **2002**, *99*, 12103–12113.

25. Manuel Rodriguez-Concepcion, Nareiso Campos, Luisa Maria Lois, Carlos Maldonado, Jean-Francois Hoeffler, Catherine Grosdemange-Billiard, Michel Rohmer and Albert Boronat, Genetic Evidence of Branching in the Isoprenoid Pathway for the Production of Isopentenyl Diphosphate and Dimethylallyl Diphosphate in *Escherichia coli*, *FEBS Letters*, **2000**, *473*, 328–332.

26. Gérard J. Martin, Sophie Lavoine-Hanneguelle, Francoise Mabon and Maryvonne L. Martin, The Fellowship of Natural Abundance 2H-isotopomers of Monoterpenes, *Phytochemistry*, **2004**, *65*, 2815–2831.

27. Wolfgang Eisenreich, Birgitta Menhard, Peter J. Haylands, Meinhart H. Zenk and Adelbert Bacher, Studies on the Biosynthesis of Taxol: The Taxane Carbon Skeleton is not of Mevalonoid Origin, *Proc. Natl. Acad. Sci.*, **1996**, *93*, 6431–6436.

28. Hans-Walter Heldt, *Plant Biochemistry and Molecular Biology*, Oxford University Press, **1997**, pp. 365–367.

29. Juraithip Wungsintaweekul and Wanchai De-Eknamkul, Biosynthesis of Plaunotol in *Croton stellatopilosus* Proceeds *via* the Deoxyxylulose Phosphate Pathway, *Tetrahedron Lett.*, **2005**, *46*, 2125–2128; and references cited therein.

30. Kyozu Ogura and Tanetoshi Koyama, Enzymatic Aspects of Chain Elongation, *Chem. Rev.*, **1996**, *96*, 1263–1276.

31. C. Dale Poulter and Hans C. Rilling, The Prenyl Transfer Reaction. Enzymatic and Mechanistic Studies of the 1′-4 Coupling Reactions in the Terpene Biosynthetic Pathway, *Acc. Chem. Res.*, **1978**, *11*, 307–313; and the references cited therein.

32. Yoshikazu Hirage, Diana I. Ito, Tetsuya Sayo, Shinji Ohta and Takayaki Suga, ^{13}C NMR Detection of Delocalized C_{10}-allylic Cation in the Biosynthesis of Farnesyl Diphosphate, *J. Chem. Soc. Chem. Commun.*, **1994**, 1057–1058.

33. Katrina Cornish, Biochemistry of Natural Rubber, a Vital Raw Material, Emphasizing Biosynthetic Rate, Molecular Weight and Compartmentalization, in Evolutionarily Divergent Plant Species, *Nat. Prod. Rep.*, **2001**, *18*, 182–189.

34. J. Retey and J. A. Robinson, Stereospecificity in Organic Chemistry on Enzymology, Verlag Chemie, Basel, **1982**, Chapter 9, pp. 209–243, (References, pp. 311–312).

35. Ludger A. Wessjohann and Wolfgang Brendt, Biosynthesis and Metabolism of Cyclopropane Rings in Natural Compounds, *Chem. Rev.*, **2003**, *103*, 1625–1647; pertinent page 1631.

36. C. Dale Poulter, Larry L. Marshs, John M. Hughes, J. Craig Argyle, Dennis M. Satterwhite, Robyn J. Goodfellow and Scott G. Moesinger, Model Studies of the Biosynthesis of Non-Head-to-Tail Terpenes. Rearrangement of the Chrysanthenyl System, *J. Am. Chem. Soc.*, **1977**, *99*, 3816–3823.

37. Michael J. S. Dewar and James M. Ruiz, Mechanism of the Biosynthesis of Squalene from Farnesyl Pyrophosphate, *Tetrahedron*, **1987**, *43*, 2661–2674.

38. W. W. Epstein and H. C. Rilling, Studies on the Mechanism of Squalene Biosynthesis, *J. Biol. Chem.*, **1970**, *245*, 4597–4605; and the cited references.
39. L. J. Mulheirn and P. J. Ramm, The Biosynthesis of Sterols, *Chem. Soc. Revs.*, **1972**, *1*, 259–291.
40. L. J. Altma, R. C. Kowerski and H. C. Rilling, Synthesis and Conversion of Presqualene Alcohol to Squalene, *J. Am. Chem Soc.*, **1971**, *93*, 1782–1783.
41. H. C. Rilling, C. Dale Poulter, W. W. Epstein and Brent Larsen, Studies on the Mechanism of Squalene Biosynthesis. Presqualene Pyrophosphate, Stereochemistry and a Mechanism for Its Conversion to Squalene, *J. Am. Chem. Soc.*, **1971**, *93*, 1783–1785
42. R. V. M. Campbell, L. Crombie and G. Pallenden, Synthesis of Presqualene Alcohol, *J. Chem. Soc. Chem. Commun.*, **1971**, 218–219
43. R.M. Coates and W.H. Robinson, Stereoselective Total Syntheses of (±) Presqualene Alcohol, *J. Am. Chem. Soc.*, **1971**, *93*, 1785–1786.
44. E. E. van Tamelen and M. A. Schwartz, Mechanism of Presqualene Pyrophosphate – Squalene Biosynthesis, *J. Am. Chem. Soc.*, **1971**, *93*, 1780–1782.

Further Reading

T.A. Geissman and D.H.G. Crout, *Organic Chemistry of Secondary Plant Metabolism*, Freeman, Cooper & Company, San Francisco, **1969**.

James B. Hendrickson, *The Molecules of Nature*, W. A. Benzamin, Inc. New York, **1965**.

Paul M. Dewick, *Medicinal Natural Products – A Biosynthetic Approach*, John Wiley & Sons, New York, **1997**, 465 pages; 2nd edn., **2001**, 520 pages; 3rd edn., **2009**, 539 pages.

J. W. Cornforth, The Logic of Working with Enzymes (Robert Robinson Lecture), *Chemical Society Reviews*, **1973**, *2*, 1-20.

J. W. Cornforth in *Structural and Functional Aspects of Enzyme Catalysis*, Eds. H. Eggerer and R. Huber, Springer-Verlag, **1981**.

Chapter 6
Monoterpenoids (C_{10})

6.1 Geranyl pyrophosphate, the Universal Precursor of Monoterpenoids

Geranyl pyrophosphate (GPP) is the first member of the elongated isoprenoids (Fig. 6.1). It is the universal precursor molecule for the biosynthetic formation of all acyclic and cyclic monoterpenes (C_{10}-terpenoids). It is formed by the combination of DMAPP and IPP, catalyzed by the appropriate prenyl transferase (see Sect. 5.5.1, Figs. 5.14 and 5.15).

6.1.1 Biosynthetic Formation of C_{10}-Acyclic Terpenes

It has been shown [1, 2] with labeled experiments that labeling always resides in the IPP portion of the labeled GPP molecule. On the basis of this observation, Battersby suggested that in the feeding experiments the externally added labeled **IPP** is trapped by the pool of endogenous DMAPP prior to IPP's isomerization to the labeled **DMAPP**.

A major problem in studying the biosynthesis of monoterpenes using labeled precursors is the difficulty in achieving the incorporation in the product. This may be explained by the fact that *the living organisms probably prevent the exogenous labeled precursors to reach the sites of synthesis.* However, with some selective cooperative plants, some biosynthetic studies on monoterpenoids have been possible. Thus, incorporation of labeled geraniol by way of secologanin in the terpene-indole alkaloids (see Sect. 6.1.6), suggested that the IPP and DMAPP part of geraniol are biosynthesized [3, 4] in the non-mevalonate (DOXP) pathway (see Sect. 5.3). Further, from the study of site-specific natural abundance of hydrogen isotopes ratio measurements [5] using deuterium NMR on several monoterpenes (e.g., geraniol) the hydrogen genealogy of the parent hydrocarbon could be elucidated. This finding will predict *the derivation of GPP from the condensation of IPP*

S.K. Talapatra and B. Talapatra, *Chemistry of Plant Natural Products*,
DOI 10.1007/978-3-642-45410-3_6, © Springer-Verlag Berlin Heidelberg 2015

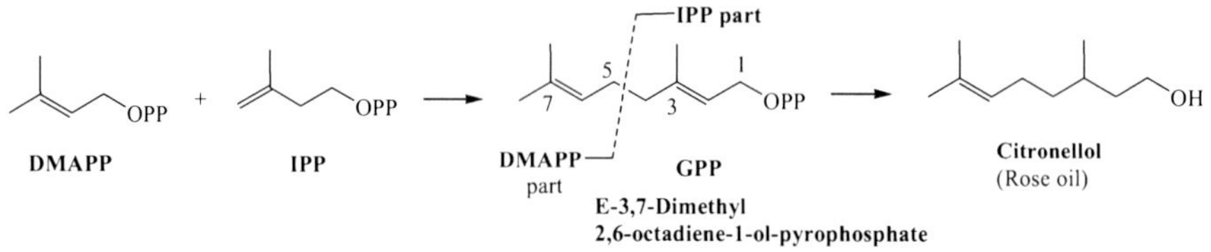

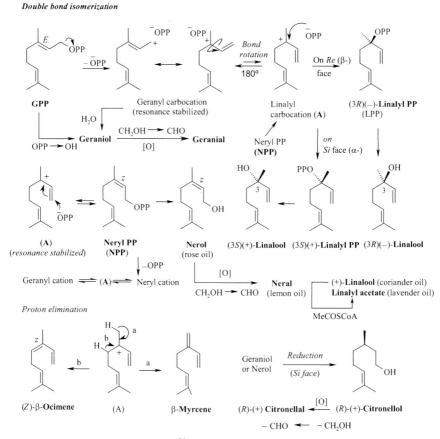

Fig. 6.1 Formation of geranyl pyrophosphate (GPP) and citronellol

Double bond isomerization

Fig. 6.2 Acyclic monoterpenes arising from geranyl pyrophosphate (GPP)

(i) *Every carbocation carries its counterpart* $^{(-)}$*OPP anion*
(ii) *The corresponding oxidation products of the alcohols like aldehydes (e.g., geranial, neral, citronellal) double bond*
     reduced products (e.g., citronellol), and proton elimination products (e.g., β-ocimene and β-myrecene) are also
     known natural products.

and independent/isomerized DMAPP and their affiliation to DOXP pathway. That is to say, "….isotope ratios at individual sites in geraniol can be traced back to the corresponding sites in GPP, then to sites of the IPP and DMAPP building blocks, then to pyruvate and glyceraldehydes-3-phosphate DOXP active molecules and finally to the carbohydrate photosynthetic precursor" [5].

The formation of acyclic monoterpenoids from GPP is delineated in Fig. 6.2.

6.1.2   Biosynthetic Formation of Cyclic Monoterpenes

The generated carbocations participate in the cation–olefin (cation-π) cyclization to yield mainly natural monocyclic and bicyclic and sometimes tricyclic monoterpenoids. The geranyl carbocation is capable of yielding several cyclic skeletal patterns under catalytic influences of suitable enzymes. Those of common occurrences are shown in Fig. 6.3. The plausible pathways for their formation are shown in the sequel (Figs. 6.4 and 6.5). Mechanistically the processes involve 1,2 anti-periplanar hydride or methyl or alkyl shift by way of a C–C bond shift to allow a new site for the carbocation and finally quenching the carbocation either through the loss of an appropriate proton or by the addition of a nucleophile like hydroxyl group. *The Wagner–Meerwein rearrangement initiated by the carbocation and stereochemically suited bond migration is almost a monopoly of terpenoid chemistry, especially of lower terpenoids. The enzymes provide compatible folding and environment for cyclization, which along with the cation chemistry is responsible for products formation.* Sometimes, a number of products in different amounts are formed by a single enzyme.

The topicity, i.e., the active site geometry, and the prochirality (*vide* Sect. 2.9) play a significant role in the biosynthesis of natural products, especially of terpenoids. The face at the location of attack and the prochirality of the hydrogen atom involved in the processes are important which in the presence of specific enzymes lead to the stereospecific products. The concerned C1 methylene hydrogen atoms in geraniol are enantiotopic which become diastereotopic in chiral environment in association with the specific enzymes or more specially in chiral cyclic monoterpenes.

6.1.2.1   Monocyclic Monoterpenes. Menthane Skeleton

The plausible pathways for the formation of some common monocyclic monoterpenes are delineated in Fig. 6.4. The significance of the double bond isomerization (E, Z) in cyclization is not obvious, as either isomer may serve as the precursor of the cyclized products, e.g., the **linalyl** carbocation (**A**) may be generated from **geranyl pyrophosphate** (E) or **neryl pyrophosphate** (Z), as shown (Fig. 6.2).

It is to be noted in Fig. 6.4 that (−)-menthol may be conceptually biosynthesized from GPP (i) via (4R)-limonyl carbocation (**B′**) or (ii) via (4S)-limomyl carbocation (**B**). The sequence of the biosynthesis of (4S)-(−)-limonene via (3S)(+)-LPP and

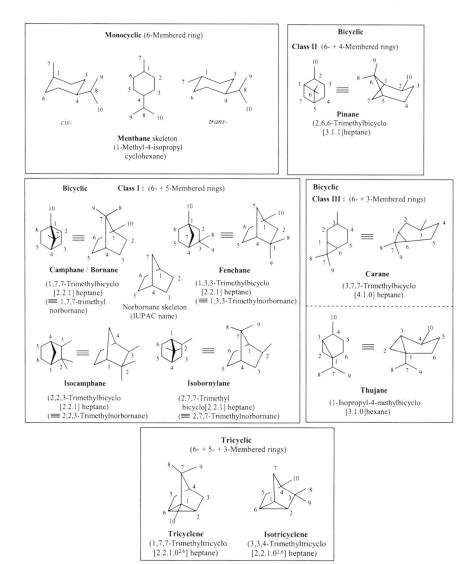

Fig. 6.3 Common skeletal patterns of cyclic monoterpenes and their nomenclatures

the sequence of the biosynthesis of (4R)-(+)-limonene via (3R)-(−)-LPP, caused by the mirror image (superposable) folding of the GPP in presence of specific enzymes, have been delineated in Fig. 6.6.

A small group of aromatic compounds, e.g., **p-cymene** and the phenol derivatives **thymol** and **carvacrol** (Fig. 6.4), occur in thyme (*Thymus vulgaris*, Labiatae). They are produced in nature from monoterpenoids, rather than via *acetate or*

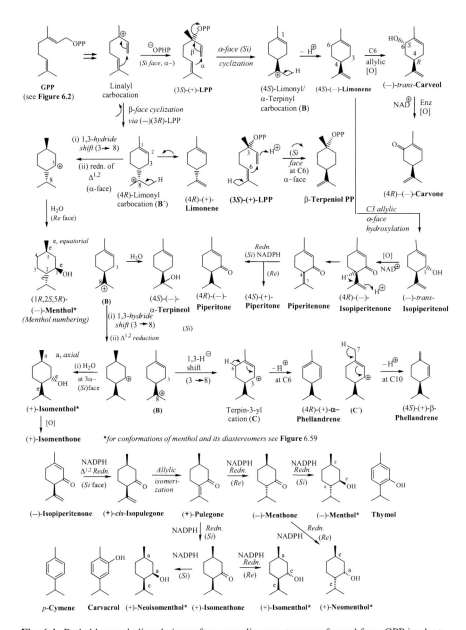

Fig. 6.4 Probable metabolic relations of monocyclic monoterpenes formed from GPP in plants

shikimate pathway, which is the more common route for aromatics. They have the same carbon skeleton as that of monocyclic monoterpenes and are structurally related to menthone, carvone—suggesting that additional dehydrogenation reactions have taken place.

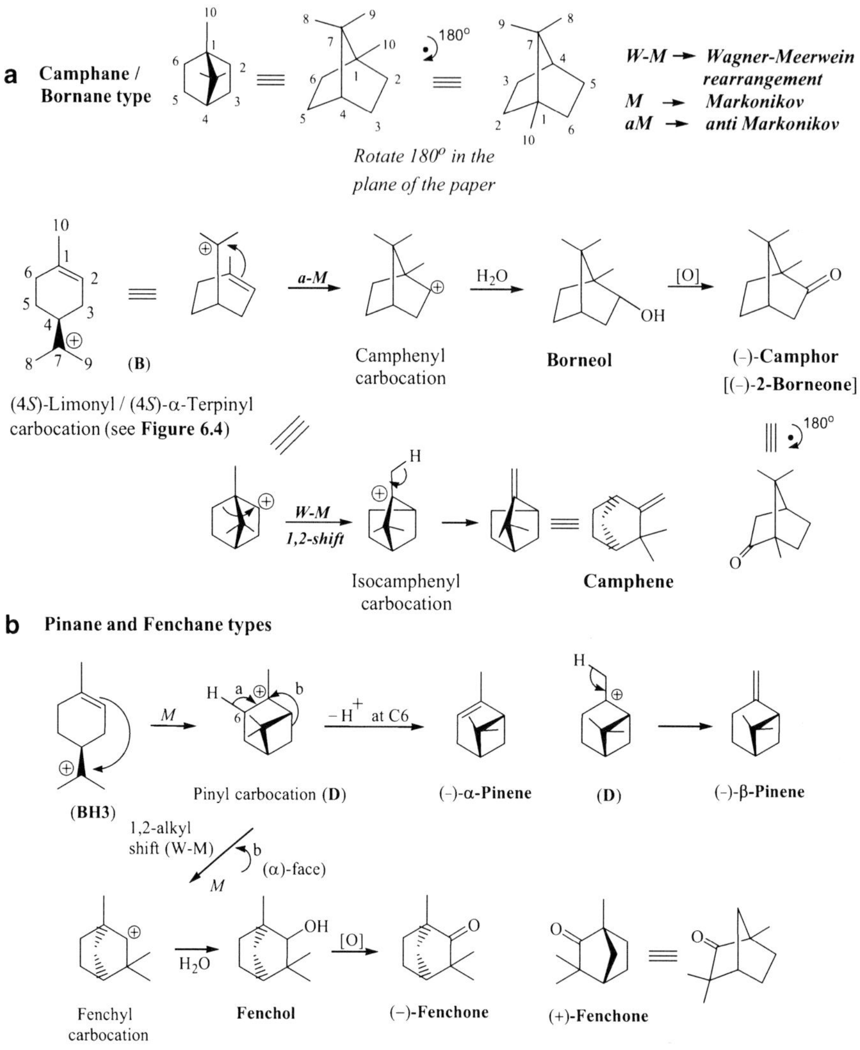

**a Camphane /
   Bornane type**

*Rotate 180° in the
plane of the paper*

W-M → *Wagner-Meerwein
       rearrangement*
M → *Markonikov*
aM → *anti Markonikov*

(B)

(4S)-Limonyl / (4S)-α-Terpinyl
carbocation (see **Figure 6.4**)

Camphenyl
carbocation

Borneol

(–)-**Camphor**
[(–)-**2-Borneone**]

W-M
1,2-shift

Isocamphenyl
carbocation

Camphene

b   Pinane and Fenchane types

(BH3)

1,2-alkyl
shift (W-M)

(α)-face)

Pinyl carbocation (**D**)

(–)-α-Pinene

(**D**)

(–)-β-Pinene

Fenchyl
carbocation

Fenchol

(–)-**Fenchone**

(+)-**Fenchone**

Fig. 6.5 (continued)

6.1.2.2   Bicyclic Monoterpenes. Formation of Some Familiar Skeletons (Camphane/Bornane, Pinane, Fenchane, Carane, and Thujane)

All these different types of bicyclic monoterpenes and their biosynthetic pathways in the presence of specific enzymes from (4S)-limonyl carbocation (**B**) are shown in Fig. 6.5. In the presence of the necessary specific enzymes the enantiomers of these bicyclic monoterpenes may be formed via the enantiomeric (4R)-limonyl carbocation (**B′**) (Fig. 6.4).

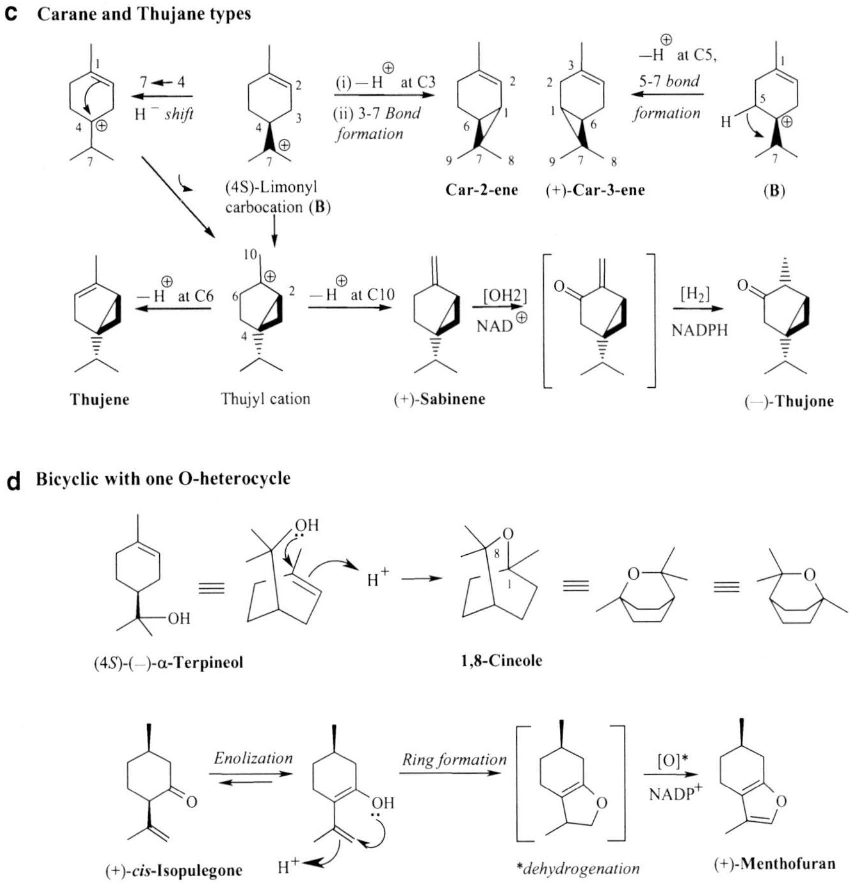

C   Carane and Thujane types

d   Bicyclic with one O-heterocycle

Fig. 6.5   Formation of bicyclic monoterpenes from limonyl carbocation

Generation of two *bicyclic monoterpenes with one O-heterocycle*, (i) from 4-S-α-terpineol by acid catalyzed ring formation (**1,8-cineole**) and (ii) from *cis*-isopulegone by acid-catalyzed ring formation followed by oxidation (**menthofuran**), are shown in Fig. 6.5.

6.1.3   Occurrence of Monoterpene Enantiomers, their Biological Responses and Biosynthesis

Individual enzyme systems present in a particular organism/plant cells control the folding of the substrate molecule, formation of a particular carbocation, and specific cyclization leading to the structure and stereochemistry of the final product. Most

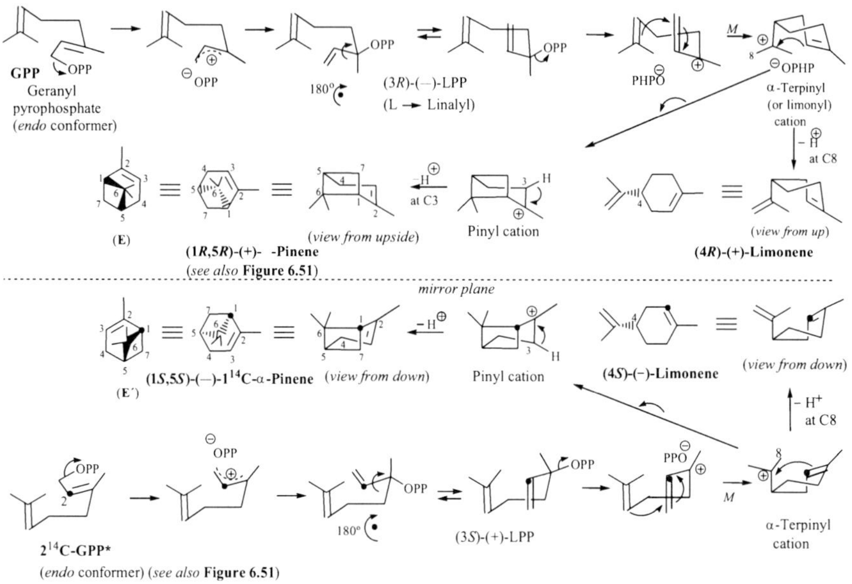

GPP
Geranyl
pyrophosphate
(*endo* conformer)

(3R)-(−)-LPP
(L → Linalyl)

PHPO

α-Terpinyl
(or limonyl)
cation

(E)

(1R,5R)-(+)- -Pinene
(*see also* **Figure 6.51**)

(*view from upside*) Pinyl cation

(4R)-(+)-Limonene (*view from up*)

mirror plane

(1S,5S)-(−)-1^{14}C-α-Pinene (*view from down*)

Pinyl cation

(4S)-(−)-Limonene (*view from down*)

(E')

2^{14}C-GPP*

(*endo* conformer) (*see also* **Figure 6.51**)

(3S)-(+)-LPP

α-Terpinyl
cation

GPP is folded and fixed by the specific enzymes in two superposable mirror image forms but nonsuperposable when complexed with the specific enzymes, thus leading to the formation of enantiomeric LPP molecules. Although the different specific enzymes generate enantiomeric monoterpenes, they themselves are not stereoisomers.
2^{14}C-Geraniol when fed in **Salvia officinalis forms 2^{14}C-GPP which is biosynthetically converted to (1S,4S)-(−)-2^{14}C-camphor via 1S,5S-(−)-1^{14}C-α-pinene (see also **Figure 6.51**)*

Fig. 6.6 Biosynthetic pathways for the stereospecific formation of enantiomers of α-pinene and limonene from GPP through enantiomers of LPP

monoterpenes are optically active. Enantiomers of the same compound have been isolated from different plants, e.g., (+)-**carvone** from caraway (*Caram carvi*, Fam. Umbelliferae) and (−)-**carvone** from spearmint (*Mentha spicata*, Labiatae) or (+)-**camphor** from sage (*Salvia officinalis*, Labiatae) and (−)-**camphor** from tansy (*Tanacetum vulgaris*, Compositae). Some enantiomeric monoterpenes are found to occur in the same plant, e.g., (+)-**limonene** and (−)-**limonene** in peppermint (*Mentha piperita*, Labiatae) and (+)-**α-pinene** and (−)-**α-pinene** in pine (*Pinus* species, Pinaceae). The enantiomers can give different biological responses, especially towards olfactory receptors in the nose. Thus characteristic odor of caraway is due to (+)-carvone whereas that of spearmint is due to (−)-carvone. (+)-Limonene smells of oranges, whereas (−)-limonene smells of lemons (see Fig. 32.2).

6.1.3.1 Biosynthesis of Pinene, Limonene, and Camphor Enantiomers

Biosynthesis of the enantiomers of limonene and α-pinene is illustrated in Fig. 6.6. It will be seen that the same achiral substrate *GPP is folded by two specific enzymes in two superposable mirror image* ways leading finally to two enantiomers. The optically pure substrates, (3S)- and (3R)-enantiomers of linalyl pyrophosphate

(LPP), catalyzed by (+)- and (−)-α-pinene synthases resulted in the predicted stereospecific transformations—(3R)-LPP to (+)-α-pinene and (3S)-LPP to (−)-α-pinene, respectively (Fig. 6.6). The reactions were demonstrated by labeling experiments and enzyme studies to proceed through *syn* isomerization of the primary allylic pyrophosphate to the tertiary allylic pyrophosphate, followed by rotation about C2–C3 single bond and *anti-endo*-cyclization [6–8]. The enantiomers (3R)-(−)-LPP and (3S)-(+)-LPP when catalyzed by different enzymes, (+)-limonene and (−)-limonene synthases present in plants, are converted into (+)- and (−)-limonenes, respectively, via loss of $H^{\oplus}$ from the corresponding enantiomeric α-terpenyl or limonyl cations (Fig. 6.6) [6–8]. The limonyl or α-terpenyl cation in the presence of (+)- and (−)-camphor synthases forms (+)-camphor and (−)-camphor, respectively, directly or through the corresponding pinyl cations. Biosynthesis of camphor has been discussed in Sect. 6.3 (Fig. 6.51).

Peppermint (*Mentha piperita*, Labiatae) produces (−)-**menthol**, with smaller amounts of the diastereoisomers, (+)-**neomenthol**, (+)-**isomenthol**, and (+)-**neoisomenthol** (Fig. 6.4). (For conformations and relative stability of these diastereomers see Fig. 6.59.) Oils from various *Mentha* species also contain significant amounts of (−)-**menthone**, (+)-**isomenthone**, (+)-**pulegone**, and (−)-**piperitone**. The metabolic relationships of these monoterpene alcohols and ketones and various other monoterpenes have been established as shown in Fig. 6.4. Enantiomers in many cases have different odors [9].

6.1.4   Occurrence of Monoterpenes in Plant Families

Monoterpenes are commonly isolated from various volatile or essential oils obtained mostly from different species of the families Rutaceae (several *Citrus* species), Labiatae (*Lavandula, Mentha, Rosmarinus, Salvia, Thymus*, etc. species), Pinaceae (*Pinus*), Myrtaceae (*Eucalyptus*), Umbelliferae (*Carum, Coriandrum, Anethum*), Lauraceae (*Cinnamomum*), Zingiberaceae (*Zingiber*), Compositae (*Chrysanthemum cinerariaefolium*). The oils are obtained by steam distillation or solvent extraction of fresh leaves, fresh or dried fruit peels/dried or fresh flowers/ripe fruit/fresh flowering tops/dried ripe berries/heartwood, etc. Useful data on volatile oils isolated from various plant materials are available in a tabular form [10].

6.1.5   Cyclopropyl Monoterpenes (C₁₀): Their Biosynthesis

The carbon skeletons of some monoterpenes, isolated from pyrethrum flower heads of a Compositae plant *Chrysanthemum cinerariaefolium*, are of irregular type, different from the regular head-to-tail coupling mechanism. They possess a cyclopropane ring (Fig. 6.7). Among them **chrysanthemic acid** and **pyrethric acid**, found as a number of esters, are especially important because of their insecticidal properties. These

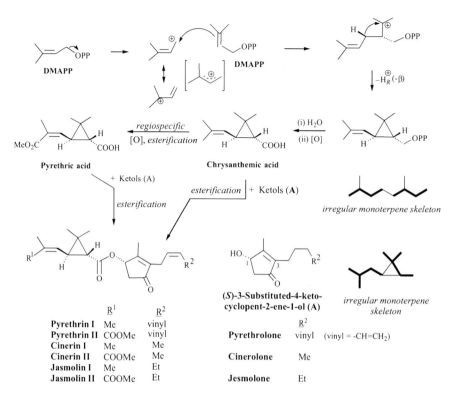

Fig. 6.7 Cyclopropyl monoterpenes: their biosynthesis from two molecules of DMAPP

cyclopropyl monoterpenes are not generated by regular head-to-tail coupling mechanism via GPP, like the regular monoterpenes discussed earlier. *They are biosynthesized from two molecules of* **DMAPP**, joining by a mechanism shown in Fig. 6.7, terminating into a cyclopropane ring formation. Chrysanthemic acid and pyrethric acid are esterified by (*S*)-4-keto-cyclopent-2-ene-1-ols (**A**), viz., +pyrethrelone, cinerolone, and jesmolone, to produce the corresponding cyclopropyl monoterpene esters. The origins of these 4-keto-1-ols are not known; it may be speculated that they are formed by cyclization of modified fatty acid derivatives.

6.1.6   Biosynthesis of Secologanin (via Loganin), the Monoterpenoid Part of Some Indole and Quinoline Alkaloids. Iridoids

The monoterpene **loganin** (Fig. 6.8), an iridoid glucoside, contains a cyclopentane ring fused to a six-membered oxygen heterocycle. It belongs to the iridoid system [11] occurring in nature, e.g., **nepetalactone** (Fig. 6.8) in *Nepeta cataria*

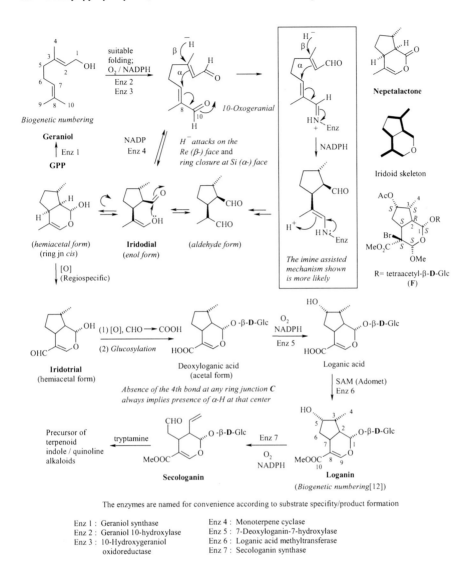

Fig. 6.8 Biosynthesis of secologanin, the monoterpene precursor of terpenoid indole and quinoline alkaloids

(Labiatae), a powerful stimulant and attractant for cats. Loganin is the key precursor for the monoterpenoid part of many complex quinoline and indole alkaloids (Chaps. 22, 25–27 and many alkaloids in Chap. 30). It is also the key intermediate in the biosynthesis of monoterpenes having iridoid structure.

It has been shown by feeding experiment that **secologanin** (Fig. 6.8) is derived from the non-mevalonate pathway [3, 6]. Hence IPP and DMAPP and, for that matter, geraniol are of triose phosphate or pyruvate origin (Sect. 5.3).

Biosynthesis of secologanin via loganin [12] involving a sequence of reactions has been delineated in Fig. 6.8. The iridoid system of loganin arises from geraniol after 10-hydroxylation and a type of folding different from the ones leading to the monoterpenoids already discussed (Figs. 6.2, 6.4, 6.5, and 6.7).

Secologanin unites with tryptamine derived from tryptophan to yield strictosidine, which rearranges to diverse indole alkaloidal skeletons in amazing fashions. The chemical structure of loganin as well as its absolute configuration has been settled by X-ray crystal structure of loganin penta-acetate monomethyl ether bromide (**F**) (Fig. 6.8) [4].

6.2   Geraniol

6.2.1   *Occurrence, Structure Determination*

Geraniol, $C_{10}H_{18}O$, b.p. 229–230°/757 mm, is a constituent of many essential oils, viz., citronella oil (20–40 %) from fresh leaves of *Cymbopogon winterianus* and *C. nardus* (Fam. Gramineae) (oil content 0.5–1.2 %)—also as acetate (~8 %) and rose oil (~20 %) from fresh flowers of *Rosa alba, R. centifolia, R. damascena*, and *R. gallica* (Rosaceae) (oil content 0.02–0.03 %). It is a colorless liquid possessing flowery rosy odor. Its structure has been determined in the classical way, i.e., derivatization, degradation, and synthesis. *The presence of two isolated double bonds, two methyls on sp^2 carbons, and an allylic methylene carrying a hydroxyl group at one of the termini made it an interesting small molecule of Nature.* Organic chemists could easily identify its chemical potential as a synthon and as a substrate for new reagents which needed the built-in structural character-istics possessed by it for studying their reagents' stereo- and regiospecificities.

Extensive work has been done on this small molecule. Both Sharpless (NL 2001) and Noyori (NL 2001) employed this molecule as an important substrate and achieved outstanding success in asymmetric synthesis with their reagents. Eventu-ally, many chemists joined the endeavor of asymmetric synthesis with different reagents using geraniol as a substrate; thus lot of new chemistry evolved (*vide infra*).

Geraniol was subjected to the reactions shown in Fig. 6.9 for obtaining structural information.

These experimental results point to its being a primary alcohol with two isolated double bonds. One double bond carries a geminal dimethyl group, and the molecule can be represented as (**1**), which has been proved by its synthesis. Identical degradation products were also obtained from its isomer nerol, b.p. 225–226°, isolated from various essential oils, e.g., neroli oil, bergamot oil, etc. Thus they must be double bond isomers at the allylic double bond. The double bond config-uration of geraniol has been settled by a simple cyclization reaction with dil. H_2SO_4

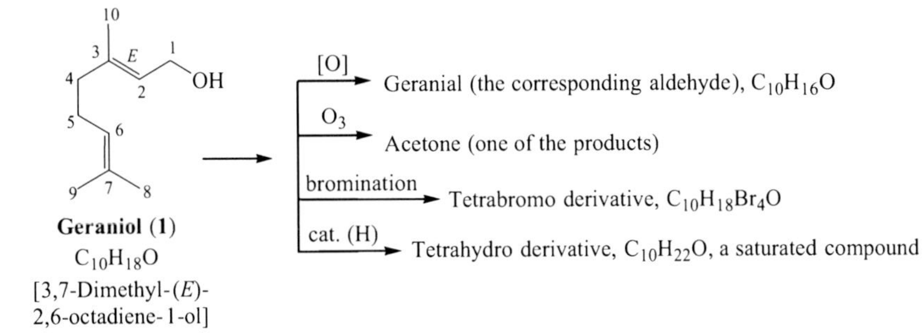

Fig. 6.9 Some reactions of geraniol

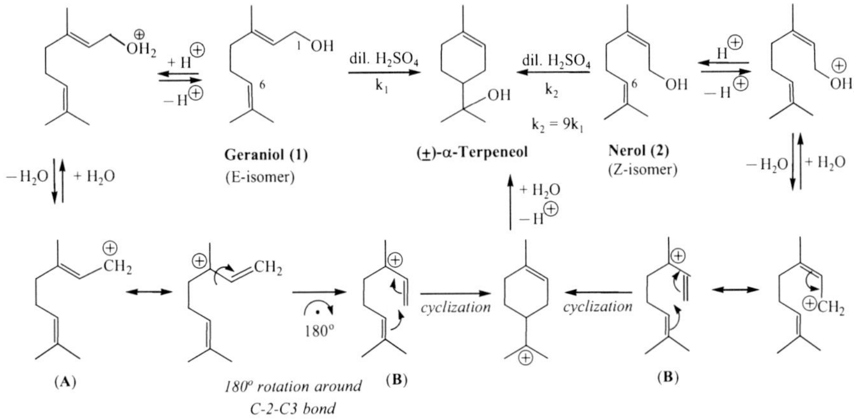

Fig. 6.10 Conversion of geraniol (**1**) and nerol (**2**) to ($\pm$)-α-terpeneol: the mechanistic sequences involved

to yield ($\pm$)-α-terpeneol, which is also obtained by cyclization of nerol with dil. H_2SO_4 (Fig. 6.10). The rate of cyclization of nerol is nine times faster than that of geraniol; this demands the proximity of the Δ^6-double bond and the hydroxyl group in nerol (**2**). On the basis of this observation *trans* (*E*) geometry is assigned to geraniol, while *cis* (*Z*) geometry to nerol (**2**).

The probable mechanistic sequence for the acid-catalyzed cyclization of geraniol and nerol involving the intermediate resonance stabilized allyl carbonium ions (**A**), and (**B**) is delineated in Fig. 6.10.

6.2.2 Spectral Properties

The IR, ^1H NMR, ^{13}C NMR, and mass spectral data of geraniol are shown in Fig. 6.11.

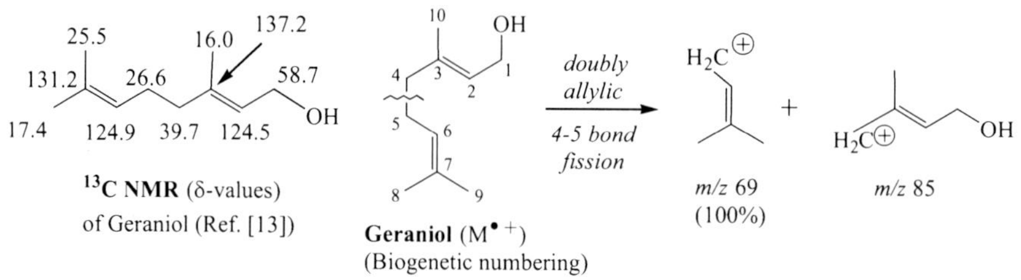
25.5
16.0
137.2
131.2
26.6
58.7
OH
17.4
124.9
39.7
124.5
13C NMR (δ-values)
of Geraniol (Ref. [13])
10
OH
4
3
1
2
5
6
7
8
9
Geraniol (M•+)
(Biogenetic numbering)
doubly
allylic
4-5 bond
fission
H2C⊕
+
H2C⊕
OH
m/z 69
(100%)
m/z 85

Fig. 6.11 IR and NMR spectral peaks and main mass fragments of geraniol

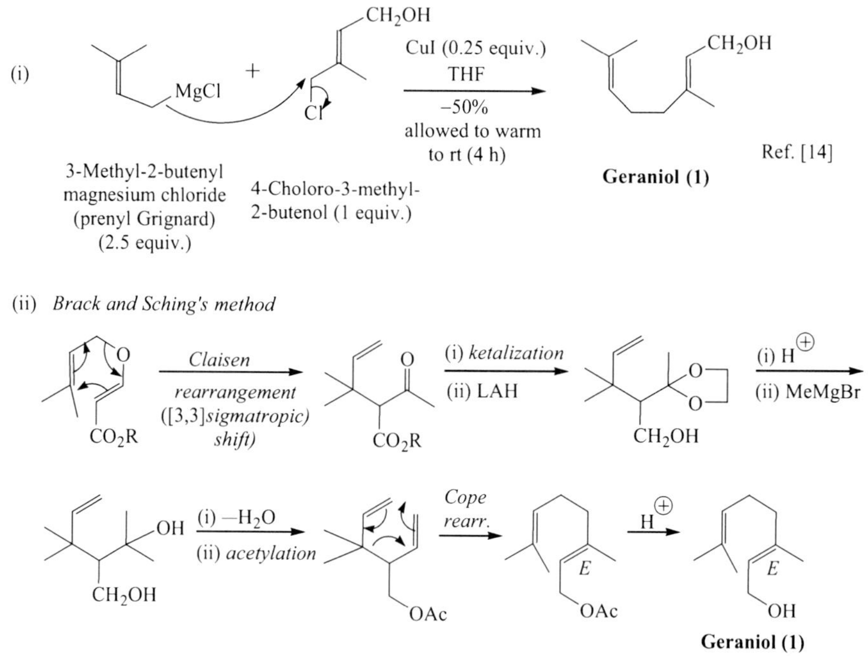
(i)
MgCl
+
CH2OH
Cl
CuI (0.25 equiv.)
THF
–50%
allowed to warm
to rt (4 h)
CH2OH
Ref. [14]
3-Methyl-2-butenyl
magnesium chloride
(prenyl Grignard)
(2.5 equiv.)
4-Choloro-3-methyl-
2-butenol (1 equiv.)
Geraniol (1)
(ii) Brack and Sching's method
O
CO2R
Claisen
rearrangement
([3,3]sigmatropic)
shift)
O
CO2R
(i) ketalization
(ii) LAH
O
O
CH2OH
(i) H⊕
(ii) MeMgBr
OH
CH2OH
(i) –H2O
(ii) acetylation
OAc
Cope
rearr.
E
OAc
H⊕
E
OH
Geraniol (1)

Fig. 6.12 Synthesis of geraniol (two methods)

6.2.3 Synthesis

Several methods of synthesis of geraniol are reported of which two are given in Fig. 6.12.

(i) One synthesis involves alkylation by an allyl Grignard reagent in the presence of CuI as a catalyst resulting in the *extensively regiospecific addition* of two C_5 units in a biosynthetic fashion [14]. The reaction is very slow in the absence of CuI [14].
(ii) Brack and Sching's method has also been delineated stepwise in this figure.

Selective reduction of geranial to geraniol has been achieved [15] by formic acid in the presence of a Ru catalyst (Fig. 6.13) in an excellent yield (99 %). Geranial has also been reduced by the transfer of a hydrogen from an alcohol diglyme, being catalyzed by a tin compound [16], in a moderate yield (67 %) (Fig. 6.13). Geranial occurs (along with neral) in lemon oil (2–3 %) obtained from the dried fruit peels of *Citrus limon* (Rutaceae) (oil content 0.1–3 %).

6.2.4 Synthesis of Chiral Geraniol-1-d

Synthesis of chiral (*S*)-(+)-geraniol-1-d (**5**) has been achieved by Noyori [17] by asymmetric reduction of geranial-1-d (**3**) with LiAlH₄-dihydroxybinaphthyl complex (**4**) in 91 % enantiomeric excess (*ee*) (Fig. 6.14). Isotopically labeled chiral geraniol and other related terpenic alcohols serve as the molecules in the studies of terpenoid biosynthesis.

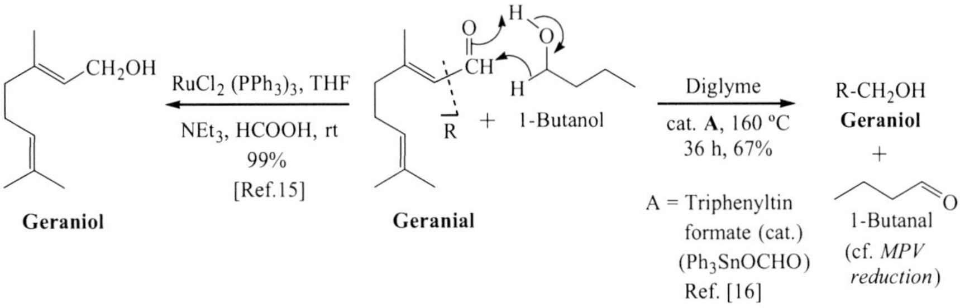

Fig. 6.13 Reduction of geranial to geraniol

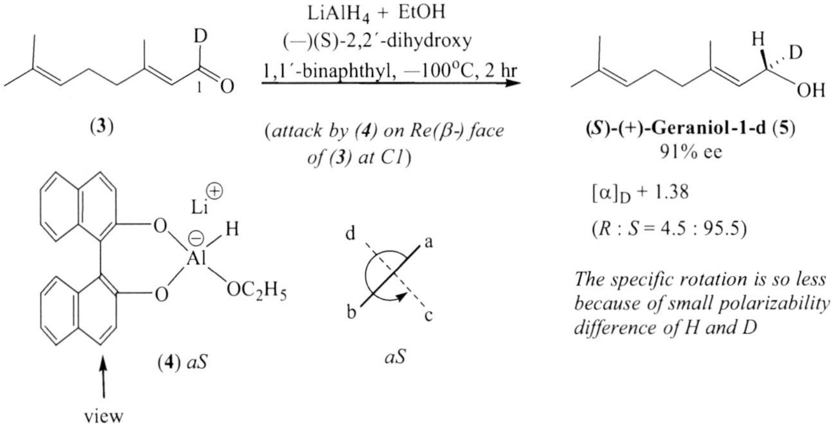

Fig. 6.14 Synthesis of (+)-(*S*)-geraniol-1-d (**5**)

6.2.5   Reactions: (i)–(x)

Several new methodologies have been developed as stated in the sequel.

(i)–(iii) **Oxidation methods.** Several reagents have been employed for the oxidation of the alcoholic OH group [18, 19] [examples (i)–(ii)] and of the vinyl methyl group in a regiospecific manner [example (iii)] [20] (Fig. 6.15).

   The success of the reaction (ii) [19] depends on the method of preparation of the reagent. Oxalyl chloride reacts violently with DMSO at room temp. The reagent is to be prepared at low temperature (−60 °C), when DMSO is successfully activated for the participation.

(iv) **Geranic acid methyl ester.** Corey designed a novel method [21] for converting geranial to the methyl ester of geranic acid. This involves the reversible formation of the intermediate cyanohydrin and its oxidation by MnO$_2$ followed by reaction with methanol to from geranic acid methyl ester in one pot. A number of aldehydes were converted to the corresponding methyl ester by this method. Geraniol also can be converted to geranic acid methyl ester via geranial in one pot (Fig. 6.16).

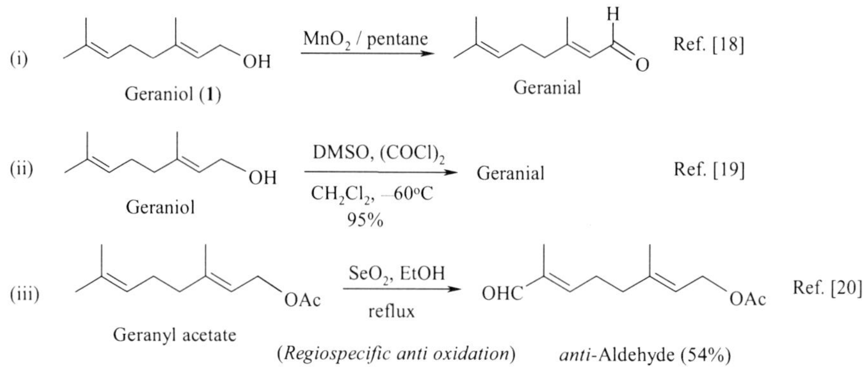

Fig. 6.15 Methods of oxidation of geraniol

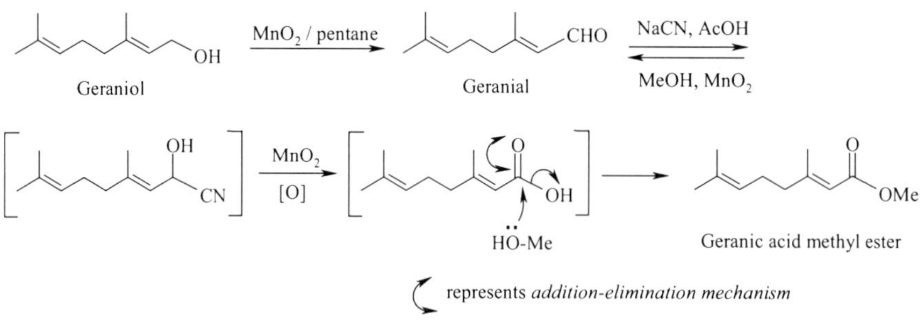

Fig. 6.16 Geranic acid methyl ester from geraniol

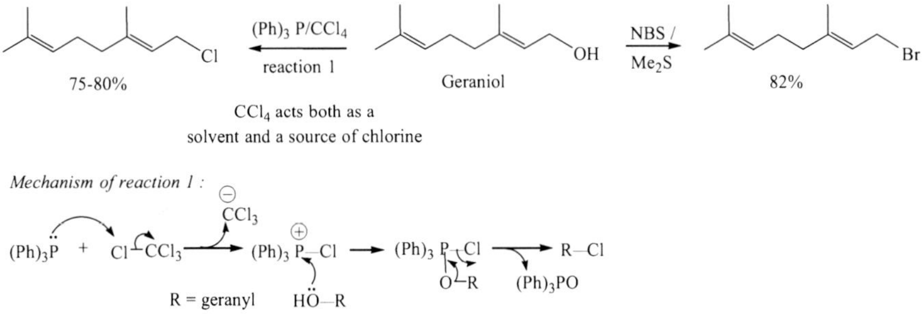

CCl$_4$ acts both as a
solvent and a source of chlorine

Mechanism of reaction 1 :

Fig. 6.17 Geranyl chloride from geraniol

Fig. 6.18 Isomerization of Δ^2 of geraniol to Δ^4

(v) ***Conversion of geraniol to geranyl chloride (or geranyl bromide)*** (Fig. 6.17)
Likewise, treatment with CBr$_4$ (instead of CCl$_4$) (Fig. 6.17) will convert
geraniol to geranyl bromide.

(vi) ***Isomerization of the double bond*** of geraniol in the presence of *N*-lithioethy-
lenediamine, a reagent discovered by Reggel et al. [22, 23], led to the
formation of a conjugated double bonded C$_{10}$-alcohol in a good yield
(Fig. 6.18).

(vii) ***Conversion of geranyl acetate into 1-alkene and 2-alkene.*** Palladium-
catalyzed hydrogenolysis of the terminal allylic acetate with formic acid-
triethylamine leads to the formation of 1-alkene. 1-Alkene and 2-alkene are
formed under different condition [24] (Fig. 6.19). A greater selectivity is
observed towards the formation of *E*-2-alkenes from allylic acetates (e.g., geranyl
acetate) using Pd°SmI$_3$ and 2-propanol as a hydrogen donor [24] (Fig. 6.19).

(viii) ***Conversion of geranial into carene*** [25] (Fig. 6.20)

(ix) ***Formation of the methyl ether by a mild method*** [26] (Fig. 6.21). The yield in
this method is better than the usual drastic one (treatment with NaH and MeI).

(x) ***Olefin metathesis.*** Tungsten hexachloride-tetramethyltin, WCl$_6$-SnMe$_4$,
catalyzes olefin metathesis in the synthesis of terpenoids with
1-methylcyclobutene as the isoprene synthon. The following interesting metathesis involving geranyl acetate and the isoprene synthon yields farnesyl acetate.
Both *E* and *Z* isomers are obtained but in very low yield (1–2 %) (Fig. 6.22) [27].

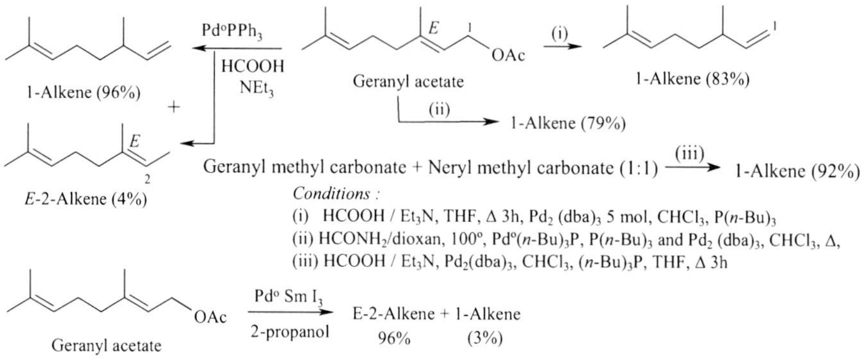

Fig. 6.19 Conversions of geranyl acetate into 1-alkene and 2-alkene

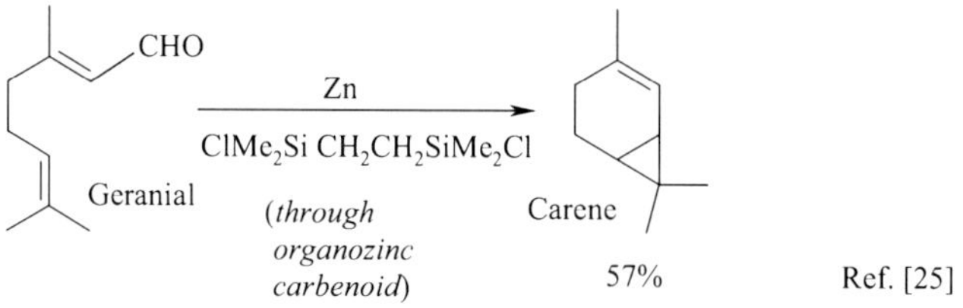

Conditions :
(i)   HCOOH / Et$_3$N, THF, Δ 3h, Pd$_2$ (dba)$_3$ 5 mol, CHCl$_3$, P(n-Bu)$_3$
(ii) HCONH$_2$/dioxan, 100°, Pd°(n-Bu)$_3$P, P(n-Bu)$_3$ and Pd$_2$ (dba)$_3$, CHCl$_3$, Δ,
(iii) HCOOH / Et$_3$N, Pd$_2$(dba)$_3$, CHCl$_3$, (n-Bu)$_3$P, THF, Δ 3h

Fig. 6.20 Conversion of geranial to carene

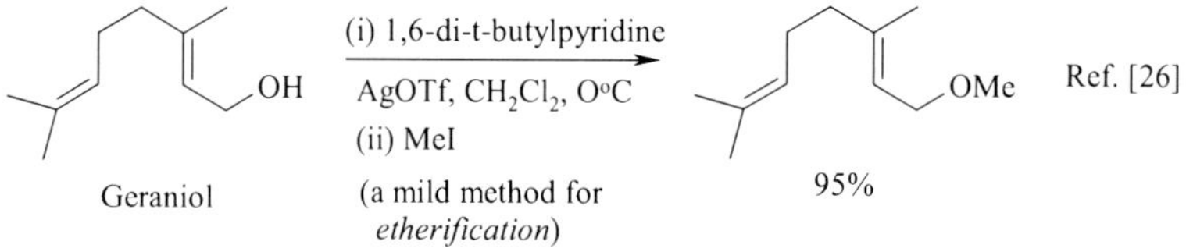

Fig. 6.21 Formation of the methyl ether of geraniol

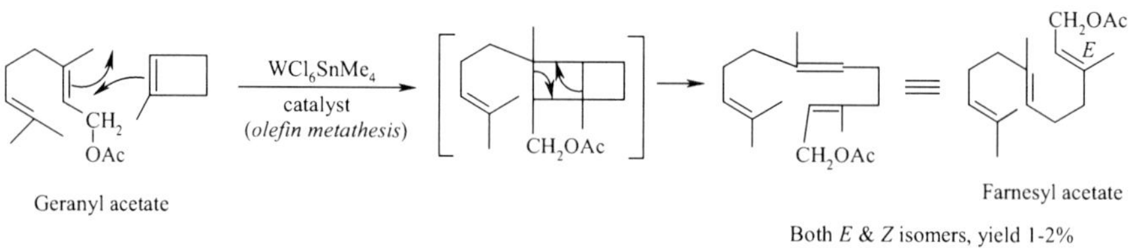

Fig. 6.22 Conversion of geranyl acetate to farnesyl acetate

6.2.6 Epoxidation methods of Geraniol (Fig. 6.23)

Five different epoxidation methods [(i)–(v)] are briefly stated.

(i) Geranyl acetate, when treated at 60 °C for 20 h with molecular oxygen in the presence of a catalyst having the structure [Fe$_3$O(piv)$_6$(MeOH)$_3$]Cl, is converted into 6,7-epoxygeranyl acetate (87 % yield) and the conversion is 82 %. Similarly neryl acetate also can be converted to the 6,7-epoxyneryl acetate (74 % yield, conversion 52 %). [28]

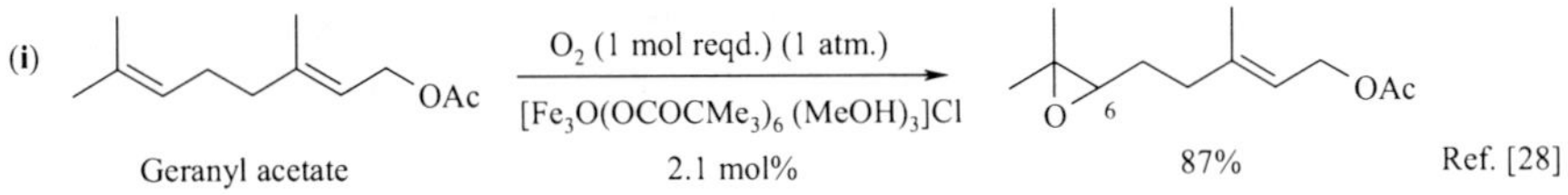

(ii) 2,3-Epoxyalcohols cannot be generally prepared from allylic alcohols with *m*-chloroperbenzoic acid (*m*-CPBA); however, geraniol-2,3-epoxide has been prepared regioselectively at the allylic alcoholic double bond with *m*-CPBA in an emulsion prepared by stirring a mixture of geraniol, *n*-hexane, *n*-octanol, water, NaOH, and dioctadecyldimethylammonium chloride [29]. Neither

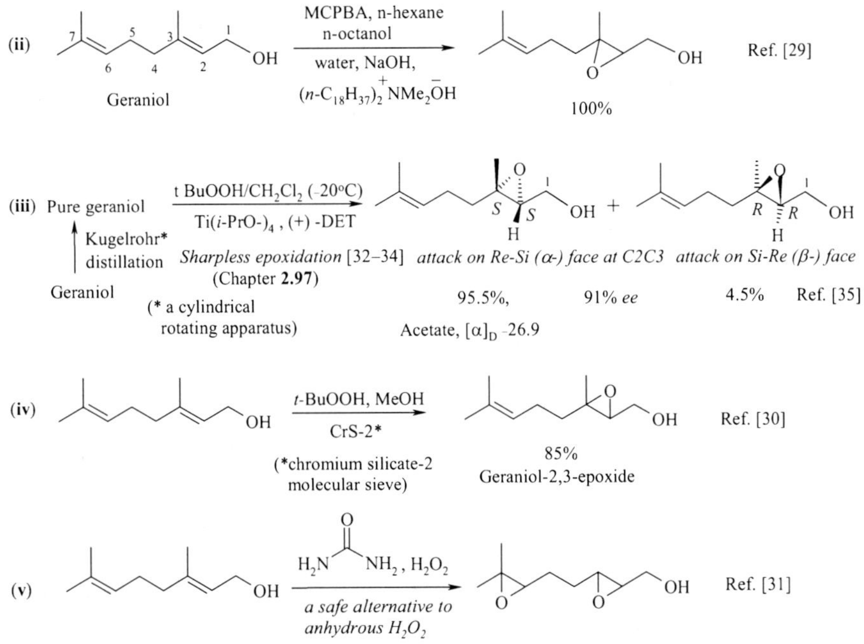

Fig. 6.23 Expoxidation of geraniol by different methods

6,7-epoxycompound nor 6,7-2,3-diepoxy-compound is formed. Such 100 % regioselectivity in epoxidation may be explained as follows: the orientation of the geraniol molecule in the emulsion system is such that due to well-known columbic attraction at micelle surface, the polar OH group will be facing the water, and *m*-CPBA will approach the substrate easily in this *cationic emulsion system*. 2,3-Double bond being close to the polar system is accessible to MCPBA for interaction, while 6,7-double bond, being away from polar head, is protected by the long alkyl chain of the surfactant or by hexane phase and hence escapes the attack by *m*-CPBA. Similarly nerol has been selectively epoxidized at 2,3-double bond in 93 % yield.

(iii) *Entirely regioselective and highly enantioselective asymmetric catalytic epoxidation of geraniol* [32–35] has been achieved by Sharpless (NL 2001, see Appendix A). Reaction of an allylic alcohol with t-butylhydroperoxide (TBHP) in the presence of Ti(*i*-PrO-)$_4$ and diethyl tartrate (DET) (*Sharpless epoxidation*) gives epoxy alcohol with high enantiomeric purity. Geraniol is epoxidized at the C2C3 double bond at the *Re–Si* face [α-,when written in the way shown in Fig. 6.23, (v)] producing the (2S,3S)-epoxide in 77 % overall yield (95 % *ee*) [35], when stoichiometric amount of titanium-(+)-DET complex is used. Use of (−)-DET leads to the preferential attack of oxygen at C2C3 double bond from the *Si–Re* face (β-) to give the enantiomeric (2R,3R)-epoxide with 90 % *ee*. By manipulating the experimental condition improvement of the overall yield to 95 % (with 95 % *ee*) has been possible. For the epoxidation of a Z-allylic alcohol (like nerol) in the presence of (+)-DET attack takes place on C2C3 double bond, again at the α-face (which is now *Re–Re* at C2C3) to produce the (2S,3R)-epoxide with ~95 % *ee. The Sharpless epoxidation has been used as the key step in the chiral syntheses of antibiotics* [33], *terpenes, carbohydrates, pheromones, and pharmaceuticals* [34]. For enantiotopic faces, see Sects. 2.9.5, 2.9.6, and for Sharpless epoxidation, see Sect. 2.11.7

6.2.7   Geraniol As a Synthon

A few examples of the use of geraniol and its functional derivatives as starting materials (synthons) for synthetic studies have already been described. Some more examples are now cited.

(i) *Synthesis of farnesol:*

    (a) Ruzicka [36] synthesized farnesol from geraniol. It constitutes the first total synthesis of farnesol (Fig. 6.24).

    (b) Corey converted geranyl acetone into farnesol through the six steps shown in Fig. 6.25 [37, 38]

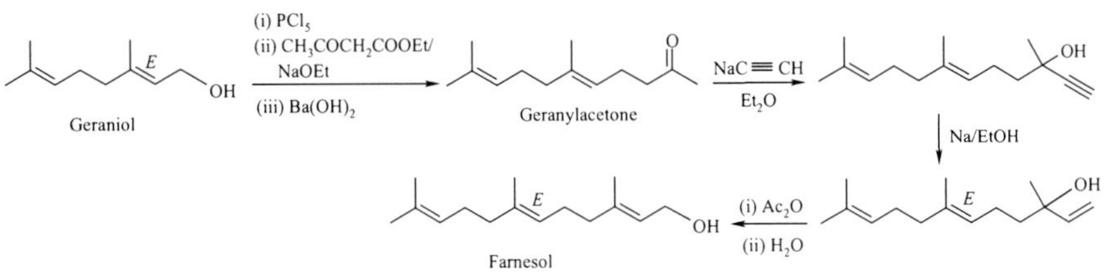

Fig. 6.24 Synthesis of farnesol from geraniol [36]

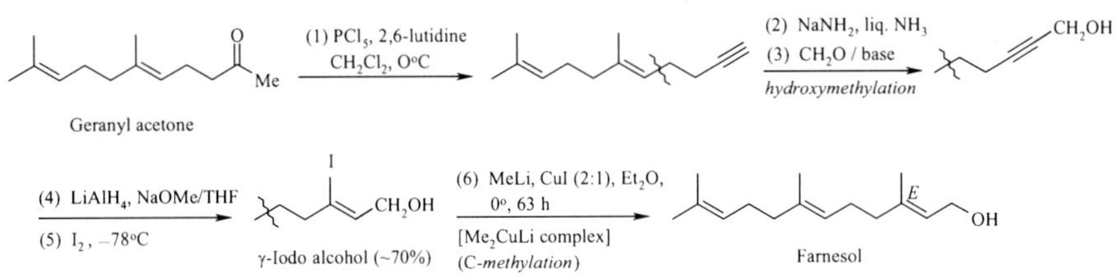

Step (1) leads to a mixture of chlorides which in step (2) undergoes dehydrochlorination to give the ethynyl derivative

Fig. 6.25 Stereospecific synthesis of farnesol [37, 38]

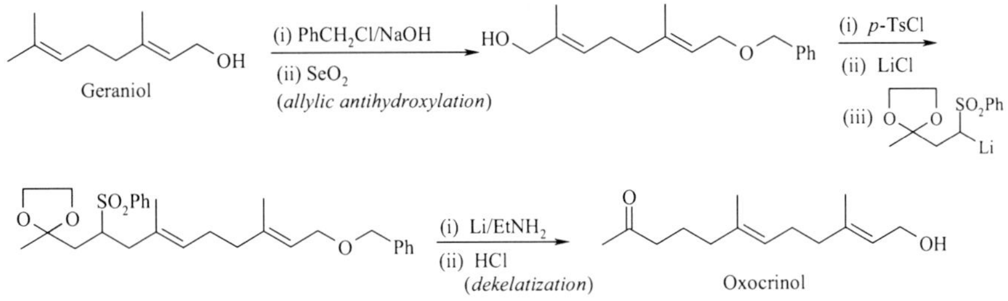

Fig. 6.26 Synthesis of oxocrinol [39]

(ii) **Synthesis of oxocrinol.** Kato et al. [39] synthesized oxocrinol, a norsesquiterpene, isolated from marine algae, using gernaiol as the starting material. The steps are quite straightforward (Fig. 6.26).

(iii) **Synthesis of Gyrinidal.** Gyrinidal [20], a defensive secretion of the whirligig water beetles, has been synthesized starting from geranyl acetate (Fig. 6.27).

(iv) **Synthesis of E,E-α-farnesene.** Chou et al. [40] have synthesized E, E-α-farnesene using geranyl bromide as an alkylating agent for an isoprene anion used in the form of 3-methyl-3-sulfolene. Thermolysis gave the product and the E-geometry of the double bond is enforced by [4 + 2] cheletropic elimination (Fig. 6.28).

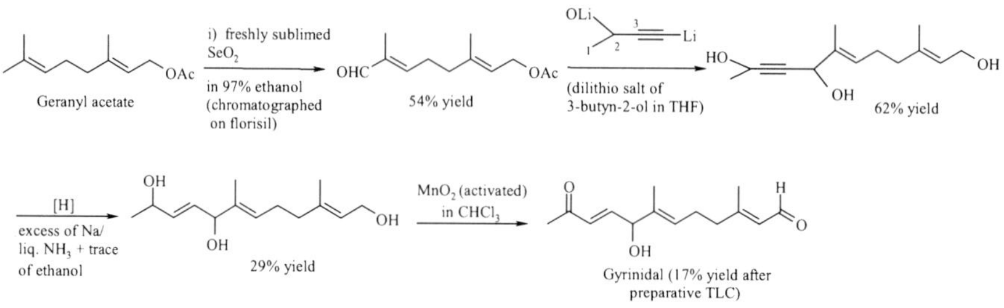

Fig. 6.27 Synthesis of gyrinidal [20]

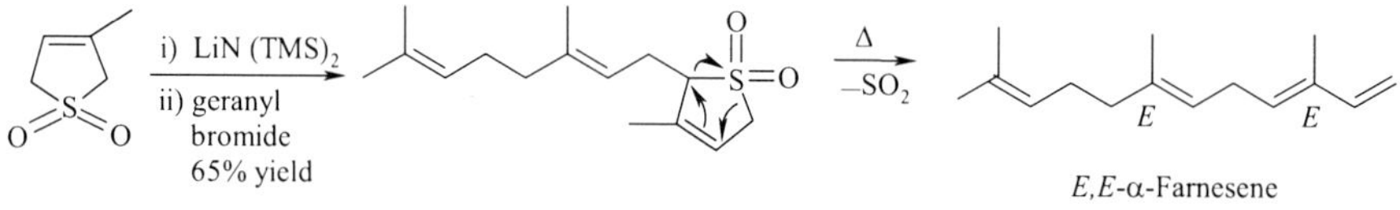

Fig. 6.28 Synthesis of *E,E*-α-farnesene [40]

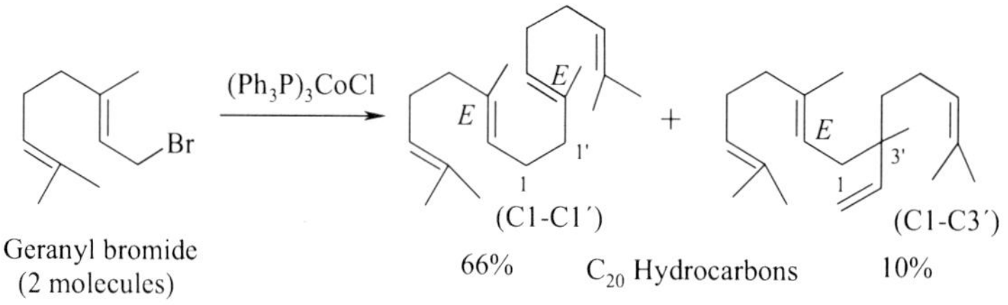

Fig. 6.29 Synthesis of C$_{20}$ hydrocarbons

(v) *Some more C–C coupling reactions:*

  (a) Geranyl bromide undergoes allylic coupling and is converted into two C$_{20}$ hydrocarbons (66 % and 10 %), formed by reductive couplings [41] of C1–C1′ and C1–C3′ of two molecules respectively; a Wurtz type of coupling takes place in the presence of chlorotris(triphenylphosphine) cobalt (Fig. 6.29).

  (b) Transition metals like Ni or Fe catalyze the S$_N$2 type coupling between a selective Grignard reagent and a primary allylic diphenylphosphate. [42] However, in the presence of a catalytic amount of CuCN. 2LiCl S$_N$2′-type coupling takes place (Fig. 6.30). Highly activated Rieke-Mg [43] is used to prepare the allyl magnesium chlorides from geranyl chloride and neryl chloride at <-100 °C in situ.

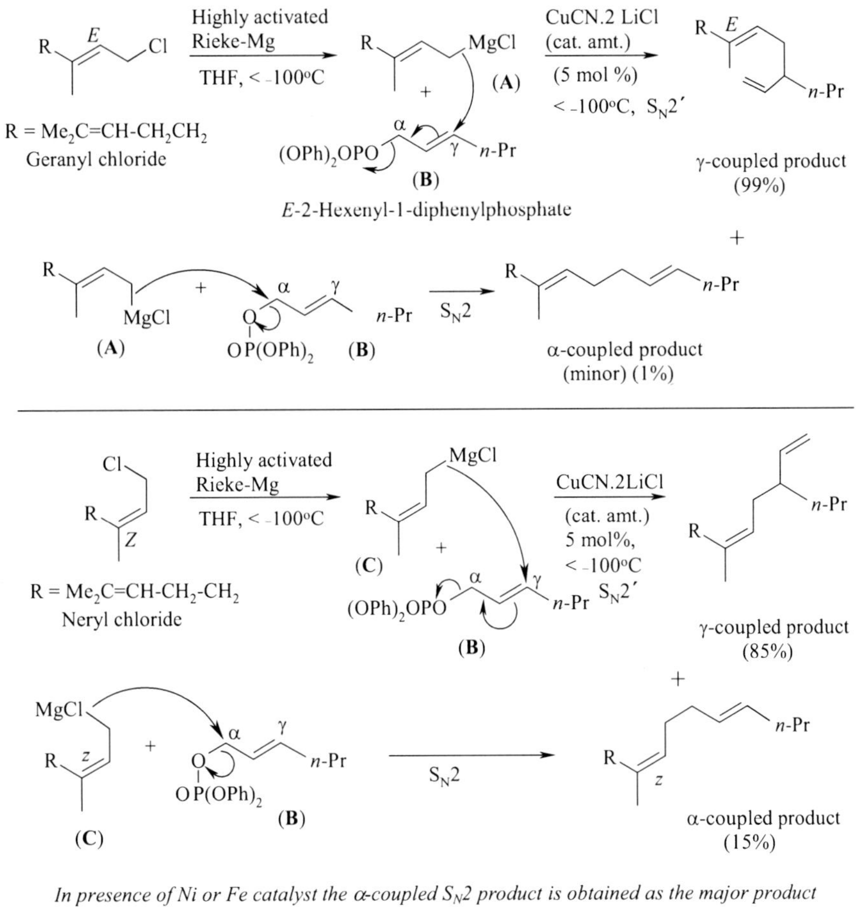

Fig. 6.30 Synthesis of α- and γ-coupled products from geranyl chloride and neryl chloride

6.2.8 Cyclic Products from Geraniol

In this subsection *five more examples of the use of geraniol as a synthon* are given.

(i) Oxidation of geranyl acetate with KMnO$_4$ proceeds stereospecifically and gives a tetrahydrofuran derivative via an intermediate 6,7-diol [44]. The stereochemistry of the product suggests the ring closure process to be suprafacial (Fig. 6.31).

(ii) OsO$_4$-catalyzed diastereoselective formation of *cis*-tetrahydrofuran derivative has been achieved using geranyl benzoate [45]. In this process the enantio-enriched 1,2-diol is formed from geranyl benzoate by Sharpless AD (asymmetric dihydroxylation). The diol-complexed Os(VI) species then undergoes an intramolecular cyclization to yield enantiopure (+)-*cis*-solamin (Fig. 6.32).

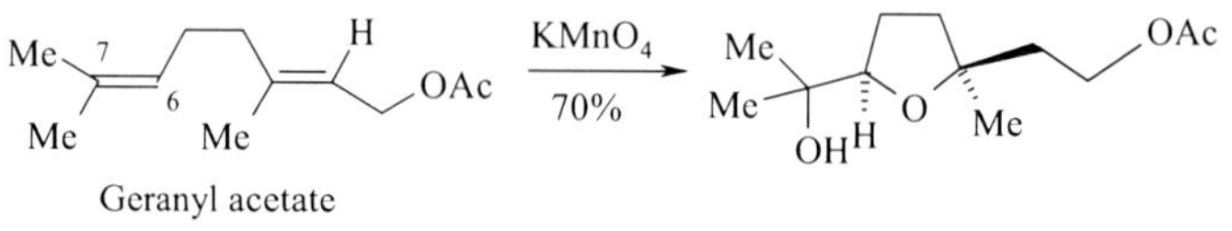

Fig. 6.31 Cyclization to a tetrahydrofuran derivative

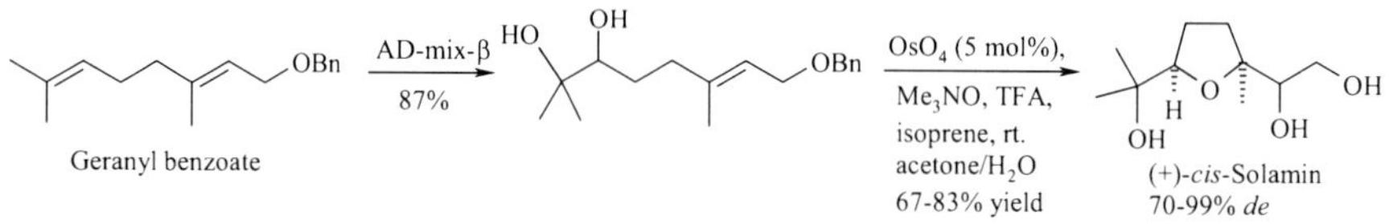

Fig. 6.32 Formal synthesis of (+)-*cis*-solamin

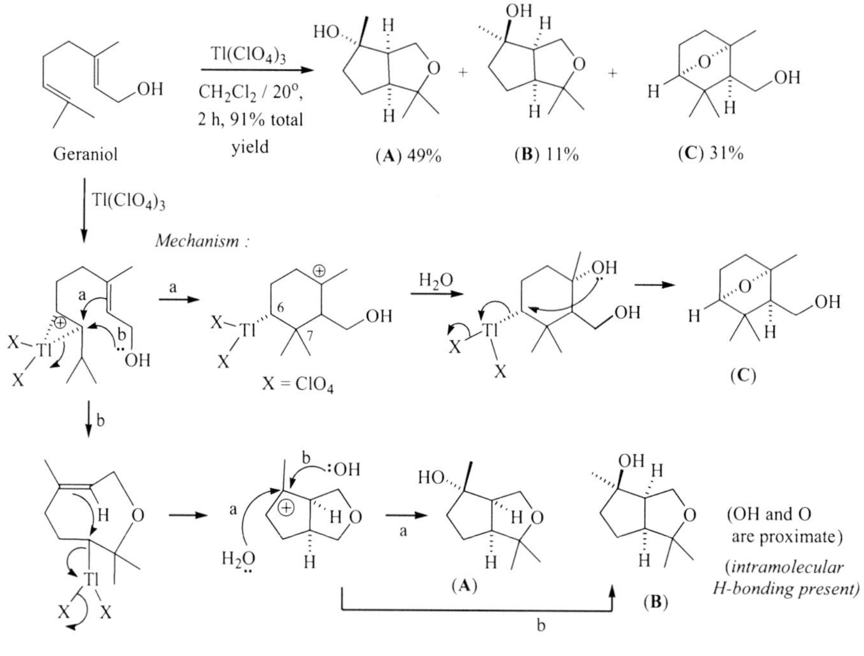

Fig. 6.33 A new type of olefinic cyclization of geraniol

(iii) A new type of olefinic cyclization of geraniol with thallium perchlorate Tl (ClO$_4$)$_3$ has been reported [46] (Fig. 6.33). The product formation (the connectivity and not the stereochemistry) can be rationalized as shown in the figure. These products possessing iridoid carbon skeleton may be useful in the synthesis and biogenesis of iridoid monoterpenes.

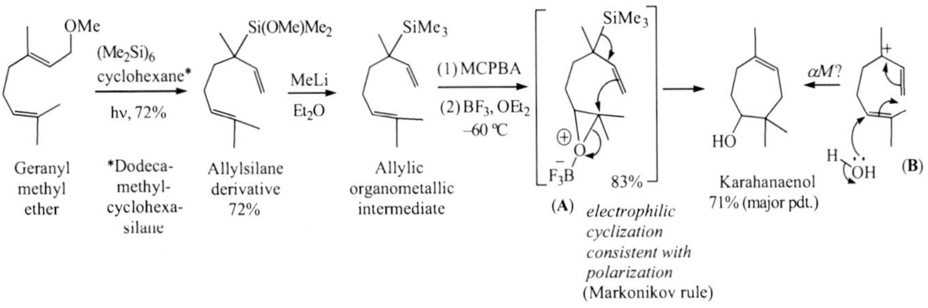

Fig. 6.34 Synthesis of karahanaenol, a seven-member ring monoterpene

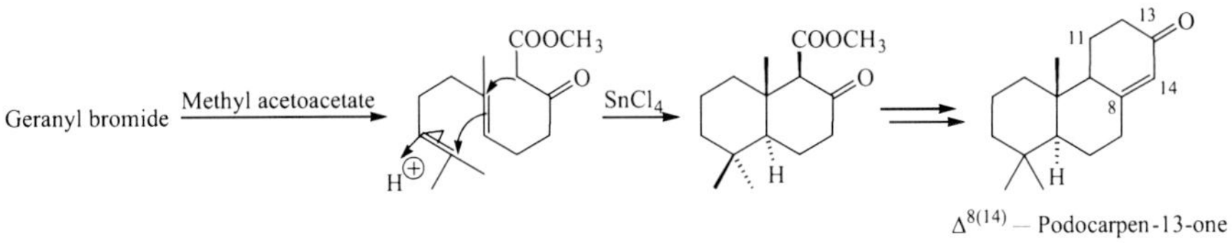

$\Delta^{8(14)}$ — Podocarpen-13-one

Fig. 6.35 Synthesis of $\Delta^{8(14)}$-podocarpen-13-one

(iv) From the methyl ether of geraniol, a seven-membered cyclic monoterpene *karahanaenol* [47] has been synthesized following the steps shown in Fig. 6.34.

This synthesis suggests the possibility that electrophilic cyclization of type (**A**) of an allylic organometallic type intermediate during enzymatic formation may account for the biogenesis of seven-membered ring monoterpnenes, karahanaenol, humulene, etc., without involving anti-Markovnikov (*aM*) cyclization like (**B**), as is usually thought.

(v) A novel synthesis of $\Delta^{8(14)}$-podocarpen-13-one [48] has been accomplished starting from geranyl bromide, as outlined in Fig. 6.35.

6.2.9 Molecular Recognition (Regio- and Stereoselective)

6.2.9.1 Molecular Recognition of Carbonyl Compounds [Examples (i)–(iii)]

(i) The α,β-unsaturated aldehydes like geranial (*E*-citral) and neral (*Z*-citral) showed regioselectivity at the γ-position when condensed with benzaldehyde in the presence of LDA and aluminum-tris(2,6-diphenylphenoxide (ATPH) (Fig. 6.36). Of the two available γ-positions with respect to carbonyl, a process of molecular recognition of carbonyl at the *anti* position seems to be operative, and the reaction takes place at two different γ-locations for two isomers [49]. Two more examples of molecular recognition of CO follow.

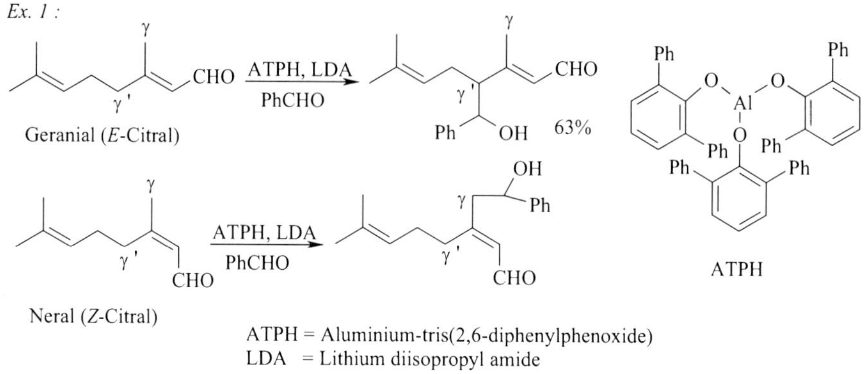

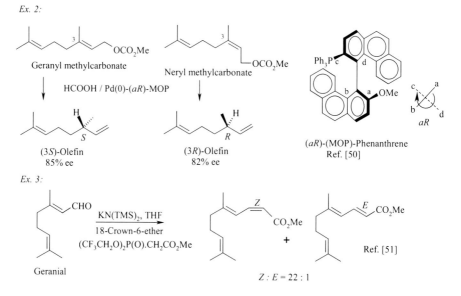

Ex. 1 :

Geranial (*E*-Citral)

Neral (*Z*-Citral)

ATPH

ATPH = Aluminium-tris(2,6-diphenylphenoxide)
LDA = Lithium diisopropyl amide

Fig. 6.36 Molecular recognition of CO group—an example [49]

Ex. 2:

Geranyl methylcarbonate

Neryl methylcarbonate

HCOOH / Pd(0)-(*aR*)-MOP

(3*S*)-Olefin
85% ee

(3*R*)-Olefin
82% ee

(*aR*)-(MOP)-Phenanthrene
Ref. [50]

Ex. 3:

Geranial

KN(TMS)$_2$, THF
18-Crown-6-ether
(CF$_3$CH$_2$O)$_2$P(O).CH$_2$CO$_2$Me

Ref. [51]

Z : *E* = 22 : 1

Fig. 6.37 Molecular recognition of CO group—two more examples

(ii) The reaction of geranyl methyl carbonate under the condition cited (Fig. 6.37) gives the reduced (3*S*)-olefin in 85 % *ee*, while the corresponding neryl methyl carbonate under the same reaction condition gives the reduced (3*R*)-olefin in 82 % *ee* [50].

(iii) Geranial is converted [51] to the *Z*:*E* olefins in the ratio 22:1 under the condition specified in Fig. 6.37.

6.2.9.2 Molecular Recognition of Prochiral Allylic Alcohols

Catalysts **A** and **A′** (Fig. 6.38) have accomplished highly enantioselective and regioselective hydrogenation of prochiral allylic alcohols [52] like geraniol and nerol. Here, the chemical multiplication of chirality is based on catalyst/substrate intermolecular *asymmetric induction*. The advantages of this unique asymmetric catalysis [53] are (i) high chemical and optical yields, (ii) regioselective reaction avoiding hydrogenation of the C6=C7 bond, (iii) high substrate–catalyst mole ratio, (iv) lack of double bond migration or E/Z isomerization, and (v) easy recovery or reusability of the catalyst. Here, both (*R*)- and (*S*)-citronellols are accessible by either variation of allylic olefin geometry or choice of handedness of the catalysts which can differentiate the C3 enantiofaces.

6.2.10 Microbial Hydroxylation

Geranyl acetate undergoes deacetylation and regiospecific hydroxylation at the terminal allylic *anti*-carbon (C8) by a strain of *Aspergillus niger* to yield 8-hydroxygeraniol (40 %) along with geraniol (50 %) (Fig. 6.39) [54]. By this method acetates of

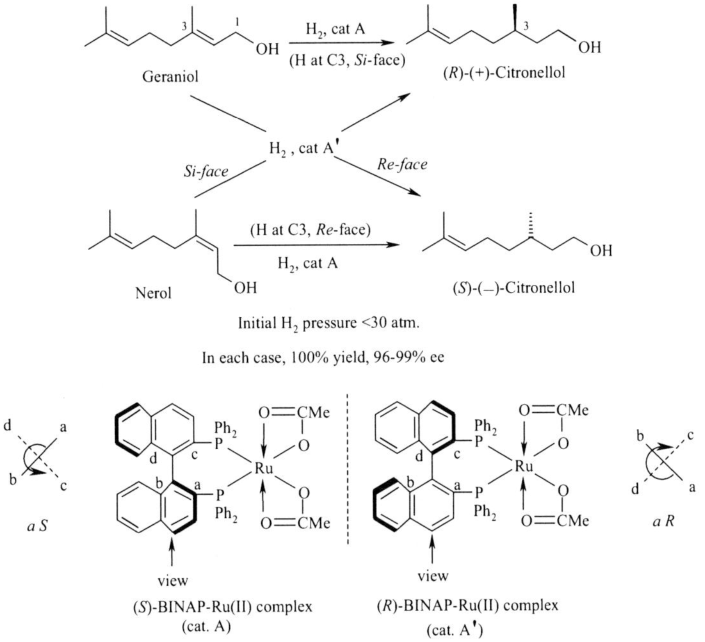

Note: *Only the BINAP part of either enantiomeric catalyst causes enantiomerism; thus, the catalysts A and A′ have aS and aR configurations respectively, as shown. The rest Ru complex part of the catalyst A or A′ possesses a horizontal plane of symmetry passing through Ru, and hence achiral.*

Fig. 6.38 Enantioselective catalytic hydrogenation of geraniol and nerol—molecular recognition

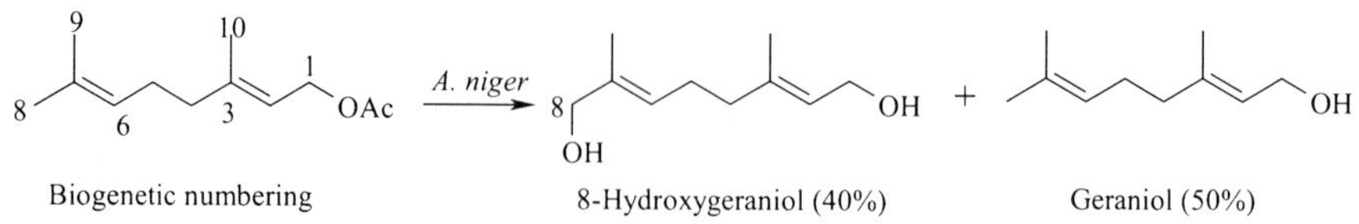

Fig. 6.39   Regiospecific hydroxylation of geraniol

citronellol (2,3-dihydrogeraniol) and linalool (3-hydroxy-2,3-dihydrogeraniol) have been regiospecifically hydroxylated at the terminal allylic *anti*-carbon (C8).

6.2.11   Metabolism of Geraniol in Grape Berry Mesocarp [55]

By using labeled geraniol (d$_6$-geraniol) it has been shown that in grape berry mesocarp of *Vitis vinifera*, geraniol acts as the precursor of the potent odorant *cis*-(2S, 4R)-rose oxide; its enantiomer [*cis*-(2R, 4S)] or two diastereomers [*trans*-(2S, 4S) and *trans*-(2R, 4R)] are not found. A scheme for metabolism of d$_6$-geraniol in grape berry mesocarp is presented in Fig. 6.40.

   This metabolism takes place during the ripening of the fruits. The concentration of the flavor in berries increases by allowing the fruits to stay on the vine for an extended period. The labeled geraniol also suffers from stereoselective reduction, E/Z isomerization, oxidation, and glycosylation to give the products delineated in Fig. 6.40 [55].

6.2.12   Bioactivity and Uses

(1) Geraniol is bioactive against mosquitoes and is used as a mosquito repellent.
(2) It has been extensively used as a synthon for the synthesis of various interesting linear as well as cyclic compounds, as already stated in the previous subsections.
(3) Because of its sweet smell and since it has no harmful effect on human system—it is used in perfumery and in the preparation of sweets (confectionery).
(4) It is also used as an odorant component in some medicines and foods.

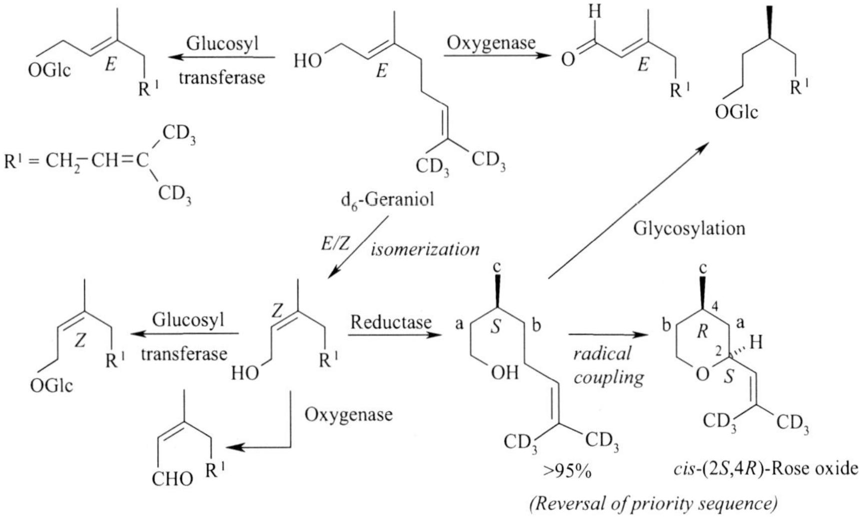

Fig. 6.40 A scheme for metabolism of d_6-geraniol in grape berry mesocarp

6.3 Camphor

6.3.1 Introduction

Camphor (Arabic, Kāfūr; Sanskrit, Korpur), $C_{10}H_{16}O$, m.p. 180°, +54.9 (EtOH), is the most well-known and classical example of biogenetically geraniol pyrophosphate-derived bicyclo[2.2.1]heptane compound. It is a volatile white crystalline solid, widely distributed in Nature, and found specially in all parts (particularly wood) of the camphor tree (*Cinnamomum camphora*, syn. *Laurus camphora*, Fam. Lauraceae) [56], a tree growing in Brazil, Japan, Borneo (hence its alternate name bornan-2-one), and some other places. It also occurs in some other related trees in the family Lauraceae, notably *Ocotea usambarensis*. The less abundant (−)-camphor, m.p. 179°, $[\alpha]_D$ −44.2 (EtOH), is present as a constituent in *Matricaria parthenium* (Fam. Compositae/Asteraceae). Camphor is present in camphor oil (25–47 %) from camphor tree and in rosemary oil (10–25 %) from fresh flowering tops of *Rosmarinus officinalis* (Fam. Labiatae/Lamiaceae), corian-der oil (~5 %) from ripe fruits of *Coriandrum sativum* (Fam. Umbelliferae/Apiaceae), and sage oil (5–22 %) from fresh flowering tops of *Salvia officinalis* (Fam. Labiatae/Lamiaceae) and many other oils. Camphor is steam volatile and also volatile at room temperature [56]. Monoterpenes are abundant in biosphere but only rarely encountered in geosphere. Camphor and borneol were found in shale (sedimentary rock) from the Precambrian Ketilidian fold of South West Greenland [57]. Camphor possesses a characteristic aromatic odor. Several compounds are reported to have camphoraceous odor, but structurally they are widely different [58] (Fig. 6.41).

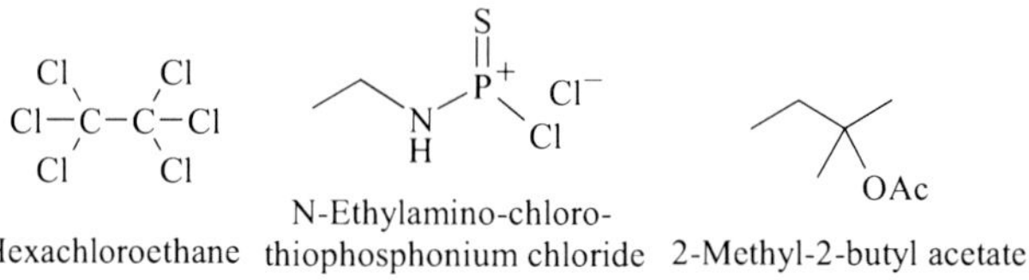

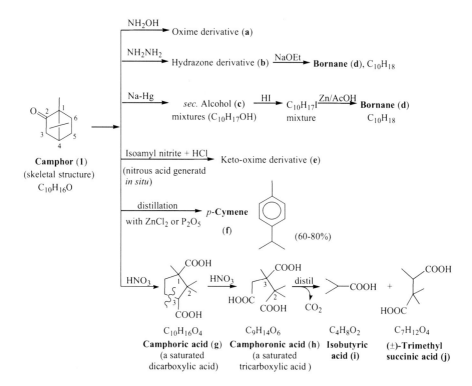

Fig. 6.41 Some compounds with camphoraceous odor

Fig. 6.42 Some degradative reactions of camphor

6.3.2   Structure Determination

The structure of camphor has been deduced from its degradative studies (done during 1894–1914) and confirmed by synthesis.

Some degradative studies are shown in Fig. 6.42. Formation of (**a**), (**b**), (**c**), and (**d**) (Fig. 6.42) suggested the presence of a ketocarbonyl group in the molecule; formation of (**e**) indicated the presence of a ketomethylene group. Further, the ketomethylene group is present as a part of a ring since the product (**g**) contains the same number of the carbon atoms as its parent compound (1). Formation of p-cymene (f) indicates the presence of a C–Me and *gem*-dimethyl moieties in (1). The products (i), (j) and CO_2 vindicated 1,2,2-trimethylcyclopentane-1,3-dicarboxylic acid and 2,2,3-trimethyl-3-carboxymethyl succinic acid structures for camphoric acid (g) and camphorinic acid (h), respectively. The ratio of carbon to hydrogen atoms of (d) ($C_{10}H_{18}$), a saturated hydrocarbon, suggested it to be a bicyclic system. All these observations pointed to the structure 1,7,7-trimethylbicyclo[2.2.1]-heptane-2-one (1) for camphor. Nearly 30 different structures were proposed for camphor prior to Bredt's correct structure (without stereochemistry) in 1893.

6.3.3 Absolute Configuration and Conformation

(+)-Camphor with two chiral centers (bridgehead carbons) has been chemically correlated to (R)-(−)-2-isopropyl-2-methylsuccinic acid of known stereochemistry (Fig. 6.43). Thus, in (+)-camphor, the (R)-configuration of C1 fixes the (R)-configuration of the other bridgehead carbon C4 also. This correlation is consistent with the results obtained by quasi-racemate method. Additionally, X-ray studies [59] of some derivatives of (+)-camphor having heavy atoms like Br or Cl (Fig. 6.44) also confirmed this absolute configuration. Thus (+)-camphor is

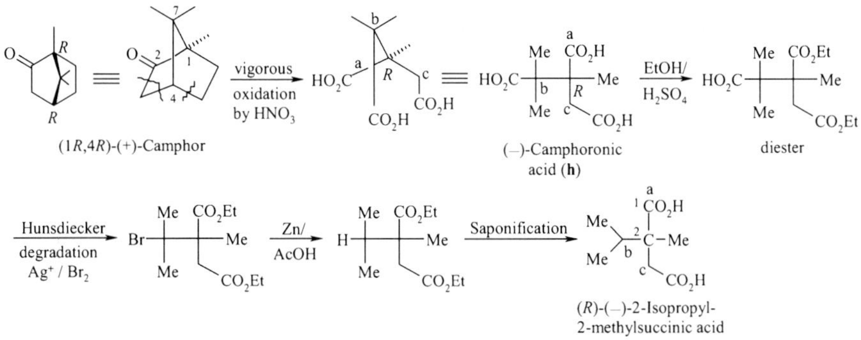

Fig. 6.43 Absolute configuration of (+)-camphor at C1 by chemical correlation

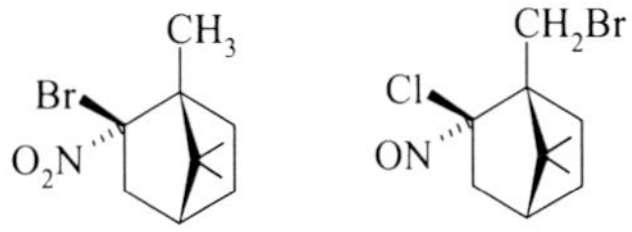

Fig. 6.44 Some (+)-camphor derivatives with Br atom (for X-ray)

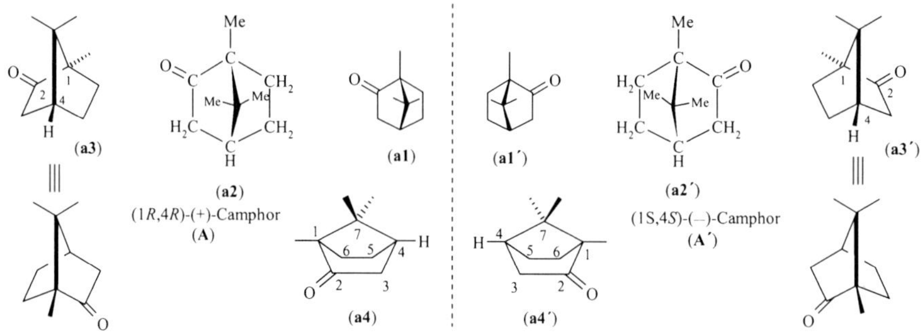

Fig. 6.45   Structural representations of (+)-camphor (**A**) and (−)-camphor (**A'**)

represented as (**A**) having (IR, $4R$) configuration, and (−)-camphor is represented as its mirror image (**A'**) having ($1S$, $4S$) configuration (Fig. 6.45).

  Conformation. The bridgehead of the bicyclo[2.2.1]heptane skeleton imparts strain to the skeletal frame of camphor and locks its molecular conformation into a rigid one in which the cyclohexanone ringattains a true boat form while the bonds C1–C7 and C4–C7 are at its flagpole positions [see (**a4**) and (**a4'**) in Fig. 6.45]. The two cyclopentane rings attain *envelope* conformations [see (**a3**) and (**a3'**) in Fig. 6.45]. The two faces of the molecule—*endo* (concave) and *exo* (convex)—are diastereotopic. In camphor the *exo* (here top or β-) face of the C=O group is sterically more hindered than the *endo* face, since one of the overhanging geminal methyls is projected towards the CO group in the *exo* face. Consequently, LiAlH$_4$/NaBH$_4$ or any other nucleophile attacks the carbonyl carbon predominantly from the *endo* face forming the *exo* or β-alcohol, isoborneol as the major product (see reaction (i), Fig. 6.51). In case of norcamphor (=norborneone) predominant attack takes place from the less hindered *exo* face producing the *endo* (or α-) alcohol, norborneol as the major product (see Fig. 6.51).

6.3.4   Meaning of Structural Representation

Interestingly, it may be mentioned that Roald Hoffmann (NL 1981) selected camphor (globally household name with a long chemical and social history) molecule to explain how "We abstract a piece of reality to show it to another person" [60]. Commonly (+)-camphor (**A**) is expressed as (**a1**), which is a graphic shorthand for the structure (**a2**), which again can be represented in three-dimensional structures (**a3**) and (**a4**) (Fig. 6.45). Roald Hoffmann then moves further by ascending the ladder of complexity in representation [60], and shows its ORTEP [61] diagram (X-ray), space-filling diagram, and finally considers the vibrations of the molecule and distribution of electrons in space which are not nailed down at static positions of the structures already represented. They are all representative models [60] of the (+)-camphor molecule. He thus elegantly explains the meaning of a structural representation of a molecule. The mirror image representations (a1'), (a2'), (a3'), and (a4') of (−)-camphor (A') molecule are also shown in Fig. 6.45.

6.3.5 Synthesis

(1) The first partial synthesis of (+)-camphor was achieved [62–64] by Albin Haller in 1896, starting from (+)-camphoric acid, a degradative product of (+)-camphor (Fig. 6.46). Komppa in 1903 completed the total synthesis of (±)-camphoric acid [65, 66] (Fig. 6.47) from which (+)-camphoric acid was obtained by resolution. So Komppa's synthesis constitutes the first formal total synthesis of (+)-camphor. It is interesting to note the following: The Na–Hg reduction step produces α-campholide (B) and not β-campholide (C) (the other possibility), as is evident from the ultimate synthesis of (+)-camphor (Fig. 6.46). In 1960 Otvös et al. have shown by using labeled $-CH_2^{14}COOH$ in homocamphoric acid that the labeled carboxyl group is lost in the pyrolysis of its calcium salt, as shown in Fig. 6.46 .

(2) (±)-Camphor has been synthesized [68] in a biomimetic fashion (Fig. 6.48) from (+)-dihydrocarvone via enol acetylation, followed by a brief treatment

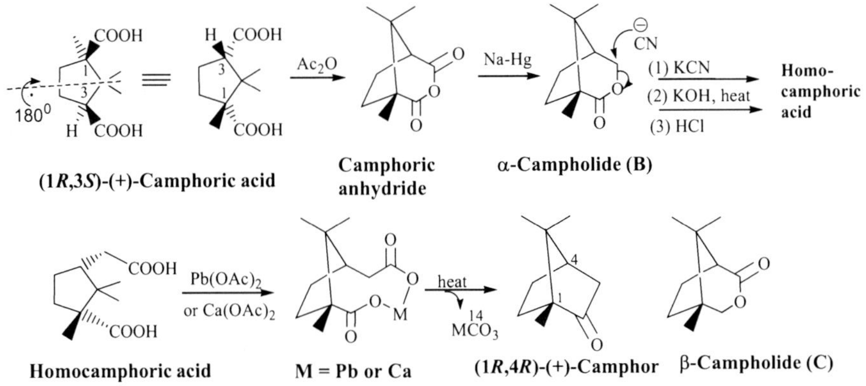

Fig. 6.46 Conversion of (+)-camphoric acid to (+)-camphor [63, 64]

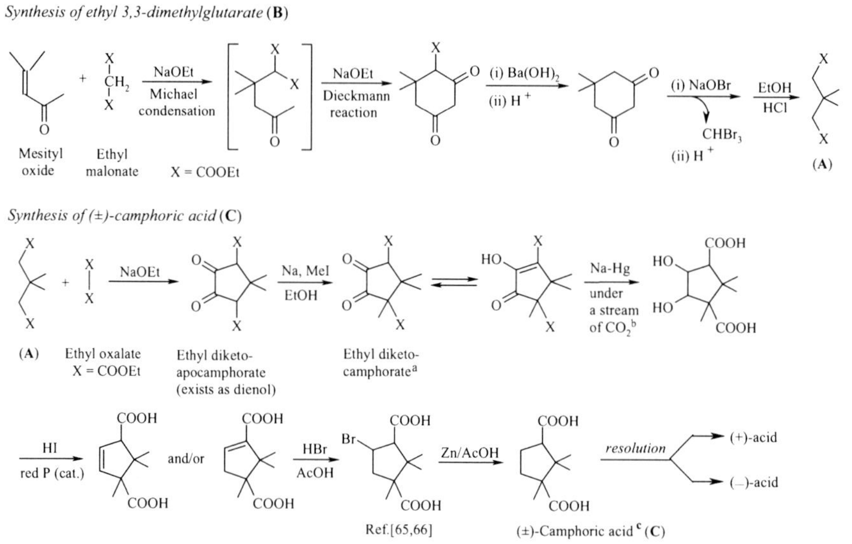

Synthesis of ethyl 3,3-dimethylglutarate (**B**)

Mesityl   Ethyl
oxide   malonate   X = COOEt   (**A**)

Synthesis of (±)-camphoric acid (**C**)

(**A**)   Ethyl oxalate   Ethyl diketo-   Ethyl diketo-
           X = COOEt   apocamphorate   camphorate^a
                                (exists as dienol)

                                                Ref.[65,66]   (±)-Camphoric acid^c (**C**)

^{a}*Exclusive C -methylation (by Na, MeI) claimed by Komppa, but disputed by chemists for several decades who believed it O-methylation, was finally settled by NMR studies in 1968 [67].*
^{b}*Double reduction of the diketone is mediated by Na -Hg, and concomitant hydrolysis of the ester groups takes place under a stream of CO$_2$.*
^{c}*Ready formation of its anhydride indicates cis-orientation of the 1,3 COOH groups.*

Fig. 6.47 Synthesis of (±)-camphoric acid [65, 66]

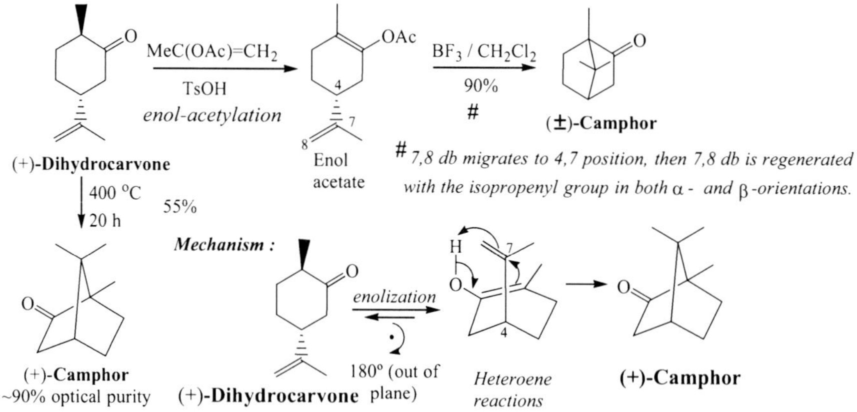

Fig. 6.48 Biomimetic synthesis of (±)-camphor [68] and (+)-camphor [69] from (+)-dihydrocarvone

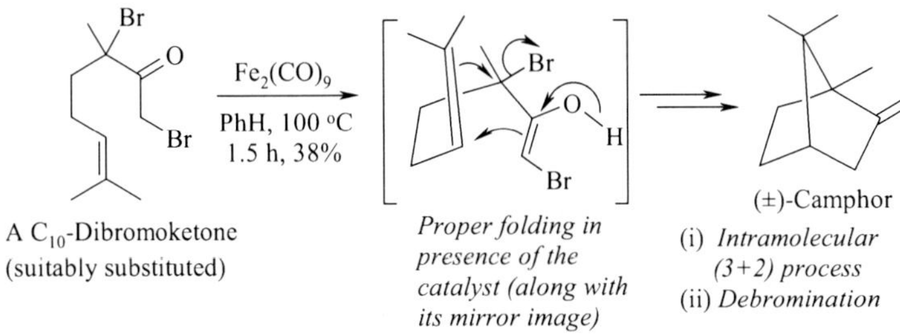

Fig. 6.49 Synthesis of (±)-camphor from a C_{10}-dibromoketone-iron carbonyl reaction [70]

with BF_3 in dilute CH_2Cl_2 solution. This reaction serves as a chemical analogy for the biosynthetic conversion of a monocyclic monoterpene into a bicyclic one. In this case racemization may be caused by the migration of the 7–8 double bond to the 4–7 position, followed by regeneration of the 7–8 double bond with the isopropenyl group in both α- and β-orientations, and formation of the bridge. However, (+)-dihydrocarvone when heated at 400 °C for 20 h gave (+)-camphor in 55 % yield (Fig. 6.48) [69]. This is an example of an *oxygen containing heteroene reaction*.

(3) (±)-Camphor has been obtained in a low yield from a C_{10}-dibromoketone (Fig. 6.49) [70].

6.3.6 Industrial Preparation of Camphor

Camphor, a very valuable material for commercial use, is now industrially prepared in thousands of tons from the readily available monoterpene α-pinene, which is abundant in the turpentine collected from pine tar (obtained from coniferous trees). The route followed involves generally Wagner–Meerwein (*W–M*) rearrangement via camphene. There are several modifications of this route [71] (Fig. 6.50) of which catalytic conversion with activated clay is the oldest one. Solid catalyst like TiO_2 has also been used. Camphene when treated with methanol in the presence of methanol-wetted cation exchange resin forms 2-methoxybornane which is subjected to catalytic oxidation by NO_2/O_2 at 10–15 °C to afford camphor. However, whatever may be the source of the catalyst, the structural framework must have been changed via W–M rearrangement (Fig. 6.50). In fact, Wagner (George Wagner, 1849–1903) first noticed the conversion of bornyl chloride and borneol to

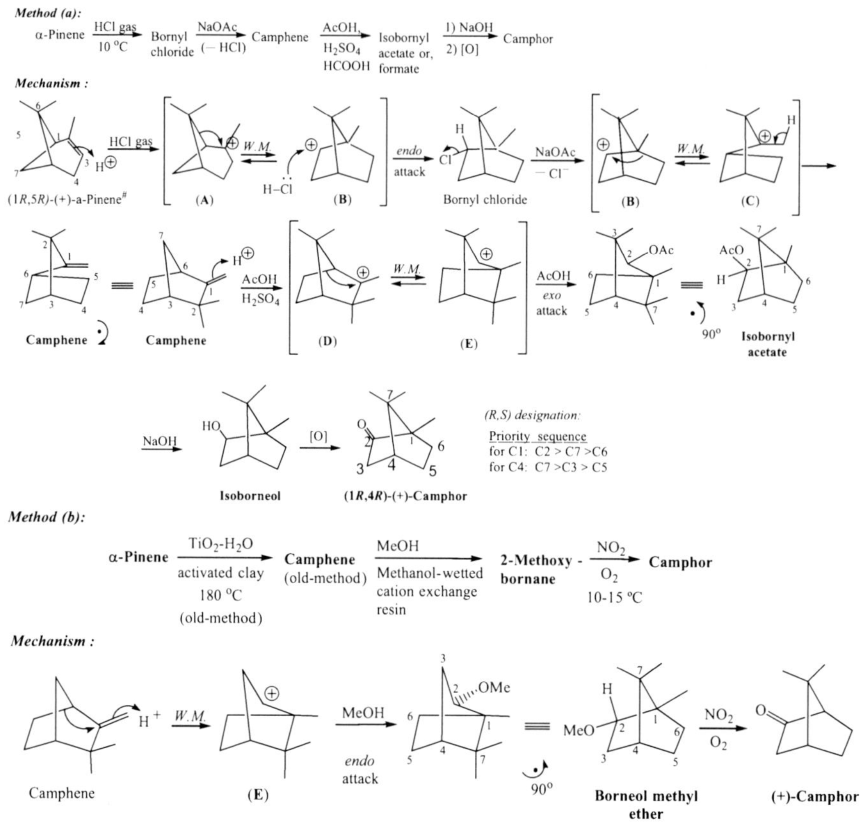

Method (a):

α-Pinene $\xrightarrow[10\,°C]{\text{HCl gas}}$ Bornyl chloride $\xrightarrow[\text{(– HCl)}]{\text{NaOAc}}$ Camphene $\xrightarrow[\substack{H_2SO_4 \\ HCOOH}]{\text{AcOH}}$ Isobornyl acetate or, formate $\xrightarrow[2)\,[O]]{1)\,\text{NaOH}}$ Camphor

Mechanism :

Camphene $\xrightarrow{\text{NaOH}}$ Isoborneol $\xrightarrow{[O]}$ (1R,4R)-(+)-Camphor

(R,S) designation:

Priority sequence
for C1: C2 > C7 > C6
for C4: C7 > C3 > C5

Method (b):

α-Pinene $\xrightarrow[\substack{\text{activated clay} \\ 180\,°C \\ \text{(old-method)}}]{TiO_2\text{-}H_2O}$ Camphene $\xrightarrow[\substack{\text{Methanol-wetted} \\ \text{cation exchange} \\ \text{resin}}]{\text{MeOH}}$ 2-Methoxy-bornane $\xrightarrow[\substack{O_2 \\ 10\text{-}15\,°C}]{NO_2}$ Camphor
(old-method)

Mechanism :

The carbocations (A), (B), (C) are in equilibrium. In presence of HCl gas (B) undergoes attack by Cl$^-$ from endo-face and forms Bornyl chloride. In presence of TiO$_2$ α-pinene forms camphene by loss of H$^\oplus$ from the carbocation (C). Cation (E), formed from (D), undergoes exo attack by AcOH (due to steric crowding of the endo face), whereas in presence of methanol-wetted cation exchange resin which perhaps blocks the exo face, it undergoes endo-attack by MeOH to form borneol methyl ether.

Fig. 6.50 Methods (a) and (b) for industrial preparation of camphor and their stepwise mechanisms

camphene, and Meerwein (Hans Lebrecht Meerwein, 1879–1965) realized the need of the rearrangement of the skeleton in such a process. *This is the genesis of Wagner–Meerwein rearrangement (W–M)* (Chap. 31). Industrially (±)-α-pinene is usually converted to (±)-camphor. In Fig. 6.50 we propose probable mechanistic pathways for the conversion of (+)-α-pinene to (+)-camphor via three Wagner–Meerwein rearrangements by two methods, for clear understanding.

6.3.7 Spectral Data of Camphor

^{13}C NMR (CCl$_4$) spectral data [72]

UV (MeOH)	λ_{max} 288 nm (ε 36) (isolated C=O)
IR (CCl$_4$)	ν_{max} 1,740 (1,760) (C=O), 1,422, 1,050 cm^{-1}
^{1}H NMR (CCl$_4$)	δ 0.83, 0.85, 0.96 (3H s each, 3t-CH$_3$'s)
MS	m/z 69, 83, 95, 108, 109 (M$^+$–CO–Me), 152 (M$^+$)
ORD and CD	1R, 4R)-(+)-Camphor exhibits the positive Cotton effect of a single isolated electronic transition.

6.3.8 Biosynthesis of Camphor [1, 2, 6]

The biosynthesis of the enantiomers (1R,5R)-(+)-α-pinene and (1S,5S)-(−)-α-pinene from GPP (endo conformer) by its specific enzyme-induced helical mirror image folding and enzyme-catalyzed six steps has already been discussed and illustrated in Fig. 6.6. Again (+)-α-pinene or (−)-α-pinene, being catalyzed by specific enzymes (+)- and (−)-camphor synthases, present in camphor producing plant cells, undergoes Wagner–Meerwein (W–M) rearrangement followed by successive nucleophilic attack of water on the incipient carbonium ion and enzymatic oxidation of the formed alcohol with the help of NAD$^+$ to generate (1R, 4R)-(+)-camphor or (1S, 4S)-(−)-camphor, respectively (Fig. 6.51).

2^{14}C-Geraniol was fed in the plant *Salvia officinalis*. 2^{14}C-Geranyl pyrophosphate (2^{14}C-GPP) formed in the cells underwent specific folding followed by the steps, shown earlier in Fig. 6.6 (lower half), and now shown with label in the lower half of Fig. 6.51, produced, as expected, (−)-2^{14}C-camphor [1, 2]. Thus incorporation of 2^{14}C-geraniol to produce 2^{14}C-camphor demonstrated the biosynthetic pathway of camphor, as shown.

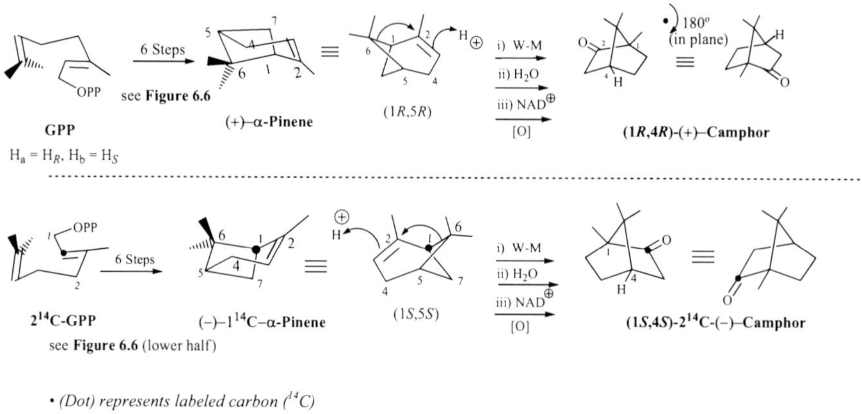

• (Dot) represents labeled carbon (^{14}C)

Fig. 6.51 Biosynthetic pathways for the stereospecific formation of (1R,4R)-(+)-camphor and (1S, 4S)-2^{14}C-(−)-camphor via (+)-α-pinene and (−)-α-pinene, respectively

6.3.9 Reactions

(i) *Stereochemistry of nucleophilic reactions* (metal hydride/Grignard reactions). A reversal of the stereochemical course of the reduction with LiAlH$_4$ or other metal hydrides, as apparent from the ratio of the products formed, is observed between norcamphor and camphor (Fig. 6.52) [73–77]. Norcamphor, lacking all three methyl groups, offers a less crowded *exo* face to the reagent and the reagent encounters one methylene bridge hydrogen and one pseudoequatorial C-3-methylene hydrogen for interactions in the transition state (TS). The *endo* face offers interactions with three pseudoaxial hydrogens in the TS. Hence the attack from the less sterically hindered *exo* face is preferred, forming the *endo* alcohol as the major product. While in camphor one of the C7-geminal methyls, the *syn* methyl, is protruded to the *exo*-face of C=O, making it crowded enough to cause preferential *endo* attack by the reagent. Hence *exo* alcohol appears as the major product in the product mixture. Reduction resulting from attack of the metal hydride from the less sterically hindered side is enhanced when the reagent is changed from lithium aluminum hydride to the more bulky lithium trimethoxya- luminium hydride, more so in a concentrated metal hydride solution indicating that such a reagent aggregates into dimeric or trimeric species [73, 74, 76, 77]. Brown (NL 1979) introduced lithium triisoamylborohydride, a new sterically hindered reagent, for the reduction of cyclic ketones with exceptional stereoselectivity [75]. Two reviews [76] and [77] are available on this topic.

Reactions [(ii)–(ix)] of camphor are delineated in Fig. 6.52 showing reaction conditions and products as well as their stereochemistry, whenever applicable, mechanism, and references.

(i) *Stereochemistry of nucleophilic reactions (metal hydride/Grignard reactions)*

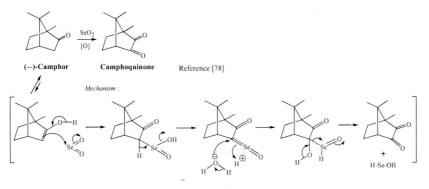

RM = LiAlH$_4$	92%	8%		9%	91%
RM = LiAl(OMe)$_3$H^c	99%	1%		2%	98%
RM = LiAl(o-t-Bu)$_3$H^c	93%	7%		7%	93%
RM=NaBH4, iPrOH (in i-PrOH)	86%	14%		14%	86%
RM = RMgX	100%	0%		3-0%	97-100%

^{a}*In case of camphor approach to the top face is strongly hindered by the overhanging syn-Me gr at C7; so attack takes place almost exclusively from the much less hindered bottom (endo) face, the more so, the more bulky the nucleophilie is.*

^{b}*In case of norcamphor there is no hindrance from the exo (top) face whereas hindrance to approach from the endo face (concave) remains, so nucleophiles approach almost exclusively or mainly from the exo face.*

^{c}*The stereoselectivity, is more in the case of the less bulky lithium trimethoxyaluminium hydride because of its tendency to aggregate into dimeric or trimeric species, whereas lithium tri-t-butoxyaluminium hydride remains monomeric and thus actually behaves like the less bulky nucleophile than the polymeric trimethoxyaluminium hydride.*

(ii) *Oxidation with SeO$_2$*

(**−**)-**Camphor**   **Camphoquinone**   Reference [78]

Mechanism :

(iii) *Wittig reaction* (Georgi Wittig, 1897-1987)   **(iv)** *Formation of a sterically hindered imine*

^{t}BuOK $^+$/ PhH
Ph$_3$ P=CH$_2$, 100°, 2h
91%

$\overset{\oplus}{H}$, Si(OH)$_4$,
H$_2$NCHPh$_2$
72%

(A') (**−**)-**Camphor**   Wittig product   **(A)** (+)-**Camphor**   A sterically hindered imine
Reference [79]   Reference [80]

Fig. 6.52 (continued)

(v) *Deblocking of the camphor ketal*

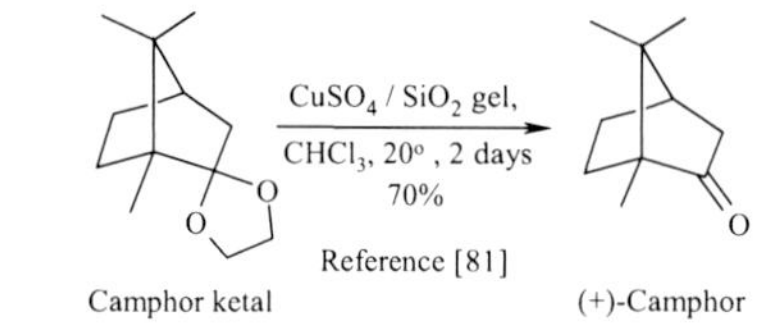

Camphor ketal

CuSO$_4$ / SiO$_2$ gel,
CHCl$_3$, 20°, 2 days
70%

Reference [81]

(+)-Camphor

(vi) *Hydrogenolysis of camphor ketal*

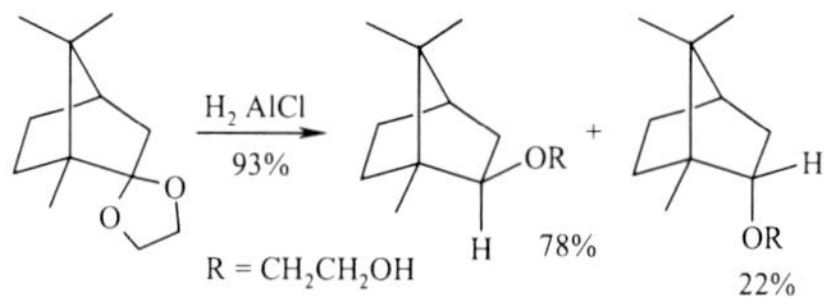

H$_2$ AlCl
93%

R = CH$_2$CH$_2$OH

OR
78%

+

OR
22%

Reference [82]

Predominantly exo reduction product is formed in conformity with the behavior of the parent ketone.

(vii) *Diastereoselective cyanation of camphor*

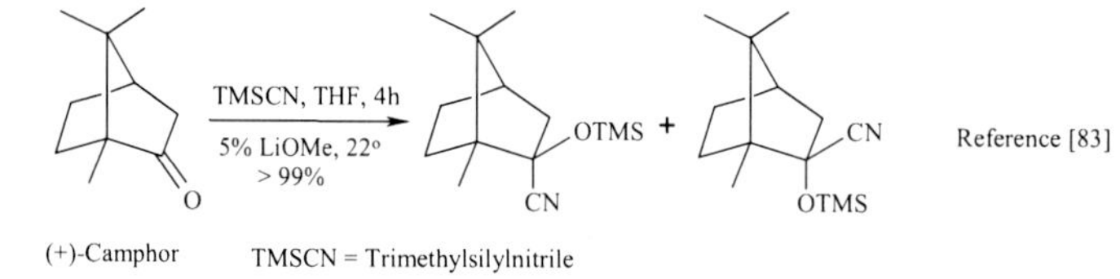

TMSCN, THF, 4h
5% LiOMe, 22°
> 99%

OTMS   +

CN

CN

OTMS

Reference [83]

(+)-Camphor   TMSCN = Trimethylsilylnitrile

(viii) *Shapiro modification of Bamford-Stevens reaction*

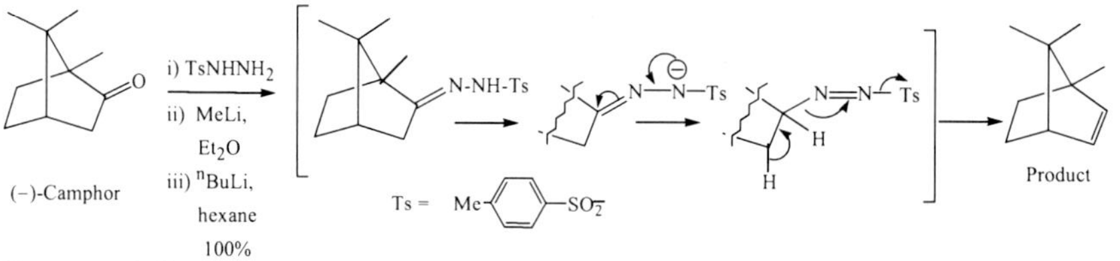

i) TsNHNH$_2$
ii) MeLi,
Et$_2$O
iii) ^{n}BuLi,
hexane
100%

(–)-Camphor

N-NH-Ts

Ts =   Me——SO$_2^-$

Product

This reaction reliably gives the alkene without skeletal rearrangement or competing insertion reaction.

(ix) *Baeyer-Villiger (Adolf-von-Baeyer and V. Villiger) oxidation*

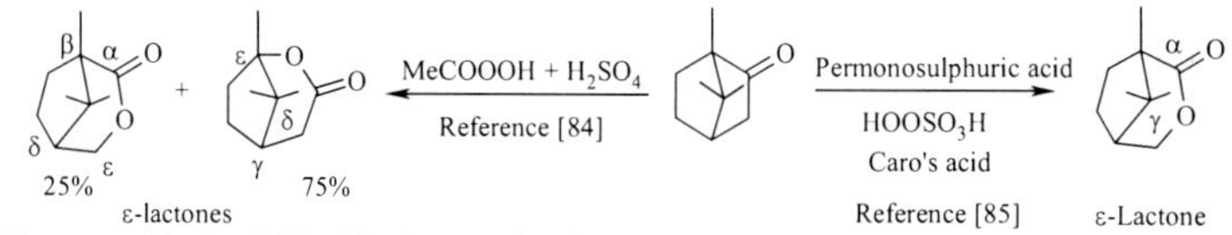

MeCOOOH + H$_2$SO$_4$

Reference [84]

25%   75%

ε-lactones

Permonosulphuric acid
HOOSO$_3$H
Caro's acid

Reference [85]   ε-Lactone

This is one of the first (1899-1900) few examples of Baeyer-Villiger oxidation
In the Baeyer-Villiger oxidation of (+)-camphor with peracetic acid a minor product (A), C$_{10}$H$_{16}$O$_4$ was obtained. The mechanistic pathway of its formation is shown. [86].

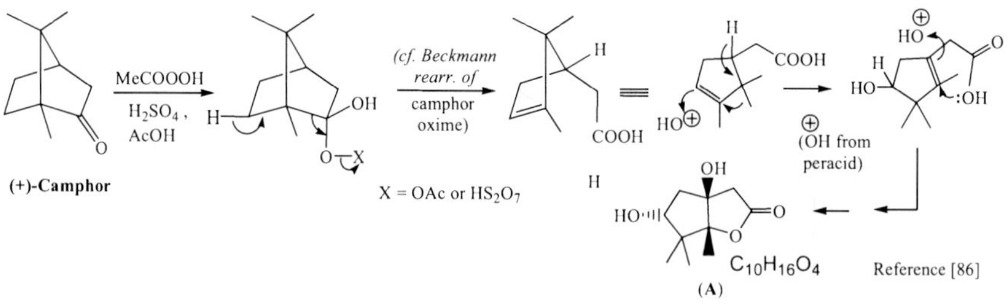

MeCOOOH
H$_2$SO$_4$,
AcOH

(+)-**Camphor**

(cf. Beckmann rearr. of camphor oxime)

X = OAc or HS$_2$O$_7$

(OH from peracid)

C$_{10}$H$_{16}$O$_4$ Reference [86]

(A)

Fig. 6.52   Reactions (i)–(ix) of camphor

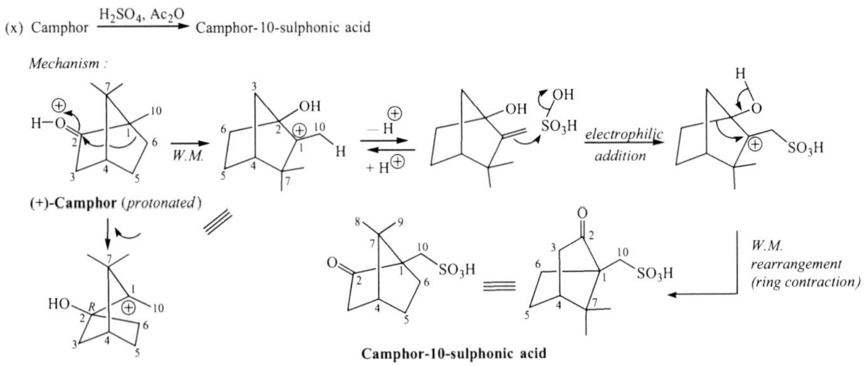

Fig. 6.53 Mechanism of formation of camphor-10-sulfonic acid from camphor

6.3.9.1 Functionalization of Camphor and Its Derivatives

Regiospecific and stereospecific (Reaction x) functionalizations of different carbon atoms of camphor and its derivatives, various bond cleavages, and rearrangements have been elegantly reviewed [87]. An example of such a stereo- and regiospecific functionalization leading to the preparation of camphor-10-sulfonic acid and the probable mechanism of the steps involved are shown in Fig. 6.53.

6.3.9.2 Ring Contraction Reaction of Camphor

Camphor on treatment with PCl_5 (or PCl_3) (Reaction xi) formed a rearranged chloro compound which upon ozonolysis and $NaBH_4$ reduction followed by treatment with NaH produced a ring contracted bicyclo[2.1.1]hexan-1-carboxaldehyde [88] (Fig. 6.54). The latter underwent Cannizaro reaction to form the corresponding primary alcohol and the carboxylic acid.

6.3.9.3 Ring Contraction by Photolysis

Photochemistry of various cyclic ketones and the elimination of CO promoted by radical formation have been reported [89]. The elimination of CO and formation of other products [89] from camphor may be explained as shown in Fig. 6.55. Photolysis of camphor is also reported to give campholenic aldehyde, a natural rearranged monoterpene [90] occurring in *Junisperus communis* [91]. The corresponding nitrile is also formed by heating the γ,γ-dimethylallyl ether of camphor oxime in 62 % yield (Fig. 6.55).

(xi)

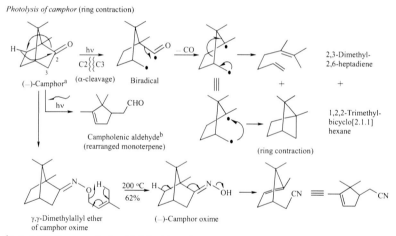

Fig. 6.54 Ring contraction reactions of camphor

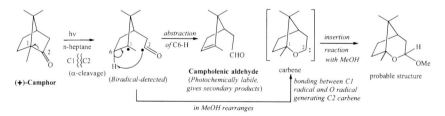

a In the lit. [90] this structure has been labeled as (+)-camphor since abs. config. was not known

*b It has been isolated as a natural product from the oil of **Junisperus communis** [91], 60 years after its identification as a photolysis product [90]*

Fig. 6.55 Photolysis products of camphor [90]

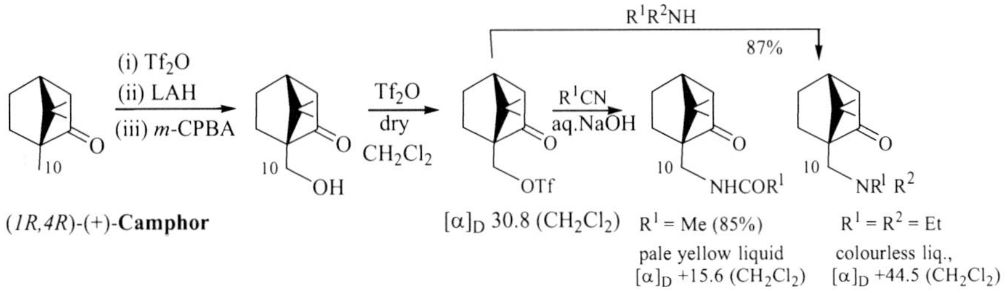

Fig. 6.56 (+)-Camphor as a chiral auxiliary

6.3.10 Camphor as a Synthon and a Chiral Auxiliary

Robert Horton in the start of his elegant route to the first synthesis [92] of taxol [93] (approved by US FDA for its use for the treatment of ovarian cancer and breast cancer, see Sect. 8.4 and Chap. 33), simultaneously with its first synthesis by Nicolaou's group, used (1S, 4S)-(−)-camphor as the cheap and readily available source of chirality to construct the eight-membered B-ring of taxol with some of its desired functionalities.

A number of enantiopure 10-aminocamphors (including amides) have been prepared in simple straightforward steps in overall good yields from (1R,4R)-(+)-camphor [94] (Fig. 6.56). They allow enantiospecific access to them and thus lead to the preparation of 10-N-substituted-camphor-based chirality transfer agents (chiral auxiliaries) (see Chap. 32) and drugs.

6.3.11 Bioactivity and Uses

Camphor has a long history of use in various religious rituals in different cultures and parts of the globe. Camphor is readily absorbed through skin and it imparts a cooling and a little anesthetic feeling like menthol. However, a large dose (>500 mg) when ingested is poisonous and can cause seizures, neuromuscular hyperactivity, irritability, and mental confusion. Generally 2 g causes toxicity and 4 g is potentially lethal.

It is extensively used as a *preservative* or *an embalming fluid* and also for *medicinal purposes*, in several cough preparations, such as "Vicks vaporub" and "Buckleys," as a cough suppressant and a tropical analgesic. Having embalming properties it is used as an active ingredient in some anti-itch gel.

Its rigid bridged structure with two distinctive faces and availability of both (+) and (−) forms make it and its various derivatives ideal molecules to function as *chiral auxiliaries* in asymmetric synthesis (Chap. 32). Modern uses include as a *plasticizer* for cellulose nitrate and as a *moth repellent* in clothes.

Recently carbon nanotubes were successfully synthesized using camphor in chemical vapor deposition process [95].

Camphor is mostly used as a flavoring agent for sweets in Asia. In ancient and medieval Europe it was widely used as an ingredient for sweets and in the dessert dish preparations in India, where it is known as "*Karpooram.*" It is available in Indian grocery stores and is labeled as "edible camphor." It is used as a scent in religious rituals. In Hindu pujas (worships) and ceremonies crude camphor is burned in a ceremonial spoon for performing "*Aarti*" or prayer.

6.4   Menthol

6.4.1   *Introduction*

(−)-Menthol, also known as levomenthol, $C_{10}H_{20}O$, m.p. 42–45 °C, $[\alpha]_D$ −50 (racemic variety, m.p. 36–38 °C), b.p. 212 °C, white or colorless crystalline solid, occurs in various mint oils, especially peppermint oil (*Mentha piperita*) (Labiatae/ Lamiaceae) along with small amounts of (−) menthone and (−)-menthyl acetate. It is one of the most used terpenoids. *Mentha arvensis* (Indian name podina/pudina), cultivated in China, is the primary species of mint used to make natural (−)-menthol crystals and flakes. This species is mainly grown in Uttar Pradesh in India.

(−)-Menthol (**1**)

(1*R*,2*S*,5*R*)-2-isopropyl-
5-methylcyclohexanol

Menthol has local anesthetic and counterirritant properties and is widely used to relieve minor throat irritation. Menthol with its characteristic odor when inhaled gives a soothing and relieving effect in nasal congestion, and hot menthol vapor as well as menthol-lozenges relieves the sore throat. In fact, plants of mint families have been used as a pain reliever for centuries. Various types of applications and uses of menthol have been stated in Sect. 6.4.8.

6.4.2   Reactions, Structure, Absolute Configuration

(−)-Menthol was subjected to the reactions shown in Fig. 6.57 for obtaining structural information. It is evident from the reactions that the gross structure of menthol is 2-isopropyl-5-methylcyclohexanol. It has been shown by correlation with D-glyceraldehyde and from a study of chemical and optical relationship and the Auwers–Skita rule that (−)-menthol has (1R,2S,5R) configuration (**1**). The structure of menthol has been confirmed by the synthesis of (±)-menthol via (±)-menthone. The absolute configuration of (−)-menthol at C1 was determined as (R)- by Prelog by applying his method (Sect. 2.11.5) [96, 97]. The phenylglyoxalate esters of (−)-menthol (**1**) and (−)-neomenthol (**3′**) [the 1-epimer of (+)-menthol (**1′**) (see Fig. 6.59)] upon reaction with methyl Grignard gave in each case after saponification predominant amount of (R)-(−)-atrolactic acid PhC(Me)OH. COOH indicating (R)-configuration of C1 in both diastereomers (2.11.5). The relative configuration of the diastereomers of menthol being known from chemical studies, the absolute configuration of (**1**) and (**3′**) at C1 determines the absolute configuration of all the diastereomers and their enantiomers at the chiral centers (at C1, C2, and C5) (see Fig. 6.59). The absolute configuration of (−)-menthol has also been established by X-ray crystallographic studies [98].

6.4.3   Spectral Data

IR: ν_{max} 3,335 (OH), 1,048, 1,020, 994, 977, 920, 876, 845 cm^{-1}

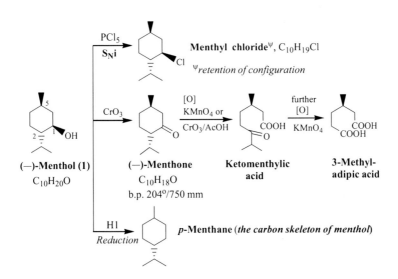

Fig. 6.57 Some reactions of menthol

^{1}H NMR (CDCl$_3$): δ 0.82 (3H, d, 5-CH$_3$), 0.90 and 0.93 (3H each, d, CH$_3$'s at C7), 3.42 [1H, m, H1 (>CHOH)]

^{13}C NMR (CDCl$_3$): δ 71.5 (C1), 50.2 (C2), 23.3 (C3), 34.6 (C4), 31.7 (C5), 45.2 (C6), 25.8 (C7), 16.1 (C8), 21.9 (C9) and 22.2 (C10)

MS: m/z 156 (M$^{\bullet+}$), 123, 95, 81, 71 (100), 55, 43, 41

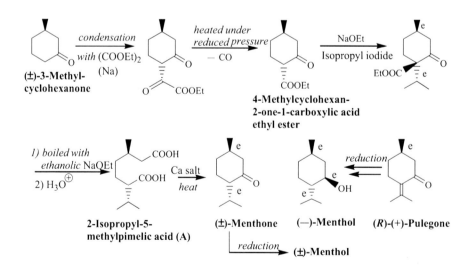

6.4.4   Synthesis of (±)-Menthone and (±)-Menthol

The structure of menthone has been confirmed by its synthesis by Kötz and Schwarz (1907) by the distillation of the calcium salt of 2-isopropyl-5-methylpimelic acid (**A**), the synthesis of which is delineated stepwise in Fig. 6.58. In each step racemic product was obtained, although the stereochemistry of each product shown corresponds to that of natural (−)-menthol.

(±)-Menthol may be obtained in good yield by reduction of (±)-menthone. This may be explained in terms of its conformation. Again (+)-pulegone of known structure and stereochemistry gives (−)-menthol on reduction.

Fig. 6.58 Synthesis of (±)-menthone

6.4.5 Stereoisomers of Menthol, Their Conformations, and Relative Stability

The projection formulae of the four diastereomers of menthol having three dissimilar chiral centers and their enantiomers are shown in Fig. 6.59. Ideally we should draw their preferred conformational structures rather than their projection formulae, and by analogy with cyclohexane we expect the six-membered cyclohexane ring to exist in the chair form. The conformation of menthol is particularly interesting because all the substituents occupy equatorial (*e*) positions, and hence it is the most stable diastereomer. All eight stereoisomers are known and their configurations are as shown in Fig. 6.59.

The most stable diastereomer (−)-menthol occurs in various *Mentha* species in more amounts than other diastereomers. Menthol possesses *anancomeric* (meaning *fixed in one conformation*—the word is derived from Greek *anankein*, which means to fix by some fate or law, common example, 4-*t*-butylcyclohexanol) or biased conformation, as shown, its flipped conformer (**1f**) possessing all three substituents

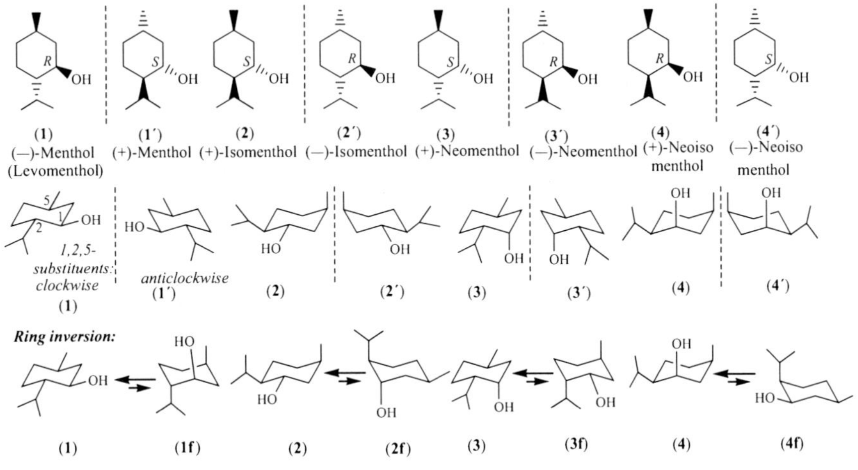

(1)	**(1′)**	**(2)**	**(2′)**	**(3)**	**(3′)**	**(4)**	**(4′)**
(−)-Menthol (Levomenthol)	(+)-Menthol	(+)-Isomenthol	(−)-Isomenthol	(+)-Neomenthol	(−)-Neomenthol	(+)-Neoiso menthol	(−)-Neoiso menthol

fa−ΔG° (approx.) 4.55   2.81   3.35   1.0

(kcal/mol)

% comp. ~100%   ~0%   ~99%   1%   ~99.5%   ~0.5%   ~15%   ~85%

*ΔG° represents the conformational energies of the diastereomers, obtained by addition of conformational energies (−ΔG° values) of OH (.6 kcal/mol), Me (1.74 kcal/mol) and iPr (2.21 kcal/mole) [4] for each conformer of any particular diastereomer. Taking the syn-axial interaction between OH and Me as 2.0 kcal/mole (approx), the −ΔG° in case of neoisomenthol (**4**) becomes ~1 kcal/mole in favor of the conformer (**4f**).*

Fig. 6.59 The configurations and conformations of menthol stereoisomers

in axial (*a*) orientation, being almost nonexistent. The isopropyl group in isomenthol and neomenthol is also equatorial (*e*) in the highly predominant preferred conformations (**2**) and (**3**), respectively, and these two diastereomers may also be regarded as anancomeric (see approximate $-\Delta G°$ values [99] and equilibrium concentrations in Fig. 6.59). The conformer (**4f**) of (+)-neoisomenthol having axial isopropyl group, and equatorial methyl and OH groups [and having no *syn*-axial interaction between OH and Me groups as in conformer (**4**)], is more stable and hence predominant (~85 %).

Thus, in view of the relative stability of the preferred conformers of the four diastereomers (in terms of their substituents) their *stability order* is expected to be as follows:

menthol > isomenthol > neomenthol > neoisomenthol

To alleviate the *syn*-axial interaction between Me and OH groups in neoisomenthol the cyclohexane ring might exist in a distorted chair conformation, decreasing the conformational energy to some extent.

6.4.5.1   Relative Rates of Esterification

It is known that the equatorial (*e*) hydroxyl undergoes esterification by acid chlorides at a faster rate than the axial (*a*) hydroxyl at the same location (epimeric molecules) (*case of steric hindrance*). The rate of esterification of a flexible molecule depends upon the specific rates and mole fractions of its (*e*) and (*a*) conformers and is expressed by the Winstein–Holness equation:

$$k = keNe + kaNa,$$

where k = specific reaction rate, *ke* and *ka* are specific rate constants of (*e*) and (*a*) conformers, and *Ne* and *Na* are the approximate mole fractions of these conformers, respectively.

Careful consideration of the approximate mole fractions (derived from % populations, Fig. 6.59) of the flipped conformers and their specific rate constants reveal that (−)-menthol (**1**), (+)-isomenthol (**2**), (+)-neoisomenthol (**4**), and (+)-neomenthol (**3**) (and their enantiomers) should follow (*e*), (*e*), (*ea*), and (*a*) pathways, respectively, in their esterification reactions. Thus, one can rationalize the following *relative rates of esterification* of these diastereomers, reported by Read et al. in 1934.

$$(-) - menthol(\mathbf{1}) : (+) - isomenthol(\mathbf{2}) : (+) - neoisomenthol(\mathbf{4})$$
$$: (+) - neomenthol(\mathbf{3}) = 16.5 : 12.3 : 3 : 1$$

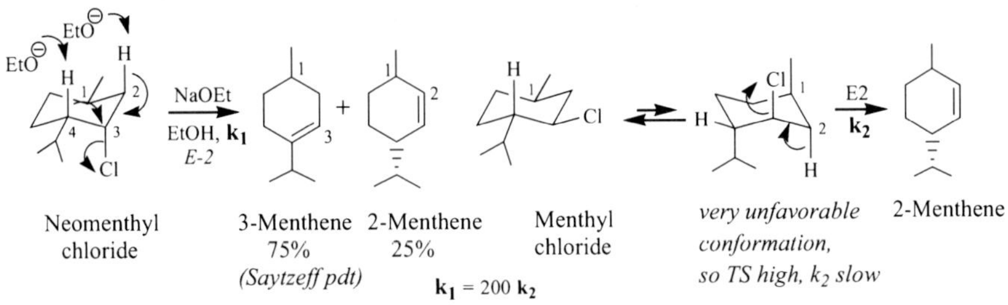

Fig. 6.60 Dehydrochlorination of neomenthyl chloride and menthyl chloride

6.4.5.2 Ionic Elimination Reactions of Menthyl Chloride and Neomenthyl Chloride

Neomenthyl chloride undergoes E2 elimination upon heating with ethanolic sodium ethoxide to produce the trisubstituted olefin, 3-menthene (75 %) along with the disubstituted olefin, 2-menthene (25 %) at a rate 200 times faster than does menthyl chloride, which under the same condition produces only 2-menthene (Fig. 6.60). These facts can be explained on the basis of the 1,2-diaxial eliminations (stereoelectronic requirement) in E2 reactions.

Neomenthyl chloride readily loses elements of hydrogen chloride at a fast rate involving axial Cl at C3 and axial H from C4 to produce predominantly 3-menthene, following the usual Saytzeff rule (double bond formed in the most substituted position); alternatively axial H from C2 may be lost to form 2-menthene, contrary to Saytzeff rule. On the contrary, menthyl chloride having equatorial chlorine can undergo dehydrochlorination only after ring inversion or flipping to an unfavorable conformation in which all three substituents including the chloride become axial. The only axial H next to the axial Cl is at C2, which is eliminated in an E2 fashion to produce the only product 2-menthene (anti Saytzeff product). Evidently, the rate in this case is much slower [99].

6.4.6 Commercial Synthesis of (−)-Menthol (Takasago Process)

Commercial synthesis of (−)-menthol is of great interest since this molecule is consumed on a massive scale in flavoring, pharmaceutical, and other applications and uses (Sect. 6.4.8). The Takasago process constitutes the major route to (−)-menthol.

6.4.6.1 Retrosynthetic Analysis and Strategy [100]

It was known that (−)-menthol (**1**) could be produced in one step through hydrogenation of isopulegol (**7**), which could arise from a Lewis acid-induced carbonyl
ene cyclization of (R)-citronellal (**6**) (Fig. 6.61). The latter reaction is particularly
productive because it simultaneously creates the six-membered ring and the two
contiguous stereogenic centers through an ordered transition state (see TS,
Fig. 6.62). Large-scale preparation of enantiomerically pure (R)-citronellal (**6**)
with a single stereogenic center is possible through hydrolysis of (R)-citronellal-
(E)-N,N-diethyl-enamine (**5**), the projected product of an enantioselective isomerization of prochiral N,N-diethylgeranylamine (**4**). Compound (**4**) could be
constructed stereoselectively from myrcene (**3**) and diethylamine through
telomerization. Myrcene could be obtained by the thermal cracking of (−)–βpinene,

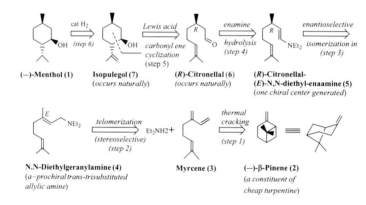

Fig. 6.61 Retrosynthetic analysis of (−)-menthol (for the Takasago process)

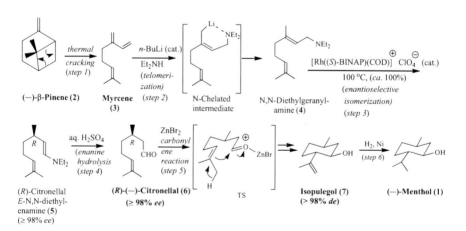

Fig. 6.62 Takasago process for the asymmetric synthesis of (−)-menthol

a major constituent of cheap turpentine. Takasago Corporation used this elegant plan to develop a highly practical and financially viable industrial process of manufacture of (−)-menthol, as delineated in Fig. 6.62.

6.4.6.2 Commercial Asymmetric Synthesis of (−)-Menthol Starting from (−)-β-Pinene (Fig. 6.62)

In this commercial synthesis [100] the *first step* constitutes thermal cracking of (−)-β-pinene (2), a constituent of cheap turpentine, to give myrcene (3). As was already known, *n*-butyllithium could catalyze the reaction of a 1,3-diene, here myrcene, with a secondary amine like diethylamine, to give regio- and stereoselectively diethylgeranylamine (4), a *trans*-trisubstituted allylic amine (*step 2*). This type of addition process is referred to as *telomerization* (Fig. 6.62).

The most important central step in Takasago process is the asymmetric double bond isomerization (*step 3*) of *N,N*-diethylgeranylamine (4) to (*R*)-citronellal *E*-diethylenamine (5) in quantitative yield (*ee* >98 %), upon treatment of the former at 100 °C with a small quantity of the catalyst [Rh((*S*)BINAP)(COD)]$^+$ClO$_4^-$. The catalysts [Rh((*S*)-BINAP)(MeOH)$_2$]$^+$ClO$_4^-$ and Rh((*S*)-BINAP)(THF)$_2$]$^+$ClO$_4^-$ are also equally effective. Process refinements permitted this reaction to be conducted on a ton scale at substrate:catalyst ratios of 8,000:1 to 10,000:1. The enantiomerically nearly pure enamine product (5) can be distilled directly from the reaction mixture at low pressure, and the (*S*)-BINAP-Rh (I) catalyst can be recycled. Using this effective catalytic asymmetric reaction as the central step the Takasago Corporation produces approximately 1,500 tons of (−)-menthol and other terpenes annually. It may be mentioned here that diethylnerylamine, the corresponding Z-isomer of (4), would also produce (5) by switching to the enantiomeric (*R*)-BINAP-Rh(I) catalyst (cf. Fig. 6.38 for enantioselective hydrogenation). This catalytic asymmetric process (*step 3*) is thus economical, efficient, and also very flexible. The remote $\Delta^{6,7}$ double bond is impervious to such asymmetric reactions.

The enamine (5) is then converted to (*R*)-citronellal (6) by the mild action of aqueous sulfuric acid (*step 4*). (*R*)-Citronellal (6) produced in this manner is of much higher enantiomeric purity (98–99 % *ee*) than the natural citronellal which is at best 80 % enantiomerically pure. In the next step (*step 5*) the neighboring C3 stereocenter having (*R*)-configuration in (−)-citronellal (6) guides the stereochemical course of the carbonyl ene cyclization to produce (−)-isopulegol (7). Thus, treatment of (*R*)-(−)-citronellal (6) with the Lewis acid ZnBr$_2$ (or ZnCl$_2$) forms (−)-isopulegol in >98 % *de*. Isopulegol (7) having all three ring substituents in equatorial orientations arises naturally from a chair-like transition state structure (TS), in which the C3 methyl group, the coordinated C1 aldehyde carbonyl, and the $\Delta^{6,7}$ double bond are all equatorial-like. This TS is expected to have lower energy than any other postulated transition state. The chemical and enantiomeric purity of

(7) is raised close to 100 % by low-temperature crystallization. Finally in the last step (*step 6*), catalytic hydrogenation of isopulegol completes the synthesis of (−)-menthol (1).

6.4.7 One-Pot Conversion of (R)-Citronellal to (−)-Menthol

Recently it has been reported [101] that cyclization of (R)-citronellal (6) over Lewis acid catalyst (step 5) may produce besides the desired (−)-isopulegol (7) three stereoisomers and other side products. One-pot full conversion of (R)-citronellal (6) to (−)-menthol (1) has been achieved [101] by treatment of (6) with beta zeotites impregnated with Ir (bifunctional catalyst) which effects consecutive acid-catalyzed cyclization and Ir-catalyzed hydrogenation at 80 °C, with 95 % selectivity for the menthol isomers, of which 75 % is the desired (−)-menthol. In this method catalyst/substrate ratio is also much less than earlier studies; 17 g (−)-menthol can be produced per g of catalyst in a single run. Presumably, this method has great industrial potential.

6.4.8 Applications and Uses

- (−)-Menthol is a substituted cyclohexanol derivative with three chiral centers and is used for the *resolution* of racemic acids via their diastereomeric esters.
- It is used as a *chiral auxiliary* in asymmetric synthesis.
  (−)-Menthol has low toxicity: Oral (rat) LD_{50}: 330 mg/kg; skin (rabbit) LD_{50}: 15,800 mg/kg. Menthol is included in many products throughout the world for various reasons as stated below:
- As a *topical analgesic* to relieve minor aches and pains such as muscle cramps, sprains, headaches, and similar conditions, alone or combined with chemicals like camphor, eucalyptus oil, or capsaicin—used as a gel or cream in Europe and as patches in the USA. Examples: *Tiger Balm* or knee/elbow *sleeves* or *IcyHot* patches.
- In *decongestants* for chest and sinuses (cream, patch, or nose inhaler). Examples: *Vicks VapoRub, Mentholatum*.
- In certain medications used to treat *sunburns* (as it provides a cooling sensation often with *Aloe*).
- As an antipruritic to reduce itching.
- Used in *oral hygiene* products and *bad-breath remedies* like mouthwashes, toothpastes, mouth and tongue-sprays and as a food flavor agent, e.g., in *chewing gum, candy*.
- In over-the-counter products for short-term relief of *minor sore throat* and *minor mouth irritation*.

- As an *additive* in *certain cigarette brands*, for flavor, to reduce throat and sinus irritation caused by smoking.
- Used in *perfumery* as menthyl esters to emphasize floral notes (especially rose).
- As a *pesticide against honey bees* or *tracheal mites*.
- In first aid products such as "mineral ice" *to produce a cooling effect* as a substitute for real ice in the absence of water or electricity (pouch, body patch, or cream).
- In some beauty products such as *hair-conditioners*, based on natural ingredients (e.g., St. Ives).
- In various patches for reducing fever, applied to children's foreheads, and in "foot patches" to relieve many ailments (more frequent in Asia, especially Japan).

References

1. A. R. Battersby, D. G. Laing and R. Ramage, Biosynthesis. Part XIX. Concerning the Biosynthesis of (–)-Camphor and (–)-Borneol in *Salvia officinalis*, *J. Chem. Soc. Perkin 1*, **1972**, 2743-2748.
2. D. V. Banthorpe and D. Baxendale, The Biosynthesis of (+)- and (–)-Camphor, *Chem. Commun.*, **1968**, 1553-1554, and the references cited.
3. Sarah E.O'Connor and Juslin J. Maresh, Chemistry and Biology of Monoterpene Indole Alkaloid Biosynthesis, *Nat. Prod. Rep.*, **2006**, *23*, 532-547, and references cited.
4. Paul J. Lentz, Jun. and Michael G. Rossmann, The Crystal Structure of Loganin Penta-acetate Monomethyl Ether Bromide, *Chem. Commun.*, **1969**, 1269.
5. Gerard J. Martin, Sophie Lavoine-Hanneguelle, Francoise Mabon and Marryvonne L. Martin, The Fellowship of Natural Abundance of ^{2}H-Isotopomers of Monoterpenes, *Phytochemistry*, **2004**, *65*, 2815-2831, and references cited.
6. Mitchell Wise and Rodney Crateau, Monoterpene Biosynthesis, in *Comprehensive Natural Products Chemistry*, (Chief Eds.: Derek Barton and Koji Nakanishi), Vol. 2, (Vol. Ed. David E. Cane), Elsevier, **1999**, pp. 97-153; pertinent pp. 128-131, and pertinent references cited.
7. David E. Cane, The Stereochemistry of Allylic Pyrophosphate Metabolism, *Tetrahedron*, **1980**, *36*, 1109-1276.
8. Rodney Croteau, Biosynthesis and Catabolism of Monoterpenoids, *Chem. Rev.*, **1987**, *87*, 929-954.
9. R. Bentley, The Nose as a Stereochemist. Enantiomers and Odor, *Chem. Rev.*, **2006**, *106*, 4099-4112.
10. Paul M., Dewick, *Medicinal Natural Products*, John Wiley, 3rd ed., **2009**, pp. 200-203.
11. B. Dinda, S. Debnath and Y. Harigaya, Naturally Occurring Iridoids. A Review, Part 1. *Chem. Pharm. Bull.*, **2007**, *55*, 159-222. Naturally Occurring Secoiridoids and Bioactivity of Naturally Occurring Iridoids and Secoiridoids. A Review, Part 2, *Chem. Pharm. Bull.*, **2007**, *55*, 689-728.
12. A. R. Battersby, R. S. Kapil and R. Southgate, Structure, Stereochemistry and Biosynthesis of Loganin, *Chem. Commun.*, **1968**, 131-133.
13. F. W. Wehrli and T. Nishida, The Use of Carbon-13 Nuclear Magnetic Resonance Spectroscopy in Natural Products Chemistry, *Fortschr .Chem. org. Naturstoffe*, **1979**, *36*, pertinent p. 24.
14. Fadila Derguini-Boumechal, Robert Lorne and Gerard Linstrumelle, Regioselective Alkylation of Allylic Grignard Reagents. A New Synthesis of Geraniol, *Tetrahedron Lett.*, **1977**, 1181-1184.

15. Bui The Khai and Antonio Arcelli, Selective Reduction of Aldehydes by a Formic acid –
    Trialkylamine-RuCl$_2$(PPh$_3$)$_3$ System, *Tetrahedron Lett.*, **1985**, *26*, 3365-3368.
16. James D. West and Boulos Zacharie, Transfer of Hydrogen from Alcohols. Catalysis by
    Compounds of Tin, *J. Org. Chem.*, **1984**, *49*, 166-168.
17. M. Nishizawa and R. Noyori, Asymmetric Synthesis of Chiral Geraniol-1-d and Related
    Terpenic Alcohols, *Tetrahedron Lett.*, **1980**, *21*, 2821-2824.
18. M. Harfenist, A. Bavley and W. A. Lazier, The Oxidation of Allyl and Benzyl Alcohols to
    Aldehydes, *J. Org. Chem.*, **1954**, *19*, 1608-1616.
19. Anthony J. Mancuso, Shui-Lung Huang and Daniel Swern, Oxidation of Long-Chain and
    Related Alcohols to Carbonyls by Dimethyl Sulphoxide "Activated" by Oxalyl Chloride,
    J. Org. Chem., **1978**, *43*, 2480-2482.
20. J. Meinwald, K. Opheim and T. Eisner, Chemical Defense Mechanisms of Arthropods
    XXXVI. Stereospecific Synthesis of Gyrinidal, A Nor-sesquiterpenoid Aldehyde from
    Gyrinid Beetles, *Tetrahedron Lett.*, **1973**, 281-284.
21. E. J. Corey, N. W. Gilman and B. E. Ganem, New Methods for the Oxidation of Aldehydes to
    Carboxylic Acids and Esters, *J. Am. Chem. Soc.*, **1968**, *90*, 5616-5617.
22. Leslie Reggel, Sideny Friedman and Irving Wender, The Lithium-Ethylenediamine System.
    II. Isomerization of Olefins and Dehydrogenation of Cyclic Dienes, *J. Org. Chem.*, **1958**, *23*,
    1136-1139.
23. Louis F. Fieser and Mary Fieser, *Reagents for Organic Synthesis*, John Wiley and Sons, New
    York, Volume 1, **1967**, p. 569.
24. J. Tsuji, I. Minami and I. Shimizu, Preparation of 1-Alkenes by the Palladium-Catalysed
    Hydrogenolysis of Terminal Allylic Carbonates and Acetates with Formic Acid –
    Triethylamine, *Synthesis*, **1986**, 623-627.
25. William B. Motherwell and L. R. Roberts, Intramolecular Cyclopropanation Reactions of
    Organozinc Carbenoids derived from Terpenoid Enals, *Tetrahedron Lett.*, **1995**, *36*, 1121-1124.
26. Robert M. Burk, Todd S. Gac and Michael B. Roof, A Mild Procedure for the Etherification
    of Alcohols with Primary Alkyl Halides in Presence of Silver Triflate, *Tetrahedron Lett.*,
    1994, *35*, 8111-8112.
27. Stephen R. Wilson and David E. Schalk, Cyclobutene Derivatives as Isoprene Equivalents in
    Terpene Synthesis. The Metathesis of 1-Methylcyclobutene, *J. Org. Chem.*, **1976**, *41*, 3928-
    3929.
28. Sotoru Ito, Koh Inoue and Masakatsu Matsumoto, [Fe$_3$O(OCOR)$_6$ L$_3$]$^+$-Catalysed Epoxida-
    tion of Olefinic Alcohol Acetates by Molecular Oxygen, *J. Am. Chem. Soc.*, **1982**, *104*, 6450-
    6452.
29. Masaki Nakamura, Nobuyasu Tsutsai, Tokuji Takeda and T. Tokoroyama, Regioselective
    Epoxidation of Geraniol with *m*-Chloroperbenzoic acid in Emulsion System. *Tetrahedron
    Lett.*, **1984**, *25*, 3231-
30. Reni Joseph, M. Sasidharan, R. Kumar, A. Sudalai and T. Ravindranathan, Chromium
    Silicate-2 (CrS-2) : An Efficient Catalyst for The Chemoselective Epoxidation of Alkenes
    with TBHP, *J. Chem. Soc., Chem. Commun.*, **1995**, 1341-1342.
31. Mark S. Cooper, Harry Heaney, Amanda J. Newbold and William R. Sanderson, Oxidation
    Reactions Using Urea-Hydrogen Peroxide; A Safe Alternative to Anhydrous Hydrogen
    Peroxide, *Synlett.*, **1990**, 533-535.
32. Tsutomu Katsuke and K. Barry Sharpless, The First Practical Method for Asymmetric
    Epoxidation, *J. Am. Chem. Soc.*, **1980**, *102*, 5974-5976.
33. Bryant E. Rossiter, Tsutomu Katsuki, and K. Barry Sharpless, Asymmetric Epoxidation
    Provides Shortest Routes to Four Chiral Epoxy Alcohols Which Are Key Intermediates in
    Syntheses of Methymycin, Erythromycin, Leukotriene C-1, and Desparlure, *J. Am. Chem.
    Soc,.* **1981**, *103*, 464-465.
34. A. Pfenninger, Asymmetric Epoxidation of Allylic Alcohols – The Sharpless Epoxidation,
    Synthesis, **1986**, 89-116.

35. Yun Gao, Robert M. Hanson, Janice M. Klunder, San Y. Ko, Hinoke Masamune and K. Barry. Sharpless, Catalytic Asymmetric Epoxidation and Kinetic Resolution. Modified Procedures Including in Situ Derivatization, *J. Am. Chem. Soc.*, **1987**, *109*, 5765-5780.

36. L. Ruzicka, Höhere Terpenverbindungen IX. Über die Total Synthese des dl-Nerolidols und des Fernesols, *Helv. Chem. Acta*, **1923**, *6*, 492-502.

37. Elias J. Corey, John A. Katzenellenbogen and Gary H. Posner, New Stereospecific Synthesis of Trisubstituted Olefins. Stereospecific Synthesis of Farnesol, *J. Am. Chem. Soc.*, **1967**, *89*, 4245-4247 and references cited.

38. Elias J. Corey and Gary H. Posner, Selective Formation of Carbon-Carbon Bonds Between Unlike Groups Using Organocopper Reagents, *J. Am. Chem. Soc.*, **1967**, *89*, 3911.

39. T. , H. Takayangi and T. Kitagawa, Synthesis of Oxocrinol, *Chem. Lett.*, **1979**, 1009.

40. Ta-Shue Chou, His-Hwa Tso and Lee-Jean Chang, Stereoselective One-step Synthesis of *trans*-β-Occimene and α-Farnesene, *J. Chem. Soc., Chem. Commun.*, **1984**, 1323-1326.

41. Den-ichi Momose, Kazuo Iguchi, Toshikazu Sugiyama and Yasuji Yamada, Reductive Coupling of Allylic Halides by Chlorotris(Triphenylphosphine)Cobalt (I), CoCl(Ph$_3$P)$_3$, *Tetrahedron Lett.*, **1983**, *24*, 921-924.

42. Akira Yanagisawa, Nobuyoshi Nomura, and Hisashi Yamamoto, Transition Metal Catalysed Substitution Reaction of Allylic Phosphate with Grignard Reagents, *Tetrahedron*, **1994**, *50*, 6017-6028.

43. T. P. Burns and R. D. Rieke, Highly Reactive Magnesium and Its Application to Organic Synthesis, *J. Org. Chem.*, **1987**, *52*, 3674-3680.

44. E. Klein and W. Rojahn, Die Permanganatoxydation von 1,5-Dienverbindungen, *Tetrahedron*, **1965**, *21*, 2353-2358.

45. D. J. Donohoe and S. Butterworth, Oxidative Cyclization of Diols Derived from 1,5-Dienes: Formation of Enantiopure *cis*-Tetrahydrofurans Using Catalytic Osmium Tetroxide: Formal Synthesis of (+)-*cis*-Solamin, *Angew. Chem.*, Int. Ed. **2005**, *44*, 4766-4768.

46. Yasuji Yamada, Hidena Sanjoh and Kazuo Iguchi, A New Type of Olefinic Cyclization of Geraniol with Thalium (III) Perchlorate, *J. Chem. Soc. Chem. Commun.*, **1976**, 997-998.

47. Dong Wang and Tak-Hang Chan, Synthesis of Karahanaenol from Geraniol *via* an Allylsilane, *J. Chem. Soc., Chem. Commun.*, **1984**, 1273-1274.

48. R. W. Skecan, G. L. Trammell and J. D. White, Cyclization of Olefinic β-Ketoesters. A Novel Synthesis of Δ$^{8(14)}$-Podocarpen-13-one. *Tetrahedron Lett.*, **1976**, 525-528.

49. S. Saito, M. Shiozawa, T. Nagahara, M. Nakadai and H. Tamamoto, Molecular Recognition of Carbonyl Compounds Using Aluminiumtris(2,6-diphenylphenoxide) (ATPH): New Regio and Stereoselective Alkylation of α,β-Unsaturated Carbonyl Compounds, *J. Am. Chem. Soc.*, **2000**, *122*, 7847-7848.

50. Tomio Hayashi, Hirashi Iwamura, Masaki Naito, Yonetatsu Matsumoto, Yasuhiro Uozumi, Misao Miki and Kazunori Yanagi, Catalytic Asymmetric Reduction of Allylic Esters with Formic Acid, Catalysed by Palladium-MOP Complexes, *J. Am. Chem. Soc.*, **1994**, *116*, 775-776.

51. W. Clark Still and Cesare Gennari, Direct Synthesis of Z-Unsaturated Esters. A Useful Modification of the Horner-Emmons Olefination, *Tetrahedron Lett.*, **1983**, *24*, 4405-4408.

52. Hidemasa Takaya, Tetsuo Ohta, Noboru Sayo, Hidenori Kumobayashi, Susumu Akutagawa, Shinichi Inoue, Isamu Kasahara and Ryoji Noyori, Enantioselective Hydrogenation of Allylic and Homoallylic Alcohols, *J. Am. Chem. Soc.*, **1987**, *109*, 1596-1597; *J. Am. Chem. Soc.* (Addition and Correction), **1987**, *109*, 4129.

53. R. Noyori, *Asymmetric Catalysis in Organic Synthesis*, Wiley Interscience, New York, **1994**, Chapter 2.

54. K. M. Madyastha and N. S. R. KrisnaMurthy, Regiospecific Hydroxylation of Acyclic Monoterpene Alcohols by *Aspergillus niger*, *Tetrahedron Lett.*, **1988**, *29*, 579-580.

55. Fang Luan, Armin Mosandl, Andreas Münch, Mathias Wüst, Metabolism of Geraniol in Grape Berry Mesocarp of *Vinitis vinifera* L. ev. Scheurebe: Demostration of Stereoselective

Reduction, E/Z Isomerization, Oxidation and Glycosylation, *Phytochemistry*, **2005**, *66*, 295-303.

56. Dictionary of Organic Compounds, Vol I, **1982**, fifth Edition, Chapman and Hall, New York, Toronto, London, p.970.

57. J. R. Maxwell, C. T. Pillinger and G. Eglinton, Organic Geochemistry, *Quart, Rev. Chem. Soc.*, **1971**, *25*, 571-628.

58. C. S. Shell, On the Unpredictibility of Odour, *Angew. Chem.*, Int. Ed. **2006**, *45*, 6254-6261 and references cited.

59. G. Ferguson, C. F. Fritchie, J. Monteath Robertson and G. M. Sim, The Crystal and molecular Structure of (+)-10-bromo-2-chloro-2-nitrosocamphane, *J. Chem. Soc.*, **1961**, 1976-1987.

60. Roald Hoffmann and Vivian Torrence, *Chemistry Imagined*, Smithsonian Institution Press, Washington and London, **1993**, pp. 24-27.

61. ORTEP stands for Oakridge (a city in Tennessee, USA with a National Laboratory) Thermal Ellipsoid Plot (Molecular Modeling)

62. K. C. Nicolaou and T. Montagnon, *Molecules That Changed the World*, Wiley-VCH, **2008**, pp. 31-32

63. Albin Haller, Sur la transformation de l'acide camphorique droit en camphre droit; synthese partitelle du camphre, *Comptes rendus*, **1896**, *122*, 446-452.

64. Albin Haller, G. Blanc, Sur la transformation de l'acide camphorique au moyen de l'acide camphorique, *Comptes rendus*, **1900**, *130*, 376-378.

65. Gustaf Komppa, Die vollständige Synthese der Apocamphersäure resp. Camphopyrsäure, *Chem. Ber.*, **1901**, *34*, 2472-2475.

66. Gustaf Komppa, Die vollständige Synthese der Apocamphersäure und Dehydrocamphersäure, *Chem. Ber.*, **1903**, *36*, 4332-4335.

67. K. Aghoramurthy and P. M. Lewis, The Komppa Synthesis of Camphoric Acid, *Tetrahedron Lett.*, **1968**, *9*, 1415-1417.

68. J. C. Fairle, G. L. Hudson and T. Money, Synthesis of (+)-Camphor, *J. Chem. Soc.*, Perkin Trans., **1973**, *1*, 2109-2112.

69. Jean M. Conia and Gordon L. Lange, Thermolysis and Photolysis of Unsaturated Ketones. 26. Preparation of Bicyclo[2.2.2]-octane-2-ones and Bicyclo[2.2.1]heptan-2-ones by Thermal Cyclization of Unsaturated Ketones. A Facile Synthesis of (+)-Camphor from (+)-Dihydrocarvone, *J. Org. Chem.*, **1978**, *43*, 564-567.

70. R. Noyori and Y. Hoyokawa, Natural Product Synthesis *via* the Polybromoketone-Iron Carbonyl Reaction, *Tetrahedron*, **1985**, *41*, 5879-5886.

71. *Total Synthesis of Natural Products*, Vol 2, Ed. John ApSimon, John Wiley Sons, New York, **1973**, p. 153-154.

72. F. W. Wehrli and T. Nishida, The Use of Carbon-13 Nuclear Magnetic Resonance Spectroscopy in Natural Products Chemistry, *Fortschritte. Chem. org. Naturst*, **1979**, *36*, 1-229.

73. Herbert C. Brown and Harold R. Deck, Selective Reductions. VIII. The Stereochemistry of Reduction of Cyclic and Bicyclic Ketones by the Alkoxy-Substituted Lithium Aluminium Hydride, *J. Am. Chem. Soc.*, **1965**, *87*, 5620-5625.

74. Herbert O. House, *Modern Synthetic Reactions*, 2nd ed., 1972, W. A. Benjamin,Inc., Menlo Park, California, pp. 62-63 and references cited.

75. S. Krishnamurthy and Herbert C. Brown, Lithium triisoamylborohydride, A New Sterically Hindered Reagent for the Reduction of Cyclic Ketones with Exceptional Stereoselectivity. *J. Am. Chem. Soc.*, **1976**, *98*, 3383-3384.

76. E. C. Ashby and J. T. Laemmle, Stereochemistry of Organometalliv Compound Addition to Ketones, *Chem. Rev.*, **1975**, *75*, 521-548.

77. J. R. Boone and E. C. Ashby, Reduction of Cyclic and Bicyclic Ketones by Complex Metal Hydrides, *Top. Stereochem.*, **1979**, *11*, 53-95.

78. R. O. C. Norman and J. M. Coxon, *Principles of Organic Synthesis*, 3rd ed., ELBS with Chapman & Hall, **1993**, pertinent pp. 602-603.

79. Lutz Fitzer and Ulrike Quabeck, The Wittig Reaction Using Potassium-tert-butoxide. High Yield Methylenation of Sterically Hindered Ketones, *Syn. Commun.*, **1985**, *15*, 855-864.

80. Brian E. Love and Jianhua Ren, Synthesis of Sterically Hindered Imines, *J. Org. Chem.*, **1993**, *58*, 5556-5557.

81. G. M. Caballero and E. G. Gros, Cleavage of Acetals Promoted by Copper (II) Sulphate Adsorbed on Silica Gel, *Syn. Commun.*, **1995**, *23*, 395-404.

82. W. W. Zajac Jr., B. Rhee and R. K. Brown, Hydrogenolysis by Lithium Aluminium Chloride of Ether Solutions of Camphor Ethylene Ketal and Norcamphor Ethylene Ketal, *Canad J. Chem.*, **1966**, *43*, 1547-1550.

83. H. Scott Wilkinson, Paul T. Grover, Charles P. Vandenbossohe, Roger P. Bakale, Nandkumar N. Bhongle, Stephen A. Wald and Chris H. Senanayake, A New Lithium Alkoxide Accelerated Diastereoselective Cyanation of Ketones, *Org. Lett.*, **2001**, *3*, 553-556.

84. R. R. Sauers and G. P. Ahearn, The importance of Steric Effects in the Baeyer-Villiger Oxidation, *J. Am. Chem. Soc.*, **1961**, *83*, 2759-2762.

85. Louis Fieser and Mary Fieser, *Advanced Organic Chemistry*, Reinhold, pp. 427-428.

86. J. D. Connolly and K. H. Overton, The Constitution and Stereochemistry of Lactone $C_{10}H_{16}O_4$, Formed in the Oxidation of Camphor by Peracids, *J. Chem. Soc.*, **1961**, 3366-3372.

87. T. Money, A Chiral Starting Material in Natural Product Synthesis, *Nat. Prod. Rep.*, **1985**, 253-289.

88. Joseph V. Paukstelis and Benoon W. Macharia, Ring Contraction of bicyclo[2.2.1]heptanes, *J. Org. Chem.*, **1973**, *38*, 646-648.

89. A. Cox and T. J. Kemp, *Introductory Photochemistry*, McGraw-Hill, London, **1971**, p. 74-75.

90. G. Ciamician and P. Sibler, Photolysis of Camphor, *Chem. Ber.*, **1910**, *43*, 1347.

91. A. F. Thomas, The Natural Occurrence of Camphonyl Skeleton, Another "Nonisoprenoid" Monoterpene System, *Helv. Chim. Acta.*, **1972**, *55*, 815-817.

92. R. A. Holton, C. Somoza, H.-B. Kim, F. Liang, R. J. Biediger, P. D. Boatman, M. Shindo, C. C. Smith, S. Kim, H. Nadizadeh, Y. Suzuki, C. Tao, P. Vu, S. Tang, P. Zhang, K. K. Murthi, L. N. Gentile and J. H. Li, First Total Synthesis of Taxol. 1. Functionalization of the B Ring, *J. Am. Chem. Soc.*, **1994**, *116*, 1597-1598.

93. K. C. Nicolaou and T. Montagnon, *Molecules That Changed the World,* Wiley-VCH, **2008**, pp. 207-218.

94. Antonio Garcia Martinez, Enrique Teso Vilar, Amelia Garcia Fraile, Santiago de la Moya Cerero and Cristina Diaz Morillo, Enantiospecific Access to 10-N-Substituted Camphors, *Tetrahedron*, **2005**, *61*, 599-601.

95. M. Kumar and Y. Ando, 'Carbon Nanotubes from Camphor: An Environmental-Friendly Nanotechnology', *J. Phys. Conf. Ser.*, **2007**, *61*, 643-646.

96. V. Prelog, Untersuchungen über asymmetrische Synthesen I. Über den sterichen Verclauf der Reaktion von α-Ketosäure-estern optisch aktiver Alkohole mit Grignard'schen Verbindungen, *Helv. Chim. Acta*, **1953**, *36*, 308-319.

97. V. Prelog and H. L. Meyer, Untersuchungen über asymmetrische Synthesen II. Über den sterichen Verclauf der Umsetzung von Phenylglyoxylsäure-estern des Menthols, Neomenthols, Borneols und Isoborneols mit Methylmagnesiumjodid, *Helv. Chim. Acta*, **1953**, *36*, 320-325.

98. J. M. Ohrt and R. Parasarathi, *Acta Cryst.* (Suppl.) **1969**, S198.

99. Ernest L. Eliel, Samuel H. Wilen and Michael P. Doyle, *Basic Organic Stereochemistry*, John Wiley & Sons, New York, **2001**, pp. 443-444.

100. K. C. Nicolaou and E. J. Sorensen, 'Menthol' in *Classics in Total Synthesis* (Targets, Strategies, Methods), VCH, Weinheim, New York, **1996**, Chapter 22, pertinent pages 354-357, and references cited.

101. Flori Iosif, Simona Coman, Vasile Pârvulescu, Paul Grange, Stephanie Delsarte, Dirk De Vos and Pierre Jacobs, Ir-Beta Zeolite as a Heterogeneous Catalyst for the One-Pot Transformation of Citronellal to Menthol, *Chem. Commun.*, **2004**, 1292-1293.

Chapter 7
Sesquiterpenoids (C_{15})

7.1 Introduction

The carbon content of monoterpenes has been taken as reference for the classification of terpenoids, like sesqui-, di-, sester-, and tri-, as they are formed by iterative condensation of C_5-isoprene units up to C_{25}. Two C_{15} units condense to form C_{30} compounds (cf. Fig. 5.15). "Sesqui" (Latin) means "one half more." Since monoterpene contains C_{10}, the next elongated chain will contain C_{15} ($C_{10}+C_5$)—thus the carbon content is one-half more compared to monoterpene and hence the prefix *sesqui* is used for C_{15}-terpenoids.

7.2 Acyclic Sesquiterpenoids: Biosynthesis

All sesquiterpenes get their skeletal carbons from *trans, trans*-farnesyl pyrophosphate (FPP). Thus, FPP serves as the fundamental molecule in the biosynthesis of sesquiterpenes. Like geranyl pyrophosphate (GPP, Sect. 6.1), the terminal (2–3) double bond undergoes $E \rightarrow Z$ isomerization, i.e., *trans, trans*-FPP (E, E) to *cis, trans*-FPP (Z, E) through successive ionization, conversion of $1°$-carbonium ion to $3°$-carbonium ion, rotation of the generated single bond, and delivery of the PP group to the carbocation (Fig. 7.1). *trans, trans*-FPP and *cis-trans*-FPP may also be expressed as FPP and *cis*-FPP, respectively.

Acyclic Sesquiterpene Hydrocarbons Acyclic sesquiterpene hydrocarbons, e.g., β-*farnesene*, *cis*-α-*farnesene*, and *trans*-α-*farnesene* are known to occur in nature. Their formation may be explained from the farnesyl carbocation (Fig. 7.2).

S.K. Talapatra and B. Talapatra, *Chemistry of Plant Natural Products*,
DOI 10.1007/978-3-642-45410-3_7, © Springer-Verlag Berlin Heidelberg 2015

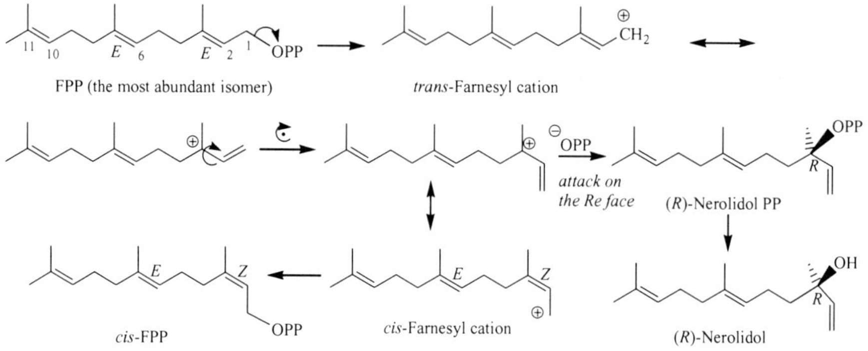

Fig. 7.1 Isomerization of FPP (E, E) to cis-FPP (Z, E) and formation of nerolidol

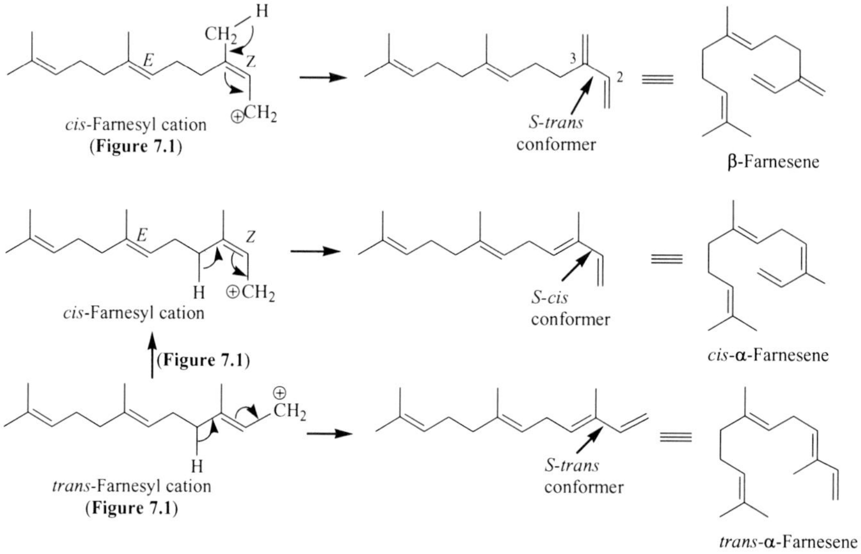

Fig. 7.2 Formation of β-farnesene, cis-α-farnesene and $trans$-α-farnesene

7.3 Cyclic Sesquiterpenoids

7.3.1 Biosynthesis. General Mechanistic Approach

Farnesyl pyrophosphate (FPP), being longer in chain length and richer in double bond content compared to geranyl pyrophosphate (GPP), can participate comparatively much more in *chemical origami* resulting into the formation of more diversified cyclic (mono-, di-, tri-, and rarely tetra-) terpenoids including fused

Fig. 7.3 *Cisoid* and *transoid* folds of FPP

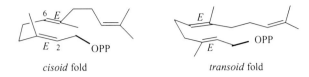

cisoid fold   transoid fold

ring systems, bridge-headed ring systems, cyclopropane, and cyclobutane systems, etc. More than 200 different types of cyclic sesquiterpenes are reported in the literature. Their formation could be conceived from FPP or its isomer *cis*-FPP via their cations involving *transoid* or *cisoid* folding (Fig. 7.3). The basic chemical reactions involved during the biosynthesis of these skeletally diversified molecules are summarized in the sequel.

All sesquiterpenes are having a common progenitor, (E, E)-farnesyl pyrophosphate (FPP). $E \rightarrow Z$ isomerization of C_2–C_3 double bond of *trans*-FPP (Fig. 7.1) is needed to generate the corresponding *cis*-farnesyl carbocation (Fig. 7.1). The *trans* carbocation reacts with the distal C10–C11 double bond prior to downstream π-facial interactions leading to the products, while *cis* farnesyl carbocation can react with either C6–C7 or C10–C11 double bond depending on the nature of folding by the relevant enzyme, prior to further transformation. Generally two different classes of *cyclases* are known—one (*transoid synthase*) is required for *trans*-farnesyl carbocation generation, while the other one (*cisoid synthase*) for *cis*-farnesyl carbocation formation—the initial requirements for subsequent conversions. Two types of fold (*cisoid* and *transoid*) of *trans*-farnesyl pyrophosphate (Fig. 7.3) may also be involved with the relevant enzymes.

The stereochemical information of the processes and hence those of the products will be dictated by a limited number of the *conformational preorganizations* of FPP and their π-cation interactions with π-facial selectivity (whenever applicable). Enzymic manipulation of various biochemical processes, e.g., hydroxylation, oxidation to various stages (e.g., $CH_3 \rightarrow CH_2OH \rightarrow CHO \rightarrow COOH$), peroxidation, dehydrogenation, reduction, lactonization, epoxidation, bond migration, etc., with total regio- and stereoselectivity involving pertinent enzymes gives rise to a plethora of diversified skeletal patterns in sesquiterpenes. Some examples are cited in Fig. 7.4.

7.3.2   Classification. Some Familiar Skeletal Patterns

Different terpene synthases recognize the compatible preorganized conformations of the cations from FPP or *cis*-FPP and direct them to different skeletal patterns. The classification of cyclic sesquiterpenes into some fully saturated skeletal patterns and their numberings are shown in the respective sections and figures given in the sequel.

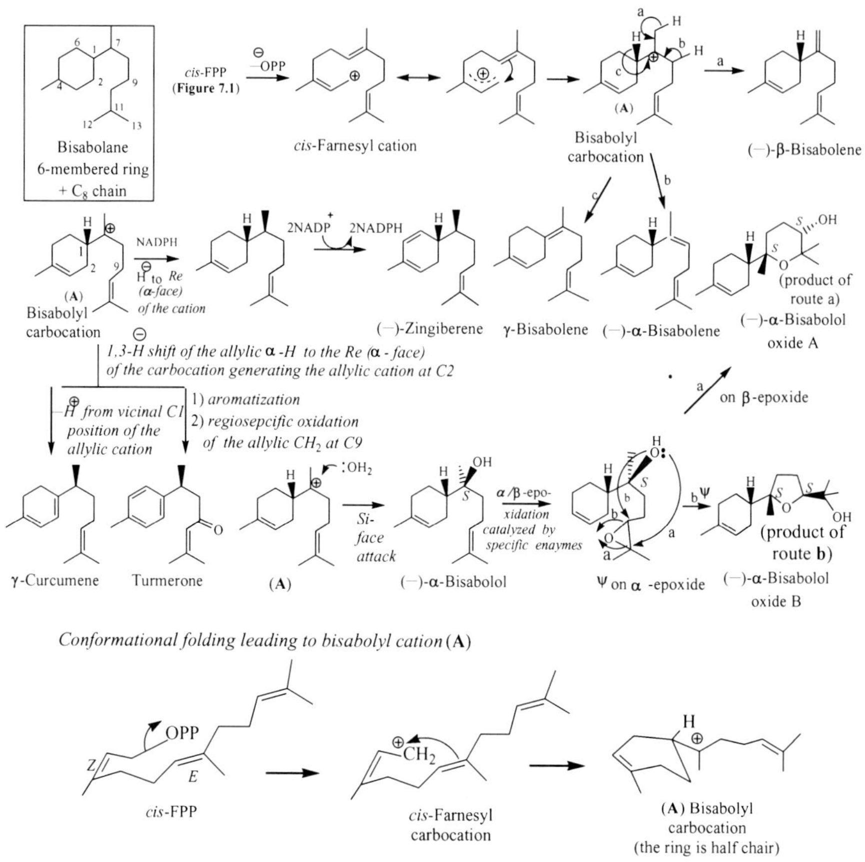

Fig. 7.4 Monocyclic sesquiterpenes derived from bisabolyl cation (**A**)

7.3.3 Monocyclic Sesquiterpenoids. Different Skeletal Patterns. Biosynthesis

Bisabolane Skeleton The plausible pathways for the biogenetic formation of some monocyclic sesquiterpenes originating from bisabolyl carbocation (**A**) are delineated in Fig. 7.4. The stereochemistry of the asymmetric centers of chiral molecules depends on the enzyme stereospecificity affecting the elimination of a particular prochiral hydrogen, or the hydride delivery on the available enantiotopic face of the olefin or carbocation.

Germacrane, Humulane, and Elemane Skeletons The probable pathways for the biosynthesis of monocyclic sesquiterpenoids possessing these skeletons are delineated in Figs. 7.5, 7.6 and 7.7 respectively.

A sesquiterpene of Humulane Skeleton. Humulene (see Sect. 7.5 also).

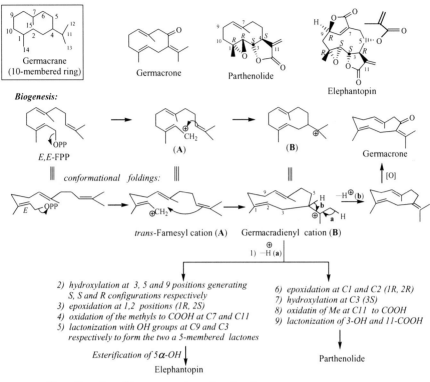

Fig. 7.5 Biosynthesis of some germacrane type monocyclic sesquiterpenes

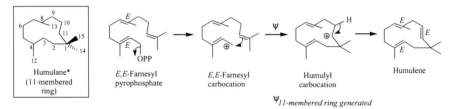

*In this numbering system the Me groups will be bonded to carbons with minimum numbers. Its name will thus be 1,1,4,8-tetramethylcyclounedecane. In any other type of nomenclature the four methyls will be bonded to carbons with higher numbers.

Fig. 7.6 Biogenesis of humulene

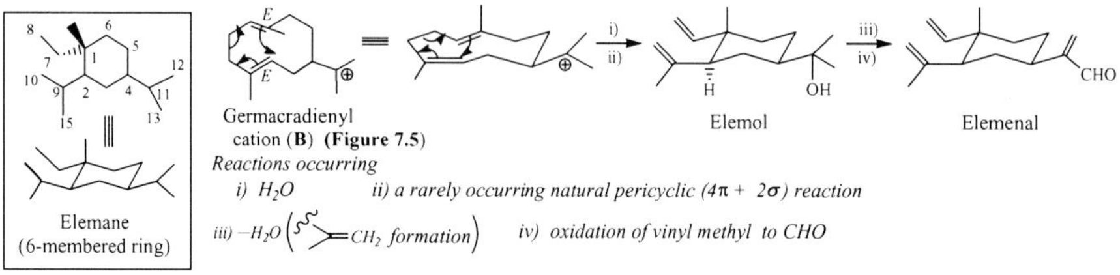

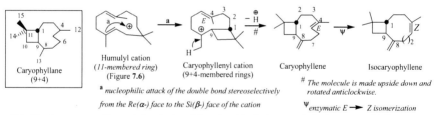

Fig. 7.7 Biogenesis of elemane type monocyclic sesquiterpenes

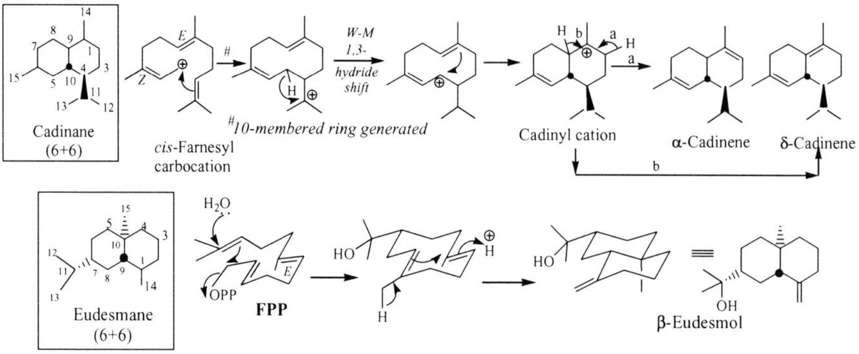

Fig. 7.8 Biosynthesis of α-cadinene, δ-cadinene, and β-eudesmol

Fig. 7.9 Biosynthesis of caryophyllene and isocaryophyllene

7.3.4 *Bicyclic Sesquiterpenoids*

The probable biosynthetic pathways of several bicyclic sesquiterpenoids possessing different skeletons, e.g., *α-cadinene*, *δ-cadinene*, and *β-eudesmol* (Fig. 7.8), *caryophyllene* and *isocaryophyllene* (Fig. 7.9), *guaiol* and *bulnesol* (Fig. 7.10), and *carotol* (Fig. 7.11) have been devised, and are delineated in the respective figures. The basic skeletons are shown in the box in each figure.

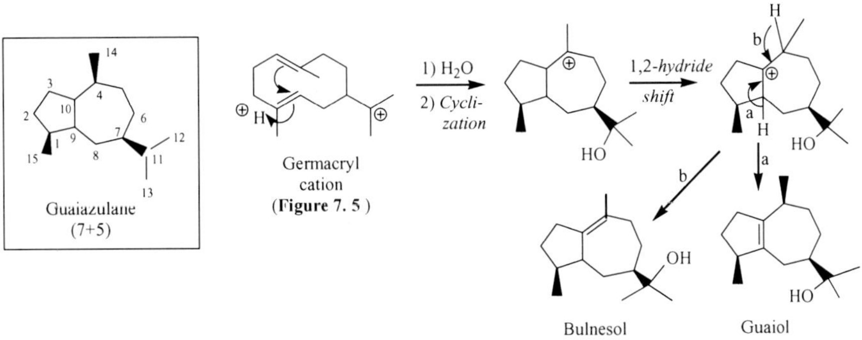

Fig. 7.10 Biosynthesis of guaiol and bulnesol

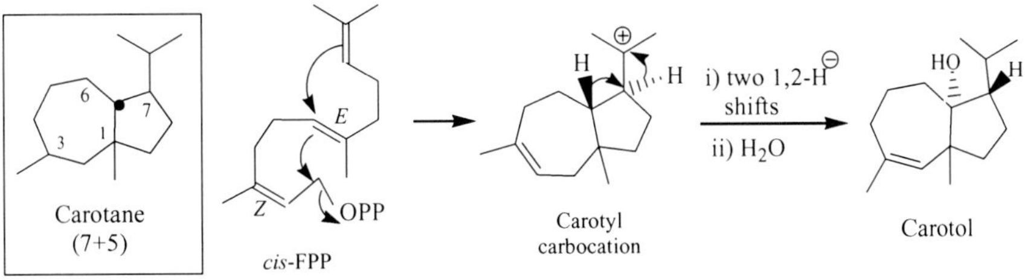

Fig. 7.11 Biosynthesis of Carotol

7.3.5 Tricyclic Sesquiterpenoids

The proposed biosynthetic sequences of sesquiterpenoids having logifolane skeleton, e.g., the longifolene enantiomers with known absolute configurations, are delineated in Fig. 7.12. Arigoni studied the biosynthesis of (+)-longifolene and achieved reasonable incorporations of activity (0.1–0.2 %) from radiolabeled 2-^{14}C-mevalonate into (+)-longifolene (Fig. 7.12), using cuttings of a young *Pinus ponderosa* tree. Applying suitable degradative methods he was able to locate the labels (Fig. 7.12). He proposed 1,3-hydride shift of the *pro-R* hydrogen at C5 of mevalonate in the humulyl cation (**A**) based on labeling experiment to form a more stable allylic carbocation, followed by two successive electrophilic attacks of the carbocations on the proximate double bonds, 1,2-migration (*W–M* type) of a proximate C–C bond to generate the carbocation (**B**). The specific enzyme-catalyzed pathways **a** or **b** is then followed to generate (+)-longifolene or (−)-longifolene. respectively. (−)-Longifolene may also be biosynthesized from the mirror image folding of the cation (**A**).

Another type of tricyclic sesquiterpene skeleton, cedrane, is known. The proposed biosynthetic sequences to the known sesquiterpenes α-*cedrene*, β-*cedrene*, *cedrol*, and *epi-cedrol* having cedrane skeleton are shown in Fig. 7.13. A densely

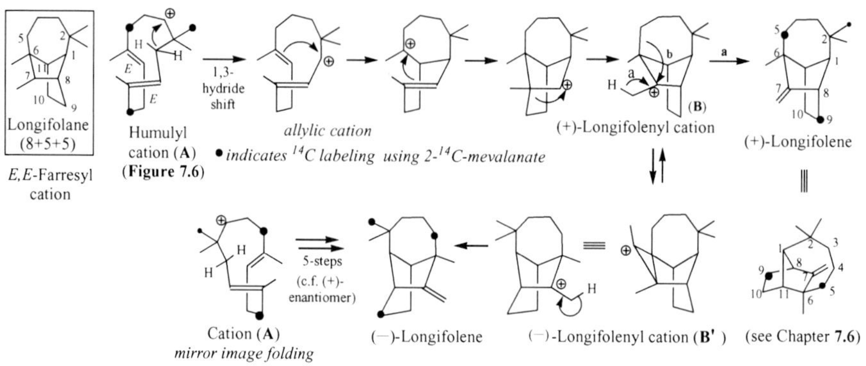

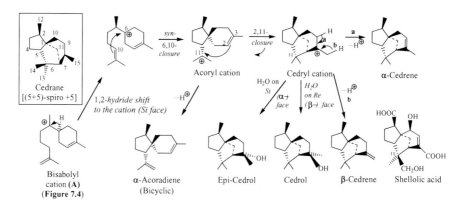

Fig. 7.12 Mechanism of the stepwise deep-seated rearrangements in the biosynthesis of (+)-longifolene and (−)-longifolene

Fig. 7.13 Biosynthesis of α-cedrene, β-cedrene, cedrol, epi-cedrol, and α-acoradiene

oxygenated tricyclic sesquiterpene *shellolic acid* is formed by oxidation of the methyls to COOH or CH$_2$OH, formation of a double bond, and oxidation of the allylic methylene to CHOH at the positions of the cedrane skeleton as shown in Fig. 7.13 by relevant enzymes. A spiro-bicyclic sesquiterpene *α-acoradiene is formed by* the collapse of the intermediate acoryl cation by loss of a proton from one of the geminal methyls (Fig. 7.13).

7.3.6   Tetracyclic Sesquiterpenoid

One example of tetracyclic sesquiterpene is *longicyclene*, derived from the longifolenyl cation (**B**) (Fig. 7.12, as shown in Fig. 7.14. Longicyclene, being devoid of any functional group or double bond, serves as the skeletal hydrocarbon for such tetracyclic sesquiterpenes.

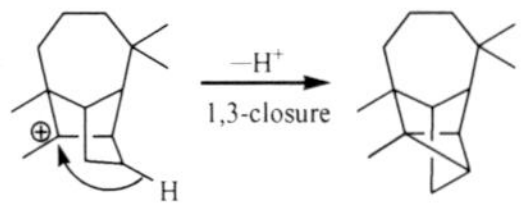

Longifolenyl
cation (B)
(Figure 7.12)

Longicyclene
(8+5+5+3)

Fig. 7.14 Biosynthesis of longicycline

7.4 Farnesol, The Parent Acyclic Sesquiterpene Alcohol

7.4.1 Introduction and Structure

Farnesol, a colorless water immiscible liquid, occurs in the oil of ambrette seeds. Abelmosk, a tropical plant (*Abelmoschus moschatus*, Fam. Malvaceae), is cultivated for musky seeds (Ambrette seeds) [1]. It also occurs in many essential oils, e.g., oils of *Acacia, Pluchea, Dioscoridis, Pitto-sporum undulatum*, and Zea Mays, possibly protecting these plants from damages induced by parasites. It is extracted from oils of rose, cyclamen (any bulbous plant belonging to the genus *Cyclamen* and primrose family with fleshy root-stock and purple, pink, and cream rose colored early blooming flowers), etc., and from the flowers of *Cassiana* species and also of *Acacia fernesiana*. The latter has been named after a noble Italian Cardinal Odoardo Farnese (1573–1626). His family was responsible for maintaining in Rome one of the first few European botanical gardens. The plant native to Caribbean and Central America was brought to the Italian Farnese Botanical Gardens and cultivated. The compound received its name from its plant source. Farnesol possesses a mild odor of flower.

The structure has been determined by Kerschbaum in 1913. Some relevant reactions pertaining to its structure are shown in Fig. 7.15.

The above reactions are consistent with the structure (**1**) of farnesol. Finally, the structure has been confirmed by its synthesis. It is the building block of possibly all sesquiterpenoids.

7.4.2 Synthesis

(i) Farnesol has been synthesized from geranyl acetate (see Fig. 6.22).
(ii) Ruzicka [2] converted nerolidol to farnesol—a case of rotation around 2,3 single bond, followed by allylic rearrangement (Fig. 7.16).
(iii) Corey synthesized farnesol from geranyl acetone in 1967 [3] (see Fig. 6.25 of Sect. 6.2).

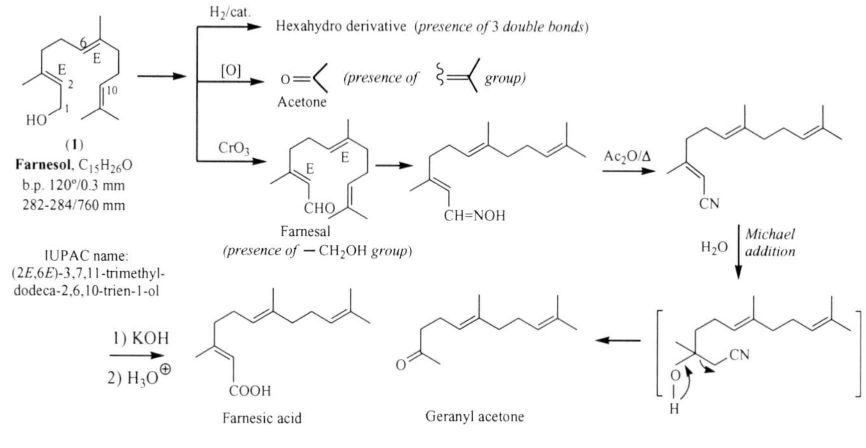

Fig. 7.15 Some reactions of farnesol

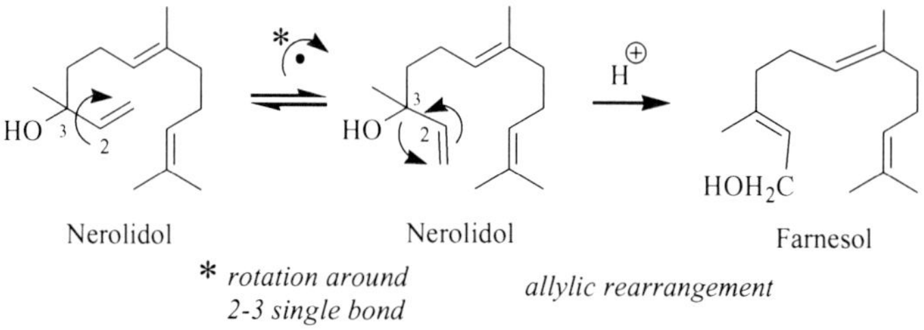

Fig. 7.16 Conversion of nerolidol to farnesol [2]

7.4.3   Biosynthesis

Farnesol pyrophosphate has been biosynthesized from one molecule of DMAPP and two molecules of IPP (Chap. 5). It undergoes hydrolysis to furnish farnesol.

7.4.4   Uses

Farnesol is a potent antimicrobial reagent. It possesses a mild sweet odor of flower and shows antibacterial properties and thus finds use in perfume industry. It helps in keeping the skin smooth and wrinkle free and in increasing skin's elasticity. It is also thought to reduce skin aging by promoting regeneration of cells and synthesis of molecules like collagen, required for healthy skin. It is, therefore, used in body lotion, shampoo, soap, etc.

7.5 Caryophyllene and Isocaryophyllene

7.5.1 Introduction

Clove oil (clove tree, *Eugenia caryophyllata, syn. Syzygium aromaticum,* Fam. Myrtaceae) serves as the major source for **caryophyllene (1)**, in which it co-occurs with **isocaryophyllene (2)** (Fig. 7.18), its biogenetic relative **humulene (1a)** (Fig. 7.17), and few other related sesquiterpenoids. Caryophyllene has been known since 1834, although the material obtained in the initial studies on clove oil was a mixture of **(1)**, **(2)**, and **(1a)**. It was obtained in pure form in 1892. Since then extensive degradative and structural studies continued. During the period 1945 to mid-1950s chemical work was done to establish its structure. Barton has been the major contributor to this field [6–9], and Corey accomplished its first total synthesis in 1963 in the racemate form [10, 11]. It has been shown to possess a bicyclo[7.2.0] undecane skeleton. Its medium-sized nine-membered ring being capable of undergoing several conformational equilibriums participates in cyclizations and rearrangements forming a large number of stereochemically interesting molecules, many of which are natural products. Formation of these natural products throws light on their probable biogenesis from caryophyllene in the plant cells.

7.5.2 Structure and Absolute Configuration

The structure of caryophyllene having a four-membered ring fused with a nine-membered ring has been deduced from the identification of its different degradation products (Fig. 7.17).

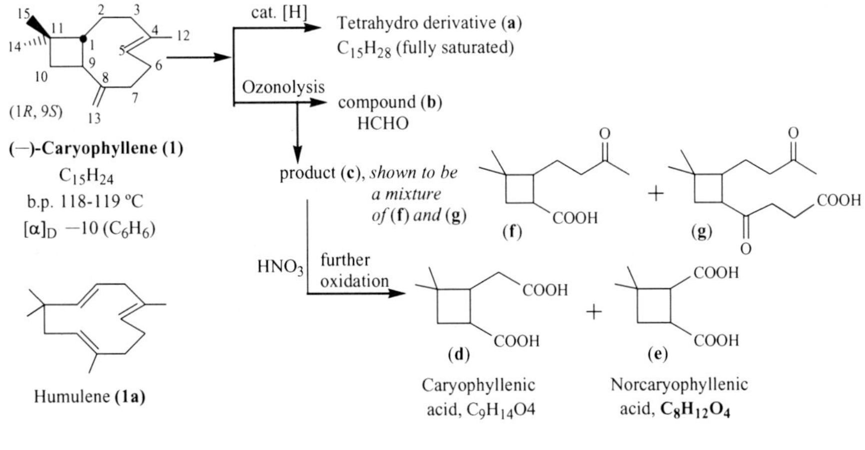

Fig. 7.17 Some degradation products of caryophyllene (1)

Formation of (a) suggests the presence of two double bonds; its hydrogen content when compared with that of the corresponding open chain saturated hydrocarbon suggests it to be a *bicyclic* system. Compound (b) (HCHO) comes from an exocyclic methylene group. Compounds (d) and (e) show the presence of a cyclobutane derivative carrying a geminal dimethyl group. The positive haloform tests of the compounds (f) and (g) require the presence of a –COCH$_3$ group in their molecules and of the –C(Me)=CH– system in the parent molecule (1). The location of the –C(Me)=CH– system in (1) follows from the structures of (f) and (g). By a consideration of the molecular rotations of some tricyclic derivatives of known structure and stereochemistry (derived from caryophyllene), the absolute configuration of the natural (–)-caryophyllene has been established by Barton [7]. Finally, the X-ray analysis of *caryophyllene chlorohydrin* [12] showed the presence of a *trans*-fused cyclobutane ring in a bicyclo[7.2.0]undecane skeleton, a *trans*-endocyclic double bond carrying a vinyl methyl group in a nine-membered ring, and an exocyclic methylene group and confirmed the structure of (–)-caryophyllene as (1) with (1*R*, 9*S*) configurations. The absolute configurations of some secondary alcohols derived from caryophyllene have been determined in 1966 by Horeau [8] by the application of his "partial resolution" method reported [9] in 1961. The results confirm the assignment of configuration of (1) made earlier [7].

7.5.3 ^{1}H NMR Spectral Data of Caryophyllene and Isocaryophyllene (Fig. 7.18)

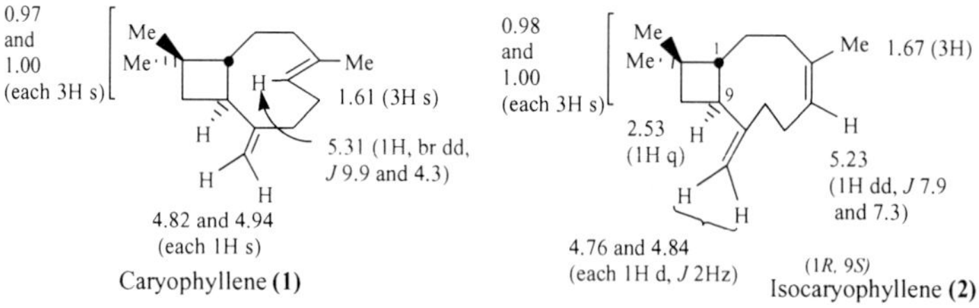

Caryophyllene (1)

Isocaryophyllene (2)

MS of (1) (EI, 70 eV, MS 9) [25]: *m/z* (% Base peak), 205 (M$^+$+1, 0.13), 204 (M$^+$, 0.78), 161 (3.5), 150 (6.8), 149 (3.4), 148 (3.3),133 (14.7), 121 (5.1), 119 (8.8), 107 (10.6), 105 (21.4), 94 (25.1), 92 (40.2), 79 (40.9), 77 (28.0), 41 (100), 40 (10.1), 39 (33.9).

Fig. 7.18 ^{1}H NMR [4] and Mass spectral [5] data of caryophyllene and isocaryophyllene

7.5.4 Synthesis of (±)-Caryophyllene

7.5.4.1 Corey's Synthesis

The first elegant total syntheses of (±)-caryophyllene (**1**) and (±)-isocaryophyllene (**2**) have been achieved by Corey and coworkers [10, 11] in 1963. The structure (**1**) represents the absolute configuration of (−)-caryophyllene, the natural enantiomer. The synthesis to be described here is not an asymmetric synthesis, and the final product is (±)-caryophyllene, i.e., racemic. For convenience, however, the structures will be drawn with the stereochemistry that will lead to (−)-caryophyllene. *At each step formation of the mirror image twin in equal amount giving the racemate is implied.*

During the chemical construction of the molecule, its following two structural features were taken care of: (i) the *trans*-fused 4-membered and 9-membered carbocyclic rings and (ii) the presence of an endocyclic *E* double bond in the 9-membered carbocyclic ring.

The endocyclic double bond of isocaryophyllene (**2**) is Z.

Though photochemical dimerization has long been known in the literature, Corey first realized the potential of this method of intermolecular addition with different olefinic substrates towards the synthesis of natural products. He put his imagination to test and successfully built the cyclobutane ring via [2+2] photocycloaddition between isobutene and cyclohex-2-enone in the desired orientation. This addition serves as the first step of the total synthesis of (±)-caryophyllene (**1**) as well as (±)-isocaryophyllene (**2**).

Two unsymmetrical alkenes can undergo [2+2] photocycloaddition to form the following possible regiochemical isomers and also stereoisomers (Fig. 7.19).

When cyclohexenone is irradiated in the presence of isobutene with UV at a longer wave length, >300 nm, the radiation will be absorbed by cyclohexenone only, while isobutene will remain in the ground state. The enone then attains a triplet state by intersystem crossing from $n-\pi^*$ excited singlet and gets polarized; the charge distribution of the enone becomes opposite to that of its ground state; consequently the β-carbon acquires δ-charge as shown in Fig. 7.20; hence, normal HOMO/LUMO interaction does not take place, and (**3a**) + (**3b**), and not (**4a**) + (**4b**), are formed.

The triplet state of the polarized enone then interacts with isobutene which is also polarized in the ground state and forms an exciplex or a charge transfer complex. Excited enone acts as the electron acceptor and electron-rich isobutene

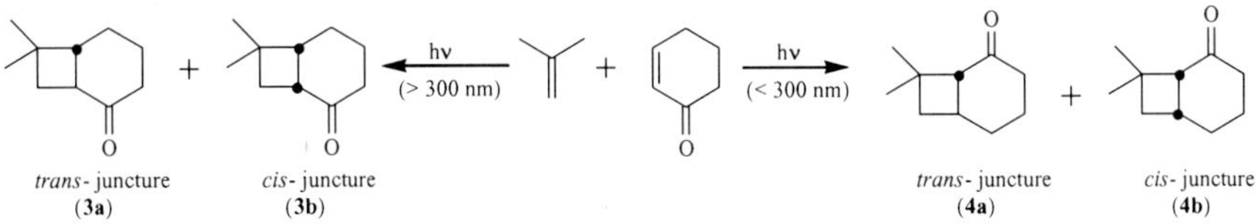

trans-juncture (**3a**)   cis-juncture (**3b**)   trans-juncture (**4a**)   cis-juncture (**4b**)

Fig. 7.19 [2+2]Photocycloaddition of isobutene and cyclohex-2-enone

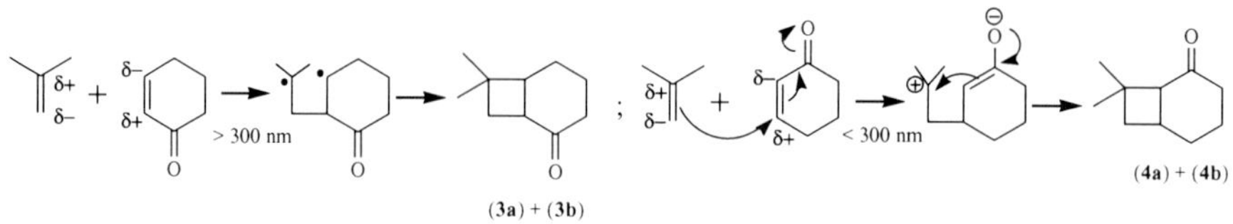

Fig. 7.20 Stereoisomers and regioisomers formed by [2+2] photocycloaddition

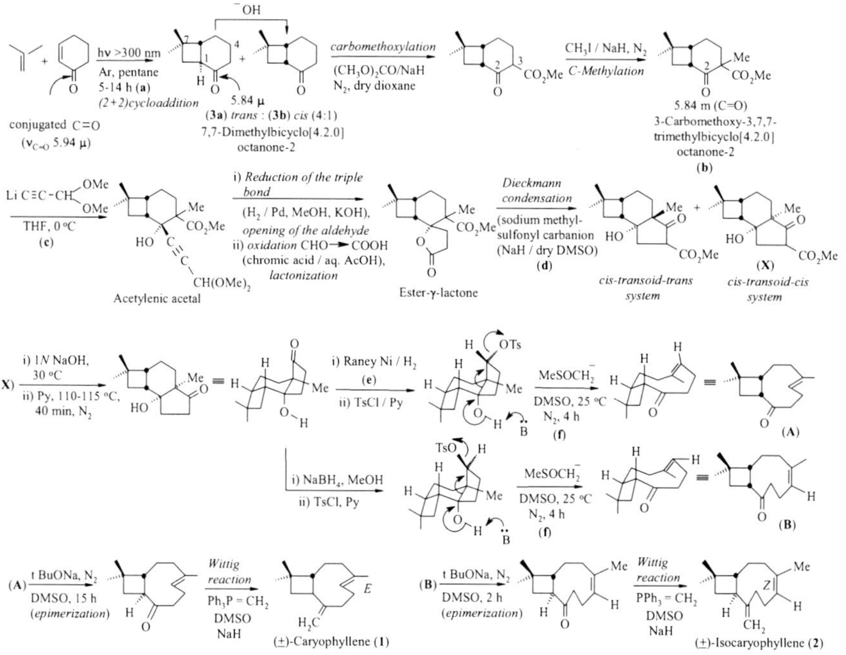

(a) Depending on the amount of the reactants, (b) Diastereoisomers are formed, the stereochemistry of the C3 chiral center is not important – as it is lost later. (c) Nucleophilic addition at the CO from the convex side (the side of the synclinal system cis to the hydrogen of the ring juncture). (d) The lactone ring opens up and probably gets esterified under the reaction condition before the Dieckmann condensation. (e) Reduction gives epimeric alcohols almost in equal amount; the desired alcohol is separated and tosylated. (f) Relative stereochemistry (cis)of the 6/5-ring juncture is of no consequence in the subsequent elimination effecting the rupture of the common bond between 6/5-membered rings. The relative stereochemistry of the angular methyl and the leaving group (OTs) confers the stereochemistry of the endocyclic double bond.

Fig. 7.21 Total Synthesis of (±)-caryophyllene and (±)-isocaryophyllene [10]

as electron donor. Finally, the nucleophilic end of isobutene attacks the electrophilic C2 of enone and then via the 1,4-diradical intermediate cis and trans isomers (3b) and (3a) are formed. The regioselectivity may be controlled by the rate of conversion of exciplex to the product. Head-to-tail regioisomers like (3a) and (3b) are the major products when electron-rich alkenes are used. In mild alkaline condition (3a) is converted to (3b).

The multistep syntheses of (±)-caryophyllene/(±)-isocaryophyllene are delineated in Fig. 7.21.

The unusual structures of caryophyllene (**1**) and isocaryophyllene (**2**), possessing fused 4- and 9-membered rings, have attracted the attention of chemists all over the world. Till 1995 five more syntheses have been reported [4, 13–18], of which we include here only two interesting syntheses.

7.5.4.2   Devaprabhakara's Synthesis

A new and improved 7-step procedure for the synthesis by Devaprabhakara et al. [14] of (±)-socaryophyllene, starting from the easily available synthon *cis–cis*-1,5-cyclooctadiene [18], is illustrated in Fig. 7.22.

7.5.4.3   Suginome's Synthesis

The newer synthesis based on a three-carbon ring expansion involving β-scission of alkoxy radicals as the key step, devised by Suginome et al. [4], is delineated in Fig. 7.23.

7.5.5   Rearrangements of Caryophyllene

The flexibility of the medium size 9-membered ring confers conformational mobility to caryophyllene. The various acid-catalyzed rearranged products of caryophyllene are competitive in their formation and are controlled by the conformation of the 9-membered ring as well as reaction condition. Two general features emerging from various rearrangements and cyclization products are the following: (i) The endocyclic *trans* 4,5 double bond of caryophyllene is more reactive than the

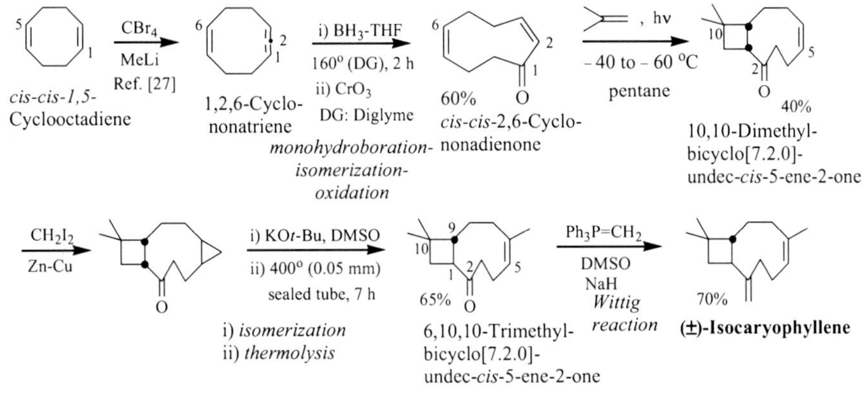

Fig. 7.22   Synthesis of (±)-isocaryophyllene by Devaprabhakara [14]

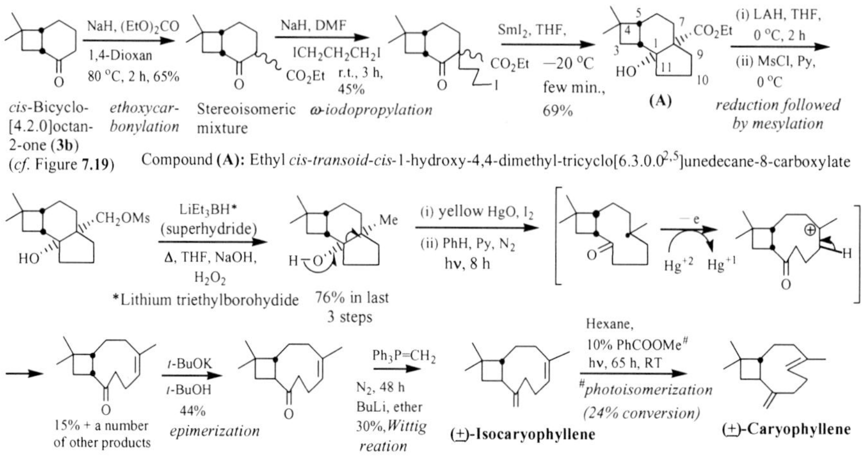

cis-Bicyclo-   ethoxycar-   Stereoisomeric   ω-iodopropylation
[4.2.0]octan- bonylation   mixture
2-one (3b)
(cf. Figure 7.19)   Compound (A): Ethyl cis-transoid-cis-1-hydroxy-4,4-dimethyl-tricyclo[6.3.0.0^{2,5}]unedecane-8-carboxylate

Fig. 7.23 Synthesis of (±)-isocaryophyllene and (±)-caryophyllene [4] (1995)

exocyclic 8(13) double bond. (ii) The system is susceptible to undergo rearrangements and cyclization reactions.

Extensive cyclization product profile analysis formed under manipulative conditions showed the formation of several natural and unnatural sesquiterpenes of much chemical interest, where the C–C bond formation takes place by way of characteristic cation-olefin addition/cyclization and 1,2-migration under strict stereoelectronic requirement providing an access to rare sesquiterpenes in fairly good quantities for their subsequent use, which otherwise would have been difficult to procure.

7.5.5.1   Conformations

Four possible conformations of (–)-caryophyllene, distinguished by the relative disposition of the exocyclic methylene and olefinic methyl groups, are shown by molecular mechanics (MM1) calculations and ^{13}C NMR studies. The predicted and experimentally determined populations of these conformations [19] are shown in Fig. 7.24, which also indicate a low inversion barrier ($\Delta G^1 = 16.25 \pm 0.11$ or 16.1 ± 0.3 kcal mol^{-1}) between the βα- and ββ-conformers. The relative populations of these conformers are reflected to some extent in the ratios of the products of various reactions of caryophyllene, *viz.*, epoxidation, hydroboration, and photooxidation, etc. [19] (*Scheme 1, p187 of* [19] *is reproduced as Fig. 7.24 with the permission of the Royal Society of Chemistry).*

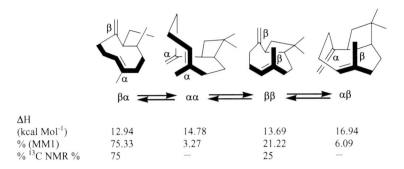

	βα ⇌	αα ⇌	ββ ⇌ αβ
ΔH			
(kcal Mol⁻¹)	12.94	14.78	13.69
% (MM1)	75.33	3.27	21.22
% ¹³C NMR %	75	–	25

ΔH (kcal Mol^{-1}): 12.94, 14.78, 13.69, 16.94
% (MM1): 75.33, 3.27, 21.22, 6.09
% ^{13}C NMR %: 75, –, 25, –

Fig. 7.24 Relative enthalpies and populations of the (–)-caryophyllene conformers in equilibrium

7.5.5.2 Rearrangements and Cyclizations

An excellent detailed account of recent advances in the chemistry of caryophyllene, including various cyclizations of caryophyllene and isocaryophyllene under different conditions, has been provided by Hanson and his collaborators in the light of their conformational mobility [19].

(i) Caryophyllene upon treatment with 3 equivalents of conc. sulfuric acid–diethyl ether at 0–20 ºC for 30 min furnishes a mixture of as many as 14 hydrocarbons and 4 alcohols. However, on keeping for three days the mixture is simplified to 3 hydrocarbons and 3 alcohols, all of which are tricyclic. The major products are **clove-2-ene (6)**, **caryan-1-ol (7)**, and **α-neoclovene (9)**. Additionally, **clove-2β-ol (5)** and **α-panasinsene (8)** are also formed in fewer amounts.

   It appears to us that the α,α-conformer generates the carbocation (A^2), which has the correct geometry to cyclize on the lower face of the molecule, followed by expansion of the cyclobutane ring to form (5) and (6) having the clovane skeleton, as shown in Fig. 7.25. In the aforesaid review [19] clove-2-ene has been shown to be formed from the β,β-conformer, which cannot be rationalized because it lacks the 4-membered ring expansion step (see Scheme 11 of [19]).

(ii) We propose that the β,β-conformer undergoes acid-catalyzed cyclization on the β-face of the molecule and a bridge is formed on the β-face of the 9-membered ring between C13 and C4; no further rearrangement takes place; only the carbocation is hydroxylated from the less hindered α-face to yield caryolan-1-ol (7) (Fig. 7.25).

In the same Scheme 11 of the aforesaid review [19] caryolan-1-ol (7) has been shown to be formed from the α,α-conformer. But a close inspection of the mechanistic sequence (Fig. 7.26) indicates that the α,α-conformer, as such, is expected to lead to a diastereomer (7a) of caryolan-1-ol (7) with the bridge formed on the α-face of the 9-membered ring between C13 and C4, followed by hydroxylation of the C8 carbocation from the β-face.

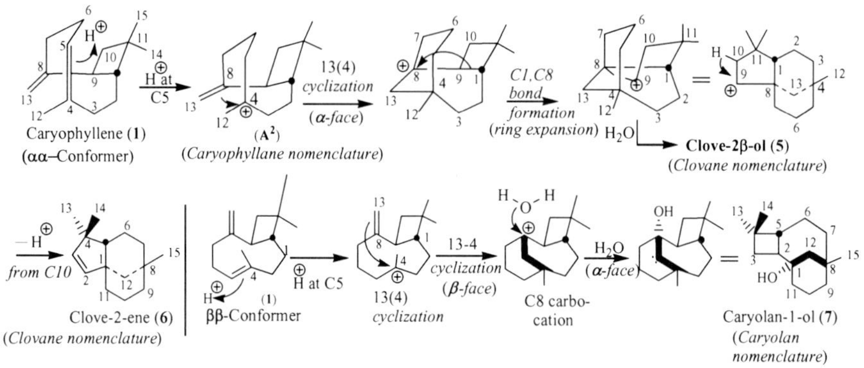

Fig. 7.25 Acid-catalyzed rearrangement of caryophyllene: formation of clove-2-ene, cloven-2β-ol, and caryolan-1-ol

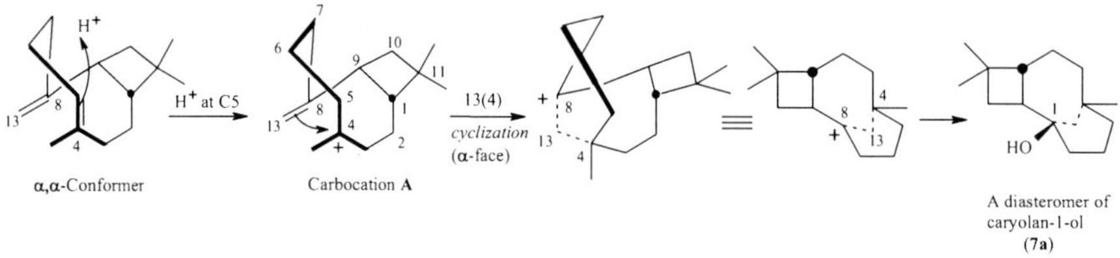

Fig. 7.26 Expected formation of the diastereomer (**7a**) of caryolan-1-ol from the α,α-conformer

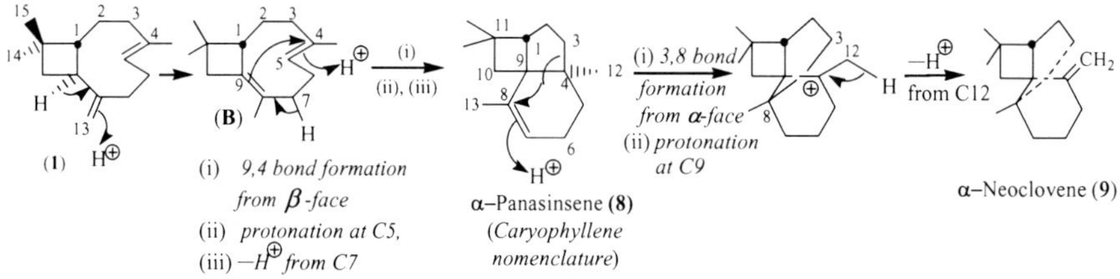

Fig. 7.27 Formation of α-panasinsene (**8**) and α-neoclovene (**9**)

The formation of the double bond isomer (**B**) of *trans*-caryophyllene (**1**) followed by cyclization yields *α-panasinsene* (**8**) (after 30 min acid treatment) which on ring expansion of the five-membered ring forms *α-neoclovene* (**9**), a product formed after 3 days (Fig. 7.27).

Solvolysis of caryophyllene derivatives leading to some tricyclic sesquiterpenes formed in a biomimetic fashion has been reported [20].

7.5.5.3 Thermal Rearrangement of (–)-Caryophyllene
to (–)-Isocaryophyllene

Pyrolysis of caryophyllene under low pressure probably causes two successive [3.3] sigmatropic rearrangements leading to the formation of isocaryophyllene [19] (Fig. 7.28).

Though terpenoids are well known for undergoing carbonium ion rearrangements, caryophyllene is regarded as the super-performer on the molecular trapeze in this regard [19].

7.5.6 Conversion of Humulene into Caryophyllene (Fig. 7.29)

The structure of humulene (**1a**) was first reported by Sukh Dev in 1951 and a full paper on humulene was published [19] in 1960. Humulene when treated with *N*-bromosuccimide (NBS) in aqueous acetone gives a mixture of a monobromo hydrocarbon (**B**) and a monobromo alcohol (**C**) (Fig. 7.29). Compound (**C**) is

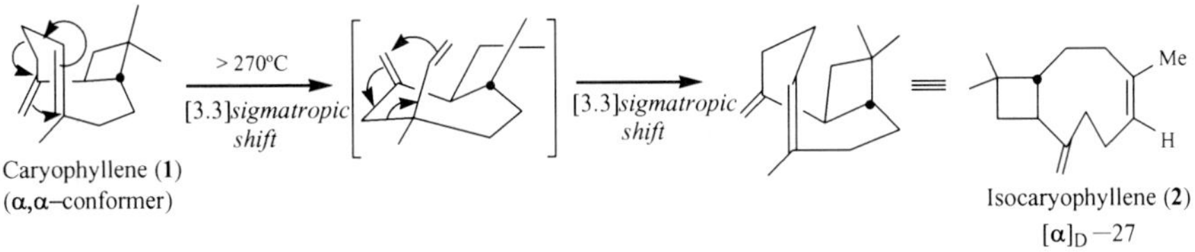

Fig. 7.28 Thermal rearrangement of caryophyllene to isocaryophyllene

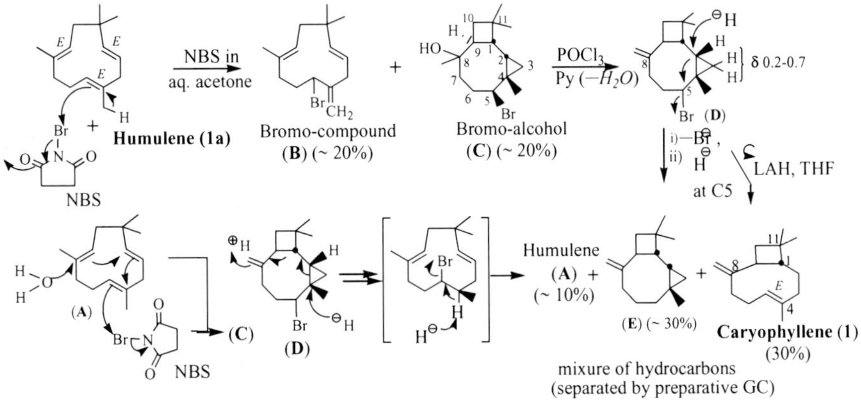

Fig. 7.29 The conversion of humulene (**1a**) into caryophyllene (**1**) [21]

separated and dehydrated with POCl$_3$ and pyridine to give compound (**D**) with a methylene moiety at C8. Compound (**D**) upon treatment with LiAlH$_4$ in THF produces a mixture of hydrocarbons caryophyllene (**1**) (30 %), the tricyclic hydrocarbon (**E**) (30, and a small amount of humulene (**A**) (~10 %)—isolated by preparative GC [21].

The probable mechanistic rationalizations (i), (ii), and (iii) for the formations of (**1**), (**E**), and (**1a**), respectively, from (**D**) are suggested, as shown in the Fig. 7.29.

(i) Rupture of the cyclopropane (inner bond) *anti* to C-Br in (**D**), expulsion of Br to form *trans* 4-5 double bond (*cf* E2 mechanism), and quenching of the generated C2 cation by H$^{(-)}$ give rise to caryophyllene (**1**).

(ii) Expulsion of Br$^-$ from C5, followed by quenching of the C5 cation by H$^{(-)}$, leads to the formation of the hydrocarbon (**E**).

(iii) Ruptures of the cyclobutane (allylic) and cyclopropane inner bonds in the cyclooctane ring of (**D**), aided by the saturation of the methylene group and attack of a hydride at the generated C4 carbocation, followed by elimination of HBr (E2 type), generate humulene (**1a**) with three *E* double bonds.

7.5.7 Apollan-11-ol: An Interesting Acid-Catalyzed Product of Humulene

Apollan-11-ol [22, 23], an acid-catalyzed (H$_2$SO$_4$/ether) optically inactive rearranged product from commercial caryophyllene, was reported in 1922 by Asahina and Tsukamoto who temporarily tagged it as "α-caryophyllene alcohol." The structure of this compound, established by Nickon [22], turned out to be a structurally very interesting symmetrical alcohol (C$_s$, plane of symmetry) and hence optically inactive. Though it was initially thought to be a product from caryophyllene, later humulene, which co-occurs in commercial caryophyllene, has been proved to be the precursor of this alcohol. This was established unambiguously by labeling experiment and spectral data of the labeled product [22]. In fact, Sukh Dev in 1951 also isolated an alcohol, C$_{15}$H$_{26}$O, m.p. 116° [24], suggested to be α-caryophyllene alcohol. That it is the skeletal rearranged product of humulene is supported by its formation in a much greater yield from humulene (**A^1**), as observed by Parker and Roberts [23]. The X-ray crystallographic studies [23] also revealed its structure (**A**). The naming of this compound, based on some fortuitous coincidental facts, is really amazing. We quote from the extraordinarily interesting book *Organic Chemistry: The Name Game* [25]: "The molecule has appealing symmetry, and a flat drawing of the ring system strikingly resembles a rocket, with side fins and exhaust tail. So Nickon dubbed the parent alkane "apollane" in timing with the Apollo 11 moon landing. By happy coincidence, proper numbering locates the –OH at C11 and thus cemented further the onomastic link with Apollo 11. In fact, Neil Armstrong's personal memorabilia include a reprint of that chemistry publication" [25]. The nomenclature and the systematic name of apollane are given in Fig. 7.30.

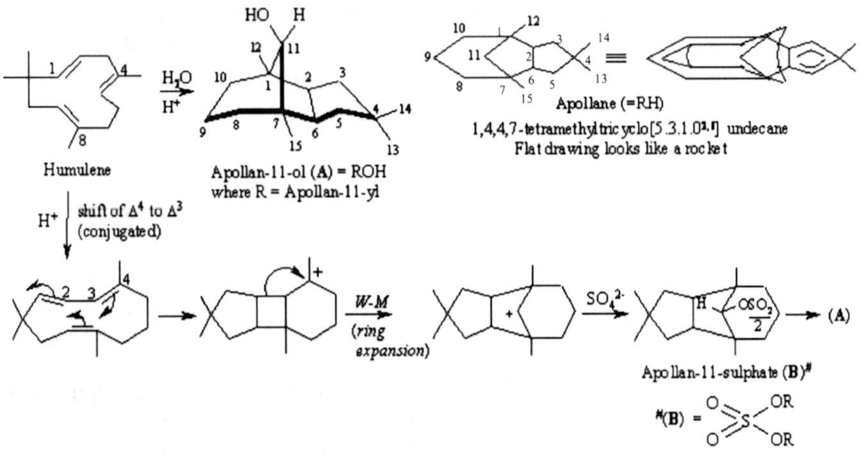

Fig. 7.30 Conversion of humulene (**1a**) into apollan-11-ol (**A**) via apollan-11-sulfate (**B**) [25]

A reasonable mechanistic explanation of the formation of "Apollan-11-ol" (**A**) from humulene is outlined in Fig. 7.30 [25]. Apollan-11-sulfate (**B**) can be converted to apollan-11-ol and the latter can be reconverted to the sulfate.

7.5.8 Biosynthesis of Caryophyllene

The probable biosynthetic sequences of caryophyllene (Fig. 7.9) and its precursor humulyl carbocation starting from farnesyl pyrophosphate (Fig. 7.6) have already been briefly outlined.

7.5.9 A Caryophyllan-Type Compound in a Sea Coral

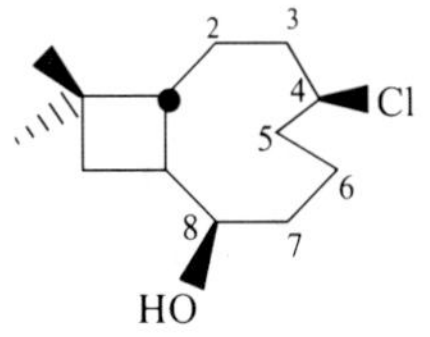

Rumphellatin A

Incidentally, it may be mentioned that *rumphellatin* A, the first chlorine-containing caryophyllane-type norsequiterpenoid ($=CH_2$ at C8 absent), has been isolated from a Formosan soft sea coral *Rumphella antipathies* [26]. Two new hemiketal norsesquiterpenoids of the same skeleton, *rumphellatin* B (having an α-Cl at C3, a β-OH at C4, and an α-hemiketal linkage involving C5 and C8 with an α-OH at C8) and *rumphellatin* C (having a β-Cl at C3, a β-OH at C4, and a β-hemiketal linkage involving C5 and C8 with a β-OH at C8) have also been reported [27] from the sea coral of the same species.

7.6   Longifolene: A Tricyclic Sesquiterpene

7.6.1   Occurrence

Longifolene (**1**), the major sesquiterpene of Indian turpentine oil, in which it occurs to the extent of 5–10 %, was first isolated by Simonsen in 1920. The turpentine oil was obtained from the oleoresin of *Pinus roxburghii* (syn. *P. longifolia*, Fam. Pinaceae). Longifolene [28–31] widely occurs also in various Pinaceae plants. It is commercially produced in hundred ton quantities. Several sesquiterpenes, e.g., *longipinene, longicyclene, caryophyllene, humulene, bisabolene* ,etc., co-occur with it though as minor constituents. Pure longifolene has [α]$_D$ +54.06 (CHCl$_3$). It is pertinent to inform that though (+)-variety is prevalent in nature and occurs mainly in the higher plants (mainly Gymnospermaceae), the (–)-form co-occurs in small amounts with (–)-sativene in *Heilanthus sativum* and *H. victoriae*. Longifolene may be looked upon as the ring-expanded *sativene*. The presence of (–)-longifolene in liverwort has also been reported [32].

7.6.2   Structure: Some Reactions of Longifolene

Extensive chemical studies on longifolene have been carried out by Simonsen during 1923–1934, and much structural information was obtained from these studies [33]. However, the complete structure of longifolene (**1**) was determined by Moffett and Rogers [34, 35] from the X-ray analysis of longifolene hydrochloride, m.p. 59–60 °C, a W–M rearranged product [34, 35]. Its formation may be rationalized as shown in Fig. 7.31.

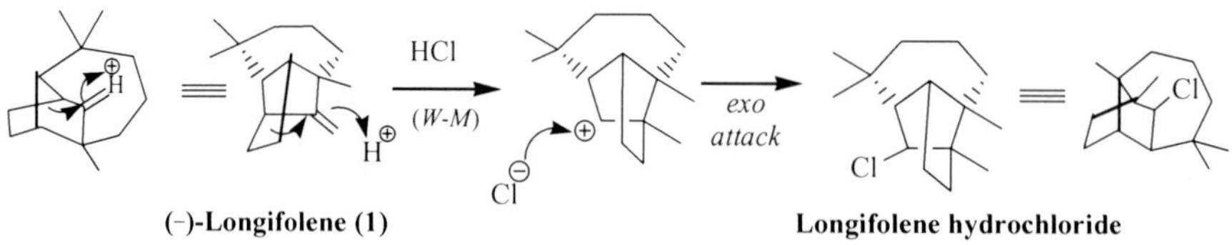

Fig. 7.31 Longifolene (**1**) to longifolene hydrochloride

−10 °C
HCl
exo
attack
Camphene
Camphene
hydrochloride
Cl
− Cl⊖
+ Cl⊖
W-M
exo
(β-face)
attack*
*due to steric crowding
of the endo face
Cl
Cl
H
Isobornyl
chloride

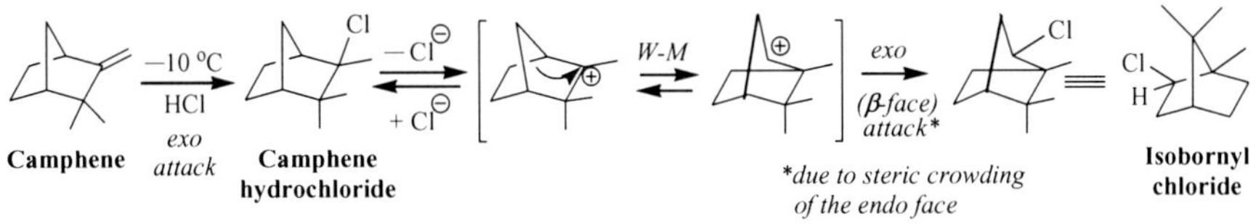

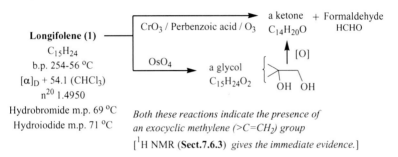

Fig. 7.32 Probable mechanism of conversion of camphene to isobornyl chloride by HCl treatment

Longifolene (1)
C15H24
b.p. 254-56 °C
[α]D + 54.1 (CHCl3)
n20 1.4950
Hydrobromide m.p. 69 °C
Hydroiodide m.p. 71 °C
CrO3 / Perbenzoic acid / O3
a ketone + Formaldehyde
C14H20O
HCHO
[O]
OsO4
a glycol
C15H24O2
OH OH
Both these reactions indicate the presence of
an exocyclic methylene (>C=CH2) group
[1H NMR (Sect.7.6.3) gives the immediate evidence.]

Fig. 7.33 Reactions indicating the presence of an exocyclic methylene group in longifolene (**1**)

This information is reminiscent of the conversion of camphene hydrochloride to isobornyl chloride by treatment with HCl, as shown in Fig. 7.32. The presence of an exocyclic methylene group in longifolene is indicated by the reactions shown in Fig. 7.33.

7.6.2.1 Conversion of Longifolene to Isolongifolene

Longifolene, like caryophyllene (**7.5**), is also a *molecular acrobat*, being very much prone to skeletal rearrangements. Several reports of its rearrangements are available in the literature [28–31]. (+)-Longifolene (**1**) on exposure to strong protic or Lewis acid undergoes a deep-seated molecular rearrangement to an isomeric tricyclic hydrocarbon, (−)-isolongifolene (**1a**) through a number of mechanistically explainable steps as shown in Fig. 7.34. However, a better gross mechanism of this stereospecific rearrangement has been elucidated by Sukh Dev, by using site specifically labeled longifolene-4,4,5,5-d_4 (Fig. 7.35) [28] (which was efficiently synthesized) to form (**D**) involving one D shift and thus has been shown to follow the pathway initially proposed by Berson et al. [31]. The less circuitous direct pathway (Fig. 7.34) would have resulted in the formation of the d_4-isolongifolene (**C**) (involving no deuterium shift), instead of (**D**). This pathway involves an *exo, exo* Me shift (Fig. 7.35) [in preference to the less circuitous *endo, endo* Me

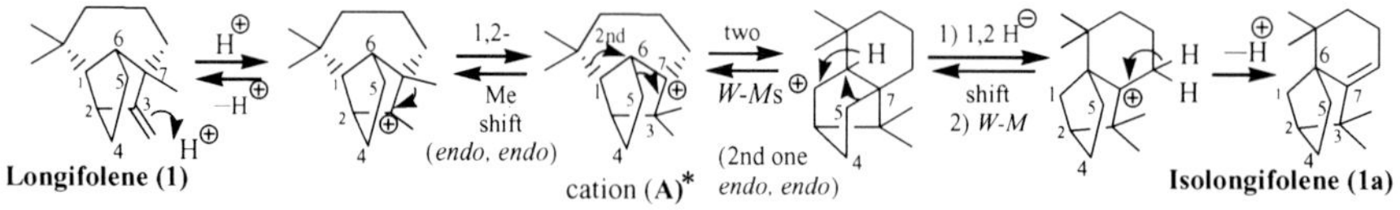

* In the carbocation (**A**) charge is located at the bridgehead of a bicyclo[4.2.1]nonane system (upper part, excluding the front two-carbon bridge). To accommodate the planar trigonal carbon in the cation (**A**) the bicyclo[2.2.1]heptane system (lower part excluding the upper four-carbon bridge) must be severely distorted, as is evident from a Dreiding model study, and hence the caion (**A**) is thermodynamically less stable relative to the caion (**B**) of (**Fig.7.35**) [29]

Fig. 7.34 Earlier proposed pathway for the acid catalyzed conversion of longifolene into isolongifolene

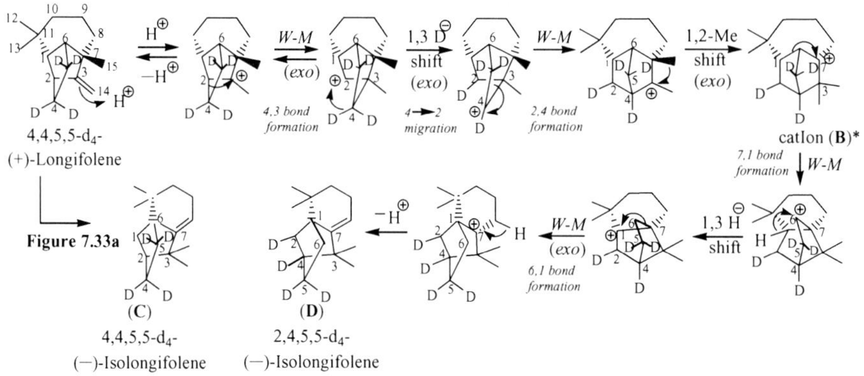

*In carbocation (**B**) charge is located at the bridgehead of a bicyclo[4.3.1]decane system, which as is observed in a Dreiding model study, can easily accommodate the planar trigonal carbon, and hence caion (**B**) is thermodynamically more stable than the caion (**A**) (**Fig. 7.34a**) [29]

Fig. 7.35 Gross mechanism for the acid catalyzed rearrangement of 4,4,5,5-d$_4$-(+)-longifolene to 2,4,5,5,-d$_4$-(−)-isolongifolene (**D**) [29]

migration route (Fig. 7.34) proposed earlier], based on the intermediacy of nonclassical carbocation, which is not tenable and acceptable. The premise that an *exo, exo shift* is energetically preferred to an *endo, endo* shift may not be the only reason here. A suitable rationale for the complete preference of the more circuitous route (Fig. 7.35) [29] has been indicated in Figs. 7.34 and 7.35 (see the arguments written against the asterisks given on the ions (**A**) and (**B**)].

7.6.3 *Spectral Data of Longifolene*

Longifolene (1)

IR: (film): ν_{max} 3,080, 1,655 and 868 cm^{-1} (exo-methylene group)

^{1}H NMR [36] (δ, CCl$_4$): 0.90 (3H, s, C4 Me, angular), 0.95 (3H, s, *exo* Me, C12), 0.99 (3H, s, *endo* Me, C11), 4.45 and 4.72 (3H, *s*, each, exocyclic methylene protons, H13a and H13b).

^{13}C NMR [36] (δ, CHCl$_3$): 47.89 (C1), 29.71 (C2), 25.48 (C3), 43.93 (C4), 43.33 (C5), 21.14 (C6), 36.39 (C7), 33.55 (C8), 167.59 (C9), 30.04 (C10), 30.51 (C11), 30.50 (C12), 98.93 (C13).

MS (EI, 70 eV, M.S.9) [37]: *m/z* (% Base peak), 205 (2.9, P+1), 204 (M$^+$, 15.0), 189 (20.7) 175 (7.4), 161 (43.4), 148 (7.4), 147 (14.7), 135 (22.3), 134 (11.2), 133 (24.8), 121 (21.8), 120 (16.0), 119 (33.5), 107 (37.4), 106 (16.1), 105 (43), 95 (34.1), 94 (42.4), 93 (43.0), 92 (17.2), 91 (52.7), 81 (23.2), 79 (35.2), 77 (30.1), 69 (22.7), 67 (23.6), 65 (17.5), 55 (43.2), 53 (24.9), 43 (21.6), **41 (100)**, 39 (33.9).

MS (EI, 70 eV, M.S.9) **of Isolongifolene** [37]: *m/z* (% Base peak) 205 (5.44, P+1), 204 (M$^+$, 31.7), 189 (31.0), 175 (61.5), **161 (100)**, 148 (42.7), 147 (17.2), 135 (9.2), 134 (10.0), 133 (41.8), 121 (13.0), 120 (11.8), 119 (38.4), 107 (21.2), 105 (43.4), 95 (13.4), 93 (22.2), 91 (39.6), 81 (11.8), 79 (20.7), 77 (25.4), 69 (16.0), 65 (13.9), 55 (25.8), 53 (21.2), 41 (74.8), 39 (41.8).

Note: The base peaks of longifolene and isolongifolene appear at *m/z* 41 and 161, respectively. Their fragmentation patterns are found to be quite similar, varying only in the relative intensities of some peaks.

7.6.4 *Synthesis of (±)-Longifolene and (+)-Longifolene by Corey's Group [38]*

The strategy for the synthesis of longifolene (Fig. 7.36) was planned by Corey [38] by exhaustive analysis of the topological properties of its carbon network having complex bridged ring system. A suitable bicycle[5.4.0]undecane system (**B**) capable of undergoing *intramolecular Michael addition* (cf. formation of santonic acid from santonin served as the precedent; see Sect. 7.8) under the influence of a base appeared to be the right intermediate to construct the desired tricyclic system with

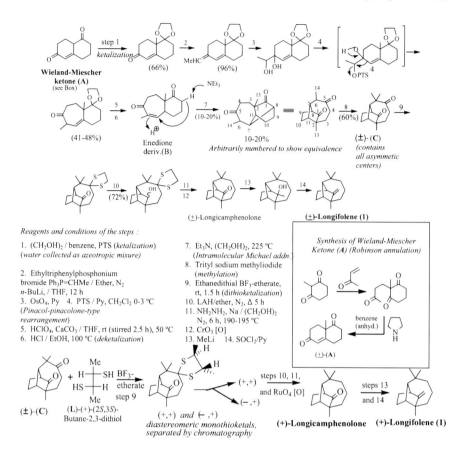

Fig. 7.36 Synthesis of (±)-and (+)-longifolene [38]

adequate functionalities and the substitution pattern of longifolene. A methyl group at the α-position of the α,β-unsaturated ketone system in the bicycle framework (**B**) is advantageous for further methylation in the subsequent step (Fig. 7.36). The Wieland–Miescher ketone (**A**) was chosen as the starting material for the synthesis of the key intermediate bicyclo[5.4.0]undecane system. The synthesis of (±)- and (+)-longifolene is delineated in Fig. 7.36. The optically active natural (+)-longifolene has been synthesized via optically active thioketal derivative using (+)-butane-2,3-dithiol and separating (+)-longicamphenolone, followed by the steps as applied to (±)-longicamphenolone with slight modification (Fig. 7.36) to obtain (+)-longifolene.

7.6.5 Synthesis of (±)-Longifolene and Some of Its Relatives by Johnson Group [39]

During the stannic chloride-catalyzed cyclization of heptynylmethylcyclopentenol (G), the expected hydroazulene system is formed along with an alcoholic compound (J) or (K), which contains the carbon network close to that of longifolene. This earlier observation [39] in Johnson's lab directed his attention to develop an appropriately substituted enynol substrate for the formation of the hydroazulene derivative, which could serve as the potential intermediate in the synthesis of longifolene (Fig. 7.37). The novel synthesis of (±)-longifolene (1) and (±)-longicamphenylone (D) (also occurring naturally) and two other relatives (B) and (C) is schematically represented in Fig. 7.37, mentioning the reaction conditions and yields [39].

Mechanism of Conversion of (M) to (Q) (Fig. 7.38) The mechanism of cyclization of the enynol (M) to form (Q) is open to question. However, it was rationalized

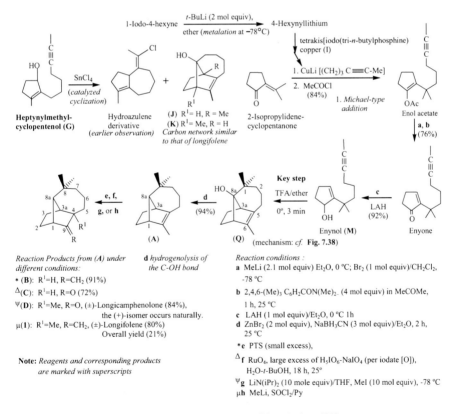

Fig. 7.37 Synthesis of (±)-longifolene (1) and some of its relatives [39]

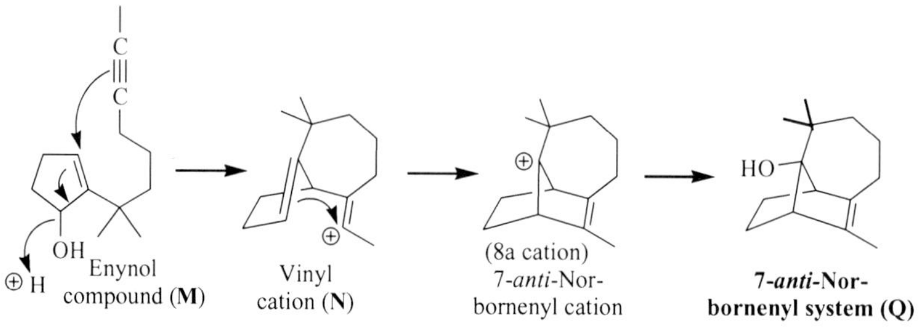

Fig. 7.38 Mechanistic rationale for the conversion of (**M**) to (**Q**) of Fig. 7.37

as follows [39]. The allylic alcohol (**M**) would first cyclize to give the vinyl cation (**N**). Intramolecular nucleophilic attack at the vinyl cation by the olefinic bond in the five-membered ring would result in further cyclization to yield the interesting cation (**8a**), which apparently may be destabilized due to violation of Bredt's rule, but it embodies the potential stabilizing characteristic of the 7-*anti*-norbonenyl cation. Nucleophilic attack of the **8a** cation (Fig. 7.38) by H$_2$O would yield the norbornenyl system (**Q**).

Other total syntheses of (±)-longifolene and (+)-longifolene have been reported by McMurry [36] and Oppolzer [40].

7.6.6 Biosynthesis

Biosynthesis of longifolene [32] has been discussed in Sect. 7.3 (Figs. 7.6 and 7.12).

7.7 Longicyclene: The First Reported Tetracyclic Sesquiterpene

7.7.1 Occurrence and Structure

Longicyclene, C$_{15}$H$_{24}$, b.p. 82 °C/2 mm, [α]$_D$+33.6 (neat), the first reported tetracyclic sesquiterpene, has been detected as a minor congener hydrocarbon of longifolene during the vapor phase chromatography (VPC/GLC) of the higher boiling portion of Indian turpentine oil obtained from the oleoresin of *Pinus longifolia* [41, 42]. This observation led to its systematic study of isolation and structure elucidation. The data in Fig. 7.39 and the spectral data for longicylene are consistent with its structure (**2**) which is confirmed by the total stereoselective synthesis of (±)-longicyclene (Fig. 7.41).

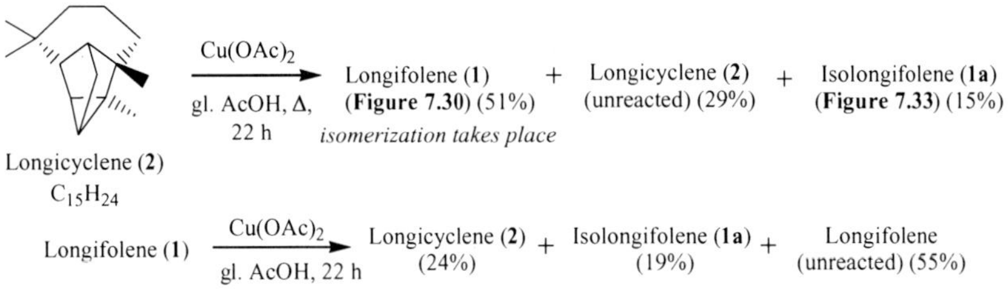

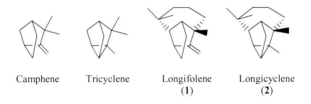

$$\text{Longicyclene (2)}$$
$$\text{C}_{15}\text{H}_{24}$$

$\xrightarrow[\substack{\text{gl. AcOH, }\Delta,\\ \text{22 h}}]{\text{Cu(OAc)}_2}$ Longifolene (1)   +   Longicyclene (2)   +   Isolongifolene (1a)
(Figure 7.30) (51%)   (unreacted) (29%)   **(Figure 7.33)** (15%)

isomerization takes place

Longifolene (1) $\xrightarrow[\text{gl. AcOH, 22 h}]{\text{Cu(OAc)}_2}$ Longicyclene (2) + Isolongifolene (1a) + Longifolene
(24%)   (19%)   (unreacted) (55%)

Note: The constituents of the mixture of the hydrocarbons were identified by GLC

Fig. 7.39 Interconversions of longicyclene and longifolene

Camphene   Tricyclene   Longifolene   Longicyclene
(1)   (2)

Fig. 7.40 Relationship of longifolene to longicyclene is like camphene to tricyclene

Longicyclene when subjected to catalytic hydrogenation or perbenzoic acid treatment remains unreacted indicating its fully saturated character. With tetranitromethane it develops a faint but distinct color indicating the presence of a cyclopropane ring [44] in its molecule. The chemical shifts of its upfield cyclopropane protons (see Sect. 7.7.2) provide definite evidence of the presence of a cyclopropane ring. The hydrogen content of its molecule, $C_{15}H_{32}$, indicates it to be a tetracyclic. Logicyclene upon prolonged refluxing with glacial acetic acid undergoes isomerization to longifolene and isologifolene (Fig. 7.39).

Interconversions of longifolene and longicyclene support the structure of longicyclene as well as constitute a formal total synthesis of ($\pm$)-longicyclene in view of the total synthesis of ($\pm$)-longifolene (Figs. 7.36 and 7.37).This new tetracyclic sesquiterpene has been named longicyclene which holds the same relationship to longifolene as tricyclene to camphene (Fig. 7.40) [42].

7.7.2   Spectral Properties [43]

IR: (CCl_4): ν_{max} 3,085 (cyclopropane type C–H stretching), 1,385; 1,370 cm^{-1} (gem-CH_3), a strong peak at 840 cm^{-1} (characteristic of a tricyclene system). In the near IR (CCl_4) it shows band at 1.640 μ, just like the absorption of tricyclene (CCl_4, 1.650 μ) assignable to the first overtone of cyclopropane ring C–H stretching [45].

^{1}H NMR (CCl_4): δ 1.04 (*s*, 3H, CH_2), 0.98 (*s*, 3H, CH_3), 0.92 (*s*, 3H, CH_3), 0.88 (*s*, 3H, CH_3). The ^{1}H NMR contains no olefinic proton; it showed the presence of four unsplit methyls. This upfield region of methyls integrated for 14H; hence 2H may be attributed to the two upfield cyclopropane protons. In fact, the spectrum also shows a 1H signal at δ 0.77, together with another minor signal at δ 0.67; thus in all probability two cyclopropane protons are indicated [42].

7.7.3   Synthesis of Longicyclene [43]

A stereoselective total synthesis of longicyclene (Fig. 7.41) has been achieved in which (−)-carvone has been converted to tetrahydroeucarvone through ring expansion. To get a 2-carboxyethyl side chain α- to the carbonyl, tetrahydroeucarvone is alkylated with 2-chloro-pent-3-ene and then trunked via oxidation. The keto acid is induced to form enol lactone. The latter on reduction with iBuAlH followed by acid rearrangement produces a bicycle[4.2.1]nonan-ketol system. The compound being unstable is immediately esterified with methane sulfonyl chloride and finally eliminates to form a bicyclic nonane compound. Sequential treatment of this compound with usual reagents led to the formation of a bicyclo-olefenic acid which is converted to the tetracyclic cyclopropyl ketone via diazoketone intermediate derived from the acid. The ketone has been reduced with diisobutylaluminium hydride and the stereochemistry of the hydroxyl group as shown in the structure is due to steric approach control—the bulky reagent should approach from the less hindered face of the molecule. Finally the OH group is removed in the conventional way.

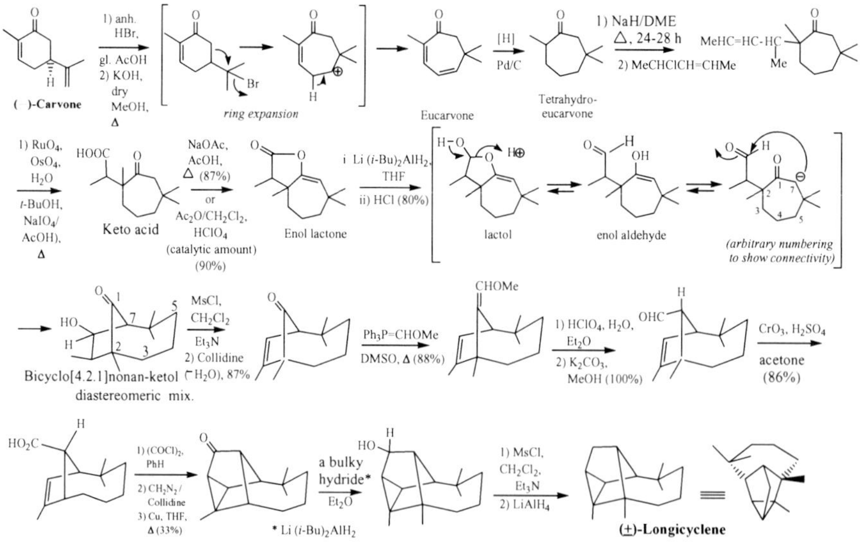

Fig. 7.41 Total synthesis of (±)-longicyclene starting from (−)-carvone [43]

7.7.4 Biosynthesis of Longicyclene

This has been discussed earlier along with that of longifolene (see Figs. 7.6, 7.12, and 7.14).

7.8 Santonin

7.8.1 Occurrence and Structure

Santonin is the principal anthelmintic component of various *Artemisia* species. It was isolated first by Kahler in 1830 from *Artemisia santonica* (Fam. Compositae/Asteraceae) and later on from various other *Artemisia* species (e.g., *A. maritima*, *A. cinica*, etc.), as well as from other related plants. *A. santonica* belongs to the family of plants known for centuries as "worm seed" [46], as it was useful in the treatment of intestinal worms.

Santonin is a very well-studied sesquiterpene lactone. Its structure (without stereochemistry) has been elucidated from its degradative studies and reactions. Some are presented here (Fig. 7.42).

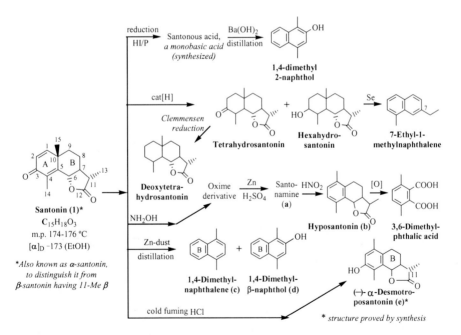

Fig. 7.42 Reactions and degradations of santonin

Earlier studies [47] on the structural elucidation involving the formation of the products (**b**), (**c**), (**d**), and (**e**) (Fig. 7.42) led to the placement of the two methyl groups at the 1 and 4 positions of the same ring in santonin molecule. However, George Clemo first proposed its correct structure (1929–1930) [48–50] with an angular methyl group on the basis of the following reactions and identification of the products (Fig. 7.43).

These reactions suggest the structure (**1**) for santonin without stereochemical description. On the basis of this structure, the formation of desmotroposantonin (**e**) and hyposantonin (**b**) (Fig. 7.42) can be explained as shown in Fig. 7.44.

The first reaction (Fig. 7.44) is known as dienone–phenol rearrangement, which finds wide application (Chap. 31) in cases of such organic molecules. Similarly, santonamine (**a**) on treatment with HNO$_2$ undergoes 1,2-methyl shift with the expulsion of N$_2$ to form hyposantonin (**b**). Likewise, 1,4-dimethyl-β-naphthol (**d**) is formed by a dienone–phenol rearrangement, and 1,4-dimethylnaphthalene (**c**) is formed by a dienol (formed by zinc dust reduction)–benzene rearrangement involving in each case a 1,2-methyl migration (Fig. 7.42). In both cases ring B gets aromatized.

The structure of santonin has been established by its synthesis (Fig. 7.45).

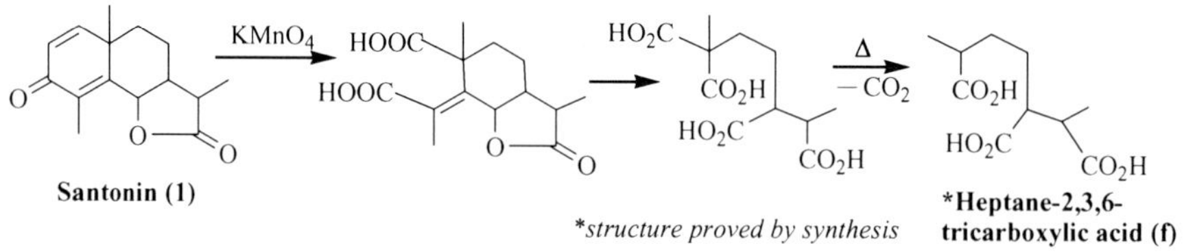

Fig. 7.43 Degradation of santonin to heptane-2,3,6-tricarboxylic acid (**f**)

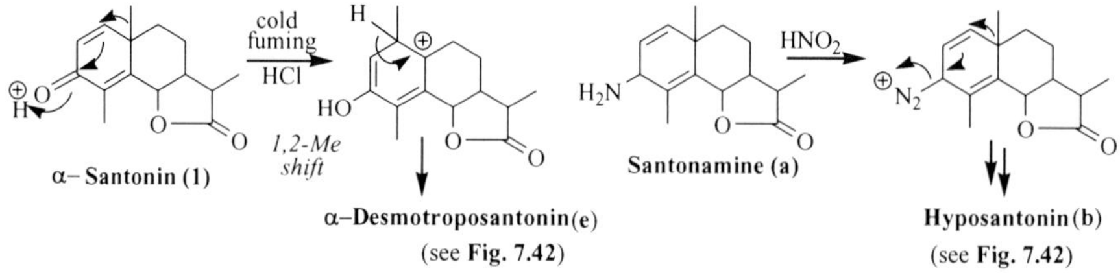

Fig. 7.44 Formation of desmotroposantonin (**e**) from santonin (**1**) and hyposantonin (**b**) from santonamine (**a**)

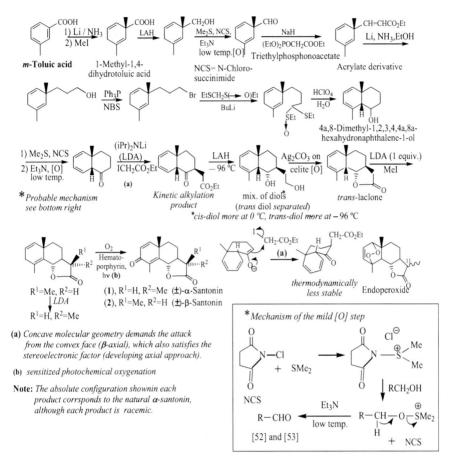

Fig. 7.45 Stereocontrolled total synthesis of (±)-α-santonin and (±)-β-santonin (1978) [51]

7.8.2 Stereocontrolled Total Synthesis of Racemic α-Santonin and β-Santonin [51]

The stereocontrolled total syntheses of α-santonin and β-santonin are delineated in Fig. 7.45. m-Toluic acid is reduced with Li in liquid ammonia (Birch reduction) followed by alkylation to obtain 1-methyl-1,4-dihydro-m-toluic acid. The COOH group is then expanded to a C_3-side chain in five steps forming n-propylbromide-substituted 1,4-dihydro-m-xylene. The latter has been alkylated with the lithium salt of the monosulfoxide of formaldehyde diethylthioacetal. Acid-catalyzed cyclization via the butanal derivative afforded 4a,8-dimethyl-1,2,3,4,4a,8a-hexahydro-naphthalen-1-ol. Oxidation, alkylation, reduction, and subsequent treatment with silver carbonate impregnated celite on the separated trans-diol yields a lactone. The lactone undergoes methylation from the convex β-face leading to the thermodynamically less stable 11-β-methyl compound. The latter could be epimerized to the more stable α-isomer. The β-isomer and the α-isomer are converted through sensitized photochemical oxygenation to β-santonin and α-santonin, respectively (Fig. 7.45). In each case, a significant amount of corresponding endoperoxide (11β-methyl derivative isolated, 11α-methyl derivative presumed to be formed also) was produced.

7.8.3 Biogenetic-Type Synthesis of Santonin [54]

The starting material in this biogenetic-type synthesis of santonin is its possible biogenetic precursor, the readily available santamarine (obtained in ample amounts from bay leaves). Its conversion to santonin [54] is shown in Fig. 7.46.

7.8.4 Absolute Configuration of α-Santonin and Related Compounds at C$_{11}$. Full Stereostructures by X-ray Studies

The absolute configuration at C11 of α-santonin has been determined to be (S) by its multistep transformation into a secondary alcohol (A) without affecting the stereochemical integrity of C11 and subsequent application of the Prelog atrolactic acid method (Sect. 2.11.5) to this compound [55, 56] as delineated in Fig. 7.47. Later, this absolute stereochemistry at C11 has also been deduced [57].

X-Ray analysis of α-santonin has firmly established (s)-configuration at both C10 and C11 and *trans* stereochemistry of the B/C ring juncture (1) [58]. Earlier, the X-ray single crystal structures of 2-bromo-α-santonin [59–61] and 2-bromo-β-santonin [62], elucidated by the usual phase-determining heavy atom

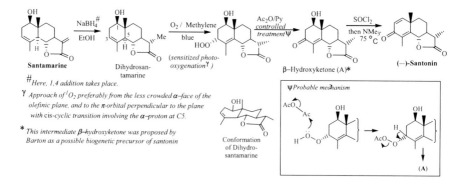

Santamarine

Dihydrosan-
tamarine

β–Hydroxyketone (A)*

(–)-Santonin

#*Here, 1,4 addition takes place.*

γ *Approach of* $^1O_2$ *preferably from the less crowded α–face of the olefinic plane, and to the π-orbital perpendicular to the plane with cis-cyclic transition involving the α-proton at C5.*

* *This intermediate β–hydroxyketone was proposed by Barton as a possible biogenetic precursor of santonin*

Conformation
of Dihydro-
santamarine

ψ *Probable mechanism*

(A)

Fig. 7.46 Biogenetic-type synthesis of (–)-santonin (1983) [54] from santamarin

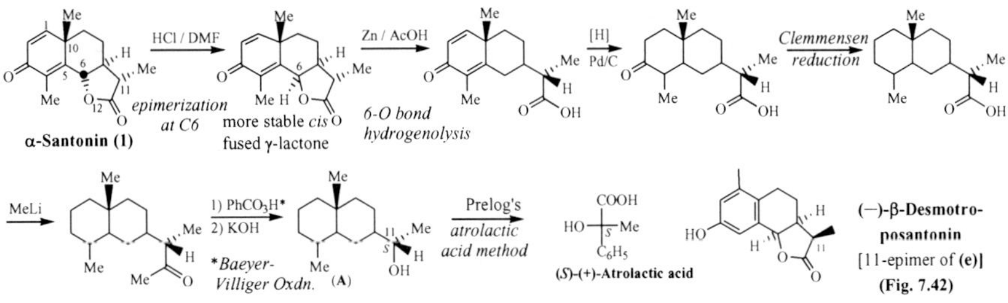

Fig. 7.47 Absolute configuration of α-santonin at C11 as (S) by Prelog atrolactic acid method [55]

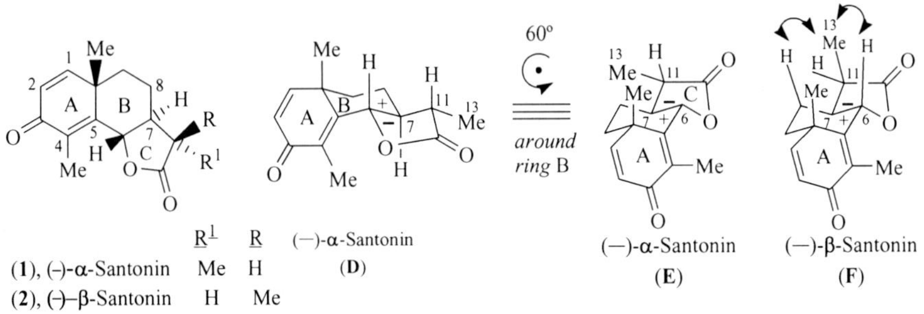

Fig. 7.48 Molecular conformations of α-santonin and β-santonin

method, revealed the full stereostructures of (−)-α-santonin and (−)-β-santonin as (1) and (2), respectively (see Fig. 7.48). The absolute stereochemistry of (−)-β-desmotroposantonin (Fig. 7.47) was established by X-ray analysis of its 2-bromoderivative [61]. It follows that (−)-α-desmotroposantonin [compound (e) of Fig. 7.42], the 11-epimer of the β-isomer, must have (S)-configuration at C11.

7.8.5 Molecular Conformations of α-Santonin and β-Santonin

The conformational features revealed from the stereostructure drawing as well as from a careful examination of a molecular model (Fischer/Dreiding) of α-santonin (1) are as follows : [cf. (D) or a different projection (E) of the same rigid conformation] (Fig. 7.48).

1. Ring A is more or less planar.
2. Ring B is in a slightly distorted chair form (cf. methylenecyclohexane moiety).
3. The γ-lactone ring C has an envelope conformation with C7 the out-of-plane atom, as has also been revealed from its X-ray crystallographic study [59–61].

4. The ring juncture B/C is *trans* with H6 and H7 in *trans* diaxial conformation with respect to the ring B, and the Me at C11 (or C13 methyl) is pseudo-equatorial (evident from X-ray studies [59–61]).
5. The torsion angle signs at the B/C ring junction is (+/–), i.e., (+) in ring B and (–) in ring C, and not (–/+) – as evident from the conformation (**D**) or (**E**) (cf. Chap. 2).

The conformation (**E**) (another representation of the molecular conformation of α-santonin) can be derived from conformation (**D**) by 60° anticlockwise (from the top) rotation of the model around an axis, passing through the center of ring **B** and vertical to it.

In the lactone ring **C** in the conformation (**D**) C6 seems to be the nonplanar atom, at a little higher position but actually—as is evident from the X-ray study of 2-bromo-α-santonin [59–61] and the molecular model analysis as well as from the conformation (**E**), C7 is the nonplanar (lower) atom. Thus it should be noted that conformational drawings do not always give the correct three-dimensional positions of all the atoms in space in a molecule. The absolute stereochemistry of β-santonin has been elucidated from the X-ray study of 2-bromo-β-santonin [62], establishing it to be C11-epimer of santonin (or α-santonin).

The less stability of β-santonin [(**2**), C11-epimer of α-santonin, (**E**)] is attributed to the steric interactions of the β-pseudo-axial C13 methyl group with the syn-axial H6 and with syn-axial-like proximate H8 [cf. conformation (**F**)] and some resulting deformations of rings B and C. ^{13}C NMR Study of β-santonin also supports the pseudo-axial orientation of its C13 methyl group (Fig. 7.48).

7.8.6   *Spectral Properties*

α–Santonin (**I**)

UV: $\lambda_{max}^{(EtOH)}$ 236 nm (ε 12,000) unsaturated conjugated ketone.

   IR: ν_{max} 1,165, 1,615, 1,645 (unsaturated ketone), 1,660, 1,780 cm^{-1} (saturated γ-lactone).

   ^{1}H NMR: δ (CDCl3) 1.25 (3H, *d*, C11–Me), 1.35 (3H, *s*, C10–Me), 2.10 (3H, *s*, C4–Me), 4.83 (1H, *d*, C6–H), 6.16 (1H, *d*, *J* 8.5 Hz, H2), 6.72 (1H, d, *J* 8.5 Hz, H1).

   ^{13}C NMR [63]: 155.1 (C1), 125.9 (C2), 186.0 (C3), 128.4 (C4), 151.5 (C5), 81.5 (C6), 54.0 (C7), 23.8 (C8), 39.3 (C9), 41.7 (C10), 41.2 (C11), 177.4 (C12), 12.5 (C13), 10.9 (C14), 25.3 (C15).

   It has been observed [63] that in case of β-santonin (**2**), the C7. C8, C9, C11, and C13 are shielded by 4.5, 3.0, 1.1, 3.0, and 2.6 p.p.m., respectively, relative to those

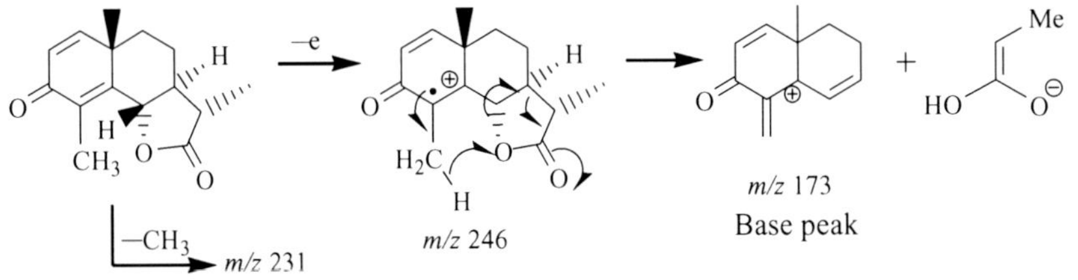

Fig. 7.49 Fragmentation of santonin to generate the base peak

of α-santonin, due to the pseudo-axial β-orientation of C13 methyl in the former, causing some distortion of the conformation of its B and C rings. All other carbons of β-santonin, excepting C12 (which is deshielded by 0.9 p.p.m.), have almost the same chemical shifts as those of α-santonin [63]. The configurational dependence of the carbon-13 chemical shifts of 6-epi-α-santonin and 6-epi-β-santonin (both having B/C rings *cis* fused) has also been discussed in detail by Randall group [63]. Inversion of C6 (*trans* to *cis* fusion) causes 5.0 p.p.m. upfield shift for the α-cases, and 3.9 p.p.m. upfield shift for the β-cases. Thus, 13C NMR studies provides with a sure method of assigning the stereochemistry of the ring juncture and the orientation of C11 methyl in santonin derivatives.

MS: The most intense peak that appeared in the mass spectrum of santonin (**1**) is at m/z 173 (M$^+$-73). The fragmentation is initiated by the removal of one electron from C4–C5 double bond when a resonance stabilized ion at m/z 173 is formed [64] (Fig. 7.49).

7.8.7   Conversion of Santonin to Santonic Acid [46, 65]

Conversion of santonin to santonic acid by prolonged vigorous boiling with concentrated alkali, discovered by Hvoslev in 1863 and studied later by Cannizzaro, remained a riddle for many decades. Woodward elucidated the structure of santonic acid [65] in 1948. His approach was based on the mechanistic way of thinking this conversion leading to its, till then, unidentifiable structure. In course of this study he discovered *an intramolecular Michael addition* as a means of a bridgehead formation via carbanion leading to a tricyclic skeleton (Fig. 7.50). Later, this type of intramolecular Michael addition has been employed during longifolene synthesis (Fig. 7.36) by Corey. Mechanistically the above conversion is explained stepwise in Fig. 7.50. The first step is the hydrolysis of the lactone to form the hydroxycarboxylate anion (**A**). The latter after loss of H$^+$ from C6, and the 3,6-diketone (**B**) formation, as shown, is converted to a resonance stabilized C7 carbanion (**C**). In the last step the carbanion (**C**) undergoes an intramolecular Michael addition to the proximate β-carbon (C1) of the α,β-unsaturated carbonyl system, in the conformation shown, followed by acidification to give santonic acid. The absolute

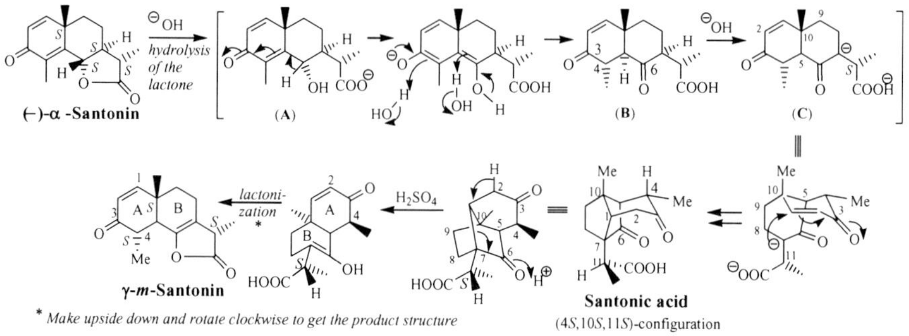

Fig. 7.50 Mechanistic rationales for the conversions of santonin to santonic acid and of the latter to γ−m-santonin

configuration of santonic acid has been firmly established in 1999 by single crystal X-ray analysis [58].

Santonic acid, on treatment with sulfuric acid, yields γ-m-santonin and its interesting mechanistic rationale is also shown in Fig. 7.50.

Thus, the structure and absolute configuration of santonin and santonic acid could be determined by single crystal X-ray analysis in a matter of days, rather than decades of chemical investigation. *At present, NMR spectroscopy and X-ray analyses are used as ultimate weapons for complicated structure determination leaving little or no room for important chemical information or intuition. However the chemical intuition and mechanistic way of thinking are indispensable for development of synthetic chemistry.*

7.8.8   Biosynthesis of Santonin

A preliminary study of the biosynthesis of santonin [66] in *A. maritime* has been carried out (Fig. 7.51). Santonin, a member of the eudesmol family of sesquiterpenoids, is derived from *E,E*-farnesyl carbocation. The latter cyclizes to a ten-membered carbocation (germacranyl carbocation) having two ethylenic linkages ideally suitable for cyclization to eudesmol. However, it has been shown through intermediate feeding in the biosynthetic study that prior to cyclization the ten-membered cation is converted into constunolide and then to dihydrocostunolide, both bearing the angular lactone and the latter a C11–Me instead of C11–C13–methylene. It then undergoes π-cation cyclization to build the eudesmol cation which quenches through the loss of a proton into 5α,6β,11β(H)-eudesm-3-en-6,12-olide. Experiments on the sequence of lactone formation suggest that the lactonization precedes the construction of the dienone group. Evidence that the ten-membered ring compound dihydrocostunolide is an intermediate has been secured. The probable sequence of the biogenesis of santonin is delineated in Fig. 7.51. It was demonstrated that the tritium-labeled compounds (marked with

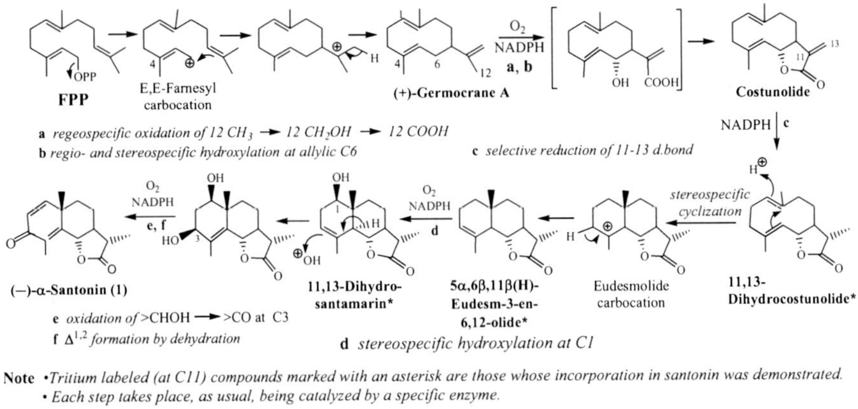
FPP
E,E-Farnesyl carbocation
(+)-Germocrane A
Costunolide
O2
NADPH
a, b
NADPH c
a regeospecific oxidation of 12 CH3 → 12 CH2OH → 12 COOH
b regio- and stereospecific hydroxylation at allylic C6
c selective reduction of 11-13 d.bond
stereospecific cyclization
(–)-α-Santonin (1)
11,13-Dihydro-santamarin*
5α,6β,11β(H)-Eudesm-3-en-6,12-olide*
Eudesmolide carbocation
11,13-Dihydrocostunolide*
e oxidation of >CHOH → >CO at C3
f Δ1,2 formation by dehydration
d stereospecific hydroxylation at C1
Note •Tritium labeled (at C11) compounds marked with an asterisk are those whose incorporation in santonin was demonstrated.
• Each step takes place, as usual, being catalyzed by a specific enzyme.

Fig. 7.51 Probable biosynthetic route to santonin (Barton) [66]

asterisks) were incorporated in santonin [66] though in small percentages (3–8 %). Such low incorporations even in alkaloid biosynthesis have often been reported. 1,2-Dihydrosantonin, a major lactonic component of *A. stellariana*, quite unexpectedly failed to act as a precursor of santonin. Zero incorporation of 1-^{3}H-FPP and minute incorporation (2.25 %) of 1-^{3}H-farnesol may be due to the difficulty of translocation of these compounds to the site of biosynthesis [66].

7.8.9 Santonin as a Synthon

Santonin has been employed as a synthon for the synthesis of quite a big number of sesquiterpenes ans their intermediates and derivatives [67–71].

Natural Norsesquiterpenes Having Nonsteroid and Steroid *cis*-Decalin Conformations Santonin has been used for the synthesis of *chamaecynone* having a novel nonsteroid *cis*-decalin conformation (–/– torsion angles at the ring junction) (see Sects. 2.15.1.2 and 2.15.1.3). It is one of the first few examples of natural acetylenic compounds of terpenoid origin [71, 72] (Fig. 7.52). It was isolated from the essential oil of the Benchi tree (*Chamaecyparis formosensis*, Fam. Cupressaceae).

The corresponding steroid conformation (+/+) torsion angles at the ring junction) of chamaecynone is highly unstable due to the steric interactions of the 4α-methyl group with 7α- and 9α-axial hydrogens (Fig. 7.52).

In this connection it is interesting to note that another natural norsesquiterpene, *isochamaecynone* [72], a C4-epimer of chamaecynone, possesses a steroid *cis*-decalin conformation (+/+) torsion angles at the ring juncture) in which 4β-methyl

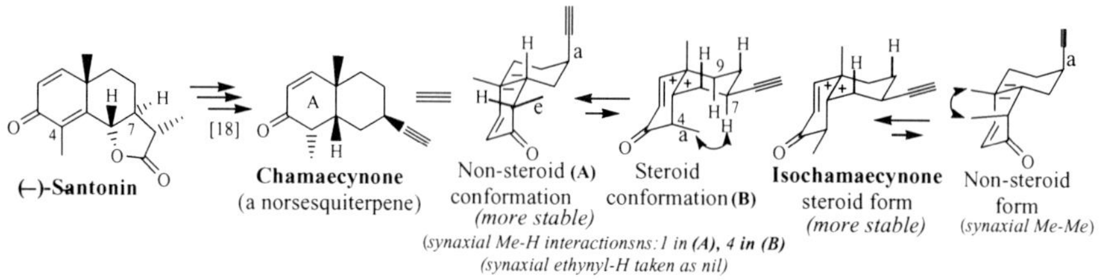

Note: *The ring A which should be of somewhat flattened half chair conformation has been drawn as chair for convenience of drawing.*

Fig. 7.52 Structure and nonsteroid conformation of chamaecynone

group being equatorial entails no destabilizing interaction with 7α- and 9α-axial hydrogens (Fig. 7.52). On the contrary, its flipped nonsteroid conformation imposes strong *syn* axial interaction between 4-CH$_3$ and 10-CH$_3$ groups, and hence it does not exist (for more details, see Sect. 2.15.1.2 and 2.15.1.3). For convenience of reproduction the ring A in both these norsesquiterpenes has been presented as chair conformation [72], although because of the α,β-unsaturated moiety ring A will exist in a flattened half-chair conformation.

7.8.10   Photochemical Transformations of Santonin

Santonin (**1**) containing a cross-conjugated ketone, as expected, should be sensitive towards light. In fact, the photochemistry of santonin has an illustrious history of the formation of various photolysis products in manifold media or conditions. The mechanisms of these photochemical transformations have been studied by many leading organic chemists [73–77]. The formation of isophotosantonic acid lactone (**E**) (having a guaianolide skeleton) as a product of photolysis of santonin in aqueous acetic acid is best rationalized in terms of Zimmerman's explanation (Fig. 7.53) [78, 79]. An initial $n \rightarrow \pi^*$ transition produces a diradical (**A**) which isomerizes to (**B**), followed by collapse (intersystem crossover, ISC) to a dipolar species (**C**). In aqueous acetic acid the anion is protonated, and the resulting carbonium ion (**D**) rearranges to the lactone (**E**). Barton showed this rearrangement to be general, and consequently it has been used in the synthesis of several perhydroazulenes of the guaianolide series [80]. This photolysis if performed in glacial acetic acid yields the corresponding acetate (**F**).

The photochemical conversion of santonin in absolute ethanol to lumisantonin (**G**) (~13 % yield) and that of lumisantonin in cold aq. AcOH at –5 to 5 °C to photosantonic acid (**J**) via an intermediate dienone (**H**) (isolated and characterized) [76, 77] are delineated in Fig. 7.54. Plausible stepwise mechanism based on the

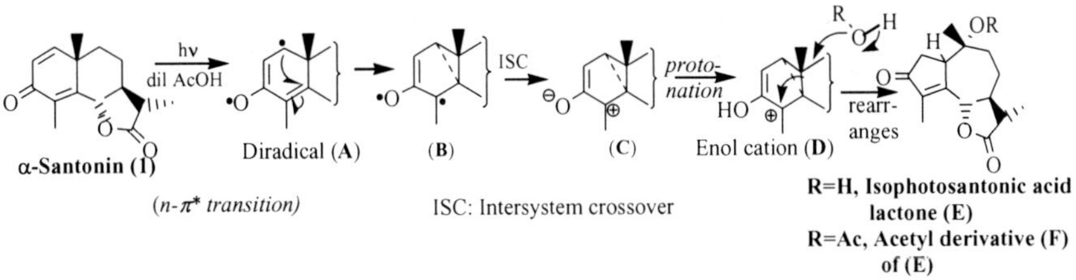

Fig. 7.53 Mechanistic rationale for the photoconversion of santonin to isophotosantonic acid

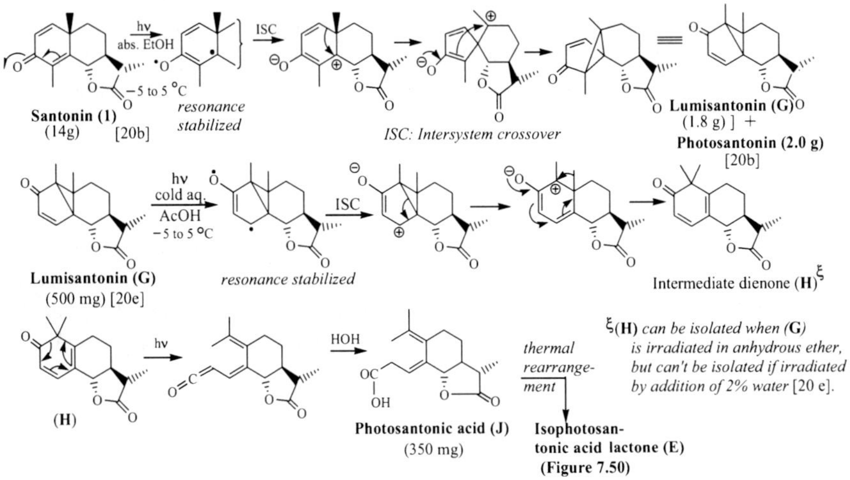

Fig. 7.54 Mechanistic rationale for the photoconversion of santonin to lumisantonin (G) and of (G) to (J) via (H)

intensive studies by Barton, Chapman, Richards, and others is shown in Figs. 7.53 and 7.54. Lumisantonin upon thermal rearrangement forms isophotosantonic acid (E). The intermediate dienone (H) is rapidly converted to photosantonic acid (J) in presence of light and water. Irradiation of lumisantonin in anhydrous ether requires 2.5 h to achieve maximum concentration of (H). Addition of 2 % water and continued irradiation under identical condition leads to complete disappearance of (H) in less than 45 min. Thus, for correct interpretation of the photochemical processes the nature and number of discrete photochemical reactions involved should be known [76].

7.9 Artemisinin: *A Sesquiterpene Lactone with an Endoperoxide Linkage and Profound Antimalarial Activity*

7.9.1 Introduction. Occurrence. Structure

Qing hao meaning "green herb" and pronounced as Ching how (*Artemisia annua*, Fam. Compositae) has been used in Chinese traditional medicine as a treatment for fever and malaria for many centuries. Its uses have appeared in several Chinese medicinal texts. The *Book of Fifty Two Prescriptions* discovered in the tomb of Mawangdui Han dynasty, dating from 168 B.C. contains the use of *qing hao* in the treatment of hemorrhoids. The antimalarial property of this drug was first described in 340 A.D. in *Zhou Hou Bei Ji Fang*, (Handbook of Prescriptions for Emergency Treatments) [81]. In 1596 Li Shizhen mentioned in the treatise *Ben Cao Gung Mu* (Compendium of Treatments) the use of *qing hao* for the treatment of shivering and fever of malaria [81–84]. In 1971 it was observed that crude ether extracts of *A. annua* produced encouraging results in mice infected with the malaria parasite *Plasmodium berghei*. In 1972 the active principle responsible for the profound antimalarial activity of *Artemisia annua* (*qing hao*) was isolated by a Chinese group of workers [81–83] as a white crystalline compound, C$_{15}$H$_{22}$O$_5$, colorless needles (from hexane), m.p. 156–157 °C, $[\alpha]_D$ + 66.3 (CHCl$_3$) from the leafy part of the plant.

7.9.2 Absolute Stereochemistry and Conformation

The compound named artemisinin (*qinghaosu*) was characterized as a sesquiterpene lactone with an unusual trioxane structure (Fig. 7.55) from its X-ray crystallographic studies by Qinghaosu Research Group of China. It was shown to possess seven stereogenic centers (*R/S designation of all chiral centers is shown in the*

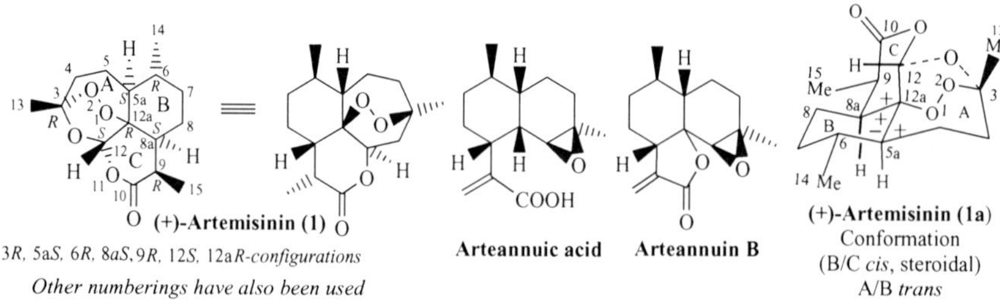

Fig. 7.55 Stereostructures of artemisinin and arteannuin B

Figure) on a tetracyclic framework (**1**). The absolute configuration and A/B *trans* and B/C *cis* ring junctures have also been settled by comparing its ORD curve with that of arteannuin B of known stereochemistry. Its rigid conformation (1a) with rings B/C in chair/chair *cis*-decalin steroidal conformation (Sect. 2.15.1.3) and A/B *trans*-decalin (Sect. 2.15.1.2)-type conformation is depicted in Fig. 7.55. The absolute configurations at all the seven stereogenic centers (Sects. 2.6.2 and 2.7.5) are shown. Chemical support to this structure was also provided [82, 83]. Artemisinin is a timely discovery since malaria caused by the protozoa of the genus *Plasmodium* (strain *Plasmodium falciparum*) became resistant to chloroquine and many other parallel synthetic drugs (Chap. 33).

7.9.3   Synthesis

The overwhelming antimalarial property of *artemisinin* with no side effect (unlike the synthetic antimalarials) and its interesting unusual structure claimed vigorous studies on it. Consequently, a number of its total as well as semi-syntheses appeared in the literature. The first total synthesis, achieved by Schmid and Hofheinz [85] of Hoffman-La Roche in Basel, displayed remarkable stereoselectivity. However, semi-syntheses [86, 87] (Figs. 7.56 and 7.57) and a recent (2010) shortest synthesis of artemisinin by Yadav et al. [88] will be presented here. Artemisinin and its derivatives have become attractive target molecules for their semisyntheses [89] and total syntheses [90]. A recent report [91, 92] on the semisynthesis of artemisinin employing biologically synthesized artemisinic acid through engineered yeast cell culture and its subsequent conversion to artemisinin using modified commercially viable chemical methodology will also be discussed briefly.

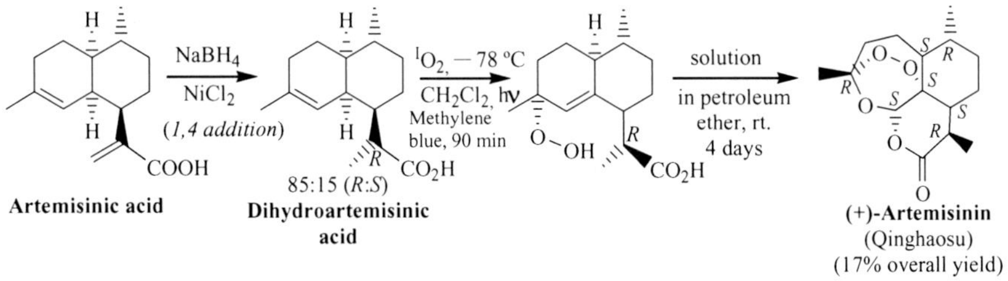

Fig. 7.56 Semisynthesis of artemisinin from artemisinic acid [86]

Application of Synthetic biology [6f]

| Strains of *Saccharomyces cerevisiae* (bakers' yeast) containing genes for expressing the enzymes as in plant *(Artemisia annua)* artemisinic acid production | → Cultural broth → | *Artemisinic Acid (good yield) |

Application of Synthetic Chemistry [6f]   : Artemisinic acid   →   Artemisinin (40-45% overall yield)

The sequence of reactions are shown below

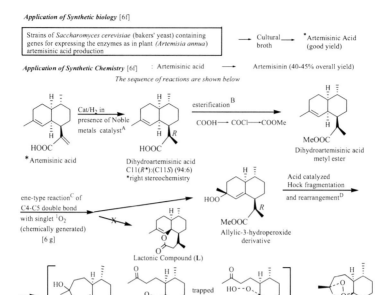

Notes: [A]   *Noble metal catalyzed hydrogenation of Δ$^{11(13)}$ gives improved yield of the right stereoisomer C11R (94%) compared to the earlier method where nickel boride (NaBH$_4$ or LiBH$_4$/NiCl$_2$) was used (C11R, 85%). Further, no significant amount of tetrahydro derivative formation observed.*

[B]   *Esterification of COOH group is needed to block it as ester and hence its participation to form the lactone (L) is arrested and the ester proceeds into the main sequence of reactions leading to artemisinin formation.*

[C]   *Chemically generated singlet oxygen is preferred since in commercial scale photosynthetic steps are rarely used.*

[D]   *Hock fragmentation [6h] is a well-known reaction for benzylic, allylic and dienylic hydroperoxides*

This reaction has been applied to the synthesis of artemisinin as shown below

In this step less expensive reagent benzenesulphonic acid/sulphonate Cu(II) – Dowex resin has been used in place of copper triflate.

[E]   *Pure oxygen has been replaced by air for safely reason*

HPLC studies show that artemisinin prepared by this method is purer than commercial artemisinin (plant source). Thus the entire chemical conversion of artemisinic acid to artemisinin showed a significant efficiency.

Fig. 7.57   Biological synthesis of artemisinic acid in yeast cell culture and its chemical conversion to artemisinin

7.9.3.1 Semisynthesis

(1a) A semisynthesis of artemisinin starting from *artemisinic acid*, a relatively abundant constituent of *Artemisia annua*, has been reported by Roth and Acton [86, 87, 89–94]. Artemisinic acid is converted to dihydroartemisinic acid with $NaBH_4$ in the presence of $NiCl_2$. Allylic hydroperoxidation is carried with singlet oxygen. On standing in air in the presence of a little acid and in a hydrocarbon solution the hydroperoxide facilitates the introduction of a second molecule of oxygen and is converted into artemisinin in 17 % overall yield (Fig. 7.56). It is a biomimetic synthesis, as artemisinic acid co-occurs with artemisinin in the same plant and acts as the precursor of artemisinin in the biosynthetic pathway (Fig. 7.60). The conversion of the hydroperoxide to artemisinin is a remarkable reaction. The mechanism of such conversion is not obvious. It involves an oxidation, ring expansion, an intramolecular nucleophilic attack, and lactonization with the competence of enzymatic reactions of its biosynthetic process [87] (*cf.* Fig. 7.60). The importance of artemisinin has also led to its semi-synthesis [89] from another naturally occurring biogenetic precursor *arteannuic acid* (Fig. 7.55).

The systematic name of artemisinin is 3,6,9-trimethyl-9,10b-epidioxyperhydropyrano[4,3,2-*ik*]benzoxepin-2-one, but this is hardly convenient. *Chemical Abstracts* adapted the name artemisinin [84].

(1b) "Yeast makes artemisinin on demand" [91]. This title refers to a research note appearing in the May 2013 issue of *Chemistry World*. It includes a summarized report on the reconstruction of the biosynthetic path of plant artemisinin in yeast cells. Jay Keasling's group at the Lawrence Berkeley National Laboratory in California, Berkeley, reported in *Nature* in 2006 about the genomically engineered yeast cells and their functions in the biosynthesis of artemisinic acid, a biogenetic precursor of artemisinin which can be chemically converted to the drug. However, genes necessary to produce the two particular biosynthetic enzymes made the yeast ill and the entire process was not competitive with plant source extraction, Recently, Chris Paddon of Amyris, Emeryville, California, developed strains of yeast cells (*Saccharomyces cerevisiae*, baker's yeast) using synthetic biology that can encode two more dehydrogenase enzymes necessary to synthesize plant artemisinic acid. Thus with full complement of five key enzymes artemisinic acid is biosynthesized in yeast cell culture in increased yield. The yields are 10 times higher than the previous reports and the yeast cells remain healthy. Chris Paddon, his group, and his collaborators reported their findings in *Nature* 2013 [92] and the title of their paper reads as "High-level semi-synthetic production of the potent antimalarial artemisinin." This method with slight difference has already been commercially used and semi-synthesized 35 metric tons of artemisinin—roughly equivalent to 70 million malaria cures [91]. The conversion of artemisinic acid to artemisinin (40–45 % overall yield) has been done in a scalable and practical way to offer a stable and affordable source for the drug [92] (Fig. 7.57).

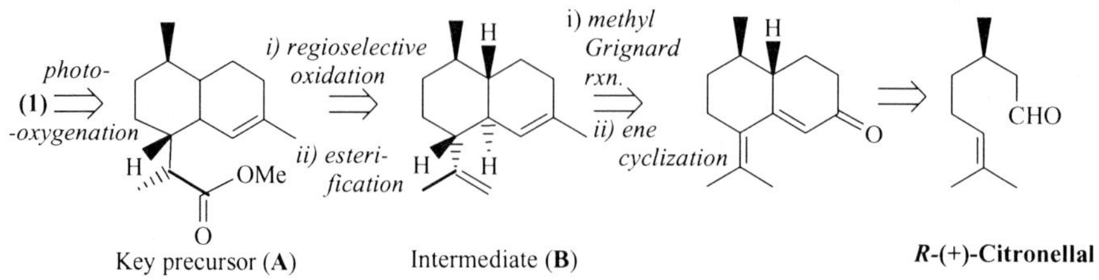

Fig. 7.58 Retrosynthetic plan for artemisinin [88]

7.9.3.2   Total Synthesis of (+)-Artemisinin by Yadav et al. [88]

A protective group-free, concise, streoselective total synthesis of (+)-artemisinin has been reported in which R-(+)-citronellal, a commercially available monoterpene, serves as the starting material. The retrosynthetic plan for this synthesis is delineated in Fig. 7.58 [88].

In the present synthesis (Fig. 7.59) R(+)-citronellal is alkylated with methyl vinyl ketone (MVK) by asymmetric 1,4-addition (Michael reaction) in the presence of a chiral proline-derived catalyst, when the product was obtained in 70 % yield with 83 % *de*. Intramolecular aldol condensation yielded an enone. The latter on Grignard reaction formed a mixture of diastereisomeric alcohols. The stereochemistry at C4 will be absent in subsequent steps. The mixture underwent SnCl$_4$-catalyzed cyclization to give the ene intermediate (**B**), a *trans*-decalin derivative. Regioselective hydroboration of (B) with 9-BBN formed the primary alcohol in 85 % yield. Swern oxidation of the latter formed the aldehyde (94 %), which upon oxidation with NaClO$_2$/NaH$_2$PO$_4$ produced the acid in 80 % yield. The corresponding methyl ester was subjected to photooxidation following Haynes protocol (DCM, rose Bengal, copper triflate) [95] to yield (+)-artemisinin (**1**) (Fig. 7.59) in 13.0 % overall yield.

7.9.4   Spectral Properties [88]

IR (KBr): ν 2,921, 2,854, 1,738, 1,115, 994 cm^{-1}.

^{1}H NMR (CDCl$_3$, 300 Mz): δ 1.00 (d, 3H, J = 5.8 Hz), 1.02–1.13 (m, 1H), 1.21 (d, 3H, J = 7.3 Hz), 1.45 (s, 3H), 1.33–1.54 (m, 4H), 1.69–1.83 (m, 2H), 1.84–1.94 (m, 1H), 1.95–2.11 (m, 2H), 2.36–2.53 (m, 1H), 3.34–3.46 (m, 1H), 5.86 (s, 1H).

^{13}C NMR (CDCl$_3$, 75 MHz): δ 12.5, 19.8, 23.4, 24.8, 25.2, 32.9, 33.6, 35.9, 37.5, 45.0, 50.0, 79.5, 93.7, 105.4, 172.0.

Mass: APCI: m/z 283 (M+H)$^+$.

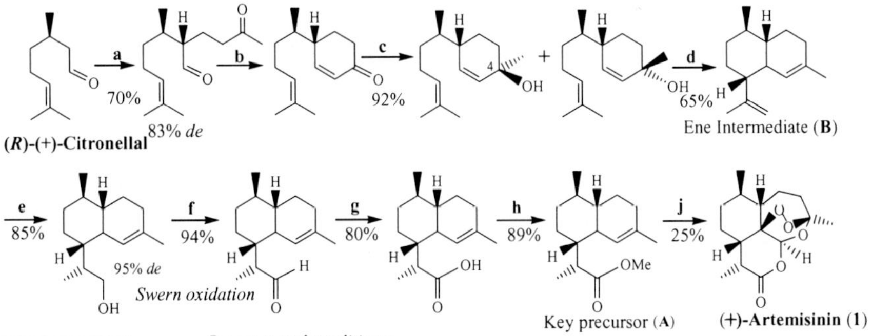

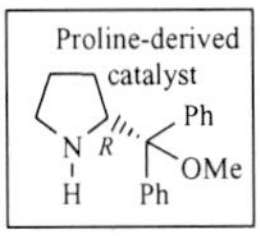

Reagents and conditions:

(**a**) MVK, catalyst (5 mol%), Ethyl 3,4-dihydroxybenzoate (20 mol%), neat, 0-4 °C, 48 h

(**b**) KOH (0.1 *N* aq, 1.0 equiv), nBu_4NOH (40% aq, cat.), Et_2O:THF:H_2O (3:1:3), reflux, 8 h

(**c**) MeMgI, Et_2O, 2 h, rt.

(**d**) $SnCl_4$, Benzene:Et_2O (4:1), 0 °C

(**e**) 9-BBN, 3*N* NaOH, H_2O_2

(**f**) *Swern oxidation* [$(COCl)_2$ (2 eq), CH_2Cl_2, Me_2SO (4 eq), − 78 0°C N_2, 30 min, add substrate, 1.5 h, − 78 °C, Et_3N, rt, 1 h]

(**g**) $NaClO_2$, NaH_2PO_4, 0°C

(**h**) MeI, K_2CO_3, acetone rt

(**j**) (i) O_2, Rose Bengal sensitizer, − 30 °C, 6 h, CH_3CN, 500 W tungsten filament lamp
(ii) O_2, $Cu(OTf)_2$, CH_3CN, − 20°C, 8 h
(iii) TsOH (cat.), CH_2Cl_2, 0 °C, 4 h, rt

Fig. 7.59 Total synthesis of (+)-artemisinin [88]

7.9.5 Biosynthesis of Artemisinin (qinghaosu) [84, 96]

A number of enzymes exhibiting sesquiterpene synthase properties have been found to occur in the leaves of *Artemisia annua*. One such enzyme, a sesquiterpene cyclase, has been isolated and sequenced and has been found to cyclize FPP to a number of products, of which amorpha-3,11-diene predominates. The latter has been shown to be the precursor of artemisinic acid which is subsequently converted to artemisinin (Fig. 7.60).

7.9.6 Uses

Uses of artemisinin and some of its bioactive derivatives have been discussed in Chap. 33.

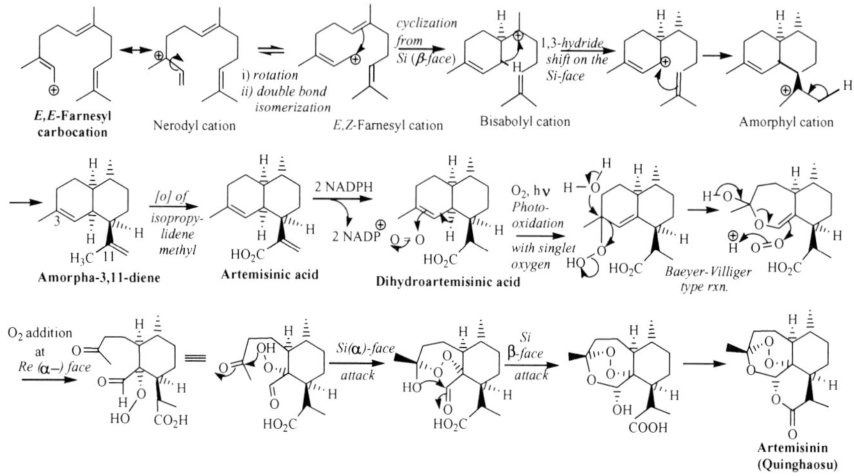

Fig. 7.60 Biosynthesis of artemisinin *via* amorpha-3,11-diene, artemisinic acid, and dihydroartemisinic acid

7.10 Abscisic Acid: A Sesquiterpene Phytohormone

7.10.1 Introduction. Occurrence

There are five major classes of phytohormones having diverse structures and functions (Fig. 7.61). One of them is *abscisic acid*, a sesquiterpene in composition with cyclofarnesane-type skeletal pattern. It regulates the water balance in plants (stomata opening and closing) and causes abscission (shedding of leaves) acceleration, bud dormancy, and growth inhibition. Such growth-inhibitory compounds are also known as "dormins." On the other hand, there are *gibberellins* (*e.g.*, GA$_3$), C$_{19}$-diterpenes (Chap. 8), which induce elongation growth of internodes, *cytokinin*—a prenylated adenine which stimulates cell division, *auxin*—an indole-derived plant hormone causing stimulation to cell elongation, and *ethylene*—derived from the methionyl moiety of *S*-adenosylmethionine, which helps in ripening the fruits. Plant hormones are structurally quite different from animal hormones.

Abscisic acid mp 160–161 °C, [α]$_D$ + 430 has been isolated in extremely low yield (0.000004 %) from the young fruits of *Gossypium hirsutum* (Fam. Malvaceae). It also occurs in "sycamore" (*Acer pseudoplatanus,* Aceraceae) [97] and birch (*Betula pubescens,* Betulaceae) leaves, rose (*Rosa arvensis,* Rosaceae), cabbage (*Brassica oleraceae,* Crucifarae), potato (*Solanum tuberosum,* Solanaceae), avocado (*Persea grantissima*), and lemon (*Citrus medica,* Rutaceae). Its structure has been deduced from degradation products, synthesis, and spectral properties. The absolute configuration of the only asymmetric center has been determined to be *S*. These are briefly discussed in the sequel.

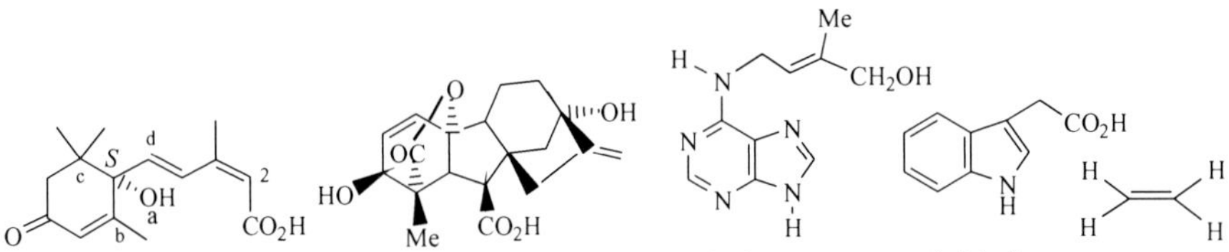

(+)-Abscisic acid (**1**) Gibberellic acid (GA_3) Zeatin (Cytokinin) Indole-3-acetic acid (auxin) Ethylene

Fig. 7.61 Some phytohormones including (+)-abscisic acid

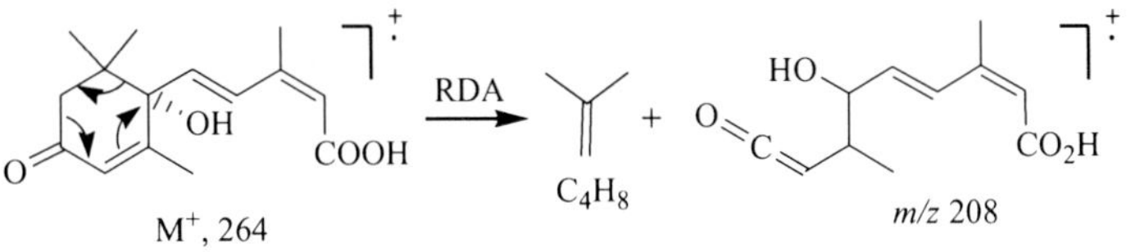

Fig. 7.62 Spectral properties of abscisic acid

7.10.2 Spectral Properties (Fig. 7.62)

UV (EtOH): λ_{max} 246 nm (ε 25,00).

IR (KBr): ν in cm^{-1} 3,405 (OH), 1,674 (acid > =O), 1,650 (α,β-unsaturated > =O), 978 (trisubstituted double bond).

PMR ($CDCl_3$): δ (J Hz) [98].

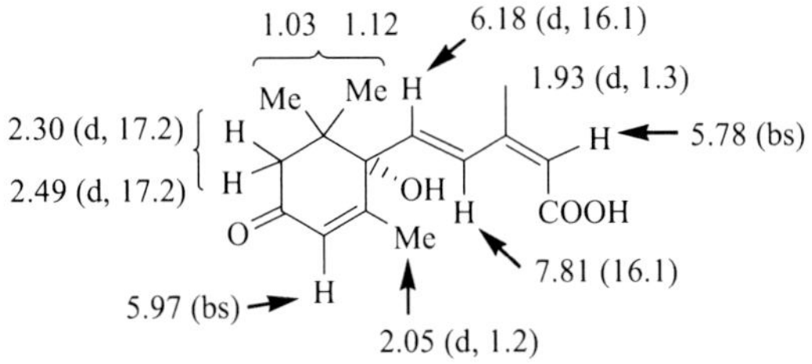

MS [98]: m/z 264 (M^+), 246 (M^+ $-H_2O$), 208 (M^+ $-C_4H_8$).

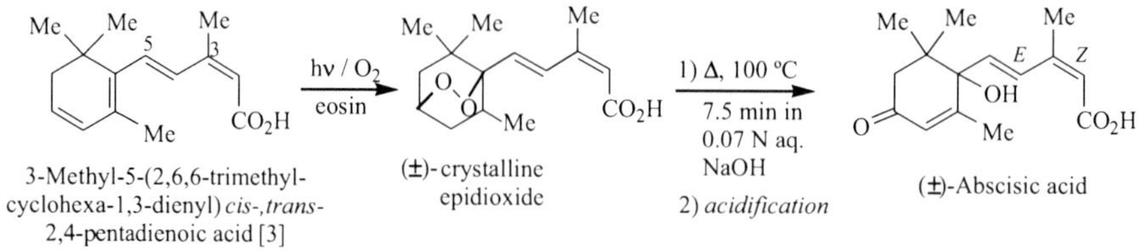

3-Methyl-5-(2,6,6-trimethyl-
cyclohexa-1,3-dienyl) *cis-,trans-*
2,4-pentadienoic acid [3]

(±)-crystalline
epidioxide

(±)-Abscisic acid

Fig. 7.63 Cornforth's synthesis of (±)-abscisic acid (1965) [99]

7.10.3 Synthesis

7.10.3.1 Cornforth's Synthesis of (±)-Abscisic Acid

The photosensitized 1,4-addition of oxygen to homoannular 1,3-diene (*epidioxidation*), a well-known reaction, has been used as the key step in the synthesis of abscisic acid by Cornforth [99]. The synthesis is outlined in Fig. 7.63.

It should be mentioned that the m.p. of the synthetic product is 188–190 °C, is around 30 °C more than the natural one and is with half activity (being *dl*) compared to the natural one, the (+)-enantiomer.

7.10.3.2 Synthesis of Optically Active Abscisic Acid

In a number of syntheses of abscisic acid (ABA) *Reformatsky reaction* has been employed conveniently for the stereospecific formation of (2Z,4E)-diene side chain responsible for its hormonal activity. Its isomerization to (2E,4E) causes the loss of the hormonal activity [98]. The orientation of the carboxylic group is thus significantly important for its activity. More than 100 abscisic acid derivatives have been synthesized to find out ABA-like strong activity and the molecular conformation needed for ABA action [100].

An ordinary synthon with an epoxy aldehyde group could not be tried for the (2Z,4E)-diene side-chain formation using Reformatsky reaction since the latter affects the epoxide also. However, Sakai et al. [98] employed a suitable epoxyaldehyde synthon, in which the expoxy group is guarded by substitution from the attack of the Reformatsky reagent, which then only reacts with the aldehyde function generating the (2Z, 4E)-diene side chain (Fig. 7.64).

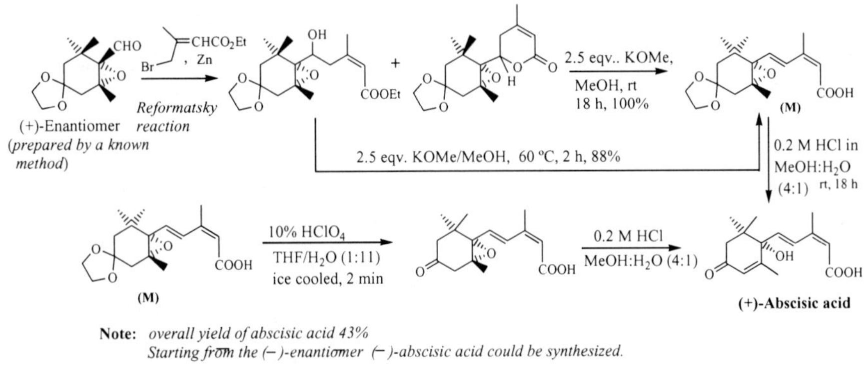

Note: *overall yield of abscisic acid 43%*
Starting from the (–)-enantiomer (–)-abscisic acid could be synthesized.

Fig. 7.64   Synthesis of (+)-abscisic acid [98]

7.10.4   Absolute Configuration

7.10.4.1   By Chemical Correlation

Violaxanthin, derived from carotenoid (see Fig. 7.69) on photooxidation, gave a number of products of which *xanthoxin*, a widespread carotenoid-derived natural growth inhibitor, is one. The latter could be converted to (+)-*trans*-abscisic acid (Fig. 7.65). Since the stereochemistry of the chiral centers of violaxanthin is known, this conversion shows (*S*) (new CIP convention, Sect. 2.4) configuration of the only chiral center of (+)-abscisic acid [103]. The reversal of configuration during this chemical conversion is extremely unlikely. Thus the absolute configuration *(S)* of C6 of (–)-*cis*-xanthoxin, which has been converted to (+)-abscisic acid, is also established.

The absolute configuration for natural (+)-abscisic acid (ABA) {6β-OH in place of 6α-OH in (**1**) or (6*R*)-configuration according to the new CIP convention)} was proposed by Cornforth et al. [104] on the basis of [*M*]$_D$-values of *cis*- and *trans*-3,6-diol esters [obtained from the (+)- and (–)-abscisic acid)] by application of Mill's empirical rule. The absolute configuration has been revised by other authors [105, 106] as (6*S*). Perhaps the diol esters are not suitable for the application of Mills' empirical rule. According to old (1956) CIP convention of (+)-abscisic acid was designated (*S*) by Cornforth [104].

7.10.4.2   By CD Studies (Exciton Chirality Method) (Sect. 2.19.6.5)

The configuration (6*S*) has been established by Nakanishi et al. [107] by CD studies (exciton chirality method) of ABA and some of it structural relatives having similarly substituted cyclohexenone ring with C6 (bearing OH) having both (*R*)-

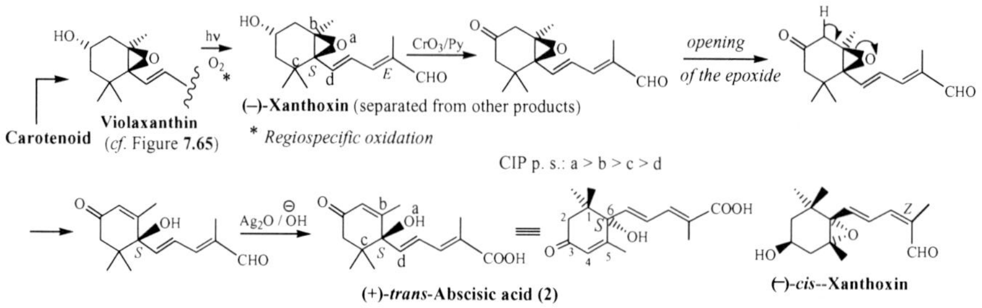

Fig. 7.65 The absolute configuration of (+)-abscisic acid

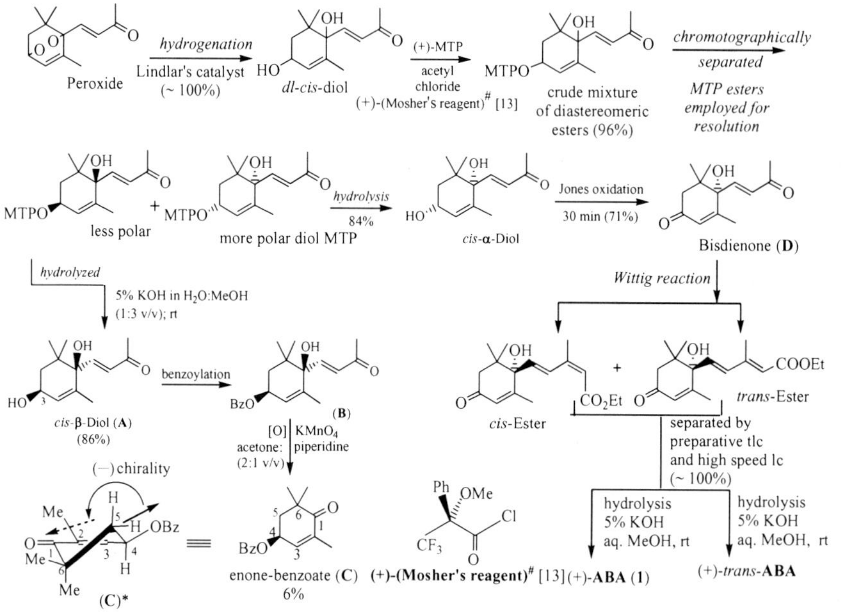

*The direction of the front electric transition moment is parallel to the C4-OBz bond, and that of the back is passing through the mid-point of C3 and the carbonyl oxygen. The chirality between the two chromophores is negative, exhibiting (−)-CE. The absolute configuration of the benzoate is hence represented by (**C**) and that of the diol by (**A**).*

(+)-MTP Acetyl chloride = (+)-1-Methoxy-1-trifluoromethyl-1-phenylacetyl chloride [(+)-**Mosher's reagent]**[#]

Fig. 7.66 Preparation of (**A**), (**B**), and (**C**) exhibiting (−)-CE, and of (**D**), (+)-**ABA,** and (+)-*trans*-ABA exhibiting (+)-CE in CD studies

and (*S*)-configurations. The preparation of the compounds (**A**) and (**B**), having (6*R*)-configuration, and also compound (**C**), and that of the compounds (**D**), (+)-ABA, and (+)-*trans*-ABA are outlined in Fig. 7.66.

The CD of the benzoate (**C**) (Table 7.1) showed a split Cotton effect due to interaction between the benzoate and the enone chromophores. The negative first Cotton effect as par the exciton chirality method indicates a negative chirality (**C1**)

Table 7.1 The CD, $\Delta\varepsilon$
(nm) in MeOH

Compound	$\pi - \pi^*$	Compound	$\pi - \pi^*$
(A)	–6.0 (232)	(D)	+ 38.4 (242)[a]
			–30.2 (208)[a]
(B)	–5.3 (230)	(+)-ABA	+34.5 (261)[a]
			–28.0 (229)[a]
(C)	–19.0 (238)[a]	(+)-*trans*-ABA	+25.5 (254)[a]
	+3.21 (219)[a]		–12.6 (221)[a]

[a]Davidov split Cotton effects centered around UV maxima

between the two axes of electric transition moments. The absolute configuration of the benzoate is hence represented by (**C**) and that of the diol by (**A**) [109].

The positive first Cotton effects of the bisdienone (**D**) and of (+)-ABA and (+)-*trans*-ABA each one showing Davidov splitting (Table 7.1) support the absolute configuration (**S**) at C6 of these compounds, as depicted which is consistent with the (+)-chirality of the two interacting chromophores.

The enantiomers (–)-ABA and (–)-*trans*-ABA were similarly prepared from the *cis*-β-diol and their CD data were also measured and found to exhibit (–)-Cotton effect (CE), like compounds (**A**), (**B**), and (**C**).

The absolute configuration (6*S*) for (+)-*trans*-abscisic acid has been established by Harada [110] by quantitative application of the exciton chirality method.

Natural (+)-abscisic acid has also been correlated with (*S*)-malic acid (*S*-COOH. CHOH.CH2.COOH), leading to the same absolute configuration (*S*) at C6 [111].

7.10.5   Molecular Conformations of (+)-ABA and (+)-trans-ABA

The cyclohexenone ring of (+)-ABA or (+)-*trans*-ABA adopts a half-chair conformation as shown in Fig. 7.67. Both types of molecules are quite flexible. The half-chair undergoes ring inversion giving rise to two possible half-chair conformers in each case, viz., (**1tr**) and (**1ftr**) and (**1ci**) and (**1fci**) for ABA (**1**). Moreover, the side-chain conformation at C8 is expected to be mainly *cisoid* or 5-*cis* (C8 and OH on the same side of the 6–7 single bond), since the side chain goes away from the cyclohexenone ring. Of the two flipped *cisoid* conformers in each case, (**1Ci**) and **2Ci** having pseudoequatonal (e′) side chain will be more stable and hence the most stable conformer of (+)-ABA and (+)-*trans*-ABA, respectively. Moreover, the 7,9-diene in (**1ci**) is *S-trans* and hence contributes to its stability, while it is *S-cis* in (**1tr**). The above arguments point out the conformers (**1ftr**) and the (**2ftr**) as the least stable ones for ABA and *trans*-ABA, respectively. Harada [110] determined the absolute configuration of (+)-*trans*-ABA by a quantitative application of the exciton chirality method for all the four conformers (**2tr**), (**2ftr**), (**2ci**), and (**2fci**). Thus quantitative application of the exciton chirality method [110] may provide a useful tool for conformational analyses of natural products.

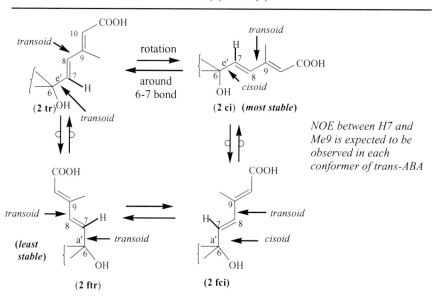

Conformers of (+)-ABA (1)

The conformations of the ring are same as the corresponding conformers of (1)

Conformers of (+)- *trans*-ABA (2)

Fig. 7.67 Conformers of (+)-ABA (**1**) and (+)-*trans*-ABA (**2**)

7.10.6   Biosynthesis [101, 102, 112]

As a fungal metabolite abscisic acid is formed from farnesyl pyrophosphate [112] (Fig. 7.68) while in higher plants it is formed by the oxidative cleavage of the carotenoid violaxanthin.

Abscisic acid is formally a sesquiterpene. However, it is also very similar to the end portion of certain carotenoids. The absolute configuration of abscisic acid is same as that of violaxanthin (isolated from orange peel) (Sect. 7.10.4.1), the pigment precursor of ABA. A reversal of configuration during the biosynthesis would be extremely unlikely.

In plants abscisic acid is derived from carotenoids (Fig. 7.69). The first committed precursor is zeaxanthin which undergoes a series of enzymatic reactions

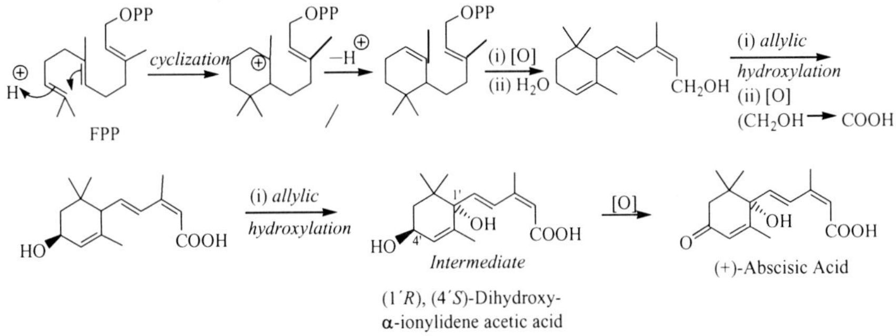

Fig. 7.68   Probable biosynthesis of abscisic acid in fungus [112]

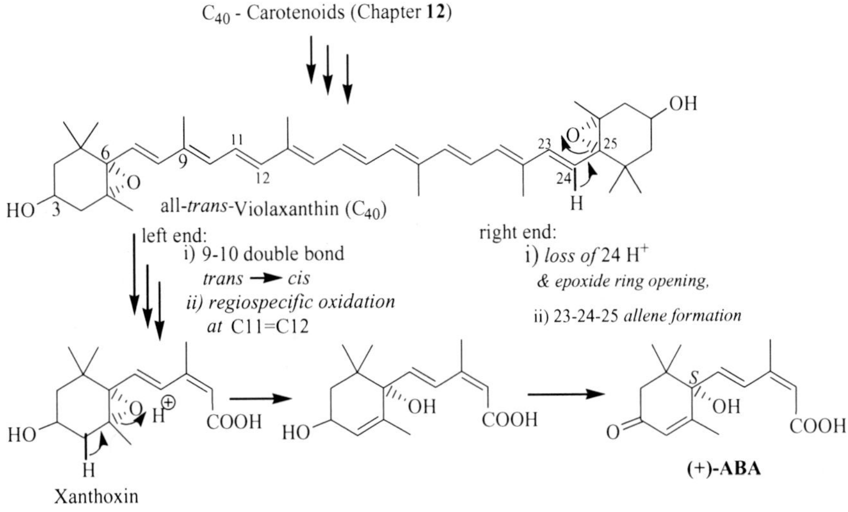

Fig. 7.69   Biosynthesis of abscisic acid in higher plants [101, 102]

involving epoxidation and cleavage of C$_{40}$ violaxanthin to xanthoxin [101], a C$_{15}$ precursor of abscisic acid. When [2-^{14}C]-xanthoxin was fed to cut shoots of bean and tomato it was converted to (+)-abscisic acid in high yield [111]. However, it was shown that [2-^{3}H]-MVA when fed to avocado pears ^{3}H-labeled ABA was found. Most ABA in plants might be produced directly from MVA. However, it appears that under different metabolism the above two different routes may be operative. The biosynthesis of (+)-abscisic acid in plants is outlined in Fig. 7.69.

7.11　Gossypol: An Interesting Dinaphthyl Bis-Sesquiterpene with Cadinane Skeletal Pattern

7.11.1　Introduction. Occurrence, Biological Activity

Gossypol, an interesting and unusual example of a bis-sesquiterpene, is a yellow-colored highly functionalized binaphthyl crystalline compound possessing axial chirality [113] (Sect. 2.16) arising out of restricted rotation around the aryl–aryl bond. Biogenetically it belongs to sesquiterpenoids with cadinane skeletal pattern. Its polyphenolic character apparently does not attest its association with the sesquiterpene class; rather it seems to be biogenetically associated with the polyketide pathway which is not true. However, the presence of isopropyl group, carbon content, and the methyls at right intervals of the basic C$_{15}$ chain certainly point to it being a dimeric sesquiterpene (cf. Sect. 7.11.5). Some monomeric naphthol and naphthaldehyde derivatives with cadinane skeletal pattern of plant origin are known. Gossypol was first isolated in 1899 from the seeds of cotton (*Gossypium hirsutum*, Fam. Malvaceae) in chemically pure but stereochemically mixture forms. It occurs in seeds of most of the cotton plants (*Gossypium* species) (*G. herbaceum*, *G. barbadense*, *G. arboretum*) (0.1–0.6 %).

Gossypol occurs in *Gossypium* species mainly as a dextrorotatory atropisomer (Sect. 2.16.5), [α]$_D$ +445 (CHCl$_3$)], and also as its (±)-form. In these species (+)-form predominates over the (−)-form. The (±)-variety may be resolved into active forms via diastereomeric Schiff's bases with various amino acid esters (*e.g.*, methyl L-phenylalanate), followed by silica gel column chromatography [114]. (±)-Gossypol complexes with acetic acid, but neither enantiomer does so. Thus (±)-gossypol can be separated from the excess of (+)-isomer by suitable treatment of cotton seed extracts. Cotton being a major plant crop, a large amount of cotton seed is obtained as a by-product, which is mainly used as a food for cattle. Cotton seed is rich in protein, but it is not used as food due to the hepatotoxic and cardiotoxic effects of gossypol in humans and animals other than ruminants. Gossypol acts as a male contraceptive; it alters sperm maturation and spermatozoid motility and inactivates the sperm enzymes necessary for fertilization. Although the antifertility effect is reversible, irreversible infertility results from the prolonged use of the drug for longer periods [115]. Some other plants of Malvaceae family, *viz.*, *Thespesia populnea* (3.3 %) and *Montezuma speciosissima* (6.1 %), are especially rich sources of almost entirely the (+)-form that does not possess contraceptive property. However, *Thespesia danis* stands as an exception; its aerial parts contain 72 % (−)-gossypol (yield ~ 0.2 %). Gossypol has also shown antimalarial property as it inhibits the enzyme essential for the parasite *Plasmodium falciferum*.

7.11.2   Absolute Configuration of Gossypol

In the mid-1980s, the resolution of enantiomeric mixture into the active forms of gossypol has been achieved in the preparative scale.

The absolute stereostructures of the enantiomers (Fig. 7.70) were elucidated [116] by Snatzke in collaboration with Huang's group in 1988 by the application of circular dichroism and exciton chirality theory (Sect. 2.18 and 2.19.6.5).

7.11.3   Synthesis

The first stereoselective synthesis of (P)-(+)-gossypol has been achieved by Meyer et al. using Ullmann coupling of the appropriate bromo monomer (Fig. 7.71). The (P),(S)-atropdiastereomer formed in 94.6 % diastereomeric excess is converted to (P)-(+)-gossypol through several appropriate steps.

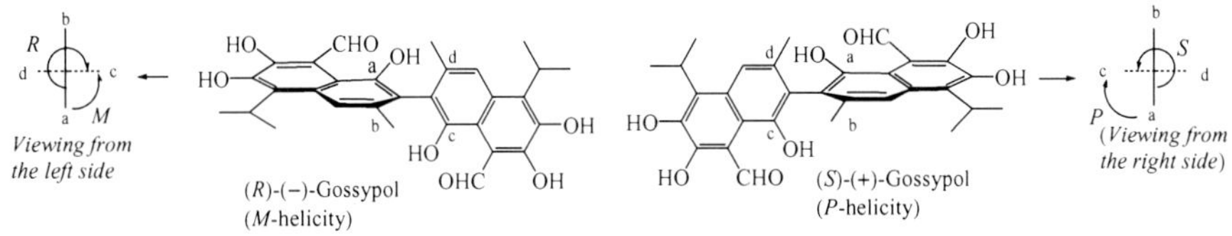

Fig. 7.70   Absolute configuration of (−)-gossypol and (+)-gossypol

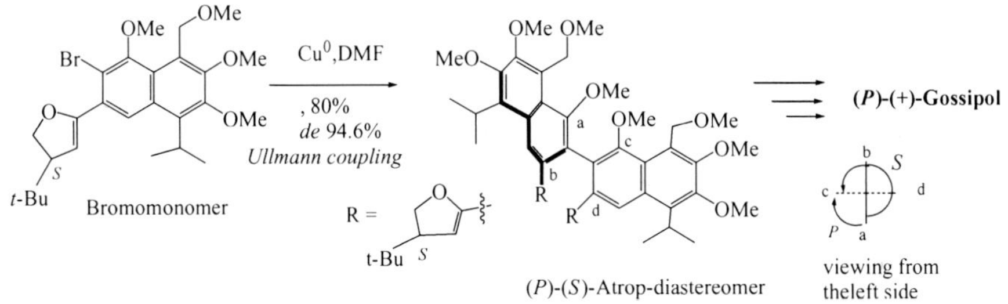

Fig. 7.71   First stereoselective synthesis of (P)-(+)-gossypol

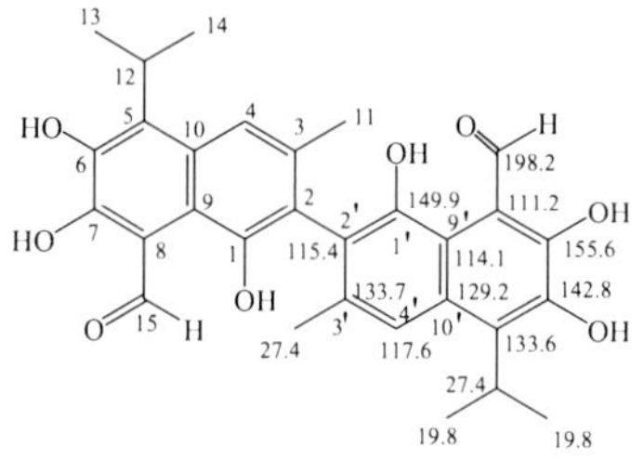

Note: *The C2 axis of the molecule is orthogonal to the pivotal C2-C2' bond and to the projection plane, and passes through its midpoint. The homotopic pairs of carbon atoms , viz., 1 & 1', 2 & 2', 3 & 3', and so on, exchangeable by the C2 operation are isotropic , and have same chemical shifts.*

Fig. 7.72 ^{13}C NMR spectral chemical shifts of gossypol

7.11.4 ^{13}C NMR Spectral Data of Gossypol [117]

^{13}**C NMR** spectral chemical shifts of gossypol having **C$_2$** symmetry are shown in Fig. 7.72.

7.11.5   Biosynthesis of Gossypol [118]

From the biosynthetic experiments for gossypol [118] with three different radioactive precursors (4-^{14}C-mevalonate, 2-^{14}C-mevalonate, and 5-^{3}H-mevalonate) labeled gossypol was isolated from the incubation flask and purified as dianilinogossypol (2-CHO → 2-CH=N-Ph). The degradative study of the latter in search of incorporation (radioactivity measurements) and locations suggested that three molecules of mevalonate are involved in the formation of the C$_{15}$-linear precursor and the compatible folding pattern (**A**) (Fig. 7.73). The formation of the hemigossypol component occurs as with other sesquiterpenes of cadinane skeleton *via* a ten-membered ring cation, formed by 1,3-hydride shift [(**B**) → (**C**)], and various oxidative stages. During hemigossypol biosynthesis *via* deoxyhemigossypol ^{3}H-equivalent from C5 is preserved in the isopropyl chain (5-^{3}H-mevalonic acid precursor) (radioactivity measurements). Diradical coupling takes place in both possible ways to form *P*-(+)-gossypol in larger amount in the said *Gossypium* species and also *M*-(−)-gossypol, depending upon the specific enzyme present.

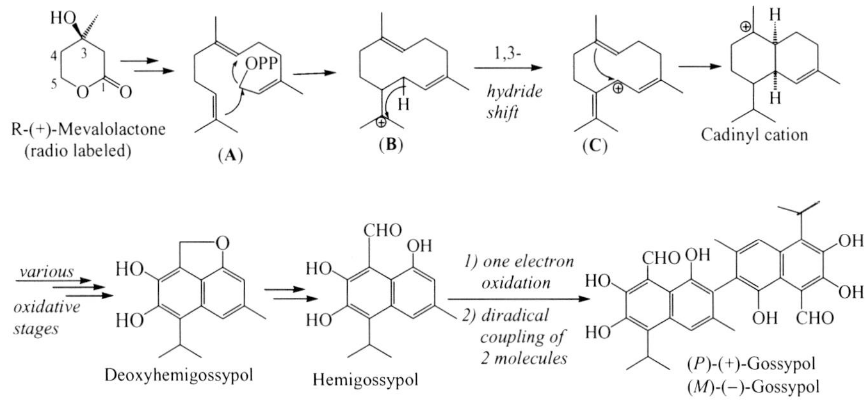

R-(+)-Mevalolactone
(radio labeled)

(A)

(B)

1,3-
hydride
shift

(C)

Cadinyl cation

various
oxidative
stages

Deoxyhemigossypol

Hemigossypol

1) one electron
oxidation

2) diradical
coupling of
2 molecules

(P)-(+)-Gossypol
(M)-(−)-Gossypol

Fig. 7.73 A scheme for the biosynthesis of gossypol

7.12   Ainsliadimer A. A Novel Sesquiterpene Lactone Dimer with a Cyclopentane Ring

7.12.1   Occurrence. Structure. Biogenesis. Bioactivity

The genus *Ainslisea* (Compositae) comprises 70 species of which 48 are indigenous to China. Some of the species are used in folk medicine. *Ainslisea* species are well known for elaborating sesquiterpenes, and recently (2008) [119] a novel sequiterpene lactone dimer ainsliadimer A (**1**) has been isolated from a chemically virgin Southwestern Chinese *Ainslisea* species (*A. macrocephala*). Its structure has been elucidated from its extensive NMR studies and single crystal X-ray analysis (ORTEP). The dimer is generated biogenetically (Fig. 7.74) from two molecules of dehydrozaluzanin C which is chemically close to the known compound anolide (**2**), a guanolide derivative.

Ainsliadimer (1), $C_{30}H_{34}O_7$ was isolated as colorless prism, $[\alpha]_D$ +47 ($CHCl_3$); (m.p. was not reported).

7.12.2   Spectral Data [119] (Fig. 7.74)

IR (KBr): $_{max}$ 3471 (OH), 1773 (five-membered keto carbonyl), 1750, 1719 (ester carbonyl), 1262 (-O-CO stretching) cm^{-1}.

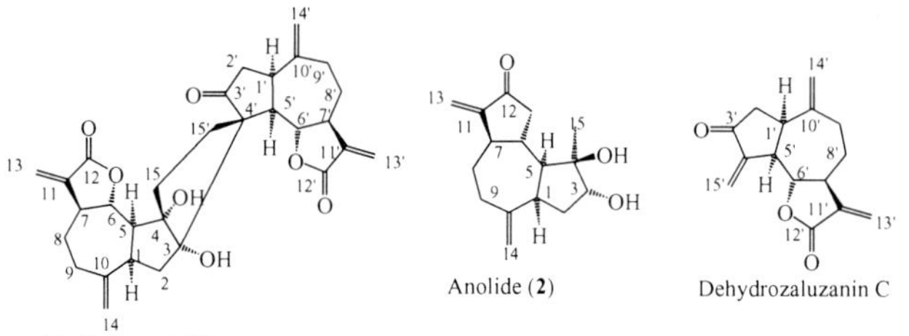

Ainsliadimer A (1)   Anolide (2)   Dehydrozaluzanin C

^{1}H NMR (CDCl$_3$) δ: *for left hand part* H1 3.15 (dd, 11.4, 12.0), H2 2.38 (dd 12.0, 13.8), H2 (1.88-1.92*), H5 2.55 (t, 11.4), H6 4.07 (dd, 9.0, 11.4), H7 2.72 (m), H8 2.22-2.24 (m), H8 1.37 (dq 6.0, 12.6) H9 2.67 (m), H9 1.88-1.92*, H-13a 6.19 (d, 3.0), H13b 5.46 (d, 3.0), H14a 5.15 (s), H14b 5.03 (s), H15a, 2.25-2.30#, H15b 2.16-2.21#   * / # overlapping

For right hand part: H1' 3.04-3.06*, H2' 2.91 (dd, 16.8, 9.6) H2' 2.45 (d 16.8), H5' 3.04-3.06* H6' 4.12 (dd 8.4, 10.2), H7' 3.04-3.06*, H8' 2.25-2.30, H8'β 1.50 (dq 3.6, 12.6) H9' 2.16-2.21, H9' 2.57 (m), H13'a 6.31 (d, 3.0), H13'b 5.61 (d, 3.0), H14'a 4.96(s), H14'b 4.59(s), H15'a 2.05 (dd, 5.4, 14.4), H15'b 2.01 (dd, 7.8, 14.4)   * overlapping.

^{13}C NMR (CDCl$_3$) δ : *for left hand part:* C1 41.8, C2 38.1, C3, 90.5, C4 89.1, C5 53.1, C6 82.9, C7 49.0,C8 31.3, C9 37.8, C10 147.5, C11 139.5, C12 170.0, C13 119.7,C14 113.8, C15 33.1.

For right hand part: C1' 50.1, C2' 46.9, C3' 224.9,C4' 62.2, C5' 38.9,C6' 84.0,C7' 43.7, C8' 32.8, C9' 38.9, C10' 150.0, C11' 138.6, C12' 169.0, C13' 122.0, C14' 113.7, C15' 26.3.

The assignments have been made on the basis of extensive ^{13}C NMR (including HMBC correlation) and comparison of some data with those reported in the literature for the monomers anolide and dehydrozaluzanin..

MS : m/z 505 2170 (calc. for C$_{30}$H$_{34}$O$_7$ 505.2226) based on HRESIMS (negative mode) [M–1].

Fig. 7.74 Spectral data of ainsliadimer A

7.12.3   Biogenesis

The biogenetic dimerization through hetro Diels–Alder reaction of dehydrozaluzanin C leading to the formation of ainsliadimer A might have taken place as shown in Fig. 7.75. The steps following the first one are different from the pathway shown in Scheme 1 of [119].

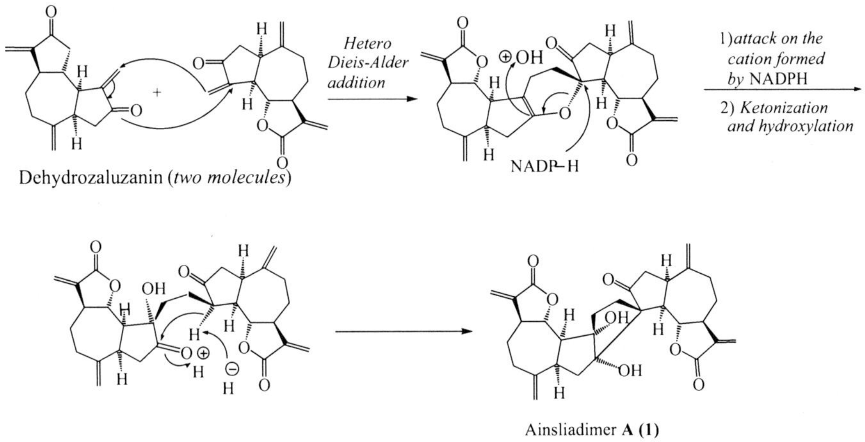

Dehydrozaluzanin (*two molecules*)

Hetero Dieis-Alder addition

NADP–H

1) *attack on the cation formed by* NADPH

2) *Ketonization and hydroxylation*

Ainsliadimer **A** (**1**)

Fig. 7.75 Probable biogenetic pathway of ainsliadimer A

7.12.4   Bioactivity

Ainsliadimer A shows remarkable inhibitory effect against NO production. NO plays an important role in the inflammatory process and hence ainsliadimer A might find therapeutic use in the inflammatory diseases.

References

1. *Webster's Encyclopedic Unabridged Dictionary of the English language*, Gramery Books, New York, **1996**, p. 4.
2. L. Ruzicka, Höhere Terpenverbindungen IX Über die Totalsynthese des d,l-Nerolidols und des Farnesols. *Helv. Chim. Acta.*, **1923**, *6*, 492-502.
3. Elias J. Corey, John A. Katzenellenbogen, and Gary H. Posner, New Stereospecific Synthesis of Trisubstituted Olefins. Stereospecific Synthesis of Farnesol, *J. Am. Chem. Soc.*, **1967**, *89*, 4245-4247.
4. Hiroshi Suginome, Takahiko Kondoh, Camelia Gogonea, Vishwakarma Singh, Hitoshi Goto and Eiji Osawa, Photoinduced Molecular Transformations. Part 155. General Synthesis of Macrocyclic Ketones Based on a Ring Expansion Involving a Selective β-Scission of Alkoxy Radicals, Its Application to a New Synthesis of (±)-Isocaryophyllene and (±)-Caryophyllene, and a Conformational Analysis of the Two Sesquiterpenes and the Radical Intermediate in the Synthesis by MM3 Calculations, *J. Chem. Soc. Perkin Trans. 1*, **1995**, 69-81, and pertinent references cited.
5. H. C. Hill, R. I. Reed, and (Miss) M. T. Robert-Lopes, Mass Spectra and Molecular Structure. Part 1. Correlation Studies and Metastable Transitions, *J. Chem. Soc.* (C), **1998**, 93-101.
6. A. Aebi, D. H. R. Barton, A. W. Burgstahler and A. S. Lindsey, Sesquiterpenoids. Part V. The Stereochemistry of the Tricyclic Derivatives of Caryophyllene, *J. Chem. Soc.* (London), **1954**, 4659-4665.
7. D. H. R. Barton and A. Nickon, The Absolute Configuration of Caryophyllene, *J. Chem. Soc.*, **1954**, 4665-4669.

8. A. Horeau and H.K. Sutherland, The Absolute Configuration of Some Caryophyllene Derivatives, *J. Chem. Soc.* (C), **1966**, 247-248.

9. Alain Horeau, Principe et Applicatioons d'une Nouvelle Methode de Determination des Configurations dite "par Dedouplement Partiel", *Tetrahedron Lett.*, **1961**, 506-512.

10. E. J. Corey, Rajat B. Mitra and Hisashi Uda, Total Synthesis of *d,l*-Caryophyllene and *d,l*-Isocaryophyllene, *J. Am. Chem. Soc.*, **1963**, *85*, 362-363.

11. E. J. Corey, Rajat B. Mitra and Hisashi Uda, Total Synthesis of *d,l*-Caryophyllene and *d,l*-Isocaryophyllene, *J. Am. Chem. Soc.*, **1964**, *86*, 485-492.

12. D. Rogers and Mazhar-ul-Haque, The Molecular and Crystal Structure of Caryophyllene Chlorohydrin, *Proc. Chem. Soc.*, **1963**, 371-372.

13. Jean L. Gras, Robert Maurin and Marcel Bertrand, Une Voie Dacces Possible au Caryophyllene – Synthese Du Dihydro-5,6-norcaryophyllene, *Tetrahedron Lett.*, **1969**, 3533-3536.

14. A. Kumar, A. Sing and D. Devaprabhakara, A Simple and Efficient Route to Caryophyllene System – Synthesis of dl-isocaryophyllene, *Tetrahedron Lett.*, **1976**, 2177-2178.

15. John E. McMurray and Dennis D. Miller, Synthesis of Isocaryophyllene by Titanium-Induced Keto Ester Cyclization, *Tetrahedron Lett.*, **1983**, 1885-1888.

16. Yasuo Ohtsuka, Setsuko Niitsuma, Hajime Tadokoro, Toshio Hayashi, and Takeshi Oishi, Medium-Ring Ketone Synthesis. Total Syntheses of (+)-Isocaryophyllene and (+)-Caryophyllene, *J. Org. Chem.*, **1984**, *49*, 2328-2332.

17. M. Bertrand and J. L Gras, Synthese Totale du (+)-Isocaryophyllene, *Tetrahedron*, **1974**, *30*, 793-796.

18. R. Vaidyanathaswami and D. Devaprabhakara, A Convenient Synthesis of *cis,cis*-1,5-Cyclononadiene, *J. Org. Chem.*, **1967**, *32*, 4143-4143.

19. Isidro G. Collado, James R. Hanson and Antonio J. Macias-Sánchez, Recent Advances in the Chemistry of Caryophyllene, *Nat. Prod. Rep.*, **1998**, 187-204 and relevant references cited.

20. Sriram Shankar and Robert M. Coates, Solvolysis of Caryophyllene-8β-yl Derivatives: Biomimetic Rearrangement – Cyclization to 12-Nor-8α-presilphiperfolan-9β-ol, *J. Org. Chem.*, **1998**, *63*, 9177-9182.

21. J. M. Greenwood, J. K. Sutherland and A. Turre, The Conversion of Humulene into Caryophyllene, *Chem. Commun.*, **1965**, 410-411.

22. A. Nickon, T. Iwadare, F. J. McGuire, J. R. Mahajan, S. A. Narang and B. Umezawa, The Structure, Stereochemistry, and Genesis of α-Caryophyllene Alcohol (Apollan-11-ol), *J. Am. Chem. Soc.*, **1970**, *92*, 1688-1696.

23. K. W. Gemmell, W. Parker, J. S. Roberts and G. A. Sim, The Structure of α-Caryophyllene Alcohol, *J. Am. Chem. Soc.*, **1964**, *86*, 1438-1439.

24. Sukh Dev, Studies in Sesquiterpenes- XVIII, The Proton Magnetic Resonance Spectra of Some Sesquiterpenes a Structure of Humulene, *Tetrahedron*, **1960**, *9*, 1-9; references cited.

25. A. Nickon and E. F. Silversmith, *Organic Chemistry: The Name Game*, Pergamon Press, New York, **1987**, 36.

26. Ping-Jyun Sung, Li-Fan Chuang, Jimmy Kuo, Tung-Yung Fan and Wan-Ping Hu, Rumphellatin A, the first chlorine-containing caryophyllan-type norsequiterpenoid from *Rumphella antipathies*, *Tetrahedron Lett.*, **2007**, *48*, 3987-3989.

27. Ping-Jyum Sung, Li-Fan Chuang, and Wan-Ping Hu, Rumphellatins B and C, Two New Caryophyllane-Type Hemiketal Norsesquiterpenoids from the Formosan Gorgonian Coral *Rumphella antipathies, Bull. Chem. Soc. Jpn.*, **2007**, *80*, 2395-2399.

28. Sukh Dev, The Chemistry of Longifolene and its Derivatives, *Fortschr. Chem. org. Naturstoffe*, **1981**, *40*, 49-104 and references cited.

29. J. S. Yadav, U. R. Nayak, and Sukh Dev, Studies in Sesquiterpenes – LV. Isolongifolene (Part 6): Mechanism of Rearrangement of Longifolene to Isolongifolene,*Tetrahedron*, **1980**, *36*, 309-315.

30. Sukh Dev, Aspects of Longifolene Chemistry. An Example of Another Facet of Natural Products Chemistry, *Acc. Chem. Res.*, **1981**, *14*, 82-88.

31. Jerome A. Berson, James H. Hammons, Arthur W. McRowe, Robert G. Bergman, Allen. Remanick, and Donald. Houston, Chemistry of Methylnorbornyl Cations. VI. The Stereochemistry of Vicinal Hydride Shift. Evidence for the Nonclassical Structure of 3-Methyl-2-norbornyl Cations, *J. Am. Chem. Soc.,* **1967**, *89*, 2590-2600.

32. D. Arigoni, Stereochemical Aspects of Sesquiterpene Biosynthesis, *Pure Appl. Chem.,* **1975**, *41*, 219-245.

33. J. S. Simonsen and D. H. R. Barton, *The Terpenes,* Vol. III, Cambridge University Press, Cambridge, England, **1952**, pp. 92-98.

34. R. H. Moffett and D. Rogers, The Molecular Configuration of Longifolene Hydrochloride, *Chem. Ind.* (London), **1953**, 916.

35. P. Naffa and G. Ourisson, Chemical Approach to the Structure of Longifolene, *Chem. and Ind.* (London), **1953**, 917-918.

36. John E. McMurry and Stephen J. Isser, Total Synthesis of Longifolene, *J. Am. Chem. Soc.,* **1972**, *94*, 7132-7137.

37. H. C. Hill, R. I. Reed, and (Miss) M. T. Robert-Lopes, Mass Spectra and Molecular Structure. Part I, Correlation Studies and Metastable Transitions, *J. Chem. Soc.* (C), **1968**, 93-101.

38. E. J. Corey, Masaji Ohno, Rajat B. Mitra, and Paul A. Vatakenchery, Total Synthesis of Longifolene, *J. Am. Chem. Soc.,* **1964**, *86*, 478-485.

39. Robert A. Volkmann, Glenn C. Andrews and William S. Johnson, A Novel Synthesis of Longifolene, *J. Am. Chem. Soc.,* **1975**, *97*, 4777-4779.

40. Wolfgang Oppolzer and Thiery Godel, A New and Efficient total Synthesis of (+)-Longifolene, *J. Am. Chem. Soc.,* **1978**, *100*, 2583-2584.

41. U. Ramdas Nayak and Sukh Dev, Longicyclene, the First Tetracyclic Sesquiterpene, *Tetrahedron Lett.,* **1963**, 243-246.

42. U. R. Nayak and Sukh Dev, Studies in Sesquiterpenes – XXXV Longicyclene, The First Tetracyclic Sesquiterpene, *Tetrahedron,* **1968**, *24*, 4099-4106.

43. Steven C. Welch and Roland L. Walters, Stereoselective Total Synthesis of (+)-Longicyclene, (+)-Longicamphor and (+)-Longiborneol. *J. Org. Chem.,* **1974**, *18*, 2665-2673.

44. D. H. R. Barton, Triterpenoids. Part III. Cycloartenone, a Triterpenoid Ketone, *J. Chem. Soc.,* **1951**, 1444.

45. P. G. Gassman and W. M. Hooker, Near-Infrared Studies. Norbornenes and Related Compounds, *J. Am. Chem. Soc.,* **1965**, *87*, 1079-1083.

46. Ludmila Birladeanu, The Stories of Santonin and Santonin Acid, *Angew. Chem. Int. Ed.,* **2003**, *42*, 1202-1208.

47. For earlier literature see J. L. Simonsen and D. H. R. Barton, *The Terpenes,* Cambridge University Press, **1952**, Vol. III.

48. George Roger Clemo, Robert Downs Haworth, and Eric Walton, The Constitution of Santonin. Part I. The Synthesis of dl-Santonous Acid, *J. Chem. Soc.,* **1929**, 2368-2387.

49. G. R. Clemo and R. D. Haworth, The Constitution of Santonin. Part II. The Synthesis of Racemic Desmotroposantonin, *J. Chem. Soc.,* **1930**, 1110-1115.

50. G. R. Clemo and R. D. Haworth, The Constitution of Santonin. Part III. Proof of the Positions of the Methyl Groups, *J. Chem. Soc.,* **1930**, 2579-2582.

51. James A. Marshall and Peter G. M. Wuts, Stereocontrolled Total Synthesis of α- and β-Santonin, *J. Org. Chem.,* **1978**, *43*, 1086-1089.

52. E. J. Corey and C. U. Kim, New and Highly Effective Method for the Oxidation of Primary and Secondary Alcohols to Carbonyl Compounds, *J. Am. Chem. Soc.,* 1972, 94, 7586-7587.

53. E. J. Corey and C. U. Kim, Improved Synthetic Routes to Prostaglandins Utilizing Sulfide-Mediated Oxidation of Primary and Secondary Alcohols, *J. Org. Chem.,* **1973**, *38*, 1233-1234.

54. Farouk S. El-Feraly, Daniel A. Benigni, and Andrew T. McPhail, Biogenetic-type Synthesis of Santonin, Chrysanolide, Dihydrochrysanolide, Tulirinol, Arbusculin-C, Tanacetin, and Artemin, *J. Chem. Soc. Perkin Trans 1,* **1983**, 355-364 and the references cited.

55. I. Abé, T. Miki, M. Sumi, and T. Toga, *Chem. Ind.* (London), **1956**, 95.

56. J. C. Fiaud and H. B. Kagan, Determination of Stereochemistry by Chemical Correlation Methods in *Stereochemistry, Fundamentals and Methods*, Ed. Henri B. Kagan, George Thieme Publishers, Stuttgart, **1977**, Vol. *3*, p. 30.

57. W. Cocker and T. B. H. McMurry, Stereochemical Relationships in the Eudesmane (Selinaceae) Group of Sesquiterpenes, *Tetrahedron*, **1960**,*8*, 181-204.

58. Andrew P. J. Brunskill, Hugh W. Thompson and Roger A. Lalancette, Santonic Acid: Catemeric Hydrogen Bonding in a γ,ε−Diketo Carboxylic Acid, *Acta Crystallogr. Sec. C*, **1999**, *55*, 566-568.

59. J. D. M. Asher and G. A. Sim, Sesquiterpenoids. Part III. Stereochemistry of Santonin: X-Ray Analysis of 2-Bromo-α-Santonin, *J. Chem. Soc.*, **1965**, 6041-6055.

60. J. D. M. Asher and G. A. Sim, Sesquiterpenoids. Part II. The Stereochemistry of Isophotosantonic Lactone: X-ray Analysis of 2-Bromodihydroisophoto-α-santonic Lactone Acetate, *J. Chem. Soc.*, **1965**, 1584-1594.

61. A. T. McPhail, B. Rimmer, J. Monteath Robertson, and G. A. Sim, Sesquiterpenoids, Part VI, The Stereochemistry of Desmotroposantonon: X-ray Analysis of 2-Bromo-(−)- β-desmotroposantonin, *J. Chem. Soc.* (B),**1967**, 101-106.

62. P. Coggon and G. A. Sim, Sesquiterpenoids. Part VIII. Stereochemistry of Santonin : X-Ray Analysis of 2-Bromo-β-Santonin, *J. Chem. Soc.* (*B*), **1969**, 237-242.

63. P. S. Pregosin, E. W. Randall, and T. B. H. McMurry, ^{13}C Fourier Studies, The Configurational Dependence of the Carbon-13 Chemical Shifts in Santonin Derivatives, *J. Chem. Soc. Perkin 1*, **1972**, 299-

64. P. Brown and C. Djerassi, Electron-Impact Induced Rearrangement Reactions of Organic Molecules, *Angew. Chem. Int. Ed..*, **1967**, *6*, 477.

65. R. B. Woodward, F. J. Brutschy, and Harold Baer, The Structure of Santonic Acid, *J. Am. Chem. Soc.*, **1948**, *70*, 4216-4221.

66. D. H. R. Barton, G. P. Moss and J. A. Whittle, Investigation on the Biosynthesis of Steroids and Terpenoids. Part 1. A Preliminary Study of the Biosynthesis of Santonin, *J. Chem. Soc.* (C), **1968**, 1813-1818.

67. D. M. Simonovic, A. Somasekar Rao and S. C. Bhattacharyya, Terpenoids XXXIX. The Synthesis of Tetrahydrosaussura lactone, *Tetrahedron*, **1963**, *19*, 1061-1071.

68. Masayoshi Ando, Ken Nanami, Toru Nakagawa, Toyonobu Asao, and Kahei Takase, Synthesis of (−)-Occidentalol and its C-7 Epimer, *Tetrahedron Lett.*, **1970**, 3891-3894.

69. Yasuo Fujimoto, Takeshi, Shimizu, and Takashi Tatsuno, Modification of α-Santonin II. Synthesis of Dihydrocostunolide, *Tetrahedron Lett.*, **1976**, 2041-2044.

70. G. Blay, L. Cardona, B. García and J. R. Pedro, The Synthesis of Bioactive Sesquiterpenes from Santonin in *Studies in Natural Products Chemistry*, Att-ur-Rahman (Ed.), **2000**, *24*, 53-129, Elsevier.

71. Tatsuo Nozoe, Toyonobu Asao, Masayoshi Ando, and Kahei Takase, The Total Synthesis of Chamaecynone, *Tetrahedron Lett.*, **1967**, 2821-2825.

72. T. Nozoe, T. S. Cheng, and T. Toda, The Structure of Chemaecynone, A Novel Norsesquiterpenoid from *Chamaecyparis formosensis* Matsum, *Tetrahedron Lett*, **1966**, 3663-3669.

73. D. H. R. Barton, P. de Mayo, and Mohammed Shafiq, The Mechanism of the Light-catalyzed Transformation of Santonin into 10-Hydroxy-3-oxogual-4-ene, *Proc. Chem. Soc.*, **1957**, 205.

74. D. H. R. Barton, P. de Mayo, and Mohammed Shafiq, Photochemical Transformations. Part II. The Constitution of Lumisantonin, *J. Chem. Soc.*, **1958**, 140-145.

75. D. Arigoni, H. Bosshard, H. Bruderer, G. Büchi, O. Jeger, and K. J. Krebaum, Uber gegenseitige Beziehungen und Umwandlungen bei Bestrahlungsprodukten des Santonins, *Helv. Chim. Acta*, **1957**, *40*, 1732.

76. O. L. Chapman and L. F. Englert, A. Mechanistically Significant Intermediate in the Lumisantonin to Photosantonic Acid Conversion, *J. Am. Chem. Soc.*, **1963**, *85*(19), 3028-3029.

77. M. H. Fisch and J. H. Richards, The Mechanism of the Photoconversion of Santonin, *J. Am. Chem. Soc.*, **1963**, *85*(19), 3030-3031.

78. H. E. Zimmerman and D. L. Schuster, A New Approach to Mechanistic Organic Photochemistry. IV. Photochemical Rearrangements of 4,4-Diphenylcyclohexadienone, *J. Am. Chem. Soc.*, **1962**, *84*, 4527-4540.

79. Also see Howard E. Zimmerman, Report on Recent Photochemical Investigations, *Pure Appl Chem.*, **1964**, *9*, 493-498.

80. D. H. R. Barton, J. E. D. Levisalles, and J. T. Pinhey, Photochemical Transformations. Part XIV. Some Analogues of Isophotosantonic Lactone, *J. Chem. Soc.*, **1962**, 3472-3482.

81. Paluther (Artemether), Product Monograph, Rhône-Paulene (India) Limited (when the world celebrated in 1997 the Centenary of the discovery (in 1897) of malaria parasite by Sir Ronald Ross, this monograph was dedicated to his memory.

82. H. Ziffer, R. J. Highet and D. L. Klayman, Artemisinin, An Endoperoxide Antimalarial from *Artemisia annua* L. *Fortschr. Chem. org. Naturstoffe*, **1997**, *72*, 121-214.

83. Daniel L. Klayman, Q*inghaosu* (Artemisinin): An Antimalarial Drug from China, *Science*, **1985**, *228*, 1049-1055.

84. Anthony R. Butler and Yu-Lin Wu, Artemisinin (Qinghaosu): A New Type of Antimalarial Drug, *Chem. Soc. Revs.*, **1992** , 85-90.

85. G. Schmid and W. Hofheinz, Total Synthesis of Qinghaosu, *J. Am. Chem. Soc.*, **1983**, *105*, 624-625.

86. Ronald J. Roth and Nancy Acton, A Simple Conversion of Artemisinic Acid into Artemisinin, *J. Nat. Prod.*, **1989**, *52*, 1183-1185.

87. Ronald J. Roth and Nancy Acton, A Facile Semisynthesis of the Antimalarial Drug Qinghaosu, *J. Chem. Educ.*, **1991**, *68*, 612-613.

88. J. S. Yadav, B. Thirupathaih and P. Srihari, A Concise Stereoselective Total Synthesis of (+)-Artemisinin, *Tetrahedron*, **2010**, *66*, 2005-2009.

89. For other semisyntheses see references 6(a), and 6(d) to 6(f) of the paper of Yadav *et al*. [7].

90. For other total syntheses of artemisinin see references 4(b) to 4(h) of the paper of Yadav *et al*. [7].

91. Yeast Makes Artemisinin on Demand, *Chemistry World*, **2013**, 10, May issue (Number 05), Research.

92. C. J. Paddon, P. J. Westfall, D. J. Pitera, K. Benjamin, K. Fisher, D. McPhee, M. D. Leavell, A. Tai, A. Main, D. Eng, D. R. Polichuk, K. H. Teoh, D. W. Reed, T. Treynor, J. Lenihan, M. Fleck, S. Bajad, G. Dang, D. Dengrove, D. Doila, G. Dorin, K. W. Ellens, S. Fickes, J. Galazzo, S. P. Gaucher, T. Geistlinger, R. Henry, M. Hepp, T. Horning, T. Iqbal, H. Jiang, L. Kizer, B. Lieu, D. Melis, N. Moss, R. Regentin, S. Secrest, H. Tsuruta, R. Vazquez, L. F. Westblade, L. Xu, M. Yu, Y. Zhang, L. Zhao, J. Lievense, P. S. Covello, J. D. Keasling, K. K. Reiling, N. S. Renninger, and J. D. Newman, High-Level Semi-Synthetic Production of the Potent, Antimalarial Artemisinin, *Nature*, **2013**, *496*, 528-532.

93. K. Boehme and H. –D Brauer, Generation of Singlet Oxygen from Hydrogen Peroxide Disproportionation Catalyzed by Molybdate Ions., *Inorg Chem.*, **1992**,*31* 3468-3471.

94. Arych A. Frimer, The Reaction of Singlet Oxygen with Olefins: The Question of Mechanism, *Chem. Rev.* **1979**, *79*, 359-387, and references cited; pertinent page 364.

95. Simone C. Vonwiller, Jacqueline A. Warner, Simon T. Mann, and Richard K. Haynes, Copper(II) Trifluoromethanesulfonate-Induced Cleavage. Oxygenation of Allylic Hydroperoxides Derived from Qinghao Acid in the Synthesis of Qinghaosu Derivatives: Evidence for the Intermediacy of Enols, *J. Am. Chem. Soc.*, **1995**, *117*, 11098-11105.

96. Richard K. Haynes and Simone C. Vonwiller, Catalysed Oxygenation of Allylic Hydroperoxides Derived from Qinghao (Artemisinic) Acid. Conversion of Qinghao Acid into Dehydroqinghaosu (Artemisitene) and Qinghaosu (Artemisinin), *J. Chem. Soc., Chem. Commun.*, **1990**, 451- 453.

97. J. W. Cornforth, B. V. Milborrow, G. Ryback, and P. F. Wareing, Chemistry and Physiology of 'Dormins' in Sycamore, *Nature*, **1965**, *205*, 1269-1270.

98. Kunikazu Sakai, Kyoko Takahashi, and Tomoko Nukano, Convenient Synthesis of Optically Active Abscisic Acid and Xanthoxin, *Tetrahedron*, **1992**, *48*, 8229-8238.

99. J. W. Cornforth, B. V. Milborrow and G. Ryback, Synthesis of ($\pm$)-Abscisin II, *Nature*, **1965**, *206*, 715.

100. Yasushi Todoroki and Nobuhiro Hiarai, Abscisic Acid Analogs for probing the Mechanism of Abscisic Acid Reception and Inactivation in *Studies in Natural Products Chemistry*, Atta-ur-Rahaman (Ed.), Elsevier, **2002**, *27*, 321-360.

101. R. S. Burden and H. F. Taylor, Xanthoxin and Abscisic Acid, *Pure & Appl. Chem.*, **1976**, *47*, 203-209, and references cited.

102. Paul M. Dewick, The Biosynthesis of C_5-C_{25} Terpenoid Compounds, *Nat. Prod. Rep.,* **1999,** *16*, 91-130; pertinent page 117.

103. R. S. Burden and H. F. Taylor, The Structure and Chemical Transformations of Xanthoxin, *Tetrahedron Lett.*, **1970**, 4071-4074.

104. J. W. Cornforth, W. Draber, B. V. Milborrow, and G. Ryback, Absolute Stereochemistry of (+)-Abscisin II, *Chem. Comm.*, **1967**, 114-116.

105. Sachihiko Isoe, Suong Be Hyeon, Shigeo Katsumura, and Takeo Sakan, Photo-oxygenation of Carotenoids. II. The Absolute Configuration of Loliolide and Dihydroactinidiolide, *Tetrahedron Lett.*, **1972**, 2517-2520.

106. Takayuki Oritani and Kyohei Yamashita, Synthesis of Optically Active Abscisic acid and Its Analogs, *ibid.*, **1972**, 2521-2624.

107. Masato Koreeda, George Weiss, and Koji Nakanishi, Absolute Configuration of Natural (+)-Abscisic Acid, *J. Am. Chem. Soc.*, **1973**, *95*, 239-240.

108. James A. Dale and Harry S. Mosher, Nuclear Magnetic Resonance Enantiomer Reagents. Configurational Correlation via Nuclear Magnetic Resonance Chemical Shifts of Diastereomeric Mandelate, O-Methyl Mandelate, and α-Methoxy-α-trifluoromethyl-phenylacetate (MTPA) Esters, *J. Am. Chem. Soc.*, **1973**, *95*, 512-519.

109. Nobuyuki Harada and Koji Nakanishi, The Exciton Chirality Method and Its Application to Configurational and Conformational Studies of Natural Products, *Accounts Chem. Res.*, **1972**, *5*, 257-263.

110. Nobuyuki Harada, Absolute Configuration of (+)-*trans*-Abscisic Acid as Determined by a Quantitative Application of the Exciton Chirality Method, *J. Am. Chem. Soc.*, **1973**, *95*, 240-242.

111. G. Rybach, Revision of the Absolute Configuration of (+)-Abscisic Acid, *J. C. S. Chem. Comm..,* **1972,** 1190-1191.

112. Masahiko Okamoto, Nobuhiro Hirai, and Koichi Koshimizu, Biosynthesis of Abscisic Acid from α-Ionylidecthanol in *Cercospora pini-densiflorae*, *Phytochmistry*, **1988**, *27*,

113. G. Bringmann, C. Gunther, M. Ochse, O. Schupp and S. Tasla, Biaryls in Nature, A Multi-facetted Class of Stereochemically, Biosynthetically and Pharmacologically Intriguing Secondary Metabolites, *Fortschr. Chem. org. Naturstoffe,* (Eds. W. Herz, H. Falk, G. W. Kirby and R. E. Moore), **2001**, *82*, 1-291, pertinent pp. 5,10, 38-40, 46.

114. R. Tyson, *Chem. Ind.* (London), **1988**, 118.

115. P. Kovacic, Mechanism of Drug and Toxic Actions of Gossypol : Focus on Reactive Oxygen Species and Electron Transfer. *Curr. Med. Chem.*, **2003**, *10*, 2711-2718.

116. Li Huang, Y-K. Si, G. Snatzke, D-K. Zheng, and J. Zhou, Absolute Configuration of Gossypol, *Coll. Czech. Chem. Commun.* **1988**, *53*, 2644.

117. Liliane Lacombe, Attribution Complete des Signaux du Spectre de RMN ^{13}C du Gossypol aL'aide des Techniques a Deux Dimensions, *J. Nat. Prod.* **1987**, *50*, 277-280.

118. Raffaello Mesciadri, Werner Angst, and Duilio Arigoni, A Revised Scheme for the Biosynthesis of Gossypol, *J. Chem Soc. Chem. Commun.*, **1985**, 1573-1574.

119. Zhi-Jun Wu, Xi-Ke Xu, Yun-Heng Shen, Juan Su, Jun-Mian Tian, Shuang Liang, Hui-Liang Li, Rui-Hui Liu, and Wei-Dong Zhang, Ainsliadimer A, A New Sesquiterpene Lactone Dimer with an Unusual Carbon Skeleton from *Ainsliasea macrocephala*, *Org Lett,* **2008**, *10*, 2397-2400 and references cited.

Chapter 8
Diterpenoids (C_{20})

8.1 Occurrence. Biosynthesis

Geranylgeranyl pyrophosphate, $C_{20}H_{33}OPP$ (GGPP) (Chap. 5), an allylic isoprene tetramer diphosphate, serves as the universal precursor of diterpenoids. Acyclic diterpenes are of rare occurrence because of the participation of its rightly disposed olefinic bonds in the cation–olefin cyclization under the influence of diterpene cyclases to yield diverse skeletal patterns. Four different types of GGPP cyclases (synthases) were known till 1999 and their functional proteins have been sequenced [1–3]. They are casbene synthase, *ent*-copalyl disphosphate synthase, taxadiene synthase, and abietadiene synthase. The basic building blocks (IPP and DMAPP) of (*E,E,E*)-GGPP are of non-mevalonate origin (deoxyxylulose pathway, Chap. 5) and are biosynthesized most probably in the plastids where mevalonate or acetate precursors for IPP are not accepted [4], while the precursors for deoxyxylulose pathway are utilized.

Plant cells thus have two distinguishable compartments where IPP could be synthesized independently. Location and precursors based generated isoprenes have been earmarked for the biosynthesis of monoterpenoids, diterpenoids, and carotenes (plastids) and for sesquiterpenes, triterpenes, and steroids (cytoplasm) [4]. Though this compartmentalization is supported by labeling experiments, some controversial experimental observations of labels in products related to precursors and use of inhibitors debated the use of exclusive compartmentalization for IPP synthesis of triterpenoids of different classes [4].

A few examples of diterpenoids, both acyclic and cyclic (mono-, di-, tri-, and tetracyclic diterpenes), and their biosynthesis are schematically presented in the Figures in the sequel.

S.K. Talapatra and B. Talapatra, *Chemistry of Plant Natural Products*,
DOI 10.1007/978-3-642-45410-3_8, © Springer-Verlag Berlin Heidelberg 2015

8.1.1   Acyclic Diterpenes

Structures of some acyclic diterpenes and acyclic diterpene derivatives are shown in Fig. 8.1.

Phytol is universally distributed in green plants as a lipophilic chain (linked as an ester) component of chlorophyll (Chap. 3). It is also present as a structural part of both vitamin E and vitamin K$_1$. The sequence of reduction of all the double bonds save the allylic one of geranylgeraniol is not well known. However, it has been shown by labeling experiment that both geranylgeraniol and geranyllinalool are converted to the phytol part of chlorophyll in corn seedling.

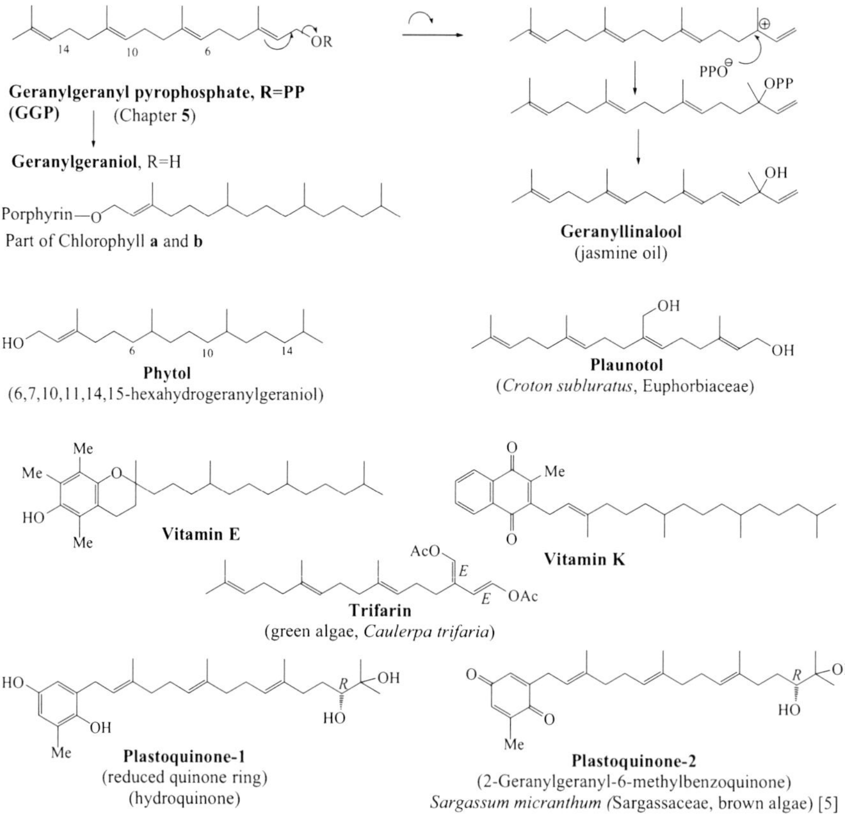

Fig. 8.1   Acyclic diterpenes and some of their derivatives

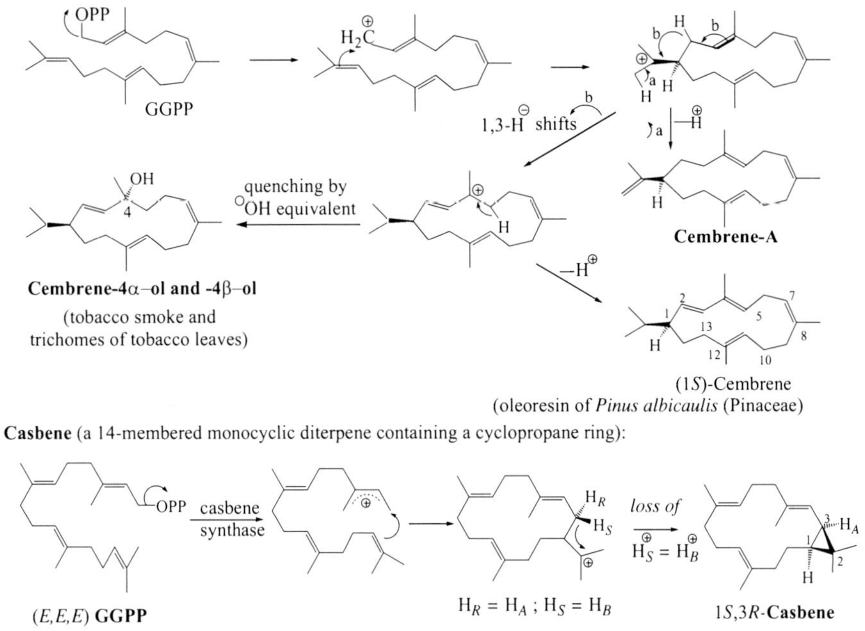

Fig. 8.2   Biosynthesis of some monocyclic diterpenes

8.1.2   Monocyclic Diterpenes

Monocyclic diterpenes occur comparatively rarely; a few of them are *cembrene* and *casbene* (14-membered carbocyclic compounds) [6] (Fig. 8.2). Casbene is a *phytoalexin*, produced by the fungus *Rhizopus stolonifer,* which grows on *Ricinus communis* (Euphorbiaceae).

8.1.3   Bicyclic Diterpenes

In case of bicyclic diterpenes decalin derivatives are formed in different stereo-chemical routes *via* preorganized chair–chair (normal and antipodal) and chair–boat (normal and antipodal) conformations of the substrate (*E,E,E*)-GGPP (Fig. 8.3). Thus four types of cyclizations give rise to four possible products, depending on the conformation of the prochiral substrate, as shown in Fig. 8.3.

   Some Examples of Bicyclic Diterpenes: Biosyntheses of some bicyclic diterpenes are outlined in Figs. 8.4 and 8.5.

   It is interesting to note that normal and enantiomeric products do occur some-times in the same plant, e.g., sclarene and ent-sclarene are found in different specimens of *Dacrydium intermedium* [8].

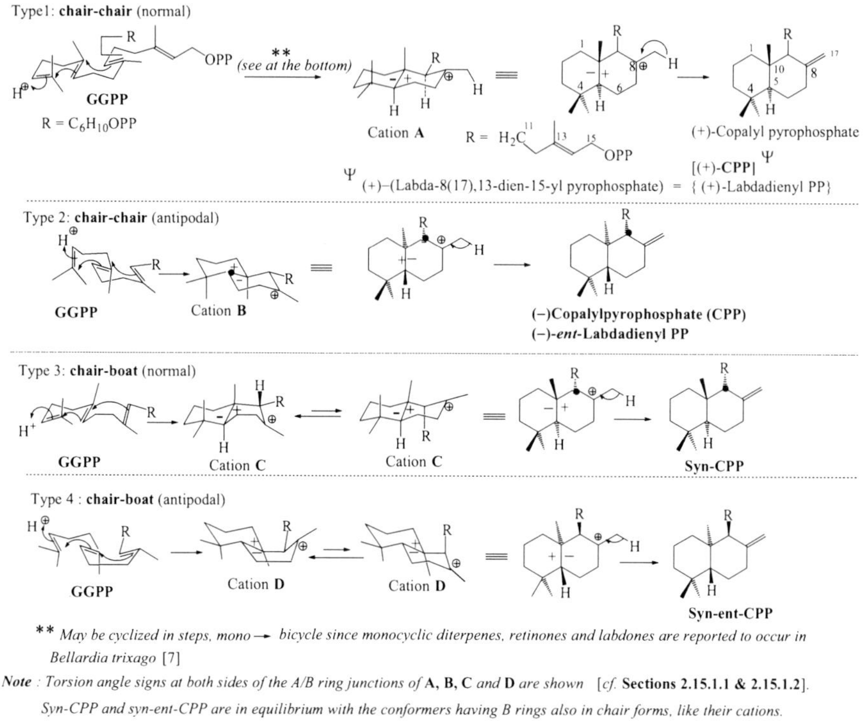

Fig. 8.3 Four types of cyclizations to form two pairs of enantiomers of two diastereomeric 6–6 bicyclic diterpenes

8.1.4 Tricyclic Diterpenes

In tricyclic diterpenes of perhydrophenanthrene-type ring orientation, A and B rings could be formed by any of the types of cyclizations presented in Fig. 8.3 . This is followed by further cation–olefin cyclization to form ring C. During ring C formation the attack on the *Re* or *Si* face of 13-14 olefinic plane will leave the stereochemical information about the chirality generated. In many cases of tetracyclic diterpenes (kaurens), the unused part of the chain undergoes further intramolecular cation–olefin cyclization of the right rotamer [caused by bringing the side chain in bonding proximity of the cation by its slow movement at angles required for the purpose clockwise/anticlockwise, preferably avoiding nonbonding steric interactions (Fig. 8.6)]. 1,2-Migration, though prevalent in triterpenes, also takes place during diterpene skeletal biosynthesis [9]. Regio- and stereospecific hydroxylation, oxidations (e.g., $-OH \rightarrow > = O$, $-CH_3 \rightarrow -CH_2OH \rightarrow -CHO \rightarrow COOH$), lactonization, etc., are the common biological events that take place at some defined/undefined steps (sequence of which may not be defined) during their biosynthesis. Through some common examples these biogenetic concepts of cyclization of tri- and tetracyclic diterpenes will be illustrated in subsequent figures (Figs. 8.6 and 8.8).

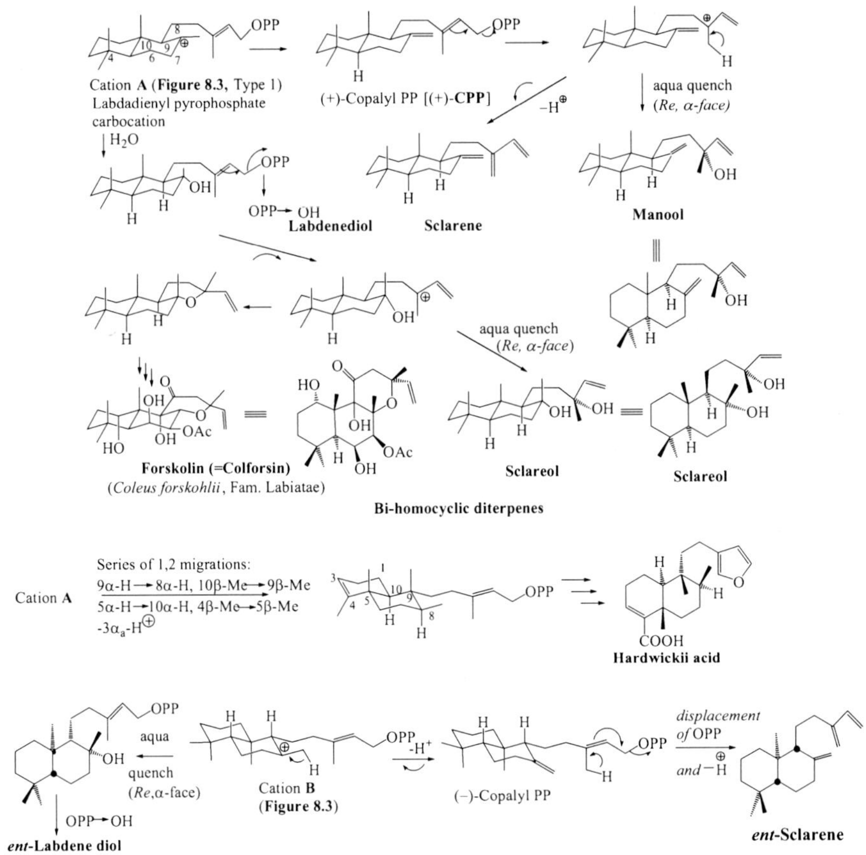

Fig. 8.4 Biosynthetic pathways of some bicyclic diterpenes

The structures of rosenonolactone and rosololactone (Fig. 8.6) have been defined by extensive chemical degradation [10]. Their absolute stereochemistry has been settled by CD and ORD studies [11, 12], as well as by X-ray crystallographic analysis [9] of the dibromo derivative of rosololactone, a close relative of rosenonolactone. Its biosynthesis, studied in detail by Arigoni [9], Birch [10], and Hanson [13, 14], is outlined in Fig. 8.6. The hydride and methyl migrations which accompany ring C formation have been documented by careful labeling studies.

Biosynthesis of abietic acid, the most important diterpene possessing abietane skeleton and abietadiene, and biosynthesis of taxol, the diterpene with unique antitumor activity, have been discussed in Sects. 8.3 and 8.4, respectively.

1,2-migrations:
9α-H → 8α-H, 10β-Me→9β-Me
5α-H→10α-H,4β-Me→5β-Me
trans-Clerodanes
Cation A (Figure 8.3)
R = H2C−CH2
Me
OPP
Cation B
9βH→ 8βH, 10αMe→9αMe
5βH→ 10βH, 4α-Me→5α-Me
-3α-H (axial)
Neoclerodiene PP
AcO
MeO2C
Salvinorin A

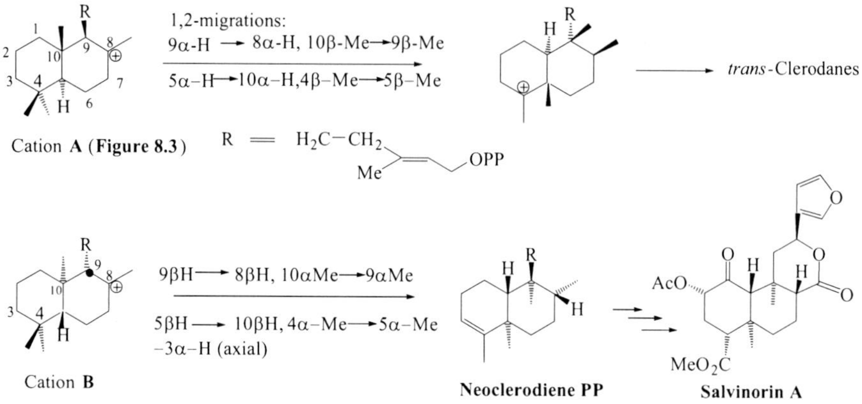

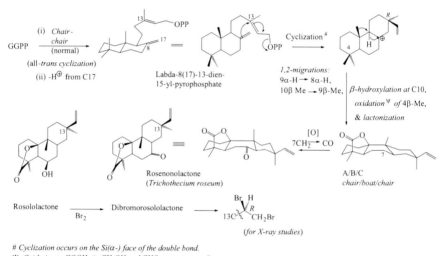

Fig. 8.5 Biosynthesis of some diterpenes derived from cations A and B

GGPP
(i) Chair-chair (normal)
(all-trans cyclization)
(ii) -H⊕ from C17
OPP
Labda-8(17)-13-dien-15-yl-pyrophosphate
Cyclization #
1,2-migrations:
9α-H → 8α-H,
10β Me → 9β-Me,
β-hydroxylation at C10,
oxidation Ψ of 4β-Me,
& lactonization
[O]
7CH2 → CO
A/B/C
chair/boat/chair
OH
Rosenonolactone
(Trichothecium roseum)
Rosololactone
Br2
Dibromorosololactone
CH2Br
(for X-ray studies)
Cyclization occurs on the Si(α-) face of the double bond.
Ψ Oxidation to COOH via CH2OH and CHO may occur earlier.

Fig. 8.6 Biosynthesis of rosenonolactone, a tricyclic diterpene lactone

8.1.4.1 Ring C Aromatized Diterpenes

A number of ring C aromatized diterpenes of abietane skeleton are known to occur in nature (Fig. 8.7).

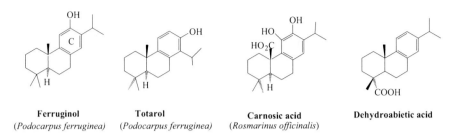

Ferruginol
(*Podocarpus ferruginea*)

Totarol
(*Podocarpus ferruginea*)

Carnosic acid
(*Rosmarinus officinalis*)

Dehydroabietic acid

Fig. 8.7 Some natural diterpenes with ring C aromatized

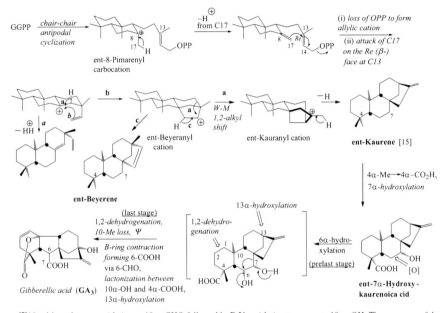

Ψ *10α- Me undergoes oxidation to 10 α-CHO followed by* **B.V.** *oxidation to generate 10 α-OH. The sequence of the steps involved in the last two stages cannot be defined.*

Fig. 8.8 Biosynthesis of gibberellic acid and some common tetracyclic diterpenes

8.1.4.2   Sandarocopimaradiene

The absolute configuration of the angular methyl and vinyl group at C13 depends on the face of attack on 13–14 double bond. In case of kaurene and sandarocopi-maradiene cyclization takes place by *Re*(β-face) attack of C17 on C13 of 13–14 double bond generating (*S*)-configuration of C13 (Fig. 8.8).

8.1.5   Tetracyclic Diterpenes

The probable biosynthetic pathways to sandaracopimaradiene, ent-7-hydroxykaurenoic acid, and gibberellic acid are delineated in Fig. 8.8. See Sect. 8.6.1 for some other information on gibberellins.

8.1.6   Ginkgolide Biosynthesis [4]

Ginkgolides are diterpene trilactones isolated from *Ginkgo biloba*. During their biosynthesis GGPP is first converted into an abietane-type tricyclic hydrocarbon intermediate which is subsequently converted to ginkgolides. The *t*-butyl group has been generated by ring A cleavage (Fig. 8.9). The compounds being lactones, oxidations take place densely and the gross carbon skeleton is a *W-M* rearranged product of abietane-type tricyclic skeleton.

By labeling experiment it has been shown that the C2–C3 bond of the dehydroabietane skeleton is preserved throughout the biosynthesis of ginkgolides. The C3-H$_2$ radical generated during C3–C2 bond cleavage picks up a neighboring hydrogen to form a methyl of the tertiary butyl group of ginkgolides.

The probable sequence of steps for the biosynthesis of the ginkgolides starting from (+)-dehydroabietane has been delineated in Fig. 8.10.

The formation of ginkgolides from primary carbocation via dehydroabietane involves a number of oxidation steps and *W-M* rearrangements (1,2-shifts) until it ends up into a densely oxygenated lactonized spiro skeleton (Fig. 8.10). (For detail [4] should be consulted.)

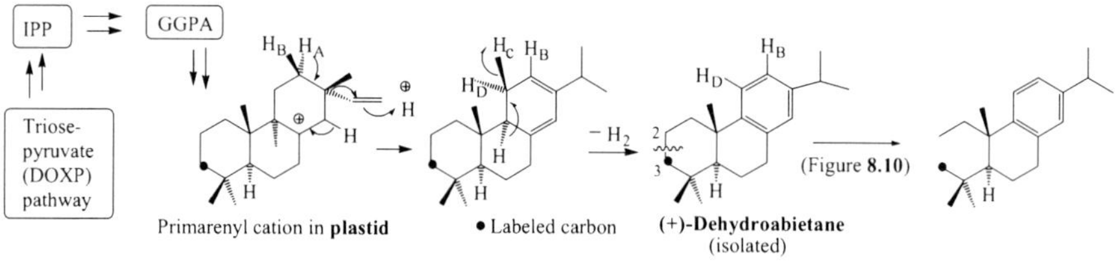

Fig. 8.9 Bioformation of the *t*-butyl group of ginkgolides

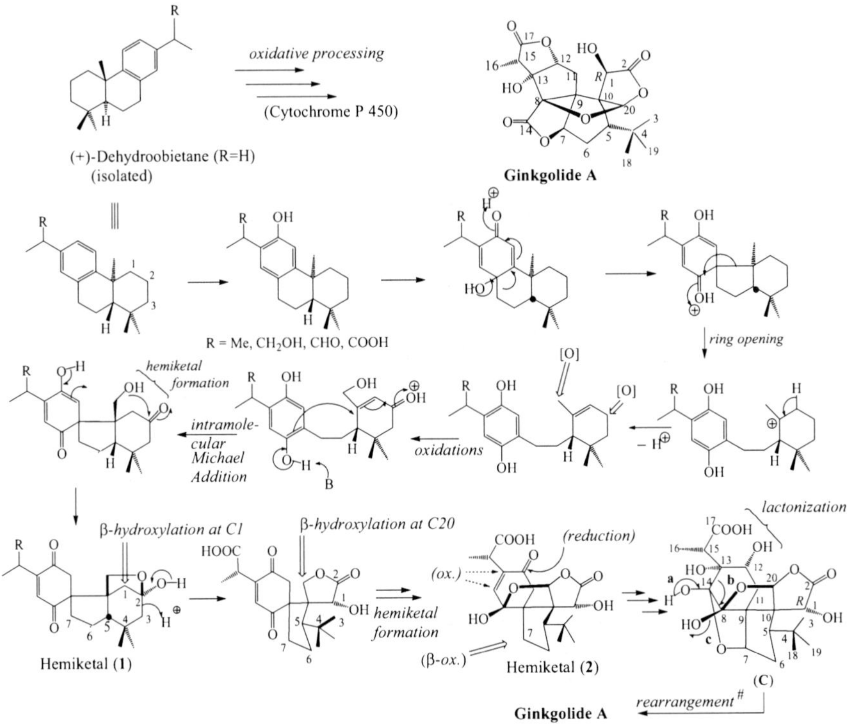

Fig. 8.10 Biosynthesis of Ginkgolides

Mechanism of rearrangement involving lactol OH at C14 in (C) (step a), ring contraction (step b) and explusion of OH from C8 (step c) has been shown in the precursor (C) of ginkgolide A

Note : Starting from the hemiketal (1) several redox steps and a well precedented rearrangement gives ginkgolide A. Most other ginkgolides (Figure 8.12) are accessible from ginkgolide A by hydroxylation at the appropriate positions

8.2 Geranylgeraniol

8.2.1 Introduction

Geranylgeraniol, $C_{20}H_{34}O$, b.p. 145 °C/0.35 mm, has been isolated and character-ized in 1967 from the wood of *Cedrela toona* (Meliaceae) [16]. It has also been isolated in 1966 from linseed oil *Linum usitatissimum* (Linaceae) [17]. In *Cedrela toona*, it occurs in the free state and also as esters of fatty acids. The pleasant aroma of the wood is due to geranylgeraniol. Its structure has been deduced in the conventional way and from spectral properties [1] of the alcohol, its acetate, and oxidized product, the aldehyde. A number of syntheses have been reported in the literature. The synthesis by Trost [18] is delineated in Fig. 8.11.

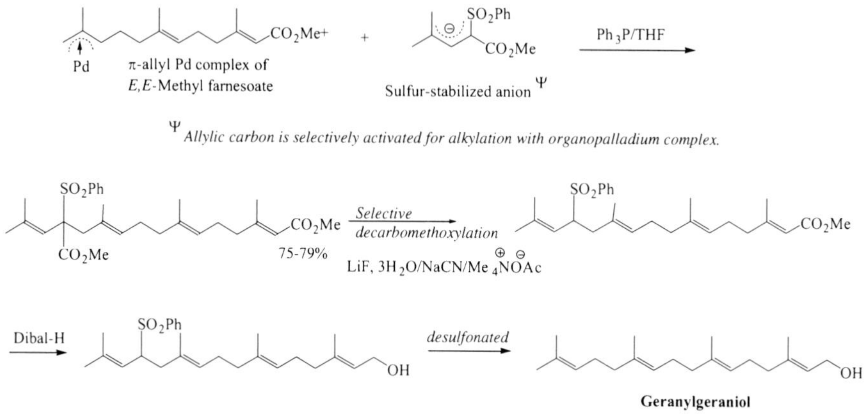

Fig. 8.11 Synthesis of Geranylgeraniol (1) [18]

8.2.2 Synthesis of Geranylgeraniol

In this synthesis [18] (Fig. 8.11) π-allylpalladium complex, which requires a soft nucleophile, is alkylated with a quite versatile sulfur-stabilized allylic carbon anion capable of undergoing selective alkylation. This is followed by selective decarboxylation, reduction of COOMe to CH$_2$OH, and subsequent desulfonation to yield the desired product. This strategy serves as a method for the synthesis of higher isoprenoids from the lower ones.

Spectral data [16]

IR: 3,200, 1,000 cm^{-1}, acetate, 1,745, 1,232 cm^{-1} oxidation product (aldehyde H > =O, 1,682 cm^{-1}).

^{1}**H NMR** (60 MHz): 9H singlet at 97 c/s, 8H singlet at 101 c/s, 6 vinylic methylenes (2 signals at 120 and 123 c/s total 12H), 2 hydroxymethine protons at 240 and 247 c/s shifted at 267 and 275 c/s in acetate, 4 vinylic protons (overlapping triplets centered ~ 307 and 324 c/s).

Mass : m/z 290 (M$^+$ 2), 217 (7), 132 (77), 104 (100), 91 (70), 69 (36).

The biosynthesis of geranylgeraniol (**1**) has been discussed in Chap. 5.

8.3 Abietic Acid and Other Resin Acids

8.3.1 Introduction. Occurrence

Abietic acid was obtained from "firs" which belongs to the genus *Abies* and hence the name. Abietic acid occurs mainly in the exudates of the pine tree bark (*Pinus pulustris, P. sylvestris* (Fam. Pinaceae)) and also in other *Pinus* species (e.g., Portuguese *Pinus pinester*, etc.). In ancient time the nonvolatile part of the exudates (*oleorosin*) was traded in Colophon [19], one of the 12 ancient Ionian cities of Asia Minor. The oleorosin (rosin) perhaps was named colophony (rosin) after the city of its trade.

Rosin is a complex mixture of resin acids of general formula $C_{19}H_{29}COOH$. Abietic acid, the best known resin acid, is a secondary acid, since it is formed from its precursor diterpene, levopimaric acid (see Fig. 8.22), by acid treatment of colophony during the process of its isolation, when double bond isomerization of levopimaric acid takes place. Abietic acid is purified as its sodium salt.

8.3.2 Structure

Its structure has been deduced from the reactions shown in Fig. 8.12 and its synthesis (see Fig. 8.15) [20–22].

The corresponding decarboxylated hydrocarbon of the saturated tetrahydroabietic acid ($C_{19}H_{33}CO_2H$) will have molecular formula $C_{19}H_{34}$ which possesses 6H less than the corresponding open chain saturated hydrocarbon (C_nH_{2n+2}, $C_{19}H_{40}$); hence this hydrocarbon and abietic acid are tricyclic. The probable part structures based on the above reactions are shown in Fig. 8.13.

Formation of (**a**), (**b**), and (**c**) and other data from Fig. 8.12 suggest (2) as a part structure for abietic acid in which two double bonds are to be placed. Formation of (**b**) claims the placement of isopropyl group on a double bond. Since a Diels–Alder product is formed under comparatively drastic condition, the double bond must be conjugated but in *S-transoid* configuration and not in Diels–Alder friendly *S-cisoid*

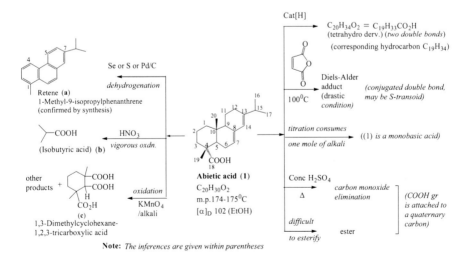

Fig. 8.12 Some reaction products of abietic acid

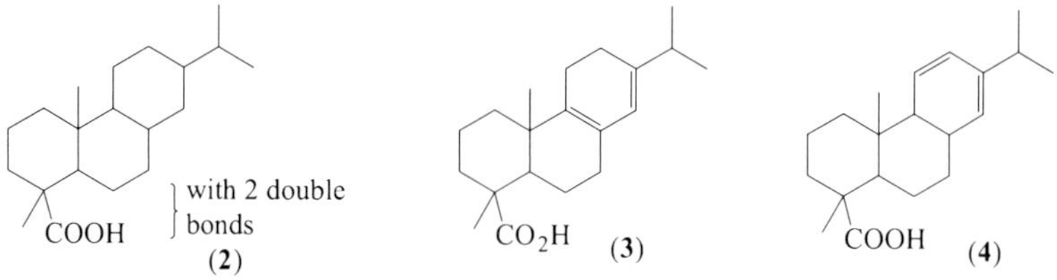

Fig. 8.13   Probable part structures for abietic acid

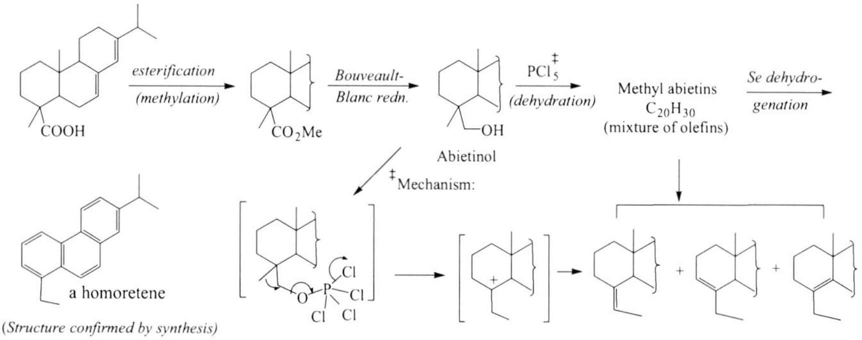

Fig. 8.14   Location of the COOH group in abietic acid

configuration. This observation excludes structures (3) and (4) and points (1) as the structure for abietic acid without the stereochemical features. The configuration of the double bonds is supported by the closeness of the calculated λ_{max} based on structure 1 and the observed value. λ_{max} calcd. [214 nm (heteroannular diene) + 5 nm × 4 (5 nm for each of the four alkyl substituents) + 5 nm (one exocyclic double bond) = 239 nm; λ_{max} observed = 241 nm] (cf. Sect. 4.2). Thus the configuration of the double bonds and their locations have been proved by some chemical reactions and are supported by UV and also by NMR spectral (see Fig. 8.17) data.

However, though the germinal disposition of the –CH_3 and –COOH groups at C4 of abietic acid is known from the formation of compound (c) (Fig. 8.12), it has been elegantly demonstrated through the formation of a homoretene (Fig. 8.14), the structure of which has been confirmed by synthesis to eliminate other isomeric homoretenes.

8.3.3 Synthesis

The structure of abietic acid was confirmed by its synthesis. The total synthesis has been carried out in two parts, as delineated in Fig. 8.15. The first part, the synthesis of (±)-dehydroabietic acid, was reported by Stork [20, 21] in 1956. The synthesis of abietic acid from dehydroabietic acid (the second part) was achieved by Burgstahler [22] in 1961.

Mechanisms with stereochemical features (when necessary) of the formation of the products in stages (a) to (g) in the synthesis of (±)-abietic acid (Fig. 8.15), involving several well-known and synthetically very useful reactions, are shown in Fig. 8.16, for immediate understanding of the students.

(±)-Dehydroabietic acid and its methyl ester have also been synthesized by Ireland [23] and Meyer [24] respectively.

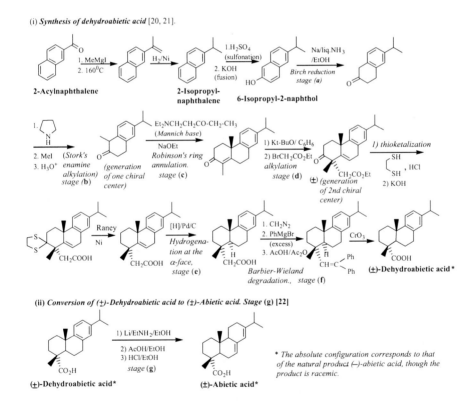

Fig. 8.15 Synthesis of (±)-dehydroabietic acid and (±)-abietic acid

Stage (a). Birch reduction

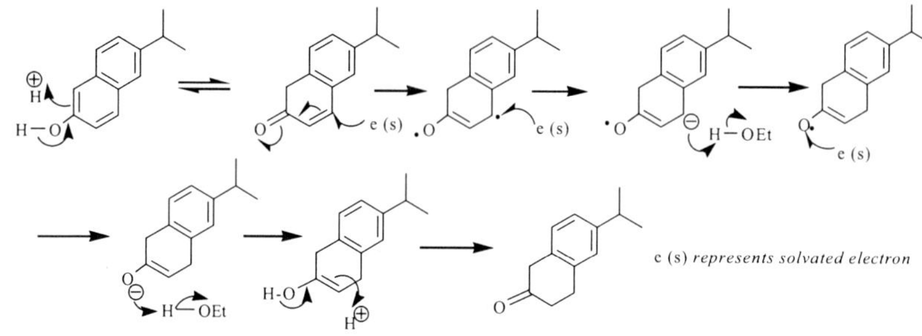

c (s) represents solvated electron

Stage (b). Stork enamine alkylation

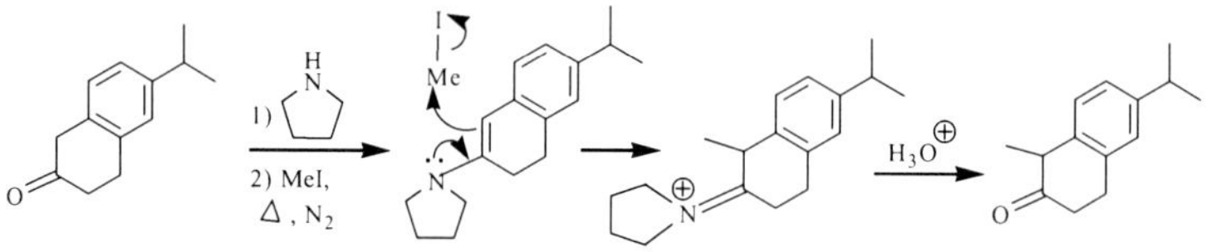

the direction of dehydration is governed by the aromatic ring
(extended conjugation)

Stage (c). Robinson annulation   (a [4+2] − condensation approach)

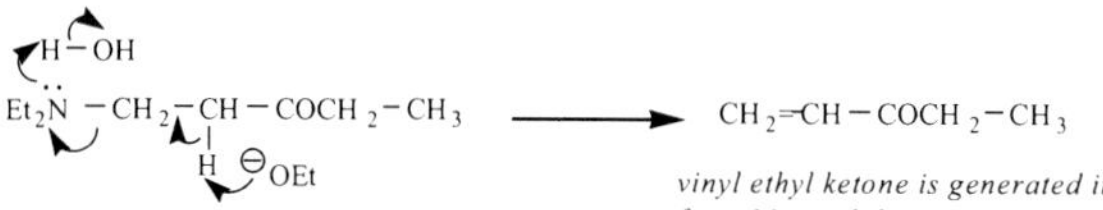

Mannich base

CH$_2$=CH−COCH$_2$−CH$_3$

vinyl ethyl ketone is generated in silu
from Mannich base

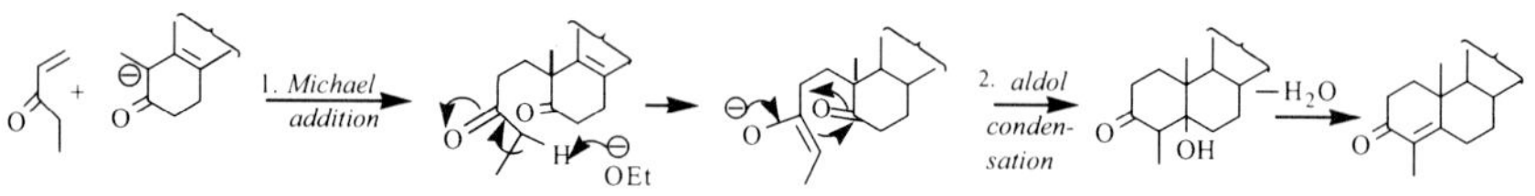

(Michael addition and aldol condensation are operative)

In step 2 the base removes the proton from the ketomethylene group. The enolate thus formed is trapped by the
carbonyl of the tetrahydronaphthalone moiety resulting into the formation of a new six-membered ring; dehydration of
the tertiary alcohol is extremely smooth and usually spontaneous leading to the formation of an α,β−unsaturated ketone;
this conjugation is the driving force of the elimination.

Stage (d). Stereochemistry of alkylation

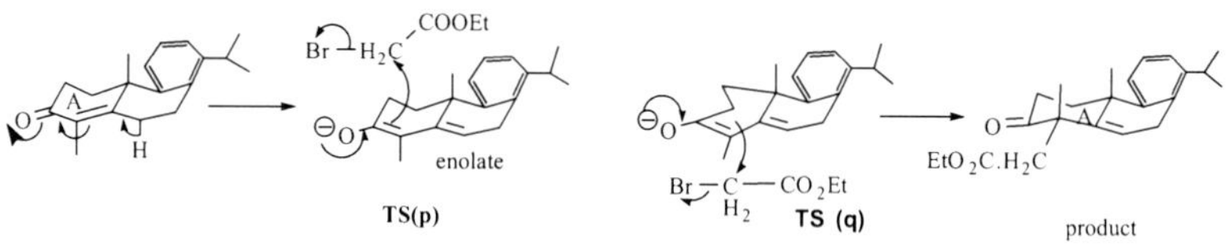

The approach of the electrophile from the β-face, orthogonal to the double bond, for head-on π-bond overlap at the
location of the attack (C-4) is sterically hindered by non-bonded 1,3-diaxial type interaction with the β−methyl of the
enolate in the **TS (p)**. However, α-axial approach in the buckled A ring attaining a half-boat type conformation will
suffer no such interaction in the **TS (q)**, and will allow π-bond overlap. Immediately after the attack the ring of the
product buckles back to the chair conformation, bringing the CH$_2$COOEt group in the equatorial orientation.

Fig. 8.16 (continued)

Stage (e). Stereochemistry of hydrogen delivery: β-face of the molecule being crowded in this conformation, the less

*hindered α-(Re-Si) face at C5=C6 (vide sects. **2.9.6 & 2.9.7, Figure 2.64**) of the molecule would be adsorbed on the catalyst surface, and thus hydrogenation would take place on the α-face making A/B ring fusion trans.*

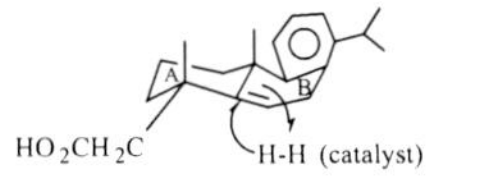

Stage (f). Barbier-Wieland degradation from higher to lower homologue of an acid

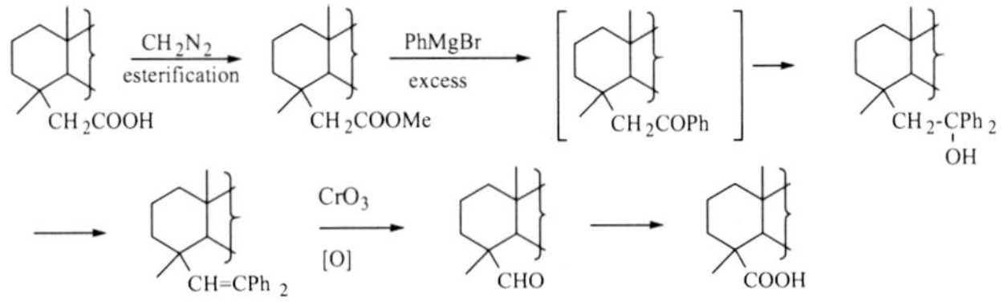

Stage (g). Conversion of (±)-Dehydroabietic acid to (±)-Abietic acid

Burgstahler and Worden [22] employed Benkeser lithium-ethylamine reduction in converting (±)-dehydroabietic acid to (±)-abietic acid.

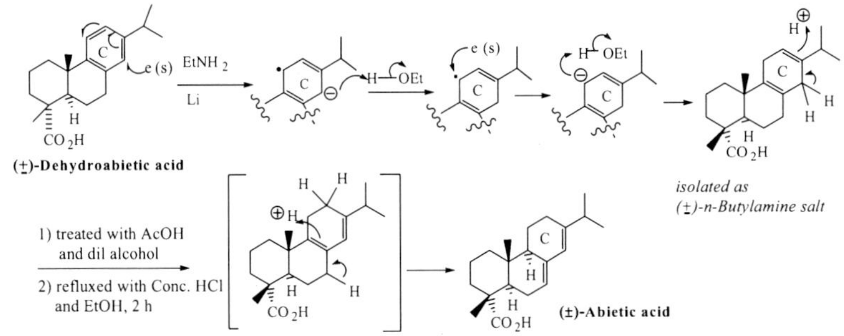

This conversion has been achieved in good over-all yield under carefully defined conditions. One such condition is to add finely-divided lithium over a period of 15 min to a rapidly stirred solution of dehydroabietic acid taken in redistilled ethylamine [22].

Fig. 8.16 Mechanistic rationalization of some stages in the formation of (±)-dehydroabietic acid and of its conversion to (±)-abietic acid

8.3.4   *Spectral Properties of Abietic Acid Methyl Ester (Fig. 8.17)*

UV: [22] (for abietic acid) λmax in nm (ε) 235 (21,500), and 241.5 (23,000)

IR: [25] λmax in cm^{-1} 1735 (ester carbonyl), 1250 (-O-CO-stretching).

^{1}H NMR: (δ) [25] (300 MH$_3$) (CDCl$_3$) δ = 0.82 (3H, s, H$_3$-20), 1.00 (d, J = 6.9 Hz, 3H, H$_3$-16 or H$_3$-17), 1.01 (d, J = 6.9 Hz, 3H, H$_3$-17 or H$_3$-16), 1.25 (s, 3H, H$_3$ -18), 2.16-2.29 (m, 1H, H15), 3.63 (s, 3H, ester C H$_3$), 5.33-5.41 (m, 1H, H7), 5.77 (s, 1H, H14).

^{13}C NMR [25]: (75.6 MHz, CDCl $_3$) : δ = 14.03 (C20), 17.01 (C19), 20.85 (C16 or C17), 21.42 (C17 or C16), 22.46 (C11), 25.68 (C6), 27.48 (C12), 34.53 (C10), 34.88 (C15), 37.12 (C3), 38.33 (C1), 45.10 (C5), 46.59 (C4), 50.94 (C9), 85 (C21 ester methyl), 120.62 (C14), 122.36 (C7), 135.53 (C8), 145.33 (C13), 179.01 (C1 8) (for numbering see structure 1).

(APC1-MS [25]: m/z (%) 317.2 (100) (M+1)$^{+}$, 315.2 (77), 313.2 (47), 257.2 (25, (M+1) –HCOOCH$_3$).

Fig. 8.17 Spectral properties of abietic acid methyl ester

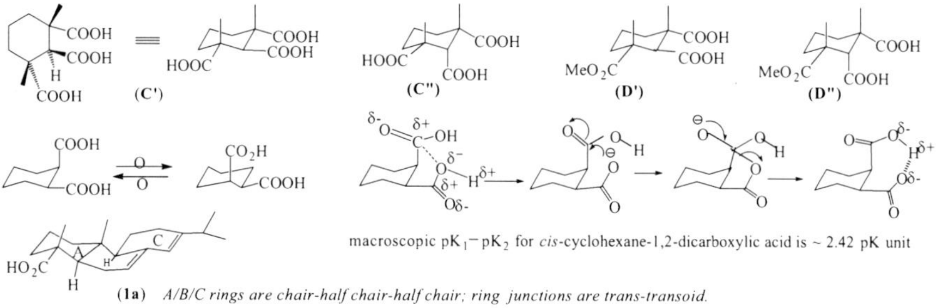

macroscopic pK$_1$– pK$_2$ for cis-cyclohexane-1,2-dicarboxylic acid is ~ 2.42 pK unit

(1a) A/B/C rings are chair-half chair-half chair; ring junctions are trans-transoid.

(–)-Abietic acid

Fig. 8.18 Field effect: dissociation constants of cis-cyclohexane-1,2-dicarboxylic acid

8.3.5 Stereochemistry and Molecular Conformation

Compound (c) (cyclohexane-1,3-dimethyl-1,2,3-tricarboxylic acid, Fig. 8.12) contains three of the four asymmetric centers of (–)-abietic acid, but has been found to be optically inactive though these three chiral centers of (–)-abietic acid were not touched during its formation. Two centers are quaternary and the third one does not have any provision for racemization. Hence in all probability, the molecule should have a plane of symmetry and should thus be expressed as (C′) or (C″) (Fig. 8.18). This is one of the major problems that gave birth to conformational analysis [26]. Barton and Schmeidler [27] were able to determine the stereochemistry of the A/B ring junction in abietic acid and hence in a number of other resin acids. They used measured values of the dissociation constants of the unsymmetrical monomethyl ester of the tricarboxylic acid (C′) or (C″). It was concluded that the difference between the "microscopic" pK$_1$ and pK$_2$ values for the asymmetric monomethyl ester (D′) or (D″) (Fig. 8.18) was 1.11 pK units. This agrees well with the analogous value of 1.15 pK units for trans-cyclohexane-1,2-dicarboxylic acid and differs from that (1.80 pK units) for cis-cyclohexane-1,2-dicarboxylic acid. The monomethyl ester should therefore be (D′) and not (D″). The secondary COOH group in the tricarboxylic acid (C′) is thus shown to be trans to the other two

COOH groups, which had earlier been shown to be *cis* to each other. Thus A/B ring junction in abietic acid and related compounds was shown to be *trans*.

Dissociation constants of cyclohexane-1,2-dicarboxylic acids. In general, the closer in space the COOH groups are, greater is the difference of the pK values; the positive dipole of one of the COOH groups eases the departure of the proton from a closely proximate OH of another COOH by a field effect. But the negative charge of the COO^- ion in the ionized acid prevents the easy departure of the second H^+ from the other proximate COOH group, as delineated in Fig. 8.18.

The relative stereochemistry in abietic acid is shown to be *trans*-transoid by its resistance to acid isomerization. Had this been a *trans*-cisoid system, it would have picked up H^+ from axial face through transient double bond isomerization to form the stereochemically comfortable conformation. The molecular conformation (−)-abietic acid may thus be represented as (**1a**) (Fig. 8.18).

8.3.6   A Few Interesting Reactions

A few interesting reactions of abietic acid will be discussed (Figs. 8.19, 8.20, and 8.21).

(i) When dehydroabietic acid is treated with conc. H_2SO_4 at low temperature, a conformationally interesting product is obtained (Fig. 8.19).

The ring size of the lactone was much debated [29]. However, Barton gave the correct structure with a five-membered lactone ring (γ-lactone). He further suggested that the reaction should proceed through an olefinic acid and the relationship of C4-COOH group and C5-H should be changed from *cis* to *trans* relationship in order to form a sterically comfortable lactone (Fig. 8.19). His

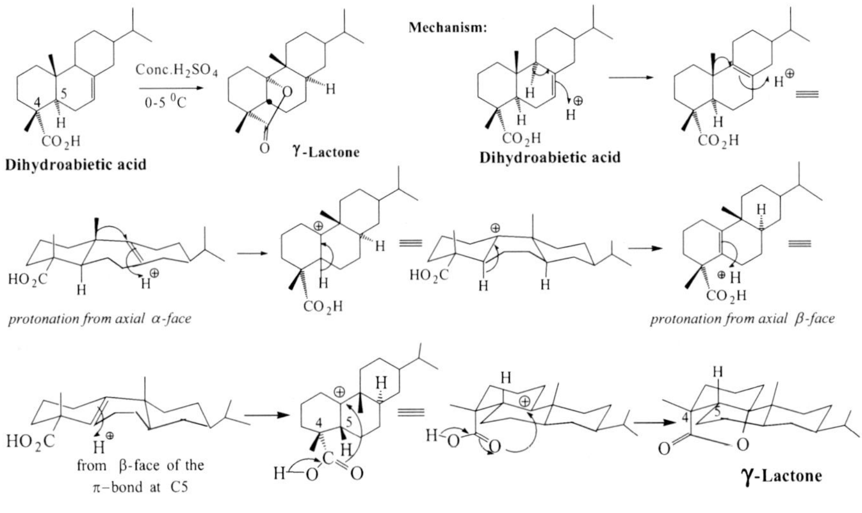

Fig. 8.19 Mechanism for the formation of γ-Lactone from dihydroabietic acid

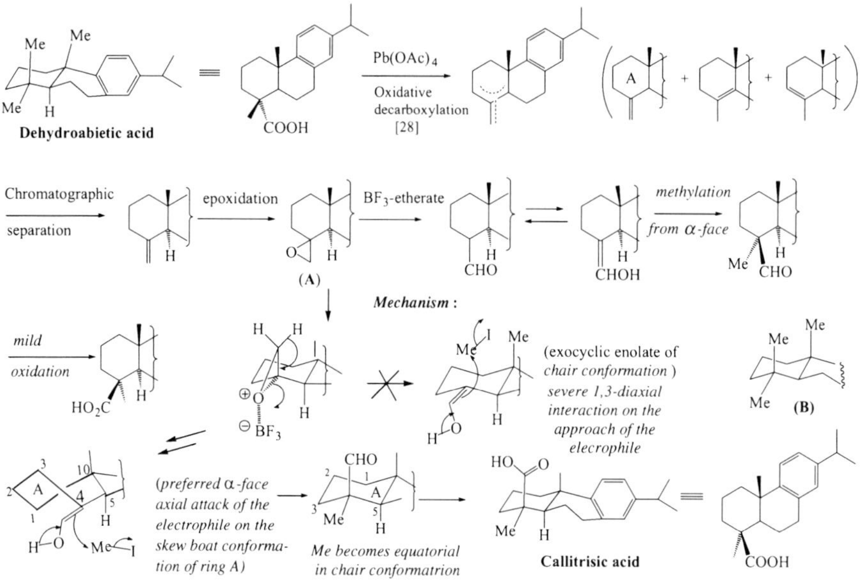

Fig. 8.20 Conversion of dehydroabietic acid to callitrisic acid

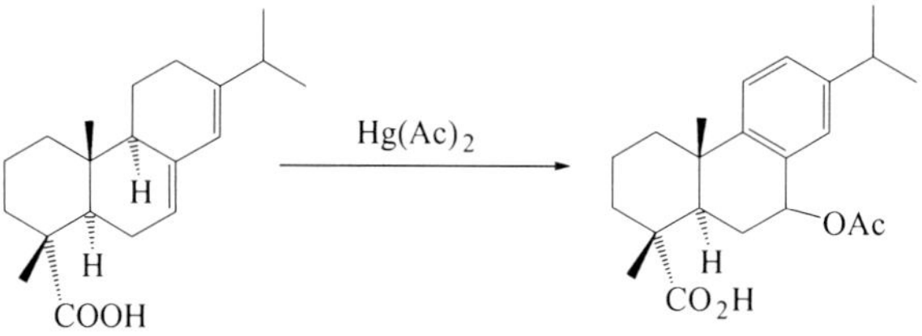

Fig. 8.21 Action of mercuric acetate on abietic acid

conjecture has been confirmed by X-ray crystallographic studies of the lactone formed from dihydroisopimaric acid under similar condition [30].

(ii) C4-epimerization of Dehydroabietic Acid. Conversion of Dehydroabietic Acid to Callitrisic Acid [31] (Fig. 8.20) .

Incidentally, it may be mentioned that in cases of C4-axial acids (like callitrisic acid) the C10-Me suffers deshielding by ~17 cps because their 1,3-diaxial relationship brings the C10-Me under the deshielding zone of the carboxyl carbonyl in their average conformation (Me comes in the plane of the carbonyl). Equatorial COOH which has no such relationship shows only a deshielding of ~4 cps (compared to **B**), thus leaving a net deshielding of ~13 cps for axial acids [32].

(iii) One interesting example is where Hg^{II} acetate incorporates allylic oxidative rearrangement and simultaneous dehydrogenation in abietic acid.

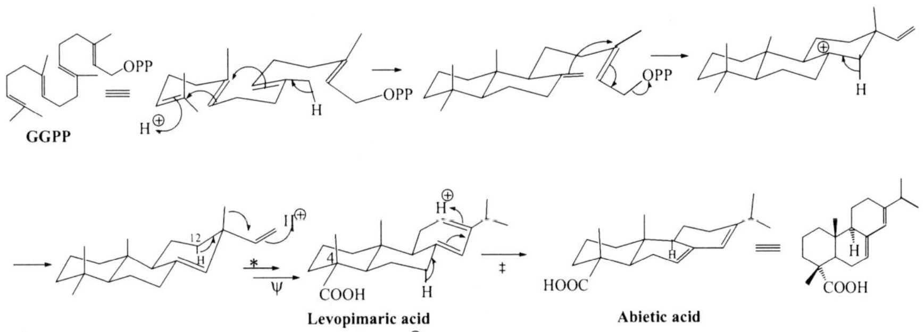

Fig. 8.22   Probable biogenesis of levopimaric acid and its isomerization to abietic acid

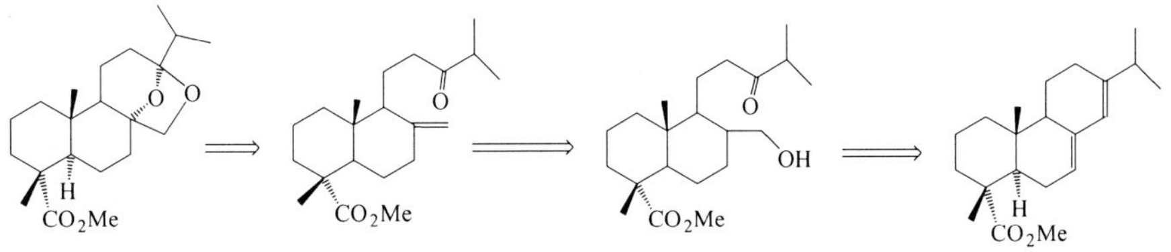

Levopimaric acid

Abietic acid

* *1,2 migration of doubly allylic Me group, followed by* $-H^{\oplus}$ *(axial) from C12*

Ψ *biological oxidation of C4 α-Me to COOH in steps at this or some earlier stage*

‡ *acid catalyzed rearrangement during isolation to form the more stable heteroannular diene system in abietic acid*

Fig. 8.23   Retrosynthesis of ambraketal

8.3.7   Biosyntheses of Abietic Acid

Levopimaric acid synthase catalyzes the stepwise formation of levopimaric acid from geranyl geranyl pyrophosphate (GGPP) folding it properly and proceeding through the steps shown in Fig. 8.22. Levopimaric acid undergoes acid-catalyzed rearrangement to produce abietic acid. The latter is considered to be a secondary resin acid and thus the isomerization of the homoannular diene to heteroannular diene may take place in vitro during isolation.

8.3.8   Uses as a Synthon

Ambergris, a metabolic product of the sperm whale, is long known as one of the most valuable animal perfumes (besides civet, musk, and castreum), and its tincture has been used as a fixative for rare perfumes. It also helps in making the fragrance of other perfumes to last long. Ambraketal (Fig. 8.24) is among the most expensive synthetic equivalents of the scarce natural ambergris. Ambraketal possesses long-lasting ambergris-type fragrance and many efforts have been made for the synthesis of ambraketal or its analogues all starting from (–)-abietic acid [25].

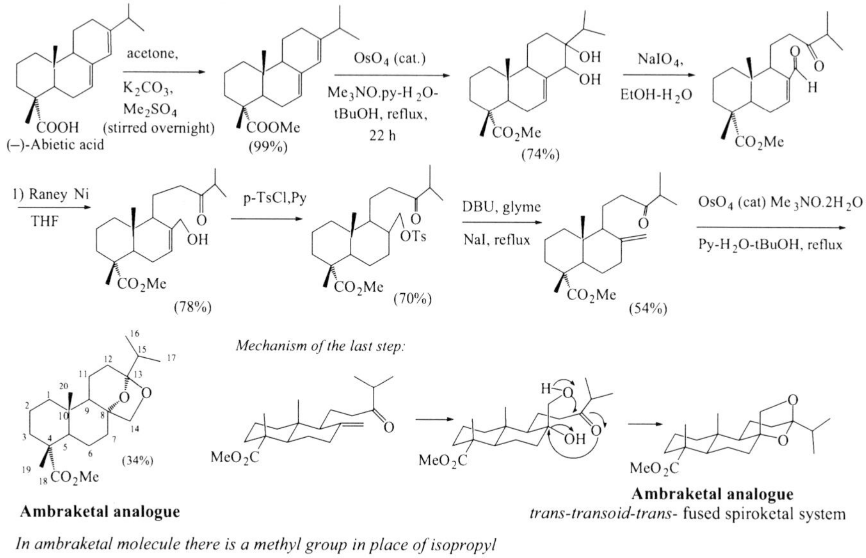

Fig. 8.24 Synthesis of an ambraketal analogue [Methyl (8R,13S)-8α,13:13,14-diepoxiabiet-19-oate] [25]

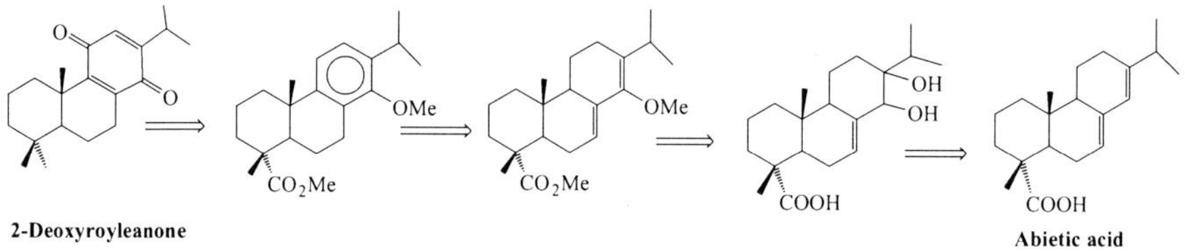

2-Deoxyroyleanone   **Abietic acid**

Fig. 8.25 Retrosynthesis of 12-deoxyroyleanone

A synthesis of an ambraketal analogue has been designed by CeuCosta et al. [25] starting from (–)-abietic acid. The synthetic strategy is shown in Fig. 8.23 and the actual synthesis is delineated in Fig. 8.24.

The first enantiospecific synthesis of the antileishmanial, 12-deoxyroyleanone (Fig. 8.25), has been achieved from (–)-abietic acid in 11 steps [33] in a 25 % overall yield. The strategy of this semisynthesis is shown in Fig. 8.25 and the actual conversion is shown in Fig. 8.26.

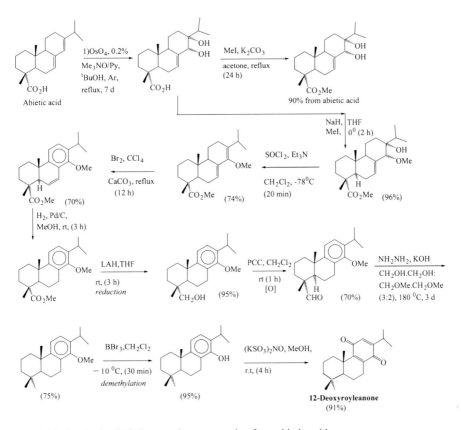

Fig. 8.26 Synthesis of 12-deoxyroyleanone starting from abietic acid

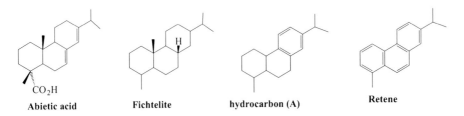

Fig. 8.27 Probable diagenetic products of abietic acid

8.3.9   Diagenetic Products of Abietic acid

Natural products which suffer chemical reactions over geological time, caused by thermal, catalysis bacterial, and other processes, are known to yield diagenetic products and the phenomenon is known as diagenesis. Abietic acid, present in the buried pine trees, results in the diagenetic products fichtelite and retene. The last two products are unknown as natural products [34].

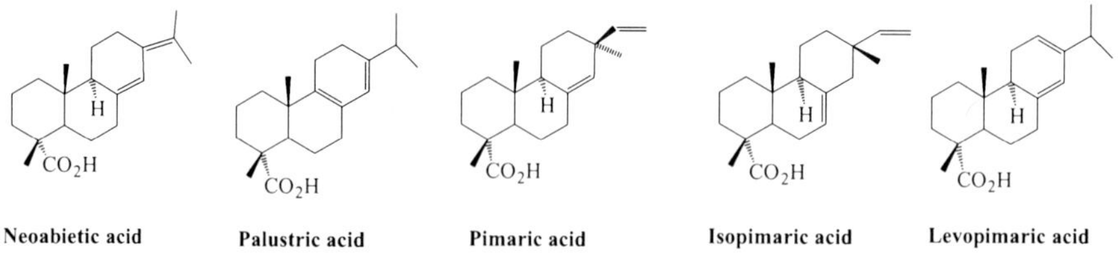

Fig. 8.28 Some resin acids related to abietic acid

It is noteworthy to mention that a hydrocarbon (A) (Fig. 8.27) has been isolated from pine forest soil. Structural similarity suggests that it could be a diagenetic intermediate in the conversion of abietic acid to retene, and hence it is suggested that compound (A) should be sought in places where fichtelite and retene are obtained (Fig. 8.27) [34].

8.3.10 Structure Diagrams of Some Related Resin Acids

Some common diterpenes resin acids structurally related to abietic acid are given in Fig. 8.28.

8.4 Taxol®: A Nitrogenous Diterpene Ester with Unique Antitumor Activity

8.4.1 Introduction. Occurrence

As a part of the screening program of the antitumor agents of plant origin, initiated in 1960 under Dr. Jonathan L. Hartwell, NCI (The United States National Cancer Institute, Bethesda, MD), extracts of bark, twigs, leaves, and fruits of *Taxus brevifolia* (Pacific Yew) (Fam. Taxaceae), one of the slowest growing trees, were examined in 1966 and identified as the most important samples [35]. The bark was first collected by a botanist Arthur S. Barclay in 1962 for chemical and biological studies [36]. Around 0.5 g of Taxol$^{(R)}$ (a registered trademark for BMS), m.p. 213–216 °C (dec.), $[\alpha]_D^{20}$ – 49 (MeOH), was isolated from 12 kg of air-dried stem and bark [35]. The name was given before the structure was known. However, it was found to contain an alcoholic OH group and hence the name is justified. All 11 species of *Taxus* contain it with varying yields; *Taxus brevifolia* ("yew") is one of the seven major species. The locations of the species are scattered and also sometimes in remote areas [37]. The production of taxol and taxane by *Taxomyces andreanae*, an endophytic fungus of Pacific yew, has been reported. The yew tree has long been known for making bows. The bark extract is recorded to be poisonous. Cativolcus, a chieftain of a Gaelic tribe, committed suicide by drinking tea

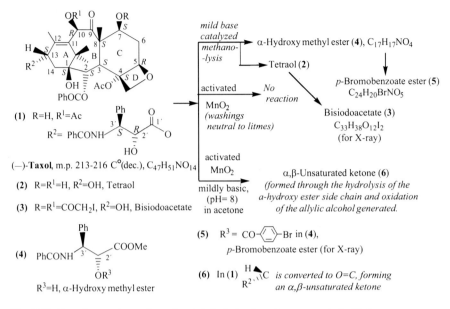

Fig. 8.29 Reaction products of taxol and their derivatives for structural elucidation

made from the bark of the yew after his defeat at the hands of Roman legions [36], as recorded by Julius Caesar in one of his books. The genus name *Taxus* perhaps had its origin from two Greek words, "toxo" (bow) and "toxikon" (poison) [36]. Taxol's remarkable antitumor activity has been discussed in Chap. 33.

8.4.2 Structure [38–40]

A high-resolution mass spectrum of taxol showed the molecular ion peak at m/z 853 and the elemental composition to be $C_{47}H_{51}NO_{14}$. It is a highly oxygenated molecule of a complex structure. Unfortunately suitable crystalline halogenated derivative could not be obtained [38] for X-ray crystallographic analysis. The structure **1** of taxol has been determined from the reactions shown in Fig. 8.29. The formation of (**6**) defined the location of the ester with nitrogen containing α-hydroxy acid at C13. Taxol was subjected to a mild base-catalyzed methanolysis at 0 °C affording a nitrogenous α-hydroxy methyl ester (**4**) and a tetraol (**2**). Their structures and absolute configuration were solved by X-ray crystallographic analysis of their *p*-bromobenzoate (**5**) and bisiodoacetate (**3**) derivatives. Its inertness to carefully washed (neutral to litmus) and activated MnO_2 indicated that the two esters were located at the allylic positions C10 and C13. The 11 stereogenic centers of taxol have 1*S*, 2*S*, 3*S*, 4*S*, 5*R*, 7*S*, 8*S*, 10*R*, 13*S*, 2'*R*, and 3'*S* configurations (Fig. 8.29), as revealed from X-ray analysis. The chemistry of taxol and related taxoids has been reviewed [39] in detail in 2003.

The spectral data of taxol (**1**) are given in Figure **8.30**

UV: λ$_{max}$ (MeOH) 227 nm (ε 29,800), 273 (1700)

IR: ν$_{max}$ (Nujol) : 3300-3500 (OH, NH), 1730 (ester), 1710 (ketone), 1650 (amide) cm^{-1}

M$^+$ at *m/z* 853 (calcd. for C$_{47}$H$_{51}$NO$_{14}$, 853)

^{1}H NMR spectral data (100 MHz, CDCl$_3$, TMS):

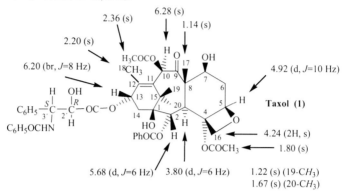

In the reference [38] the absolute configuration of C3' has been drawn to have (R)configuration by oversight.

Fig. 8.30 The spectral data of taxol

It is worth mentioning that Halsall [41] reported in 1970 the isolation (from *Taxus baccata*) of the diterpene baccatin-V, a naturally occurring oxetane similar to the tetraol (**2**), differing only in the configuration of the OH group at C7, i.e., the 7-epimer of the tetraol (**2**). A few other taxane derivatives have been reported from the same plant [42] in 1969.

8.4.3 Spectral Data [35, 38]

The spectral data of taxol (**1**) are given in Fig. 8.30.

8.4.4 Conformation of Taxol (1C)

The following characteristic features of the preferred conformation (**1C**) (Fig. 8.31) of the (−)-taxol molecule are evident from a careful analysis of its Dreiding model.

The 6-membered ring A assumes a half-chair-like conformation. The C13α–O bond of the side chain is having a pseudoaxial type (a′) orientation in ring A. The carbon atoms C15, C11, C12, and C13 are almost planar, while C1 and C14 are nonplanar.

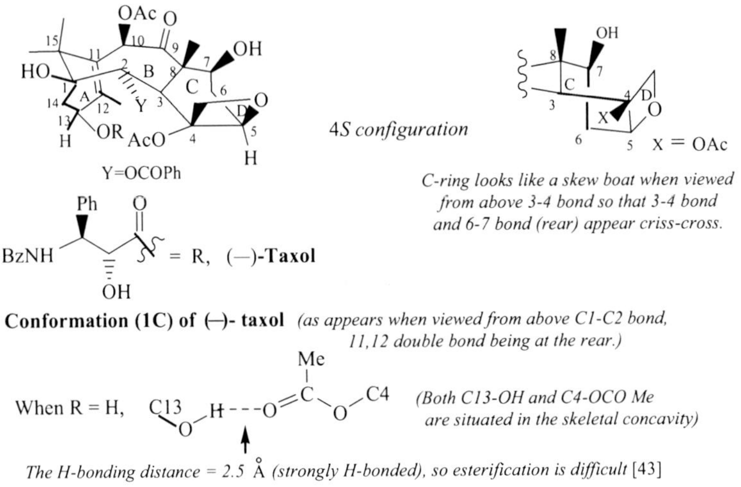

4S configuration

X = OAc

C-ring looks like a skew boat when viewed from above 3-4 bond so that 3-4 bond and 6-7 bond (rear) appear criss-cross.

Y=OCOPh

= R, (−)-Taxol

Conformation (1C) of (−)- taxol *(as appears when viewed from above C1-C2 bond, 11,12 double bond being at the rear.)*

When R = H,   *(Both C13-OH and C4-OCO Me are situated in the skeletal concavity)*

The H-bonding distance = 2.5 Å (strongly H-bonded), so esterification is difficult [43]

Fig. 8.31   Conformation of the taxol molecule

1. The 8-membred ring B is quite flexible. However, its preferred conformation is shown (one perspective view). In it the nonbonded interaction between 8-β-methyl and 15β-methyl is completely avoided. This interaction becomes severe in some other conformations of the flexible ring B.
2. In ring B, in the preferred conformation (**1C**) the carbon atoms C3, C8, C9, C10, and C11 come almost in one plane, while C2, C1, and C15 are above this plane.
3. The ring C looks like a skew boat (Fig. 8.31) when viewed from a particular direction.
4. The oxetane ring D is almost perpendicular to the puckered plane of ring C.

Thus the upper half of the taxol molecule forms its convex, more exposed surface, while the lower half constitutes its concave, more crowded face. Among other factors the shape of the taxol molecule or its any analogue presumably has an important role in its bioactivity (cf. Sect. 33.6.3 and [90] and [91]).

8.4.5   Synthesis of Taxol

The remarkable antitumor activity of taxol and its dense substitutional decoration along the periphery of the gross tricyclo[9.3.1.0^{3,8}]pentadecane skeleton involving seven oxygen atoms directly hooked to the skeleton as C-O bonds with their configurational identity and one oxygen atom as C=O throw a great challenge to the synthetic organic chemists community. Nearly 30 or more groups were involved in that hot race [36] and it became a very crowded field for accomplishing the same goal. Nearly 23 years (1971–1994) after its reported isolation and structure elucidation [38] its total synthesis first appeared in the literature [36, 44–49]. Of the first

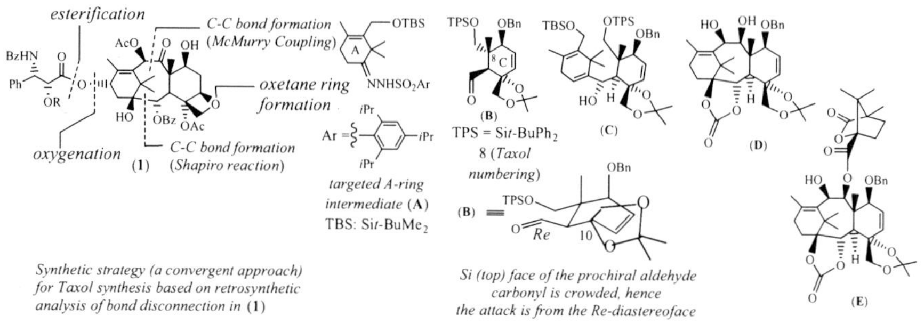

Fig. 8.32 Synthetic strategy for taxol (K. C. Nicolaou) [44]

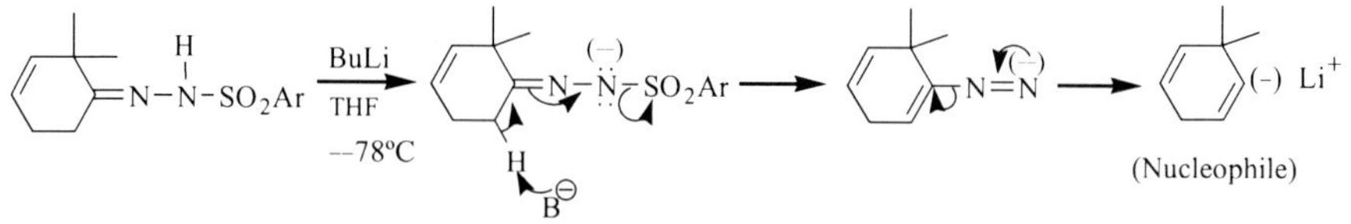

Fig. 8.33 The formation of the nucleophile used in Shapiro reaction

two syntheses of taxol the one by Nicolaou et al. [44–47] will be discussed briefly. The other synthesis by Holton et al. appeared in the literature within a span of few days only [48, 49]. Later on, many more syntheses of taxol have been reported. The ecstasy and agony during the conquest of taxol have been elegantly narrated by Nicolaou and Guy [36], which is an inspirational story to the research students.

Strategy for the Synthesis of Taxol by Nicolaou: The synthetic strategy as given and followed by the authors [36, 44–47] is shown in structure (**1**) (Fig. 8.32). Intermediates are identified chemically and by X-ray analysis when needed. The *chronology of the strategic events in the synthesis of taxol* is summarized below:

(1) Construction of appropriately functionalized ring A (**A**)
(2) Synthesis of appropriately functionalized ring C (**B**), another targeted inter-
    mediate (use of Narasaka protocol of phenylboronic acid)
(3) Union of ring A and ring C precursors [(**A**) and (**B**)], Fig. 8.32) to form an
    intermediate (**C**) using *Shapiro reaction*, in which ring A hydrazone behaves as
    a latent nucleophile (Fig. 8.33). The latter could react with the aldehyde group
    of ring C precursor. The formation of only one stereoisomer is caused by the
    attack of the nucleophile from the less hindered *Re* (*β* -diastereoface of the
    chelated aldehyde (see Fig. 8.34, Strategic Event 3, 1st row). The 8-methyl
    offers hindrance to the approach of the nucleophile from *Si* (α)-diastereoface.
    The alcoholic OH at C2 being allylic is useful in subsequent functionalization
    at C1.
(4) Prior to the C9–C10 coupling a hydroxyl group is generated at C1 and blocked
    as carbonate with C2-OH. The carbonate will serve as the potential source of

Strategic Event 1.

Synthesis of Adequately Substituted Ring A [Compound (A)]

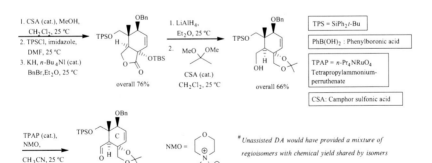

Strategic Event 2.

Synthesis of Appropriately Substituted Ring C [Compound (B)] (Application of Narasaka reaction using boronate as a template to tether the diene and dienophile in a specific orientation (endo) for allowing rigorously regiochemically controlled intramolecular Diels-Alder regioisomer formation)

Fig. 8.34 (continued)

Strategic Events 3-7.

Union between Ring A and Ring C Precursors [(A) and (B)] *(Application of Shapiro Reaction) (reaction of C1, C2, C3, three contiguous asymmetric carbon atoms with desired relative stereochemistry of taxol, and resolution of McMurry coupling product)*

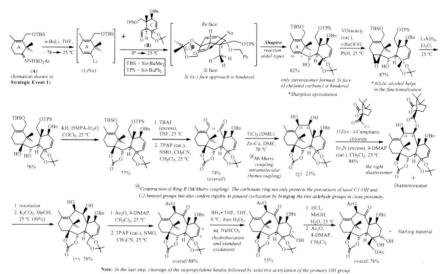

@*Construction of Ring B (McMurry coupling). The carbonate ring not only protects the precursors of taxol C1-OH and C2-benzoyl groups but also confers rigidity to pinacol cyclization by bringing the two aldehyde groups in close proximity.*

Note: *In the last step, cleavage of the isopropylidene ketals followed by selective acetylation of the primary OH group.*

The Remaining Strategic Events

Formation of hydroxyoxetane system. Generation of hydroxyl at C13(S) and its subsequent esterification with N-benzoyl β-lactam derivative

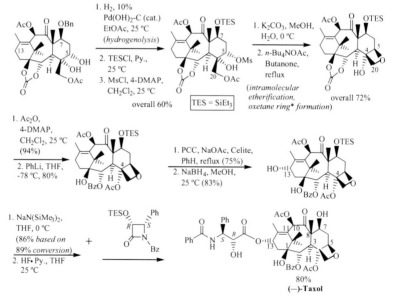

The oxetane ring is generated by displacement of C5-mesylate by the C20-OH. The latter is formed by the selective deacetylation of C20-acetate. During this intramolecular 4-membered cyclic ether formation the configuration at C5 is reversed (SN^2 displacement).

Fig. 8.34 Total synthesis of taxol (Nicolaou) [44]

C1-OH and C2-benzoate of taxol. This carbonate will further bring C9 and C10 closer to effect in a better way the subsequent McMurry coupling (single electron transfer—intramolecular homocoupling).

(5) Ring B is thus generated by McMurry coupling between C9 and C10 of the carbonate derivative (±)-(**D**). At every step racemates are formed.

(6) The resolution of (±)-(**D**) by (1S)-(–)-camphanic acid chloride and the stereoisomerically pure (+)-enantiomer has been used for subsequent steps. An interesting observation is the formation of the ester (**E**) with (–)-camphanic acid at hindered C9-OH and not with C10-OH while with Ac_2O the acetyl derivative of C10-OH and not of C9-OH is formed. These two selective esterifications are not obvious and do not offer immediate rationalization.

(7) Formation of 9-keto-10 acetate as in natural taxol.

(8) Generation of an α-OH at C5 and the formation of an oxetane ring using mesylate protocol.

(9) The other steps follow the usual rational sequence.

The entire synthetic reaction strategy is delineated in Fig. 8.32.

The formation of an organolithium base from a tosylhydrazone of a ketone is shown in Fig. 8.33. The lithiated anion acts as a nucleophile in Shapiro reaction.

Total synthesis of taxol by Nicolaou by the convergent strategic events already defined is delineated in Fig. 8.34. These strategies opened a chemical pathway for the production of not only the natural product itself but also a variety of designed taxoids [44].

8.4.6   Search for Commercial Sources for Taxol

In most cases of the drugs of plant origin, the plants cannot serve as the source for commercial supply of the drugs to the users. The yield of taxol is low and the main source plant, *Taxus brevifolia*, is a slowly growing tree, which takes nearly 200 years to mature. For drug extraction, the bark is to be stripped and the full course of the drug per patient needs more than three plants [50]. This is indeed a threat to the species. Though taxol occurs in all the *Taxus* species, the yields are not encouraging. Further, time-consuming multistep synthesis of taxol will not be economically viable for commercial supply. The endeavor for the synthesis of such a fantastic molecule has always been a great challenge, intellectually rewarding, and chemically explorative, but could not be a means to meet the market demand. Hence other ways of preparation of this anticancer drug of great demand have been explored.

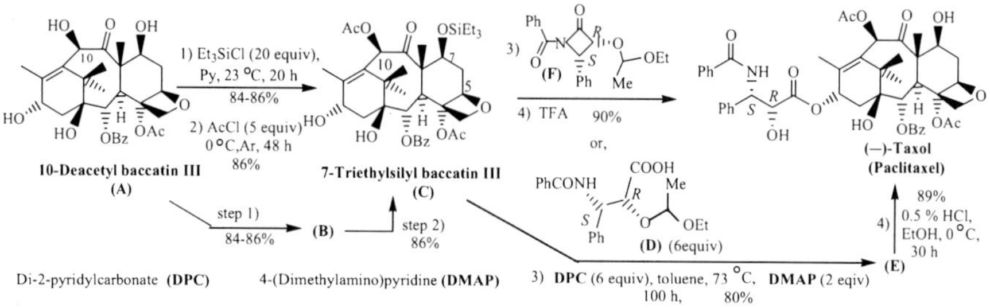

Fig. 8.35 Semisynthesis of (–)-taxol (paclitaxel)

8.4.6.1 Semisynthesis [43, 51]

The reported yields of taxol from the bark of various American, European, and Asian *Taxus* species (yew) range from 40 to 165 mg/kg. However, the yew is one of the slowest growing trees in the world. The several multistep syntheses of this complex molecule taxol have already been achieved. They are of high academic interest, but of little practical value. So an efficient partial synthesis (semisynthesis) of taxol from an easily and abundantly available taxol congener would be much useful for getting sufficient supply of this exceptionally promising cancer chemotherapeutic agent having a broad spectrum of antileukemic and tumor-inhibiting activity.

10-Deacetylbaccatin III (**A**) is a congener of taxol but is far less active than taxol. However, it is obtained in ~1 g/kg of fresh leaves [43], without harming the tree. A highly efficient 4-step conversion of 10-deacetylbaccatin III [43] into taxol is delineated in Fig. 8.35. In this methodology 10-deacetylbaccatin III (**A**) was triethylsilylated under carefully controlled condition to give 7-triethylsilyl-10-deacetylbaccatin III (**B**) in 84–86 % yield, keeping 10-OH unaffected. Thus, 7-OH is protected. Acetylation of (**A**) gives the 7-acetyl derivative and not baccatin III. Acetylation of (**B**) under controlled condition furnished 7-triethylsilyl-baccatin III (**C**). C13-OH Group is situated in the skeletal concavity of (**C**) and is able to form a stable H-bond with the C4 acetate (C13.OH...O = C(Me).O.C4 distance = 2.5 Å, as measured by a Dreiding model and molecular mechanics [43]) (conformation 1C, Fig. 8.31). Hence its esterification is not possible. The esterification with optically pure (2R,3S)-N-benzoyl-O-(1-ethoxyethyl)-3-phenylisoserine (**D**) under the reaction condition shown produced (**E**) in 80 % yield [43]. The carefully chosen protecting group could be removed by treatment with 0.5 % HCl to give (–)-taxol (paclitaxel) in 38 % overall yield [43]. Further improvement was made acylating with a less sterically demanding acylating agent, the β-lactam (**F**), to give (–)-taxol in up to 90 % yield [51]. Obviously this latter methodology (steps 3 and 4) may be the preferred one for its commercial production in future. The semisynthesis is expected to serve to alleviate the shortage of taxol and thus greatly facilitate its application in cancer chemotherapy.

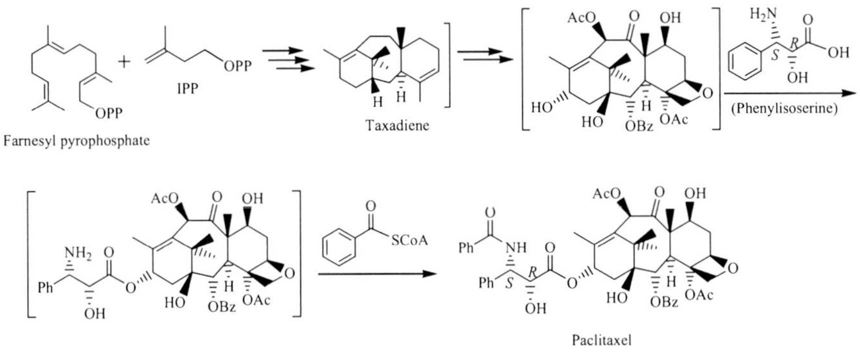

Fig. 8.36 Fermentation process for taxol formation

8.4.6.2   Application of Biocatalysis. Fermentation Process [52]

A fermentation process has been developed using *Taxus* cells for the isolation of taxol from the culture media. The entire process is catalyzed by a string of enzymes dedicated to specific steps. In this process isopentenyl diphosphate and farnesyl diphosphate are allowed to combine to form geranylgeranyl PP (GGPP), the universal molecule in the diterpene biosynthesis, catalyzed by geranylgernanyl diphosphate synthetase. The GGPP thus formed subsequently cyclizes to baccatin through the participation of a series of enzymes which cause hydroxylation, acetylation, oxidation, and the generation of the oxetane ring. Because of the specificity of enzymes these enzymic transformations follow their sequence according to the substrates they are meant for. The side chain at C13 is attached through enzymic transfer of phenylisoserine and then benzoylation (Fig. 8.36). Methyl jasmonate acts as elicitor (compounds that can enhance the production of secondary metabolites in an enzymic process) and enhances the production of paclitaxel ($110 \ mg \ L^{-1}$) in cell-suspension culture media.

The plant cell fermentation process is a clean process, and it eliminates all steps of the semisynthesis.

Microbes thriving on plants sometimes produce the same compound as their hosts. This property of the microbes has been thought to be a clean process for the production of pharmacologically useful compounds. With this idea, search for yew-associated microbes that could produce taxol has been made. In this endeavor, 25 *Taxus brevifolia* species from different locations have been collected and nearly 200 microbes associated with these plants have been studied, of which only *Taxomyces andreanae*, an endophytic fungus showed the ability of producing taxol [37, 53]. These microbes are collected and made free from associated traces of taxol that have accompanied from the plant. They are allowed to grow in a suitable culture medium, and after 21 days' incubation, the culture is filtered, extracted with CH_2Cl_2 or $CHCl_3/MeOH$. Removal of the solvent gave taxol which is purified. When compared with yew-taxol, the fungal taxol showed identical spectral properties. The yield is low. However, efforts are going on to improve

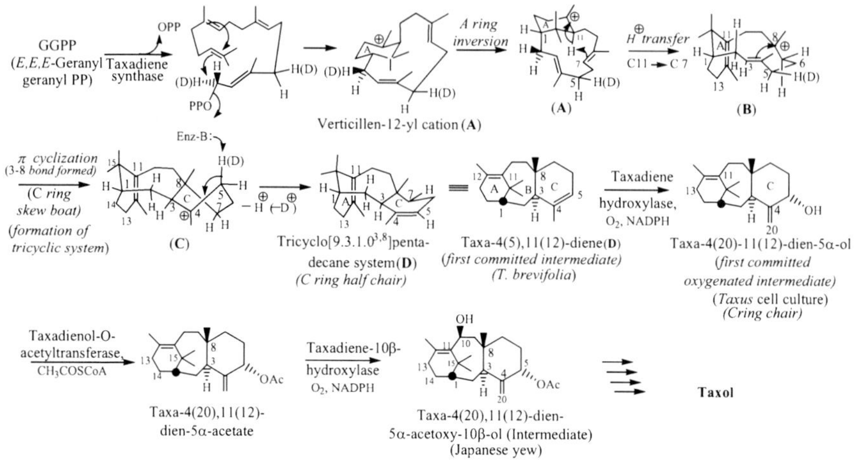

Note: Ring A of cation (**A**) is generated from the particular folding of GGPP, as shown, made by the specific enzyme taxadiene synthase.

Fig. 8.37 Biosynthesis of taxol

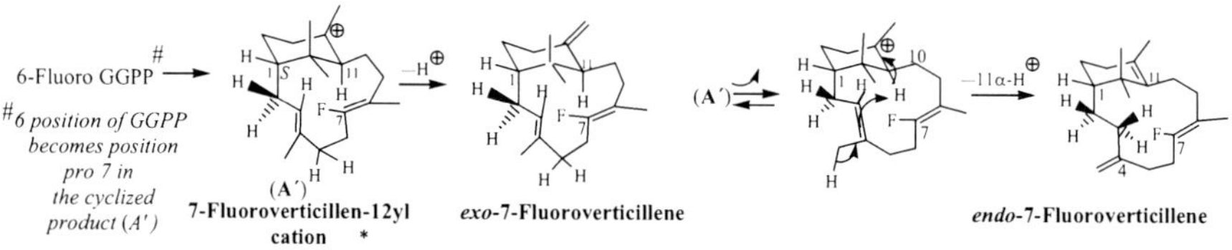

*Proton transfer from C11 to C7, like in (**A**) (**Figure 8.37**) is arrested in (**A'**) because of the decreased basicity of the π-bond by the F substituent.

Fig. 8.38 Chemical identity of the intermediates and the nonformation of the tricyclic system

the yield with the application of genetic engineering and to make the process commercially viable.

In the fermentation process, taxol is generated from the genetically engineered *Taxus* species cell culture media following the biosynthetic pathway [52].

8.4.7   Biosynthesis of Taxol

The study of the biosynthesis of taxol showed that the taxane skeleton is not of mevalonoid origin (labeling experiment) [54] and that the first committed intermediate in the biosynthetic pathway has been taxa-4(5),11(12)-diene, which has also been isolated from *T. brevifolia* [54]. Soon after the isolation of 4(20),11(12)-dien-5α-ol (Fig. 8.37) from *Taxus* cell culture media it has been proved to be the first oxygenated intermediate in this biosynthetic pathway [55]. These intermediates have been synthesized with appropriate labels and biosynthetic experiments were carried out to show their intermediacy in the process. Further identification of other intermediates will map the entire biosynthetic path to taxol. Another intermediate taxa-4(20),11(12)-dien-5α-acetoxy-10β-ol (Fig. 8.37), isolated from Japanese yew, has been shown to be an intermediate in taxol biosynthesis [56].

During the cyclization of GGPP to taxadiene, a bicyclo[9,3.1]-pentadeca-3,7-diene precursor (verticillen-12yl carbocation) is formed (Fig. 8.37). This is followed by a proton transfer from C11 to C7, prior to further cyclization to tricycle [9.3.1.0^{3,8}] pentadecane skeleton (cf taxane). The event of hydrogen transfer from C11 to C7 [(A) to (B) in Fig. 8.37] has been elegantly shown (Fig. 8.38) [57] using 6-fluorogeranylgeranyl PP. Further cyclization of the latter has been arrested at verticillene stage as the fluorine substituent at C7 decreases its π basicity sufficiently to prevent proton transfer at this position to form tricyclo[9.3.1.0^{3,8}] pentadecane system like (C). Thus 7-fluoroverticellen-12-yl cation faces other fates and undergoes exocyclic or endocyclic elimination of proton to form a few verticillene derivatives (Fig. 8.37), of which exo-fluoroverticillene is important. The stereochemical description of these derivatives is obtained from their extensive NMR spectral studies [57]. The ring juncture B/C in (D) is *trans* with (3R)-H3-α and (8S)-Me8-β configurations (cf taxadiene).

8.4.8   Uses

Taxol is used for an important first-line treatment of ovarian and breast cancer and head and neck cancer and has potential applications in the treatment of other cancers. Some studies on the structure–activity relationships of taxol have been reported [40].

The antitumor and antileukemic activity and structure–activity relationship of taxol and its derivatives/analogues have been briefly discussed in Chap. 33 (Sect. 33.6.3).

8.5   Gibberellins

8.5.1   Introduction. Biological Activity

Gibberellins belong to an odd class of diterpenes. It was discovered in late 1820s that a rice plant (*Oryza sativa*, Gramineae) disease was caused by a fungus identified later as *Gibberella fujikuroi*. The disease known as *bakanae* in Japan causes the elongation of the rice seedlings more rapidly than the normal seedlings when ultimately they wilt and die. This observation was published in 1828 [58, 59]. A compound thought to be responsible for such elongation was isolated in 1938 and named gibberellin by Yabuta after the name of the fungus. Later it was shown by Takahashi [60] to be a mixture of three compounds GA$_1$, GA$_2$, and GA$_3$. All compounds possessing this type of biological activity with gibbane skeleton are called gibberellins and are represented by the general formula GA$_n$ (n represents a digit given according to the chronological order of the discovery) [61]. 112 Gibberellins have been characterized till 1997 [62, 63]. Gibberellins regulate a number of different rate-determining steps in plant metabolism, e.g., breaking the dormancy, inhibition of senescence, a primary stimulus to germination (especially when the pericarp is removed), overcoming juvenility, and determining the developmental sequence of sex organs of flowers.

In short, gibberellins play an important role in almost all phases of plant growth. Of all the bioactive gibberellins GA$_3$ has been the most important one and is of wide occurrence. It is named Gibberellic acid. Gibberellins also occur in higher plants, e.g., *Citrus reticulate*, *Phaseolus vulgaris*, and *P. multiflorus*.

8.5.2   Structure. Synthesis

Unlike other tricyclic and tetracyclic diterpenes, gibberellins do not yield phenanthrene derivatives on dehydrogenation but yield fluorene derivatives.

The gross structure of gibberellic acid (GA$_3$) was elucidated as (**1**) [60]. Its complete stereostructure was subsequently determined by X-ray diffraction studies [64] of the di-*p*-bromobenzoate of the methyl ester of gibberellic acid (GA$_3$).

It has high density of functional groups attached to the strained gibbane skeleton (**2**), and it is labile toward many reagents. Its synthesis [61, 64] needed a lot of manipulation for the introduction of functional groups that become ultimately the integral part of its structure and the construction of the carbon framework. Corey who synthesized gibberellic acid [65, 66] commented [67].

(**1**) GA$_3$ Gibbane (**2**)

m.p. 233-235 ^{0}, $[\alpha]_D$ +52(EtOH)

"The plant bioregulator gibberellic acid resisted synthesis for more than two decades, because it abounds in all the elements contributing to complexity, including reactive and dense functionality, in an usually forbidding arrangement."

8.5.3 Biosynthesis

Biosynthesis of GA_3 has been discussed in Sect. 8.1.5. Both C_{20} Gibberellins and C_{19} (in which C20 is lost) Gibberellins are known to occur [62].

8.5.4 Uses

Gibberellic acid is used in agriculture to stimulate the swelling of fruits, e.g., grapes and tomatoes. It is also used to break the dormancy of seeds, e.g., lettuce and peach, to accelerate the germination of barley, and also for various other agricultural purposes. For detailed studies on Gibberellins, see [62].

8.6 Ginkgolides

8.6.1 Introduction. Occurrence

The plant *Ginkgo biloba* has been regrown in the center of Hiroshima from her charred womb caused by the heat that resides only in the heart of Stars [68], within 1 year of the beastliest day (August 6, 1945) in Man's history, perhaps as a symbol of Hiroshima's strong will to live and our humble vow "Peace in this World" [68].

In 1923, a devastating earthquake shook Tokyo and a great fire broke out which engulfed everything on its way; however, a temple surrounded by *Gingko* trees survived. Perhaps the liberated chemicals from the trees acted as a fire retardant [69].

Ginkgo biloba (order, Ginkgoales) is a mono-species genus with longest survival history (250 million years) with unbelievable endurance ignoring the Nature's evolution of millions of years and Her environmental pressures and changes. *Ginkgo biloba* was found in the fossil, in the love-lyrics of Johann Wolfgang von Goethe (1749–1832) and in Charles Robert Darwin's (1809–1882) interest who nicknamed it as "living fossil" in 1859 since the species has not changed over millions of years [69]. The beautiful bipartite-shaped leaves of the plant add beauty to its gait. The painters use the tree and the leaves as their subjects and the beautiful shape of the leaves is copied in the hairstyle of the brides and sumo-wrestlers—all have top knots in the style of Ginkgo leaves. The trees are now planted as roadside ornamental trees in oriental cities. They are precious collections in various museum premises and parks all over the globe.

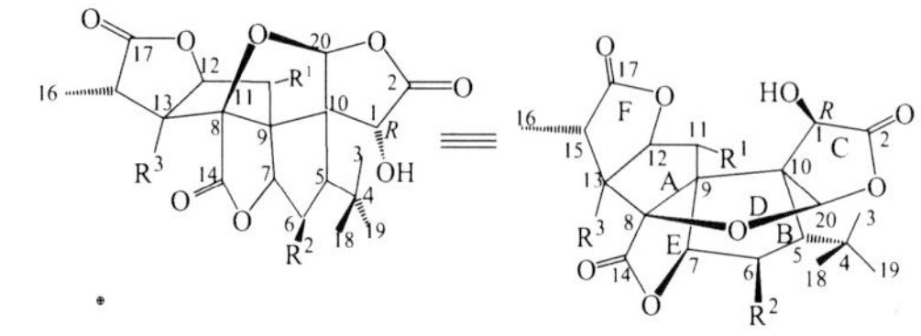

Structure	Name	R^1	R^2	R^3	
(1)	Ginkgolide A	H	H	OH	$C_{20}H_{24}O_9$
(2)	Ginkgolide B	OH	H	OH	$C_{20}H_{24}O_{10}$
(3)	Ginkgolide C	OH	OH	OH	$C_{20}H_{24}O_{11}$
(4)	Ginkgolide J	H	OH	OH	$C_{20}H_{24}O_{10}$
(5)	Ginkgolide M	OH	OH	H	$C_{20}H_{24}O_{10}$

Fig. 8.39 Stereostructures of ginkgolides

8.6.2   Chemical Constituents. Ginkgolides; Biosynthesis. Synthesis

The chemical constituents, called ginkgolides, a family of polycyclic terpenoids were first isolated by Furukawa in 1932 and the structures were elucidated by Nakanishi in a series of papers [70–73], based on chemical and NMR spectroscopic studies [structures (1)–(5)] (Fig. 8.39). Finally, the correctness of the structures was fully supported by X-ray crystallographic analysis [74]. They all belong to diterpenes with heavily oxygenated functionalities and the ratio of carbon:oxygen ~ 10:4.5 → 10:5.5, and other ginkgolides discovered differ in the positions of hydroxyls. The ginkgolides (1)–(5) have three lactones in the same positions. Later the biosynthesis of ginkgolides has been extensively studied by Arigoni [75], which is already outlined briefly in Fig. 8.10. Serious attempts have been made to synthesize ginkgolides as their complex scaffolds offer a great challenge to the synthetic organic chemists. Corey has elegantly synthesized ginkgolide B [76]. It is a therapeutic agent of potentially wide application—a potent antagonist of platelet-activating factor, and also an insect antifeedant.

8.7   Forskolin

8.7.1   Occurrence, Stereostructure

Forskolin (1), $C_{22}H_{34}O_7$, m.p. 230–238 °C, $[\alpha]_D$ −26.9, a diterpene of labdane skeleton having an additional heterocyclic ring (Fig. 8.40), has been isolated from the roots of *Coleus forskohlii* (Labiatae) [77, 78]. Its structure has been shown to be 7β-acetoxy-8,13-epoxy-1α,6β,9α-trihydroxy-lab-14-ene-11-one (1) from its reactions, synthesis, spectral data, and X-ray crystallographic analysis. Coleonol, independently isolated from the same plant, was assigned the 7-epimeric stereostructure of (1), based on degradative and spectroscopic studies [78]. Later on it has been found to be identical with forskolin (1) having 7-(S)-β-acetoxy group [79, 80]. A series of compounds possessing the basic skeleton of forskolin with different oxygenation patterns and also with 7-epimeric hydroxy/acetoxy compounds have

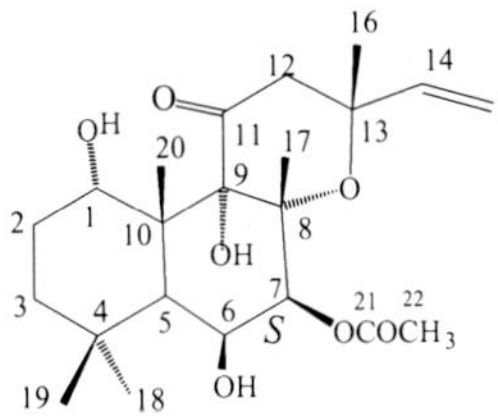

Forskolin (1)
(≡ Coleonol)

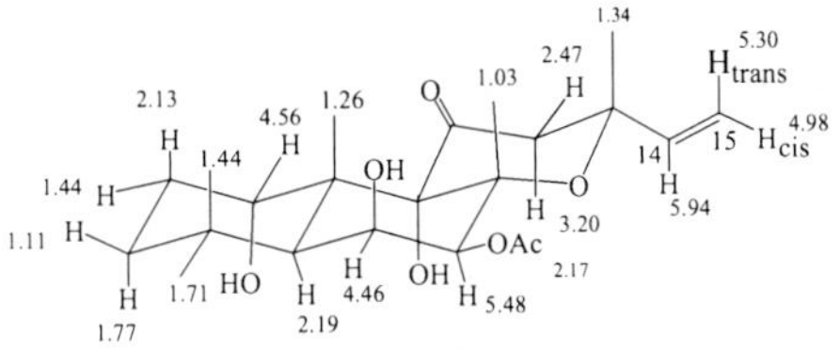

Conformation of **Forskolin** .
^{1}H NMR chemical shift values

^{1}H NMR : δ (CDCl$_3$) (J in Hz), 4.56 (dd, J 6.0, 3.4, H1), 1.44 (m, H2-eq), 2.13 (m, H2-ax), 1.11 (m, H3-eq), 1.77 (m, H3-ax), 2.19 (d, J, 1.9, H5), 4.46 (dd, 1.9, 3.9, H6), 5.48 (d, J 3.9, H7), 2.47 (d, J$_{gem}$ 17.1, H12-eq), 3.20 (d, J$_{gem}$ 12.1, H12-ax), 5.94 (dd, J 17.1, 10.2, H14), 4.98 (d, J 10.2, H15-cis), 5.30 (d, J 17.1, H15-trans), 1.34, 1.03, 1.71, 1.44 and 1.26 (s each, H$_3$ 16, H$_3$ 17, H$_3$ 18, H$_3$ 19, H$_3$ 20 respectively), 2.17 (H$_3$ 22).

^{13}C NMR : δ (CDCl$_3$), 74.5 (C1), 26.6 (C2), 36.1 (C3), 34.4 (C4), 42.9 (C5), 69.9 (C6), 76.7 (C7), 81.4 (C8), 75.0 (C9), 43.1 (C10), 205.3 (C11), 48.8 (C12), 82.7 (C13), 146.4 (C14), 110.7 (C15), 31.5 (C16), 33 (C17), 23.6 (C18), 19.8 (C19), 24.3 (C20), 169.6 (C21), 21.0 (C22).

Fig. 8.40 ^{1}H & ^{13}C NMR Spectral data of forskolin (= coleonol)

been isolated (see [81]). Biogenetic relationship of the polyoxygenated diterpenes in *Coleus forskohlü* has been studied [81].

8.7.2 NMR Spectral Data [78]

^{1}H NMR spectral assignments for coleonol (forskolin) (**1**) (Fig. 8.40) have been made by two-dimensional methods. The 7-OAc group is assigned equatorial β-configuration from the coupling constants of H7 : $J_{7,6} = J_{ae} = 3.9$ Hz. ^{13}C NMR spectral assignments (Fig. 8.40) have been made with the help of DEPT and ^{1}H-^{13}C chemical shift correlation studies.

8.7.3 Bioactivity

Forskolin possesses unique antihypertensive, intraocular pressure lowering, and platelet-inhibiting properties and is also a unique cardiotonic [77, 78] (Chap. 33). A few forskoditerpenosides have been isolated possessing relaxation effects in vitro [82].

References

1. Jake MacMillan and Michael H. Beale, Diterpene Biosynthesis in *Comprehensive Natural Products Chemistry*, Editor-in-Chief, Sir Derek Barton and Koji Nakanishi, Vol.2, Volume Editor,David E.Cane **1999**, Elsevier, pp. 217-243.
2. Paul M. Dewrick, The Biosynthesis of C5-C25 Terpenoid Compounds, *Nat. Prod. Rep.*, **1999**, *16*, 97-130.
3. Paul M. Dewrick, The Biosynthesis of C5-C25 Terpenoid Compounds, *Nat. Prod. Rep.*, **2002**, *19*, 181-222.
4. Matthias Schwarz and Duilio Arigoni, Ginkgolide Biosynthesis in *Comprehensive Natural Products Chemistry*, Editors -in-Chief, Sir Derek Barton and Koji Nakanishi, Vol 2, Volume Editor, David E. Cane, **1999**, Elsevier, pp. 367-400.
5. Makoto Iwashima, Jun Mori, Xiang Ting, Takayuki Matsunaga, Kyoko Hayashi, Daishi Shinoda, Haruo Saito, Ushio Sankawa, and Toshimitsu Hayashi, Antioxidant and Antiviral Activities of Plastoquinones from the Brown Alga *Sargassam micracanthum,* and a New Chromene Derivative Converted from the Plastoquinones, *Biol. Pharm. Bull.* **2005**, *28*, 374-377.
6. William G. Dauben, William E. Thiessen, and Paul R. Resnick, Cembrene, a Fourteen-Membered Ring Diterpene Hydrocarbon, *J. Org. Chem.*, **1965**, *30*, 1693-1698.
7. Alejandro F. Barrero, Juan F. Sanchez, and Felix G. Cuenca, Dramatic Variation in Diterpenoids of Different Populations of *Bellardia trixago*, *Phytochemistry*, **1988**, *27*, 3676-3678.
8. Nigel B. Perry and Rex T. Weavers, Foliage Diterpenes of *Dacrydium intermedium*: Identification, Variation and Biosynthesis, *Phytochemistry*, **1985**, *24*, 2899-2904.
9. A. I. Scott, S. A. Sutherland, D. W. Young, L. Guglielmetti, D. Arigoni, and G. A. Sim, The Structure and Absolute Configuration of Rosololactone and Related Diterpenoid Lactones, *Proc. Chem. Soc.,* **1964**, 19-21, and references cited.
10. A. J. Birch, R. W. Richards, H. Smith, A. Harris, and W. B. Whalley, Studies in Relation to Biosynthesis-XXI. Rosenonolactone and Gibberellic Acid, *Tetrahedron*, **1959**, *7*, 241-251.
11. W. B. Whalley, B. Green, D. Arigoni, J. J. Britt, and Carl Djerassi, The Absolute Configuration of Rosenonolactone and Related Diterpenoids, *J. Am. Chem. Soc.* **1959**, *81*, 5520-5521.
12. Carl Djerassi, B. Green, W. B. Whalley, and (in part) C. G. DeGrazia, The Chemistry of Fungi. Part III. [Optical Rotatory Dispersion Studies. Part CV]. The Absolute Configuration of Rosenono- and Rosolo-Lactone *J. Chem. Soc.* (C), **1966**, 624-627.
13. B. Achilladelis and J. R. Hanson, The Rearrangement During Rosenonolactone Biosynthesis, *Chem. Commun.,* **1969**, 488-489.
14. Brian Dockerill and James R. Hanson, Studies in Terpenoid Biosynthesis . Part 19. Formation of Pimara-8(9),15-diene by Trichothecium roseum, *J. Chem. Soc.* Perkin Trans, **1977**, *1*, 324-327.
15. David E. Cane, The Stereochemistry of Allylic Pyrophosphate Metabolism, *Tetrahedron* Report No.82, **1980**, *36*, 1109-1159 and references cited.
16. B. A. Nagasampagi, L. Yankov and Sukh Dev, Isolation and characterization of Geranylgeraniol, *Tetrahedron Lett.*, **1967**, 189-192.
17. E. Fedeli, P. Capella, M. Cirimela, and G. Jacini, *J. Lipid Res.,* **1966**, *7*, 437.
18. Barry M. Trost, Lother Weber, Paul Strege, Terry J. Fullerton, and Thomas J. Dietsche, Allylic Alkylation. Nature of the Nucleophile and Application to Prenylation, *J. Am. Chem. Soc.*, **1978**, *100*, 3426-3435.
19. Stefan Berger and Dicter Sicker, *Classics in Spectroscopy*, Wiley VCH, **2009**, p.459-480, pertinent page 461.
20. Gilbert Stork and John W. Schulenberg, The Total Synthesis of dl-Dehydroabietic acid, *J. Am. Chem. Soc.* **1956**, *78*, 250-251.
21. Gilbert Stork and John W. Schulenberg, The Total Synthesis of dl-Dehydroabietic acid, *J. Am. Chem. Soc.* **1962**, *84*, 284-292

22. Albert W. Burgstahler and Leonard R. Worden, The Synthesis of Abietic Acid from Dehydroabietic acid, *J. Am. Chem. Soc.,* **1961**, *83*, 2587-2588.

23. Robert E. Ireland and Roger C. Kierstead, Experiments Directed toward the Total Synthesis of Terpenes. X. A Sterioselective Scheme for Diterpenoid Resin Acid Synthesis, *J. Org. Chem.,* **1966**, *31*, 2543-2559.

24. Walter I. Meyer and Carl W. Siegel, Diterpenoid Total Synthesis, an A→B→ C Approach 11. C-Ring Deoxy Aromatic Systems. Total Synthesis of Methyl (±)-Dehydroabietate, *J .Org. Chem.,* **1977**, *42*, 2769-2771.

25. M. Ceu Costa, S. P. Alves, M. Eduarda Correia, and M. Joäo Marcelo-Curto, Synthesis of an Ambergris-Type Ketal from Abietic Acid, *Synthesis*, **2006**, 1171-1175.

26. D. H. R. Barton, Some Recollection of Gap Jumping, Profiles, Pathways and Dreams, Autobiographies of Eminent Chemists, Jeffrey I. Seeman, Series Editor, American Chemical Society, Washington D C, **1991**.

27. D. H. R. Barton and (in part) G. A. Schmeidler, The Application of the Method of Electrostatic Energy Difference Part I. Stereochemistry of Diterpenoid Resin Acids, *J. Chem. Soc.* (London), **1948**, 1197-1203.

28. Herbert O. House*, Modern Synthetic Reactions*, 2nd Edn, W.A. Benjamin, Inc., California, London, **1972**, 372-374.

29. John W. ApSimon, Andrew M. Holmes, Helmut Beierbeck, and John K. Saunders, Revised Structure for the Lactone Derived from the Acid Treatment of Dihydroisopimaric Acid, *Can. J. Chem.,* **1976**, 54, 418-422.

30. Werner Herz and John F. Blount, X-ray Structure of the γ-Lactone Formed by Acid Treatment of Dihydroisopimaric Acid, *J. Org. Chem.,* **1979**, *44*, 1172-1173.

31. S. W. Pelletier and D. L. Herold Jun. Inverssion of C-4 Substituents in Dehydroabietic acid. Synthesis of (±)-Callistrisic Acid, *Chem. Commun,* **1971**, 10.

32. C. R. Narayanan and N. K. Venkatasubramanian, Stereochemical Studies by PMR Spectroscopy – III, Axial and Equatorial Acids, *Tetrahedron Lett,* **1965**, 3639-3646.

33. E. J. Alveaez-Manzaneda Roldán, R. Chahboum, F. Bentaleb, E. Cabrera Torres, E. Alvarez, A. Haidour, J. M. Ramos López, R. Alverez-Manzaneda Roldán, and S. EI Houssame, First Enantiospecific Synthesis of Antileshmanial 12-Deoxyroyleanone from Abietic Acid, *Synlett*, **2004**, 2701-2704.

34. J. R. Maxwell, C. T. Pillinger, and G. Eglinton, Organic Geochemistry, *Quaterly Reviews*, **1971**, *25*, 571-628; pertinent pages 594, 602.

35. Monroe E. Wall, Camptothecin and Taxol: Discovery to Clinic, *Med. Res. Rev.*, **1998**, *18*, 299-314.

36. Kyriacos C. Nicolaou and Rodney K. Guy, The Conquest of Taxol, *Angew. Chem. Int. Ed.*, **1995**, *34*, 2079-2090.

37. Andrea Stierle, Gary Strobel, and Donald Stierle, Taxol and Taxane Production by *Taxomyces andreanae*, an Edophytic Fungus of Pacific Yew. *Science*, **1993**, *260*, 214-216.

38. M. C. Wani, H. L. Taylor, Monroe E. Wall, P. Coggon, and A. T. McPhail, Plant Antitumor Agents. VI. The Isolation and Structure of Taxol, A Novel Antileukemic and Antitumor Agent from *Taxus brevifolia. J. Am. Chem. Soc.*, **1971**, *93*, 2325-2327.

39. D. G. I. Kingston, P. G. Jagtap, H. Yuan, and Samala, The Chemistry of Taxol and Related Taxoids, *Fortschr. Chem. org. Naturstoffe*, **2003**, *84*, 53-225, and references cited.

40. D. G. I. Kingston, Studies on the Chemistry of Taxol, *Pure Appl. Chem.*, **1998**, *70*, 331-334.

41. D. P. Della Casa de Marcano, T. G. Halsall, E. Castellano, and O. J. R. Hodder, Crystallographic Structure Determination of the Diterpenoid Baccatin-V, a Naturally Occurring Oxetan with a Taxane Skeleton, *Chem. Commun.*, **1970**, 1382-1383.

42. D. P. Della Casa de Marcano and T. G. Halsall, The Isolation of Seven New Taxane Derivatives from the Heartwood of Yew (*Taxus baccata* L.) *Chem. Commun.*, **1969**, 1282-1283.

43. Jean-Noel Denis, Andrew E. Greene, Daniel Guenard, Francoise Gueritte-Voegelei, Lydie Mangatal, and Pierre Potier, A Highly Efficient, Practical Approach to Natural Taxol, *J. Am. Chem. Soc,* **1988**, *110*, 5917-5919

44. K. C. Nicolaou, Z. Yang, J. J. Liu, H. Ueno, P. G. Nantermet, R. K. Guy, C. F. Claiborne, J. Renaud, E. A. Couladouros, K. Paulvannan, and E. J. Sorensen, Total Synthesis of Taxol, *Nature*, **1994**, *367*, 630-634.

45. K. C. Nicolaou and E. J. Sorensen, Taxol in *Classics in Total Synthesis*, VCH, New York, 1996, pp. 655-671.

46. K. C. Nicolaou, C. R. H. Hale, C. Nilewsky, and H. A. Loanmidou, Constructive Molecular Complexity and Diversity. Total Synthesis of Natural Products of Biological and Medicinal Importance, *Chem. Soc. Rev.*, **2012**, *41*,5185-5138,- a review article.

47. K. C. Nicolaou, Christopher R. H. Hale, and Christian Nilewski, A Total Synthesis Trilogy: Calicheamicin γ$_1$, Taxol®, and Brevetoxin A, *Chem. Rec.*, **2012**, *12*, 407-441,—for a detailed behind-the-scenes accounts.

48. Robert A. Holton, Carmen Somoza, Hyeong-Baik Kim, Feng Liang, Ronald J. Biediger, P. Douglas Boatman, Mitsuru Shindo, Chase C. Smith, Soekchan Kim, Hossain Nadizadeh, Yukio Suzuki, Chunlin Tao, Phong Vu, Suhan Tang, Pingsheng Zhang, Krishna K. Murthi, Lisa N. Gentile, and Jyanwei H. Liu, First Total Synthesis of Taxol, 1. Funtionalization of the B Ring, *J. Am. Chem. Soc.*, **1994**, *116*, 1597-1598.

49. Robert A. Holton and the same (18) other authors, First total Synthesis of Taxol 2. Completion of the C and D Rings, *J. Am. Chem. Soc.*, **1994**, *116*, 1599-1600.

50. K. C. Nicolaou and T. Montagnon, Molecules That Changed the World, Wiley-VCH, 2008, pp. 208-215 pertinent p. 212

51. Iwao Ojima, Ivan Habus, Mangzhu Zhao, Martine Zucco, Young Hoon Park, Chung Mingh Sun, and Thierry Brigaud, New and Efficient Approaches to the Semisynthesis of Taxol and Its C-13 Side Chain Analogs by Means of β-Lactam Synthon Method, *Tetrahedron*, **1992**, *48*, 6985-7012.

52. Ningqing Ran, Lishan Zhao, Zheaming Chen, and Junhua Tao, Recent Applications of Biocatalysis in Developing Green Chemistry for Chemical Synthesis at the Industrial Scale, *Green Chemistry*, **2008**, *10*, 361-372.

53. Andrea Stierle, Gary Strobel, Donald Stierle, Paul Grothaus, and Gary Bignami, The Search for a Taxol-Producing Microorganism among the Endophytic Fungi of the Pacific Yew, *Taxus brevifolia*, *J. Nat. Prod.*, **1995**, *58*, 1315-1324.

54. Wolfgang Eisenreich, Birgitta Menhard, Peter J. Hylands, Meinhart H. Zenk, and Adelbert Bacher, Studies on the biosynthesis of Taxol: The Taxane Carbon Skeleton is Not of Mevalonoid Origin, *Proc. Natl. Acad. Sci. USA*, **1996**, *93*, 6431-6436.

55. Alfredo Vázquez and Robert M. William, Studies on the Biosynthesis of Taxol. Synthesis of Taxa-4(20),11(12)-diene-2α,5α-diol, *J. Org. Chem.*, **2000**, *65*, 7865-7869.

56. Tohru Horiguchi, Christopher D. Rithner, Rodney Croteau, and Robert M. Williams, Studies on Taxol Biosynthesis. Preparation of Taxa-4(20),11(12)-dien-5α-acetoxy-10β-ol by Deoxygenation of a Taxadiene Tetraacetate Obtained from Japanese Yew, *J. Org. Chem.*, **2002**, *67*, 4901-4903.

57. Yinghua Jin, David C. Williams, Rodney Croteau, and Robert M. Coates, Taxadiene-Synthase Catalyzed Cyclization of 6-Fluorogeranylgeranyl Diphosphate to 7-Fluoroverticillenes, *J. Am. Chem. Soc.*, **2005**, *127*, 7834-7842.

58. B. B. Stowe and T. Yamaki, The History and the Physiological Action of the Gibberellins, *Ann. Rev. Plant Physiol.* **1957**, *8*, 181-216.

59. Historical Introduction in *Gibberellins* and *Plant Growth*, Etd. By H.N. Krishnamoorthy, Wiley Eastern Limited, New Delhi, **1975**, p., XIII-XV.

60. N. Takahashi, H. Kitamura, A. Kawarada, Y. Seta, M. Takai, S. Tamura and Y. Sumiki, Biochemical Studies on 'bakanae' Fungus, Part 34. Isolation of Gibberellin and their Properties, *Bull. Arg. Chem. Soc., Japan*, **1955**, *19*, 267-277.

61. Lewis N. Mander, New Strategies for the Construction of Highly Functionalized Organic Molecules : Applications to C$_{19}$ Gibberellin Synthesis, *Acc. Chem. Res.* **1983**, *16*, 48-54 ref. [62] (pertinent ref.).

62. Noboru Murofushi, Hisakazu Yamane, Youji Sakagami, Hidemasa Imaseki, Yuji Kamiya, Hajime Iwamura, Noboru Hirai, Hideo Tsuji, Takao Yokota and Junichi Ueda, Plant Hormones in *Comprehensive Natural Products Chemistry*, Editor-in-chief.Sir Derek Barton and Koji Nakanishi, 1999,Vol.**8**, Vol. Editor Kenji Mori, Pergamon Press, Elsevier, pp.19-136, Pertinent pp.35-57.

63. B. E. Cross, J. F. Grove, J. MacMillan, J. S. Moffatt, T. P. C. Mulholland, J. C. Seaton, and N. Sheppard, Proc. Chem. Soc. London, **1959**, 302-303.

64. Jean A. Hartsuck and Willam N. Lipscomb, Molecular and Crystal Structure of the Di-*p*-bromobenzoate of the Methyl Ester of Gibberellic Acid, *J. Am. Chem. Soc.*, **1963**, *85*, 3414-3419.

65. E. J. Corey, Rick L. Danheiser, Srinivasan Chandrasekharan, Patrice Siret, Gary E. Keck, and Jean Louis Grass, Stereospecific Total Synthesis of Gibberellic Acid. A Key Tricyclic Intermediate, *J. Am. Chem. Soc.,* **1978**, *100*, 8031-8034.

66. E. J. Corey, Rick L. Danheiser, Srinivasan Chandrasekharan, Gary E. Keck, B. Gopalan, Samuel D. Larsen, Patrice Siret, and Jean Louis Grass, Stereospecific total synthesis of Gibberellic Acid, *J. Am. Chem. Soc.,* **1978**, *100*, 8034-8036.

67. E. J. Corey and Xue-Min Cheng, *The Logic of Chemical Synthesis,* John Wiley & Sons, New York, Singapore, p. 84.

68. M. K. Paul, *Forever free,* Sarat Book Distributors, Kolkata, **2003**, p. 76.

69. K. C. Nicolaou and T. Montagnon*, Molecules That Changed the World*, Wiley VCH, **2008**, p.171.

70. M. Maruyama, A. Terahara, Y. Itagaki, and K. Nakanishi, The Ginkgolides, 'Stereochemistry of the Ginkgolides I. Isolation and Characterization of Various Groups *Tetrahedron Lett.,* **1967**, 299-302.

71. M. Maruyama, A. Terahara, Y. Itagaki, and K. Nakanishi, The Ginkgolides. II. Derivatives of Partial Structures, *Tetrahedron Lett.*, **1967**, 303-308.

72. M. Maruyama, A. Terahara, Y. Nakadaira, M. C. Woods, and K. Nakanishi, The Ginkgolides. III. The Structures of Ginkgolides, *Tetrahedron Lett.*, **1967**, 309-315.

73. M. Maruyama, A. Terahara, Y. Nakadaira, M. C. Woods, Y. Takagi, and K. Nakamishi, The Ginkgolides. IV. Stereochemistry of the Ginkgolides. IV. Stereochemistry of the Ginkgolides, *Tetrahedron Lett.*, **1967**, 315-319.

74. Noriyoshi Sakabe, Sasumu Takada, and Kei Okabe, The Structure of Ginkgolide A, A Novel diterpenoid Trilactone, *Chem. Commun.* 1967, 259-261.

75. Matthias Schwarz and Duilio Arigoni, Ginkgolide Biosynthesis, in *Comprehensive Natural Products Chemistry*, Editor-in-Chief Sir Derek Barton, and Koji Nakanishi, Vol 2, (Volume Editor, David E. Cane), Elsevier, **1999**, pp. 367-400.

76. E. J. Corey, M.-C. Kang, M. C. Desai, A. K. Ghosh, and I. N. Houpis, Total Synthesis of (±) Ginkgolide B, *J. Am. Chem. Soc.*, **1988**, *110*, 649-651.

77. S. V. Bhat, B. S. Bajwa, H. Dornauer, N. J. De Souza, and H. -W. Fehlhaber, Structure and Stereochemistry of New Labdane Diterpenoids from *Coleus forskohlii Briq., Tetrahedron Lett.,* **1977**, 1669-1672.

78. J. S. Tandon, M. M. Dhar, S. Ramkumar, and K. Venkatesan, *Indian J. Chem.* **1977**, *15B*, 88.

79. Anil K. Saksena, Michael J. Green, Ho-Jane Shue, Jesse K. Wong, and Andrew T. McPhail, Identity of Coleonol with Forskolin. Structure Revision of a Base-Catalysed Rearrangment Product, *Tetrahedron Lett.*, **1985**, *26,* 551-554.

80. Om Prakash, Raja Roy, and Monojit M. Dhar, A Nuclear Magnetic Resonance Study of Coleonol, *J. Chem. Soc. Perkin Trans. II*, **1986**, *11*, 1779-1783.

81. Anand Akhila, Kumkum Rani, and Raghunath S. Thakur, Biogenetic Relationship of Polyoxygenated Diterpenes in *Coleus forskohlii, Phytochemistry,* **1990**, *29*, 821-824 and references cited.

82. Gautam Brahmachari, *Handbook of Pharmaceutical Natural Products*, Wiley-VCH, **2010**, Vol.1, 323.

Further Reading

K. C. Nicolaou, W.-M. Dai, and R. K. Guy, Chemistry and Biology of Taxol, *Angew. Chem. Int. Ed. Engl.* **1994,** *33*, 15-44.

K. C. Nicolaou and R. K. Guy, The Conquest of Taxol, *Angew. Chem. Int. Ed. Engl.*, **1995**, *34*, 2079-2090.

K. C. Nicolaou, R. K. Guy, and P. Potier, Taxoids: New Weapons Against Cancer, *Sci. Am..* **1996,** *274*, 94-98.

David G. I. Kingston, The Shape of Things to Come: Structural and Synthetic Studies of Taxol and Related Compounds, *Phytochemistry*, **2007**, *68*, 1844-1854.

J. Mann, Natural Products in Cancer Chemotherapy: Past, Present and Future, *Nat. Rev. Cancer*, **2002,** *2*, 143-148.

Chapter 9
Sesterterpenoids (C_{25})

9.1 Introduction. Occurrence. Structure

As per Ruzicka's biogenetic hypothesis of isoprenoids, the sesterterpenoids (C_{25}-isoprenoids) are expected to occur in nature. However, a long sought sesterterpene was not discovered until 1958 when a sesterterpene was isolated as a metabolic product from the cultural broth of the plant pathogenic fungus *Ophiobolus miyabeanus* and named *ophiobolin*. The stereostructure of ophiobolin has been established [1] in 1965 as (1) on the basis of ^{1}H NMR spectral data of ophiobolin and its bromomethoxy derivative and an X-ray crystallographic analysis of the latter. The skeletal pattern of ophiobolin was unknown at the time of its structural elucidation.

9.2 Spectral Data of Ophiobolin (1) (Fig. 9.1)

Since the report of the isolation and structure elucidation of ophiobolin, *a large number of sesterterpenes with ophiobolin, epiophiobolin, and other skeletal patterns have been reported from pathogenic fungi, insect secretions, and marine sponges* [2, 3]. Himalayan lichens elaborate sesterterpenoids having interesting skeletal patterns. Quite a good number of sesterterpenoids have been obtained from sea sponges, while their occurrence in plants, especially in higher plants, is rare. All these compounds originate biogenetically from geranyl farnesyl pyrophosphate (GFPP). It has been shown by labeling experiments that in plant pathogens GFPP is formed from MVA [4, 5]. Geranylfarnesol (GF), the hydrolyzed product of GFPP, has been isolated from the insect wax of *Ceroplastes albolineatus* [6], while geranylnerolidol and also geranylfarnesol have been found in fungus *Cochliobolus heterostrophus* [3].

S.K. Talapatra and B. Talapatra, *Chemistry of Plant Natural Products*,
DOI 10.1007/978-3-642-45410-3_9, © Springer-Verlag Berlin Heidelberg 2015

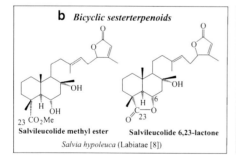

Ophiobolin (1)   C$_{25}$H$_{36}$O$_4$, m.p.182^{0}C
[a]$_D^{29}$ +270

^{1}H NMR: δ 0.82 (s, 3H, 22Me), 1.08 (d, J = 6.5 Hz, 3H, 23Me), 1.3 (s, 3H, 20Me), 1.7 (s, 6H, 24Me, 25Me), 4.43 (q, J = 8 Hz, 1H, H17), 7.13 (t, J = 8 Hz, 1H, H8), 9.26 (s, 1H, H21), 5.15 (d, J = 8 Hz, H18). Additional signals at 2.46 and 2.77 (AB type q, J = 20, 2 Hz) and at 3.20 (d, J = 11, 1H) are attributed to H$_2$4 and H6 protons adjacent to >C=O group.

UV: λ$_{max}$ (EtOH): 238 nm (ε, 13,800)

IR: υ$_{max}$ (CHCl$_3$): 3500 (OH), 1743 (5-membered ketone), 1674 (α,β-unsaturated carbonyl) and 1633 cm^{-1}

Fig. 9.1   Spectral data of ophiobolin

a   *Partly saturated acyclic C25-isoprenyl alcohol*

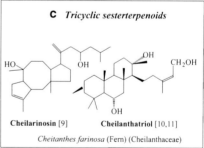

(2Z,6E)-3,7,11,15,19-Pentamethylicosa-2,6-dien-1-ol
(*Solanum tuberoscum,* Solanaceae, potato leaves/lipids) [7]

b   *Bicyclic sesterterpenoids*

Salvileucolide methyl ester   **Salvileucolide 6,23-lactone**

Salvia hypoleuca (Labiatae [8])

c   *Tricyclic sesterterpenoids*

Cheilarinosin [9]   **Cheilanthatriol** [10,11]

Cheitanthes farinosa (Fern) (Cheilanthaceae)

Fig. 9.2   Sesterterpenoids of plant origin

9.3   Sesterterpenoids of Plant Origin

Only few plant-based sesterterpenoids are known. Some examples with structures and sources are shown in Fig. 9.2.

(1) Ophiobolin-type cyclization

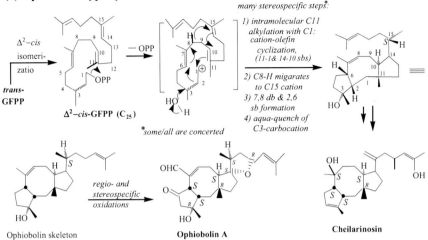

Ophiobolin skeleton

regio- and
stereospecific
oxidations

Ophiobolin A

Cheilarinosin

(2) Labdane-type cyclization and aqua-quench leading to bicyclic sesterterpines

cyclization and
aquaquench

Δ^2-*cis*-GFPP
(C$_{25}$)

Oxidations
(regio- and
stereospecific)
& lactonization

and

**Salvileucolide
methyl ester**

Salvileucolide lactone

(3) Cyclization and aquaquench leading to a tricyclic sesterterpene

oxidation
(revgio- and
stereospecific)

Δ^2-*cis*-GFPP (C$_{25}$)

Cheilanthatriol

Fig. 9.3 Biogenesis of sesterterpenes having different skeletons

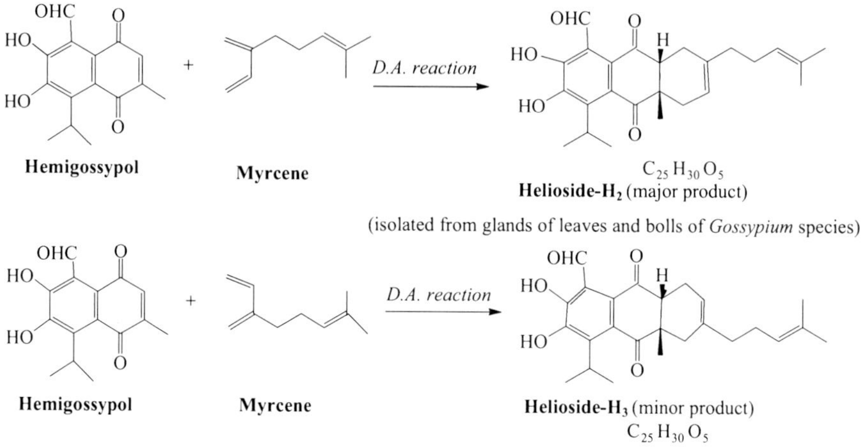

Fig. 9.4 C$_{25}$ Natural compounds but biogenetically not related to GFPP

9.4 Biosynthesis of Some Sesterterpenoids (Fig. 9.3)

The biogenetic cyclizations of some sesterterpene skeletal patterns are delineated in Fig. 9.3.

9.5 Natural C$_{25}$ Compounds Biogenetically Not Related to Geranylfarnesyl PP

From the molecular composition the heliosides (Fig. 9.4) appear to be sesterterpene derivatives. But biogenetically they are not derived from geranylfarnesyl pyrophosphate; rather, they are formed from the Diels–Alder reaction between hemigossypol and myrcene (Fig. 9.4). Both are co-occurring in the same plant. In fact in the laboratory the Diels–Alder reaction between the two synthons forms helioside-H$_2$ as the major product, while helioside-H$_3$ has been obtained from repeated chromatography of the mother liquor of helioside-H$_2$ [12, 13].

References

1. Shigeo Nozoe, Masuo Morisaki, Kyosuke Tsuda, Yoichi Iitaka, Nobutaka Takahashi, Saburo Tamura, Keijiro Ishibashi, Makoto Shirasaka, The Structure of Ophiobolin, a C$_{25}$ Terpenoid Having a Novel Skeleton, *J. Am. Chem. Soc.*, **1965**, *87*, 4968-4970

2. Fumio Sugawara, Nobutaka Takahashi, Gary Strobel, Choong-Hyo Yun, George Gray, Yali Fu and Jon Clardy, Some New Phytotoxic Ophiobolins Produced by *Drechslera oryzae*, *J. Org. Chem.*, **1988**, *53*, 2170-2172.

3. J. R. Hanson, Sesterterpenoids, *Nat. Prod. Rep.*, **1986**, 123-132.

4. Shigeo Nozoe and Masuo Morisaki, Enzymic Formation of a Tricyclic Sesterterpene Alcohol from Mevalonic Acid and all-*trans*-Geranylfarnesyl Pyrophosphate, *Chem. Commun.*, **1969**, 1319-1320.

5. L. Canonica, A. Fiecchi, M. Galli Kienle, B. M. Ranzi and A. Scala, The Biosynthesis of Cochlibolins A and B., *Tetrahedron Lett*, **1966**, 3035-3039.

6. Tirso Rios and S. Pérez C, Geranylfarnesol, A New Acyclic C_{25} Isoprenoid Alcohol Isolated from Insect Wax, *Chem Commun.*, **1969**, 214-215.

7. Masahi Toyoda, Masako Asahina, Hideaki Fukawa, and Toru Shimizu, Isolation of a New Acyclic C_{25}-Isoprenyl Alcohol from Potato Leaves, *Tetrahedron Lett*, **1969**, 4879-4882.

8. Abdolhossein Rustaiyan, Akram Niknejad, Lilly Nazarians, Jasmin Jakupovic and Ferdinand Bohlmann, Sesterterpenes from *Salvia hypoleuca, Phytochemistry*, **1982**, *21*, 1812-.1813

9. R. Thanu Iyer, K. N. N. Ayengar and S. Rangaswami, Structure of Cheilarinosin, A New Sesterterpene from *Cheilanthes farinose, Indian J. Chem.*, **1972**, *10*, 482-484

10. Hafizullah Khan, Asif Zaman, G. L. Chetty, A. S. Gupta and Sukh Dev, Cheilanthatriol – A New Fundamental Type in Sesterterpenes, *Tetrahedron Lett*, **1971**, 4443-4446.

11. Ram P. Rastogi and B. N. Mehrotra, *Compendium of Indian Medicinal Plants* Vol. 2, 1970-1979, Ram P. Rastogi (Ed.), CDRI, Lucknow and Publications & Information Directorate, New Delhi, **1991**, p173.

12. Robert D. Stipanovic, Alois A. Bell, Daniel H. O'Brien and Maurice J. Lukefahr, Helioside H, An Insecticidal Terpenoid from *Gossypium hirsutum, Phytochemistry*, **1978**, *17*, 151-152.

13. Alois A. Bell, Robert D. Stipanoivic, Daniel H. O'Brien and Paul A. Frysxell, Sesquiterpenoid Aldehyde Quinones and Derivatives in Pigment Glands of *Gossypium, Phytochemistry*, **1978**, *17*, 1297-1305.

Chapter 10
Triterpenes (C_{30})

10.1 Introduction. Biogenesis. Functions of Enzymes

Triterpenoids belong to a large and structurally diverse class of natural products which occur abundantly in plant kingdom, especially in higher plants in cases of pentacyclic and tetracyclic triterpenes, while mono-, bi-, and tricyclic triterpenes are prevalent in ferns and nonflowering lower plants (cryptogams). Some of them were found in fossils as such and also as their diagenetic products.

The biogenetic precursor of this class of compounds is squalene, an isoprene-derived C_{30} linear polyene hydrocarbon with C_2-symmetry (Chap. 5). Under the influence of different enzymes, collectively called tripterpene synthases, nearly 200 different triterpene skeletons have been generated [1] from *squalene, oxidosqualene*, and *bisoxidosqualene*. Till 2007, more than 30 *oxidosqualene cyclases* (OSCs) have been cloned and characterized [2–4]. β-*Amyrin synthase* and *lupeol synthase* are the first two synthases to be cloned. Mutagenesis studies on these synthases showed that exchange of only one amino acid dramatically changes the enzyme's own specificity and endowed them with different product specificity (cf. Chap. 11, Ref. [32]). Some of the synthases are monofunctional (e.g., β-amyrin synthase). They elaborate single triterpene each, whereas multifunctional triterpene synthases are capable of producing more than one triterpene, e.g., *mixed amyrin synthase* isolated from peas, *Pisum sativum* (Fam. Papilionaceae). Another multifunctional synthase from *Arabidopsis thaliana* (Fam. Brassicaceae) is capable of yielding as many as nine different triterpenes.

Enzymes responsible for producing 20 different triterpene skeletal patterns have been identified so far, and the rest enzymes remain unidentified. Identification of all the involving enzymes would have uncovered the origin and diversity of the triterpene skeletal patterns. However, all the triterpenes are formed by mono- and multifunctional triterpene synthases, according to specificity.

The majority of the tetracyclic triterpenes are having 6-6-6-5 cyclopentanoperhydrophenanthrene skeleton while pentacyclic skeletal patterns possess 6-6-6-6-5 cyclopentanoperhydrochrysine and 6-6-6-6-6 perhydropicene carbon skeletal

S.K. Talapatra and B. Talapatra, *Chemistry of Plant Natural Products*,
DOI 10.1007/978-3-642-45410-3_10, © Springer-Verlag Berlin Heidelberg 2015

frames. In addition to the above prevalent triterpene ring systems, acyclic, mono-cyclic, bicyclic, tricyclic, and even hexacyclic triterpenes are known to occur and have been isolated and characterized.

The experimentally explored/unexplored but mechanistically reasonable modes of 2,3(S)-oxidosqualene/squalene/2,3(S),22(S),23-bis-oxidosqualene cyclization in the presence of specific enzymes to generate the mono-, di-, tri-, tetra-, and pentacyclic triterpenes are depicted in Figs. 10.1–10.16 that follow.

Unlike mono-, sesqui-, and diterpene cyclases, triterpene cyclases never use pyrophosphate (PP) substrates. The mother molecule squalene does not contain pyrophosphate group. Further, ionization through the loss of PP group in most of the former cases (Chaps. 6, 7 and 8) causes the initiation of the cyclization process, while in triterpene biosynthesis the cyclization is initiated by the electrophilic addition of a proton at the terminal double bond (squalene) or at the epoxy oxygen of oxidosqualene/bisoxidosqualene [1, 5]. Other steps are more or less similar to those for mono-, sesqui-, and diterpene bioformation. Unlike sesquiterpenes (Chap. 7) where 1,3-hydride shift is a common phenomenon, in triterpene biosyn-thesis, especially in tetracyclic and pentacyclic systems, antiparallel 1,2-hydride as

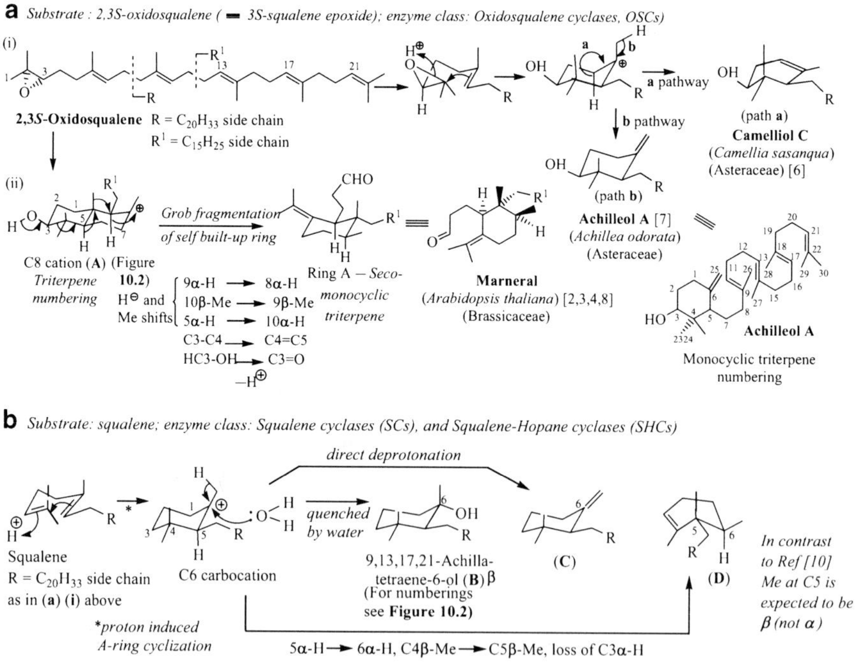

a *Substrate : 2,3S-oxidosqualene (▬ 3S-squalene epoxide); enzyme class: Oxidosqualene cyclases, OSCs)*

b *Substrate: squalene; enzyme class: Squalene cyclases (SCs), and Squalene-Hopane cyclases (SHCs)*

$^{\beta}$*Compound (B) occurs in a fern [9]*

• *Alicyclobacillus acidocaldarius (a bacterium) SHC mutants have been shown to generate (B) and the related monocyclic triterpenes (C) and (D) [10]*

Fig. 10.1 Biosynthesis of some monocyclic triterpenes originating from 2,3S-oxidosqualene and squalene

well as 1,2-alkyl (methyl) migrations are prevalent. Oxidosqualene (2,3S-oxo) is the obligatory precursor for all 3β-hydroxyl (in ring A) of cyclic triterpenes and also for some 3-oxo compounds (e.g., friedelin). Prior to protonation or initiation of cyclization, the triterpene synthases, as per their specificity, fold the linear substrate into a compatible conformation (which does not change much during cyclization) to bring the reactive sites within bonding distances. Proton addition at the terminal double bond or opening of the protonated epoxide triggers the cyclization when cation–olefin cyclization preferably in Markovnikov fashion (each time a low energy carbocation is generated) starts following the stereoelectronic requirement of the processes. The cation reaches its home carbon depending on the specificity of the cyclase in operation, and a specific proton is eliminated to quench the carbocation as olefin; sometimes the carbocation is quenched by some nucleophile, mainly water (aqua quench). All the above events will be mechanistically (organic chemistry based) explained through the formation of some selected triterpenes of different skeletal patterns (Figs. 10.1–10.16). Whether cyclization to the different carbocations and the host of 1,2 migrations is concerted or nonconcerted still remains to be settled; however, the stereospecificity in these processes is maintained by the specific enzymatic folding of the substrate molecules in definite conformations for cyclizations giving rise to innumerable skeletal patterns of triterpenes through the intermediacy of various carbocations. It is to be noted that squalene numbering remains unchanged as long as the molecule is acyclic; the numbering changes when cyclization takes place, since C5 of squalene after cyclization of ring A becomes C1 and the starting gem dimethyl groups of squalene are numbered only at the end. Numbering also depends upon the number of consecutive fused rings—as will be applicable for mono-, bi-, tri-, tetra-, and pentacyclic triterpenes (Sects. 10.1.1–10.1.5).

10.1.1   Monocyclic Triterpenes

10.1.2   Bicyclic Triterpenes

Bicyclic triterpenes are mostly *trans*-decalin derivatives. Electrophilic proton addition at the epoxy or terminal double bond of epoxysqualene [*cf.*, (a)] or squalene [*cf.*, (b)] and ring A and B formations are concerted followed by quenching of carbocations when direct cyclization leads to products formation. Sometimes migrations of hydrides and methyls take place prior to final carbocation quenching leading to the products (Fig. 10.2). All the bicyclic triterpenes of Fig. 10.2 are *trans*-decalin derivatives, the conformations of which are shown in the figure.

   c Lansic acid [16, 17]. The major constituent of the latex of the fruit peels of *Lansium domesticum* Jack v. duku (Meliaceae) serves as an example of a bicyclic triterpene which unlike the common bicyclic triterpene does not contain a decalin ring system. Its structure revealed the presence of two substituted six-membered

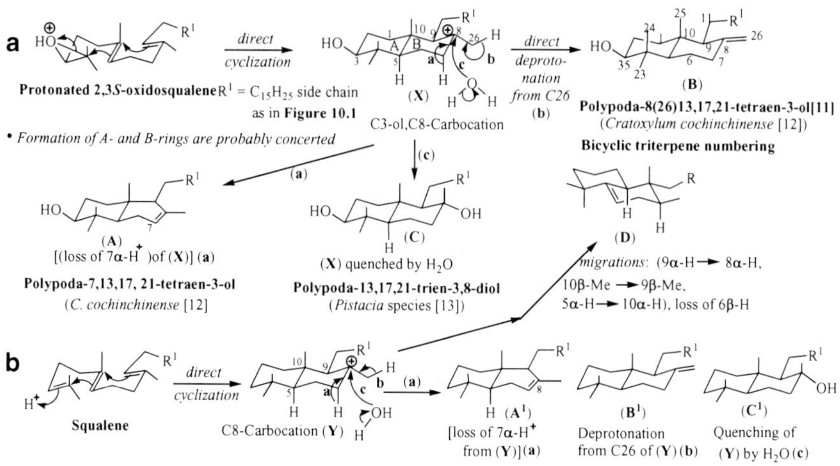

- For R and R[1] see the first structure of **Figure 10.1**
- *The bicyclic triterpenes* (**A**[1]) *and* (**B**[1]) *occur in the fern genus* ***Polystichum*** [14] *and have also been obtained from A.* ***acidocaldarius*** *SHC mutants.*
- *The triterpene alcohol* (**C**[1]) *is from the fern Polypodiodes and the triterpene* (**D**) *has been generated by A. acidocaldarius SHC mutants but is not found in nature* [15]

Fig. 10.2 Biosynthesis of some bicyclic triterpenes originating from 2,3*S*-oxidosqualene and squalene

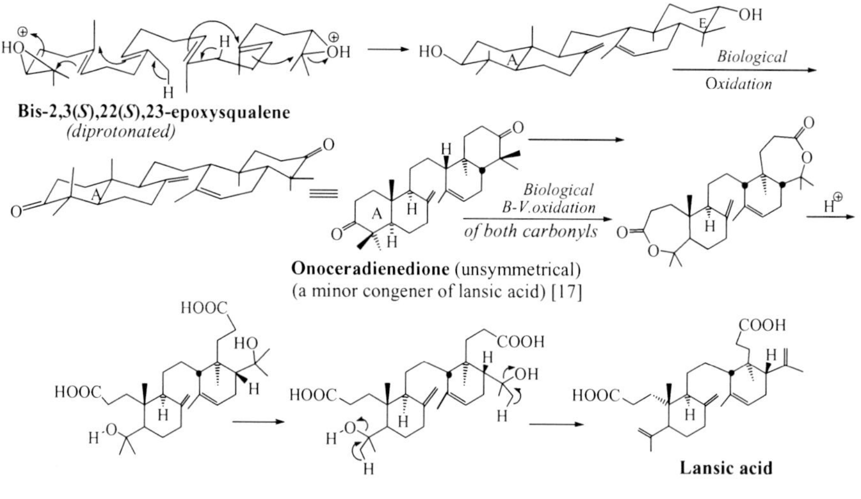

Fig. 10.3 Biogenetic formation of lansic acid

rings linked with a C_2-linker. It has been chemically correlated with α-onocerin. Biogenetically it may be derived from unsymmetrical onoceradienedione (a minor congener of lansic acid from *Lansium domesticum* [17]), as outlined in Fig. 10.3.

10.1.3 Tricyclic Triterpenes

(a) **Malabaricol** The first tricyclic triterpene, malabaricol, has been isolated by Sukh Dev [18] from *Ailanthus malabarica* (Simaroubaceae). Malabaricol is proposed to be biogenetically formed from 2,3S-oxidosqualene as shown in Fig. 10.4. The stereochemistry of (+)-malabaricol has been settled by correlating it with (+)-ambreinolide of known structure and stereochemistry [19]. Malabaricanediol, a precursor of malabaricol, is structurally consistent with cyclization from 2,3-epoxy-18,19-dihydrosqualene-18,19-diol.

Later, Bohlmann et al. isolated malabarica-14(27),17,21-trien-3-ol (**B**) [20] and an aqua-quenched product (**C**) has been generated in the laboratory. Incidentally, it may be mentioned that isomalabaricane has been isolated from a sea sponge [21].

Sharpless [22] carried out the conversion of an epoxydiol of squalene to dl-malabaricane diol (7 % yield) using picric acid (Fig. 10.5). This conversion serves as the first examples of a nonenzymatic catalyzed conversion of squalene into a natural product. The conventional acids used ($SnCl_4$, $SnCl_3$, BF_3-etherate) for polyene cyclization affect not only epoxy group but also other

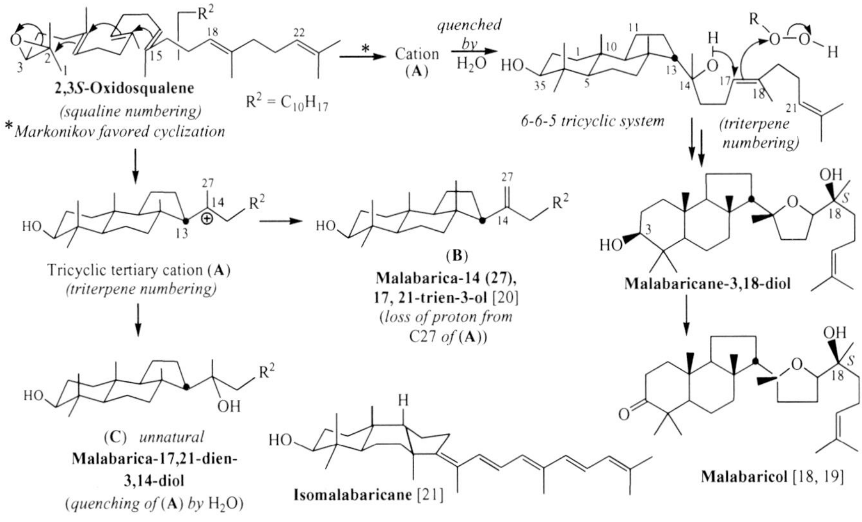

Fig. 10.4 Biosynthesis of malabaricol and related tricyclic triterpenes

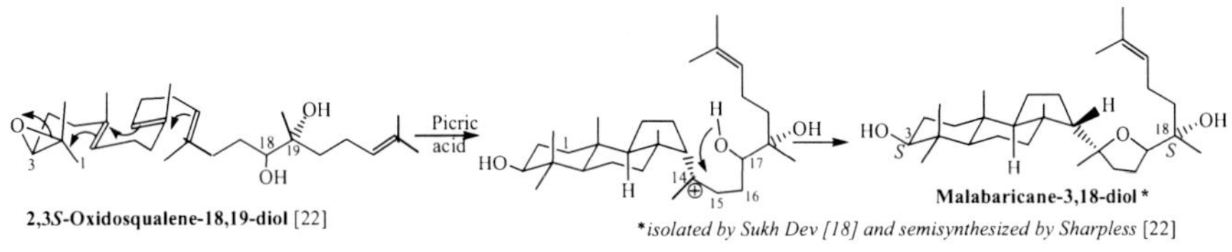

Fig. 10.5 Semisynthesis of malabaricane-3,18-diol

vulnerable centers yielding different cyclic products. The selection of picric acid, used for such cyclization, is the first case and is based on its acidity (nucleophilicity) as well as bulk which makes it convenient for the purpose. In nonenzymatic cyclization oxidosqualene gave thermodynamically favored tertiary carbocation with 6.6.5-fused tricyclic products (Markovnikov addition) as found in natural malabaricol [23]. Such a product has also been obtained from 18,19-dihydro-2,3-oxidosqualene [24].

Biogenetic pathways of some tricyclic triterpenes derived from squalene are delineated in Fig. 10.6.

(b) **Lansioside A** (Fig. 10.7) [27] isolated from the fruit peels of *Lansium domesticum* Jack v. duku (Meliaceae) is an aminosugar glycoside. The

Squalene
(folded in this manner by the specific enzyme)
$R^2 = C_{10}H_{17}$
or
migrations:
13α-H → 14α-H
8β-Me → 13β-Me
loss of a 27-proton
13β-14(27),17,21-malabaricatriene [25] (Fern)
13α-14(27), 17,21-Malabaricatriene [25] (Fern)
loss of 9α-H
loss of 7α-proton
8,17,21-Podiadatriene [26]
7,17,21-Podiodatriene [26] (Fern)

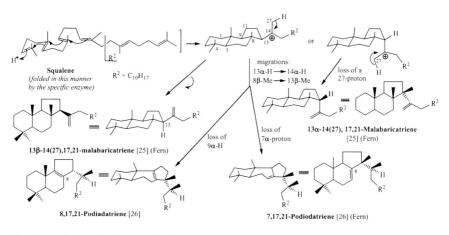

Fig. 10.6 Biogenesis of some tricyclic triterpenes derived from squalene

1) Regioselective reduction of Ring A C=O from Re face
2) Biological B–V oxidationof ring E C=O group
Steps (i) and (ii), stated below
Onoceradienedione (Ring C seco)
(a minor congener of lansic acid)
[See Figure.10.3]
Step (i) Glycosylation of ring A hydroxyl group with N-acetylated aminosugar
Step (ii) Hydrolysis of the lactonic group - generation of a COOH group and a hydroxypropyl group. From the latter elements of water are eliminated to generate a propylidene group (cf. Figure 10.3). However, the chronology of the sequence of reactions cannot be predicted.
Lansioside A

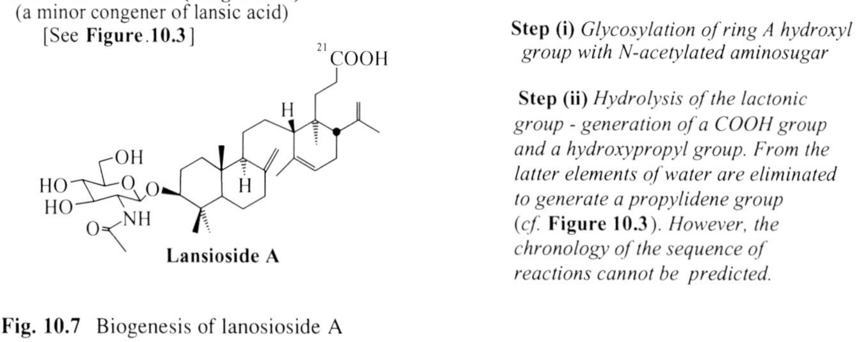

Fig. 10.7 Biogenesis of lanosioside A

aglycone part of it is a tricyclic triterpene having secoonocerane skeleton. Biogenetically it may be derived from α-onoceradione (Fig. 10.7)

Subsequently, two more tricyclic triterpenes, **lanosioside B** and **C**, have been isolated and their structures were established.

10.1.4   Tetracyclic Triterpenes. Substrates: (a) Oxidosqualene, (b) Squalene

(a) OSCs (*oxidosqualene cyclases*) can generate a large number of tetracyclic triterpenes from 2,3*S*-oxidosqualene (**1**) via two major intermediate carbocations (Fig. 10.8) as follows. (i) The 17β-*protosteryl carbocation* (**C**) is derived from (**1**) folded with rings A and C in *pro*-chair conformation and ring B in *pro*-boat conformation (**A**), prior to cyclization and 1,2 migrations, and biosynthesizes 6-6-6-5 tetracyclic triterpenes as shown in Fig. 10.9. The most well-known compounds that are formed from the carbocation (**C**) are lanosterol and cycloartenol. Lanosterol is responsible for the biosynthesis of cholesterol in mammals and ergosterol in fungi, while cycloartenol is responsible for the biosynthesis of phytosterols including cholesterol (Chap. 11). (ii) The 17β-*dammarenyl carbocation* (**D**) is derived from all *pro*-chair conformation (**B**) of 2,3*S*-oxidosqualene and yields a large variety of 6,6,6,5 tetracyclic triterpene alcohols from the 17β-dammarenyl cation, the detailed pathways for the bioformation of which have been outlined in Fig. 10.10. Further, a number of pentacyclic triterpenes are generated from the dammarenyl carbocation as stated in Sect. 10.1.5.

(b) **Cyclization of Squalene:** The enzymatic cyclization of squalene differs in several respects from that of oxidosqualene. The formation of all chair (A–B–C rings) (**F**) carbocation is the characteristic of squalene, while

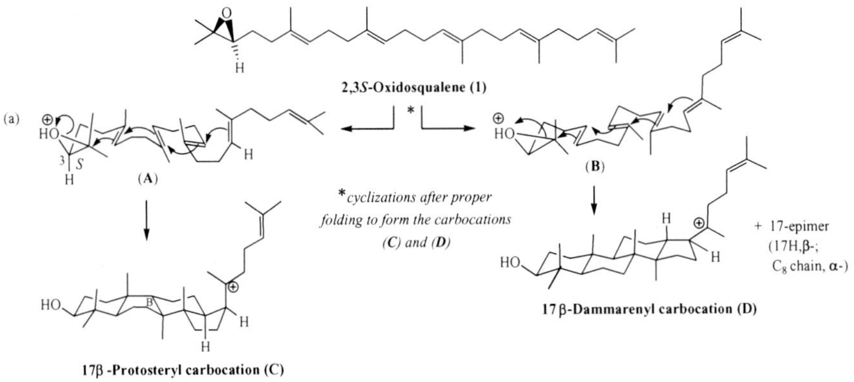

Fig. 10.8 Bioformation of tetracyclic cations (**C**) and (**D**) from 2.3*S*-oxidosqualene

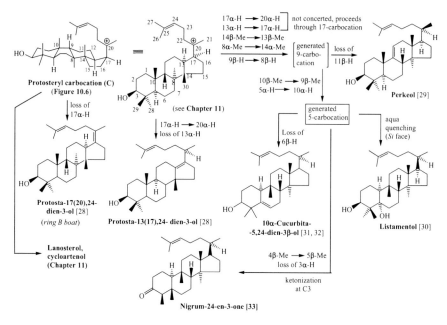

Fig. 10.9   Bioformation of 6-6-6-5 tetracyclic triterpenes derived from protosteryl carbocation (**C**)

oxidosqualene can form chair–boat–chair (A–B–C) cation (protosteryl carbocation) as well as chair–chair–chair (C–C–C) cation (dammarenyl carbocation) (Fig. 10.8) after cyclization [1, 2–4].

In case of tetracyclic triterpenes from squalene the latter will form the C–C–C all chair tetracyclic carbocation (**F**), i.e., the deoxy analog of dammarenyl cation. The carbocation (**F**) gives rise to some 6-6-6-6 or 6-6-6-5 tetracyclic triterpenes (Fig. 10.11) A particular way of folding of squalene and cyclization giving rise to 8(26),14(17)-onoceradiene, a C2-symmetric tetracyclic triterpene is also shown in Fig. 10.11.

10.1.5   Pentacyclic Triterpenes

Pentacyclic triterpenes are biosynthesized from the dammarenyl carbocation (**D**) (Fig. 10.8) or protosteryl carbocation (**C**). Formation of various isomeric carbocations due to 1,2-hydride and/or methyl migrations and ring D expansion and subsequent quenching of the carbocations through the loss of adequate protons or quenching by hydroxyl equivalent (e.g., H$_2$O) lead to the formation of a plethora of pentacyclic triterpenoids of diverse skeletal patterns. A few common compounds have been taken as examples and their biosynthetic pathways derived from lupanyl

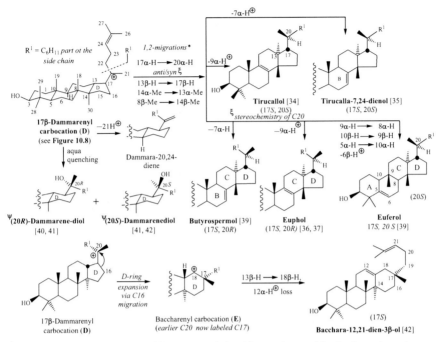

Migrations are always suprafacial. In a cyclohexane system only the axial group migrates and the migration terminus undergoes inversion of configuration, if an axial group from that center also migrates.

ᵠ*C20(R) and C20(S) products are formed by quenching the carbocation (D) (planar at C20) with the side-chain R' anti to the C17-C13 bond) with an hydroxide equivalent from the lower (α-) face and upper (β- face) respectively. They may also be formed by attack of the planar carbocation with R' syn to C17-C13 bond from the upper face and the lower face respectively.*

ξ *1,2-Migration of 17α-H to the α-face of the C20 carbocation in the rotamer with R' anti to the C17-C13 bond; followed by the 1,2-migrations and deprotonation shown produces tirucallol, or euferol with (20S)-configuration, whereas when the rotamer with R' syn to the C17-C13 bond is involved the final product is butyrospermol or euphol with (20R)-configuration.*

Fig. 10.10 Some tetracyclic 6-6-6-5 and 6-6-6-6 triterpenes derived from the dammarenyl cation (**D**) and bachharenyl cation (**E**)

cation (**G**) (Fig. 10.12), 18α-oleanyl cation (**H**) (Fig. 10.13), and hopyl cation (**J**) (Fig. 10.14) are delineated .

The pathway shown for the biosynthesis of eupacannol (Fig. 10.13) from bauerenol is a speculative mechanistic possibility, explaining the axial β-hydroxylation at C7. All the anti-migrations following the migration of 8β-methyl to 14-β-methyl may be continued without the involvement of bauerenol, hydroxylation taking place at any stage, even before cyclization of squalene epoxide. *Alternatively, eupacannol might be biosynthesized from squalene* (rather than from squalene epoxide), through the intermediacy of 3-deoxy-α-amyrinyl-8-cation (= deoxy-J cation). The latter undergoes migration of 14α-Me → 13α-Me, 8β-Me → 14β-Me and loss of 7α-proton to form 3-deoxybaurenol, followed by enzymatic β-axial hydroxylation at C7, and a series of methyl and hydride transfers in the stereospecific trans manner (cf. Fig. 10.13), and loss of C3 proton, and finally

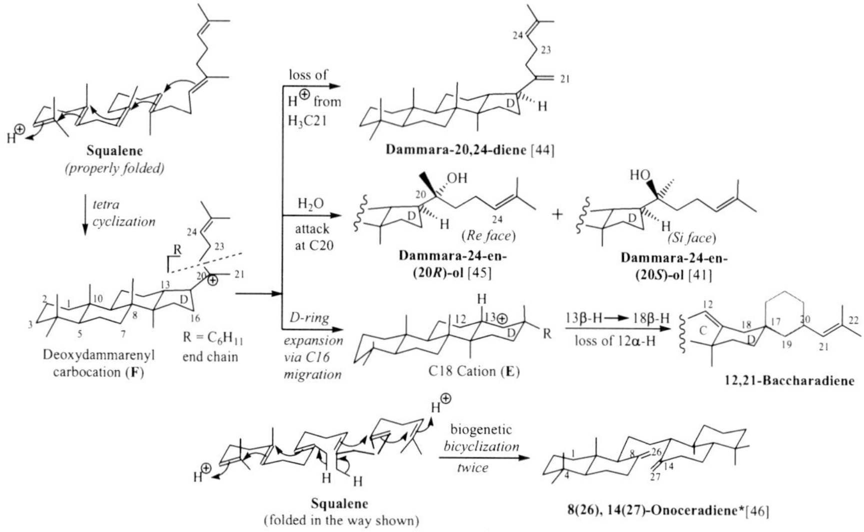

Since squalene is symmetrical, acid catalyzed cyclization may start from either end in succession or from both ends simultaneously

Fig. 10.11 Bioformation of some 6-6-6-5 and 6-6-6-6 tetracyclic and a C_2-symmetric 6-6-$(CH_2)_2$-6-6 tetracyclic triterpenes from squalene

enzymatic reduction of the generated C3 = C4 double bond to push C4-Me to β-equatorial position (For molecular conformation see Fig. 10.33). The allylic hydroxylation at C9 in squalene itself, prior to its cyclization, might also take place, catalyzed by specific enzymes [51]. The substrate specificity of most enzymes still remains to be defined.

10.1.6 Cyclization of Bis-Oxidosqualene

The oxidation state of several triterpenoids indicates that they arise from a bis-oxidosqualene substrate. Two general modes of bis-oxidosqualene cyclizations that have been established experimentally are shown in Fig. 10.15. Squalene is epoxidized by squalene epoxidase to 2,3(*S*)-oxidosqualene, the cyclization modes of which have been outlined in Fig. 10.8. Under some conditions, when lanosterol synthase is present in restricted amount, the distal terminus of 2,3(*S*)-oxidosqualene can enter into reepoxidation by squalene epoxidase to form 2,3(*S*)-22(*S*),23-bisoxidosqualene (Fig. 10.15). The latter under the influence of different specific enzymes undergoes the conversions to a series of varied types of tetracyclic and pentacyclic triterpenes. The bioformations of a few such triterpenes, for example, reissantenol [57], 24,25(*S*)-epoxycycloartan-3(*S*)-ol [58, 59], α-onocerin [60], and serratenediol [61], derived from 2,3(*S*)-22(*S*),23-bisoxidosqulene, are delineated in Fig. 10.15.

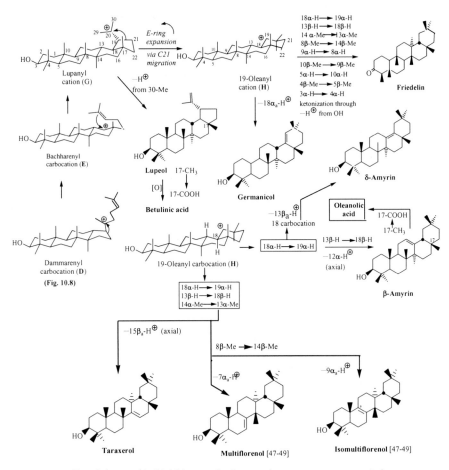

Fig. 10.12 Biosynthesis of lupeol (a 6-6-6-6-5 pentacyclic triterpene) and of some common 6-6-6-6-6 pentacyclic triterpenes derived from dammarenyl cation (**D**) *via* lupanyl cation (**G**)

A tetracyclic triterpene 3β-cabraleadiol [62] containing a substituted tetrahydro-furan moiety linked at 17β position would originate from 2,3-(*S*)-22-(*S*),23-bis-oxidosqualene following the probable biogenetic route depicted in Fig. 10.16 .

10.1.7   *Sesquiterpene–Nortriterpene Adduct (C$_{44}$)*

Cheiloclines A-I, nine octacylic C$_{44}$-terpenoids reported from the root bark of *Cheiloclinium hippocratoides* (Fam. Celastraceae) [63], represent first examples

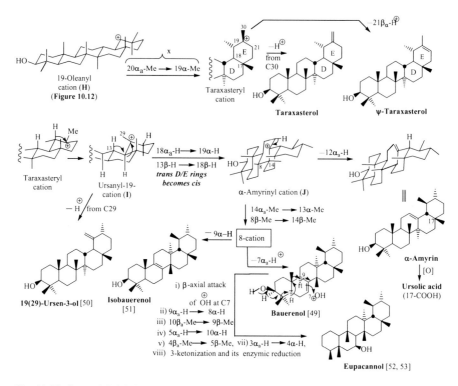

Fig. 10.13 Some 6-6-6-6-6 pentacyclic triterpenes derived from 19-oleanyl cation (**H**)

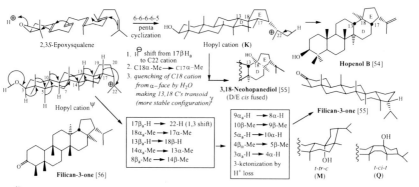

$^{\gamma}$ The generated **trans-transoid-cis** configuration (**M**) of C/D/E rings is much more stable than their **trans-cisoid-trans** configuration (**Q**), (in which ring D attains. true boat conformation), resulting from quenching by H$_2$O from the β-face (see also Chapter 2).

$^{\Psi}$ The hopyl cation undergoes a 1-3H$^{\ominus}$ shift followed by a series of suprafacial1,2 anti migration and finally 3-ketonization occurs. to produce filican-3-one.

Fig. 10.14 Biosynthesis of 6-6-6-6-5 pentacyclic triterpenes derived from 2,3S-epoxy squalene *via* hopyl cation (**K**)

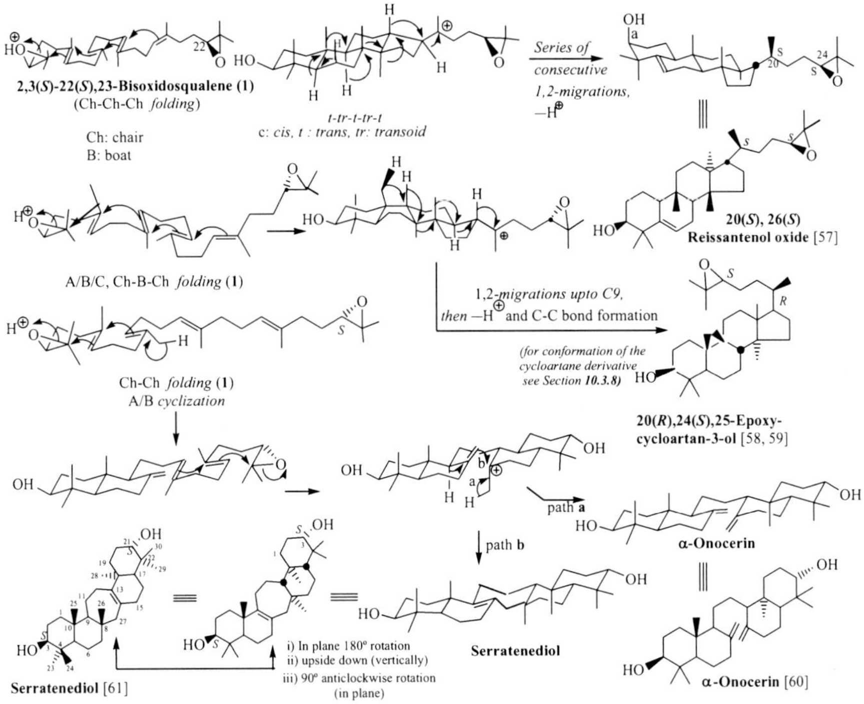

Fig. 10.15 Bioformation of tetracyclic and pentacyclic triterpenes derived from 2,3(S)-22(S),23-bisoxidosqualene

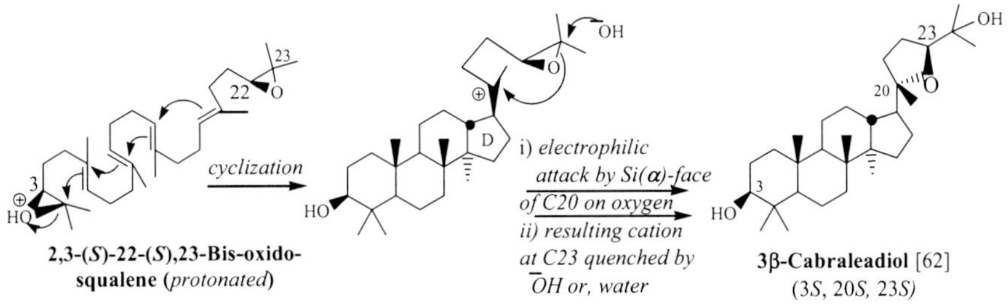

Fig. 10.16 Biosynthesis of 3β-cabraleadiol

of hetero-Diels–Alder adducts between a nortriterpenequinone and a sesquiterpene. Their structures were elucidated on the basis of COSY, ROESY, HSQC, and HMBC NMR experiments. They are composed of one aromatic triterpene unit derived from pristimerin (D:A-friedo-nor-oleanane skeleton) and one sesquiterpene unit derived from guaia-1(5),3(4), 11(12)-triene (Fig. 10.17). One such terpenoid, cheilocline A, is shown in this figure.

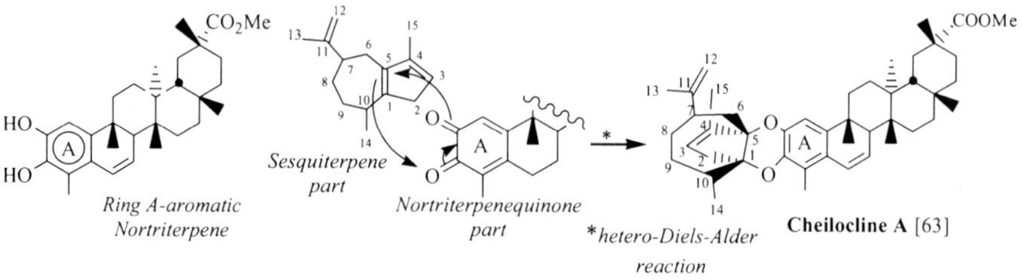

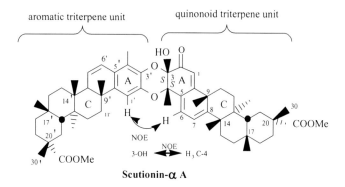

Fig. 10.17 Possible biogenetic formation of cheilocline A

aromatic triterpene unit   quinonoid triterpene unit

Scutionin-α A

Fig. 10.18 Structure of a triterpene dimer scutionin-α A

10.1.8   Triterpene Dimers (C$_{60}$) and Triterpene Trimers

Triterpene dimers and triterpene trimers are also known to occur in nature. Seven dimeric compounds have been isolated [64] from the root bark of the subtropical shrub *Maytenus scutioides* (Celastraceae), distributed in North of Argentina and South of Paraguay and Bolivia. They were found to be composed of one quinoid-type triterpene, derived from co-occurring pristimerin or 7,8-dihydropristimerin and one aromatic triterpene, derived from pristimerin or 6-hydroxypristimerin linked together by two ether linkages between the two A rings. One of them scutionin-α A is shown in Fig. 10.18. The other triterpene dimers are 7,8-dihydroscutionin-αB, 7,8-dihydroscutionine-βB, scutidin-αA, the structures and absolute configurations of which have been elucidated [64].

González and his group [65] and Itokawa and his collaborators [66] have isolated more triterpene dimers and González et al. [67] have also reported triterpene trimers from Celastraceae plants.

10.2   Squalene, the Universal Precursor of Triterpenoids and Steroids

10.2.1   Occurrence. Biogenesis

Squalene, $C_{30}H_{50}$, has been named after its first source of isolation—the shark (*Squalus* species) liver oil. Later, it has been found to be ubiquitously distributed. It has been biosynthesized from two molecules of all *trans*-FPP. The puzzling mechanism of this coupling involves the alkylation of C2–C3 bond of one molecule of FPP by the second molecule of FPP and the formation of a cyclopropane intermediate (*presqualene*) followed by its opening, and the carbocation thus formed receives a hydride from NADPH (labeling experiment) serving as the H_R of the central methylene group of squalene. Its biogenesis has been exemplified in a sequence of steps in Chap. 5 (Fig. 5.20). Its structure has been settled in the conventional way by degradation and synthesis.

10.2.2   Synthesis of Squalene [68]

Cornforth applied their method of highly stereoselective olefin synthesis [69] to the synthesis of squalene in which no separation of geometrical isomers was necessary before the final step.

The C_{30} carbon skeleton is constructed according to the carbon skeletal linkage of the following three units.

$C_{11} + C_8 + C_{11} = C_{30}$ (Fig. 10.19).

Many other syntheses of squalene have been reported later.

10.3   β-Amyrin

10.3.1   Occurrence. Structural Elucidation

β-Amyrin, the parent compound of the oleanane family, is the major component of Manila elemi resin and is quite prevalent in higher plants. It sometimes occurs as the major constituent as a single triterpene or in a mixture, either in the free state or as its ester like acetate, palmitate, etc. It was isolated from the latex of rubber trees and from *Erythroxylum coca* (Erythroxylaceae) [70]. Its structure has been decided on the basis of its reactions and degradation studies, some of which are delineated in Fig. 10.20. The reaction products have been identified and the elimination of other possibilities of the location of the double bond and the generated CO group and carboxylic groups during oxidative studies [71, 72] are not discussed. The

(i) *Synthesis of the terminal C$_{11}$-unit (g)*

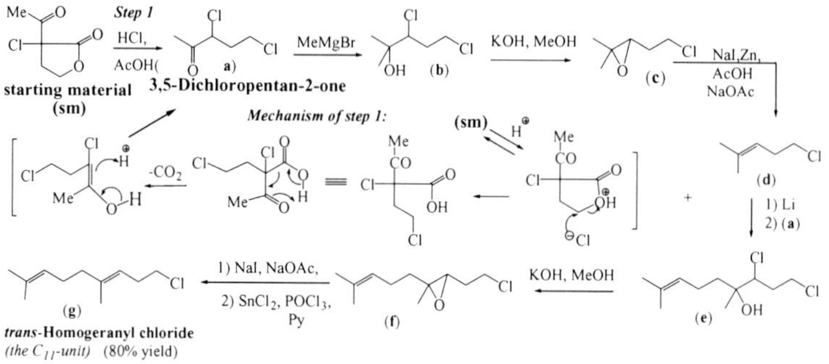

starting material (sm)   3,5-Dichloropentan-2-one

Mechanism of step 1:

(g)
trans-Homogeranyl chloride
(the C$_{11}$-unit)   (80% yield)

Mechanism of some steps:

The addition of the lithium alkyl [from (d)] to (a) takes place on the rotamer in which two vicinal dipolar groups
>C$^{\delta+}$=O$^{\delta}$ and >C$^{\delta'}$-Cl$^{\delta}$ are as far as possible and antiparallel and the addition follows Cram's rule. Now, this
addition can be better explained by Felkin model, as shown below.

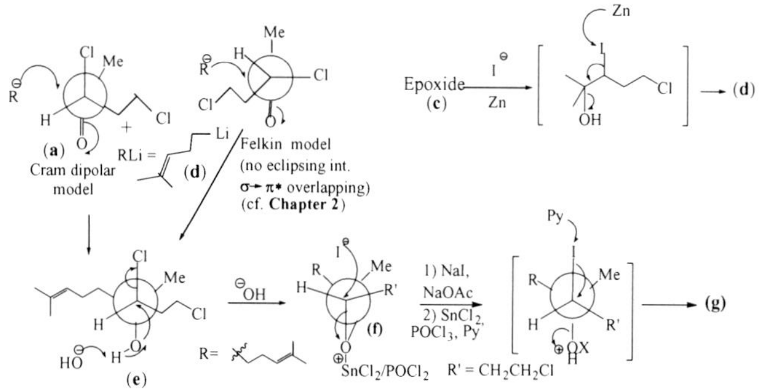

This sequence of reactions constitutes a general stereoselective synthesis of trisubstituted olefins.

(ii) *Synthesis of the central C$_{8}$-unit*

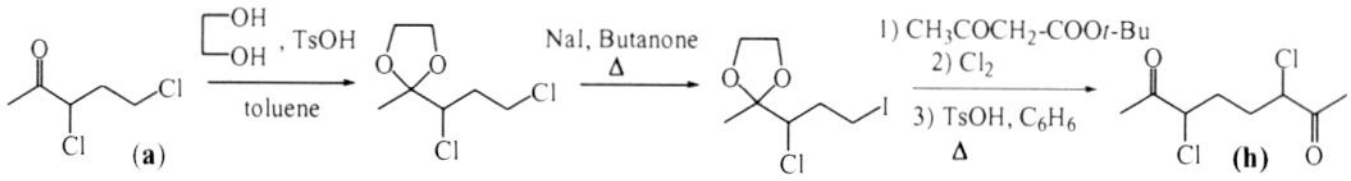

(iii) *Building of the C$_{30}$-chain*

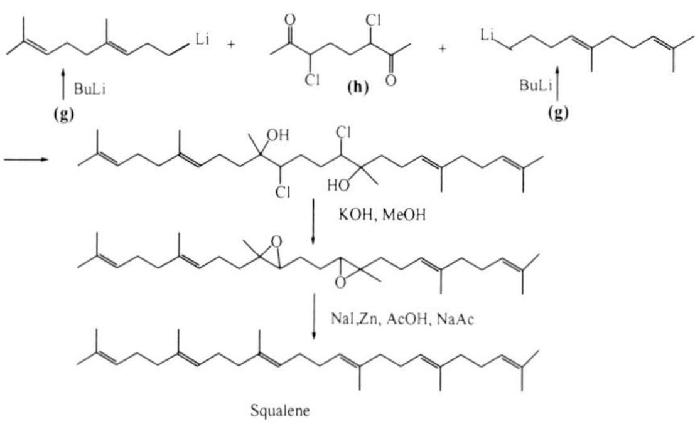

Squalene

Fig. 10.19   First synthesis of squalene by Cornforth [68] (1959)

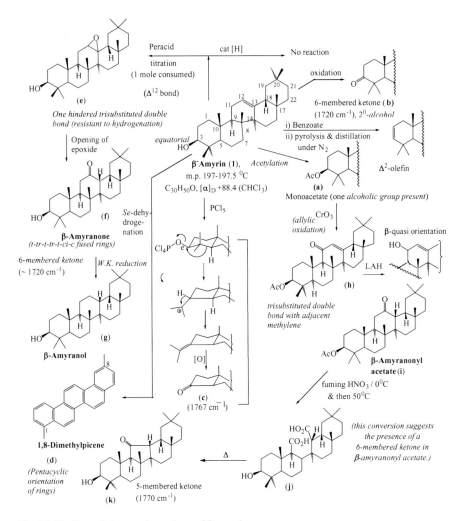

Fig. 10.20 Degradations and reactions of β-amyrin

formation and structures of the products which led to the correct structure of β-amyrin [71, 72] are exemplified in the figure.

Formation of (**a**) and (**b**) indicates the presence of a secondary OH group. Formation of (**c**) and (**d**) indicates that the secondary OH group is equatorial and is in the ring A of a perhydropicene skeleton.

It is resistant to hydrogenation; it forms an epoxide (**e**), which opens up to a ketone (**f**) and the latter could be reduced to (**g**). Further, (**a**) could be oxidized to an enone (**h**).

All these compounds suggest the presence of one hindered trisubstituted double bond with an adjacent methylene group. The compound (**i**) [≡ (**f**)-acetate] could be oxidized to a dicarboxylic acid (**j**) which on heating gives a five-membered ketone (**k**). These indicate the presence of a six-membered ketone in (**i**).

10.3.2   Stereochemistry

Stereochemistry at C17 and C18. β-Amyrin has perhydropicene skeleton with a trisubstituted double bond in ring C. The relation of β-amyrin with oleanolic acid is shown in Fig. 10.21, which elucidates the stereochemistry at C17 and C18 [73].

Barton [74] has elegantly shown that C17 COOH in oleanolic acid is β-axial in ring D, and D/E rings are *cis*-fused as is evident from the following facts: the C17-CO and C13-O bonds in the lactone (**A**) are in 1,3-diaxial relationship in ring D (Fig. 10.21). However, opening of the lactone ring to form the thermodynamically more stable (**B**) having the 13,18 double bond is not possible due to the presence of C18–H and C13–O bonds (*e* and *a*, respectively, in ring D) in *cis* orientation. Thus, C18-H and C17-COOH in oleanolic acid must be in *e* and *a* orientation in ring D. This experiment showed that the D/E ring junction in β-amyrin and its derivatives is *cis*-fused. Had this ring junction been *trans* [(having the conformation (**C**)] (Fig. 10.22), opening of the lactone ring to form the $\Delta^{13(18)}$ compound (**B**) would have been stereochemically feasible (Fig. 10.22).

The stereochemistry at C18 and C9 has been deduced by base-catalyzed isomerization studies of methyl 11-oxo-oleanolate 3-acetate (**D**) (Fig. 10.22) [obtained by CrO$_3$ oxidation of acetyl oleanolic acid]. Epimerization at either or both the centers C9 and C18 of (**D**) (being activated by keto or conjugated keto group at C11) is possible. That the configuration at C9 remains unchanged has been decided by converting the epimeric product (**E**) to the 11,13(18)-diene (**F**) of

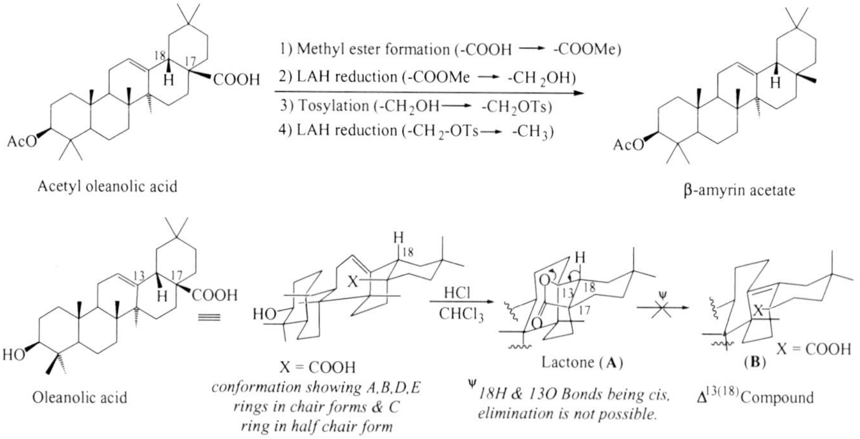

Fig. 10.21 Conversion of acetyl oleanolic acid to β-amyrin acetate and formation of the lactone from oleanolic acid

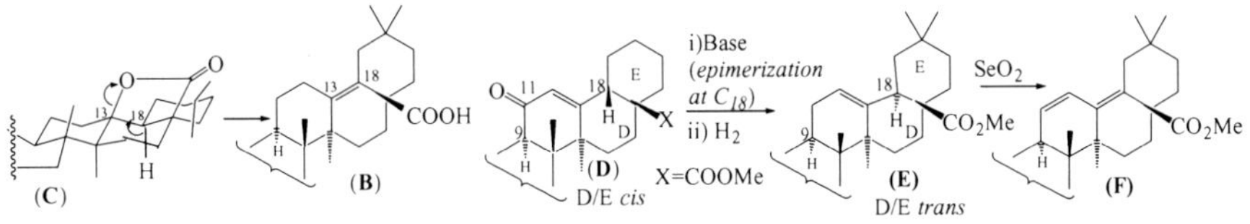

Fig. 10.22 Elucidation of stereochemistry at C18 and C9

known structure, in which stereochemistry at C18 is destroyed. Its C9 configuration remains unchanged indicating that B/C rings are *trans*-fused, which cannot be changed to the less stable *cis* fusion.

The configurations of A/B and B/C ring junctures have been settled by correlating oleanolic acid with other compounds of oleanane series, and by consideration of the stability of various perhydrophenanthrene diastereomers, of which the *trans-transoid-trans* is the most stable one (*vide* Chap. 2). In oleanane series the A–B–C system possesses the *trans-transoid-trans* stereochemistry. The synthesis of (**A**), the left-hand part precursor for the synthesis of β-amyrin from ambreinolide of known structure and stereochemistry (Fig. 10.26), establishes the stereochemistry of the A/B rings as *trans*. The absolute stereochemistries of β-amyrin and other triterpenes of oleanane series follow from their biosyntheses as described in Figs. 10.8 and 10.12.

10.3.3 Spectral Data of β-Amyrin

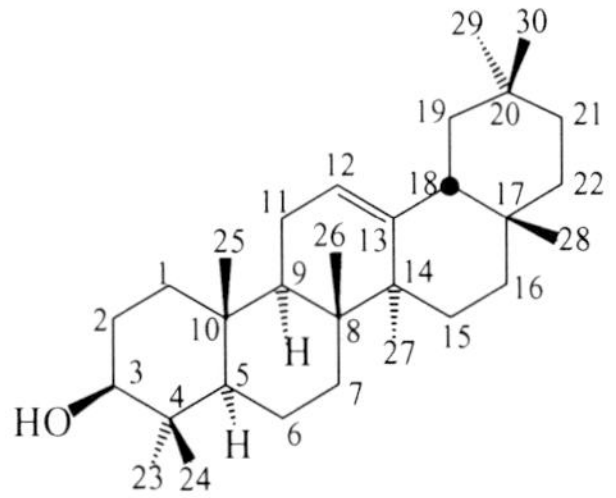

IR (KBr): 3350 (OH), 1392, 1368, 1140, 1048 cm^{-1}.

^{1}**H NMR** (CDCl$_3$; 400 MHz, δ) [70] 5.18 (t, J = 3.2 Hz, 1H), 3.27-3.18 (*m*, 1H), 1.13 (*s*, 3H), 1.00 (*s*, 3H), 0.96 (*s*, 3H), 0.93 (*s*, 3H), 0.86 (*s*, 3H), 0.83 (*s*, 3H), 0.79 (*s*, 3H), and very complex methylene envelope.

^{13}**C NMR:** (CDCl$_3$; 400 MHz, δ) C1 (38.5), C2 (27.0), C3 (78.9), C4 (38.7), C5 (55.1), C6 (18.3), C7 (32.7), C8 (39.7), C9 (47.6), C10 (37.0), C11 (23.5), C12 (121.9), C13 (145), C14 (41.7), C15 (26.2), C16 (27.3), C17 (32.5), C18 (47.2), C19 (46.8), C20 (31.1), C21 (34.8), C22 (37.2), C23 (28.3), C24 (15.4), C25 (15.5), C26 (16.8), C27 (26.0), C28 (28.8), C29 (33.2), C30 (23.6).

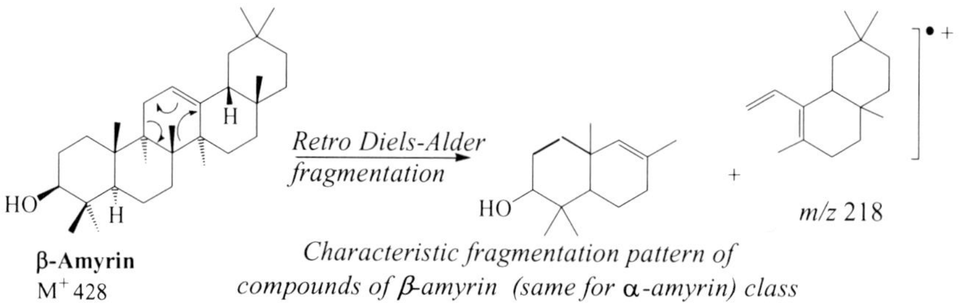

Fig. 10.23 Main mass fragmentation of β-amyrin

Mass Spectrometry

The characteristic mass fragmentation pattern [74] of β-amyrin is shown in Fig. 10.23.

10.3.4 Synthesis of β-Amyrin

Total synthesis of pentacyclic triterpenes with ring arrangement like β-amyrin by stepwise ring extension (linear synthesis) could offer great difficulty specially for having two adjacent quaternary carbon atoms at C8 and C14. Halsall and Thomas [75] suggested that this difficulty could be overcome, and a pentacyclic triterpene could be synthesized by constructing a suitable tetracyclic system with A,B and D,E rings, followed by ring closure to generate ring C. This suggestion received support from the acid-catalyzed cyclization of α-onocerin to γ-onocerin by Barton and Overton (Fig. 10.24).

Based on the above strategy Corey synthesized [76] a β-amyrin derivative, olean-11,13(18)-diene via cation–olefin cyclization of a tetracyclic intermediate containing A,B and D,E rings. During the preparation of the pentacyclic interme-diate a precursor containing the A and B rings is joined to a monocyclic interme-diate with a preformed E ring, and ring D is generated by an internal aldol condensation. Finally ring C is constructed by internal cation–olefin cyclization (Fig. 10.25). The two precursors (**A**) and (**B**) of the oleanane skeleton are shown in Fig. 10.25.

The pentacyclic skeleton was later on converted to β-amyrin by Barton [78]. Thus composite efforts of two groups of workers led to the total synthesis β-amyrin. The objective of Corey's work was to test the possibility of construction of a pentacyclic nucleus (perhydropicene) by cyclizing a tetracyclic intermediate with built-in rings A, B, D, and E via cation–olefin cyclization when ring C is formed (Fig. 10.26).

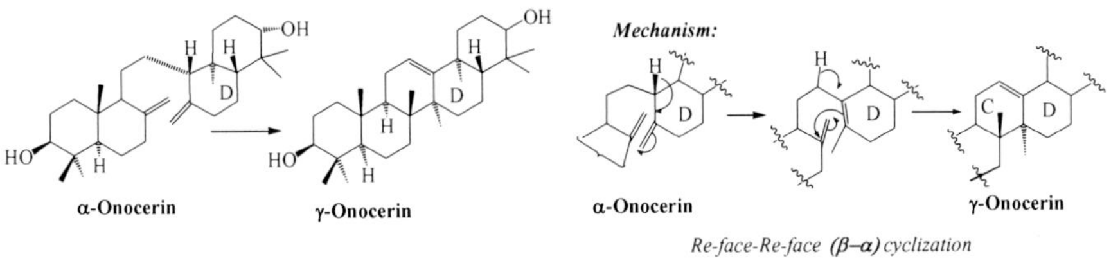

Fig. 10.24 Acid-catalyzed cyclization of α-onocerin to γ-onocerin to form the ring-C

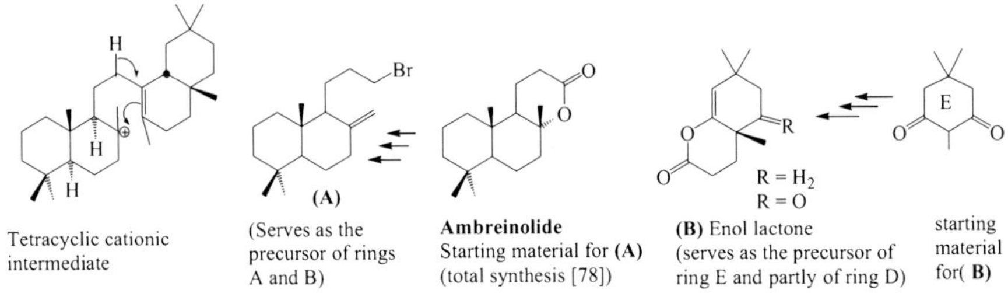

Fig. 10.25 Strategy for the synthesis of β-amyrin

10.3.5 Formal Syntheses of β-Amyrin by Polyene Cyclization

In 1972 van Tamelen has shown that stannic chloride-nitromethane-induced cyclization of an epoxide (A) leads to δ-amyrin in 4 % yield [79]. The latter has been converted to 18α-olean-12-ene [B] by Brownlie and coworkers [80]. Compound [B] has been converted to β-amyrin by Barton et al. [78] (Fig. 10.27). Thus van Tamelen's synthesis of δ-amyrin constitutes the formal synthesis of β-amyrin.

Tori and coworkers reported [81] the acid-catalyzed backbone rearrangement of 3β,4β-epoxyfriedelane (C) to β-amyrin (Fig. 10.28), accompanied by other products. Friedelin has been synthesized by Ireland [82] and Kametani [83]. 3β,4-β-Epoxyfriedelane (C) has been derived from friedelin. Hence Tori's conversion constitutes another formal synthesis of β-amyrin. β-Amyrin has also been obtained as a backbone rearrangement product of taraxerol and multiflorenol (Fig. 10.28) [84].

10.3.6 Johnson's Total Synthesis of β-Amyrin by Polyene Cyclization

In 1993 W.S. Johnson et al. reported [70] a total synthesis of dl-β-amyrin (Fig. 10.29) in ~0.2 % overall yield by biomimetic polyene cyclization using fluorine atom as a cation-stabilizing (C-S) auxiliary. An appropriate polyene (D)

(i) Synthesis of (A)

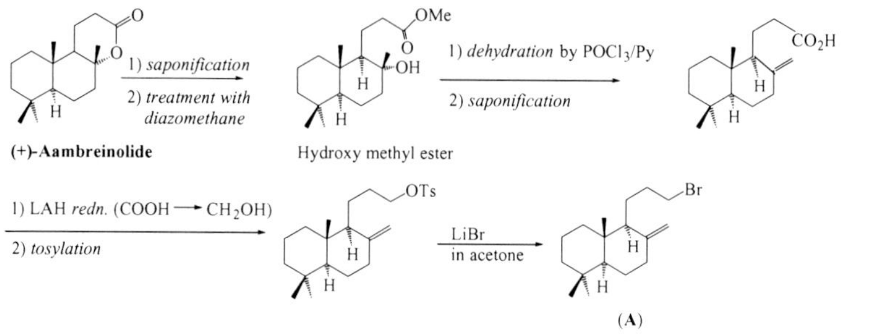

(+)-**Aambreinolide**   Hydroxy methyl ester

(ii) Synthesis of (B)

2,5,5-**Trimethyl-cyclohexane-1,3-dione**

(±)-**Enol lactone**   **(B)**

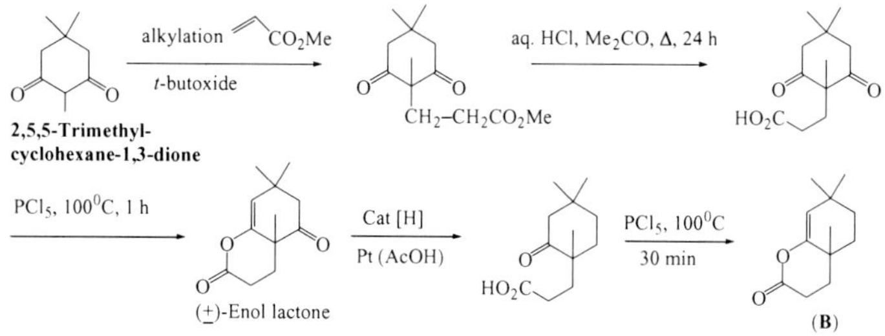

(iii) Coupling of (A) and (B), two precursors of oleanane ring system

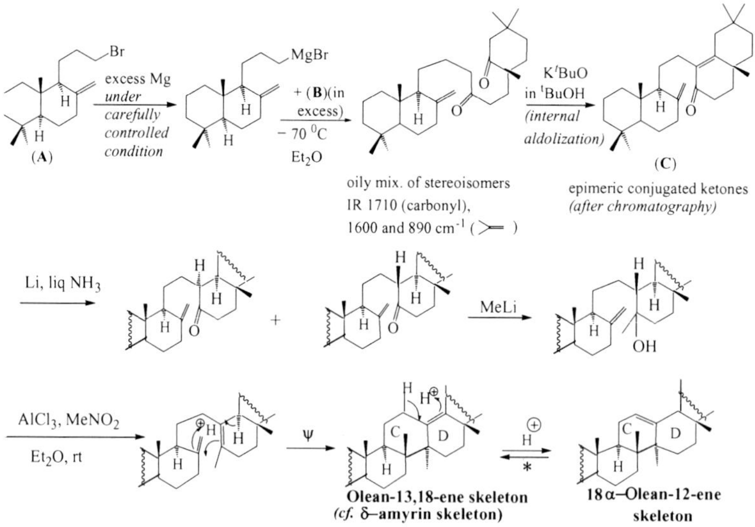

(A)

oily mix. of stereoisomers
IR 1710 (carbonyl),
1600 and 890 cm^{-1} (⟩═)

epimeric conjugated ketones
(after chromatography)

(C)

Olean-13,18-ene skeleton
(cf. δ–amyrin skeleton)

18α–Olean-12-ene
skeleton

Notes: Ψ *Direction of cation-olefin cyclization is important. In principle, the ring C can be formed*
by addition of AB ring cation to DE ring unsaturation.

* *This acid catalyzed equilibrium leans more towards tetra-substituted 13-18 double band.*

Fig. 10.26 (continued)

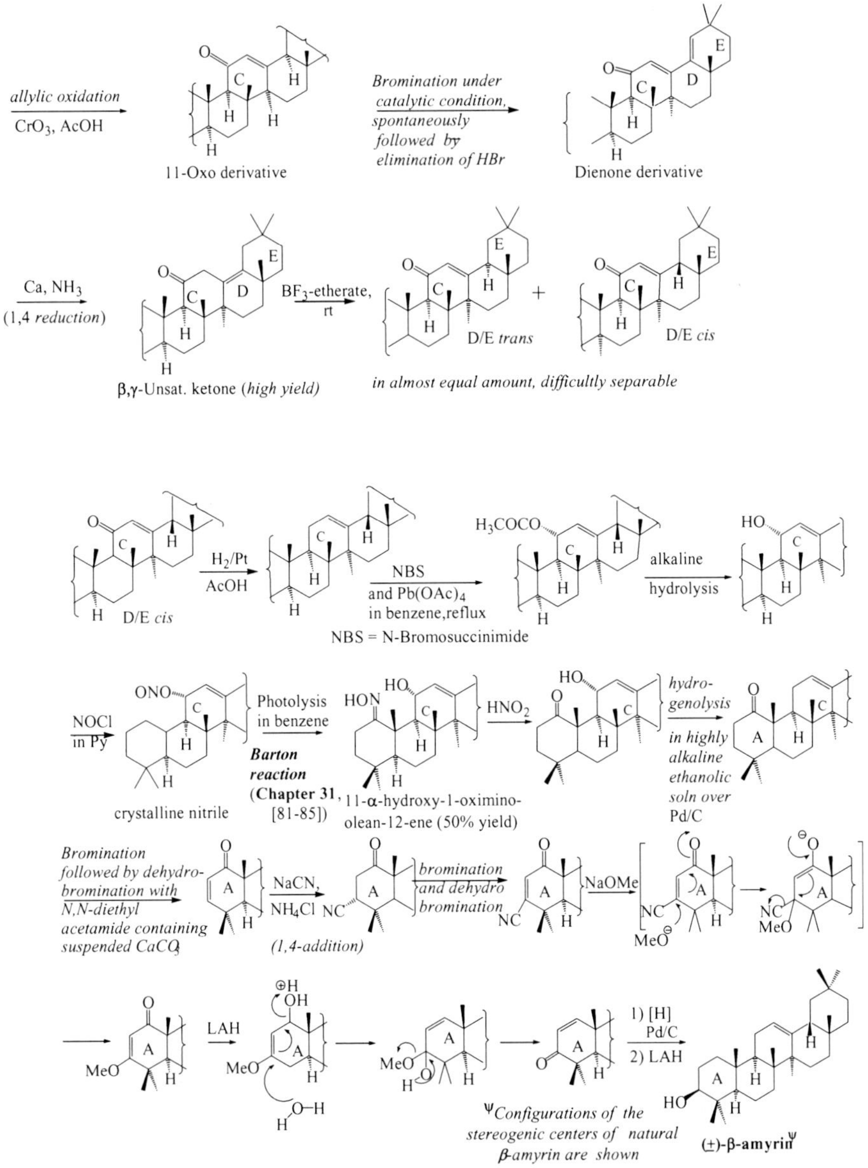

Fig. 10.26 Synthesis of (±)-β-amyrin

with tetrasubstituted *trans* fluoroalkene bond (at pro-C13) and having 3-*trans*, 7-*trans*,11-*cis*-alkene (alkene nomenclature) stereochemistry and a propargylsilane terminating group attached to a pentenoid system has been synthesized [70]. The latter was cyclized in CF$_3$COOH taken in CH$_2$Cl$_2$ at −70 °C to give a pentacyclic system (**E**) containing six out of the eight stereogenic centers of β-amyrin with

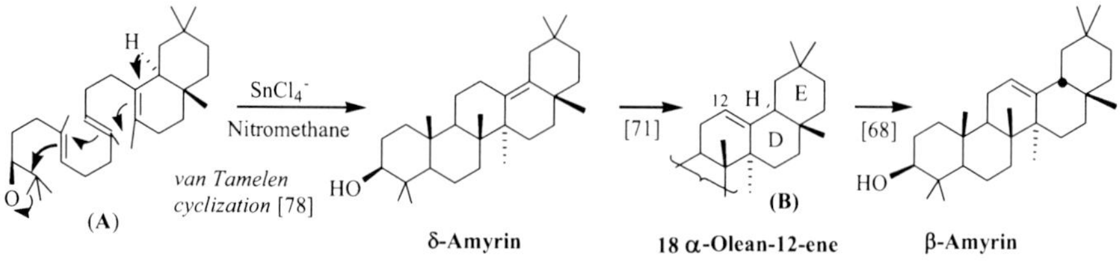

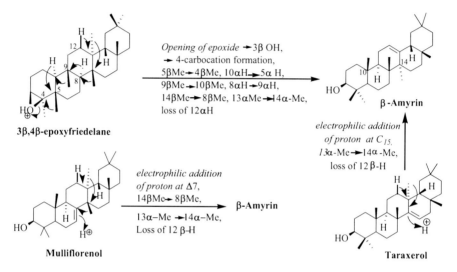

Fig. 10.27 Formal synthesis of β-amyrin via δ-amyrin by polyene cyclization

Fig. 10.28 Acid-catalyzed conversions of 3β,4β-epoxyfriedelane, taraxerol, and multiflorenol to β-amyrin

correct relative stereochemistry. Compound (**E**) was subjected to a number of steps delineated in Fig. 10.29 to finally give *dl*-β-amyrin [70].

10.3.7   *3-Deoxy-β-Amyrin by Backbone Rearrangement of 3β-Fridelanol*

Friedelin, a commonly occurring triterpene ketone on reduction, produces 3β-friedelanol. The latter on acid treatment produces 3-deoxy-β-amyrin. The spectacular backbone rearrangement is triggered by elimination of –OH as H$_2$O and is followed by a series of concerted stereospecific migration of axial H or CH$_3$ with electron pair on the same face as the migrating species and *anti* to the species migrating from that terminus, as delineated in Fig. 10.30.

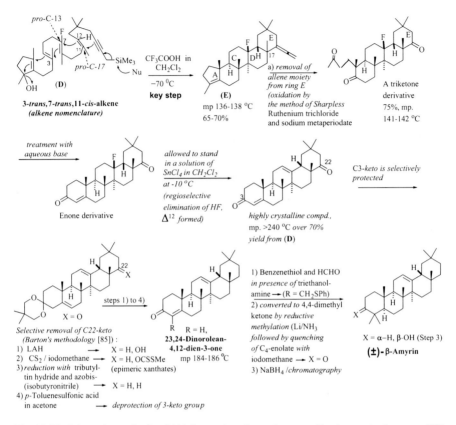

Fig. 10.29 Johnson's synthesis of (±)-β-amyrin using polyene cyclization as the key step [70] (1993)

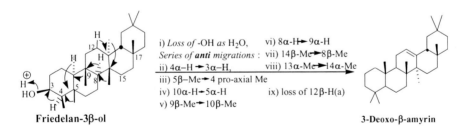

Fig. 10.30 A spectacular backbone rearrangement of friedelan-3β-ol to 3-deoxy-β-amyrin

10.3.8 Biosynthesis

For biosynthesis of β-amyrin see Fig. 10.12.

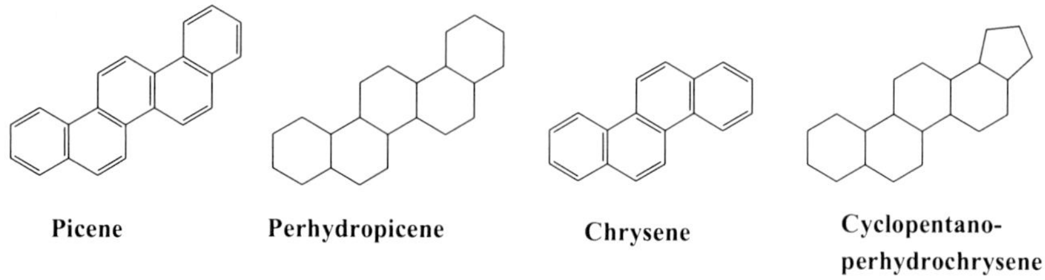

| Picene | Perhydropicene | Chrysene | Cyclopentano-perhydrochrysene |

Fig. 10.31 Picene, chrysene, and their perhydroderivatives

10.4   Analysis of Molecular Conformations of Some Common Pentacyclic Triterpenes

The beauty of the polycyclic terpenoid biosynthesis lies in the conversion of a single substrate into different skeletal patterns under the influence of different specific cyclases. In cases of triterpenoids the individual substrates are squalene, 2,3(S)-oxidosqualene, and 2,3(S),22(S),23-bisoxidosqualene. Each of them cyclizes to yield a large number of monocyclic and/or polycyclic triterpenoids of different skeletal patterns (Sects. 10.1–10.3) under the influence of specific enzymes. The basic carbocyclic rings are formed from a substrate undergoing a particular folding by a single enzyme-controlled asymmetric reaction. *The final configurations of the stereogenic centers are dictated by the extent of the antiperiplanar 1,2-hydride/alkyl shifts (migrations) and quenching of the terminal carbocation thus formed on the way towards the final product formation.*

The common fused ring systems with the pentacyclic triterpene framework belong to 6-6-6-6-6-fused cyclohexanes in the array of perhydropicene and to 6-6-6-6-5-fused cyclohexanes with a cyclopentane—in the way of cyclopentano-perhydrochrysene (Fig. 10.31).

Molecular conformations of some common pentacyclic triterpenes will be illustrated by reference of the above carbocyclic systems as revealed by careful examination of their Dreiding molecular models. The rings attain chair, half-chair, boat, or twist-boat conformations depending on the nature of ring fusion, presence or absence of endocyclic sp^2 carbons (as C=C), nonbonded interactions through space, steric crowding, etc. The preferred molecular conformation is the one which is energetically most comfortable.

10.4.1   Conformation of β-Amyrin

The molecular conformation of β-amyrin (**A**) having A/B, B/C, and C/D *trans*-fused and D/E *cis*-fused (steroid conformation) may be expressed as **A**^{1} or **A**^{2} or **A**^{3}. In **A**^{1} β-bonds are going up and α-bonds are going down (Fig. 10.32). These conformations are equivalent. The conformation **A**^{1} is first obtained by constructing

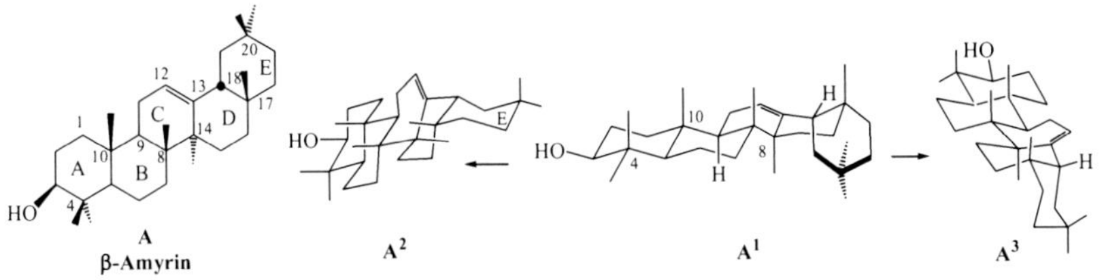

A[1]: The molecular model of **A** is held horizontally at the eye-level of the viewer in such a way that looking at the upper face of the molecule in a slanting manner (tangentially) the ring A, B, C, and D appear as chair, chair, half-chair, and chair (little distorted because of the sp[2] carbon at 13. In this perspective E ring will be viewed from the lower face.

A[2]:   A[1] is held at the eye-level, looking from the side of the ring E rotated 90° anticlockwise about the puckered plane of the molecule and viewed tangentially on the upper face in a slightly slanting position of the model to get the conformation as in **A**[2]; the ring E also appears as a chair form.

A[3]:   If A[1] is rotated about 60° clockwise in the puckered plane of the molecule (which looking from the side of the ring E) and viewed in a slanting manner (tangentially) on the upper face, the molecular conformation like A[3] will be seen.

Fig. 10.32 Conformation of β-amyrin, different representations, and the manner of getting the perspective formulas

the molecule by Dreiding model and keeping the correct absolute stereochemistry of each chiral centre as in **A**. Rings A, B, D and E are in chair form, while ring C is in half chair form. Now projecting the model in different ways on the projection plane one can get the different representations of the same molecular conformation. This has been indicated in the Fig. 10.32.

β-Amyrin. The Conformations of Rings in β-Amyrin The molecule possesses chair–chair–half-chair–chair–chair conformations of rings A, B, C, D, and E, respectively, and thus involves 1,3-*syn* diaxial nonbonded interactions. To alleviate such interactions to some extent ring A may have slightly distorted chair conformation to minimize the energy content contrary to a report [86]. The corresponding boat conformation (lacking 4β-Me and 10β-Me *syn-a*xial interaction), if attained, will involve increase of energy due to flagpole–bowsprit (fp–bs) interaction between 3β-OH and 10β-Me (Fig. 10.33, Ic) in addition to the increase of energy by >5 kcal/mol due to the boat/skew-boat conformation.

10.4.2   Molecular Conformation of α-Amyrin

α-Amyrin possesses the same stereoconfiguration of β-amyrin and has 19β- and 20α-methyl groups (both equatorial in ring E chair) instead of 20-gem-dimethyl groups in molecular conformation of α-amyrin (vide Fig. 10.13) and is thus same as that of β-amyrin.

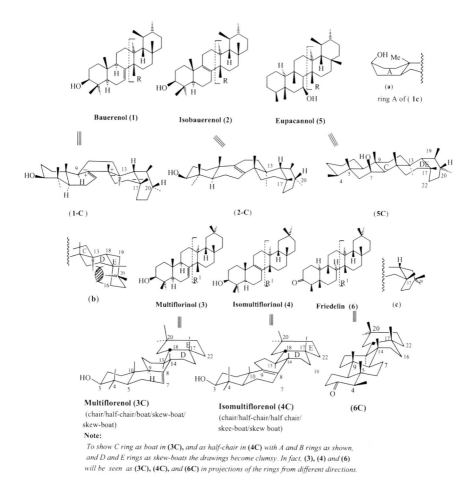

Fig. 10.33 Structures and molecular conformations of some pentacyclic triterpenes

10.4.3   Conformations of Bauerenol, Isobauerenol, Multiflorenol, and Isomultiflorenol

10.4.3.1   Conformation of Ring A

In 1968 the present authors have shown [51] from careful analysis of Dreiding models of bauerenol and isobauerenol (Fig. 10.33) that although ring A is free to assume boat form the associated extra strain will be enhanced by a strong fp–bs interaction between 10β-Me and 3β-OH(**a**), as already stated in cases of α-amyrin and β-amyrin. Contrary to the literature report [86] on bauerenol *a slightly distorted chair conformation is, therefore, preferred in the ground state for all polycyclic triterpenes having 10β-Me and 4-gem-dimethyl groups, as in bauerenol*(1), *isobauerenol* (**2**), *multiflorenol* (**3**), *and isomultiflorenol* (**4**) (Fig. 10.33).

10.4.3.2   Conformations of Rings B and C

As expected, bauerenol (**1C**) will have rings B and C in half-chair and true boat conformation, respectively, to maintain the correct stereochemistry at C9, C13 andC14. The opposite torsion angle signs at 9-8 and 14-13 bonds (1st and 3rd bonds in ring C) also dictate that ring C must possess boat conformation (*cf.* Sect. 2.14.2).

10.4.3.3   Conformations of Rings D and E

The nonbonded interactions existing in different possible conformations of the *cis*-fused nonsteroid forms of D/E rings of bauerenol and multiflorenol, as revealed from careful analysis of their Dreiding models, have been recorded in Table 10.1, in their possible decreasing order of enthalpy (or increasing order of stability). Careful analysis of the additional nonbonded interactions present in the four possible conformations in case of bauerenol (**1**) [(i) to (iv)] and multiflorenol (**3**) [(v) to (viii)] leads to the increasing order of relative stability of the different conformers in each case, as shown in the Table, and also to the facts that conformer (iv) with *cis*-fused D/E chair/skew-boat form and the conformer (viii) with *cis*-fused D/E skew-boat/skew-boat form are the most stable and hence the predominant conformers of bauerenol and multiflorenol, respectively. The enthalpy difference measured qualitatively also indicates that the conformations (iv) and (viii) of the D/E rings of bauerenol and multiflorenol are expected to be their only predominant conformations.

Again the torsion angle signs $(-/-)$ at the D/E ring junction are consistent with the *cis*-fused D/E rings of non-steroid form (**1C**) (*cf.* Sect. 2.15.1.3). In isobauerenol (**2**) [51] both the rings B and C are in half-chair conformation. Like Bauerenol D/E rings of isobauerenol are *cis*-fused and of nonsteroid form. In both bauerenol and isobauerenol ring D prefers chair form, but ring E must exist in a boat (or skew-boat) conformation with 17β-Me and 20β-H at the fp–bs positions, since a severe steric interference between 13α-Me and 20α-Me in addition to 1,3-*syn* diaxial interaction between 17β-Me and 19β-Me makes it impossible for rings D and E to attain chair conformation as shown in (**b**) [51] (Fig. 10.33).

Chemical Support Facile $SeO_2/AcOH$ oxidation of bauerenyl acetate to form the corresponding 7,9(11)-hetereoannular diene and inertness of isobauerenyl acetate to $SeO_2/AcOH$ even under drastic condition have been explained [51] on the basis of their proposed conformations. These facts reflect the relative instability of bauerenyl acetate due to its nonbonded interactions, already mentioned, and

Table 10.1 Nonbonded interactions existing in different conformations of D/E rings of bauerenol (1) and multiflorenol (3)

Conformations of rings D/E	Nonbonded interspace interactions + additional interactions over chair-chair conformations
Bauerenol (in order of decreasing enthalpy/increasing stability)	
(i) Chair/chair	13α-Me↔20α-Me (v. strong), cannot be attained, 17β-Me↔19β-Me (*syn* axial), 13α-Me↔17-20 bond, 14β-Me (axial) in ring D, 19β-Me(*a*) in ring E
(ii) Skew boat/skew-boat	13α-Me↔16α-H (*fp↔bs*), 14β-Me↔17β-Me (fp↔bs alleviated), additional enthalpy for two skew-boats.
(iii) Boat/chair	14β-Me↔17β-Me (*fp↔bs*) (v. strong) + additional enthalpy for true boat
(iv) Chair/skew-boat	17β-Me↔20β-H(*fp↔bs*) (alleviated in skew-boat), 13α-Me(*a*) (ring D) ↔ 17-22bond (*syn*-diaxial type) + (additional enthalpy for skew-boat);
Multiflorenol (in order of decreasing enthalpy/increasing stability)	
(v) Boat/chair	14β-Me ↔ 17β-Me (*fp↔bs*), + additional enthalpy for true boat
(vi) Chair/chair	13α-Me (a)(ring D) ↔ 20α-Me (*a*)(ring E)(v.strong) *does not allow this conformation*, 13α-Me ↔ 17-22 (bond), 17β-Me↔20α-Me (axial in ring E)
(vii) Chair/skew-boat	17β-Me ↔ 20β-Me (fp ↔ bs, alleviated in skew-boat), 13α-Me ↔ 21α-H (*fp ↔ bs*, alleviated in alternative skew boat, 14β-Me (*a* in ring D, additional enthalpy for skew boat.
(viii) Skew-boat/skew-boat (worked through 15-16 bond in eye level)	13α-Me ↔ 16α-H (*fp ↔ bs*), + additional enthalpy for two skew-boats

Note The enthalpy due to the remaining part (A/B/C-fused rings) is same for all possible conformations of D/E rings of bauerenol and multiflorenol

involvement of low activation energy of this reaction (involving release of strain). On the contrary, isobauerenyl acetate, which is free from any severe nonbonded interaction has to involve high-energy TS of boat conformation of ring C and severe nonbonded interaction for undergoing this reaction, remains unchanged. These facts have been explained mechanistically [51].

Multiflorenol (3) and Isomultiflorenol (4) It has been suggested [51] that multiflorenol (3) and isomultiflorenol (4)—both having geminal methyls at C20—will have conformations of A, B, C ring identical with those of bauerenol and isobauerenol, respectively. Both will have rings D and E in the boat or skew-boat form, as shown in (3C) and (4C) (Fig. 10.34), to avoid strong fp–bs interaction between 17-Me and 20β-Me groups in case; rings D and E assume chair and boat conformation respectively as shown in (C).

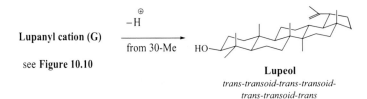

Lupanyl cation (G)

see **Figure 10.10**

$\xrightarrow[\text{from 30-Me}]{-\text{H}^{\oplus}}$

Lupeol

trans-transoid-trans-transoid-
trans-transoid-trans

Fig. 10.34 Molecular conformation of lupeol

10.4.4   Molecular Conformations of Eupacannol, Friedelin, and Derivatives

From the structures of bauerenol (**1**), isobauerenol (**2**), and eupacannol (**5**), it is evident that their **R** part is common (Fig. 10.34); hence the conformations of D and E rings will be similar in these cases [*cf.* (**1C**), (**2C**) and (**5C**)]. Thus, in eupacannol [52, 53] also *cis*-fused D/E rings are in chair/skew-boat form (**5C**). R′ part is common in multiflorenol (**3**), isomultiflorenol (**4**), and friedelin (**6**); hence the *cis*-fused D and E rings will be similarly adopting skew-boat/skew-boat conformations.

From the conformational analyses (with the Dreiding model) of multiflorenol, bauerenol and their isomers it is well understood that during the biosynthesis of pentacyclic triterpenes (perhydropicene type) in plant cells, when the 1,2 *anti*-migrations advance to such an extent that the terminal carbonium ion (prior to quenching) gets a home beyond C13 in ring D or C or B or A, all triterpenoids (olefin or alcohol) derived from that carbonium ion will have D/E *cis*-fused rings. If the carbonium ion gets a home in ring B or A the *cis*-fused D/E rings will be in *skew-boat/skew-boat* conformation with *geminal methyls at C20*, while with *C19, C20 vicinal methyls* in *chair-skew boat* conformation.

Our conjecture made in 1968 [51] has been shown to be correct by detection in 1983 of D/E boat conformations in friedelin by methyl-to-methyl nuclear Overhauser enhancement studies [87], and also by ^{1}H NMR data of some friedelin derivatives isolated from the roots of *Martonia diffusa* (Celastraceae) [88]—with special reference to the in influence of 20α-Me and its natural oxidation products and also of the oxidation of 13α-Me to CH_2OH, CHO and COOH. The bulk requirement of CH_2OH (sp^3), CHO (sp^2), and COOH (sp^2) at C20 and that of Me (sp^3) at C13-α position play an important role in acquiring the conformations of rings D and E as follows: 13α-Me and 20α-Me or 13α-Me and 20α-CH_2OH (*cis* D/E boat–boat) and 13α-Me and 20α-CHO or 13α-Me and 20α-COOH (*cis* D/E chair–chair). These results confirmed our conjecture proposed in 1968 [51]. These are well explained in terms of their interspace nonbonded interactions. Further support came from the X-ray analysis of a co-crystal of friedelin-3β-ol and friedelin (0.75/0.25), in which three cyclohexane rings adopt chair conformation, one adopts a twist-boat conformation, and the fifth is in boat conformation [89], consistent with our assignment from Dreiding model analysis.

10.4.5   Molecular Conformation of Lupeol

Lupeol, a very common pentacyclic triterpene of higher plants, is having a cyclopentanoperhydrochrysene skeleton. The molecular conformation of Lupeol is straightforward. It attains chair–chair–chair–chair five-membered cyclopentane conformation; here the lupanyl carbocation (**G**) has been quenched as an olefin forming an isopropylidene group (Fig. 10.34).

Interestingly, it has been shown that the absolute configurations of all the eight chiral centers of lupeol isolated from a fossil have been found to be identical with those of lupeol isolated from higher plants. The specimens of lupeol isolated from these two sources of a gap of millions of years are identical in all respects (and not diastereomeric—being epimeric at any one or more chiral center/s [90]. This is an amazing finding, since the molecule escapes the pressure of a large period of evolution.

10.5   Conclusion

Careful analyses by means of Dreiding models of the nonbonded interactions present in different conformations of polycyclic molecules containing three, four, or five rings (six-membered) having terminal cis-fused rings may lead to their preferred conformations which might be confirmed by suitable X-ray and/or NMR studies.

10.5.1   Diagenetic Product of β-Amyrin

1,8-Dimethylpicene [Fig. 10.20, structure (**d**)] is the major diagenetic product of β-amyrin.

References

1. Ran Xu, Gia C. Fazio and Seiichi P. T. Matsuda, On the Origins of Triterpenoid Skeletal Diversity, *Phytochemistry*, **2004**, *65*, 261-291, and references cited.
2. Masaaki Shibuya, Ting Xiang, Yuji Katsube, Miyuki Otsuka, Hong Zhang, and Yutaka Ebizuka, Origin of Structural Diversity in Natural Triterpenes: Direct Synthesis of *seco*-Triterpene Skeletons by Oxidosqualene Cyclase, *J. Am. Chem. Soc.*, **2007**, *129*, 1450-1455.
3. Albert Eschemoser and Dulio Arigoni, Revisited after 50 years: The Stereochemical Interpretation of the Biogenetic Isoprene Rule for the Triterpenes, *Helv. Chim. Acta*, 2005, *88*, 3011-3050.
4. Ikuro Abe, Michel Rohmer and Glen D. Prestwitch, Enzymatic Cyclization of Squalene and Oxidosqualene to Sterols and Triterpenes, *Chem. Rev.*, 1993, *93*, 2189-2206.

5. Karl Poralla, Cycloartenol and Other Triterpene Cyclases in *Comprehensive Natural Products Chemistry*, Editor-in-Chief: Sir Derek Barton and Koji Nakanishi, Pergamon Press, Elsevier, **1999**, vol 2, (Volume Editor David E. Cane), pp 299-319.

6. T. Akihisa, K. Arai, Y. Ki-mura, K. Koike, W. C. M. C. Kokke, T. Shibata, and T. Nikaido, Camelliols A-C, Three Novel Incompletely Cyclized Triterpene Alcohols from *Sasanqua* Oil (*Camellia sasanqua*), *J. Nat. Prod.*, **1999**, *62*, 265-268.

7. A. F. Barrero, E. J. Alvarez-Manzaneda, and R. Alvarez-Manzaneda, Achilleol A: A New Monocyclic Triterpene Skeleton from *Achillea odorata* L; *Tetrahedron Lett.*, **1989**, *30*, 3351-3352.

8. Quanbo Xiong, William K. Wilson, and Seiichi P. T. Matsuda, An *Arabidopsis* Oxidosqualene Cyclase Catalyzes Iridal Skeleton Formation by Grob Fragmentation, *Angew. Chem. Int. Ed.*, **2006**, *45*, 1285-1288.

9. Y. Arai, M. Hirohara, H. Ageta and H. Y. Hsu, Fern Constituents: Two New Triterpenoid Alchols with Mono and BIcyclic Skeletons, Isolated from *Polypodiodes formosana*, *Tetrahedron Lett.*, **1992**, *33*, 1325-1328.

10. T. Hoshino and T. Sato, Squalene-Hopane Cyclase: Catalytic Mechanism and Substrate Recognition, *Chem. Commun.*, **2002**, 291-301.

11. G. J. Bennett, L. J. Harrison, G.-L. Sia, and K.-Y. Sim, Triterpenoids Tocotrienols and Xanthones from the Bark of *Cratoxylum cochinchinense*, *Phytochemistry*, **1993**, *32*, 1245-1251.

12. L. H. D. Nguyen and L. J. Harrison, Terpenoid and Xanthone Constituents of *Cratoxylum cochinchinense*, *Phytochemistry*, **1998**, *50*, 471-476.

13. R. B. Boar, L. A. Couchman, A. J. Jaques, and M. J. Perkins, Isolation from *Pistacia* Resins of a Bicyclic Triterpenoid Representing an Apparent Trapped Intermediate of Squalene-2,3-epoxide Cyclization, *J. Am. Chem. Soc.*, **1984**, *106*, 2476-2477.

14. K. Shiojima, Y. Arai, K. Masuda, T. Kamada, and H. Ageta, Fern Constitutents: Polypoda-tetraenes, Novel Bicyclic Triterpenoids, Isolated from Polypodiaceous and Aspidiaceous Plants. *Tetrahedron Lett.*, **1983**, *24*, 5733-5736.

15. T. Sato and T. Hoshino, Catalytic Function of the Residues of Phenylalanine and Tyrosine Conserved in Squalen-Hopene Cyclases, *Biosci. Biotechnol. Biochem.*, **2001**, *65*, 2233-2242.

16. A. K. Kiang, E. L. Tan, F. Y. Lim, K. Habaguchi, K. Nakanishi, L. Fachen, and G. Ourisson, Lansic Acid, A Bicyclic Triterpene, *Tetrahedron Lett.*, **1967**, 3571-3574.

17. K. Habaguchi, M. Watanabe, Y. Nakadaira, and K. Nakanishi, The Full Structure of Lansic Acid and Its Minor Congenor, An Unsymmetric Onoceradienedione, *Tetrahedron Lett.* 1968, 3731-3734.

18. A. Chawla and Sukh Dev, A New Class of Triterpenoids from *Ailanthus malabarica* DC. Derivatives of Malabaricane, *Tetrahedron Lett.*, **1967**, 4837-4843.

19. R. R. Sobti and Sukh Dev, A Direct Correlation of (+)-Malabaricol with (+)-Ambreinolide, *Tetrahedron Lett.*, **1968**, 2215-2217.

20. J. Jakupovic, F. Eid, F. Bohlmann, and S. El-Dahmy, Malabaricane Derivatives from *Pyrethrum santolinoides*, *Phytochemistry*, **1987**, *26*, 1536-1538.

21. T. McCabe, J. Clardy, L. Minale, C. Pizza, F. Zollo, and R. Riccio, A Triterpenoid Pigment with the Isomalabaricane Skeleton from the Marine Sponge *Stelletta* sp., *Tetrahedron Lett.*, **1982**, *23*, 3307-3310.

22. K. B. Sharpless, d.l.-Malabaricanediol. The First Cyclic Natural Product Derived from Squalene in a Nonenzymic Process, *J. Am. Chem. Soc.*, **1970**, *92*, 6999-7001.

23. E. E. van Tamelen, J. Willet, M. Schwartz, and R. Nadeau, Nonenzymic Laboratory Cyclization of Squalene 2,3-Oxide. *J. Am. Chem. Soc.*, **1966**, *88*, 5937-5938.

24. Eugene E. van Tamelen, K. B. Sharpless, R. P. Hanzlik, Raymond B. Clayton, Alma L. Burlingame, and P. C. Wszolek, Enzymic Cyclization of *trans,trans,trans*-18,19-Dehydrosqualene-2,3-Oxide, *J. Am. Chem. Soc.*, **1967**, *89*, 7150-7151.

25. K. Masuda, K. Shiojima, and H. Ageta, Fern Constituents-2. New Malabaricatrienes Isolated from *Lemmaphyllum microphyllum* var *obovatum*, *Chem. Pharm. Bull.*, **1989**, *37*, 1140-1142.

26. Y. Arai, M. Hirohara and H. Ageta, Fern Constituents: Three New Skeletal Triterpenoid Hydrocarbons Isolated from *Polypodiodes niponica, Tetrahedron Lett.*, **1989**, *30*, 7209-7212.

27. Mugio Nishizawa, Hisaya Nishide, and Yugi Hayashi, The Structure of Lansioside A: A Novel Triterpene Glycoside with Amino-Sugar from *Lansium domesticum, Tetrahedron Lett.*, **1992**,1349-1350

28. H. Hattori, H. Igarashi, S. Iwasaki, and S. Okuda, Isolation of 3β-Hydroxy-4β-methylfusida-17 (20)[16,21)-*cis*]24-diene, (3β-Hydroxy-protosta-17(20)[16,21-*cis*]24-diene, and a Related Triterpene Alcohol, *Tetrahedron Lett.*, **1969**, *13*, 1023-1026.

29. K. Schreiber and G. Osske, Sterine und Triterpenoide-IV. Darstellung von Parkeol aus Cycloartenol, *Tetrahedron*, **1964**, *20*, 1803-1805.

30. T. R. Govindachari, N. Viswanathan, and P.A. Mohamed, Structure of Litsomentol, A New Tetracyclic Triterpene, *Tetrahedron*, **1971**, *27*, 4991-5009.

31. T. Ito, T. Tamura, T. M. Jeong, T. Tamura, and T. Matsumoto, 10α-Cucurbita-5,24-dien-3β-ol from gourd seed oil, *Lipids*, **1980**, *15*, 122-123.

32. G. Balliano, O. Caputo, F. Viola, L. Delprino, and L. Cattel, Biosynthesis of Cucurbitacins. Part 2. Cyclization of Squalene 2,3-epoxide to 10α-Cucurbita-5,24-dien-3β-ol by Microsomes from *Cucurbita maxima* Seedlings, *Phytochemistry*, **1983**, *22*, 915-921.

33. C. Toiron, A. Rumbero, E. Wollenweber, F. J. Arriaga, and M. Bruix, A New Skeletal Triterpenoid Isolated from *Empetrum nigrum, Tetrahedron Lett.*, **1995**, *36*, 6559-6562.

34. D. Arigoni, O. Jeger, und L. Ruzicka, Über die Konstitution und Konfiguration von Tirucallol, Euphorbol und Elemadienolsäure *Helv. Chim. Acta*, **1955**, *38*, 222-230.

35. T. Itoh, T. Tamura, and T. Matsumoto, Tirucalla-7,24-dienol: A New Triterpene Alcohol from Tea Seed Oil, *Lipids*, **1976**, *11*, 434-441.

36. D. Arigoni, R. Viterbo, M. Dünnenberger, O. Jeger, und L. Ruzicka, Zur Kenntnis der Triterpene. Konstitution und Konfiguration von Euphol und iso-Euphenol, *Helv. Chim. Acta*, **1954**, *37*, 2306-2322.

37. D. H. R. Barton, J. F. McGhie, S. K. Pradhan, and S. A. Knight, The Constitution and Stereochemistry of Euphol, *Chem. Ind.* (London), **1954**, 1325-1327

38. *Idem*, same title, *J. Chem. Soc.*, **1955**, 876-886.

39. D. S. Irvine, W. Lawrie, A. S. McNab, and F. S. Spring, Triterpenoids. Part L. The Constitution of Butyrospermol, *J. Chem. Soc.*, **1956**, 2029-2033.

40. M.-J. U. Ferreira, A. M. Lobo, C. A. O'Mahoney, D. J. Williams, and H. Wyler, Euferol and Melliferol: Two Novel Triterpenoids from *Euphorbia mellifera, J. Chem. Soc., Perkin Trans. 1*, **1990**, 185-187.

41. Hiroshi Yamashita, Kazuo Masuda, Tomomi Kobayashi, Hiroyuki Ageta, and Keniji Shiojima, Dammarane Triterpenoids from Rhizomes of *Pyrrosia lingua, Phytochemistry*, **1998**, *49*, 2461-2466.

42. J. S. Mills, The Constitution of the Neutral Tetracyclic Triterpenes of Dammar Resin, *J. Chem. Soc.*, **1956**, 2196-2202.

43. T. Akihisa, Y. Kimura, and T. Tamura, Bacchara-12,21-dien-3-ol from the Seeds of *Glycine max, J. Chem. Soc.*, **1955**, 876-886.

44. K. Masuda, K. Shiojema, and H. Ageta, Fern Constitutents-6. Tetracyclic Triterpenod Hydrocarbons Having Different Carbon Skeletons, Isolated from *Lemmaphyllum microphyllum* var *obovatum, Chem. Pharm. Bull.*, **1983**, *31*, 2530-2533.

45. C. H. Baker, E. J. L. Barreiro, and B. Gilbert, Tetracyclic Triterpenes of *Barbacenia bicolor, Phytochemistry*, **1976**, *15*, 785-787.

46. Reference 1., p 276, structure number **109**.

47. H. N. Khastgir, and P. Sengupta, Structure of Multiflorenol, *Chem. Ind.*, **1961**, 1077-1078.

48. P. Sengupta, and H. N. Khastgir, Banerenol and Multiflorenol from *Gelonium multiflorum* A. Juss, The Structure of Multiflorenol, *Tetrahedron*, 1963, *19*, 123-132.

49. F. N. Lahey and M. V. Leeding, A New Triterpene Alcohol, Bauerenol, *Proc. Chem. Soc.* (London), **1958**, 342-343.

50. K. K. Bhutani, D. K. Gupta, and R. S. Kapil, Occurrence of D/E *trans* Stereochemistry Isomeric to Ursane (*cis*) Series in a New Pentacyclic Triterpene from *Cslotropis procera, Tetrahedron Lett.,* **1992**, *33*, 7593-7596.

51. Sunil K. Talapatra, Subrata Sengupta and Bani Talapatra, A New Pentacyclic Triterpene Alcohol from *Evodia fraxinifolia* Hook F., *Tetrahedron Lett.,* **1968**, 5963-5968.

52. Sunil Kumar Talapatra, Durga Sankar Bhar, Ramaprasad Chakraborty, and Bani Talapatra, Eupacannol, a Pentacyclic Triterpene of a New Skeletal Type, and other Chemical Constituents from *Eupatorium cannabinum, J. Indian Chem. Soc.,* **2003**, *80*, 1209 1216 [Professor S. M. Mukherji Commemoration Volume (Part II)].

53. S. K. Talapatra, D. S. Bhar, and B. Talapatra, Eupacannol, A New Skeletal Type in Pentacyclic Triterpenes, *9 th IUPAC Symposium on the Chemistry of Natural Products,* Ottawa, Canada, **1974**, *Abstracts,* p 39B.

54. S. Matsunaga, and R. Morita, Hopenol-B, a Triterpene Alcohol from *Euphorbiua supina, Phytochemistry,* **1983**, *22*, 605-606.

55. B. Achari, A. Pal, and S. C. Pakrashi, Indian Medicinal Plants. XXXVI. New D,E-*cis*-fused Neohopane derivatives from *Alangium lamarckii, Tetrahedron Lett.,* **1975**, 4275-4278.

56. R. Verpoorte, On the Occurrence of Filican-3-one in *Strychnos dolichothyrsa, Phytochemistry,* **1978**, *17*, 817-818.

57. C. B. Gamlath, A. A. L. Gunatilaka, and S. Subramaniam, Studies on Terpenoids and Steroids. Part 19. Structures of three Novel 19(10- >9) abeo-8α,9β,10α-euphane triterpenoids from *Reissantia indica* (Celestraceae), *J. Chem. Soc.,* Perkin Trans. 1, **1989**, 2259-2267.

58. J. De Pascual Teresa, J. G. Urones, I. S. Marcos, P. Basabe, M. J. Cuadrado, and R. Fernandez Moro, Triterpenes from *Euphorbia broteri, Phytochemistry,* **1987**, *26*, 1767-1776.

59. M. Della Greca, A. Florentino, P. Monaco, and L. Privitera, Cycloartene Triterpenes from *Juncus effuses, Phytochemistry,* **1994**, *35*, 1017-1022.

60. D. H. R. Barton and K. H. Overton, Triterpenoids. Part XX, The Constitution and Stereochemistry of a Novel Tetracyclic Triterpenoid, *J. Chem. Soc.,* **1955**, 2639-2652.

61. Y. Inubshi, T. Sano, and Y. Tsuda, Serratenediol – A New Skeletal Triterpenoid Containing A Seven-Membered Ring, *Tetrahedron Lett.,* **1964**, 1303-1310.

62. J. De Pascual Teresa, J. G. Urones, P. Basabe, and F. Granell, Components of *Cistus bourgeanus* Coss. *An. Quim,* **1979**, *75*, 131-134.

63. Dulce Mesa-Siverio, Haydee Chávez, Ana Estévez-Braun, and Angel G. Ravelo, Cheiloclines A-I. First Examples of Octacyclic Sesquiterpene-Triterpene Hetero-Diels-Alder Adducts, *Tetrahedron,* **2005**, *61*, 429-436.

64. Antonio G. González, Nelson L. Alvarenga, Ana Estévez-Braun, Angei G. Ravelo, Isabel L. Bázzocchi, and Laila Moujir, Structure and Absolute Configuration of Triterpene Dimers from *Maytenus scutioides, Tetrahedron,* **1996**, *52*, 9597-9608.

65. Antonio G. González, Maria L. Kennedy, Felix M. Rodriguez, Isabel L. Bazzocchi, Ignacio A. Jiménez, Angel G. Ravelo, and Laila Moujir, Absolute Configuration of Triterpene Dimers from *Maytenus* Species (Celastraceae), *Tetrahedron,* **2001**, *57*, 1283-1287.

66. Osamu Shirota, Hiroshi Morita, Koichi Takeya, and Hideji Itokawa, New Geometric and Stereoisomeric Triterpene Dimers from *Maytenus chuchuhuasca, Chem. Pharm. Bull.,* **1998**, *46*, 102-106 and references cited.

67. Antonio G. González, NNelson L. Alvarenga, Isabel L. Bazzocchi, Angel G. Ravelo, and Laila Moujir, Triterpenes Trimers from *Maytenus scutioides*: Cycloaddition Compounds? *J. Nat. Prod.,* **1999**, *62*, 1185-1187.

68. J. W. Cornforth, (Mrs.) R. H. Cornforth, and K. K. Mathew, A Stereoselective Synthesis of Squalene, *J. Chem. Soc.,* **1959**, 2539 – 2547

69. J. W. Cornforth, (Mrs.) R. H. Cornforth and K. K. Mathew, A General Stereoselective Synthesis of Olefins, *J. Chem. Soc.,* **1959**, 112 –127

70. William S. Johnson, Mark S. Plummer, S. Pulla Reddy, and William R. Bartlett, The Fluorine Atom as a Cation–Stabilizing Auxilliary in Biomimetic Polyene Cyclizations. 4. Total Synthesis of *dl*-β-Amyrin, *J. Am. Chem. Soc.,* **1993**, *115*, 515-521.

71. Armin Meyer, O. Jeger, V. Prelog und L. Ruzicka, Zur Kenntnis der Triterpene, Zur Konstitution des Ringes C der pentacylischen Triterpene, *Helv. Chim. Acta,* **1951,** *34,* 747-755.

72. H. Menard und O. Jeger, Zur Kenttnis der Triterpene, 177 Mitteilung, Zur Gliederzahl des Ringes C der pentacyclischen Triterpene, *Helv. Chim. Acta,* 1953, *36,* 335-336.

73. D. H. R. Barton and N. J. Holness, Triterpenoids. Part v, Some Relative Configurations in Rings C, D, and E of β-Amyrin and Lupeol Group of Triterpenoids, *J. Chem.Soc.*(London), **1952,** 78-92.

74. H. Budzikiewicz, I. M. Wilson and Carl Djerassi, Mass Spectrometry in Structural and Stereochemical Problems. XXXII, PentacyclicTriterpenes, *J. Am. Chem. Soc.,* **1963,** *85,* 3688-3699.

75. T. G. Halsall and D. B. Thomas, An Approach to the Total Synthesis of Triterpenes, Part 1, *J. Chem. Soc.,* **1956,** 2431-2443.

76. E. J. Corey, Hans- Jürgen Hess, and Stephen Proskow, Synthesis of a β-Amyrin Derivative, Olean-11,12; 13,18-diene, *J. Am. Chem. Soc.,* **1963,** *85,* 3979-3983.

77. D H R Barton, E. F. Lier, and J. F. Mcghie, The Synthesis of β-Amyrin, *J. Chem. Soc.* (c), 1968, 1031-1040].

78. P. Dietrich et E. Lederer, Synthèse totale de íambrèinolide racémique et de quelques – uns de ses dérivés, *Helv. Chim.Acta,* 1952, *35,* 1148.

79. E. E. van Tamelen, M. P. Seiler and W. Wierenga, Biogenetic-type Total Synthesis. δ-Amyrin, β-Amyrin and Germanicol, *J. Am. Chem. Soc.,* **1972,** 94, 8229-8231.

80. George Brownlie, M. B. E. Fayez, F. S. Spring, Robert Stevenson, and W. S. Strachan, Triterpenoids. Part XLVIII. Olean-13(18)-ene. Isomerisation of Olean-12-ne and Related Hydrocarbons with Mineral Acid *J. Chem. Soc.* **1956,** 1377-1384.

81. M. Tori, T. Tsuyuki, and T. Takahashi, Bull. Chem. Soc. Jpn., **1977,** *50,* 3381-3383.

82. Robert E. Ireland and David M. Walba, The Total Synthesis of (+)- Friedelin, an Unsymmetrical Pentacyclic Triterpene, *Tetrahedron Lett.,* **1976,** 1071-1074.

83. Tetsuji Kametani, Yoshiro Hirai, Yuichi Shiratori, Keiichiro Fukumoto, and Fumio Satoh, Convenient and Stereoselective Route to Basic Frameworks for Synthesis of Unsymmetrical Pentacyclic Triterpenes, *J. Am. Chem. Soc.,* **1978,** *100,* 554-560.

84. K. Nakanishi, T. Goto, S. Ito, S. Natori and S. Nozoe, *Natural Products Chemistry,* Academic Press, Inc. New York, Vol.1, **1974.**

85. Derek H. R. Barton and Stuart W. McCombie, A New Method for the Deoxygenation of Secondary Alcohols, *J. Chem. Soc., Perkin Trans. 1,* **1975,** 1574-1585.

86. L. Ramachandra Row and C. Sankara Rao, Chemical Examination of *Diospyros* Species-Part V : A None Aromatization of Ring B and Other Reactions of Bauerenol *Tetrahedron Lett.,* **1967,** 4845-4852.

87. Francisco Radler de Aquino Neto and Jeremy K. M. Sanders, Detection of Boat Conformations in the Triterpene Friedelin by Methyl-to-Methyl Nuclear Overhauser Enhancements, *J. Chem. Soc., Perkin Trans 1,* **1983,** 181-184.

88. Mariano Martinez V., Miguel Munoz Corona, Christina Sanchez Velcz, Lydia Rodrigucz-Hann and Pedro Joseph-Nathan, Terpenoids from *Martonia diffusa, J. Nat. Prod.,* **1988,** *51,* 793-796.

89. H. K. Fun, N. Boonnak and S. Chantrapromma, A Cocrystal of Friedelan-3β-ol and Friedelin, *Acta Cryst.* **2007,** 02014-02016.

90. G. Ourisson, Personal Communication during the Pre-IUPAC Symposium on Natural Products, held in Calcutta, **1990.**

Chapter 11
Steroids: Cholesterol and Other Phytosterols

11.1 Introduction

Steroids belong to a class of organic compounds having a common cyclopentano-perhydrophenanthrene nucleus, a fused four-ring system made up of 17 carbon atoms. Most steroids carry a side chain at C17, varying in composition. Sometimes the side chain remains attached to C16 and C17 in the form of a spiroketal ring. The steroids contain a hydroxyl group at C3 (a few contains C3-keto) and extra hydroxyl/s at different position/s, a double bond at C5–C6, and often with aromatized ring A. The term steroid has been first used in 1827. Steroids upon heating with selenium at ~360 °C in a sealed tube generally yield *Diels hydrocarbon*, shown to be 3′-methyl-1,2-cyclopentenophenanthrene derivative (last structure in Fig. 11.1). Steroids are bioactive compounds; many of them behave as hormones and prohormones and drugs and prodrugs and control many important physiological actions. They include *bile acids* in which the terminal carbon atom of the truncated side chain has been oxidized to a carboxylic group. Some examples of common steroidal compounds of plant (P) and animal (A) origins are shown in Fig. 11.1.

11.2 Cholesterol

11.2.1 Introduction: Functions in Human System

Cholesterol and some other sterols, e.g., cholic acid and ergosterol, were isolated in the nineteenth century. However, their structural elucidations were achieved in the twentieth century when chemical knowledge came out of its infancy and was ripe enough for the structural work. In course of time (1936–1992) a number of steroidal sex hormones and more than 30 steroids were isolated. The synthesis of cortisone by Sarett and Kendall and the report by Philip Hench on the anti-inflammatory

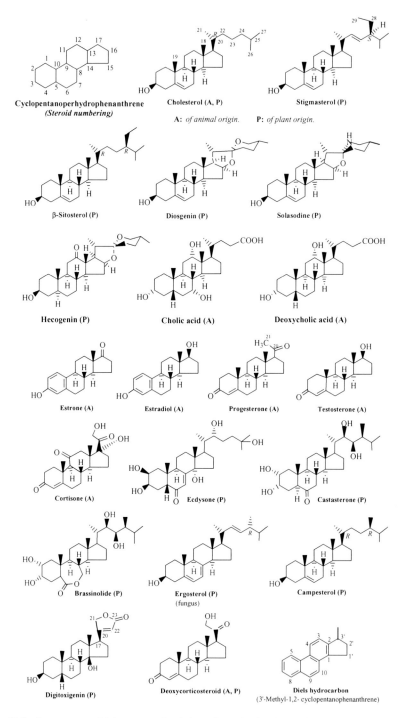

Fig. 11.1 Common steroidal compounds of plant (P) and animal (A) origin

property of cortisone in the treatment of rheumatoid arthritis made the whole field of steroid an attractive area for chemical and biochemical research [1].

"A string of Nobel Prizes signaled contemporary appreciation of the significance of steroid discoveries during these decades. Wieland (1927), Windaus (1928), Butenandt, and Ruzicka (1939) were awarded the Chemistry Prize, while Hench, Kendall and Reichstein shared the Nobel Prize in Physiology or Medicine in 1950 for their pioneering work on cortisone. The same Prize went to Konrad Bloch in 1964. The classic early structural studies of the 1920s and 1930s were followed by the elucidation of steroid stereochemistry, which took a new turn following the introduction of conformational analysis by Derek Barton in his classic *Experientia* paper in 1950. Barton and Odd Hassel were awarded the 1969 Nobel Prize in Chemistry for their work in this area." [1, see also 2 and 3]

Industries also got involved in the discovery and synthesis of steroidal drugs having profound bioactivity and high sensitivity as well as high specificity.

Of all the bioactive steroids mostly occurring in animals cholesterol is the most familiar and abundant one. It is also the first steroid isolated from human gall stones by Conradi in 1775 and thus was the first member of the steroid family to be discovered (see [4]). It derived its name from the Greek version of gallstone— *Chole* (gall bile) and *stereo* (solid) and *ol* for alcohol. Cholesterol is found in tissues of mammalian body, particularly in the spinal chord and brain. Mammals are capable of absorbing cholesterol from dietary sources. It is also synthesized in the body. It serves as the precursor of various steroidal hormones and bile acids. The latter as the sodium salt can act as detergent-like substances that emulsify fats and oils to make them easily digestible.

Improper balance by way of a marked change in the cholesterol content in our system is responsible for some serious diseases. When cholesterol content of the bile acids increases to the extent of making its supersaturated bile acid solution, cholesterol is thrown out of the solution as crystals and forms gallstones in which it remains as a major component.

Defects in maintaining proper cholesterol level in blood serum have serious consequences. Elevated level of serum cholesterol causes atherosclerosis (hardening of arteries due to cholesterol deposits) and coronary artery disease. *When the cholesterol deposits break loose and the resulting thrombus blocks the flow of blood in the heart it causes a heart attack and causes a stroke when the flow of blood in the brain is blocked.*

In order to reduce the cholesterol content of the body system, some inhibitors have been designed which interrupt certain step/s in the biosynthesis of cholesterol. But cholesterol being the precursor of various steroidal hormones is necessary to maintain the homeostasis in the organisms; the desirability of such restriction or "full-scale assault" in its biosynthetic pathway is questioned. Bloch wrote [5]: "Researchers in the field refer to cholesterol as the paradigm of a Janus-faced substance, a villain causing cardiovascular disease on the one hand, yet at the same time, a molecule essential for bodily function."

11.2.1.1 Functions of LDL and HDL

Cholesterol and its esters are transported in the plasma in the form of water-miscible lipoprotein particles. They are characterized by their density as (i) *very low-density lipoprotein* (VLDL, *highest lipid content*), (ii) *low-density lipoprotein* (LDL, *intermediate lipid content*), and (iii) *high-density lipoprotein* (HDL, *least lipid content*). LDL transports cholesterol and its esters from hepatocyte to extrahepatocyte tissues, i.e., from liver to other tissues. An elevated LDL level will transport more cholesterol especially to cell walls of blood vessels (arteries) and forms *atherosclerotic plaques* causing narrowness of the vessel. Therefore, LDL is commonly known as *bad cholesterol*. High-density lipoprotein (HDL) binds and esterifies cholesterol released from peripheral tissues and then transfers cholesteryl esters to hepatocyte tissues, i.e., from the blood vessel walls to the liver or to tissues that use cholesterol to synthesize steroid hormones. Hence, HDL is usually known as *good cholesterol*. Thus, the words good cholesterol and bad cholesterol, signifying HDL and LDL, respectively, in our system, are justified. For a healthy person the HDL/LDL ratio is ~3.5.

11.2.1.2 Occurrence in Animals and Plants

The occurrence and function of cholesterol in animal, especially in human tissues, have been a much discussed topic; thus, most people, even some scientists, believe that cholesterol does not occur in plants. In view of such a belief and also in view of the absence of adequate mention in the biochemistry text books and in the *Dictionary of Natural Products* [6] of the presence of cholesterol in plants, Behrman and Gopalan [7] wrote a paragraph for the next generation of biochemistry texts as stated below. "But plants also contain cholesterol both free and esterified. Cholesterol occurs as a component of plant membranes and as a part of the surface lipids of leaves where it is sometimes the major sterol. The quantity of cholesterol is generally small when expressed as percent of total lipid. While cholesterol averages perhaps 50 mg/kg total lipid in plants, it can be as high as 5 g/kg (or more) in animals."

It is pertinent to point out that biosynthetic proofs were available for the formation of cholesterol from oxidosqualene via cycloartenol in plants and via lanosterol in animals and fungi. Since its first isolation from plants (algae) by Tsuda [8], reports on the occurrence of cholesterol from various plants are known in the literature [9, 10]. Further, since mammals are capable of absorbing cholesterol from dietary sources, people become more concerned about the intake of vegetable and non-vegetable diets, more especially of **vegetable oils** which contain cholesterol (Chap. 34).

11.2.2 Structural Elucidation. Relative and Absolute Stereochemistry. Conformation

The structural elucidation of cholesterol (**1**) has a long chemical history [11] which includes the brilliant and extensive work of a number of outstanding chemists of those days (1903–1933), of which special mention may be made of Adolf Windaus (dating from 1903), Otto Diels, H.O. Wieland (dating from 1912), E. Abderhalden, Otto Rosenheim, and Harold King. X-ray crystallographic work on steroids by J. D. Bernal led to the revision of the structure initially proposed by Windaus and Wieland for cholesterol.

Cholesterol, $C_{27}H_{46}O$, is a laevorotatory crystalline solid, m.p. 149°, $[\alpha]_D$ –39 (EtOH), –31.8 (Et$_2$O). The stereostructure (**1**) and its molecular conformation (**1a**) have been established on the basis of the findings shown schematically in Fig. 11.2, stereochemical analysis, and finally by its total synthesis. The steroidal nomenclature is shown in (**1**). A huge amount of experimental data, observations, and often

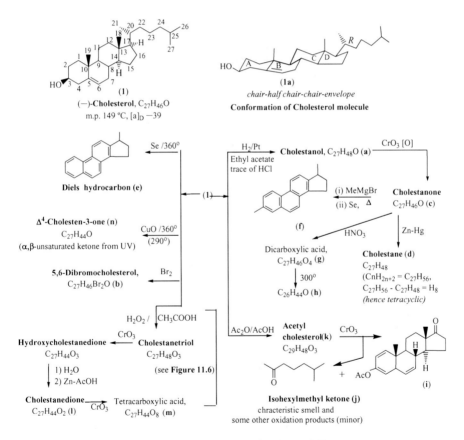

Fig. 11.2 Schematic representation of the reactions of cholesterol (**1**)

debated conclusions have accumulated during many years of hard work. We have presented here only a very small part of it, helpful in developing the structure.

(i) The nature and location of the hydroxyl group, (ii) presence of a double bond, (iii) the basic carbocyclic skeleton, and (iv) the presence of an isooctyl side chain and its site of attachment in the steroid skeleton have been deduced through the formation of the products (Fig. 11.2).
(ii) Derivatives (a) and (b) indicate the presence of a double bond.
(iii) Derivatives (c), (f), (k), and (n) suggest the presence of a secondary OH group and (f) indicates its presence at C3.
(iv) Derivatives (d), (e), (f), and (i) suggest that the tetracyclic skeleton of cholesterol has the cyclopentanoperhydrophenanthrene orientation as shown.
(v) Derivatives (i) and (j) suggest the presence of an isooctyl C_8 side chain located at C17 of the steroid nucleus.

The partial structure of cholesterol may thus be represented as (1') (Fig. 11.3) which takes care of all the carbon atoms, barring two methyls and their locations. Since (e) and (f) lack these two methyls, they must have been knocked off at the demand of aromatization and hence *they must be angular in disposition*. The methyls in (e) and (f) originate from the side chain and the Grignard reaction.

Locations of the angular methyls have been decided by the sequence of reactions shown in Figs. 11.4 and 11.5. The keto acid (p) (Fig. 11.4) is resistant to esterification, and eliminates CO on treatment with conc. H_2SO_4. These observations are indicative of the tertiary nature of the CO_2H. *Thus the methyl group*

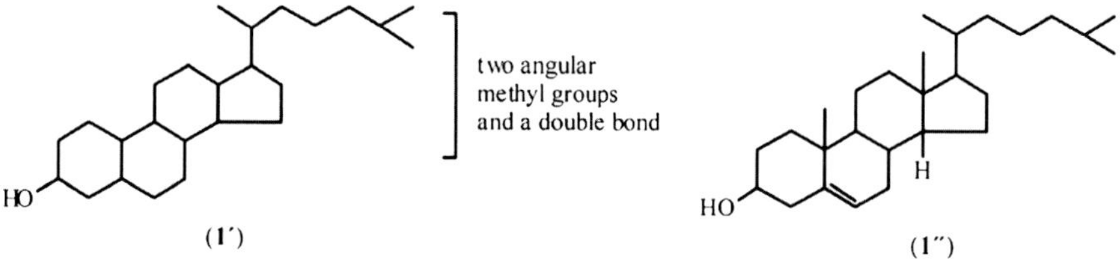

two angular methyl groups and a double bond

(1')   (1'')

Fig. 11.3 Structures (1') and (1'')

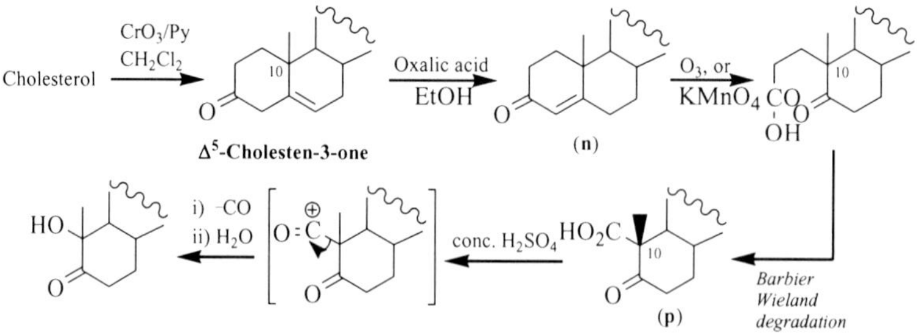

Fig. 11.4 The location of a methyl group at C10

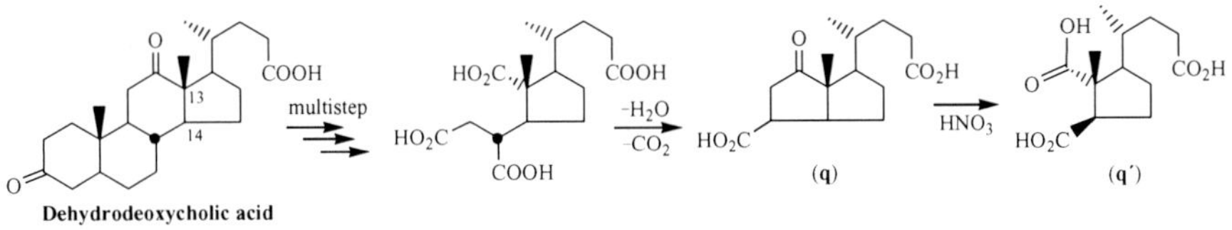

Fig. 11.5 The position of a methyl group at C13

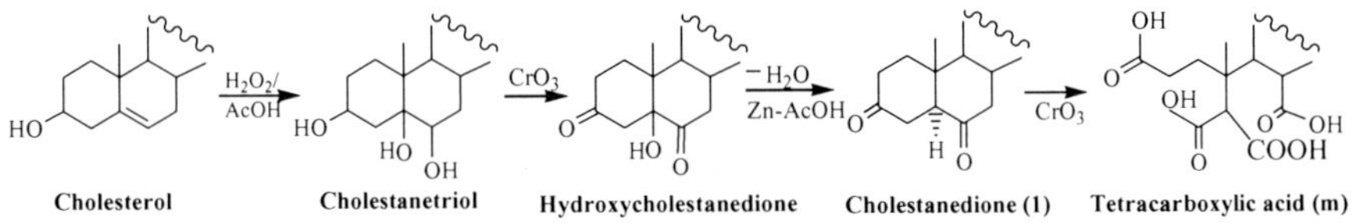

Fig. 11.6 The location of the double bond

must be geminal to COOH, and hence angular, and is attached to C10 of the skeleton.

The position of the second tertiary methyl group has been determined from the degradative methods shown in Fig. 11.5 in which dehydrodeoxycholic acid has been used, being known to be structurally related to cholesterol. Of the three COOH groups of (**q′**), one COOH group is tertiary in nature, being resistant to esterification, suggesting the geminal disposition of the COOH and the methyl group. The methyl may thus be attached either at C13 or C14. The COOH groups at these centers are *trans*, as no anhydride is formed on vacuum distillation. The C/D rings are thus *trans* fused. The *cis* 5,5-bicyclic system is known to be energetically more comfortable than the *trans* system in (**q**). Had there been a hydrogen instead of a methyl at C13, the C13 could have epimerized via enolization with C12–CO group to produce the *cis* system, which was not the case. *Hence a methyl group must be at C13.*

11.2.2.1 Location of the Double Bond

The formation of cholestanedione (**l**) and its subsequent conversion to a tetracarboxylic acid (**m**) suggest the presence of the two keto groups in two different rings and the genesis of (**l**) can be well explained (Fig. 11.6) if *the double bond is present at C5–C6*. Thus cholesterol's non-stereostructure may be expressed as (**1″**) (Fig. 11.3).

11.2.2.2 Configuration at C3

The reactions carried out by Shoppee (1948), as delineated in Fig. 11.7, suggest that 3-OH and 5-COOH must be *cis* disposed in the oxidation product (**r**). The

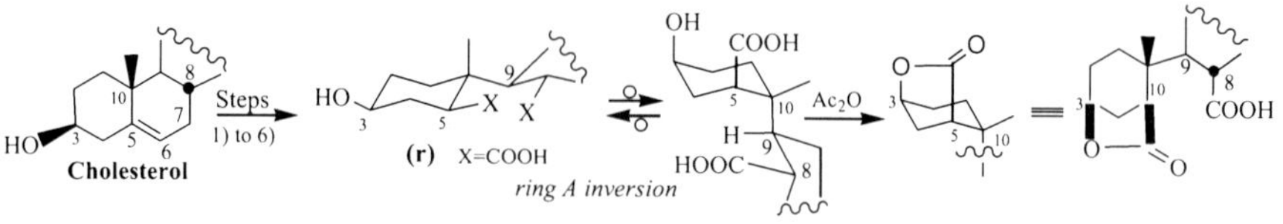

Fig. 11.7 Determination of configuration of cholesterol at C3

conversion of cholesterol to the compound (**r**) involved six straightforward successive steps, viz, treatments with (1) HNO_3 to form the 6-nitro derivative,
(2) Zn/AcOH to get the reduced product 5α-6-imino compound via the 6-amino
derivative, (3) acid to give the 5α-6-ketone, (4) Br_2, (5) $AgNO_3$ in pyridine to give
the 5α-6,7-diketo derivative, and (6) $NaOH/H_2O_2$ to get oxidized to the dicarboxylic acid (**r**). The ring A flips to enable the 1,3-diequatorial relationship to become
1,3-diaxial to form the lactone. Thus, 3-OH group is β-equatorial in this system and
A/B rings are *trans* fused.

Thus the structure and the relative configuration of cholesterol were elucidated
in pre-NMR age by ingenious planning/methods.

11.2.2.3 A/B-, B/C-, and C/D-Ring Fusions. Absolute Configuration

In the 5,6-dihydroderivatives of cholesterol the ring junction A/B may be *trans
fused* or *cis fused;* the former is *cholestanol* (5α-steroid), and the latter is
coprostanol (5β-steroid). The *trans fused* B/C rings offer conformational stability
to 5β-steroids due to conformational mobility of ring A. In most of the steroids like
sterols, bile acids, and related compounds the C/D rings are *cis fused.* The absolute
stereochemistry of cholesterol (**1**) and some other steroids has been established
by X-ray crystallographic studies. The absolute configurations of different
5α-cholestanones and 5β-coprostanones are in conformity with the observed Cotton
effects in their ORD curves (see Fig. 2.171).

11.2.3 Synthesis of Cholesterol

The total synthesis of cholesterol [4, 12, 13] by Woodward et al. constitutes the first
total synthesis of a non-aromatic natural steroid. The synthetic schemes contain the
great qualities of the master revealing imagination, planning, and their successful
implementation. On the whole, they offer a beautiful microcosm of organic chemistry *securing a permanent place in the history of the synthetic organic chemistry.*
The synthesis of cholesterol by Woodward et al., involving more than 30 steps, is
outlined in Fig. 11.8, showing mechanistic rationalization, when necessary.

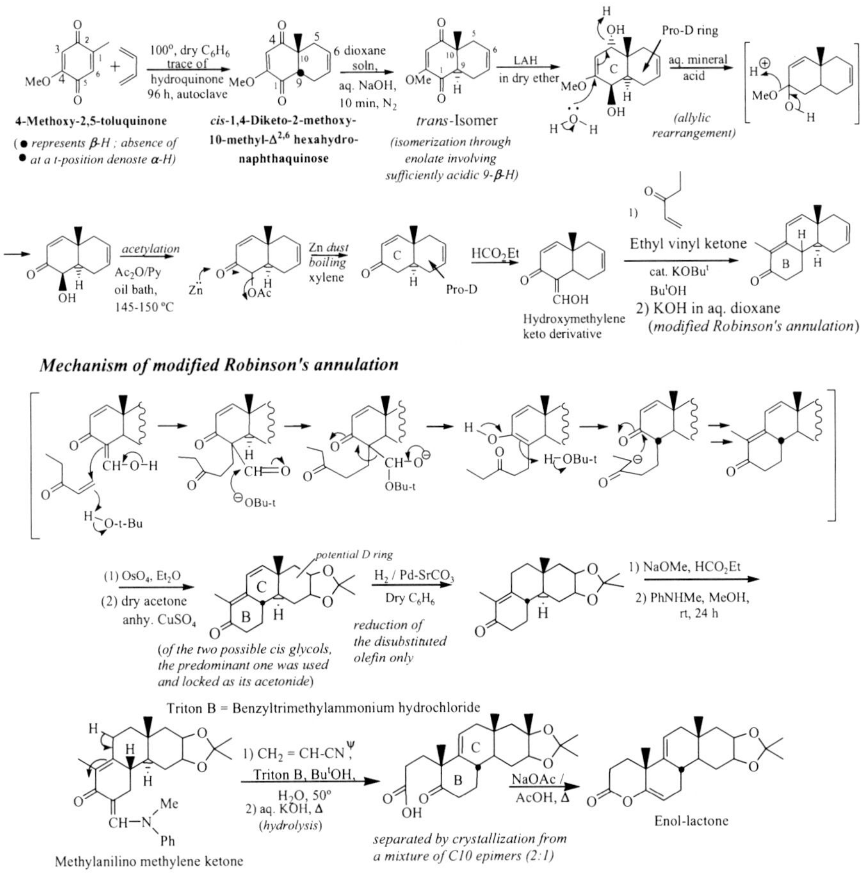

100°, dry C6H6
trace of
hydroquinone
96 h, autoclave
4-Methoxy-2,5-toluquinone
(● represents β-H ; absence of
● at a t-position denoste α-H)
cis-1,4-Diketo-2-methoxy-
10-methyl-Δ2,6 hexahydro-
naphthaquinose
dioxane
soln,
aq. NaOH,
10 min, N2
trans-Isomer
(isomerization through
enolate involving
sufficiently acidic 9-β-H)
LAH
in dry ether
Pro-D ring
aq. mineral
acid
(allylic
rearrangement)
acetylation
Ac2O/Py
oil bath,
145-150 °C
Zn dust
boiling
xylene
Pro-D
HCO2Et
Hydroxymethylene
keto derivative
1)
Ethyl vinyl ketone
cat. KOBut
ButOH
2) KOH in aq. dioxane
(modified Robinson's annulation)
Mechanism of modified Robinson's annulation
H-O-t-Bu
OBu-t
H-OBu-t
potential D ring
(1) OsO4, Et2O
(2) dry acetone
anhy. CuSO4
(of the two possible cis glycols,
the predominant one was used
and locked as its acetonide)
H2 / Pd-SrCO3
Dry C6H6
reduction of
the disubstituted
olefin only
1) NaOMe, HCO2Et
2) PhNHMe, MeOH,
rt, 24 h
Triton B = Benzyltrimethylammonium hydrochloride
1) CH2 = CH-CNΨ,
Triton B, ButOH,
H2O, 50°
2) aq. KOH, Δ
(hydrolysis)
Methylanilino methylene ketone
separated by crystallization from
a mixture of C10 epimers (2:1)
NaOAc /
AcOH, Δ
Enol-lactone

Ψ*Preferential attack of the electrophile from the axial β-face in the low energy pre-chair TS leading to the undesired α-Me epimer (~66%). The desired β-methyl epimer is formed by the attack of the electrophile from the α-face in the higher energy pre-twist TS in less amount (~38%)*

Fig. 11.8 (continued)

Since Woodward did not achieve an asymmetric synthesis, both the natural (−)-cholesterol and its enantiomer, unnatural (+)-cholesterol, are formed in equal amounts leading to the formation of racemic (±)-cholesterol.

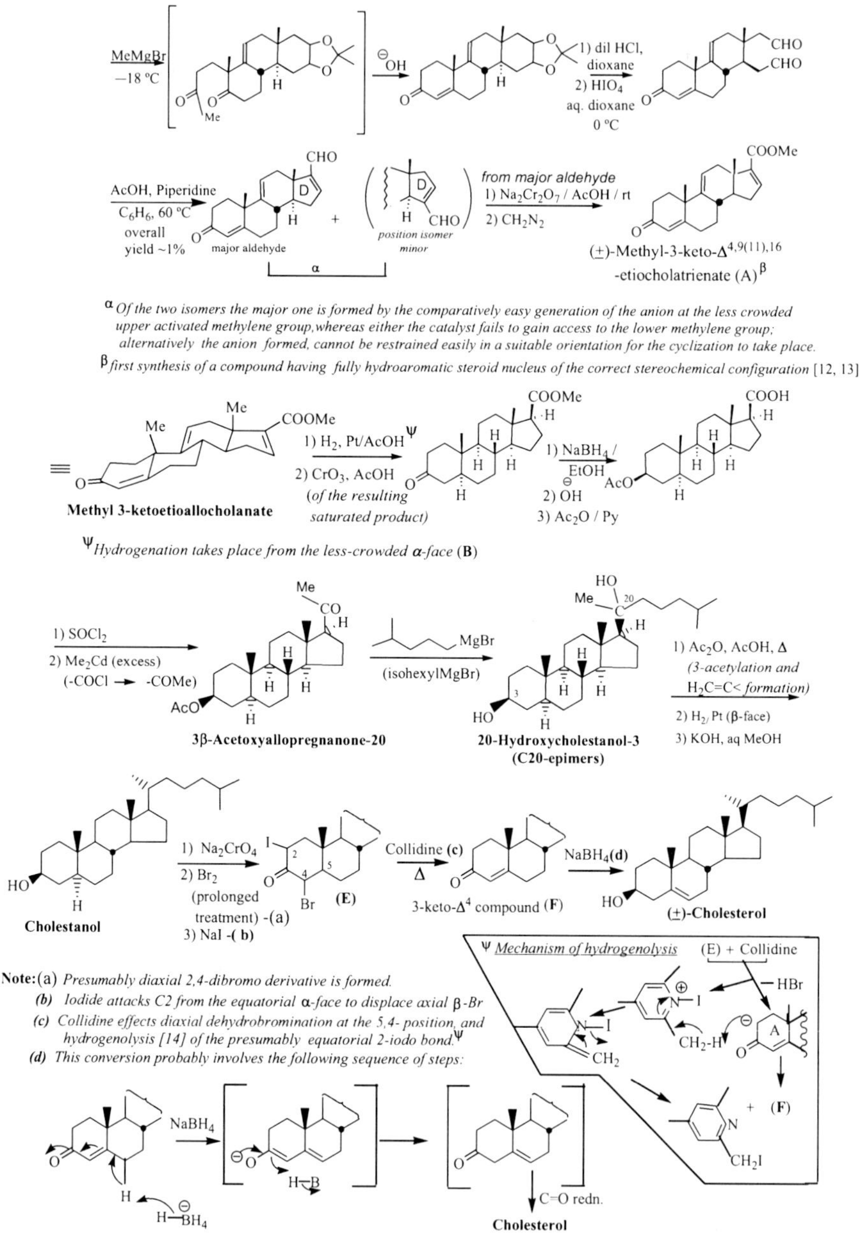

MeMgBr
−18 °C
Me
OH
1) dil HCl, dioxane
2) HIO4 aq. dioxane 0 °C
CHO
CHO
AcOH, Piperidine
C6H6, 60 °C
overall yield ~1%
major aldehyde
position isomer minor
from major aldehyde
1) Na2Cr2O7 / AcOH / rt
2) CH2N2
COOMe
(±)-Methyl-3-keto-Δ4,9(11),16 -etiocholatrienate (A)β
α Of the two isomers the major one is formed by the comparatively easy generation of the anion at the less crowded upper activated methylene group, whereas either the catalyst fails to gain access to the lower methylene group; alternatively the anion formed, cannot be restrained easily in a suitable orientation for the cyclization to take place.
β first synthesis of a compound having fully hydroaromatic steroid nucleus of the correct stereochemical configuration [12, 13]
Methyl 3-ketoetioallocholanate
1) H2, Pt/AcOH ψ
2) CrO3, AcOH (of the resulting saturated product)
1) NaBH4 / EtOH
2) OH
3) Ac2O / Py
COOH
ψ Hydrogenation takes place from the less-crowded α-face (B)
1) SOCl2
2) Me2Cd (excess) (-COCl → -COMe)
MgBr
(isohexylMgBr)
3β-Acetoxyallopregnanone-20
20-Hydroxycholestanol-3 (C20-epimers)
1) Ac2O, AcOH, Δ (3-acetylation and H2C=C< formation)
2) H2/ Pt (β-face)
3) KOH, aq MeOH
Cholestanol
1) Na2CrO4
2) Br2 (prolonged treatment) -(a)
3) NaI -(b)
(E)
Collidine (c)
Δ
3-keto-Δ4 compound (F)
NaBH4(d)
(±)-Cholesterol
ψ Mechanism of hydrogenolysis
(E) + Collidine
− HBr
Note: (a) Presumably diaxial 2,4-dibromo derivative is formed.
(b) Iodide attacks C2 from the equatorial α-face to displace axial β-Br
(c) Collidine effects diaxial dehydrobromination at the 5,4-position, and hydrogenolysis [14] of the presumably equatorial 2-iodo bond.ψ
(d) This conversion probably involves the following sequence of steps:
NaBH4
C=O redn.
Cholesterol

Fig. 11.8 Total synthesis of (±)-cholesterol (Woodward) (1951–1952) [4, 12, 13]

Further, during an achiral synthesis whenever a chiral center is generated, the absolute configuration of the natural molecule is shown, though in each step, the corresponding enantiomer is formed in equal amount. The tabletop construction of cholesterol requires the building up of the carbocyclic system, cyclopentanoperhydrophenanthrene, the placement of the substituents, and the double bond in the later stages. The rings are constructed in the reverse order C → CD → BCD → ABCD. The angular methyls fall on the right places during ring construction. Strategy wise this is a linear multistep synthesis in which each new part is added stepwise leading to the formation of the final product. Even though the yields are good to excellent in individual steps, the overall yield is low.

(3S, 8S,
9S, 10R,
13R, 14S,
17R, 20R)

(1)

The ring D was initially kept as a six-membered one, but at a later stage was converted to the desired 5-membered ring with a handle via the seco derivative (i.e., D-ring opening). Through the handle by chemical manipulation a C8-hydrocarbon side chain with a chiral center was constructed.

The reactions leading to the total synthesis of (±)-cholesterol as per the above strategy involved more than 30 steps using common laboratory chemicals. In this method *cis-trans* equilibrium is shifted more to *trans* by seedling with the pure *trans* isomer obtained from previous batch and purified through crystallization. A mechanism of the formation of the 3-keto-Δ^4 compound, as shown in Fig. 11.8, has been suggested by Carl Djerassi [14].

Woodward et al. achieved a total synthesis [15] cholestanol from the intermediate (B) and another intermediate methyl 3α-acetoxy-$\Delta^{9(11)}$ etiocholenate (C) from (A) by partial hydrogenation and oxidation. From these intermediates, the paths to progesterone, desoxycorticosterone, testosterone, androsterone, and cortisone have been described previously by other investigators. The saturated keto-ester (B) has previously been converted to the Δ^4-compound. The corresponding acid was previously converted to the hormones desoxycorticosterone and progesterone. Thus this reaction sequence constitutes their total syntheses [12, 13].

Another classical synthesis of (±)-cholesterol was simultaneously reported in 1951 by Robinson's group [16–18]. It consists essentially of the addition of ring D onto a saturated A-B-C tricyclic diketone derived from 1-methyl-5-methoxy-2-tetralone [16–18].

11.2.3.1 Specification of the Chiral Centers. Conformation

Cholesterol molecule (**1**) contains eight dissimilar stereogenic (chiral) centers and accordingly it may have $2^8 = 256$ stereoisomers. Only few including cholesterol exist in nature due to biogenic feasibility, which also avoids steric encumbrance. The *trans fusions* of B/C and C/D rings confer conformational rigidity to such skeletons, and the *axial* and *equatorial* substituents can be differentiated in chemical reactions. The absolute configurations of the chiral centers of (−)-cholesterol are specified according to CIP rules as 3*S*, 8*S*, 9*S*, 10*R*, 13*R*, 14*S*, 17*R*, and 20*R* (steroid numbering) (structure **1**). This is in conformity with its biosynthesis in the presence of specific enzymes to be discussed in the sequel. The architecture of the ring scaffold, the stereochemistry of the fusion sites and other chiral centers, the specified absolute configuration of each stereogenic center, and the unequivocal conformation of cholesterol are shown in structures (**1**) above and (**1a**) in Fig. 11.2.

11.2.4 NMR Spectral Data of Cholesterol

¹H NMR: (CDCl₃, δ-values)

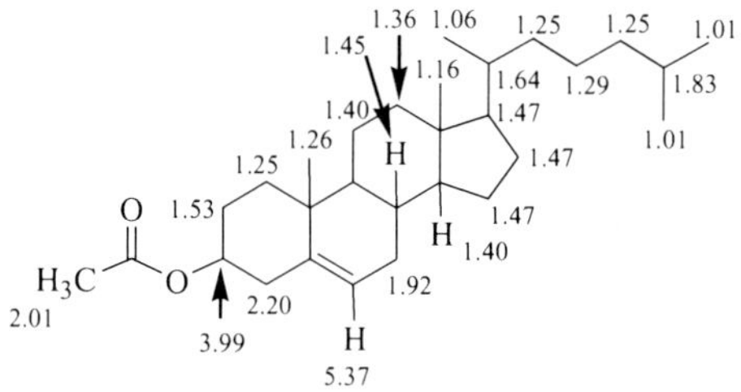

¹³CNMR: (CDCl₃, δ-values)

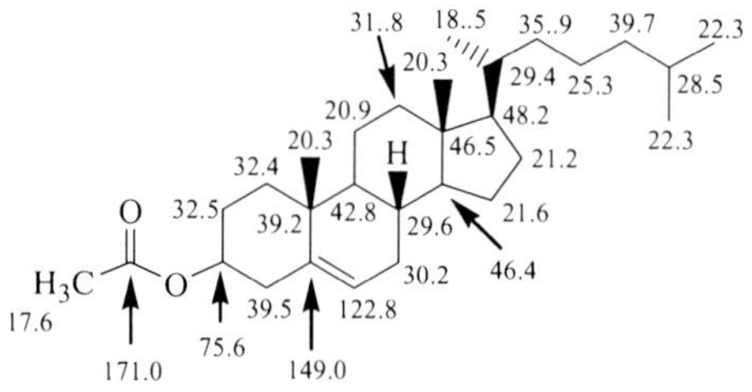

11.2.5 Biosynthesis of Cholesterol in Animals, Fungi, and Plants

The biosynthesis of cholesterol in animals has been discussed along with its biosynthesis in plants to show the similar and dissimilar steps involved in the processes.

The biosynthesis of cholesterol may be expressed in the following compressed form (Fig. 11.9).

Steps (a). The conversion of R-(+)-mevalonic acid into squalene has been discussed in detail in Chap. 5.

Step (b). "The discovery of Elias J. Corey and independently of Eugene van Tamelen of squalene epoxidation prior to cyclization was another key element for formulating cholesterol pathway" [5]. The enzyme squalene epoxidase (squalene monooxygenase) converts squalene into $3(S)$-2,3-oxidosqualene [4] by a regio- and stereospecific epoxidation of the former. The absolute stereochemistry of the latter has been established [19]. It has been shown that oxidosqualene cyclase shows substrate specificity on (2) and not on its $3(R)$-enantiomer.

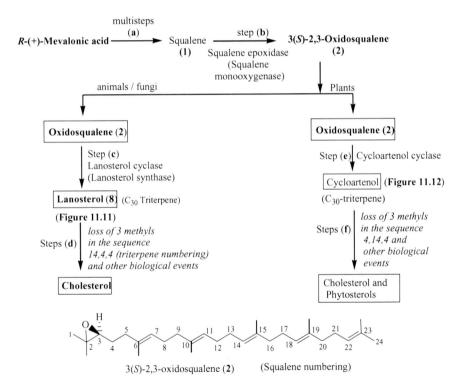

Fig. 11.9 Biosynthetic routes for formation of cholesterol in animals/fungi and plants and phytosterols in plants

11.2.5.1   Formation of Lanosterol in Animals and Fungi

Lanosterol (**8**) is the experimentally supported essential precursor of cholesterol in animal and fungi, and the study of its biosynthesis in the cell has evolved much interest. As a consequence of intensive studies several parallel proposals with thoughts in similar lines as well as some reviews [5, 20–24] came out in print [**Steps (c)**]. The enzymatic folding of (**2**) is assumed to attain preformed chair–boat–chair–open conformation (Fig. 11.10). During cyclization the conformational change is very less and the cyclization is initiated by enzymatic electrophilic addition of a proton to the epoxide which opens up to generate a transient carbocation. The latter is then added to the suitably disposed π bond, and the cation–olefin cyclization takes place. For each cyclization a new carbocation is formed and the substrate being a polyene, cation–olefin cyclization is repeated until it reaches a stopping point which is protosterol C20 carbocation (**4**). Subsequently a series of 1,2 shifts of hydrides and methyls result in the formation of lanosterol C8 carbocation (**4″**); the latter is quenched through the loss of C9-H as proton to yield lanosterol (**8**).

Extensive study on this conversion (**Step c**) (Fig. 11.10) provides a remarkable example of polyene cyclization, revealing a comprehensive stereochemical course of electrophilic antiparallel addition of carbocation to the olefinic double bond, followed by Wagner–Meerwein type 1,2-migrations of hydrides and methyls, and finally quenching of the carbocation through the loss of an appropriate proton. Incidentally, it should be mentioned that the brilliant work of some stalwarts [25–27] will always have a lasting influence on such studies.

Initially the cyclization was thought to proceed in a concerted nonstop fashion as per Ruziicka's "isoprene rule," while van Tamelen suggested that the cyclization proceeds through discrete carbocation intermediates based on some experimental observations and entropy considerations. However, Corey [28] suggested that during cyclization to form A, B, and C rings, ring C appears first as a

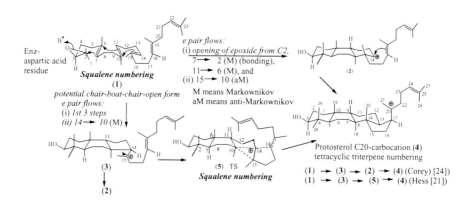

Fig. 11.10   Formation of protosteryl C20 carbocation (**4**)

5-membered ring (**3**) to avoid energetically disfavored anti-Markovnikov addition needed to effect six-membered ring formation via a secondary carbocation. The five-membered ring thus formed expands to 6-membered ring [6.6.5 (A.B.Pro C (**3**) → 6.6.6 A.B.C) (**2**)] (Fig. 11.10). The ring expansion is then followed by ring D (five membered) formation and a cascade of 1,2-hydride and methyl migrations, and lanosterol (**8**) is finally formed through the loss of H-9 as proton causing the quenching of carbocation (**4″**) (Fig. 11.11).

At this stage it should be pointed out that Hess Jr. [22] suggested that potential ring C (Pro C, being 5-membered) and ring D are formed in concert (**5**) (Fig. 11.10) in TS prior to the expansion of ring C in the biosynthesis of lanosterol and thus avoids energetically unfavorable ring expansion through a secondary carbocation in preference to a tertiary carbocation.

The absolute configuration at C20 of natural lanosterol is R. Had this side chain at C17 been α- and pseudo-equatorial, it would have necessitated ~120° rotation of the side chain around C17–C20 bond prior to H17 migration at C20 to confer R configuration at C20. However, this rotation of C17–C20 bond with C_8 side chain is disfavored because of 14β-Me group (severe nonbonding interactions). Further, the rate of hydride migration from C17 to C20 is faster than the rate of bond rotation. All these observations point towards the β- and pseudo-axial orientation of the side chain. This conjecture is supported by trapping the protosteryl carbocation analogue (**7**) and establishment of its structure by synthesis by Corey [28] from (**6**). This configuration of the side chain at C17 allows more stereochemical control of the enzyme on C20 carbocation (**4**) compared to the α-pseudo-equatorial orientation (**4′**).

The points revealed from the above studies on lanosterol biosynthesis are

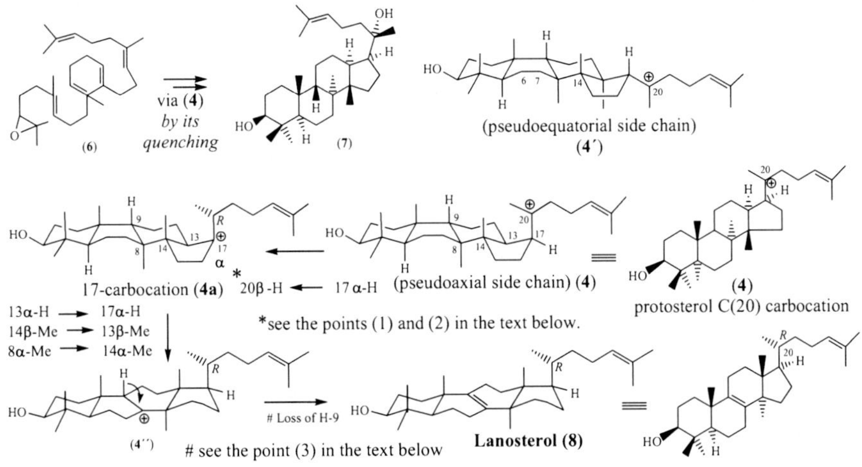

Fig. 11.11 Probable pathway for the formation of lanosterol in animals and fungi

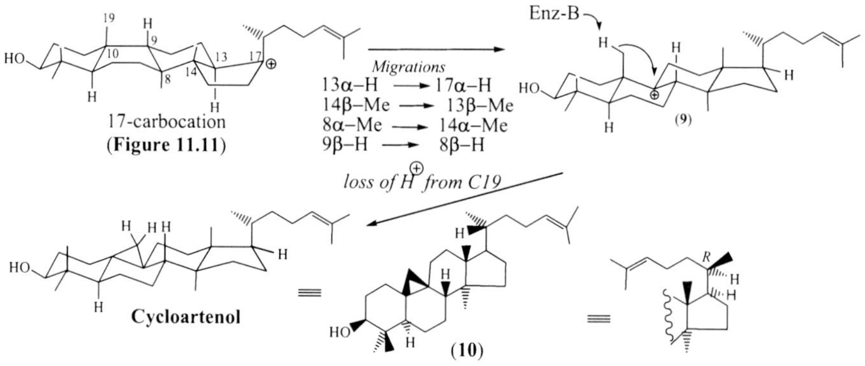

Fig. 11.12 Formation of cycloartenol in plants

(1) The enzyme protects the C20 carbocation from quenching by a nucleophile like water and holds the conformation of the cationic side chain in such a way that on delivery of the hydride the configuration at C20 becomes *R*.

(2) C17-H or H17) migrates first to C20 carbocation prior to other 1,2 spontaneous concerted migrations of hydrides and methyls leading to lanosterol C8 carbocation (**4′**).

(3) Enzyme plays an important role in the elimination of H9 as proton to quench C8 carbocation, completing the biosynthetic process of lanosterol.

11.2.5.2 Formation of Cycloartenol in Plants

In the plant cell in the presence of the enzyme cycloartenol cyclase, the carbocation moves in concert one carbon atom further, compared to lanosterol C8 carbocation (**4″**) and gets a new home at C9 (**9**) (**Steps e**). The latter is then quenched through the loss of one of the hydrogens as proton from an enzyme-activated C19-Me to form a cyclopropane ring [29]—a characteristic structural feature of cycloartenol (**10**). The enzyme plays a vital role in eliminating such proton (Fig. 11.12).

11.2.5.3 Lanosterol to Cholesterol and Cycloartenol to Cholesterol Conversions (Fig. 11.13)

The biosynthetic conversion (**Steps d**) (Fig. 11.9) requires

(1) *Loss of three angular methyls*

(2) *Change in the following existing structural features :*

   (a) *Reduction of the side chain double bond*

   (b) *Introduction of a new $\Delta^{5,6}$ double bond in place of $\Delta^{8,9}$ double bond.*

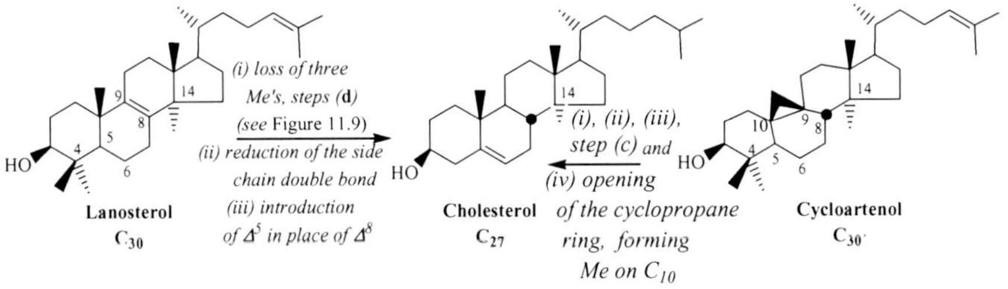

Fig. 11.13 Stages in the bioconversion of lanosterol into cholesterol (in animals and fungi), and of cycloartenol into cholesterol (in plants)

The biogenetic conversion path (**e**) (Fig. 11.9) followed in plants requires in total

(a) *Loss of the three angular methyls*
(b) *Opening of the cyclopropane ring generation C_{19}-Me on C_{10}*
(c) *Reduction of the side chain double bond and*
(d) *Creation of a double bond at $\Delta^{5,6}$.*

The three angular methyl groups are lost in the sequence 14, 4, 4 from lanosterol and 4, 14, 4 from cycloartenol. In between, phenomena like reduction, isomerization, and introduction of a new double bond by desaturation, etc., occur. The sequence of demethylation has been settled in both the cases through the isolation of selectively demethylated or oxygenated methyl sterols or products. In phytosterols the extra carbon atom/s are delivered by AdoMet (SAM) (S-adenosyl-1-methionine) at C24 and the π-facial diastereoselectively will leave the stereochemical information at the newly generated chiral center.

11.2.5.4   Biosynthetic Conversion of Lanosterol to Cholesterol

Biosynthetic conversion of lanosterol to cholesterol (Fig. 11.14) in animals and fungi starts with 14α-Me demethylation catalyzed by Lanosterol-14-demethylase (cytochrome-P-450 monooxygenase). By two successive oxidation reactions ($CH_3 \rightarrow CH_2OH$–CHO) and finally by Baeyer–Villiger type of oxidation, the original 14α Me is eliminated as HCOOH. Elimination of other methyls takes place as COOH $\rightarrow CO_2$ (with the help to 3 keto (3-OH $\rightarrow$ 3-CO) group, which is then reduced to 3β-OH.

The established association of high plasma cholesterol level, especially of LDL with plaque formation and development of artery blockage and coronary disease, stimulated the interest in the biosynthesis of cholesterol with a view to intercepting the formation of biosynthetic intermediate/s by identifying the relevant enzymes and inhibiting the enzyme/s to control the endogenous formation of cholesterol. In this effort the inhibitor/s of HMG-CoA reductase (Chap. 5, footnote 1), squalene epoxide cyclase, and demethylating enzymes of lanosterol have been discovered.

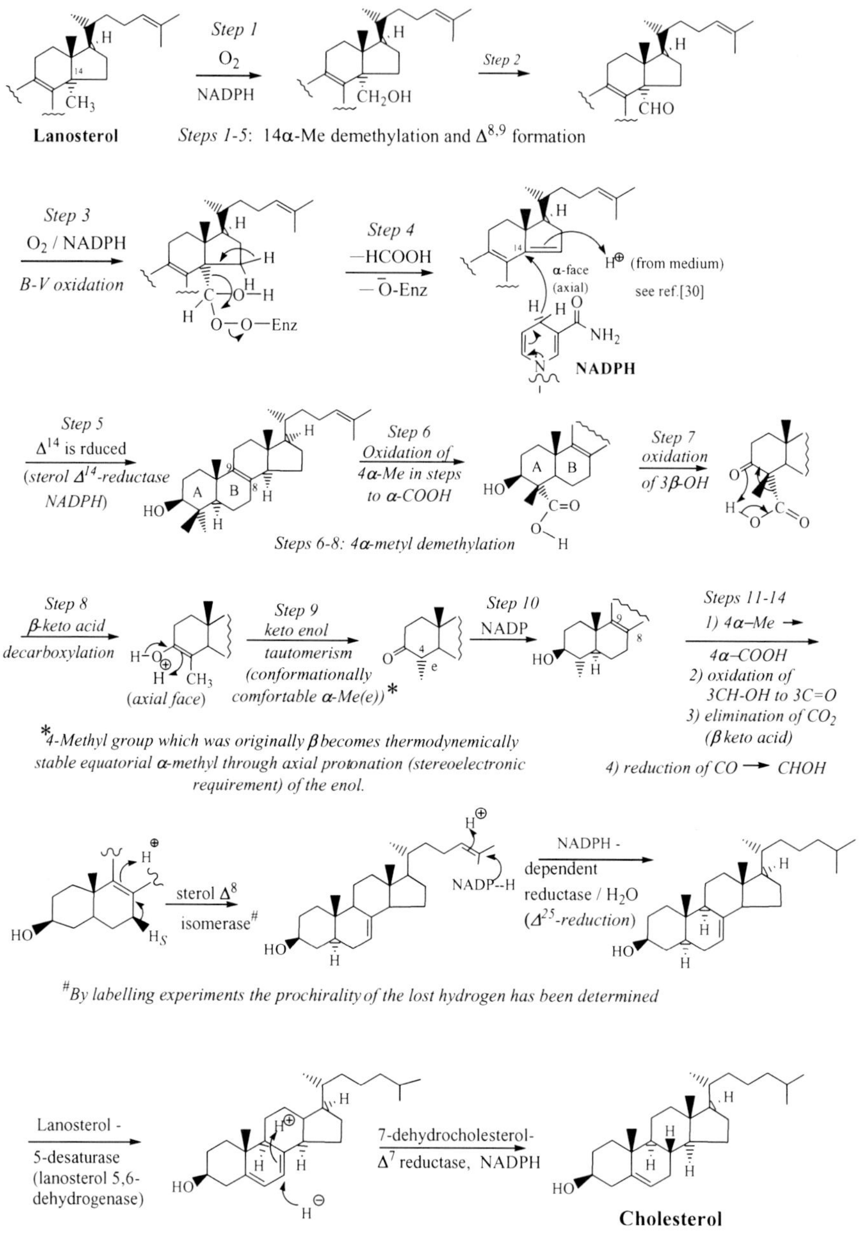

Lanosterol   *Steps 1-5*: 14α-Me demethylation and Δ^{8,9} formation

Step 3
O$_2$ / NADPH

B-V oxidation

Step 4
—HCOOH
—O-Enz

α-face
(axial)

H$^{\oplus}$ (from medium)

see ref.[30]

NADPH

Step 5
Δ^{14} is rduced
(sterol Δ^{14}-reductase
NADPH)

Step 6
Oxidation of
4α-Me in steps
to α-COOH

Step 7
oxidation
of 3β-OH

Steps 6-8: 4α-metyl demethylation

Step 8
β-keto acid
decarboxylation

Step 9
keto enol
tautomerism
(conformationally
comfortable α-Me(e))*

(axial face)

Step 10
NADP

Steps 11-14
1) 4α–Me →
4α–COOH
2) oxidation of
3CH-OH to 3C=O
3) elimination of CO$_2$
(β keto acid)

4) reduction of CO → CHOH

*4-Methyl group which was originally β becomes thermodynemically
stable equatorial α-methyl through axial protonation (stereoelectronic
requirement) of the enol.*

sterol Δ^8
isomerase#

NADP--H

NADPH -
dependent
reductase / H$_2$O
(Δ^{25}-reduction)

#*By labelling experiments the prochirality of the lost hydrogen has been determined*

Lanosterol -

5-desaturase
(lanosterol 5,6-
dehydrogenase)

7-dehydrocholesterol-
Δ^7 reductase, NADPH

Cholesterol

Fig. 11.14 Biogenetic conversion of lanosterol to cholesterol (in animal and fungi)

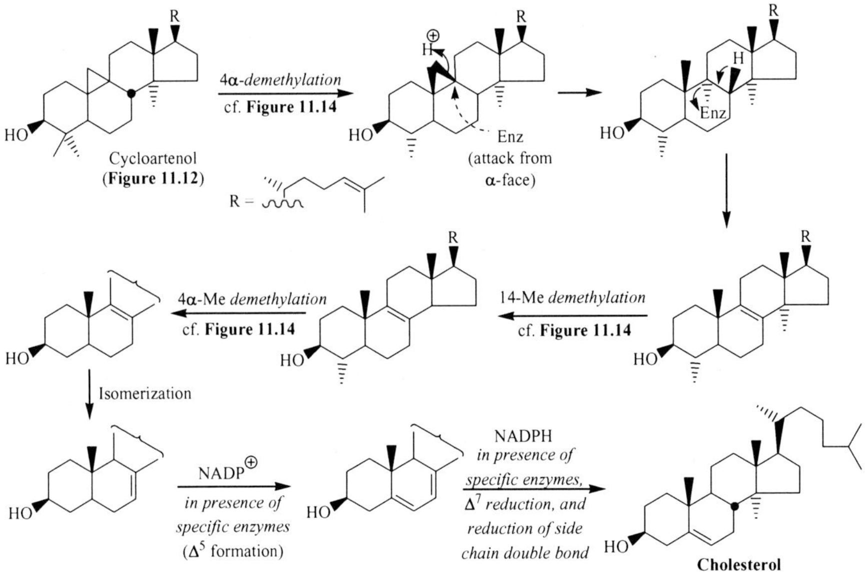

Fig. 11.15   Probable sequence of biogenesis of cholesterol from cycloartenol in plants

11.2.5.5   Biosynthesis of Cholesterol from Cycloartenol in Plants

Probable sequence of steps in the biogenesis of cholesterol from cycloartenol is outlined in Fig. 11.15. This proposal based on the present knowledge of biosynthesis of cholesterol via lanosterol in animals and fungi (Figs. 11.11 and 11.14) and phytosterols from cycloartenol is delineated in these Figures. However, it seems from the literature that chronology of the steps for plant cholesterol, in the absence of any circumstantial evidence, cannot be defined in definite terms.

11.2.5.6   Biogenetic Conversion of Cycloartenol to Other Phytosterols (Fig. 11.16)

Unlike the animal membranes, the plant membranes contain a mixture of sterols, some of which are sitosterols $(24R)$-24-ethylcholesterol, stigmasterol$(24S)$-24-ethylcholesta-5,22-dienol, $(24R)$-24-methylcholesterol, 24-epicampesterol, and campesterol. Sitosterols are involved in membrane reinforcement.

The biosynthesis of phytosterol (Fig. 11.17) starts with the methylation at Δ^{24} of the side chain. The latter is considered to be the obligatory location for first alkylation. The addition of the methyl group is governed by the enzyme which dictates the π-facial diastereoselectivity of alkylation and plays a vital role for being nucleophilic toward electrophilic SAM.

Lanosterol is nowhere present in the plant system. The side chain variation using Δ^{24} is shown in Fig. 11.17. Biogenesis of these phytosterols has been studied [31–33] and the enzymes have been identified.

*Ψ Concerted opening of the β-cyclopropane ring through the elimination of 8β–H proton is stereoelectronically unfavourable being in cis relationship. Hence the ring opens through the nucleophilic attack of an enzyme followed by **trans** elimination.*

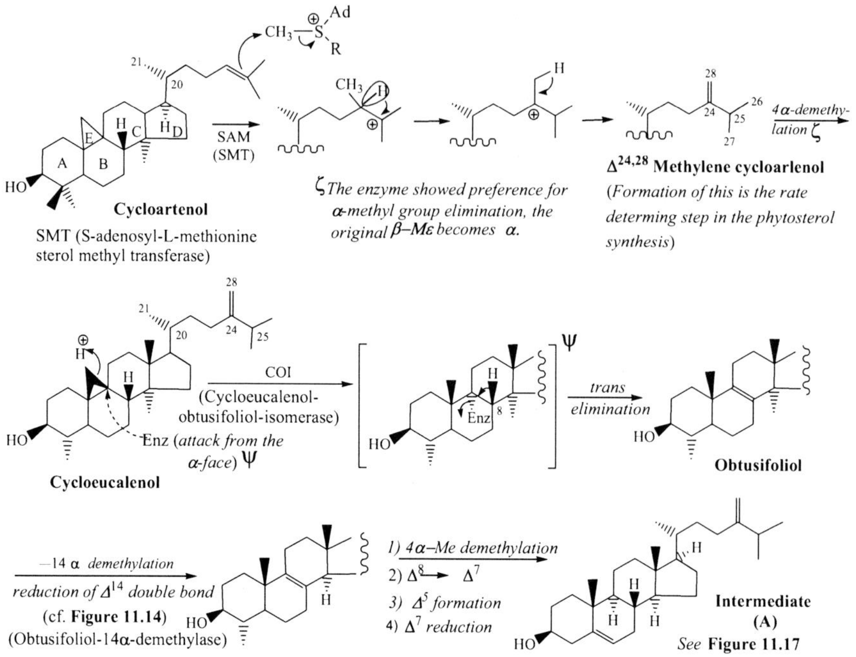

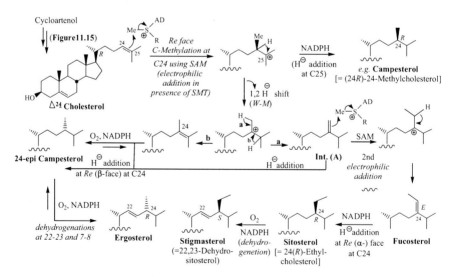

Fig. 11.16 Biogenetic modifications of the side chain and cyclopropane ring of cycloartenol to form phytosterols

Note: *For full structures of the phytosterols see* **Figure 11.18**

Fig. 11.17 Bioformation of different phytosterols by side chain variation using Δ^{24}-cholesterol

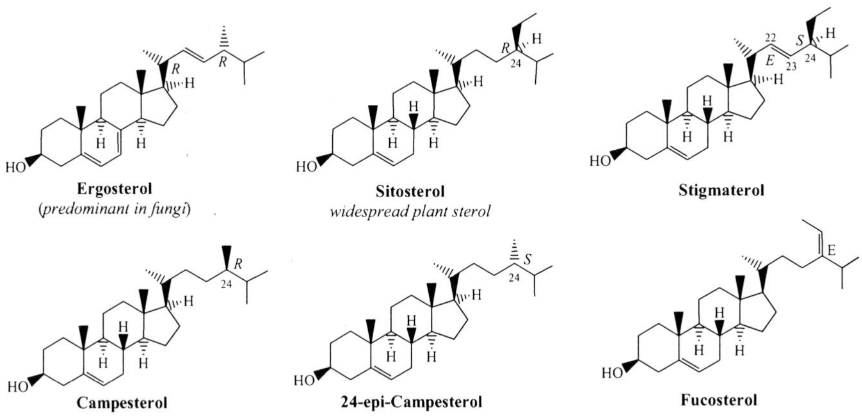

Fig. 11.18 Some phytosterols

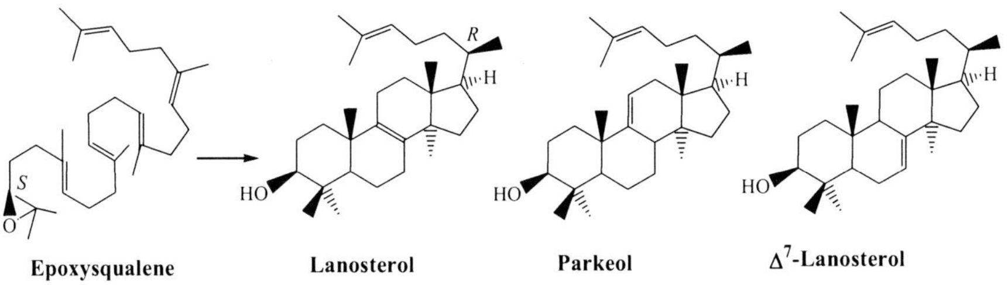

Fig. 11.19 Mutation of cycloartenol synthase lanosterol and allied compounds

Plant cell membranes produce phytosterols elaborating different variations of side chains (Fig. 11.17), which include campesterol [=(24R)-24-methylcholesterol], 24-epicampesterol, ergosterol [=(24R)-24-methylcholesta-5,22-dienol], sitosterol [=(24R)-24-ethylcholesterol) and stigmasterol [=(24S)-24-ethylcholesta-5,22-dienol] [31], and fucosterol (=24E-ethyledenecholesterol). The structures of these phytosterols are shown in Fig. 11.18. The biosynthesis of sitosterol (=24-R-ethylcholesterol) and stigmasterol (=22,23-dehydrositosterol) in plant cells has demonstrated that their isoprene units are supplied exclusively from the mevalonate (MVA) pathway [31, 32]; MVA is located in the cytoplasm.

The specific deprotonation by these two different enzymes (lanosterol synthase and cycloartenol cyclase) is responsible for different pathways for cholesterol formation in animals and plants. However, recent studies [34] showed that mutagenesis at specific site of cycloartenol synthase allows lanosterol biosynthesis (Fig. 11.19). A novel site-specific mutant His477Asn was uncovered that produces 88 % lanosterol and 12 % parkeol, while His477Gln mutant produces 73 % parkeol, 22 % lanosterol, and 5 % Δ^7-lanosterol; however, in plants no lanosterol biosynthesis takes place in the absence of lanosterol synthase.

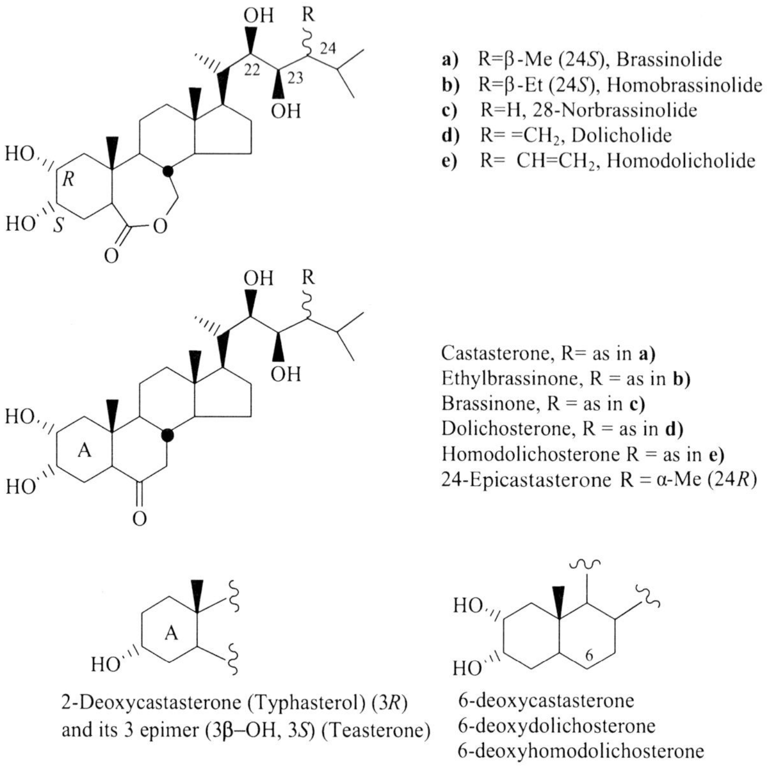

a) R=β-Me (24S), Brassinolide
b) R=β-Et (24S), Homobrassinolide
c) R=H, 28-Norbrassinolide
d) R= =CH₂, Dolicholide
e) R= CH=CH₂, Homodolicholide

Castasterone, R= as in **a)**
Ethylbrassinone, R = as in **b)**
Brassinone, R = as in **c)**
Dolichosterone, R = as in **d)**
Homodolichosterone R = as in **e)**
24-Epicastasterone R = α-Me (24R)

2-Deoxycastasterone (Typhasterol) (3R)
and its 3 epimer (3β–OH, 3S) (Teasterone)

6-deoxycastasterone
6-deoxydolichosterone
6-deoxyhomodolichosterone

Fig. 11.20 Some natural brassinosteroids

11.3   Brassinosteroids [35–40]

11.3.1   Introduction. Some Brassinosteroids

Brassinosteroids represent a new class of phytohormones and its first member named brassinolide **1** has been isolated in extremely small yield (4 mg from 40 kg bee collected pollens) of rape seed plant (*Brassica napus*, Cruciferae/Brassicaceae) [35, 36]. Isolation of brassinolide confirms the role of steroids as hormones in plants and demonstrates an evolutionary conservation between plants and animals. It is a powerful growth-promoting agent and the increase in the growth is due to both cell elongation and cell division. For this duel effect of brassinolide, its discovery appears to be more important to plant physiologists and biochemists. It contains an unprecedented seven-membered B-ring lactone and its structure has been elucidated as (2R,3S,22R,23R,24S)-2,3,22,23-tetrahydroxy-24-methyl-B-homo-7-oxa-5-α-cholestan-6-one (Fig. 11.20) from chemical, spectral, and X-ray analysis. Till 1991, more than 60 brassinosteroids have been isolated from a wide variety of plants [37, 38]. A few representatives of brassinosteroid family are shown to possess the following structural characteristics [36, 39, 40] (Fig. 11.20).

All the brassinosteroids are (2R,3S,22R,23R)-tetrols except in *typhasterol* and *teasterone* where only one 3R or 3S OH group is present. Brassinosteroids carrying *lide* in the suffix contain a 7-membered-B-ring lactone with differently substituted side chain. Compounds of this class with 6-oxo group also contain the same type of side chains.

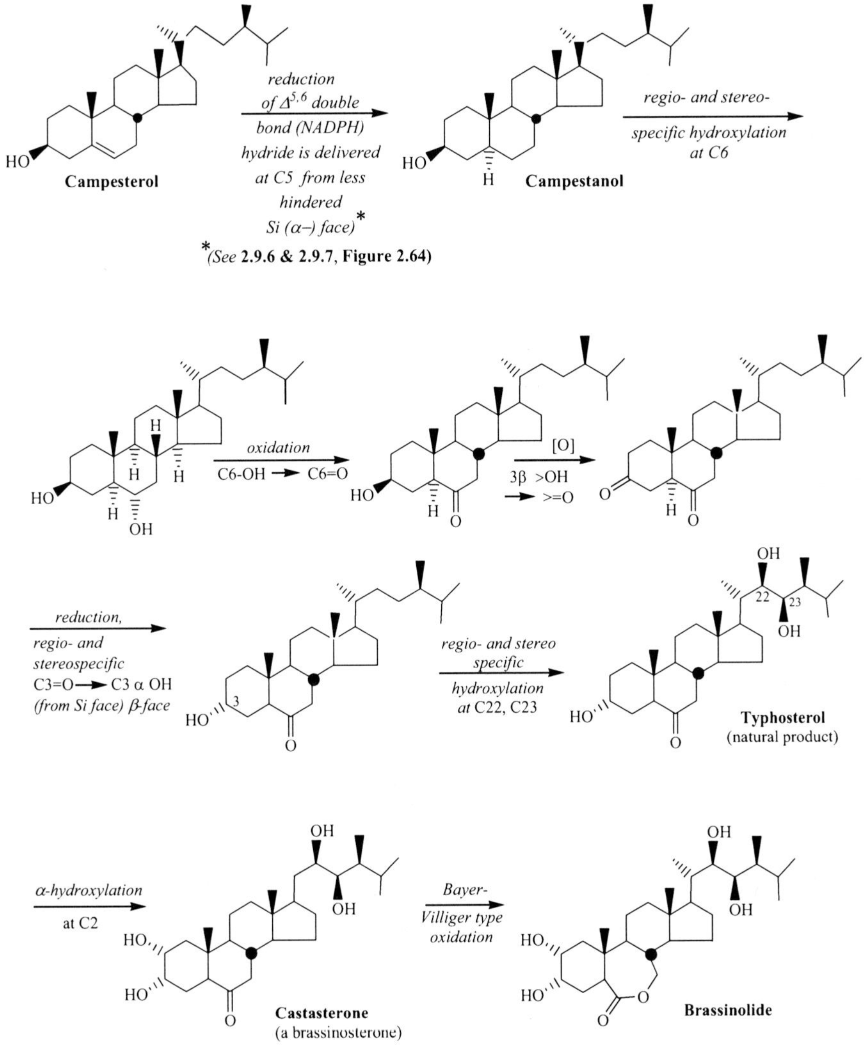

Fig. 11.21 Biosynthesis of brassinosterone and brassinolide

11.3.2   Biosynthesis

From the relationship of brassinolides and the brassinosterones, it is obvious that in the final step Baeyer–Villigar oxidation takes place and brassinosterones are the immediate precursors of the corresponding brassinolides. From label experiments it has been shown that campesterol is converted to a brassinosterone and finally to a brassinolide (Fig. 11.21). Further the side chain functionalization takes place during the formation of the 6-oxo-compound.

In view of their low natural content in plants, and their importance as agricultural agents, much effort has been directed toward the total and semisynthesis of brassinosteroids. However these are multistep processes and the yields are not encouraging for commercial use. The challenging stereostructures of brassinosteroids have attracted the attention of synthetic chemists and a number of syntheses have been reported. The first synthesis of brassinolide has been reported in 1980 by two groups of workers [41, 42].

11.3.3   Spectral Data of Brassinolide [35, 36, 38]

Brassinolide

$C_{28}H_{48}O_6$, mp. 274–275 °C, $[\alpha]_D$ + 16

IR (KBr) ν_{max} 3450, 1,690 cm^{-1}, (CH$_2$Cl$_2$), 1,720 cm^{-1}

^{1}H NMR (C$_5$D$_5$N): δ 0.72 (s, 3H), 1.03 (d, 6H, J = 6.3 Hz), 1.04 (s, 3H), 1.10 (d, 3H, J = 6.4 Hz), 1.13 (d, 3H, J = 6.5 Hz)

^{13}C NMR (CD$_2$Cl$_2$–CD$_3$OD) (9:1): δ 10.3, 11.9, 12.1, 15.6, 20.8, 21.0 (CH$_3$), 68.3, 68.4, 71.0, 73.7, 74.8 (4CHOH, 1 CH$_2$–O–), 177.6 (lactone carbonyl)

MS (FD): m/e 463 (M+1 – H$_2$O)$^+$, 445 (M+1 – 2H$_2$O), 409, 379, 361, 349, 101, 71 (side chain fragmented species)

The structure of brassinolide was completely established by its X-ray crystallographic analysis [35, 36].

11.4 Other Bioactive Steroidal Compounds

11.4.1 Ecdysones [43, 44]

11.4.1.1 Introduction. Structures

Ecdysones (Fig. 11.22) are a group of moulting hormones necessary for the larval development (cf. *Sect.* 1.6.2.2) of the insects. Insects cannot synthesize sterols, the biogenetic precursors of ecdysones. They take food which supplies the ecdysones or sterols (cholesterol, sitosterols, etc.). The sterols are converted into ecdysones in their system. Thus they are of dietary origin. Ecdysis (Greek, shedding), the entomological term for moulting, has been used to name this class of moulting hormones. Higher plants elaborating ecdysones play important role in the metabolism of phytophagous insects.

11.4.1.2 Biogenesis

The probable events (1–7) in the biogenesis of ecdysones from cholesterol are outlined in Fig. 11.23.

The hydroxylation is a very common biosynthetic phenomenon in ecdysones. The sequence of the biological events (1–7) is speculative and may be different. It depends on the enzyme specificities. The side chain containing an extra methyl, e.g., β-sitosterol chain, may be modified. Other sterol side chains may also undergo similar hydroxylations during their conversion to ecdysones.

Some other examples of ecdysones [43, 44] are given in Fig. 11.24.

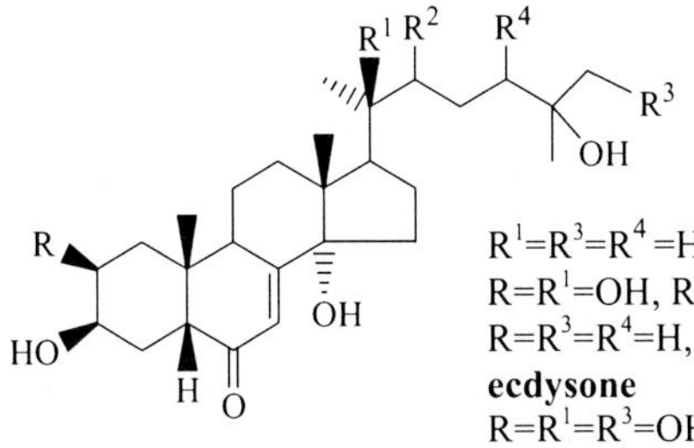

$R^1=R^3=R^4=H$, $R=OH$, $R^2=\alpha$-OH, Ecdysone (**α-Ecdysone**)
$R=R^1=OH$, $R^2=\alpha$-OH, $R^3=R^4=H$, Ecdysterone (**β-Ecdysone**)
$R=R^3=R^4=H$, $R^1=OH$, $R^2=\alpha$-OH, **2-Deoxy-20-hydroxyecdysone**
$R=R^1=R^3=OH$, $R^2=\alpha$-OH, $R^4=H$, **Inokosterone**
$R=R^1=OH$, $R^3=H$, $R^2=\beta OH$, $R^4=Me$, **Makisterone-A**

Fig. 11.22 Structures of some ecdysones

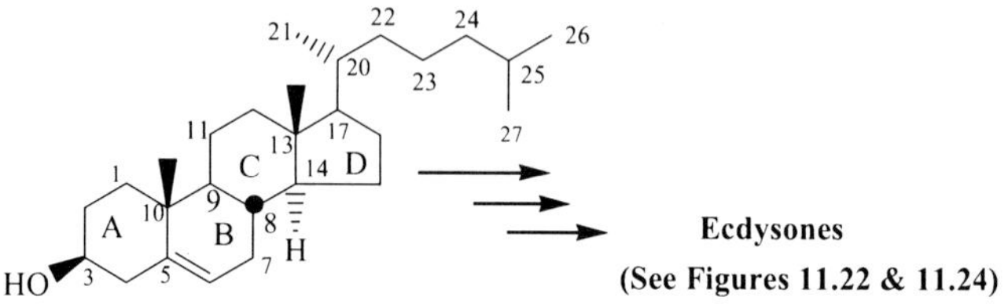

Cholesterol

Ecdysones
(See Figures 11.22 & 11.24)

(1)   Reduction of $\Delta^{5,6}$ bond to yield cis fused A/B rings

(2)   Hydroxylation at C6→ regiospecific oxidation to C6>=O or

(3)   α-epoxidation at $\Delta^{5,6}$, trans opening – delivery of hydride from β-face making A/B cis fused, followed by regiospecific
oxidation of C6>CHOH; operations (1-2) or 3, converts A/B rings to (X).

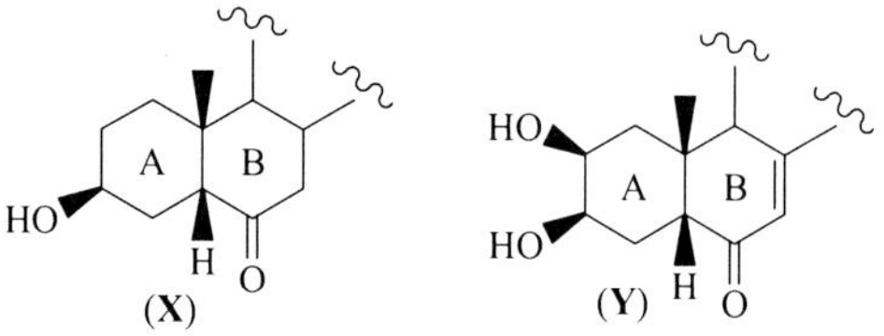

(4)   Dehydrogenation at C6-C7 (2NADP$^+$ → 2NADPH)

(5)   In most of the cases β-hydroxylation at C2; after operations 4-5 A/B rings are converted to (Y).

(6)   Selective hydroxylation/s at C20, C22, C25, C26/C27 elaborating the following side chains is ecdysones,

(7)   A compulsory α-hydroxylation at C14, and hydroxylation at C25 in most ecdysones

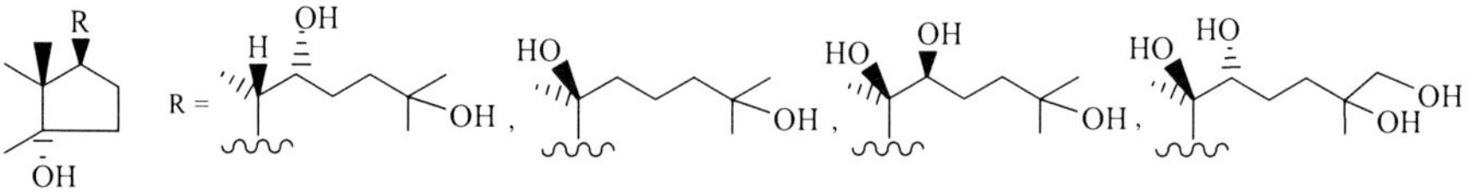

Fig. 11.23 Probable steps in the biogenesis of some ecdysones from cholesterol and hydroxylation patterns in the side chain

11.4.2   Diosgenin: Diosgenin-Derived Steroidal Drugs

Diosgenin (Fig. 11.25), $C_{27}H_{42}O_3$, m.p. 204–207, $[\alpha]_D$ + 129 (CHCl$_3$), an important steroidal derivative, occurs mainly as the glycosides and rarely in the free state in the rhizomes of the plants belonging to the genus *Dioscorea* (Dioscoreaceae). Dioscine, a common glycoside, is diosgenin-rhamno-rhamno glucoside. Other sources of diosgenin where it occurs in the free state are *Costus speciosus* (Zingiberaceae), *Kallstroemia pubescens*, and *Trigonella-foenum-graecum* (Leguminosae). Diosgenin has been the steroid drug precursor of choice. It is converted to 16-DPA (16-dehydropregnenolone acetate and then into various steroidal drugs (Fig. 11.25)—corticosteroids. Mexico is the largest producer of diosgenin.

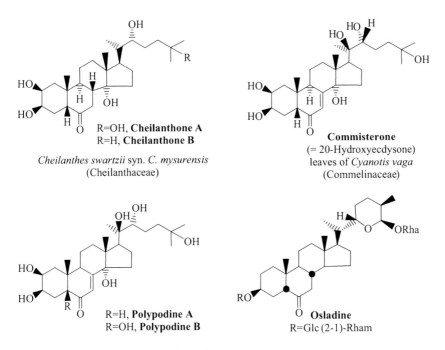

R=OH, **Cheilanthone A**
R=H, **Cheilanthone B**

Cheilanthes swartzii syn. *C. mysurensis*
(Cheilanthaceae)

Commisterone
(= 20-Hydroxyecdysone)
leaves of *Cyanotis vaga*
(Commelinaceae)

R=H, **Polypodine A**
R=OH, **Polypodine B**

Osladine
R=Glc (2-1)-Rham

Fig. 11.24   The structures of some other ecdysones

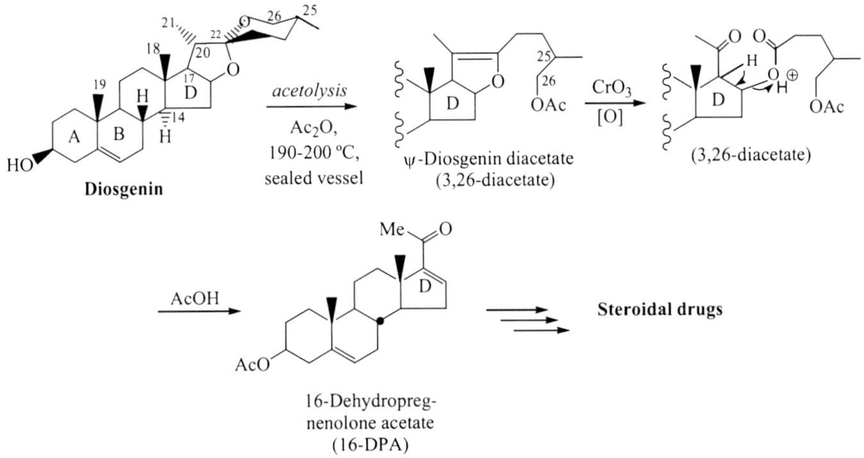

Diosgenin

acetolysis
Ac₂O,
190-200 °C,
sealed vessel

ψ-Diosgenin diacetate
(3,26-diacetate)

CrO₃
[O]

(3,26-diacetate)

AcOH

16-Dehydropreg-
nenolone acetate
(16-DPA)

Steroidal drugs

Fig. 11.25   Diosgenin and derived steroidal drugs

11.4.2.1 Spectral Data of Diosgenin

IR (KBr) ν_{max} 3,460 (OH)
^{1}H NMR (CDCl$_3$) δ: 0.79 (*s*, C27 Me and C18-Me), 0.98 (*d*, J = 6.4 Hz, C21 Me), 1.03 (*s*, C19-Me), 4.40 (*q*, C16 H), 5.31 (brd, C6-H)
^{13}C NMR (CDCl$_3$) δ: C1 (37.3), C2 (31.6), C3 (71.6), C4 (42.3), C5 (140.6), C6 (121.3), C7 (32.6), C8 (31.4), C9 (8.501), C10 (36.6), C10 (36.6), C11 (20.4), C12 (34.8), C13 (40.3), C14 (56.5), C15 (31.8), C16 (80.7), C17 (62.1), C18 (16.3), C19 (19.4), C20 (14.6), C21 (14.5), C22 (109.3), C23 (31.4), C24 (28.8), C25 (30.3), C26 (66.8)
MS: *m/z* 414 (M$^+$), 342, 139, 115.

11.4.3 Cardioactive Glycosides

The aglycone part of the cardioactive glycosides is having steroidal structures and is characterized by A/B and C/D *cis* fused rings; sugar molecules forming glycosidic bonds with 3β-OH group; a β-hydroxyl at C14; and an unsaturated γ- or δ-lactone on C17β. The most well known is **digitoxin** which occurs in the genus *Digitalis*. In Hellebrigenin having the δ-lactone on C17β., an additional β-OH at C5 and more oxidized C10-Me as C10-CHO are present (Fig. 11.26). The aglycone part improves the action of the heart. The sugar part of the glycoside is responsible for the solubility of the drug. These basic structures arise biosynthetically by metabolism of cholesterol effecting modifications of side chains.

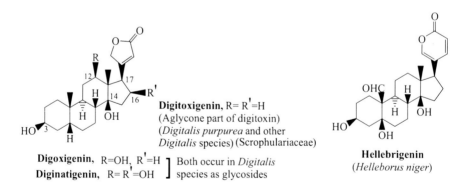

Digitoxigenin, R= R′=H
(Aglycone part of digitoxin)
(*Digitalis purpurea* and other
Digitalis species) (Scrophulariaceae)

Digoxigenin, R=OH, R′=H ⎤ Both occur in *Digitalis*
Diginatigenin, R= R′=OH ⎦ species as glycosides

Hellebrigenin
(*Helleborus niger*)

Fig. 11.26 Structures of the aglycone parts of two cardiac glycosides, digitoxin, and hellebrigenin

References

1. Leon Gortler and Jeffrey L. Sturchio, 'Introduction' in *Steroids*, **1992**, *57*, 356, Butterworth-Heinemann.
2. D. H. R. Barton, Conformation of Steroid Nucleus, *Experientia*, **1950**, 316-319.
3. D. H. R. Barton and O. Hassel, Conformational Analysis – The Fundamental Contributions of D. H. R. Barton and O. Hassel, *Top. Stereochem*, **1971**, *6*, 1–17.
4. R. B. Woodward, Franz Sondheimer and David Taub, The Total Synthesis of Cholesterol, *J. Am. Chem. Soc.*, **1951**, *73*, 3548.
5. Konrad Bloch, Steroid Molecule: Structure, Biosynthesis, and Function, *Steroids*, **1992**, *57*, 378-383.
6. Buckingham, Ed., *Dictionary of Natural Products* **1994**, Vol. 1, p. 1011 (C-01218), Chapman and Hall.
7. E. J. Behrman and Venkat Gopalan, Cholesterol and Plants, *J. Chem. Educ.*, **2005**, *82*, 1791-1793
8. Kyosuke Tsuda, Saburo Akagi and Yukichi Kishida, Steroid Studies, VIII. Cholesterol in Some Algae, *Chem. Pharm. Bull.*, **1958**, *6*, 101-104.
9. R. P. Rastogi and B. N. Mehrotra, *Compendium of Indian Medicinal Plants,* Vols. 1-4, Publication & Information Directorate, CSIR, New Delhi, **1991**.
10. A. M. Rizk, *Phytochemistry of the Flora of Qatar,* Scientific & Applied Research Centre, University of Qatar, Published by Kingprint of Richmond, **1986**.
11. Louis F. Fieser, and Mary Fieser, *Steroids*, Reinhold Publishing Corporation, New York, Amsterdam, London, **1959**, 2nd print, **1967**.
12. R. B. Woodward, Franz Sondheimer and David Taub, The Total Synthesis of Some Naturally Occurring Steroids, *J. Am. Chem. Soc.*, **1951**, *73*, 3547-3548
13. R. B. Woodward, Franz Sondheimer, David Taub, Karl Heusler and W. M. McLamore, The Total Synthesis of a Steroid, *J. Am. Chem. Soc.*, **1951**, *73*, 2403-2404.
14. G. Rosenkranz, O. Mancera, J. Gatica and Carl Djerassi, Steroids, IV. α-Iodoketones. A Method for the Conversion of Allosteroids into Δ^4-3-ketosteroids, *J. Am. Chem. Soc.*, **1950**, *72*, 4077-4080.
15. R. B. Woodward, Franz Sondheimer, David Taub, Karl Hansler and W. M. McLamore, The Total Synthesis of Steroids, *J. Am. Chem. Soc.*, **1952**, *74*, 4223-4251, and references cited.
16. H. M. E. Cardwell, J. W. Cornforth, S. R. Duff, Hugo Holtermann and Sir Robert Robinson, Total Synthesis of Androgenic Hormones *Chem. Ind (London).*, **1951**, 389-390
17. H. M. E. Cardwell, J. W. Cornforth, S. R. Duff, Hugo Holtermann and Sir Robert Robinson, Experiments on the Synthesis Related to the Steroids Part LI. Completion of the Synthesis of Androgenic Hormones and of the Cholesterol Group of Sterols *J. Chem. Soc.*, **1953**, 361-384.
18. J. W. Cornforth, *Progress Org. Chem.*, **1955**, *3*, 21.
19. Derek H. R. Barton, Trevor. R. Jarman, Keith. C. Watson, David. A. Widdowson, Robin. B. Boar and Kathleen. Damps, Investigation on the Biosynthesis of Steroids and Terpenoids, Part XII. Biosynthesis of 3- hydroxyl terpenoids and Steroids from (3S)-2,3 –Epoxy-2.3-dihydrosqualene *J. Chem. Soc., Perkin Trans*. 1, **1975**, 1134-1138.
20. Ikuro Abe, Michel Rohmer and Glenn D. Prestwich, Enzymatic Cyclization of Squalene and Oxidosqualene to Sterols and Triterpenes, *Chem. Rev.*, **1993**, *93*, 2189-2260 and the references cited.
21. K. Ultrich Wendt, George E. Schulz, Elias J. Corey and David R. Liu, Enzyme Mechanism for Polycyclic Triterpene Formation, *Angew. Chem., Intl. Ed.*, **2000**, *39*, 2812-2833.
22. B. Andes Hess, Jr., Formation of C Ring in the Lanosterol Biosynthesis from Squalene, *Org. Lett.*, **2003**, *5*, 165-167.
23. Geoffrey D. Brown, The Biosynthesis of Steroids and Triterpenoids, *Nat. Prod. Rep.*, **1998**, 653-696, and references cited.
24. L. J. Mulheirn and P. L. Ramm, Biosynthesis of Sterols, *Chemical Society Reviews*, **1972**, *1*, 259-291.

25. G. L. Stork and A. W. Burgsthaler, Stereochemistry of Polyene Cyclization, and the Prophetic Concept of 'isoprene rule' by L. Ruzicka *et al.*, *J. Am. Chem. Soc.,* **1955**, *77*, 5068-5077

26. A. Eschenmoser, L. Ruzicka, O. Jeger, and D. Arigoni, Eine Stereochemische Interpretation der Biogenetischen Isoprene Regel bei den Triterpen, *Helv. Chim. Acta*, **1955**, *36*, 1611-1614.

27. Albert Eschenmoser and Dullio Arigoni, Revisited after 50 Years: The 'Stereochemical Interpretation of the Biogenetic Isoprene Rule for the Triterpenes, *Helv. Chim. Acta*, **2005**, *88*, 3011-3049.

28. R. Ulrich Wendt, George E. Schulz, Elias J. Corey and David R. Liu, Enzyme Mechanisms for Polycyclic Triterpene Formation, *Angew. Chem. Int. Ed.*, **2000**, *39*, 2812-283.

29. Ludger A. Wessjohann and Wolfgang Brandt, Biosynthesis and Metabolism of Cyclopropane Rings in Natural Products, *Chem. Rev.*, **2003**, *103*, 1625-1647.

30. M. Akhtar, A. D. Rahuntula and D. C. Welton, The Incorporation of a 15β-Hydrogen Atom from the Medium in Cholesterol Biosynthesis, *J. C. S. Chem. Comm.*, **1969**, 1278-1260 (1969).

31. D. Arigoni and M. Schwarz, in *Comprehensive Natural Product Chemistry,* D. Cane, Volume Ed., Pergamon, Oxford, **1999**, Vol . 2, pp 45-68.

32. Damrong Kongduang, Juraithip Wungsintaweekul, and Wanchai De-Eknamkul, Biosynthesis of β-Sitosterol and Stigmasterol Proceeds Exclusively *via* the Mevalonate Pathway in Cell Suspension Cultures of *Croton stellatopilosus,*, *Tetrahedron Lett.*, **2008**, 4067-4072.

33. Edgar Lederer, Some Problems Concerning Biological C-Alkylation Reaction and Phytosterol Biosynthesis, *Quart. Rev.*, The Chemical Society, **1969**, *23*, 453-481.

34. Michael J. R. Segura, Silvia Lodeiro, Michelle M. Meyer, Akash J. Patel, and Seiichi P. T. Matsuda, Directed Evolution Experiments Reveal Mutations at Cycloartenol Synthase Residue His477 that Dramatically Alter Catalysis, *Org. Lett.*, **2002**, *4*, 4459-4462.

35. Michael D. Grove, Gayland F. Spencr, William K. Rohwedder, Nagabhushanam Mandava, Joseph F. Worley, J. David Warthen Jr., George L. Steffens, Judith L. Flippen-Anderson and J. Carter Cook Jr., Brassinolide, a Plant-Growth Promoting Steroid Isolated from *Brassica napus* Pollen, *Nature*, **1979**, *281*, 216-221.

36. G.Adam, J. Schmidt and B. Schneider, Brassinosteroids, hem *Fortschritte.Organischer Naturstoffe*, **1999**,*78*, 1-46.

37. S.-K. Kim in *Natural Occurrence of Brassinosteroids,* Eds. H. G. Cutler, T. Yokota and G. Adam, *ACS Symp. Ser.* 474, **1991**, Chapter **3**.

38. Noboru Murofushi, Hisakazu Yamane, Youji Sakagami, Hidemasa Imaseki, Yuji Kamiya, Hajime Iwamura, Noboru Hirai, Hideo Tsiyi, Takao Yokota and Juniehi Ueda, Plant Hormones in *Comprehensive Natural Products Chemistry*, Editor-in-Chief Sir Derck Barton and Koji Nakanishi, Vol **8**, Volume Editor Kenji Mori, Pergamon Press, Elsevier, **1999**, pp. 19-136; pertinent pp. 99-107.

39. G. Adam and Y. Marquardt, Review Article Number 19, Brassinosteroids, *Phytochemisty*, **1986**, *25*, 1787-1799.

40. Braja G. Hazra and Vandana Pore, Brassinoids, New Class of Phytohormones, *J. Indian Chem. Soc.* (Professor Sukh Dev 75th Birthday Commemoration Issue), **1998**, *75*, 746-757.

41. Steven Fung and John B. Siddall, Stereoselective Synthesis of Brassinolide: a Plant-Growth Promoting Steroidal Lactone, *J. Am. Chem. Soc.*, **1980**, *102*, 6580-6581.

42. Masaji Ishiguro, Suguru Takatsuto, Masuo Morisaki, and Nobuo Ikekawa, Synthesis of Brassinolide, a Steroidal Lactone with Plant-growth Promoting Activity, *J. Chem. Soc., Chem. Comm.*, **1980**, 962-964.

43. M. Bathory, I. Toth, K. Szendrei, and J. Reisch, Ecdysteroids in *Spinacia oleracea* and *Chenopodium bonus-henricus*, *Phytochemistry*, **1982**, *21,* 236-238.

44. J. Jizba, V. Herout, and F. Sorm, Polypodine B – a Novel Ecdysone-like Substance from Plant Material, *Tetrahedron Lett.*, **1967**, 5139-5143.

General References

Louis F. Fieser, and Mary Fieser, *Steroids*, Reinhold Publishing Corporation, New York, Amsterdam, London, **1959**, 2nd print, **1967.**

I. L. Finar, *Organic Chemistry, Volume 2: Stereochemistry and the Chemistry of Natural Products*, Fifth Edition, **ELBS**, Longman, **1975**.

Geoffrey D. Brown, The Biosynthesis of Steroids and Triterpenoids, *Nat. Prod. Rep.*, **1998**, pp 653-696.

Christopher J. Coulson, Steroid Biosynthesis and Action, in *Molecular Mechanism of Drug Action*, Taylor & Francis, London, New York, **1988**, Chapter 6, pp 75-98.

Chapter 12
Carotenoids: GGPP-Derived Polyisoprenoid (C_{40}) Coloring Pigments

12.1 Introduction

Carotenoids (carotenes, tetraterpenoids) are biogenetically GGPP derived widely distributed and highly conjugated coloring pigments which absorb light between 400 and 500 nm. The color of carotenoids ranges from deep red to light yellow and sometimes even extends to purple, depending on the environment of the extended conjugated system. The name carotene is derived from carrots (edible roots) *Daucus carota* (Umbelliferae) in which these polyene pigments were first found by Wackenroder in 1832 [1], and the word termination "ene" was proposed by Hofmann in 1866 for unsaturated compounds [2]. Carotenes impart yellow to reddish color to carrots. They are widespread and distributed in flowers (daffodils, marigold), fruits (orange, tomato, pumpkin, red pepper or paprika), algae, fungi, photosynthetic bacteria, fall-coloration of deciduous plants, and also in animals— imparting natural coloration to birds, reptiles, amphibians, fishes, and various invertebrates [3]. During the ripening of fruits and bright yellow-orange coloration of leaves in autumn (fall color), the chlorophyll pigments break down in chloroplast thylakoid membrane, and the otherwise masked color of carotenoids starts revealing their brilliant bright yellow to orange colors. Further, during ripening biosynthesis of some new carotenoids also occurs.

Chlorophylls are the primary light harvesting pigments, and carotenoids are important accessory light-harvesting molecules that transfer energy to the reaction center during photosynthesis. They protect plants from the damage caused by oxygen especially during fall, when chlorophyll starts degrading and is not able to absorb light energy. Hence, carotenes minimize or stop the photooxidative damage. They quench the triplet excited state of photosynthesizers as well as the excited singlet state of oxygen. Thus, plants lacking carotenoids are damaged and killed quickly on exposure to light and oxygen, compared to the plants having their presence. The light harvesting molecules chlorophyll **a**, chlorophyll **b**, and carotenoids remain arranged in the highly organized way around the reaction center. These accessory pigments increase the efficiency of the system by transferring the energy to the

S.K. Talapatra and B. Talapatra, *Chemistry of Plant Natural Products*,
DOI 10.1007/978-3-642-45410-3_12, © Springer-Verlag Berlin Heidelberg 2015

reaction center through a mechanism of resonance energy transfer. Similar light harvesting combination also exists in photosynthetic bacteria as energy donors. Carotenoids being fat soluble are also known as lipochromes or chromolipids.

12.2   Structures of Carotenoids

Currently more than 800 structurally unique natural carotenoids are known, each of which can form further *cis-trans* isomers [4]. Carotenoids containing oxygen functions such as OH, OMe, epoxy, carboxy, aldehyde, etc. are collectively known as xanthophylls. Apocarotenoids are structurally close to carotenoids but contain carbon atoms less than C$_{40}$. Dietary vitamin A deficiency is a severe nutritional problem for children in developing and underdeveloped countries. β-Carotene and lycopene are precursors of vitamin A. Attempts have, therefore, been made to genetically introduce these carotenoids or their precursor (*15Z*)-*phytoene* in seed tissues of transgenic rice plants. Carotenoids with C$_{45}$, C$_{50}$ units are also known. Structures of a few familiar carotenoids are shown in Fig. 12.1.

Most natural acyclic carotenoids are having *all-E*-configuration of the double bonds (X-ray). The presence of some specific Z isomers has also been established. Some examples (**2**)–(**6**) are cited in Fig. 12.2. Prolycopene (**6**), the poly-Z-isomer of lycopene and the main pigment of Tangarine tomato fruits (*Lycopersicon esculentum* var. "*Tangella*" (Solanaceae)), has been shown to possess (7Z, 9Z, 7′Z, 9′Z) stereochemistry from spectroscopic data and comparison with synthetic model compounds. In the same way the other co-occurring pigments—the poly-Z-isomeric carotenoids (**2**)–(**5**) have been shown to possess (15Z) (15Z, 9′Z), (9Z, 9′Z), (9Z, 7′Z, 9′Z) stereochemistries, respectively [5] (Fig. 12.2). The remaining double bonds in the pigments have the *E*-geometry. Sterically hindered as well as *cis* isomers are found in much lesser number (Fig. 12.2).

12.3   Spectral Properties

IR ν_{max} Diagnostic peaks for (Z)-double bond (~730–780 cm^{-1}) and (E)-double bond (~960 cm^{-1}).

NMR, especially ^{13}C NMR spectra reveal the detailed structural information of carotenoids. The number of peaks in noise-decoupled spectra suggests the homogeneity as well as whether the molecule is symmetrical (C$_2$) in the event of which half of the carbon atoms will appear in the noise-decoupled ^{13}C NMR spectra of the compound under investigation. Since the accurate assignments of the sp^2 carbons are not always easy, the advantage of the distinct ^{13}C NMR peaks of vinyl methyls and vinyl methylenes on (Z) and (E) configuration has been used [6]. To generalize these data, many open chain polyenes having C$_{15}$, C$_{20}$, and C$_{30}$ chains with (E) and (Z) as well as (E, Z) stereochemistries of the double bonds have been synthesized

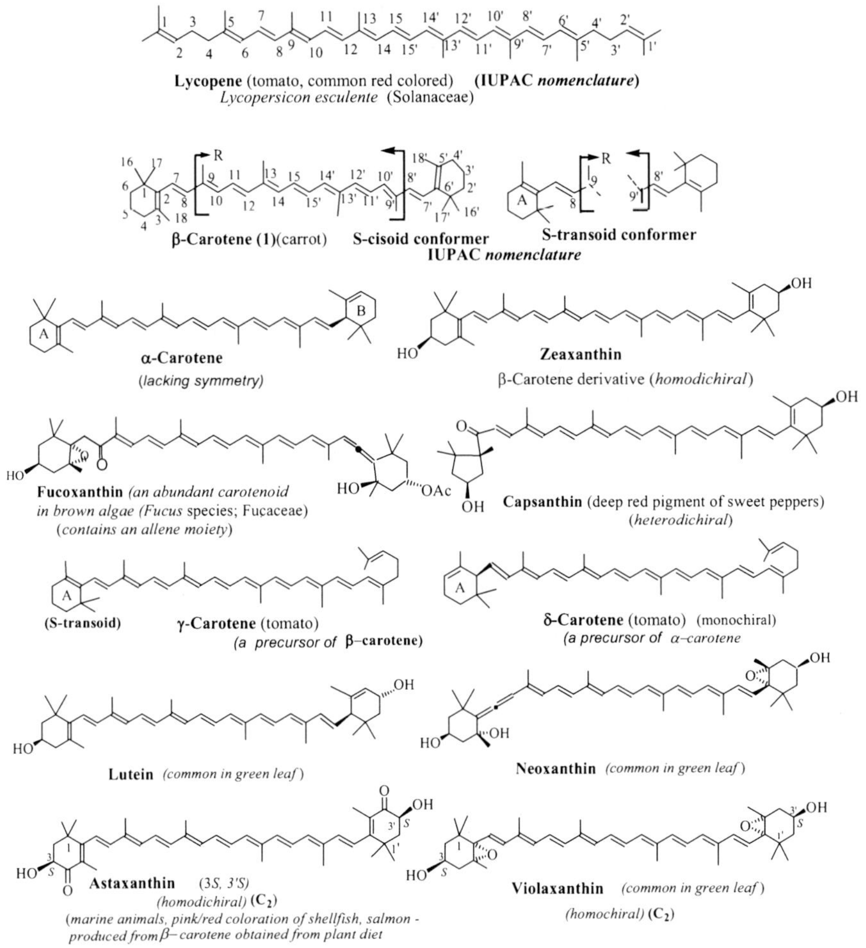

Lycopene (tomato, common red colored)   **(IUPAC nomenclature)**
Lycopersicon esculente (Solanaceae)

β-Carotene (1)(carrot)   **S-cisoid conformer**   **S-transoid conformer**
IUPAC *nomenclature*

α-Carotene   **Zeaxanthin**
(lacking symmetry)   β-Carotene derivative *(homodichiral)*

Fucoxanthin *(an abundant carotenoid*   **Capsanthin** (deep red pigment of sweet peppers)
in brown algae (Fucus species; Fucaceae)   *(heterodichiral)*
(contains an allene moiety)

(S-transoid)   **γ-Carotene** (tomato)   **δ-Carotene** (tomato) (monochiral)
(a precursor of β-carotene)   *(a precursor of α-carotene*

Lutein *(common in green leaf)*   **Neoxanthin** *(common in green leaf)*

Astaxanthin   *(3S, 3'S)*   **Violaxanthin**   *(common in green leaf)*
(homodichiral) **(C₂)**   *(homochiral)* **(C₂)**
(marine animals, pink/red coloration of shellfish, salmon -
produced from β-carotene obtained from plant diet

Fig. 12.1 Structures of some *all–E*-carotenoids

and the ^{13}C NMR data have been recorded and generalized [6]. Some data with
structures have been given for natural carotenoids in Fig. 12.2 and for model
synthetic compounds (Fig. 12.3).

12.4 β-Carotene and Lycopene

β-Carotene and lycopene are most familiar of all carotenoids. The structure of
β-carotene (**1**) (a symmetrical C40 polyolefin), based on its degradation products
(Fig. 12.4), its synthesis (Figs. 12.5 and 12.6), and the synthesis of lycopene
(Fig. 12.7) (another symmetrical C40 polyolefin) have been discussed in brief.

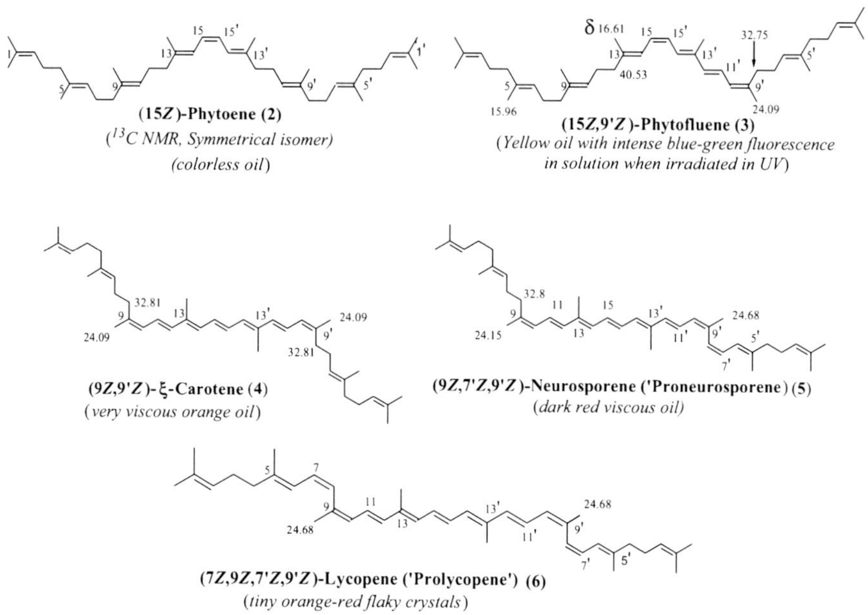

(15Z)-Phytoene (2)
(^{13}C NMR, Symmetrical isomer)
(colorless oil)

(15Z,9'Z)-Phytofluene (3)
(Yellow oil with intense blue-green fluorescence
in solution when irradiated in UV)

(9Z,9'Z)-ξ-Carotene (4)
(very viscous orange oil)

(9Z,7'Z,9'Z)-Neurosporene ('Proneurosporene') (5)
(dark red viscous oil)

(7Z,9Z,7'Z,9'Z)-Lycopene ('Prolycopene') (6)
(tiny orange-red flaky crystals)

Note : The system of numbering carotenoids is recommended by IUPAC

Fig. 12.2 Carotenoids with Z-configurations at specific location from tangerine, tomato fruits and some of their diagnostic ^{13}C peaks (Sect. 12.3) [5]

Based on these degradation products, the structure of β-carotene has been shown to be (1) (Fig. 12.1) which has been confirmed by its synthesis. Several syntheses of β-carotene have been reported. The synthesis [7] by Isler et al. (1957) which gave a high yield has been outlined later (see Fig. 12.5).

β-Carotene, lycopene, zeaxanthin, astaxanthin, and violaxanthin, shown in Fig. 12.1, and the carotenoids (4), and (6), shown in Fig. 12.2, belong to the C$_2$ point group possessing a C$_2$ axis passing through the midpoint of C15-C15′ bond and orthogonal to the polyolefin chain. The carotenoid (2) (Fig. 12.2) also belong to C$_2$ point group; however, its C$_2$ axis passing through the midpoint of C15–C15′ bond lies in the plane of the polyolefin chain.

12.5   Synthesis of β-Carotene

The synthesis by Isler (1957) is outlined later (see Fig. 12.6).

There are other methods of joining two or three smaller molecules to form the C40 skeleton, e.g., C16+C8+C16→C40 (symmetric synthesis), whereas C25+C15→C40 (unsymmetrical synthesis). Retinal (C20 aldehyde)

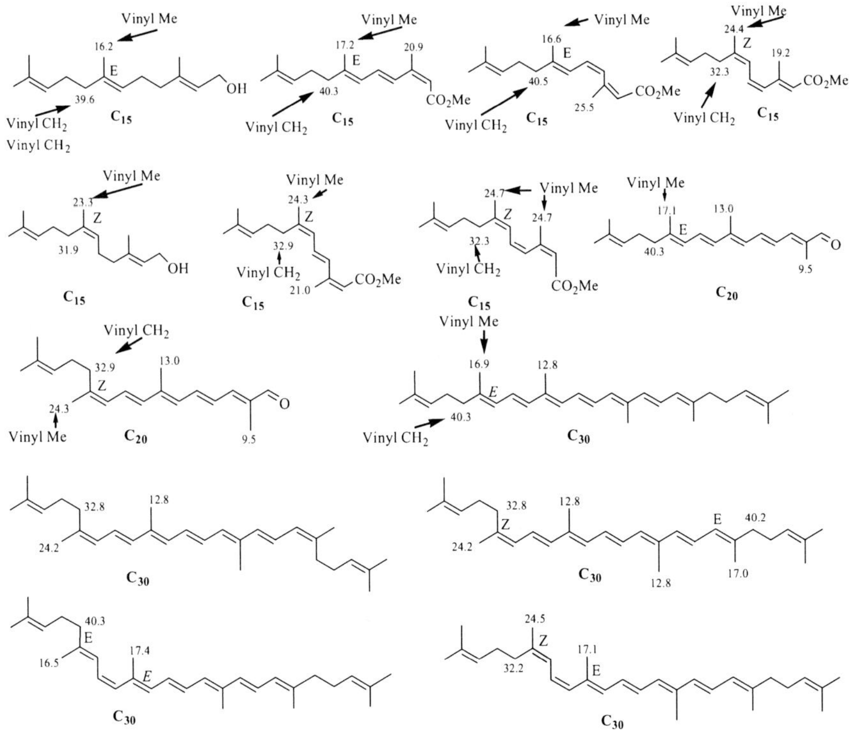

Diagnostic peaks [6]

Vinyl **Me** (***Z***) ~ 24.0 Vinyl **Me** (***E***) ~ δ 17

Vinyl $\mathbf{CH_2}$ (***Z***) ~ δ 33.0 Vinyl $\mathbf{CH_2}$ (***E***) ~ δ 40

Fig. 12.3 Diagnostic peaks for vinyl methyl and vinyl methylene carbons of model compounds [6]

(see Fig. 12.6), when subjected to intermolecular McMurry coupling by treatment with $LiAlH_4$–$TiCl_3$ underwent reductive dimerization (C20+C20) (symmetric reaction) to yield β-carotene in 85 % yield [8].

Industrial syntheses [7, 9] of β-carotene have been accomplished by (i) BASF by coupling of β-retinyltriphenylphosphonium chloride (C20) and retinal (C20) (*cf.* Figs. 12.6 and 34.8), and (ii) Roche by condensing two units of β-C19-aldehyde with acetylenedimagnesium bromide (C2 unit) (C19+C2+C19) and conversion of the diol formed to the all-*trans* β-carotene. (iii) Both (i) and (ii) involve symmetric synthesis.

12.6 Conversion of Vitamin A to β-Carotene

β-Carotene is generally prepared by a multistep route from vitamin A, as outlined in Fig. 12.6.

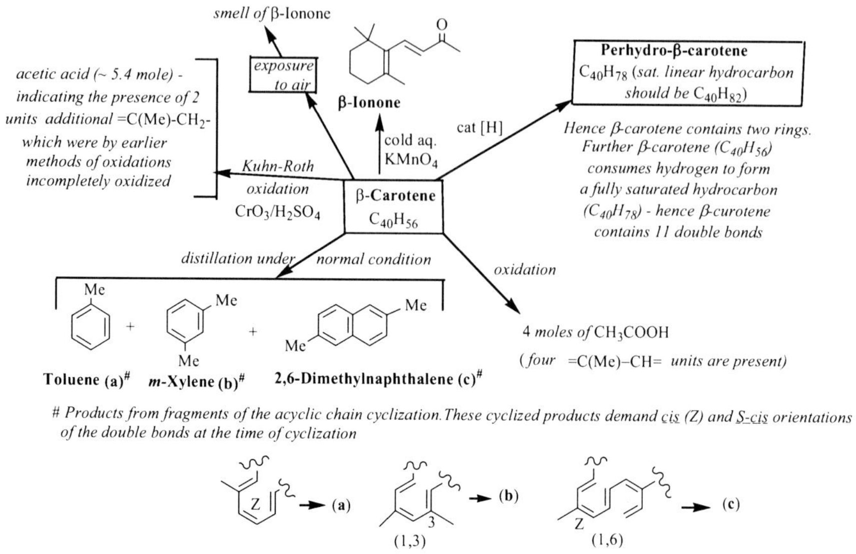

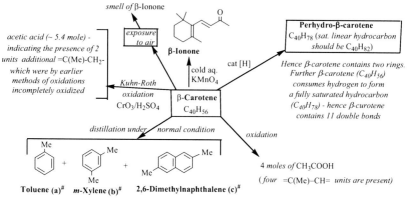

Fig. 12.4 Degradation products of β-carotene

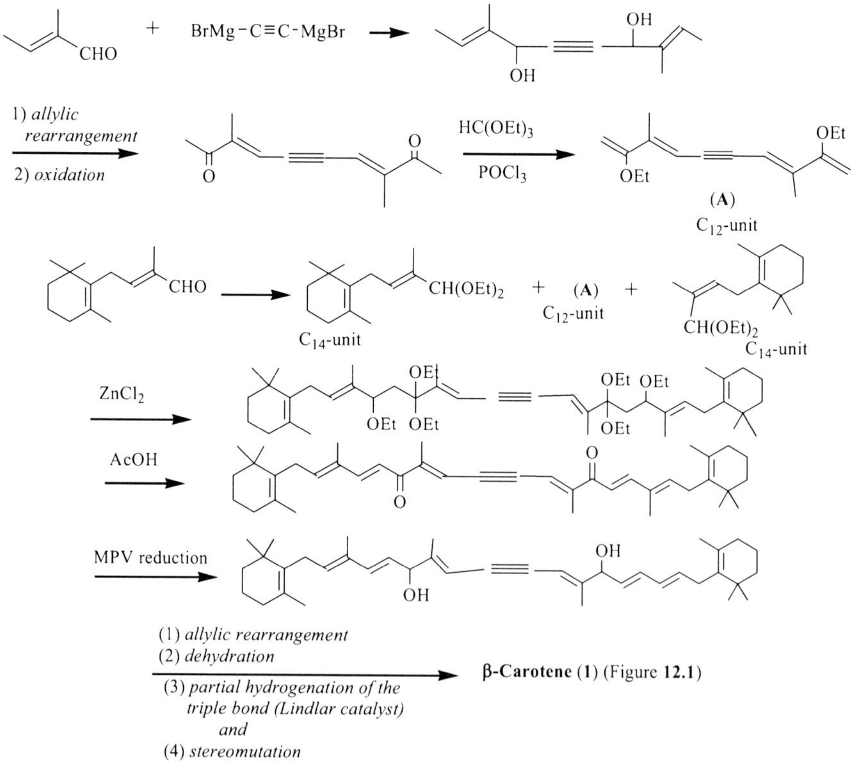

Fig. 12.5 Synthesis of β-carotene (1) by Isler (1957)

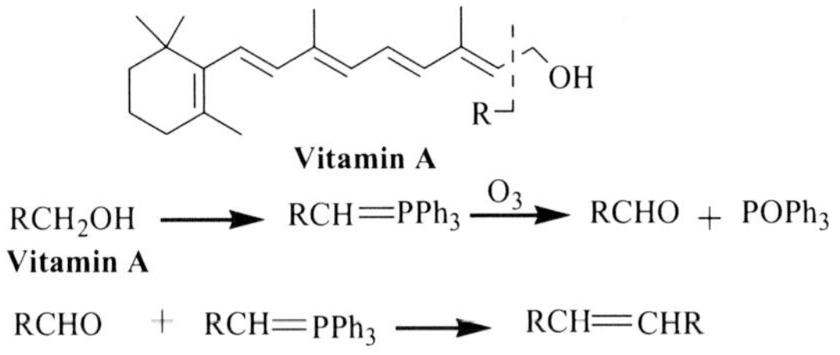

Vitamin A

$RCH_2OH \longrightarrow RCH=PPh_3 \xrightarrow{O_3} RCHO + POPh_3$

Vitamin A

$RCHO + RCH=PPh_3 \longrightarrow RCH=CHR$

Fig. 12.6 Conversion of vitamin A to β-carotene

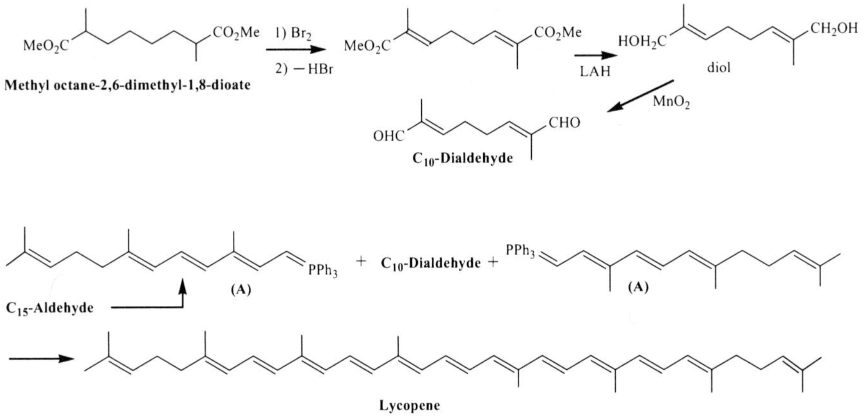

Fig. 12.7 Synthesis of lycopene

12.7 Synthesis of Lycopene

Of the several syntheses of lycopene (the red tomato pigment), the synthesis by Weedon (1965) by employing Wittig's reaction is outlined in Fig. 12.7.

Synthesis of a dialdehyde (C_{10}-unit):

12.8 Biosynthesis [10]

Most carotenoids are tetraterpenes (C_{40}). Two C_{20} units originating from geranylgeraniol are joined by tail-to-tail to form a chain of 32 carbon atoms bearing 8 methyl side chains. Often the basic C_{40} skeleton is retained or only slightly modified by cyclization at one or both ends (Figs. 12.1 and 12.2). Biosynthesis of squalene from FPP via presqualene has been discussed in Chap. 5 (Fig. 5.20). Biosynthesis of phytoene (C_{40}) from GGPP via prephytoene follows analogous mechanism like squalene from FPP. It then undergoes desaturation in stages, each

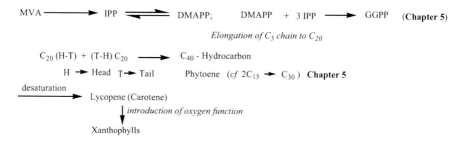

Fig. 12.8 Schematic expression of carotenoid biosynthesis

of which removes two hydrogen atoms and extends the conjugation of polyene by two double bonds on either side. Four such desaturations of two H atoms convert phytoene to lycopene. The latter undergoes cyclization at both or at any one terminal to form cyclocarotenes (β, α, γ, δ-carotenes) (Fig. 12.1). Epoxidation and hydroxylation occur at later stages to yield a special class of carotenes called xanthophylls. The steps in the process could be shown schematically in Fig. 12.8.

The genesis of C_{40}-linear chain from two molecules of GGPP is shown in Fig. 12.9.

Unlike squalene in which at the terminal step the carbocation is quenched by the donation of a hydride from NADPH, the carbocation in carotenoid biosynthesis is quenched through the loss of the adjacent prochiral hydrogen. Label experiment with $(5R)$-$(5$-$^{3}H_{1})$-MVA suggests the formation of $(15Z)$-phytoene as the true intermediate in the biosynthesis of carotenoids. Since carotenoids with all E-double bonds are formed from $(15Z)$-phytoene, $(15Z \rightarrow 15E)$ isomerization is an essential step in lycopene (*all E*) biosynthesis. Further, during desaturation the prochiral hydrogen that is lost has also been identified. The exact mechanism of desaturation is not known but each newly generated double bond holds *trans* relationship. The stepwise desaturation reactions take place in the following sequence: phytoene $\rightarrow$ phytofluorene (7,8,11,12,7',8' hexahydro-ψ,ψ-carotene) $\rightarrow$ ζ-carotene (7,8,7',8'-tetrahedro ψ,ψ-carotene) $\rightarrow$ and neurosporene (7,8-dihydro-ψ,ψ-carotene) $\rightarrow$ lycopene. Because of the geometry of the molecule and long extended conjugation, the scope for cyclization, involving polyene, is not likely except at the termini. Here, squalene-like folding behavior under influence of different cyclases is not expected. Hence, the main chain of the carotenoids remains linear, and the terminal parts undergo cyclizations and oxidations to generate different carotenoids, as outlined in Fig. 12.10.

12.9   Uses

Natural carotenoids are precursors of vitamin A and also important photosynthetic light-harvesting pigments. β-Carotene is commercially important as a food coloring agent. In 1995 commercial preparation of β-carotene to the scale of 500 tons per year was planned [11].

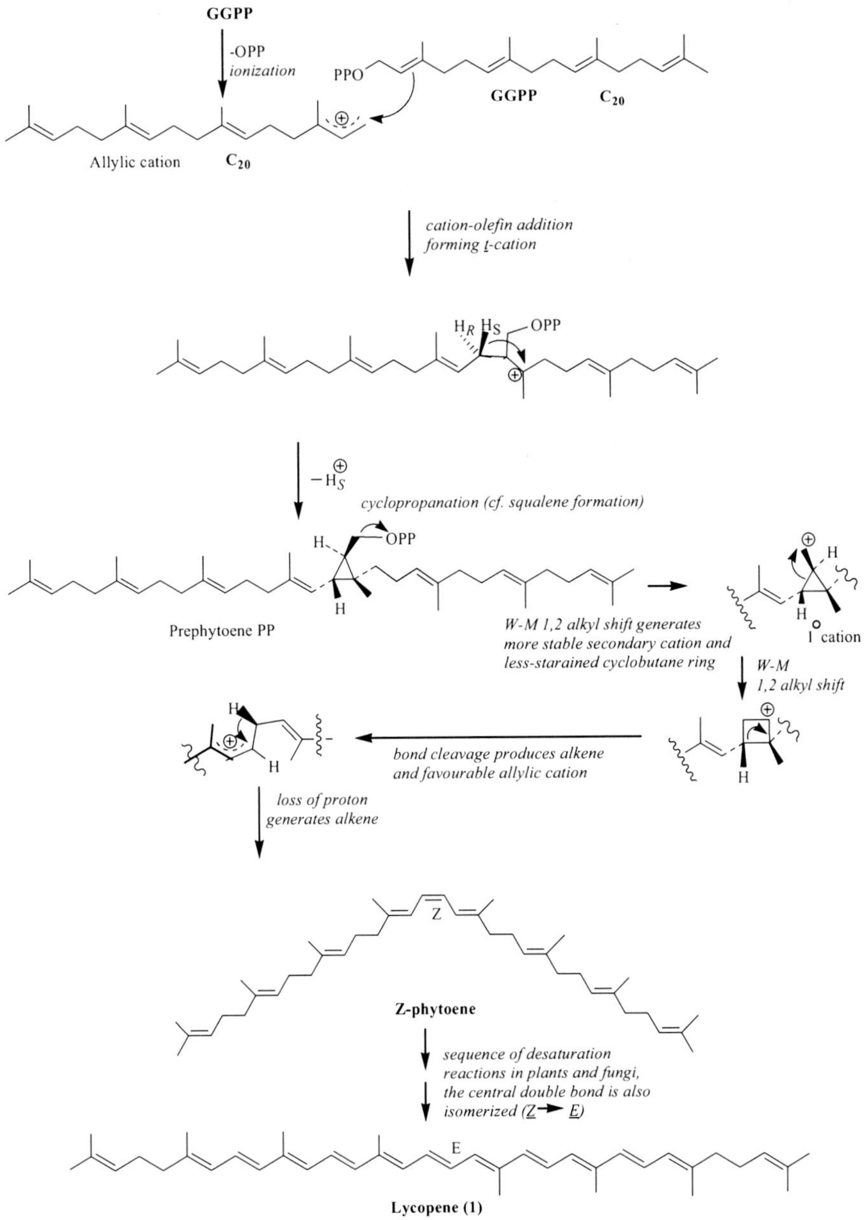

Fig. 12.9 Biosynthesis of the C_{40}-polyene lycopene

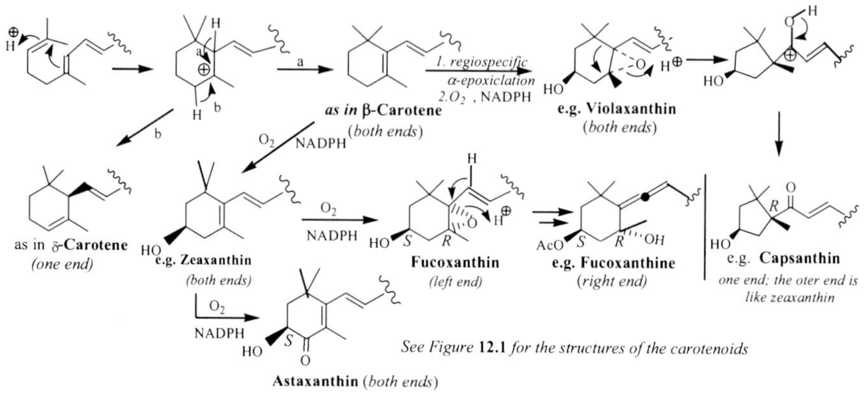

Fig. 12.10 Formation of various types of terminal rings of carotenoids

References

1. Stanley C. Bevan, S. John Greeg and Angela Rosseinsky, *Concise Etymological Dictionary of Chemistry*, Applied Science Publishers Ltd., London, **1976**, p. 37.
2. Stanley C. Bevan, S. John Greeg and Angela Rosseinsky, *Concise Etymological Dictionary of Chemistry*, Applied Science Publishers Ltd., London, **1976**, p. 57.
3. Gregory Armstrong, Carotenoid Genetics and Biochemistry, in *Comprehensive Natural Products Chemistry,* Editors-in-chief, Sir Derek Barton and Koji Nakanishi Vol 2, **1999**, Volume Editor David E.Cane, Pergamon Press, pp. 321-350.
4. Hugo Scheer, Chlorophylls and Carotenoids in *Encyclopedia of Biological Chemistry*, Elsevier, Volume 1, **2004,** 430-437.
5. John M. Clough and Gerald Pattenden, Stereochemical Assignment of Prolycopene and Other Poly-Z-Isomeric Carotenoids in Fruits of the Tangerine Tomato Lycopersicon esculentum var. 'Tangella', *J. Chem. Soc.* Perkin Trans.1, **1983**, 3011-3018.
6. Linda Cary, John M. Clough, and Gerald Pattenden, Application of Nuclear Magnetic Resonance Spectroscopy in the Stereochemical Assignment of Poly-Z-Isomeric Conjugated Polyene Isoprenoids, *J. Chem. Soc., Perkin Trans.*1, **1983**, 3005-3009.
7. For a listing of β-carotene syntheses see Otto Isler, Ed. *Carotenoids,* Birkhauser Verlag, Basel, **1971**, Chapter 6.
8. John E. McMurry and Michael P. Fleming, A New Method for the Reductive Coupling of Carbonyls to Olefines. Synthesis of β-Carotene, *J. Am. Chem. Soc.*, **1974**, 96, 4708-4709.
9. Website of β-carotene syntheses.
10. G. Britton, Stereochemistry of Carotenoid Biosynthesis, in *Studies in Natural Products Chemistry*, Vol.7 (Part A), Ed. Atta-ur-Rahman, **1990**, 317-367 and references cited.
11. H. K. Chopra and P. S. Panesar, *Food Chemistry*, Narosa Publishing House, **2010**, pertinent page 407.

Plant Index[1]

A

Abelmoschus moschatus (Malvaceae[2]), 411

Abies grandis (Pinaceae), 309t

Acacia catechu (Leguminosae[3]), 695–696

  fernesiana, 411

  species, 411f

Acer pseudoplatanus (Aceraceae), 450

  pseudoplatanus, 450

Achillea odorata (Asteraceae), 518

  ochroleuca, 643f

  pseudopectinata, 643f

Aconitum heterophyllum (Ranunculaceae), 916f

  nepelles, 916f

Acronychia baueri (Bauerella baueri) (Rutaceae), 917f

Adenostemna fasciculatum (Compositae/ Asteraceae), 15

Aegle marmelos (Rutaceae), 246t, 643f, 667, 719

Ailanthus malabaricus (Simarubaceae) malabarica, 521

Ainsliaea macrocephala (Compositae), 461

Alangium lamarckii (Alangiaceae), 528f, 916f

Alchornea floribunda (Euphorbiaceae), 916

Allium cepa (Liliaceae/Alliaceae), 1028

  sativum, 1027–1028

Alnus japonica (Betulaceae), 688f

Aloe barbadensis, 1004

officinali, 1004

vulgaris, 1004

Aloe vera (Liliaceae), 1004

Alseodaphne semecarpifolia (Lauraceae), 719t

Alstonia scholaris (Apocynaceae), 245, 246, 917

  constricta, 875

Amaryllis belladonna (Amaryllidaceae), 719

Amni majus (Umbelliferae+), 670

Ananas comosus (Bromeliaceae), 1013t

  sativa, 1013t

  sativus, 1013t

Andrographis paniculata (Acanthaceae), 983t

Anemone raddeana (Ranunculaceae), 997

Angelica anomala (Umbelliferae+), 311

  decursiva, 643f

  gigas, 643f

  japonica, 643f

Anhalonium lewinii (Cataceae), 309

Aniba megaphylla (Lauraceae), 638t

  lancifolia, 638t

Annona reticulata (Annonaceae), 719t

Apium graveolens (Apiaceae), 669

Arabidopsis thaliana (Brassicaceae/ Cruciferae), 517

Araucaria augustifolia (Araucariaceae), 638t

Archangelica officinalis (Apiaceae), 643f

Arctostaphylos glandulosa (Ericaceae), 15

Argemone bocconia (Papaveraecae), 917

---

[1] Note: Page numbers followed by f denote figures and page numbers followed by t denote tables.

[2] The family of each genus is written in parenthesis against a species of that genus, appearing first when there are more than one species of that genus in this list.

[3] The families Leguminosae = Fabaceae; Umbelliferae = Apiaceae; Compositae = Asteraceae; Liliaceae = Alliaceae; Brassicaceae = Cruciferae; Cornaceae = Nyssaceae

S.K. Talapatra and B. Talapatra, *Chemistry of Plant Natural Products*,
DOI 10.1007/978-3-642-45410-3, © Springer-Verlag Berlin Heidelberg 2015

Subject Index[1]

---

[1] *Note: Page numbers followed by f denote figures and page numbers followed by t denote tables.*

S.K. Talapatra and B. Talapatra, *Chemistry of Plant Natural Products*,
DOI 10.1007/978-3-642-45410-3, © Springer-Verlag Berlin Heidelberg 2015